Phenolics in Food and Nutraceuticals

Fereidoon Shahidi and Marian Naczk

Phenolics in Food and Nutraceuticals

CRC Press is an imprint of the
Taylor & Francis Group, an **informa** business

CRC Press
Taylor & Francis Group
6000 Broken Sound Parkway NW, Suite 300
Boca Raton, FL 33487-2742

First issued in paperback 2019

ISBN-13: 978-1-58716-138-4 (hbk)
ISBN-13: 978-0-367-39509-4 (pbk)

Library of Congress Cataloging-in-Publication Data

Shahidi, Fereidoon, 1951-
Phenolics in food and nutraceuticals / Fereidoon Shahidi and Marian Naczk
p. cm.
Includes bibliographical references and index.
ISBN 1-58716-138-9 (alk. paper)
1. Phenols. 2. Food industry and trade. I. Naczk, Marian. II. Title.

TP453.P45S55 2003
664'.001'547632—dc21

2003043977

Library of Congress Card Number 2003043977

Visit the Taylor & Francis Web site at
http://www.taylorandfrancis.com

and the CRC Press Web site at
http://www.crcpress.com

Preface

Nutraceuticals and functional foods are of considerable interest to scientists, manufacturers and consumers because they assume protective and preventive roles in the pathogenesis of certain types of cancer and several other chronic diseases. The benefits of such products may be based on epidemiological, *in vitro* or *in vivo* cellular, organ, or whole organ assembly assays and could possibly include preclinical and human clinical trials. Thus, control of certain types of cancer, cardiovascular disease, autoimmune disorders and the process of aging may benefit from consumption of foods, ingredients and nutraceuticals capable of neutralizing reactive oxygen species and other oxidants in the body.

Phenolic and polyphenolic compounds in food and nutraceuticals represent the most widely distributed plant secondary metabolites exerting their beneficial effects as free radical scavengers and chelators of pro-oxidant metals and thus preventing low-density lipoprotein oxidation and DNA strand scission or enhancing immune function. This book reports the classification and nomenclature of phenolics as well as their occurrence in food and nutraceuticals, their chemistry and applications, and their nutritional and health effects. Food scientists, nutritionists, chemists, biochemists and health professionals will be interested in this book, which describes methods of analysis and quantification and antioxidant activity of phenolics in food and nutraceuticals.

Phenolics in Food and Nutraceuticals can be used as a text for advanced undergraduate or graduate courses on its subject. Extensive references have been provided to facilitate further reading of the original reports.

Fereidoon Shahidi
Marian Naczk

Authors

Fereidoon Shahidi, Ph.D., FACS, FCIC, FCIFST, FRSC, has reached the highest academic level, university research professor, in the department of biochemistry at Memorial University of Newfoundland (MUN). He is also cross-appointed to the department of biology, ocean sciences centre and the aquaculture program at MUN. Dr. Shahidi is the author of nearly 500 research papers and book chapters and has authored or edited over 30 books and made over 300 presentations at scientific conferences. His research contributions have led to several industrial developments around the globe. Dr. Shahidi's current research interests include different areas of nutraceuticals and functional foods as well as marine foods and natural antioxidants.

Dr. Shahidi serves as the editor-in-chief of the *Journal of Food Lipids* and is an editorial board member of *Food Chemistry, Journal of Food Science, Journal of Agricultural and Food Chemistry, Nutraceuticals and Food*, and the *International Journal of Food Properties*. He was the recipient of the 1996 William J. Eva Award from the Canadian Institute of Food Science and Technology in recognition of his outstanding contributions to food science in Canada through research and service, and also the 1998 Earl P. McFee Award from the Atlantic Fisheries Technological Society in recognition of his exemplary contributions in the seafood area and their global impact. He has also been recognized as one of the most highly cited authors in the world in the discipline of agriculture, plant and animal sciences and was the recipient of the 2002 ADM Award from the American Oil Chemists' Society.

Dr. Shahidi is the current chairperson of the nutraceuticals and functional foods division of the Institute of Food Technologists and the chair of Lipid Oxidation and Quality of the American Oil Chemists' Society. He is also the chair of the agricultural and food chemistry division of the American Chemical Society. Dr. Shahidi serves as a member of the Expert Advisory Panel of Health Canada on Standards of Evidence for Health Claims for Foods, Standards Council of Canada on Fats and Oils, Advisory Group of Agriculture and Agri-Food Canada on Plant Products and the Nutraceutical Network of Canada. He is a member of the Washington-based Council of Agricultural Science and Technology on Nutraceuticals.

Marian Naczk, Ph.D., is professor of food science and chair of the department of human nutrition at St. Francis Xavier University, Antigonish, Nova Scotia, Canada.

Dr. Naczk received his M. Eng. in technology of edible meat products and his Ph.D. in food technology in the department of food preservation and technical microbiology at the Technical University of Gdańsk, Gdańsk, Poland. He held teaching and research associate and university teacher positions in this department from 1969 to 1982. Dr. Naczk worked as visiting scientist in the department of chemical engineering and applied chemistry in the food engineering group of the University of Toronto between 1982 and 1985. In 1985 he joined the department of food science and nutrition at the University of Minnesota as research specialist. From 1987 to 1989 he worked as a senior research associate in the department of biochemistry of the Memorial University of Newfoundland, St. John's. Dr. Naczk joined the department of human nutrition (formerly the department of nutrition and consumer studies) at the St. Francis Xavier University in 1989 as assistant professor and has served as chair since 1998.

Dr. Naczk is the author of over 100 research papers and book chapters and has presented over 50 papers in different national and international conferences. His current research interests include plant phenolic–protein interactions and the antioxidant properties of plant phenolics; these are financed through grants from the Natural Sciences and Engineering Research Council of Canada.

Table of Contents

1 Biosynthesis, Classification, and Nomenclature of Phenolics in Food and Nutraceuticals

Phenolic compounds in food and nutraceuticals originate from one of the main classes of secondary metabolites in plants derived from phenylalanine and, to a lesser extent in some plants, also from tyrosine (Figure 1.1) (van Sumere, 1989; Shahidi, 2000, 2002). Chemically, phenolics can be defined as substances possessing an aromatic ring bearing one or more hydroxyl groups, including their functional derivatives. Their occurrence in animal tissues and nonplant materials is generally due to the ingestion of plant foods. Synthetic phenolics may also enter the food system via their intentional incorporation in order to prevent oxidation of their lipid components.

Plants and foods contain a large variety of phenolic derivatives including simple phenols, phenylpropanoids, benzoic acid derivatives, flavonoids, stilbenes, tannins, lignans and lignins. Together with long-chain carboxylic acids, phenolics are also components of suberin and cutin. These rather varied substances are essential for the growth and reproduction of plants and also act as antifeedant and antipathogens (Butler, 1992). The contribution of phenolics to the pigmentation of plant foods is also well recognized. In addition, phenolics function as antibiotics, natural pesticides, signal substances for establishment of symbiosis with rhizobia, attractants for pollinators, protective agents against ultraviolet (UV) light, insulating materials to make cell walls impermeable to gas and water and as structural materials to give plants stability.

Many properties of plant products are associated with the presence, type and content of their phenolic compounds. The astringency of foods, the beneficial health effects of certain phenolics or their potential antinutritional properties when present in large quantities are significant to producers and consumers of foods. Furthermore, anthocyanins are distributed widely in foods, especially in fruits, and also in floral tissues (Harborne and Williams, 2001). They may also be used as nutraceuticals in the dried and powderized form from certain fruit or fruit by-product sources. These anthocyanins are responsible for the red, blue, violet and purple colors of most plant species and their fruits and products. For example, after their extraction,

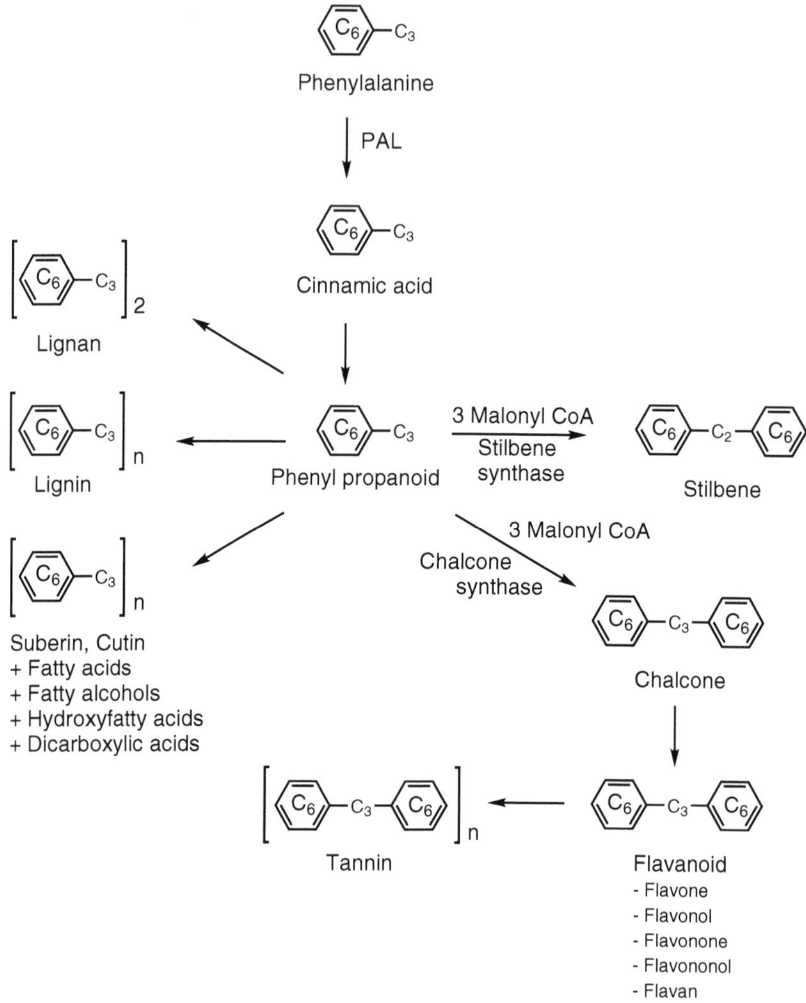

FIGURE 1.1 Production of phenylpropanoids, stilbenes, lignans, lignins, suberins, cutins, flavonoids and tannins from phenylalanine. PAL denotes phenylalanine ammonia lyase.

colorants produced from the skin of grapes may be used by the food industry (Francis, 1993). Methods of analysis of different classes of phenolics and polyphenolics have recently appeared (Hurst, 2002).

BIOSYNTHESIS, CLASSIFICATION AND NOMENCLATURE

Cinnamic and Benzoic Acid Derivatives and Simple Phenols

Phenylalanine ammonia lyase (PAL) catalyzes the release of ammonia from phenylalanine and leads to the formation of a carbon–carbon double bond, yielding *trans*-cinnamic acid. In some plants and grasses tyrosine is converted into 4-hydroxycinnamic acid via the action of tyrosine ammonia lyase (TAL). Introduction of a hydroxyl group into the *para* position of the phenyl ring of cinnamic acid proceeds via catalysis by monooxygenase utilizing cytochrome P_{450} as the oxygen binding site. The *p*-coumaric acid formed may be hydroxylated further in positions 3 and 5 by hydroxylases and possibly methylated via O-methyl transferase with S-adenosylmethionine as methyl donor; this leads to the formation of caffeic, ferulic and sinapic acids (Figure 1.2 and Figure 1.3). These compounds possess a phenyl ring (C6) and a C3 side chain and are thus collectively termed phenylpropanoids, which serve as precursors for the synthesis of lignins and many other compounds.

Benzoic acid derivatives are produced via the loss of a two-carbon moiety from phenylpropanoids. Salicylic acid is a benzoic acid derivative that acts as a signal substance (Raskin, 1992). After infection or UV irradiation, many plants increase their salicylic acid content, which may induce the biosynthesis of defense substances. Aspirin, the acetyl ester of salicylic acid, was first isolated from the bark of the willow tree. Similar to phenylpropanoid series, hydroxylation and possibly methylation of hydroxybenzoic acid leads to the formation of dihydroxybenzoic acid (protocatechuic acid), vanillic acid, syringic acid and gallic acid (Figure 1.3). Hydroxybenzoic acids are commonly present in the bound form in foods and are often the component of a complex structure like lignins and hydrolyzable tannins. They are also found in the form of organic acids and as sugar derivatives (Schuster and Herrmann, 1985). However, exceptions in which they are present mainly in the free form do exist (Mosel and Herrmann, 1974a, b; Schmidtlein and Herrmann, 1975; Stöhr and Herrmann, 1975a, b).

Conventionally, phenylpropanoids (i.e., cinnamic acid family) and benzoic acid derivatives are collectively termed "phenolic acids" in the food science literature. However, it should be noted that this nomenclature is not necessarily correct from a chemical and structural viewpoint. Nonetheless, such compounds are referred to as phenolic acids in the remainder of this book.

Decarboxylation of benzoic acid and phenylpropanoid derivatives leads to the formation of simple phenols (Figure 1.2). Thermal degradation of lignin or microbial transformation may also produce simple phenols in foods (Maga, 1978); thus, vinyl-substituted phenols may be produced by decarboxylation of hydroxycinnamic acids (Pyysalo et al., 1977). However, exposure of 4-vinylguaiacol to oxygen leads to the formation of vanillin (Fiddler et al., 1967). A number of simple phenols, namely, phenol, *o*-cresol, 4-ethylphenol, guaiacol, 4-vinylguaiacol and eugenol, are found in foods of plant origin. Reduction products of phenylpropanoids also yield phenolic compounds such as sinapyl alcohol, coniferyl alcohol, and coumaryl alcohol, among others.

Coumarins

Coumarins are lactones of *cis-O*-hydroxycinnamic acid derivatives that are present in certain foods of plant origin and exist in the free form or as glycosides. The most important of these are simple coumarins, furanocoumarins (also called psoralens) and pyranocoumarines, which are present in free and glycosidic forms in foods. This

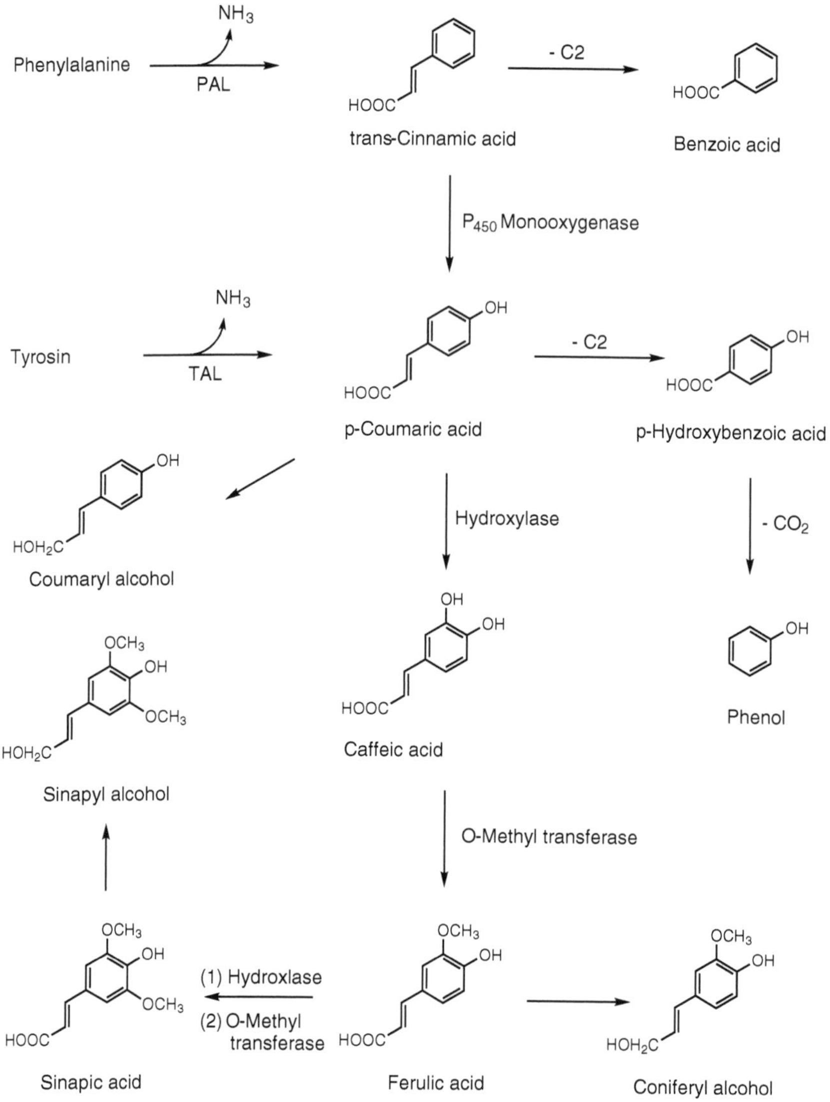

FIGURE 1.2 Formation of phenylpropanoids of cinnamic acid family as well as benzoic acid derivatives and corresponding alcohols from phenylalanine and tyrosine. PAL denotes phenylalanine ammonia lyase and TAL denotes tyrosine ammonia lyase.

group of compounds also acts as phytoalexins in response to disease infection. It is worth noting that most studies on coumarins have been carried out on Umbelliferae and Rataceae families (Murray et al., 1982). Presence of psoralens in different parts of such fruits and vegetables has been confined because high levels of it are found only in parsnip roots (Murray et al., 1982). The chemical structures of some naturally occurring coumarins are shown in Figure 1.4.

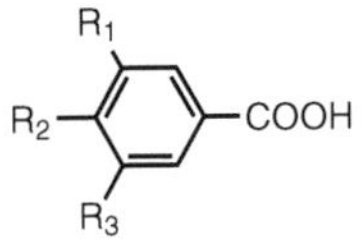

Acid		R_1	R_2	R_3
p-Hydroxybenzoic	4-Hydroxybenzoic	H	OH	H
Protocatechuic	3,4-Dihydroxybenzoic	OH	OH	H
Vanillic	4-Hydroxy-3-methoxybenzoic	OCH_3	OH	H
Syringic	3, 5-Dimethoxybenzoic	OCH_3	OH	OCH_3
Gallic	3,4,5-Trihydroxybenzoic	OH	OH	OH

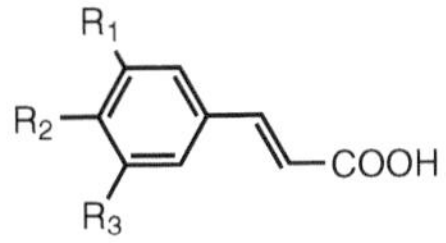

Acid		R_1	R_2	R_3
p-Coumaric	4-Hydroxycinnamic	H	OH	H
Caffeic	3,4-Dihydroxycinnamic	OH	OH	H
Ferulic	4-Hydroxy-3-methoxycinnamic	OCH_3	OH	H
Sinapic	4-Hydroxy-3,5-dimethoxycinnamic	OCH_3	OH	OCH_3

FIGURE 1.3 Phenolic acids of cinnamic and benzoic acid families found in food and nutraceuticals.

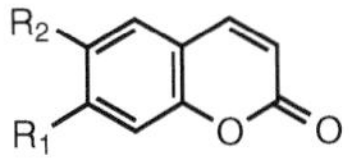

Simple coumarins	R_1	R_2
Coumarin	H	H
Umbelliferone	OH	H
Aesculetin	OH	OH
Scopoletin	OH	OCH_3

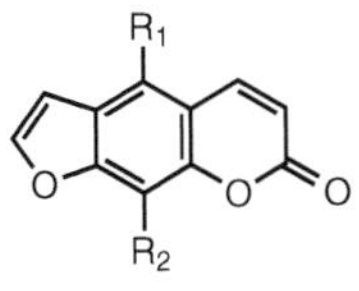

Furanocoumarins	R_1	R_2
Psoralen	H	H
Xanthaxin	H	OCH_3
Bergapten	OCH_3	H
Bergaptol	OH	H

FIGURE 1.4 Chemical structures of some coumarins found in food and nutraceuticals.

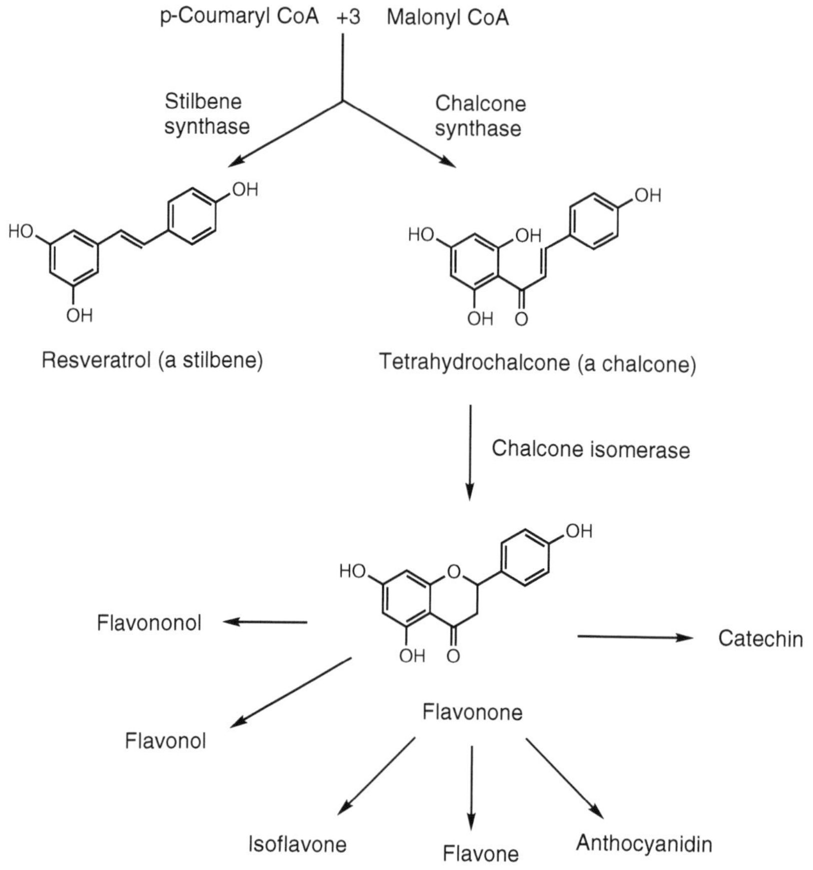

FIGURE 1.5 Production of flavonoids and stilbenes from phenylpropanoid (coumaryl CoA) and malonyl CoA.

Flavonoids and Stilbenes

The flavonoids, including flavones, isoflavones and anthocyanidins, are formed via condensation of a phenylpropane (C6–C3) compound via participation of three molecules of malonyl coenzyme A, which leads to the formation of chalcones that subsequently cyclize under acidic conditions (Figure 1.5). Thus, flavonoids have the basic skeleton of diphenylpropanes (C_6–C_3–C_6) with different oxidation level of the central pyran ring. This also applies to stilbenes, but in this case, after introduction of the second phenyl moiety, one carbon atom of the phenylpropane is split off. Stilbenes are potent fungicides in plants, e.g., viniferin of grapevine.

In the case of flavonoids and isoflavonoids, depending on the substitution and unsaturation patterns, flavones, flavonones, flavonols and flavononols, as well as flavan-3-ols and related compounds, may be formed (Figure 1.6). Flavones and flavonols occur as aglycones in foods; approximately 200 flavonols and some 100 flavones have been identified in plants. These compounds possess a double bond between C-2 and C-3. Flavonols are different from flavones in that they possess a hydroxyl group in the 3-position and can be regarded as 3-hydroxyflavones. On the other hand, flavones may be referred to as 3-deoxyflavonols. Within each class, individual flavonoids may vary in the number and distribution of hydroxyl groups as well as in their degree of alkylation or glycosylation. The formation of flavonol and flavone glycosides depends on the action of light; therefore, they are found mainly in leaves and fruit skins with only trace amounts in parts of plants below the soil surface (Herrmann, 1967). Flavonols are present mainly as mono-, di- and triglycosides. The monoglycosides occur mainly as 3-*O*-glycosides. Glycosylation in positions 5, 7, 3′ and 4′ is rarely reported in flavonols of fruits and vegetables.

In the case of diglycosides, the two sugar moieties may be linked to the same or two different carbons. Of these, 3-*O*-diglycosides and 3,7-di-*O*-glycosides are found most frequently; 3-rutinosides, with a rhamnose and a glucose molecule joined by a 1→6 bond, are the most widely distributed flavonol diglycosides (Macheix et al., 1990). Rutin is an example of a diglycoside found in a number of fruits and vegetables. Triglycosides, meanwhile, are the least frequently reported flavonol glycosides. The sugar moiety of glycosides is usually composed of D-glucose, D-galactose, L-rhamnose, L-arabinose, D-apose and D-glucuronic acid or their combinations. A number of flavonol glycosides acylated with phenolic acids such as *p*-coumaric, ferulic, caffeic, *p*-hydroxybenzoic and gallic acids have been found in fruits. Of these, kaempferol 3-(*p*-coumaryl) glycosides are most widely distributed in plants (Harborne and Williams, 1975, 2001). Meanwhile, flavonones and flavononols are characterized by the presence of a saturated C_2–C_3 bond and an oxygen atom (carbonyl group) in the 4-position. Thus, flavonones may be referred to as dihydroflavones. Flavononols differ from flavonones by having a hydroxyl group in the 3-position and are often referred to as 3-hydroxyflavonones or dihydroflavonols. Flavonones have one center of asymmetry in the 2-position, while flavononols possess a second center of asymmetry in the 3-position (Figure 1.5 and Figure 1.6). Presence of the phenyl ring in the 3-position instead of the 2-position affords isoflavones. The flavonoids include many phytoalexins, e.g., the isoflavone medicarpin from alfalfa. Flavonoids are usually present as glycosides with a sugar moiety linked through an OH group (known as *O*-glycosylflavonoid) or through a carbon–carbon bond (known as C-glycosylflavonoid).

The flavonones in plants are often glycosylated in the 7-position with disaccharides rutinose and neohesperoside. Both of these disaccharides are made of rhamnose and glucose and differ only in their linkage type: 1→6 for rutinose and 1→2 for neohesperoside. It is worth noting that flavononol glycosides are good fungistatic and fungitoxic substances (Kefford and Chandler, 1970). Meanwhile, flavonol glycosides may also occur in the skin of certain fruits (Trousdale and Singleton, 1983).

Among flavonoids, anthocyanins (Figure 1.7) and catechins (Figure 1.8), known collectively as flavans because of lack of the carbonyl group in the 3-position, are important; flavan-3-ols and flavan-3,4-diols belong to this category. Anthocyanins are glycosidically bound anthocyanidins present in many flowers and fruits. While chalcones and flavones are yellow, anthocyanins are water-soluble pigments responsible for the bright red, blue and violet colors of fruits and other foods (Mazza and Miniati, 1994). Thus, the bright red skins of radishes, the red skins of potatoes and

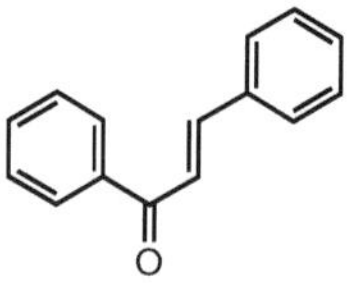

Chalcone

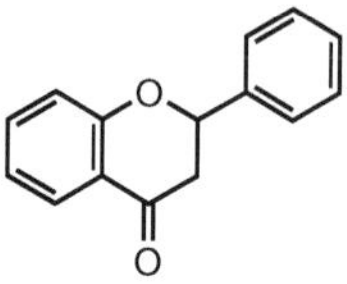

Flavonone

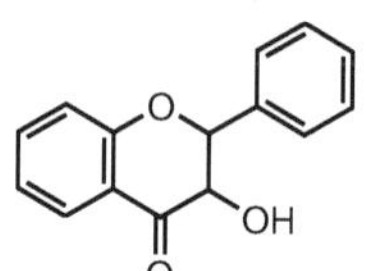

Flavononol

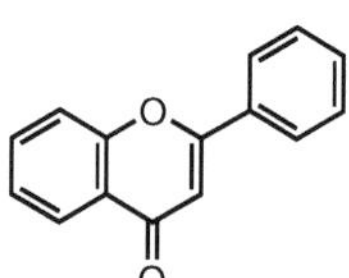

Flavone

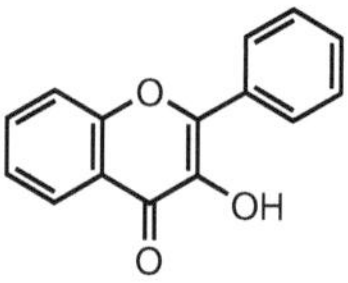

Flavonol

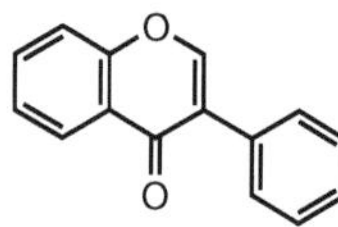

Isoflavone

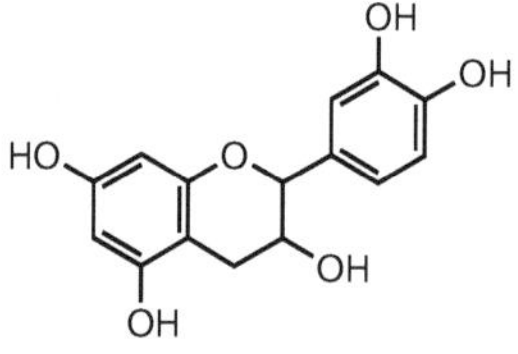

Catechin

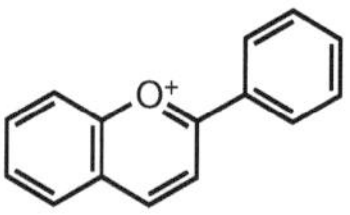

Anthocyanidin
(Flavylium cation)

FIGURE 1.6 Chemical structures of selected C_6–C_3–C_6 compounds found in food and nutraceuticals.

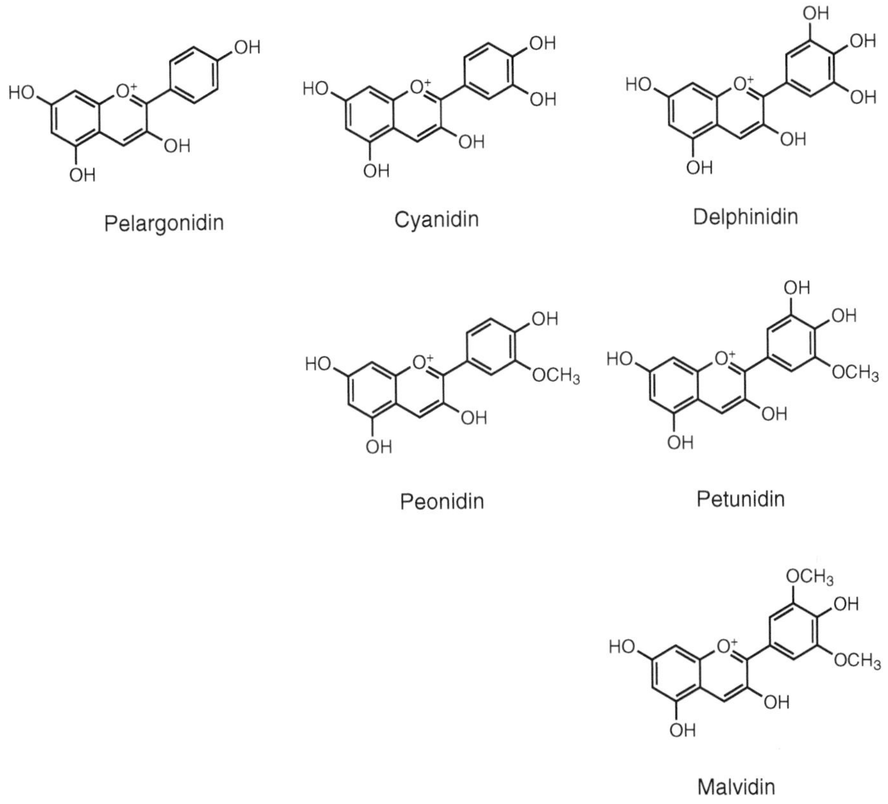

FIGURE 1.7 Chemical structures of anthocyanins.

the dark skins of eggplants are due to the presence of anthocyanins. Blackberries, red and black raspberries, blueberries, cherries, currants, concord and other red grapes, pomegranate, ripe gooseberries, cranberries and plums all contain anthocyanins. Almost 200 different anthocyanins have been identified in plants. The anthocyanin pelargonin contains pelargonidin as its chromophore. Introduction of one or more hydroxyl groups in 3′ and 5′ positions of the phenyl residue and their successive methylation yields five additional anthocyanidins that differ in color (Figure 1.7). Most anthocyanins occur as monoglycosides and diglycosides of pelargonidin, cyanidin, peonidin, delphinidin, petunidin and malvidin. Meanwhile, anthocyanins assume different colors when subjected to pH variation in solutions. In addition, catechins and epicatechins, found in different plants and in high amounts in green tea leaves, are similar in their structures to anthocyanidins; however, they are colorless (Figure 1.9).

Examples of some naturally occurring flavonoids in foods and food components are provided in Figure 1.9. The substitution pattern in the ring structure is also given.

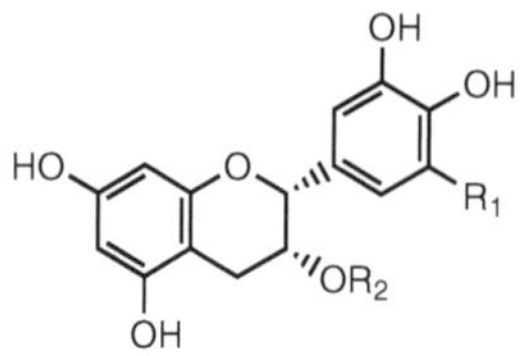

OH
OH
HO
O
R1
OR2
OH

Catechin	R_1	R_2
(-)-Epicatechin (EC)	H	H
(-)-Epicatechin gallate (ECG)	H	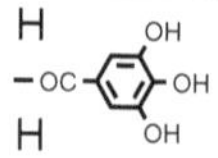 -OC- (OH, OH, OH)
(-)-Epigallocatechin (EGC)	OH	H
(-)-Epigallocatechin gallate (EGCG)	OH	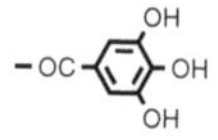 -OC- (OH, OH, OH)

FIGURE 1.8 Chemical structures of catechins.

Lignans and Lignins

By definition, lignans are dimers of phenylpropanoid (C6–C3) units linked by the central carbons of their side chains. According to their oxygenation, four major groups of linear lignans are found, namely, lignans or butane derivatives, lignolides or butanolide derivatives, monoepoxylignans or tetrahydrofuran derivatives and biepoxylignans or cyclolignans or derivatives of 3,7-dioxabicyclo-(3,3,0)-octane. Cyclized lignans or cyclolignans occur as tetrahydronaphthalene or naphthalene derivatives. Structures of some lignans are provided in Figure 1.10.

In plants, lignans and their higher oligomers act as defensive substances (Davin and Lewis, 1992). The lignan pinoresinol is formed when the plant is wounded and is toxic to microorganisms. Malognol inhibits the growth of bacteria and fungi. The pharmacological effects of lignans related to their antiviral activity have also been demonstrated (Davin and Lewis, 1992).

Lignins are formed via polymerization of a mixture of the three monolignols *p*-coumaryl, sinapyl and coniferyl alcohols (Lewis and Yamamato, 1990; Douglas, 1996). The composition of lignin varies greatly in different plants; lignin of conifers, for example, has a high coniferyl alcohol content. Lignin is covalently bound to cellulose in the cell walls. Liquified cell walls may be compared to reinforced concrete in which cellulose fibers serve as the steel and lignin the concrete; this makes plants difficult for herbivores to digest. In addition, lignin inhibits the growth of pathogenic microorganisms.

Suberins and Cutins

Suberins are another group of compounds composed of polymeric phenylpropanoids, long chain fatty acids and fatty alcohols with 18 to 30 carbon atoms (Davin and Lewis, 1992). Hydroxyfatty acids and dicarboxylic acids with 14 to 30 carbon atoms might also be present. In suberin, as in lignin, phenylpropanoids are partly linked

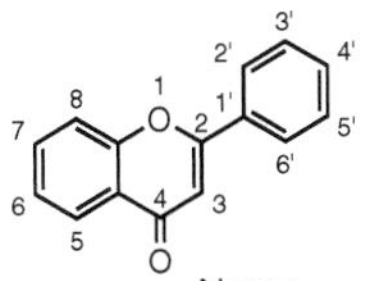

Class	Name	Substitutions	Dietary Source
Chalcone (no 1-2 bond)	Butein	2, 4, 3', 4' - OH	Miscellaneous
	Okanin	2, 3, 4, 3', 4' - OH	Miscellaneous
Flavone	Chrysin	5, 7 - OH	Fruit skins
	Apigenin	5, 7, 4' - OH	Parsley, celery
	Rutin	5, 7, 3', 4' - OH; 3 - rutinose	Buckwheat, citrus, red pepper, red wine, tomato skin
Flavonone	Naringin	5, 4', - OH; 7 - rhamnoglucose	Citrus, grapefruit
	Naringenin	5, 7, 4' - OH	Citrus
	Taxifolin	3, 5, 7, 3', 4', - OH	Citrus
	Eriodictyol	5, 7, 3', 4' - OH	Lemons
	Hesperidin	3, 5, 3' - OH, 4' - OMe; 7 - rutinose	Oranges
	Isosakuranetin	5, 7 -OH; 4' - OMe	Citrus
Flavonol	Kaempferol	3, 5, 7, 4' - OH	Leek, broccoli, endives, grapefruit, black tea
	Quercetin	3, 5, 7, 3', 4' - OH	Onion, lettuce, broccoli, tomato, tea, berries, apples, olive oil
Flavononol	Engeletin	3, 5, 7, 4' - OH; 3 - O-rhamnose	White grapeskin
	Astilbin	3, 5, 7, 5', 4' - OH; 3 - O-rhamnose	White grapeskin
	Genistin	5, 4' - OH; 7 - glucose	Soybean
	Taxifolin	3, 5, 7, 3', 4' - OH	Fruits
Isoflavone	Genistein	5, 7, 4' - OH	Soybean
	Daidzin	4' - OH, 7 - glucose	Soybean
	Daidzein	4', 7 - OH	Soybean
Flavanol	(+)-Catechin	3, 5, 7, 3', 4' - OH	Tea
	(+)-Gallocatechin	3, 5, 7, 3', 4', 5' - OH	Tea
	(-)-Epicatechin	3, 5, 7, 5', 4' - OH	Tea
	(-)-Epigallocatechin	3, 5, 7, 3', 4' - OH	Tea
	(-)-Epicatechin gallate	3, 5, 7, 3', 4' - OH; 3 - gallate	Tea
Anthocyanidin	Epigenidin	5, 7, 4' - OH	Stored fruits
	Cyanidin	3, 5, 7, 4' - OH; 3, 5 - OMe	Cherry, raspberry, strawberry
	Delphinium	3, 5, 7, 3', 4', 5' - OH	Dark fruits
	Pelargonidin	3, 5, 7, 4' - OH	Dark Fruits

FIGURE 1.9 Different classes of flavonoids, their substitution patterns and dietary sources.

with each other. Suberin is a cell wall constituent that forms gas- and water-tight layers. Cutin is a polymer similar to suberin, but with a relatively small proportion of phenylpropanoids and dicarboxylic acids. Consisting mainly of esterified hydroxyfatty acids with 16 to 18 carbon atoms, cutin impregnates the cell walls of leaves and shoot organs.

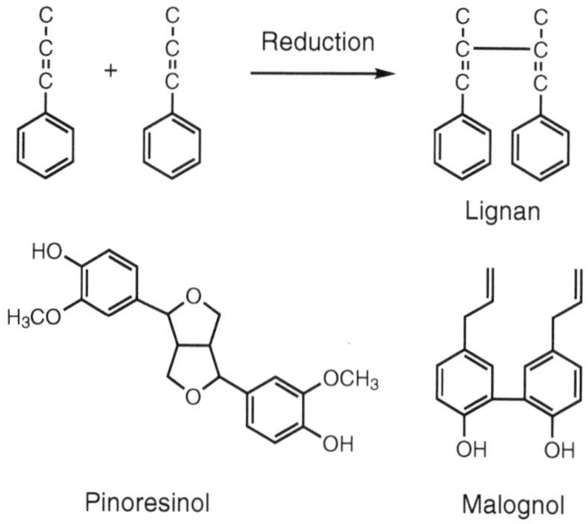

FIGURE 1.10 Chemical structures of selected lignans.

TANNINS

Tannin is a collective term used to describe a variety of plant polyphenols used in tanning raw hides to produce leather. This group of compounds includes oligomeric and polymeric constituents. Tannins are polyphenols capable of precipitating proteins from aqueous solutions and are widely distributed in plants. They are found in especially high amounts in the bark of certain trees such as oak and in galls. Tannins also form complexes with certain types of polysaccharides, nucleic acids and alkaloids (Ozawa et al., 1987). Depending on their structures, tannins are defined as hydrolyzable or condensed, the latter term known also as proanthocyanidin. The condensed tannins are oligomers and polymers of flavonoids, specifically flavan-3-ols, while hydrolyzable tannins are glycosylated gallic acids (Ferreira and Li, 2000; Khanbabaee and van Ree, 2001). Many of these gallic acids are linked to sugar molecules. The molecular weights of hydrolyzable tannins range from 500 to 2800 Da or more (Haddock et al., 1982). Based on their hydrolysis products, hydrolyzable tannins include gallotannins and elagitannins. Condensed tannins include dimers, oligomers and polymers of flavan-3-ols. Approximately 50 procyanidins ranging from dimers to hexamers have been identified (Hemingway, 1989). Based on the hydroxylation pattern of A- and B-rings, proanthocyanidins can be divided into procyanidins, propelargonidins and prodelphinidins. The consecutive units of condensed tannins are linked through the interflavonoid bond between C-4 and C-8 or C-6 (Hemingway, 1989). The molecular weight of proanthocyanidins isolated from fruits ranges between 2000 and 4000 Da (Macheix et al., 1990) (Figure 1.11).

The phenolic groups of tannins bind very tightly with the –NH groups of peptides and proteins and prevent their hydrolysis and digestion in the stomach; thus, they are known to be antinutritional in nature. In the tanning process tannins bind to the collagen of the animal hide and thus produce leather that can withstand attack by the degrading microorganisms. Because they inactivate aggressive enzymes, tannins also protect plants against attack by microorganisms.

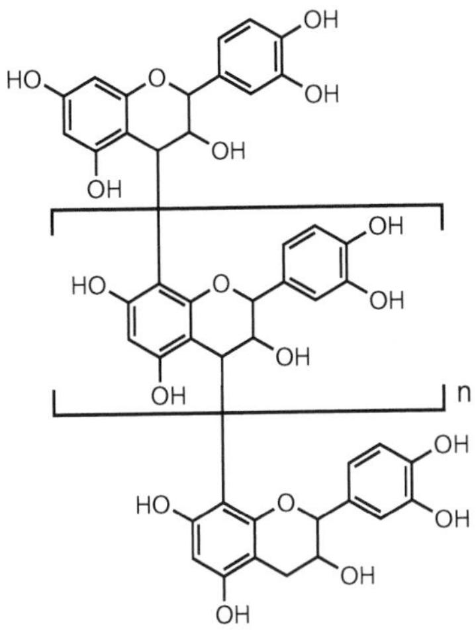

Condensed Tannin

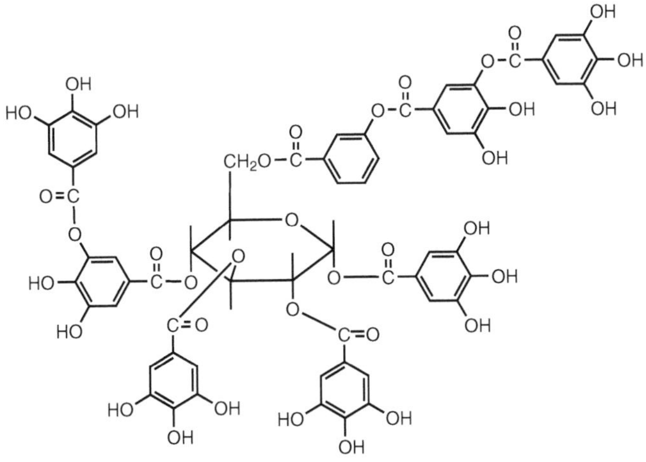

Hydrolyzable Tannin

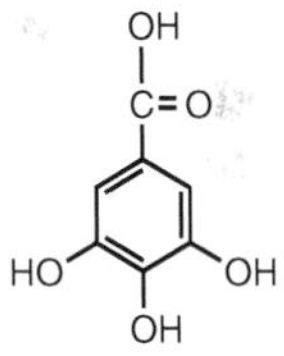

Gallic Acid

FIGURE 1.11 Chemical structures of tannins.

Tocopherols and Tocotrienols

Another class of plant phenolics often encountered in foods is tocols. These monophenolic and lipophylic compounds occur widely in plant tissues. The eight different tocols with vitamin E activity are divided into two groups known as tocopherols and tocotrienols; each of these is classified into α, β, γ and δ, depending on the number and position of methyl groups on the chromane ring. (See Figure 1.12.) The α-isomer is a 5,7,8-trimethyl; β, a 5,8-dimethyl; γ, a 7,8-dimethyl; and δ, an

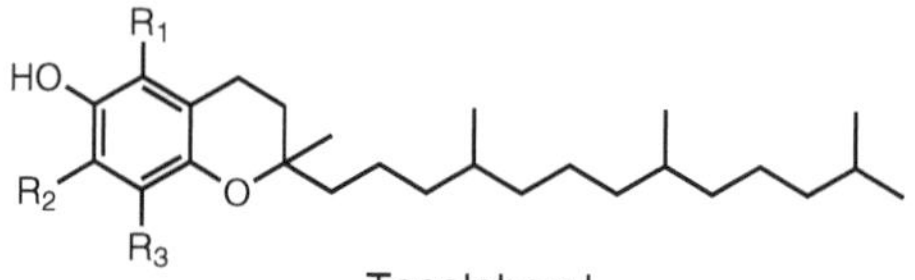

	R_1	R_2	R_3
α	CH_3	CH_3	CH_3
β	CH_3	H	CH_3
γ	H	CH_3	CH_3
δ	H	H	CH_3

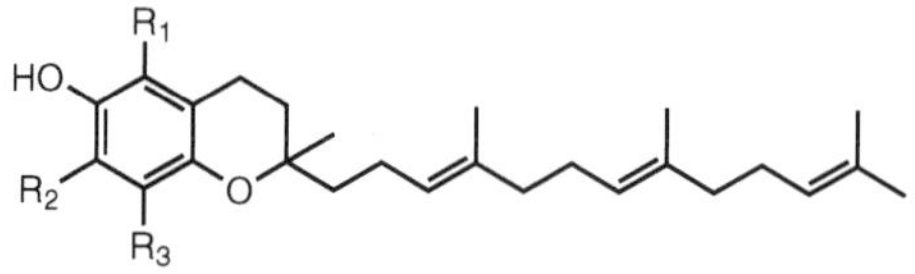

FIGURE 1.12 Chemical structures of tocopherols and tocotrienols.

8-methyl derivative of the parent compound. Furthermore, the side chain in tocopherols is saturated, although it is unsaturated in tocotrienols (Shahidi, 2000).

The occurrence of tocols in vegetable oils is diverse, but animal fats generally contain α-tocopherol and, in some cases, γ-tocopherol. The β-isomer is generally less prevalent. Tocols are by-products of edible oil processing and are isolated as distillates of the deodorization process.

REFERENCES

Butler, L.G. 1992. Protein-polyphenol interactions: nutritional aspects, in *Proc. 16th Int. Conf. Groupe Polyphenols,* 16, Part II, 11–18.

Davin, L.B. and Lewis, N.G. 1992. Phenylpropanoid metabolism: biosynthesis of monolignols, lignans and neolignans, lignins and suberins, in *Phenolic Metabolism in Plants*, Staffod, H.A. and Ibrabim, R.K., Eds., Plenum Press, New York, 325–375.

Douglas, C.J. 1996. Phenylpropanoid metabolism and lignin biosynthesis: from weeds to trees. *Trends Plant Sci*., 1, 171–178.

Ferreira, D. and Li, X-C. 2000. Oligomeric proanthocyanidins: naturally occurring O-heterocycles. *Nat. Prod. Rep*., 17, 193–212.

Fiddler, W., Parker, W.E., Wasserman, A.E., and Doerr, R.C. 1967. Thermal decomposition of ferulic acid. *J. Agric. Food Chem*., 15, 757–761.

Francis, F.J. 1993. Polyphenols as natural colorants, in *Polyphenolic Phenomena,* Scalbert, A., Ed., Institut National de la Rocherche Agronomique, Paris, 209–220.

Haddock, E.A., Gupta, R.K., Al-Shafi, S.M.K., Layden, K., Haslam, E., and Magnaloto, D. 1982. The metabolism of gallic acid and hexahydroxydiphenic acid in plants: biogenetic and molecular taxonomic considerations. *Phytochemistry*, 21, 1049–1051.

Harborne, J.B. 1967. *Comparative Biochemistry of the Flavonoids*. Academic Press, London.

Harborne, J.B. and Williams, C.A. 1975. Flavone and flavonol glycosides, in *The Flavonoids*, Harborne, J.B., Mabry, T.J., and Mabry, H., Eds., Chapman and Hall, London, 376–441.

Harborne, J.B. and Williams, C.A. 2001. Anthocyanins and flavonoids. *Nat. Prod. Rep.*, 18, 310–333.

Hemingway, R.W. 1989. Structural variations in proanthocyanidins and their derivatives, in *Chemistry and Significance of Condensed Tannins*. Hemingway, R.W. and Karchesy, J.J., Eds., Plenum Press, New York, 83–98.

Hurst, W.J. 2002. *Methods of Analysis for Functional Foods and Nutraceuticals*. CRC Press, Boca Raton, FL.

Kefford, J.B. and Chandler, B.V. 1970. *The Chemical Constituents of Citrus Fruits*. Academic Press, New York.

Khanbabaee, K. and Van Ree, T. 2001. Tannins: classification and definition. *Nat. Prod. Rep.*, 18, 641–649.

Lewis, N.G. and Yamamato, E. 1990. Lignin: occurrence, biogenesis and degradation. *Ann. Rev. Plant Physiol. Plant Mol. Biol.*, 41, 455–496.

Macheix, J-J., Fleuriet, A., and Billot, J. 1990. *Fruit Phenolics*. CRC Press, Boca Raton, FL.

Maga, J.A. 1978. Simple phenol and phenolic compounds in food flavor. *CRC Crit. Rev. Food Sci. Nutr.*, 10, 323–348.

Mazza, G. and Miniati, E. 1994. *Anthocyanins in Fruits, Vegetables and Grains*. CRC Press, Boca Raton, FL.

Mosel, H.D. and Herrmann, K. 1974a. The phenolics of fruits. III. The contents of catechins and hydroxycinnamic acids in pome and stone fruits. *Z. Lebensm. Unters. Forsch.*, 154, 6–10.

Mosel, H.D. and Herrmann, K. 1974b. Die phenolischen Inhalsstoffe der Obstes. IV. Die phenolischen Inhalsstoffe der Brombeeren und Himbeeren und deren deren Veranderun gen wahrend Wachstum und Reife der Fruchte. *Z. Lebensm. Unters. Forsch.*, 154, 324–328.

Murray, R.D.H., Mendez, J., and Brown, S.A. 1982. *The Natural Coumarins: Occurrence, Chemistry and Biochemistry*. John Wiley-Interscience, Chichester, U.K.

Ozawa, T., Lilley, T.H., and Haslam, E. 1987. Polyphenol interactions: astringency and the loss of astringency in ripening fruit. *Phytochemistry*, 26, 2937–2942.

Pyysalo, T., Torkkewli, H., and Honkanen, E. 1977. The thermal decarboxylation of some substituted cinnamic acids. *Lebensm.-Wiss.u.-Technol.*, 10, 145–149.

Raskin, J. 1992. Protein–polyphenol interactions: nutritional aspects, in *Proc. 16th Int. Conf. Groupe Polyphenols,* 16, Part II, 11–18.

Schmidtlein, H. and Herrmann, K. 1975. On the phenolic acids of vegetables. IV. Hydroxycinnamic acids and hydroxybenzoic acids of brassica species and leaves of other cruciferae. *Z. Lebensm. Unters. Forsch.*, 159, 139–141.

Schuster, B. and Herrmann, K. 1985. Hydroxybenzoic and hydroxycinnamic acid derivatives in soft fruits. *Phytochemistry*, 24, 2761–2764.

Shahidi, F. 2000. Antioxidants in food and food antioxidants. *Nahrung*, 44, 158–163.

Shahidi, F. 2002. Antioxidants in plants and oleaginous seeds, in *Free Radicals in Food: Chemistry, Nutrition, and Health Effects,* Morello, M.J., Shahidi, F., and Ho, C-T., Eds., ACS Symposium Series 807. American Chemical Society. Washington, D.C., 162–175.

Stöhr, M. and Herrmann, K. 1975a. Die phenolischen Inhaltsstoffe des Obstes. VI. Die phenolischen Inhaltsstoffe der Johannisbeeren, Stachelbeeren und Kulturheidelbeeren Veranderungen der Phenolsauren und Catechine wahrend Wachstum und Reife von Swarzen Johannisbeeren. *Z. Lebensm. Unters. Forsch.*, 159, 31–33.

Stöhr, M. and Herrmann, K. 1975b. The phenols of fruits.V. The phenols of strawberries and their changes during development and ripeness of fruits. *Z. Lebensm. Unters. Forsch.*, 159, 341–343.

Trousdale, E.K. and Singleton, V.L. 1983. Astilbin and engeletin in grapes and wines. *Phytochemistry*, 22, 619–620.

van Sumere, C.F. 1989. Phenols and phenolic acids, in *Methods in Plant Biochemistry, Volume 1, Plant Phenolics*. Harborne, J.B., Ed., Academic Press, London, 29–74.

2 Cereals, Legumes, and Nuts

INTRODUCTION

Phenolic compounds are ubiquitous in cereals, legumes, and nuts. Different phenolics belonging to the benzoic acid, cinnamic acid, flavonoid and tannin class of compounds may be present in free, esterified/etherified or insoluble bound forms (Ribereau-Gayon, 1972; Salunkhe et al., 1982). Thus, ferulic acid may be linked to polysaccharides (Faulds and Williamson, 1999; Hartley et al., 1976; Hatfield et al., 1999; Iiyama et al., 1990; Ishii, 1997; Ishii and Hiroi, 1990; Markwalder and Neukom, 1976), lignins (Higuchi et al., 1967; Iiyama et al., 1990) and suberin (Riley and Kolattukudy, 1975); *p*-coumaric acid may be linked to polysaccharides (Hartley et al., 1976), lignins (Grabber et al., 2000; Nakamura and Higuchi, 1978), and cutin (Kolattukudy et al., 1981). In aleurone layer of cereal grains, ferulic acid is mainly linked to polysaccharides via an ester bond at the *O*-5 position of arabinofuranose (Ishii, 1997). Ferulic acid may also undergo dimerization by oxidative coupling catalyzed by peroxidase (Brett et al., 1999; Geissmann and Neukom, 1973). Dehydrodimers of ferulic acid have been proposed to form cross links between arabinoxylan chains (Markwalder and Neukom, 1976; Waldron et al., 1996). The main dehydrodimers identified in plant material are shown in Figure 2.1.

A number of phenols present in cereal grains, especially roasted products, may be formed as by-products of enzymatic or thermal degradation of substituted benzoic and cinnamic acids. Tressl et al. (1976) studied the thermal degradation of cinnamic and hydroxycinnamic acids under conditions simulating roasting of cereal grains. The 4-vinyl derivatives of these acids were the major thermal degradation products. However, some phenols with extended alkyl side chains, such as 5-n-alkyl-resorcinols, occur naturally in cereal grains (Collins, 1986).

The oligomeric flavanols, products of condensation of (+)-catechin, (+)-gallocatechin and their epimers, are also found in many plant products (Thompson et al., 1972). However, in cereal grains the presence of oligomeric flavanols has been reported only in sorghum (Strumeyer and Malin, 1975) and barley (Pollock et al., 1960); in barley brewers often call them anthocyanogens (Gramshaw, 1968). Simple flavanols such as monomers, dimers and trimers are found in developing sorghum grains (Glennie et al., 1981), while highly polymerized flavanols (tannins) occur in mature sorghum seeds. Simple flavanols are generally considered precursors of haze in beer (Gramshaw, 1969) because they easily undergo acid-catalyzed and autoxidative polymerization (Goldstein and Swain, 1963) to products that may form insoluble complexes with proteins.

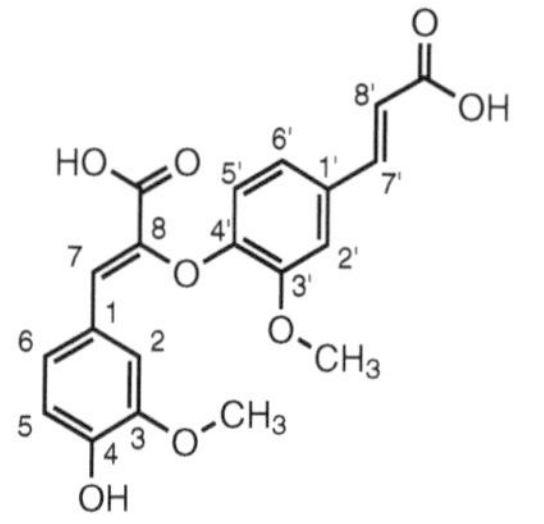

8-O-4'-DiFA

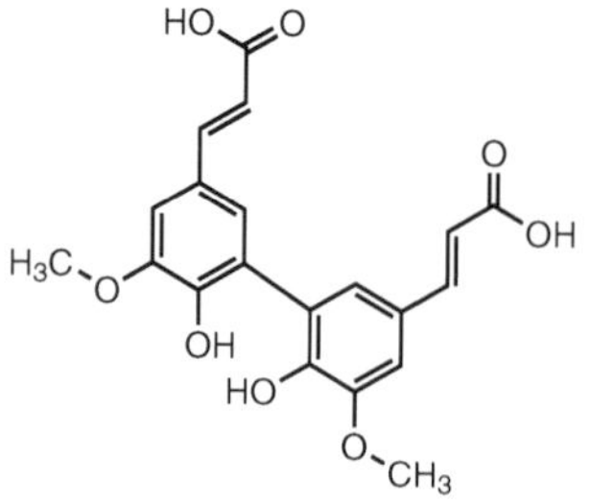

5, 5'-DiFA

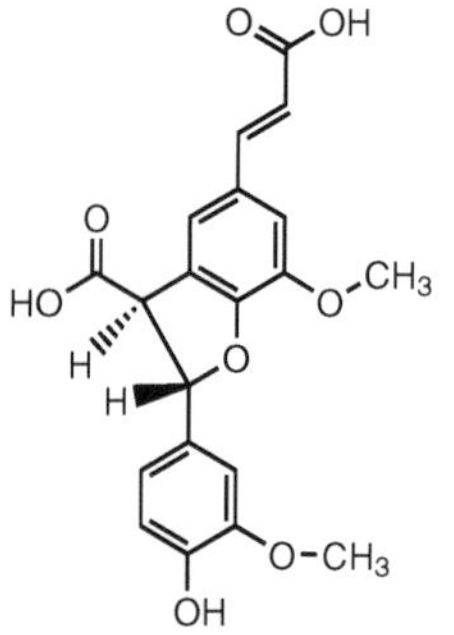

8, 5'-DiFA - benzofuran form

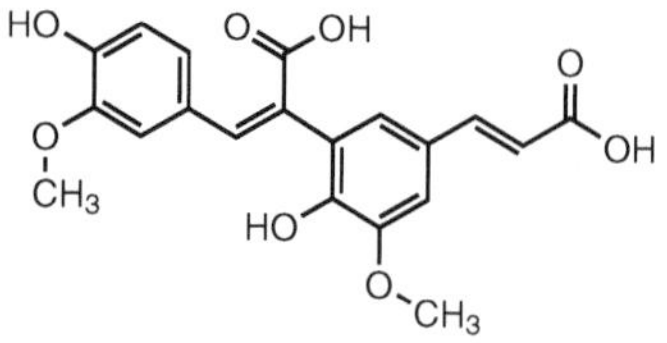

8, 5'-DiFA - open form

FIGURE 2.1 Chemical structures of some dehydrodimers of ferulic acid (DiFA).

BEANS AND PULSES

Dry beans and peas are important components of traditional diet in many Asian, African, and Central and South American countries. Faba bean, also known as broad bean, has been grown in Egypt, Europe and Asia for centuries. Cowpeas (black-eyed peas) and winged beans are grown in a number of tropical countries. In addition, pigeon peas, lima beans, horse gram, hyacinth beans, moth beans, and black and green gram are also widely harvested for human consumption in Asia and a number of other countries (Duke, 1981). Dry beans and peas serve as a good source of protein and carbohydrate, although they contain a number of antinutritional factors, including phytates, enzyme inhibitors, lectins, and tannins, among others. Flavonol glycosides, tannins and anthocyanins are responsible for the color of seed coat in dry beans (Beninger and Hosfield, 1999; Beninger et al., 1998, 1999; Takeoka et al., 1997). Tannins in beans and pulses have received considerable attention due to their possible nutritional and physiological implications, biochemical functions, and influence on the aesthetic qualities of foods (Deshpande and Cheryan, 1987; Reddy et al., 1985).

Phenolic Acids

Sosulski and Dabrowski (1984) reported that defatted flours of 10 legumes contain only soluble esters of *trans*-ferulic, *trans-p*-coumaric and syringic acids. Of these legumes, mung bean, field pea, faba bean, lentil and pigeon pea contain 18 to 31 mg of total phenolic acids per kilogram, while in Navy bean, lupine, lima bean, chickpea and cowpea, the total content of phenolics ranges from 55 to 163 mg/kg. On the other hand, Garcia et al. (1998) reported the presence of caffeic, *p*-coumaric, sinapic and ferulic acids in dehulled soft and hard-to-cook beans (*Phaseolus vulgaris*). The dehulled soft beans contained 45 times more methanol soluble esters of phenolic acids than hard-to-cook beans, while the content of phenolic acids bound to water-soluble fraction of pectins was two times higher in hard-to-cook beans than in dehulled soft beans.

Flavonoids

Hertog et al. (1992) reported that after hydrolysis of all glycosides, French green beans contain 39 mg of quercetin per kilogram and <12 mg of kaempferol per kilogram of sample. Later, Hempel and Bohm (1996) reported that 3-*O*-glucuronides and 3-*O*-rutinosides of quercetin and kaempferol are the main flavonoid glycosides in six varieties of yellow and green French beans. These authors reported that the total content of quercetin- and kaempferol-3-*O*-glycosides was 19.1 to 183.5 and 5.6 to 14.8 mg/kg of sample, respectively. Figure 2.2 shows two new isoflavonoid derivatives, mutabilin (3′-methoxy-5-hydroxy-7-*O*-β-D-glucosylisoflavone) and mutabilein (3′-methoxy-5, 7-dihydroxyisoflavone) that have been identified in seeds of *Lupinus mutabilis* (Fabaceae) (Dini et al., 1998). Beninger et al. (1998) reported that two flavonol glycosides, namely, kaempferol 3-*O*-β-D-glucopyranoside (astragalin; 499 mg/kg of fresh whole bean weight) and kaempferol 3-*O*-β-D-glucopyranoside-(2→1)-*O*-β-D-xylopyranoside (585 mg/kg of fresh whole bean weight), are responsible for imparting yellow color to the seed coat of 'Prim' variety of Manteca-type dry beans (Figure 2.3).

Later, Beninger and Hosfield (1999) identified three flavonol glycosides contributing to the color of seed coat of commercial dark red bean, cultivar Montcalm astragalin, quercetin 3-*O*-β-D-glucopyranoside-(2→1)-*O*-β-D-xylopyranoside, and quercetin 3-*O*-β-D-glucopyranoside. In addition, malvidin 3-glucoside, petunidin 3-glucoside, delphinidin 3-glucoside and 3,5-diglucoside have been detected in black-violet beans (Feenstra, 1960). Subsequently, Stanton and Francis (1966) reported that delphinidin 3-glucoside is the major anthocyanin in beans of Canadian Wonder cultivar, while cyanidin 3-diglucoside, 3,5-diglucoside, pelargonidin 3-glucoside and

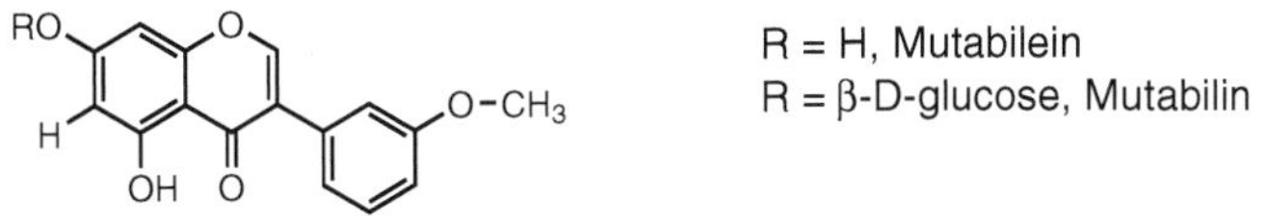

R = H, Mutabilein
R = β-D-glucose, Mutabilin

FIGURE 2.2 Chemical structures of isoflavonoid derivatives found in *Lupinus*.

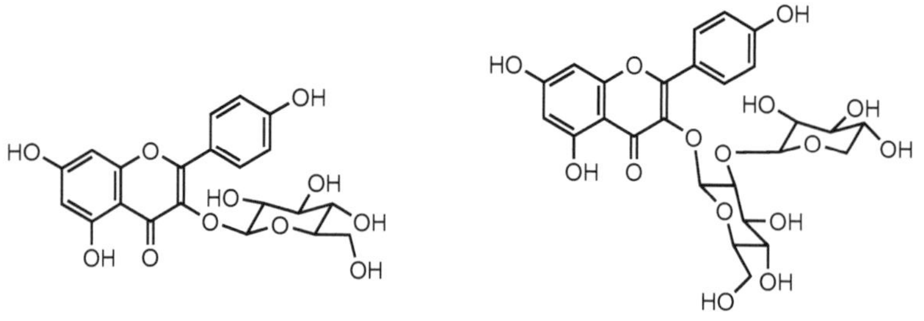

(1) (2)

FIGURE 2.3 Chemical structures of (1) kaempferol 3-*O*-β-D-glucoside and (2) kaempferol 3-*O*-β-D-glucosyl-(2→1)-β-D-xyloside found in dry beans.

3,5-diglucoside are the minor anthocyanins. Later, Nozzolillo (1972, 1973) and Nozzolillo and MacNeill (1985) identified only malvidin glucosides in three black bean cultivars, namely, Black Wax, Royalty, and Black Valentine. Recently, Takeoka et al. (1997) reported that delphinidin 3-*O*-glucoside, malvidin 3-*O*-glucoside, and petunidin 3-*O*-glucoside are the main anthocyanins responsible for the color of black and purple seed coat beans.

Condensed Tannins

Located mainly in the seed coat or testa of beans and peas, condensed tannins are often considered antinutritional factors. The content of tannins in beans and peas is up to 2.0% as expressed in (+)-catechin or tannic acid equivalents (Table 2.1). The average content of tannins in eight faba bean cultivars is 4.3% (Marquardt et al., 1978); the amount in cowpeas ranges from 0 to 0.7% and in chickpeas between 0 and 0.2% (Price et al., 1980). On the other hand, as shown in Table 2.1 and Table 2.2, cotyledons contain fewer tannins than whole seeds do (Bressani and Elias, 1980; Elias et al., 1979). Recently, Wang et al. (1998) estimated the content of condensed tannins in 17 field pea cultivars and 9 grass pea lines grown in western Canada. These authors found barely detectable levels of condensed tannins in peas, but a content of 890 to 5180 mg (+)-catechin equivalent per kilogram in grass peas.

Tannins in beans are linear polymers of flavan-3-ol (catechin and gallocatechin) and flavan-3 to 4-diol (leucocyanidin and leucodelphinidin) units. These units are linked together by carbon–carbon bonds formed between carbon-4 of one unit and carbon-6 or -8 of another. Each chain of tannin polymer contains a flavan-3-ol unit at the terminal ends (Martin-Tanguy et al., 1977). Based on the difference in their solubility in methanol and ether, bean tannins can be fractionated into four groups depending on their degree of polymerization (Cansfield et al., 1980). On the other hand, separation of condensed tannins on a Sephadex LH-20 column yielded two fractions: one containing low molecular weight polyphenolics and another containing soluble condensed tannins (Marquardt et al., 1977).

TABLE 2.1
Tannin Content of Dry Beans[a]

Bean	Whole Bean	Cotyledon
Tannin Content Expressed as Catechin Equivalents		
Black gram (*Phaseolus mungo* L.)	5.40–11.97	0.16–0.33
Chick pea (*Cicer arietimum*)	0.78–2.72	0.16–0.38
Cowpea (*Vigna sinensis* L.)	1.75–5.90	0.28
Green gram (*Phaseolus aureus* L.)	4.37–7.99	0.21–0.39
Kidney bean (*Dolichos lablab*)	10.24	0.73
Pigeon pea (*Cajanus cajan*)	3.80–17.10	0.22–0.43
Tannin Content Expressed as Tannic Acid Equivalents		
Faba bean (*Vicia faba* L.)	7.50–20.00	7.40–9.10
Horse gram (*Dolichos biflorus* L.)	16.00	—
Lima bean (*Phaseolus lunatus*)	6.50–9.30	—
Moth bean (*Phaseolus aconitifolius*)	13.00	—
Peas (*Pisum sativum* L.)	5.00–10.50	4.60–5.60
Winged bean (*Psophocarpus tetragonolobus* L.)	1.58–7.50	1.08

[a] Grams per kilogram.

Source: Adapted from Reddy, N.R. et al., 1985, *J. Am. Oil Chem. Soc.*, 62:541–549.

TABLE 2.2
Tannin Content in Bean Samples (*Phaseolus vulgaris*) with Different Seed Coat Color[a]

Sample	Seed Coat Color	Whole Seeds	Cotyledons	Seed Coat	Whole Cooked Seeds
NEPI mutant	White	3.8	3.8	1.3	1.8
Normal	White	3.9	4.5	1.3	2.1
Normal	Red	9.3	5.0	38.0	4.1
Normal	Black	9.2	4.6	42.0	4.7
San Fernando	Black	6.7	5.9	43.0	5.8

[a] Grams of tannic acid per kilogram.

Source: Adapted from Elias, L.G. et al., 1979, *J. Food Sci.*, 44:524–527.

There are differences in the content of condensed tannins of beans depending on the color of seed coats. The white varieties of beans usually contain lower concentrations of tannins than those with red, black or bronze seed coats (Table 2.1 and Table 2.2) (Bressani and Elias, 1980; Elias et al., 1979). However, Marquardt et al. (1978) reported that white varieties of faba beans have higher concentrations of tannins than those with dark testa. Moreover, the variability in the content of tannins is much greater in colored varieties of beans than in white-coated beans. Breeding studies show that the genes of low-tannin beans dominate those of high-tannins seeds; thus low-tannin colored beans can be obtained through hybridization techniques (Ma Yu and Bliss, 1978). Furthermore, Ma Yu and Bliss (1978) and Elias et al. (1979) did not find any correlation between the color of seed coat or seed size and tannin content, although a positive correlation exists between tannin concentration in seed coats and trypsin inhibitor activity (Elias et al., 1979).

Tannin content of beans decreases during seed maturation. Mature winged beans (80 to 85 days after flowering) contain about 50% fewer tannins than those harvested 40 days after flowering (Kadam et al., 1982). These changes in tannin content may be attributed to the polymerization of polyphenolic compounds to high molecular weight insoluble polymers.

Effect of Processing

Storing beans at high temperatures and humidity produces "hard-to-cook" seeds, so a longer cooking time is required for cotyledon softening. A number of hypotheses have been proposed to explain the cause of this defect; these include lipid oxidation and polymerization (Muneta, 1964), catabolism of phytin and pectin demethylation with subsequent formation of insoluble pectate (Jones and Boulter, 1983a, b) and formation of protein–polyphenol complexes as well as polymerization of polyphenols (Rozo, 1982). A highly significant negative correlation exists between tannin content and the development of hardness (Mejia, 1979). According to Hincks and Stanley (1986), hardening of beans involves phytate loss during initial storage and phenol metabolism during extended storage.

In addition, Srisuma et al. (1989) reported that storage induced "hard-to-cook" beans that contained higher levels of hydroxycinnamic acids than the control beans. These phenolic acids provide cross-linking sites to pectin in middle lamella and proteins — interactions that may result in textural bean defects. It has been reported that storing dry beans at elevated temperatures and humidity results in lower protein quality (Molina et al., 1975), possibly due to the interaction of proteins with tannins (Bressani, 1975). Sievwright and Shipe (1986) suggested that the adverse effect of tannins on the quality of bean proteins depends on the age of the beans and their storage conditions. They found that polyphenol oxidase enhances protein–tannin interaction and that the incidences of hard-to-cook phenomenon in beans can be reduced by soaking the beans in salt solutions. It has been postulated that salt is involved in breaking down the hydrogen bonds between condensed tannins and proteins.

Heat treatment and cooking of beans improve their texture and acceptability as a result of destruction of their antinutritional factors. Soaking beans before cooking is a common practice. A number of studies have been carried out to determine the effect of cooking on the antinutritional factors of beans. It has been reported that cooking reduces the content of tannins by 30 to 40% (de Espana, 1977); in this study the cooking water was not discarded. Therefore, it has been suggested that changes

TABLE 2.3
Effects of Cooking on Distribution of Tannins in *Phaseolus* Beans

Cooking Fraction	White Beans		Black Beans		Red Beans	
	(g/kg)	(%)	(g/kg)	(%)	(g/kg)	(%)
Raw seed	3.6	100	9.0	100	14.7	100
Cooked bean	2.4	66.7	5.44	60.4	5.5	37.4
Cooking water	0.56	15.5	1.72	19.1	13.7	11.7
Loss or bound	0.64	17.6	1.84	20.5	7.5	50.9

Note: Bean to water ratio of 1:3; beans were cooked 4 h at atmospheric pressure.

Source: Adapted from Bressani, R. et al., 1982, *J. Plant Foods*, 4:43–55.

in the polyphenol content may result from binding of phenolics with other organic materials, or alteration of chemical structure of polyphenols, enabling them to give a color reaction with Folin-Denis reagent (Bressani and Elias, 1980). This postulate was later confirmed by the results of studies of Bressani et al. (1982) on partition of polyphenols during cooking of beans. The values shown in Table 2.3 indicate that 37.4 to 66.7% of the total polyphenols of raw beans remained in cooked beans and that the cooking water contained less than 20% of total polyphenols. However, 17.8 to 50.9% of total polyphenols was not accounted for in this study; these polyphenols were probably involved in the formation of insoluble complexes with proteins or other compounds.

The content of tannins of beans may be reduced by the removal of hulls or soaking. Soaking winged beans (Chimbu cultivars) in water reduces the content of tannins by 34 and 67.5% after 6 h (de Lumen and Salamat, 1980) and 24 h of treatment, respectively (Sathe and Salunkhe, 1981). Additional removal of seed coat after water soaking further reduces the tannin content in winged beans (Sathe and Salunkhe, 1981) and dry beans (*Phaseolus vulgaris* L.) (Table 2.4) (Deshpande et al., 1982). Substantial reduction in tannin content may be obtained by chemical treatment of beans: soaking winged beans in a sodium hydroxide solution for 6 h reduces their tannin content by 63% (de Lumen and Salamat, 1980), whereas 24-h treatment of beans with 2% aqueous KOH removes approximately 87.2% of their tannins (Sathe and Salunkhe, 1981).

The content of polyphenolic compounds can be reduced substantially during protein isolate production. Aqueous extraction and membrane processes have been considered and are expected to lead to industrial developments. For example, the micellization process developed by Murray et al. (1978) for preparation of protein isolates from faba beans markedly reduced the content of phenolics in the protein isolates produced. This technique consists of salt extraction of proteins followed by precipitation upon dilution with cold water, which affords a protein isolate in which the proteins have a micellar structure. About 46% of phenolics are eliminated during the salt extraction step, possibly due to the removal of phenols associated with hulls as well as low recovery of proteins associated with phenols. Further reduction in

TABLE 2.4
Tannin Content of Whole and Dehulled Dry Beans (*Phaseolus vulgaris* L.)[a]

Cultivar	Whole Beans	Dehulled Beans	(%) Reduction on Dehulling
Cranberry	763	100	86.9
Viva Pink	1221	104	91.5
Pinto	2647	142	94.6
Light Red Kidney	1522	222	85.4
Dark Red Kidney	1053	287	72.7
Small Red	2828	190	93.3
Black Beauty	337	108	68.0

Note: Values are mean of triplicate determinations on dry weight basis; dry beans were dehulled after 12 h soaking in water at 21°C.

[a] Milligrams of catechin equivalent per kilogram.

Source: Adapted from Deshpande, S.S. et al., 1982, *J. Food Sci.*, 47:1846–1850.

the content of phenolics has been achieved during the dilution steps. The isolate contained only 3.5% of phenolics originally present in the concentrate used for its preparation (Arntfield et al., 1985). According to Murray et al. (1981), the precipitation by dilution favors protein–protein rather than protein–phenolic interactions. The isolate obtained using isoelectric precipitation contains a level of phenolics comparable to that found in the isolate prepared by a micellization technique.

CEREALS

Barley

Barley is grown in many parts of the world for food and feed uses. In the U.S., it is mostly a feed crop, as well as a grain for brewing and producing ethanol. Only small quantities of barley crop are used in the form of flour and grit as a component of such food products as soups, dressings, baby foods and specialty items (Shellenberger, 1980). Table 2.5 shows the total content of phenolic compounds present in barley; their content depends on the cultivar and the cultivation location (Bendelow and La Berge, 1979). Phenolics present in barley grains range from tyrosine, tyramine and its derivatives, phenolic acids and their esters and glycosides, and anthocyanins responsible for the blue and red color of barley tissue to lignans and substances related to lignin (Briggs, 1978; Salomonsson et al., 1980).

Barley-malt contributes phenolic and polyphenolic compounds to early stages of the brewing process. A correlation coefficient of 0.90 was found between the content of phenolics in barley and malt (Bendelow and La Berge, 1979). Upon processing, the malt phenolics polymerize to give rise to beer polyphenolics that furnish color, impart astringent taste, serve as a browning substrate, and participate in precipitation of poorly coagulable beer proteins. On the other hand, proanthocyanidins from barley

TABLE 2.5
Content of Polyphenols in Barley and Malt as Affected by Cultivar[a]

Station in Western Canada	Cultivar	Grain	Malt
Brandon	TR 201	3.29	3.55
	TR 907	3.00	2.96
	TR 424	2.87	2.78
	Betzes	2.83	2.84
	Klages	2.58	2.74
Acme	TR 201	3.64	3.81
	TR 907	3.72	3.89
	TR 424	3.40	3.52
	Betzes	3.04	3.03
	Klages	3.27	3.38

[a] Grams of gallic acid equivalent per kilogram.

Source: Adapted from Bendelow, V.M. and La Berge, D.E., 1979, *J. Am. Soc. Brew. Chem.*, 37:89–90.

malt influence the development of haze in beer because 80% of these phenolics present in regular beer are derived from barley malt.

Phenolic Acids

The free forms of salicylic, *p*-hydroxybenzoic, vanillic, protocatechuic, *o*-, *m*- and *p*-coumaric, syringic, ferulic and sinapic acids have been identified in barley grains. Recently, Yu et al. (2001) reported the presence of chlorogenic and protocatechuic acids in barley. The highest content of these acids is found in the grains between 19 and 31 days of anthesis. Ferulic acid is the predominant free phenolic acid in barley seeds (Nordkvist et al., 1984) and barley brans (Renger and Steinhart, 2000). The concentration of ferulic acid in 18 barley cultivars ranges from 365 to 605 mg/kg of sample (Zupfer et al., 1998). The content of all phenolic acids in unripe seeds is three to six times higher than that in mature seeds on the day of harvest (Table 2.6). The highest content of ethanol-soluble bound phenolic acids is also detected at 19 to 31 days of anthesis (Slominski, 1980). The content of free chlorogenic and protocatechuic acids in 30 varieties of barley grown in Canada ranges from 3.21 to 16.28 mg/kg of dry matter and from trace to 2.88 mg/kg of dry matter, respectively (Yu et al., 2001).

A number of bound phenolic acids, including ferulic, *p*-coumaric, vanillic, sinapic and *p*-hydroxybenzoic acids, have been found in barley after alkaline hydrolysis (Van Sumere et al., 1972). Of these, ferulic and *p*-coumaric acids are the predominant phenolic acids in alkaline hydrolyzates of barley. Recently, Yu et al. (2001) reported the presence of protocatechuic, vanillic, chlorogenic and ferulic acids in acid hydrolyzate of barley. In addition, *p*-hydroxybenzoic, caffeic, and *p*-coumaric acids have been found in the hydrolyzates obtained following treatment of barley

TABLE 2.6
Content of Free Phenolic Acids in Developing Barley Seeds[a]

Phenolic Acid	Development after Anthesis (Days)		
	19	31	42
Salicylic	0.85	0.28	0.06
p-Hydroxybenzoic	5.65	2.76	1.82
Vanillic	2.11	0.56	0.11
o-Coumaric	0.31	0.12	0.11
Protocatechuic	0.62	0.23	0.30
m-Coumaric	0.76	0.21	0.12
Syringic	0.92	1.72	0.36
p-Coumaric	1.86	0.84	0.23
Ferulic	14.80	12.72	2.58
Sinapic	0.28	0.13	0.05

[a] Grams per kilogram of dry weight.

Source: Adapted from Slominski, B.A., 1980, *J. Sci. Food Agric.*, 31:1007–1010.

TABLE 2.7
Percentage Distribution of Ferulic and *p*-Coumaric Acids and DiFA[a] in Barley Grain Fractions Separated Mechanically

Variety	Grain Fraction	Fraction	Ferulic Acid[b]	*p*-Coumaric Acid[b]	DiFA[a,b]
Boira	I[c]	47.5	77.7	78.0	83.8
	II[d]	52.5	22.3	22.0	16.2
Iranis	I	54.4	78.7	81.4	81.1
	II	45.6	21.3	18.6	18.9
Volga	I	47.6	82.3	86.3	86.0
	II	52.4	17.7	13.7	14.0

[a] Ferulic acid dehydrodimers.
[b] Percent = (grams in fraction/gram in grain) × 100.
[c] Husk and outer layer.
[d] Combined intermediate and endosperm fractions.

Source: Adapted from Hernanz, D. et al., 2001, *J. Agric. Food Chem.*, 49:4884–4888.

with acid and α-amylase and also in hydrolyzates obtained after treatment of barley with acid, amylase, and cellulase. The phenolic acids are mainly located in the outer layers of the grain (Nordkvist et al., 1984), which contain 77.7 to 82.3 and 79.2 to 86.8% of the total amounts of ferulic and *p*-coumaric acids in barley grain, respectively (Table 2.7) (Hernanz et al., 2001).

Barley bran contains the same phenolic acids together with syringic and *o*-coumaric acids (Van Sumere et al., 1958). On the other hand, Nordkvist et al. (1984) reported the presence of vanillic, *p*-coumaric, ferulic and diferulic acids in husk, aleurone layer and endosperm of barley. The highest content of total insoluble bound phenolic acids (0.6 to 0.9%) is found in husk, testa and aleurone cells of barley. The endosperm contains only trace amounts of insoluble bound phenolic acids. Ferulic acid is the predominant insoluble bound phenolic acid found in abraded fraction of barley grain enriched with aleurone layer; *p*-coumaric acid is dominant in the fraction containing a high level of husk. The walls of aleurone cells are rich in arabinoxylans and ferulic acid is known to be linked to this cell wall constituent (Mueller-Harvey et al., 1986; Smith et al., 1981). On the other hand, *p*-coumaric acid is known to form linkages with lignins (Higuchi et al., 1967) and cell walls of husk are considerably more lignified than other parts of the kernel (Salomonsson et al., 1980). Furthermore, *p*-coumaryl conjugate of agmatine (decarboxylated arginine) has been identified in germinated barley grains (Stoessl, 1965).

The content of ferulic acid, as determined in 11 barley varieties, ranges from 359 to 624 mg/kg of dry weight, while the level of *p*-coumaric acid is between 79 and 260 mg/kg of dry weight (Hernanz et al., 2001). Barley bran contains 6401 mg/kg of ferulic acid (free and bound) and 151 mg/kg of *p*-coumaric acid (Renger and Steinhart, 2000).

A number of ferulic acid dehydrodimers (DiFA) have also been identified in barley. These include 8-*O*-4′-DiFA, 8,8′-DiFA, 8,8′-aryl DiFA, 5,5′-DiFA, 8, 5′-DiFA benzofuran form and 8, 5′-DiFA open form (Renger and Steinhart, 2000; Stewart et al., 1994). Of these, 8-*O*-4′-DiFA is most abundant in the grain (73 to 118 mg/kg of dry weight) (Hernanz et al., 2001). The husk and outer layer contain from 79.2 to 86.8% of the total DiFA amount in barley grain (Table 2.7) (Hernanz et al., 2001). According to Renger and Steinhart (2000), 8,8′-aryl DiFA (178 mg/kg), 5,5′-DiFA (197 mg/kg), 8-*O*-4′-DiFA (163 mg/kg) and 8,5′-DiFA (135 mg/kg) are the main DiFA in barley bran (Table 2.8) (Figure 2.1). The distributions of DiFA in barley grain fractions are similar to those reported for wheat (Lempereur et al., 1997; Renger and Steinhart, 2000) and rye (Andreasen et al., 2000a, b; Renger and Steinhart, 2000). In plant material, ferulic acid dehydrodimers provide cross links between cell wall polymers (Ishii and Hiroi, 1990) that may contribute to cell wall extensibility (Kamisaka et al., 1990) and cell–cell adhesion (Waldron et al., 1997). In addition, ether glucosides, namely, *p*-hydroxybenzoic acid 4-*O*-β-glucoside, vanillic acid 4-*O*-β-glycoside, *o*-coumaric acid 2-*O*-β glucoside and ferulic acid 4-*O*-β-glucoside have been found in mature barley grains (Van Sumere et al., 1972).

Other Phenolics

The colored barley grains contain at least eight pigmented compounds that have been indentified as derivatives of cyanidin, delphinidin, and pelargonidin, as well as cyanidin-3-arabinoside and, possibly, cyanidin-3-glucoside (Briggs, 1978). Barley grains also contain a range of flavanols from monomeric, dimeric, and trimeric proanthocyadins to higher molecular weight condensed tannins. These include (+)-catechin, (–)-epicatechin, dimeric prodelphinidin B3 and procyanidin B3, as well

TABLE 2.8
Content of Ferulic Acid (FA) and Its Dehydrodimers (DiFA) in Cereal Brans[a,b]

Compound	Barley	Maize	Oats	Rice	Rye	Millet	Wheat
FA	6289	9440	3282	3823	9193	18047	18884
8–8′aryl DiFA	178	279	96	57	162	1313	147
8–8′DiFA	71	67	47	15	188	393	127
8–5′DiFA	135	175	84	32	117	322	146
8-O-4'DiFA	163	255	107	41	124	374	182
5–5′DiFA	197	239	132	36	123	281	205

[a] After alkaline treatment with 4 mol/L NaOH (ester + ether linked).
[b] Milligrams per kilogram.

Source: Adapted from Renger, A. and Steinhart, H., 2000, *Eur. Food Res. Technol.*, 211:422–428.

TABLE 2.9
Flavanol Content in Three Barley Varieties Grown in Ireland in 1981 (mg/kg)[a]

Barley Variety	(+)-Catechin	Procyanidin B3	Prodelphinidin B3	Trimers[b]	Total Polyphenols[c]
Ark Royal	41	202	281	464	1460
Emma	29	130	218	388	1220
Triumph	25	138	186	413	1300

[a] Milligrams per kilogram.
[b] Sum of four components separated by HPLC.
[c] Measured with Folin-Ciocalteu reagent.

Source: Adapted from McMurrough, I. et al., 1983, *J. Sci. Food Agric.*, 34:62–72.

as trimeric procyanidin C2 and three trimeric prodelphinidins (Goupy et al., 1999; McMurrough et al., 1983, 1996). The oligomeric proanthocyanidins are made of linked units of (+)-catechin and/or (+)-gallocatechin (Delcour and Tuytens, 1984; McMurrough et al., 1996).

Table 2.9 shows the content of flavanols in three barley varieties grown in Ireland, following their extraction into acetone-water. The monomeric, dimeric and trimeric flavanols collectively account for 58 to 68% of the total polyphenolic content (McMurrough et al., 1983). On the other hand, Griffith and Wayne (1982) reported that the content of water-extracted flavanols in 21 winter and spring barley varieties ranges from 0.11 to 0.17% and also found that prolonged storage of barley grains does not affect the content of flavanols. In addition, Overland et al. (1994) reported no detectable level of proanthocyanidin in the proanthocyanidin-free mutants of barley varieties, although the corresponding parent barley varieties contain from 0.19 to 0.27% of proanthocyanidins as determined by the vanillin-HCl assay. This mutation does not alter the chemical composition of grains significantly.

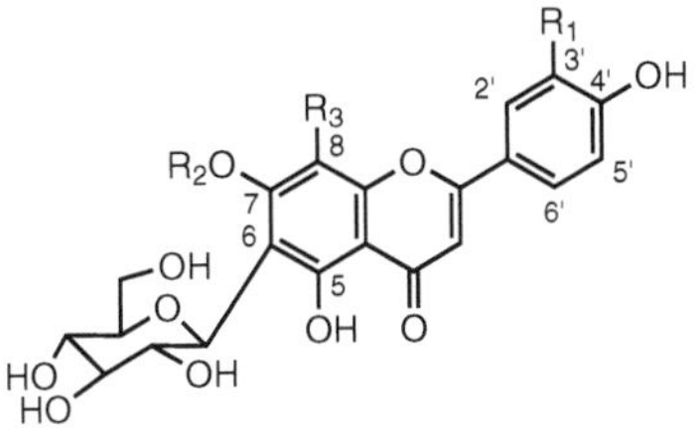

Compound	R_1	R_2	R_3
Isoscoparin-7-*O*-β-D-glucoside	OCH_3	β-glucoside	H
Carlinoside	OH	H	α-arabinoside
Schaftoside	H	H	α-arabinoside
Isoorientin-7-*O*-β-D-glucoside	OH	β-glucoside	H
Isovitexin-7-*O*-β-D-glucoside	H	β-glucoside	H

FIGURE 2.4 Chemical structures of some flavone C-glycosides isolated from barley leaves.

Osawa et al. (1992) and Kitta et al. (1992) isolated a novel flavone C-glycoside, 2″-*O*-glycosylisovitexin, from green barley leaves and reported that its antioxidative activity is similar to that exhibited by α-tocopherol. Later, several more flavone C-glycosides, namely, isoscoparin 7-*O*-β–D-glucoside, carlinoside, shaftoside and 7-*O*-β–D-glucosides of isoorientin and isovitexin (Figure 2.4), were identified in young barley leaves (Norbaek et al., 2000). According to Doll et al. (1994) and Christensen et al. (1998), these phenolics may contribute to resistance against leaf diseases.

Effect of Processing

Roasted barley has been used as a component of daily beverage in China for many centuries. Roasting barley for 1 min at 327 to 342°C reduces the content of catechins in the grain by more than 65%. An increase in the roasting temperature from 327 to 342°C does not have any significant effect on the content of catechins (Duh et al., 2001).

A novel pigment, hordeumin, is produced as a secondary product during ethanol fermentation of uncooked barley (Deguchi et al., 2000a; Ohba et al., 1992). Hordeumin is a product of oxidative polymerization of anthocyanins and tannins (Deguchi et al., 1999). According to Ohba et al. (1992), it contains cyanidin and delphidin. Recently, Deguchi et al. (2000b) demonstrated that hordeumin possesses antimutagenic activity.

Brewer's spent grain, a solid barley residue from the brewing process (after mashing and filtration), contains five times more ferulic and *p*-coumaric acids, as well as from two to five times more ferulic acid dehydrodimers (DiFA), than unprocessed grains. However, the ratio of *p*-coumaric to ferulic acids in spent grain is similar to that of unprocessed barley. The spent grain contains 1860 to 1948 mg ferulic acid and 565 to 794 mg *p*-coumaric acid per kilogram of sample (Hernanz et al., 2001). Also, a marked increase in the ratio of DiFA to ferulic acid has been observed, thus indicating that some monomer forms of ferulic acid are solubilized during the mashing process. Bartolome et al. (1997) evaluated spent grains of barley as a potential source of ferulic acid and reported that the treatment of spent grain with esterase from *Aspergillus niger* (FAE-III) releases only 3.3% of the total ferulic acid present. The addition of xylanase from *Trichoderma viride* to the reaction mixture increases this amount to 30%.

Buckwheat

Buckwheat (*Fagopyrum esculentum* Moench) belongs to the *Polygonaceae* family that includes plants such as rhubarb and Japanese indigo (Watanabe, 1998). Buckwheat is a minor crop compared with other cereals. In western Canada, buckwheat is considered an important alternative crop and is marketed to Japan to be used for production of soba noodles (Mazza, 1993). In the U.S., buckwheat endosperm is used for to manufacture pancake flour. Some buckwheat is also sold in the form of groats for use as breakfast cereal. The production of groats involves dehulling seeds that are adjusted to a moisture content of 22% and then subjected to 10- to 20-min heating at 150 to 164°C. The groats can be additionally roasted at 100 to 150°C for 1 to 2 h (Dietrych-Szostak and Oleszek, 1999).

Buckwheat has been considered a good dietary source of rutin (Ohara et al., 1989; Ohsawa and Tsutumi, 1995). The daily intake of 100 g of buckwheat would provide about 10% of a therapeutic dose of rutin of 180 to 350 mg/day (Schiller et al., 1990). The reported health effects of rutin include antiinflammatory, anticarcinogenic, estrogen receptor binding, smooth muscle relaxation activities (Pisha and Pezzuto, 1994) and ability to lower the fragility of blood vessels (Iwata et al., 1990; Yildzogle-Ari et al., 1991). According to Deschner (1992), dietary rutin and quercetin, under low dietary fat intake, may significantly suppress colonic neoplasia.

Phenolic Acids

The chemistry of the common phenolics of buckwheat seed and product differs markedly from that of other cereal products. In buckwheat, the content of ferulic and hydroxycinnamic acids is low. Bran-aleurone fraction of buckwheat contains bound syringic, *p*-hydroxybenzoic, vanillic, and *p*-coumaric acids. These acids can be liberated by alkaline or acid hydrolysis, thus indicating the possible presence of phenolic acids in the form of esters and glycosides (Durkee, 1977). Zadernowski et al. (1992) have identified 20 and 14 phenolic acids in buckwheat groats and hulls, respectively. Of these, *p*-coumaric (4.6 mg/100 g), vanillic (1.7 mg/100 g), *p*-hydroxybenzoic (1.7 mg/100 g), and caffeic (1.3 mg/100 g) acids are the predominant phenolic acids in the groats; *p*-coumaric (3.6 mg/100 g), vanillic (1.65 mg/100 g), sinapic (1.4 mg/100 g) and gentisic (1.1 mg/100 g) acids are the major phenolic acids in the hulls. Recently, Watanabe et al. (1997) isolated, by preparative HPLC, and identified protocatechuic acid (13.4 mg/100 g) and 3,4-dihydroxybenzaldehyde (6.1 mg/100 g) in ethanolic extracts of buckwheat hulls.

(a)

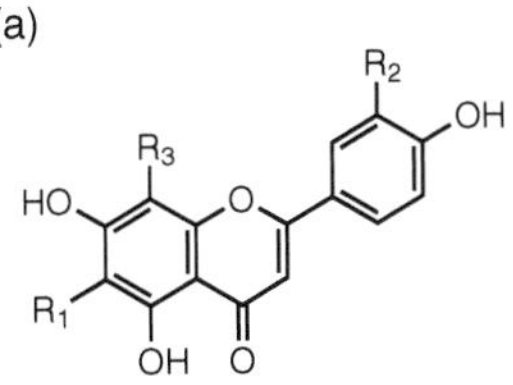

Compound	R_1	R_2	R_3
Vitexin	H	H	C-glucosyl
Orientin	H	OH	C-glucosyl
Isoorientin	C-glucosyl	OH	H
Isovitexin	C-glucosyl	H	H
Glucosylvitexin	H	H	C-glucoglucosyl
Glucosylorientin	H	OH	C-glucoglucosyl

(b)

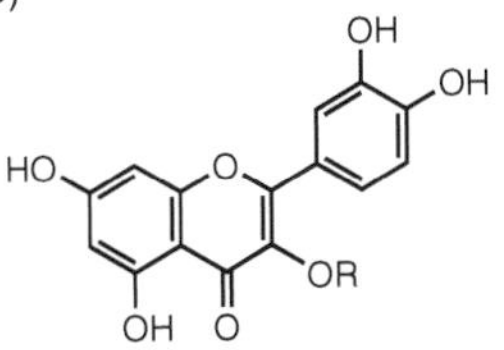

Compound	R
Quercetin	H
Rutin	rutinose
Quercetrin	rhamnose
Hyperin	galactose

FIGURE 2.5 Chemical structures of (A) flavone and (B) flavonol derivatives.

Flavonoids

Buckwheat seeds serve as a rich source of flavonoids (Ohara et al., 1989; Ohsawa and Tsutsumi, 1995; Oomah and Mazza, 1996; Oomah et al., 1996). The seeds and hulls of Canadian buckwheat varieties contain, on average, 387 and 1314 mg/100 g of flavonoids, respectively (Oomah and Mazza, 1996). On the other hand, the total flavonoid content in the seeds and hulls of Polish buckwheat variety are 18.8 and 74 mg/100 g of dry matter, respectively (Dietrych-Szostak and Oleszek, 1999). Flavonols such as rutin, hyperin, quercitrin and quercetin (Sato and Sakamura, 1975) and flavones such as vitexin, isovitexin, orientin, and isoorientin have been identified in immature buckwheat seeds (Margna and Margna, 1982). Only six flavonoids have been identified in mature seeds and their hulls. Rutin and isovitexin are the only flavonoids of seeds, although the hulls contain rutin, quercetin, orientin, vitexin, isovitexin and isoorientin (Figure 2.5) (Dietrych-Szostak and Oleszek, 1999). In addition, Watanabe (1998) isolated and identified four catechins, namely, (–)-epicatechin, (+)-catechin 7-*O*-β-D-glucopyranoside, (–)-epicatechin 3-*O*-*p*-hydroxybenzoeate, and (–)-epicatechin 3-*O*-(3,4-di-*O*-methyl)gallate in the ethanolic extracts of buckwheat groats (Figure 2.6).

The concentration of flavonoids in buckwheat is affected by several factors, including location, growth conditions and variety. According to Oomah and Mazza (1996) and Oomah et al. (1996), geographic location is the major factor responsible for variation in the total flavonoid and rutin contents in seeds; however, growing conditions also have a significant impact on the total content of flavonoids in the hulls. The hull fraction comprises 25.6 to 34% of the seed weight (Oomah and Mazza, 1996). The content of rutin in buckwheat seed may range from 12.6 to 51.1 mg/100 g (Kitabayashi et al., 1995; Oomah and Mazza, 1996), while in the hulls it may range from 4.3 to 85.3 mg/100 g (Dietrych-Szostak and Oleszek, 1999;

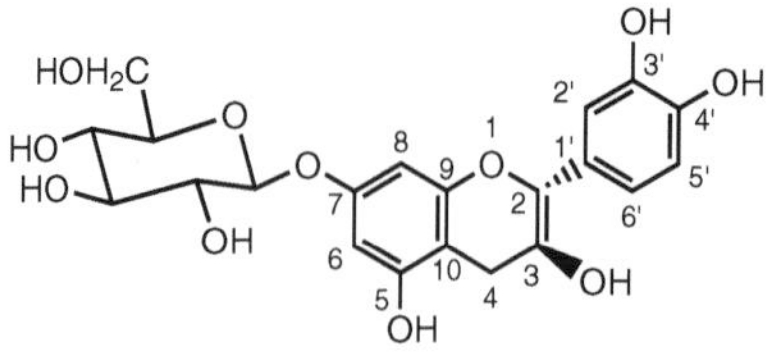

(+)-Catechin 7-*O*-β-D-glucopyranoside

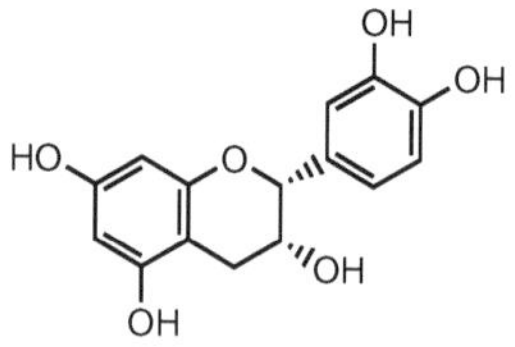

(-)-Epicatechin

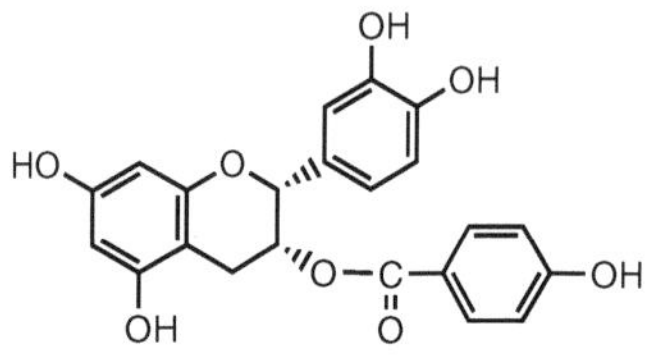

(-)-Epicatechin 3-*O*-*p*-hydroxybenzoate

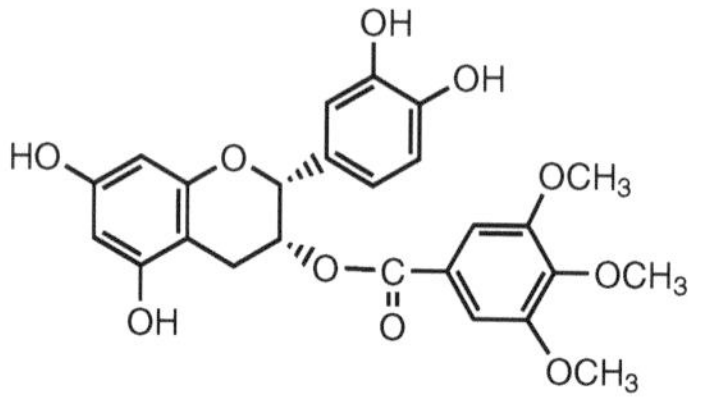

(-)-Epicatechin 3-*O*-(3,4-di-*O*-methyl)gallate

FIGURE 2.6 Chemical structures of catechins found in buckwheat groats.

Oomah and Mazza, 1996; Watanabe et al., 1997). The content of hyperin and quercetin in buckwheat hulls is 5.0 and 2.5 mg/100 g, respectively (Figure 2.5).

Soluble condensed tannins of buckwheat are based on pelargonidin and cyanidin structures (Durkee, 1977). The presence of condensed tannins in buckwheat has also been confirmed by Luthar and Kreft (1996). Later, Watanabe et al. (1997) detected the presence of condensed tannins in two fractions of buckwheat phenolics separated on a Sephadex LH-20 column. Gel permeation chromatography of acetyl derivitaves of these fractions indicated that buckwheat condensed tannins are a mixture of proanthocyanidins with various degrees of polymerization.

Effect of Processing

Kreft et al. (1999) determined the content of rutin in the husk, bran and flour milling fractions of Siva variety of buckwheat seeds grown in Slovenia. The bran fraction contains 13.1 to 47.6 mg rutin/100 g, the flour fraction 1.9 to 16.8 mg rutin/100 g and the husk fraction 2.9 mg rutin/100 g. Heat treatment of seeds prior to dehulling affects the content of rutin in the groats considerably. Untreated groats contain 18.8 mg rutin/100 g, while the content of rutin in heat-treated groats ranges from 17.76 mg/100 g (10 min at 150°C) to 11.6 mg/100 g (2 h and 10 min at 150°C). Steaming the seeds for 20 min at 164°C, followed by 50-min treatment with steam at 150°C, reduces the rutin content by almost 75% (Dietrych-Szostak and Oleszek, 1999).

Corn

Production of corn in North America constitutes more than half of the world's supply (Benson and Pearce, 1987). In the U.S. and Canada corn is primarily used as a feed

TABLE 2.10
Total Phenolic Acid Content in Some Cereal Flours[a]

Phenolic Acid	Wheat	Rice	Oats	Corn
p-Hydroxybenzoic	Trace	5.0	1.4	1.3
(*p*-Hydroxyphenyl) acetic	—	Trace	0.4	1.1
Vanillic	3.6	2.1	4.2	3.7
Protocatechuic	Trace	—	0.5	3.0
Syringic	4.2	0.2	5.3	11.5
Cis-p-Coumaric	—	Trace	—	Trace
Trans-p-Coumaric	Trace	1.3	2.0	18.9
Cis-Ferulic	Trace	1.9	2.6	6.5
Trans-Ferulic	63.6	75.1	63.7	258.6
Caffeic	Trace	—	2.6	4.5
Cis-Sinapic	Trace	Trace	Trace	—
Trans-Sinapic	Trace	Trace	4.3	Trace
Total	71.4	85.6	87.0	309.1

[a] Milligrams per kilogram.

Source: Adapted from Sosulski, F. et al., 1982, *J. Agric. Food Chem.*, 30:337–340.

crop, but it can also be a food crop, especially in the developing countries. A number of food ingredients are produced from corn, among them corn grits, corn meal and corn flour. Corn flour is an ingredient of ready-to-eat breakfast cereals as well as a component in numerous dry mixes such as those for pancakes, muffins and doughnuts (Rooney and Serna-Saldivar, 1987).

Phenolic Acids

Yellow dent corn flour contains 309.1 mg/100 g of phenolic acids (Sosulski et al., 1982). Corn flour contains three times more phenolic acids than cereal flours obtained from rice, wheat and oats (Table 2.10). Phenolic acids of corn are in the free, esterified and insoluble bound forms. Of these, the insoluble bound phenolic acids are the predominant fraction, constituting 69.2% of the total amount of phenolic acids (Grabber et al., 2000; Sosulski et al., 1982).

Some phenolic acids may be linked covalently to amine functionalities through a "pseudo peptide" bond (Figure 2.7). Presence of feruoylputrescine, diferuloylputrescine, *p*-coumarylspermidine, diferuloylspermidine and diferuloylspermine has been reported in the embryo and endosperm of corn (Martin-Tanguy et al., 1982). Furthermore, *p*-coumaryltryptamine and feruloyltryptamine have been isolated from whole corn grains (Ehmann, 1974). Certain physiological roles have been ascribed to some of these conjugates. According to Martin-Tanguy et al. (1978, 1982), phenolic acids coupled to alkyl amines may exhibit antiviral activity and may also serve as biochemical markers closely associated with pollen and ovule fertility. Three novel ferulic acid glycosides have been isolated from acid hydrolysis of corn hulls. The structures of these feruloylated disaccharides have been

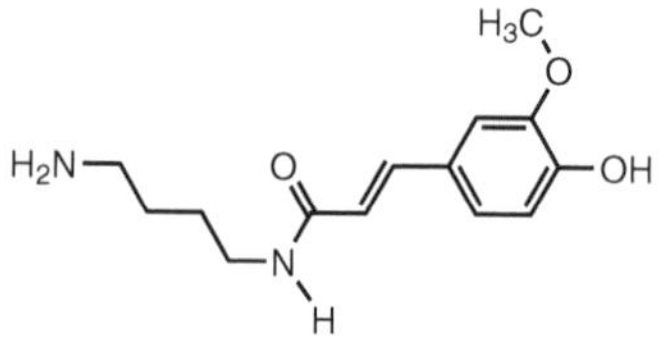

FIGURE 2.7 Chemical structures of some phenolic acid conjugates of amines found in corn.

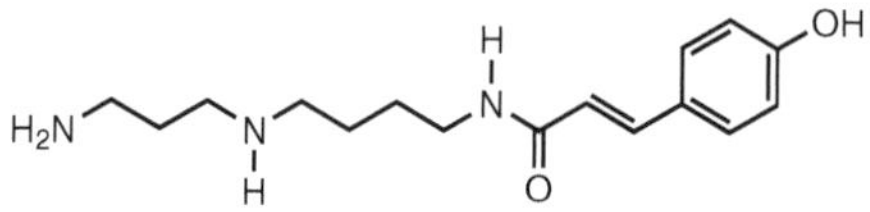

identified as *O*-(2′-*O*-*trans*-feruloyl-α-L-arabinofuranosyl-(1→3)-β-D-xylopyranose, *O*-[2′-*O*-methoxy-5′-*O*-(E)-feruloyl]- α-L-arabino-furanosyl-(1→3)-β-D-xylopyranose, and *O*-[2′-*O*-methoxy-5′-*O*-(Z)-feruloyl]- αL-arabinofuranosyl-(1→3)-β-D-xylopyranose (Hosny and Rosazza, 1997).

Stanols and sterols ferulates such as sitostanyl and campestanyl ferulates as well as sitosteryl and campesteryl ferulates have been found in corn; sitostanyl ferulate is the predominant ferulate in corn (Figure 2.8). The total concentration of these ferulates in six samples of corn ranges from 31 to 70 mg/kg. These compounds are mostly located in the inner pericarp (Seitz, 1989). Norton (1994) isolated and identified 16 derivatives of sterols esterified to *p*-coumaric and ferulic acids; the content of these esters in corn bran was between 70 and 540 mg/kg of bran and ranged from 180 to 8600 mg/kg of unrefined corn oil (Norton, 1995).

Other Phenolics

Phlobaphene and anthocyanidin pigments are commonly found in maize (Styles and Ceska, 1975); pelargonidin-3-glucoside and cyanidin-3-glucoside have been found in the aleurone tissue of maize (Harborne and Gavazzi, 1969). On the other hand, maize flowers contain cyanidin 3-glucoside, cyanidin 3-(6″-malonylglucoside), cyanidin 3-(3″,6″-dimalonylglucoside), peonidin 3-glucoside, peonidin 3-(6″-malonylglucoside), and peonidin 3-(dimalonylglucoside) (Fossen et al., 2001). Phlobaphene pigments are located mostly in the cob and pericarp tissues, whereas anthocyanidins are found in all tissues. Synthesis of phlobaphenes in maize plants is controlled genetically by P locus (Styles and Ceska, 1977). Moreover, monomeric flavan-3,4-diols, namely, leucopelargonidin and leucocyanidin, have been isolated from the aleurone tissue (Reddy, 1964). Presence of two flavonols, kaempferol and quercetin, has also been reported in hydrolyzates of aleurone tissue of corn (Kirby and Styles, 1970). In addition, 4-vinylguaiacol and 4-vinylphenol have been identified in volatile components from tortilla chips. Of these, only the 4-vinylguaiacol (1.5 mg/kg) is

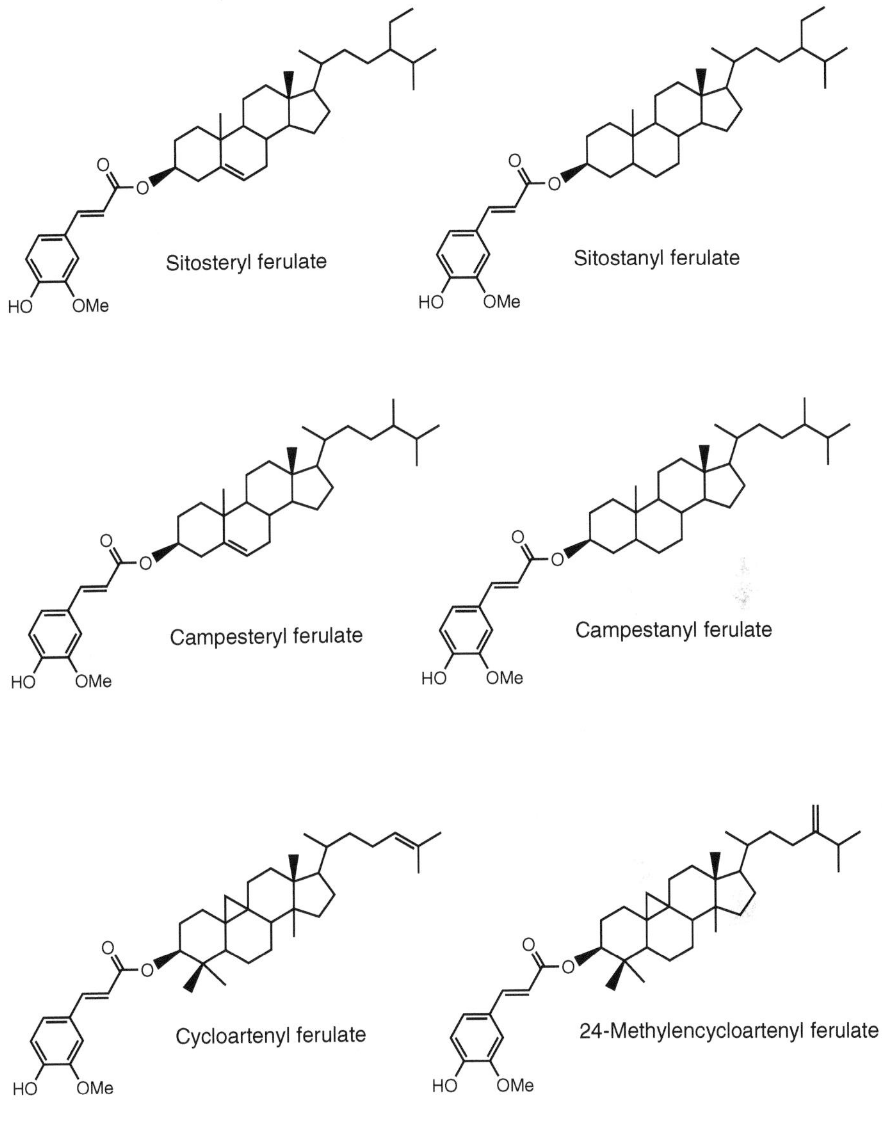

Sitosteryl ferulate

Sitostanyl ferulate

Campesteryl ferulate

Campestanyl ferulate

Cycloartenyl ferulate

24-Methylencycloartenyl ferulate

FIGURE 2.8 Chemical structures of some ferulates of cycloartenols, stanols, and sterols.

present at levels exceeding the threshold value, and therefore may contribute to the flavor of tortilla chips (Buttery and Ling, 1998).

Millets

The term millet is used for any small-seeded annual grass that is of minor importance in the West but regarded as a staple food in Africa and Asia (Schery, 1963). Five millets, namely, *Setaria italica, Pennisetum glaucum* (also known as *Pennisetum typhoideum*), *Eleusine coracana*, *Echinochloa frumentacea*, and *Panicum miliaceum*, are common. *Panicum miliaceum* is also known as Proso millet, *Pennisetum typhoideum* as pearl millet and *Eleusine coracana* as finger millet.

Pearl millet is a major source of protein, essential minerals and calories in many developing countries. It is widely grown in Africa and Asia as a food grain (Desai and Zende, 1979); however, considerable quantities of phytic acid and polyphenols found in millet (Mahajan and Chauhan, 1987) lower the digestibility of millet proteins and starches (Khetarpaul, 1988) as well as the availability of its minerals (Mahajan and Chauhan, 1988). Polyphenols are also responsible for the gray pigmentation of grains, which may be objectionable (Reichert et al., 1980b). The total content of polyphenols in the millet grain may range from 0.5 to about 0.8% (Deokar, 1987; Khetarpaul and Chauhan, 1990). The content of ferulic acid dehydrodimers (DiFA) (Figure 2.1) in millet bran is up to 10 times higher than that of those found in rice, maize, oat, rye and wheat bran. The 8,8′-aryl DiFA (1313 mg/kg) is the most abundant DiFA, comprising about 50% of the total content of DiFA in millet bran (Table 2.8) (Renger and Steinhart, 2000).

Millet pigments are methanol soluble and pH sensitive. The color of millet flour paste can be changed reversibly from gray to yellow-green at alkaline pH and partially and reversibly from gray to creamy white in the presence of acid. Glucosylvitexin, glucosylorientin, and vitexin, a flavonoid compound, (Figure 2.5) have been found to be responsible for the yellow-green discoloration of millet flour paste in the presence of base. These pigments are also known as C-glycosylflavones and are characterized by having a sugar bound via carbon–carbon bond to the flavonoid nucleus. This sugar group is not hydrozylable by normal acid and enzymatic hydrolysis procedures. Depending on the variety, the whole millet may contain from 87 to 259 mg/100 g of C-glycosylflavones expressed as glycosylvitexin equivalents (Reichert et al., 1980b). It has also been found that glucosylvitexin, glucosylorientin and vitexin are present in pearl millet in ratios of 29:11:4, respectively (Reichert, 1979).

Pearl millet varieties also contain substantial quantities of alkali-labile ferulic acid (ALFA) ranging from 133 to 241 mg/100 g. It has been suggested that ferulic acid is esterified to glucose, quinic acid or other sugars or amino acids and it may be responsible for pH-sensitivity of methanol-extracted pearl millet flours (Reichert et al., 1980b).

The concentration of C-glucosylflavones and ALFA not only depends on the pearl millet variety but is also affected by dehulling processes. The greater the degree of abrasive dehulling of millet grain the lower is the content of these phenolics (Figure 2.9) in the dehulled millets (Reichert et al., 1980b).

The gray pigmentation of pearl millet is probably due to more complex interactions involving chelation of phenolics *in vivo* with ions of Cu, Fe, Al, or other metals. In addition, copigmentation effects that enhance colors may be due to changes in the degree of ionization of the phenolics (Reichert, 1979).

Proso millet may be used as puffed or cooked breakfast cereal or as a replacement for wheat flour in certain baked products (Hinze, 1972). It contains 0.055 to 0.178% tannins, expressed as (+)-catechin equivalents. Dark-colored seeds have the highest content of tannins. The dehulling process reduces the content of tannins by 65 to

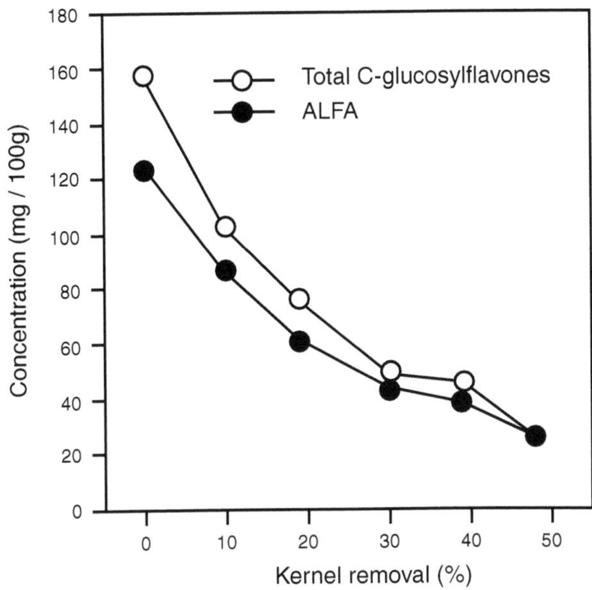

FIGURE 2.9 Concentrations of phenolics in dehulled, methanol-extracted millet flour. (Adapted from Reichert, R.D., 1979, *Cereal Chem.*, 56:291–294.)

80% because the hulls contain 15 to 40 times more tannins than the dehulled grains (Lorenz, 1983).

Finger millet is an important food crop in India and Africa. Protocatechuic, gallic, and caffeic acids were the predominant free phenolic acids, while ferulic, caffeic, and coumaric acids were the major bound phenolic acids found in finger millets (Subba Rao and Muralikrishna, 2002). This cereal grain contains a much higher level of tannins than proso millets. Of the 19 Indian and 10 African finger millet varieties examined, two Indian cultivars contained approximately 1% tannins while two African cultivars had a tannin content of over 3%. High levels of tannins in finger millets with dark seed coat have been linked with poor protein digestibility; these tannins have been found mostly in the glutelin fraction of finger millet proteins (Ramachandra et al., 1977). The dehulling process removes 88.1 to 100% of tannins originally present in grains, depending on the seed variety (Ramachandra et al., 1977). On the other hand, 24-h fermentation of finger millet flour with endogenous grain microflora reduces the content of phenols and tannins by 20 and 52%, respectively (Antony and Chandra, 1998); while 96-h of malting of finger millet seeds brought about 3-fold and 2-fold decrease in protocatechuic and bound phenolic acid content, respectively (Subba Rao and Muralikrishna, 2002).

Japanese barnyard millet is a traditional food in colder parts of Japan. It is a rich source of protein, lipid, vitamins B_1 and B_2, and nicotinic acid (Watanabe, 1999). This cereal is low in phenolics and tannins (Suman et al., 1992). Recently, Watanabe (1999) identified three phenolic compounds, namely, N-(*p*-coumaroyl)serotonin, luteolin and tricin, in the ethanolic extracts of the Japanese barnyard millet. Of these, N-(*p*-coumaroyl)serotonin exhibited antiinflammatory activity (Kawashima et al., 1998).

Oats

Oats are produced worldwide and differ from other cereals in that the bran is not separated from the endosperm during milling and that they are used primarily as breakfast cereal or as feed for animals. Oat products are also used in granola bars, infant foods, and components of breads and cookies, as well as thickening agents in soups, gravies, and sauces. However, despite broad uses of oats and their products in foods, little work has been done to characterize oat grain phenolics. Recently, an excellent review on phenolics in oat groats and hulls reported the presence of derivatives of benzoic and cinnamic acids as well as quinones, flavones, flavonols, chalcones, flavanones, anthocyanidins and amino phenolics (Peterson, 2001; Peterson et al., 2001). The structures and functionality of some conjugated phenolics of oats were summarized by Collins (1986).

Phenolic Acids

Durkee and Thiverge (1977) reported that ferulic acid is a major phenolic acid present in the soluble bound and insoluble bound phenolic acid fractions of oats. These authors also identified traces of free *p*-coumaric, ferulic and vanillic acids in oats. Presence of vanillic, sinapic, *p*-coumaric, *p*-hydroxyphenylacetic, caffeic, protocatechuic, syringic and *p*-hydroxybenzoic acids in bound forms has also been reported (Durkee and Thiverge, 1977; Sosulski et al., 1982). Later, Zadernowski et al. (1992) detected *o*-coumaric and *p*-phenyllactic acids as well as traces of veratric acid. According to Sosulski et al. (1982), oat flour (Harmon variety) contains approximately 87 mg/kg of total phenolic acids (Table 2.10). Of these, 66.3% are in the insoluble bound form and 20.6% as soluble bound phenolic acids.

Bound phenolic acids of oat may be coupled to long chain ω-hydroxy fatty acid, long-chain alcohols, glycerol, sugars, polysaccharides, lignins and amines (Collins, 1986). For example, 30% of phenolic acid diesters found in oat flour are the derivatives of octacosan-1, 28-diol (Daniels and Martin, 1965). Of these, esterified phenolic acids coupled to n-alkanols possess significant antioxidant activities due to their lipophilic properties (Daniels and Martin, 1961). Approximately 36% of a mixture of oat antioxidants was glycerol esters of phenolic acids (Daniels and Martin, 1968). Oat hulls and groats may also contain catechol, coniferyl alcohol, gallic acid, *p*-hydroxyphenzaldehyde, salicylic acid, and vanillin (Emmons and Peterson, 1999; Xing and White, 1997).

Avenanthramides

Oat groats contain at least 25 avenanthramides and N-acylanthranilate alkaloids, while oat hull extracts contain only 20 of these compounds (Figure 2.10). These compounds are anionically substituted cinnamic acid conjugates (Collins, 1989; Collins and Mullin, 1988). The complete chemical structures of five avenanthramides have been elucidated and identified as avenanthramides. The content of avenanthramides in dehulled oats is 200 to 800 mg/kg (Collins, 1986). Oats contain from 40 to 132 mg/kg of avenanthramide B (Dimberg et al., 1993); oat groats have 25 to 47 mg/kg of avenanthramide A, 21 to 43 mg/kg of avenanthramide B and 28 to

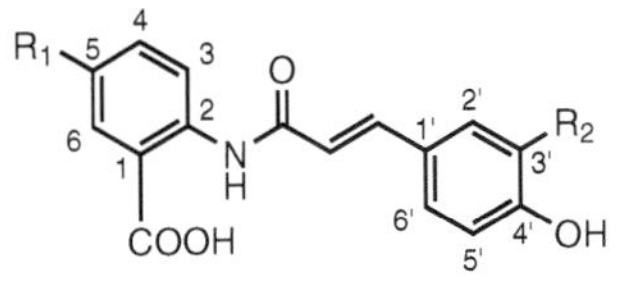

	R_1	R_2
A	OH	H
B	OH	HCH_3
C	OH	OH
D	H	H

FIGURE 2.10 Chemical structures of some avenanthramides.

62 mg/kg of avenanthramide C (Dimberg et al., 1996). In a review, Peterson (2001) recently reported some unpublished results from his laboratory: oat groats contain 54, 36 and 52 mg/kg and hulls 25, 17 and 25 mg/kg of avenanthramides A, B and C, respectively. Presence of nine novel bound phenolic acids in oats was reported by Collins et al. (1991), three of which were identified as avenalumic acid and its 3′-hydroxy and 3′-methoxy derivatives. These phenolics are ethylenic homologues of *p*-coumaric, caffeic and ferulic acids covalently bound to the amine group of hydroxy-substituted *o*-aminobenzoic acid.

Resistant to acid hydrolysis, avenanthramides are slowly hydrolyzed in the presence of alkali, with slight decomposition, to the corresponding substituted cinnamic and anthranilic acids (Collins, 1989). Two avenanthramides, namely, N-(4′-hydroxy-3′-methoxy-(E)-cinnamoyl)-5-hydroxyanthranilic acid (avenanthramide B) and N-(4′-hydroxy-3’methoxy-(E)-cinnamoyl)-5-hydroxy-4-methoxyanthranilic acid (a novel avenanthramide A_2), have been found to be heat stable (Dimberg et al., 1993). Some N-cinnamoylanthranilic acids have been described in the literature to possess antiallergic, antihistaminic and antiasthmatic activities (Devlin and Hargrave, 1985).

Other Phenolics

American varieties of oats contain from 19 to 30.3 mg of total tocols per kilogram (Peterson and Qureshi, 1993), while those grown in Hungary have 15 to 48 mg of total tocols per kilogram (Lasztity et al., 1980). Alpha-tocopherol and α-tocotrienol are the predominant tocols found in oats, contributing 86 to 91% to the total amount (Peterson and Qureshi, 1993). Apigenin, luteolin, and tricin have been identified in oat flour based on their UV spectra, co-chromatography with authentic compounds, mass spectroscopy and effects of diagnostic shift reagents (Collins, 1986).

Effect of Processing

Heating oats with hulls (at 100°C for 10 min) significantly increases the content of vanillic acid, vanillin, *p*-coumaric acid, *p*-hydroxybenzaldehyde and coniferyl alcohol in groats. This increase (25 to 900%) in phenolic content may be due to their release from hulls and their diffussion into the groats during processing (Dimberg et al., 1996).

Emmons et al. (1999) evaluated the content of phenolics in various milling fractions of oats. The total content of phenolics in pearlings and dried oat groat flours ranged from 274 to 342 and from 89 to 156 mg/kg, respectively. Pearling fractions contained higher amounts of avenanthramides (137 to 185 mg/kg) than oat groat flours (39 to 76 mg/kg). The higher levels of phenolics in pearling samples might indicate that more phenolics are located in the seed coat, aleurone and sub-aluerone layers. The presence of avenanthramides in oat hulls has also been reported (Emmons and Peterson, 1999). On the other hand, extrusion of ground oat grains brings about a 50% loss in their content of soluble phenolic compounds (Zadernowski et al., 1999).

Rice

Grown and harvested in more than 100 countries (Luh, 1980), rice is a staple food for over half of the world's population. The milling of rice includes removing hulls to produce brown rice and, following this, removing bran and polishing kernels.

Phenolic Acids

Rice flour contains 856 mg/kg of phenolic acids and in this respect is similar to oat and wheat flour (Table 2.10). Ferulic acid is the major phenolic acid present in rice flour. Approximately 74.2% of the total phenolics in rice are in the insoluble bound form; therefore, phenolic acids have little effect on the color or flavor of aqueous slurries of rice. However, wet processing, especially under alkaline conditions, and cooking or baking may bring about release of some bound phenolics, which will increase their contribution to sensory characteristics of the product (Sosulski et al., 1982). Table 2.8 shows that insoluble bound phenolics have been found to be cell wall components (Harris and Hartley, 1976; Renger and Steinhart, 2000). Moreover, the acids are considered to be present as phenolic–carbohydrate esters because they are released by alkaline solvents. The alkaline extracts of rice endosperm cell walls contain 9.1, 2.5 and 0.56 mg/g of ferulic acid, *p*-coumaric acid and dehydrodimers of ferulic acid, respectively. Dehydrodimers of ferulic acid, a dimeric product of ferulic acid (Figure 2.1), are produced by oxidative coupling (Shibuya, 1984). Rice bran contains a number of sugar–amino acid–phenolic acid conjugates, of which seven contain ferulic acid and at last one an acidic amino acid (Yoshizawa et al., 1970).

Cooked rice has a highly desirable and characteristic odor; however, the specific odor of rice bran is considered an obstacle in its efficient utilization as a food component. Commercially available polished rice contains trace quantities of rice bran, which may affect the flavor characteristics of cooked rice. This effect depends on the quantity of residual bran in rice before cooking. The volatiles of rice bran contain alcohols, carbonyl compounds, carboxylic acids and phenols. Seven phenols (namely, 4-vinylphenol, 4-vinylguaiacol, guaiacol, phenol, *o*-cresol, *p*-cresol and 3,5-xylenol) have been identified in steam volatile concentrate of rice bran. Of these, 4-vinylphenol, with a characteristic unpleasant flavor, is the predominant phenolic component present. Products of thermal decarboxylation of ferulic and *p*-coumaric acids, respectively, 4-vinylphenol and 4-vinylguaiacol can be produced during cooking of rice (Fujimaki et al., 1977).

During the traditional production of parboiled rice, paddy grains are soaked in water consecutively for 2 to 3 days, resulting in a considerable loss of dry matter.

TABLE 2.11
Changes in Phenol Content in Rice, Husk and Bran prior to and after Soaking for 72 h[a]

Rice/Component	Before Soaking	After Soaking
Paddy rice	0.408	0.478
Brown rice	0.276	0.322
Milled rice	0.148	0.418
Husk	1.284	1.361
Bran	2.722	3.792

[a] Grams of phloroglucinol equivalent per kilogram on dry weight basis.

Source: Adapted from Singaravadivel, K. and Raj, S.A., 1979, *J. Food Sci. Technol.*, 16:77–78.

Phenolics are among the components leached out from the grains during soaking. The phenol content of the water in which the grains are soaked increases markedly with the increase in soaking time. This increase is relatively higher in milled and brown rice than in paddy grains. Thus, the intact testa and pericarp might act as barriers for permeation of phenols from whole paddy grains. On the other hand, soaking rice in water also augments the level of phenolic constituents in the grain (Table 2.11) due to their liberation from conjugated phenolic glycosides by the action of β-glucosidase (Singaravadivel and Raj, 1979).

Other Phenolics

Isovitexin, a flavonoid compound also known as C-glucosylflavone, has been reported in rice hulls (Figure 2.5). This flavonoid shows good antioxidant activity and plays an important role in postharvest storage of rice by providing a chemical defense for rice germs against oxidative deterioration (Osawa et al., 1985; Ramarathnam et al., 1988, 1989). Moreover, anisole, *m*-hydroxybenzaldehyde, vanillin (4-hydroxy-3-methoxybenzaldehyde), and syringaldehyde (4-hydroxy-3,5-dimethoxybenzaldehyde) have been identified in methanolic extracts of wild rice hulls (Asamarai et al., 1996). Shin and Godberg (1996) reported that rice bran obtained from mixed long-grain varieties (Tebonnet and Lemont) contained 3.1 mg of oryzanol per kilogram of sample and that irradiation of rice bran at 10 kGy reduced its oryzanol content to 2.4 mg/kg.

Rice bran also contains feruloyl esters of triterpene alcohols and sterols (Figure 2.8) (Akihisa et al., 2000; Diack and Saska, 1994; Norton, 1995). According to Norton (1995), the total content of feruoyl esters (ferulates of cycloartenol, 24-methylenecycloartanol, campesterol, and sitosterol) of rice bran is 3.4 g/kg of rice bran. Recently, Akihisa et al. (2000) identified six novel feruloyl esters of triterpene alcohols and sterols in the methanolic extract of rice bran: cycloeucalenol *trans*-ferulate, 24-methylencholesterol *trans*-ferulate, cycloartenyl *cis*-ferulate, 24-methylenecycloartanyl *cis*-ferulate, sitosteryl *cis*-ferulate and 24-methylcholesterol *cis*-ferulate. These ferulates display good antiinflammatory activities against 12-*O*-tetradecanoylphorbol-13-acetate (TPA)-induced ear inflammation in mice (Akihisa et al., 2000). Moreover, cycloartenol *trans*-ferulate exhibits a strong inhibitory effect against the tumor-promoting activity of TPA in mice skin carcinogenesis initiated by 7,12-dimethylbenz[a]anthracene (DMBA) (Akihisa et al., 1997; Yasukawa and Akihisa, 1997; Yasukawa et al., 1998).

Rye

Because it can be grown under various conditions, rye (*Secale cereale* L.) is one the most widely distributed cereal grains. Rye plant can be grown in sandy soil of low fertility, at high altitudes and in the cool temperate zones, as well as in semiarid regions near deserts. Germany, Poland and Russia are the main rye-producing countries; the rye crop is used as flour for bread and for animal feed (Bushuk, 1976).

Weidner et al. (1999) reported that grains of Polish rye cultivars contained 14.6 to 15.0 mg/kg of free phenolic acids, 89.9 to 151.9 mg/kg of phenolic acids liberated from esters and 16.9 to 17.4 mg/kg of phenolic acids liberated from soluble glycosides. Ferulic, caffeic, *p*-coumaric and sinapic acids are the predominant phenolic acids in rye grains. Andreasen et al. (1999) also reported that ferulic acid (1000 mg/kg of dry matter) is the most abundant phenolic acid in rye grains (cv. Esprit) followed by sinapic (130 mg/kg of dry matter) and *p*-coumaric acids (60 mg/kg of dry matter). Caffeic, vanillic, protocatechuic and *p*-hydroxybenzoic acids are present in rye in minor quantities (<20 mg/kg of dry matter).

Recently, Andreasen et al. (2000b) measured the content of phenolic acids in the kernels of 17 rye varieties grown in one location in Denmark during 1997 and 1998. Their content of ferulic acid ranged from 900 to 1170 mg/kg, sinapic acid between 70 and 140 mg/kg and *p*-coumaric acid from 40 to 70 mg/kg. Andreasen et al. (2001) have also reported that 8-O-4′- DiFA is the predominant ferulic acid dehydrodimer in rye kernel, followed by 8,5′-DiFA benzofuran and 5,5′-DiFA (Figure 2.1). The total content of ferulic acid dehydrodimers in whole rye grain ranges from 241 to 409 mg/kg of dry matter (Andreasen et al., 2000b). Rye bran contains 10 to 15 times more dehydrodimers of ferulic acid than the corresponding flour fraction (Andreasen et al., 2000a). Subsequently, Renger and Steinhart (2000) reported that 8,8′-DiFA is the most abundant dehydrodimer of ferulic acid in rye bran, followed by 8,8′ aryl DiFA, 8-*O*-4′-DiFA and 5,5′-DiFA (Table 2.8). In addition, vanillin has been identified as a contributor to the flavor of sour dough rye bread crumbs (Kirchhoff and Schieberle, 2001). Subsequently, Hakala et al. (2002) investigated the distribution of steryl ferulates in rye. The grain used in this study contained 55 to 64 mg/kg of total steryl ferulates, which are predominantly located in rye bran (150 to 251 mg/kg of total steryl ferulates). Campestanyl and sitostanyl ferulates are the predominant ferulates in rye (Figure 2.8).

Sorghum

Sorghum is an important grain for human consumption in parts of Asia and Africa; in the U.S., it is used primarily as a source of animal feed (Strumeyer and Malin,

TABLE 2.12
Distribution of Tannins in Different Protein Fractions of High-Tannin Sorghum Seeds (IS 2825)

Grain Component	Tannins (%)	Proportion of Total Tannins (%)
Albumin and globulin	0.79	22.96
Prolamin	1.05	30.52
Glutelin	0.50	14.54
Total extraction	2.34	—
Tannin content	3.44	—
Recovery of tannin in protein fractions	68.02	—

Source: Adapted from Chavan, J.K. et al., 1979, *J. Food Biochem.*, 3:13–20.

1975). Sorghum varieties have been subdivided into three groups according to their level of phenolics, which are primarily proanthocyanidins (Butler and Price, 1977; Cummings and Axtell, 1973). Group I sorghums do not have pigmented testa, but contain a low level of phenols and no tannins. Groups II and III sorghums have pigmented testa. Although tannins of group II sorghums may be extracted into acidic methanol (1% HCl) and not into methanol alone, tannins of group III sorghums may be extracted into methanol and acidic methanol. The color and pigmentation of sorghum pericarp is known to be controlled by R, Y, B_1, B_2, and S genes (Nip and Burns, 1969, 1971; Rooney and Miller, 1982). Nearly all sorghums, regardless of their color, contain pigments or pigment precursors. The content and color of these pigments changes with the previously mentioned genes. The R and Y genes determine the color of the pericarp and the B_1 and B_2 genes control the presence of pigmented testa. Dominant genes R and Y greatly increase the levels of anthocyanidin pigments in groups I and II sorghum (Hahn and Rooney, 1986).

Polyphenolic compounds in sorghum include mainly flavonoids, anthocyanidins, flavanols (tannins) and phenolic acids (Butler, 1989; Hahn et al., 1984; Woodhead, 1981). However, sorghum grains are tannin free (Hoseney et al., 1981) or contain trace quantities of hydrolyzable tannins (Bullard and Elias, 1980). The condensed tannins (also called proanthocyanidins) are found in some bird-resistant varieties of sorghum (Butler, 1982, 1989, 1992).

Sorghum tannins readily associate with sorghum proteins. About 68% of total tannins are recovered from albumins and globulins, prolamins and glutelins fractions (Table 2.12) (Chavan et al., 1979b). However, it remains unknown whether such types of tannin–protein complexes exist *in situ* or are formed only during the extraction process.

Phenolic Acids

Tentative identification of free and insoluble bound phenolic acids of *Sorghum bicolor* (L.) Moench has been reported by Hahn et al. (1983). Free phenolic acids

TABLE 2.13
Free and Bound Phenolic Acid Composition of Several Sorghum Varieties[a]

	SC0748		IS2327		SC0719	
Phenolic acid	**Free**	**Bound**	**Free**	**Bound**	**Free**	**Bound**
Gallic	—	13.2	—	—	—	26.1
Protocatechuic	11.0	11.5	7.0	98.4	8.0	15.8
p-Hydroxybenzoic	10.1	23.7	2.3	23.6	9.3	24.2
Vanillic	15.5	—	126.7	—	23.3	27.4
Caffeic	6.0	44.6	6.3	17.0	8.7	26.8
p-Coumaric	109.1	123.0	47.4	53.5	6.4	79.9
Ferulic	74.0	213.0	18.2	139.0	26.0	91.9
Cinnamic	4.7	—	2.0	—	—	9.4

[a] Milligrams per kilogram, dry weight basis.

Source: Adapted from Hahn, D.H. et al., 1983, *Cereal Chem.*, 60:255–259.

have been extracted from sorghum grain with methanol and the residue then subjected to acid hydrolysis in order to liberate bound phenolics. Gallic, protocatechuic, *p*-coumaric, *p*-hydroxybenzoic, vanillic, caffeic, ferulic and cinnamic acids have been identified in sorghum extracts (Table 2.13). Moreover, 12 other peaks have been separated, but not identified.

Protocatechuic, *p*-hydroxybenzoic, caffeic, *p*-coumaric, and ferulic acids are found in free and bound forms. However, cinnamic and vanillic acids are found in free and/or bound forms only in some sorghum varieties. Gallic acid is present only in the bound form at an average content of 23.6 mg/kg (Hahn et al., 1983).

Polyphenolic Composition

A number of proanthocyanidins and leucoanthocyanidins have been identified in sorghum grains. Proanthocyanidins are oligomers of flavan-3-ols; flavan-3,4-diols and flavan-4-ols are also known as leucoanthocyanidins (Figure 2.11). Of these, profisetinidins have been isolated and identified in the pericarp of Martin sorghum (Blessin et al., 1963) and propelargonidins in the pericarp of a "commercial" sorghum (Yasumatsu et al., 1965). Later, luteoforolor or proluteolinidin (3′,4′,5,7-tetrahydroxyflavan-4-ol) was found to be a principal proanthocyanidin of Kafir variety of sorghum (Bate-Smith and Rasper, 1969). Luteolinidin is a major component of red or red-pink pigments as green seeds began to ripen (Figure 2.12) (Bate-Smith and Rasper, 1969; Haslam ,1977). According to Strumeyer and Malin (1975), tannins isolated from Leoti and Georgia varieties are procyanidin in nature. These researchers also showed that these tannins consist of a series of oligomers of flavan-3-ols that, upon hydrolysis, depolymerize to afford cyanidin.

Apiferol (4′,5,7- trihydroxyflavan-4-ol) has been tentatively identified in the leaves and seeds of forage sorghum (Figure 2.12) (Watterson and Butler, 1983). This flavan-4-ol is suspected to be a precursor of red coloration observed in leaf tissue

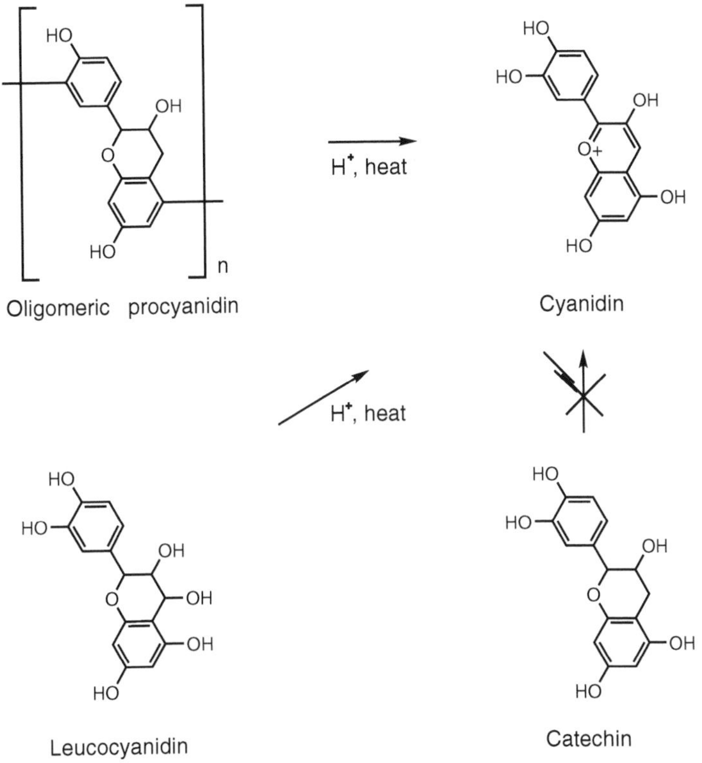

FIGURE 2.11 Conversion oligomeric flavan-3-ols and flavons 3,4-diols to anthocyanidins. (Adapted from Watterson, J.J. and Butler, L.G., 1983, *J. Agric. Food Chem.*, 31:41–45.)

as a result of injury or physiological stress. Gupta and Haslam (1978) identified (+)-catechin, procyanidin dimer B-1, and procyanidin polymer corresponding to (–)-epicatechin units linked to (+)-catechin as a chain termination unit in sorghum. On the other hand, methanolic extracts of a Hungarian sorghum (Szegedioerpe) contain four procyanidins with the basic formula epicatechin-(epicatechin)$_n$-catechin and one procyanidin trimer corresponding to epicatechin-catechin-epicatechin (Figure 2.13) (Gujer et al., 1986).

The pigmentation of sorghum pericarp is mostly due to the presence of anthocyanins and anthocyanidins as well as other flavonoids. Two glucosides of apigenidin (both yellow) and two glucosides of luteolinidin (both orange) were isolated from white pericarp of six sorghum varieties (Nip and Burns, 1971). Eriodictyol (a chalcone) was found to be a predominant pigment in the lemon-yellow pericarp (Kambal and Bate-Smith, 1976), while luteoferol was present in the red pericarp of sorghum (Nip and Burns, 1969). The red sorghum pericarp contained also two yellow pigments identified as apigeninidin-5-glucoside and kaempferol-3-rutinoside-7-glucuronide, one orange pigment only identified as anthocyanin (Nip and Burns, 1969), and two 3-deoxyanthocyanidins: apigeninidin and luteolinidin (Melake-Berhan et al., 1996).

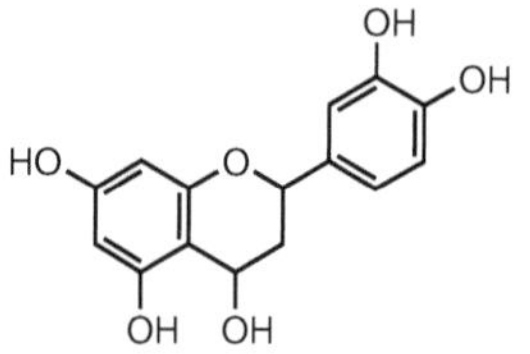

Luteoferol

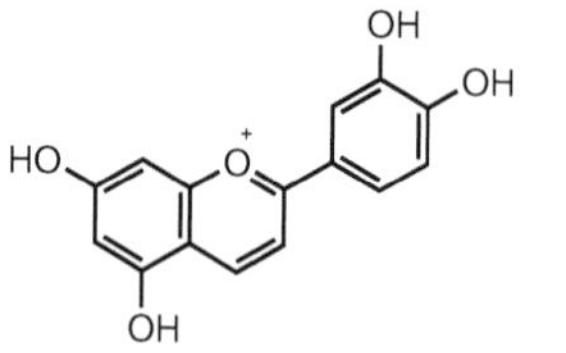

Luteolinidin

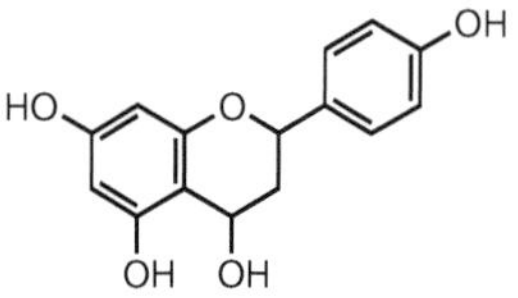

Apiferol

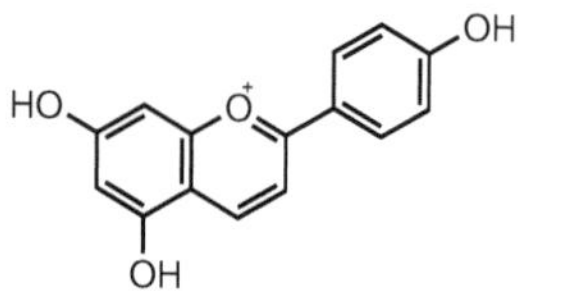

Apigenidin

FIGURE 2.12 Chemical structures of some sorghum flavan-4-ols and corresponding anthocyanidins. (Adapted from Watterson, J.J. and Butler, L.G., 1983, *J. Agric. Food Chem.*, 31:41–45.)

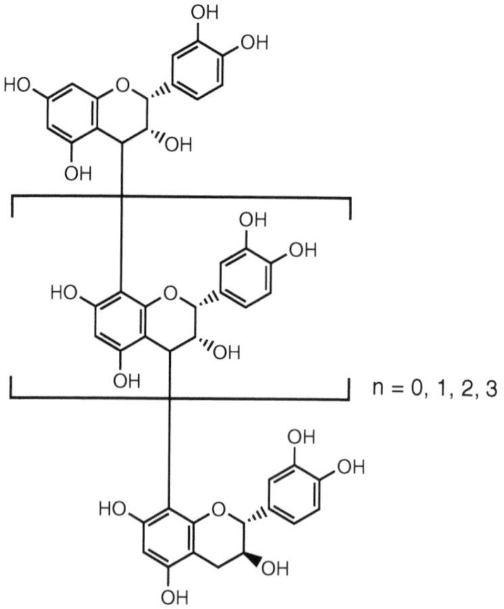

n = 0, 1, 2, 3

FIGURE 2.13 Chemical structures of procyanidins identified in sorghum grains.

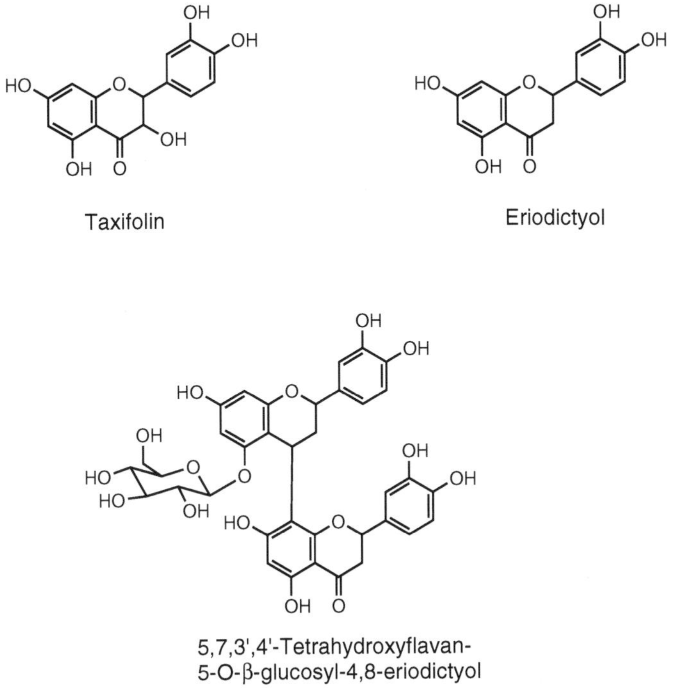

FIGURE 2.14 Chemical structures of some flavonoids isolated from sorghum grains. (Adapted from Gujer, R. et al., 1986, *Prog. Clin. Biol. Res.*, 213:159–161.)

A number of monomeric and oligomeric flavonoids linked by 4,8-interflavan bond have also been found in sorghum grains. The low molecular weight phenolics have been tentatively identified as (+)-catechin, naringenin and eriodictyol (+)-taxifolin, along with three monomeric glycosides (Figure 2.14). Dimeric flavonoids have been identified as 5,7,3′,4′-tetrahydroxyflavan-5-*O*-α-glucosyl-4,8-eriodictyol and 5,7,3′,4′-tetrahydroxyflavan-5-O-α-glucosyl-4,8-eriodictyol-5-O-α-glucoside. A trimeric flavonoid has been identified as 5,7,3′,4′-tetrahydroxyflavan-5-*O*-α-glucosyl-4,8–5,7,3′,4′-tetrahydroxy flavan-5-O-α-glucosyl-4,8-eriodictyol (Gujer et al., 1986). Recently, Krueger et al. (2003) have detected a number of heteropoly-flavan-3-ols linked by A-type and B-type interflavan bonds in Ruby Red sorghum (*Sorghum bicolor* L. Moench). In addition, a series of glucosylated heteropolyflavans containing a flavonone (eriodictoyl or eriodictoyl-5-*O*-β-glucoside) were also identified.

Polyphenolic Content

The content of polyphenolic compounds in sorghum grains depends on a number of factors, including plant type, cultivar, age of plant or plant part, stage of development and environmental conditions (Table 2.13 and Table 2.14). The content of condensed tannins in sorghum, estimated as (+)-catechin equivalents by different research groups, ranged from 3.6 to 10.2% (Harris and Burns, 1970), 2.69 to 6.88% (Jambunathan and Mertz, 1973), 0.11 to 4.25% (Maxson et al., 1972), 0.1 to 8.0%

TABLE 2.14
Content and Distribution of Tannins in Different Parts of Sorghum Grain

Grain Component	Content (g/kg)	Proportion of Total Tannins (%)
Testa	224	81.4
Pericarp	42	15.1
Endosperm and germ	0.8	3.3
Whole grain	36.3	—

Source: Adapted from Reichert, R.D. et al., 1980b, in *Polyphenols in Cereals and Legumes*, International Development Research Centre, Ottawa, Canada, 50–60.

(McMillan et al., 1972), 0 to 3.88% (Elkin et al., 1996) and 0 to 1.97% (Dicko et al., 2002). However, values expressed as tannic acid equivalents varied from 0.31 to 1.5% (Maxson et al., 1972), 0.20 to 0.76% (Maxson and Rooney, 1972), 0.37 to 1.57% (Rostango et al., 1973b), and 0.33 to 1.41% (Rostango et al., 1973a). Moreover, the mold-resistant sorghum genotypes contain higher levels of proanthocyanidins than those of the susceptible lines (Melake-Berhan et al., 1996).

Maturation of sorghum grains brings about changes in their polyphenolic content and composition. These changes may be due to the metabolism or polymerization of phenolic compounds. The content of condensed tannins increases during the early stages of grain development and then remains constant or declines as the grain approaches harvest time (Bullard and Elias, 1980; Chavan et al., 1979b; Davis and Hoseney, 1979).

The period of time at which the maximum level of polyphenolics is produced in the grain depends on the sorghum variety. The variety GA-615 has the maximum level of (+)-catechin equivalents at 16 days, while for double dwarf Feretita, a significant increase in (+)-catechin equivalents is observed at 32 days (Figure 2.15) (Rooney et al., 1980). Price et al. (1979a) found that, for tannin-containing varieties, the maximum values per seed are reached between 25 and 40 days after half anthesis. The decrease in tannin content is attributed to increased polymerization or association with other cellular components (Bullard et al., 1980; Price et al., 1979a). However, Butler (1982) reported that drying seeds at any stage of maturation has a more pronounced effect on the relative degree of tannin polymerization than maturation by itself. On the other hand, depending on the sorghum cultivar, maturation of grain also brings about different changes in polyphenolic compounds. Thus, the content of assayable tannins of some cultivars from group III sorghum is reduced greatly during seed maturation, while those present in the seeds of some cultivars from group II remain unchanged (Asquith et al., 1983).

Effect of Processing

Various attempts have been made to reduce the detrimental effects of sorghum polyphenolics on digestibility of proteins and enzyme inhibition. Sequential

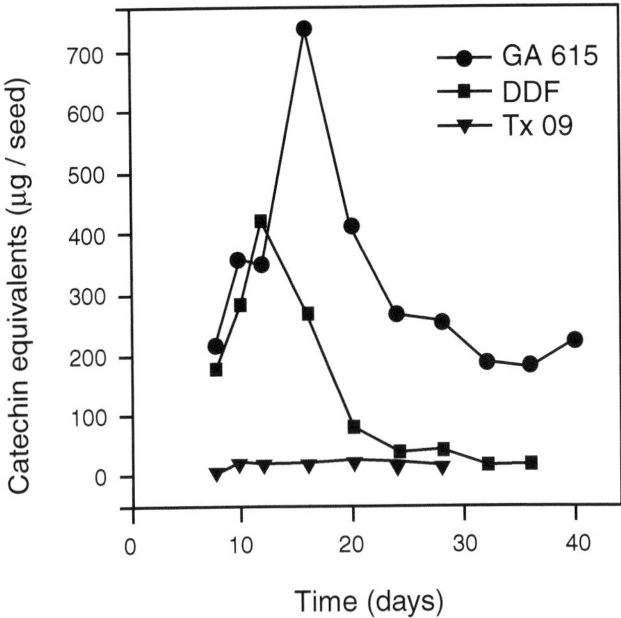

FIGURE 2.15 Variation in catechin content of kernels from three sorghum varieties from anthesis to maturity. The catechin equivalents were determined with the vanillin method on 1% HCl methanol extracts of the grain. Catechin equivalents were calculated subtracting a blank and after subtracting a blank; thus, two curves exist for each variety. (Adapted from Rooney, L.W. et al., 1980, in *Polyphenols in Cereal and Legumes*, International Development Research Centre, Ottawa, Canada, 25–35.)

dehulling of sorghum grains results in the removal of up to 98% of their tannins; however, this process brings about up to 45% protein loss as well as adversely affecting the nutritive value of proteins. The proteins in dehulled grains have a lower content of lysine, histidine and arginine than those in whole grains (Chibber et al., 1978). Dry heat has little effect on sorghum tannins, whereas moist heat decreases the content of assayable tannins from 0. 4 to 2.6% (Price et al., 1978). On the other hand, soaking whole seeds in water or alkali also reduces the content of tannins.

Extraction at higher temperatures reduces the time needed for soaking grains. Although 24-h soaking of sorghum grains in distilled water at 30°C is required for 31% removal of tannins, 20 min at 100°C is necessary to achieve a similar removal of tannins. The treatment of sorghum with 0.005 to 0.05 M sodium hydroxide at 30°C for 24 h results in greater tannin removal (from 36 to 84%) than using distilled water. At 100°C, alkali treatment gives similar results after 20 min of soaking. Replacing sodium hydroxide with sodium carbonate or potassium hydroxide produces similar effects on tannins (Chavan et al., 1979a). Steeping sorghum grain in 0.3% NaOH (w/v) solution for 8 h at 25°C is more effective in reducing the tannin level than soaking in water or 0.9% (v/v) HCl solution (Beta et al., 2000). Alkali treatment is expected to convert tannins into insoluble, nutritionally unreactive phlobaphenes (Swain, 1965), which may result in formation of indigestible tannin–protein complexes (Price et al., 1979b; Salunkhe et al., 1982) and promote oxidative polymerization of condensed tannins (Porter, 1992).

WHEAT

Wheat varieties differ from one another with respect to optimal growing conditions, kernel color, hardness and composition. Varieties planted in late summer or fall for maturation the following summer are called winter wheats; these include red and white as well as hard and soft varieties. Wheat varieties planted in the spring for harvesting in late summer are known as spring varieties and include durums, as well as hard red and hard and soft white varieties (Campbell, 1992).

Phenolic Acids

Wheat kernels have been reported to contain free phenolic compounds as well as their derivatives (El-Basyouni and Towers, 1964). The total content of phenolic acids in six wheat cultivars grown at four locations in Canada ranged from 1608 to 2687 mg of ferulic acids equivalents/kg (Abdel-Aal et al., 2001). Weidner et al. (1999) reported that grains of Polish wheat cultivars contain 3.44 to 3.55 mg/kg of free phenolic acids, 42.9 to 69.9 mg/kg of phenolic acids liberated from esters and 3.68 to 9.83 mg/kg of phenolic acids liberated from soluble glycosides. Ferulic, vanillic, gentisic, caffeic, salicylic, syringic, *p*-coumaric and sinapic acids, as well as vanillin and syringaldehyde, have been identified in wheat kernels (Table 2.10 and Table 2.15) (McKeehan et al., 1999; Sosulski et al., 1982; Weidner et al., 1999).

TABLE 2.15
Phenolic Acids Obtained on Alkaline Hydrolysis of Wheat[a,b]

	Ungerminated	Germinated (48 h)	
Phenolic Acid	Grains	Grains	Embryos
Ferulic acid			
Ethanol-soluble	3	26	310
Ethanol insoluble	ND[b]	181	1200
p-Coumaric acid			
Ethanol-soluble	1	8	261
Ethanol-insoluble	ND[b]	0	0
Sinapic acid			
Ethanol-soluble	0	7	95
Syringic acid			
Ethanol-soluble	2	8	197
Vanillic acid			
Ethanol-soluble	14	10	292

[a] Milligrams per kilogram, dry weight basis.
[b] ND = not determined.

Source: Adapted from El-Basyouni, S. and Towers, G.H.N., 1964, *Can. J. Biochem.*, 42:203–210.

Of these, ferulic acid is the primary phenolic acid in the grain during all stages of development and is present in seeds in free and esterified forms (McCallum and Walker, 1991; McKeehan et al., 1999; Rybka et al., 1993).

Ferulic acid is known to occur in high quantities in the aleurone cell walls of kernels and to a lesser extent in the seed coats and embryos (Fulcher, 1982; Fulcher, et al., 1972; Pussayanawin et al., 1988). This acid is esterified to arabinose in the pentosan fraction (Faurot et al., 1995; Izydorczyk et al., 1991), stanol and sterol (Seitz, 1989) and glucose (Herrmann, 1989). In mature grain the content of ferulic acid may range from 535 to 783 mg/kg of grain (Abdel-Aal et al., 2001). In addition, Weidner et al. (1999) found that, besides ferulic acid, caffeic, *p*-coumaric and sinapic acids are the main phenolic acids in Polish wheat cultivars. Renger and Steinhart (2000) have found the total amounts of ferulic and *p*-coumaric acids in wheat bran subjected to treatment with 4 M NaOH to be 18.8 and 0.34 g/kg, respectively.

Wheat bran also contains 808 mg/kg of total ferulic acid dehydrodimers (DiFA), of which 8-*O*-4′-DiFA is the dominating esterified DiFA and 8,5′-DiFA the most abundant esterified DiFA in wheat bran (Table 2.8) (Figure 2.1) (Garcia-Conesa et al., 1997; Lempereur et al., 1998; Renger and Steinhart, 2000). Ferulic acid and *p*-coumaric acid are the main phenolic acid linked to arabinoxylans, pectic substances or xyloglucans in the cell walls of Gramineae (Hartley et al., 1990; Iiyama et al., 1994; Lam et al., 1992; McDougall, 1993; Scalbert et al., 1985). Régnier and Macheix (1996) reported that bound ferulic acid and bound *p*-coumaric acid reach a maximum level in cell walls of durum wheat during the hydrical step (associated with phase II of grain development) and then decrease rapidly during grain dehydration.

Germination of grains significantly increases the content of ethanol-soluble phenolic acids (Table 2.15); the maximum concentration is reached about 9 days after germination. Thereafter, the content of phenolic acids starts to decline and reaches a minimum 4 to 5 weeks later (El-Basyouni and Towers, 1964). McKeehan et al. (1999) reported that the level of ferulic acid decreases by almost 50% during grain ripening. This decrease is thought to be due to

- The increase in rate of endosperm development surpassing the rate of synthesis of the outer grain covering
- The decrease of phenylalanine ammonia-lyase activity (an enzyme that catalyzes the reductive deamination of phenylalanine to cinnamic acid)
- The formation of covalent cross links between grain phenolics and cell wall components

Other Phenolics

Presence of alkylresorcinols in wheat grains has been reported. A series of n-alkylphenols has been found in wheat bran in which n-alkyl side chains containing 17, 19, 21, 23 and 25 carbons are coupled to a resorcinol ring at the 5 position (Wenkert et al., 1964). The total content of these alkylphenols, expressed as 5-n-pentadecylresorcinol equivalents on a dry basis, ranges from 657 to 927 mg/kg for wheat and from 633 to 863 mg/kg for triticale. Wheat bran has the highest content of alkylresorcinols (2110 mg/kg); corresponding flours contain only 380 mg/kg (Musehold, 1978).

TABLE 2.16
Content of Flavonoids in Wheat Bran[a]

Class and Variety	Total Flavonoids
Hard Red Spring	
Len	149.1
Alex	245.4
Stoa	289.0
Hard Red Winter	
Roughrider	288.2
Agassiz	271.1
Durum Wheat	
Rugby	405.7
Lloyd	358.4
Vic	291.2

[a] Milligrams per kilogram.

Source: Adapted from Feng, Y. and McDonald, C.E., 1989, *Cereal Chem.*, 66:516–518.

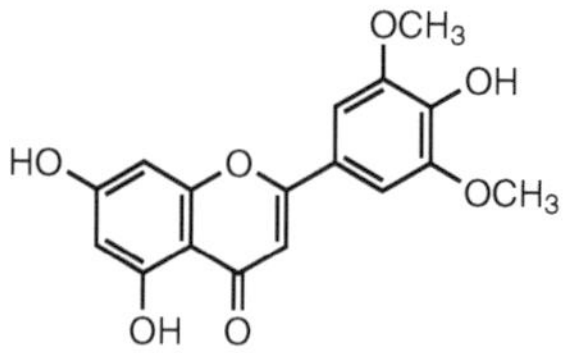

FIGURE 2.16 Chemical structure of tricin, a dominant flavone pigment in wheat.

The bran of hard wheat and commercial wheat germs contains a number of flavonoid pigments. Depending on the wheat class, the content of flavonoids in brans ranges from 149.1 to 405.7 mg/kg (Table 2.16) (Feng and McDonald, 1989). The wheat germs contain the same level of flavonoids as those found in the bran. Tricin (5,7,4′-trihydroxy 3′,5′-dimethyoxy flavone) is the dominant flavone pigment in cultivated and wild wheats (Figure 2.16). Anderson and Perkin (1931) first isolated this compound from ether extracts of wheat. According to Chen and Geddes (1945), about one third of the total content of flavones in wheat, expressed as tricin equivalents, is contained in the endosperm and two thirds in the bran and embryo. Two other flavonoids have also been isolated from wheat germ and identified as apigenin glycosides (King, 1962). On the other hand, 6-C-pentosyl-8-C-hexosylapigenin and 6-C-hexosyl-8-C-pentosylapigenin have been isolated and identified by spectroscopic methods (Feng et al., 1988; Feng and McDonald, 1989).

Two phenolic acid diacylglycerols (containing two ferulic acids or ferulic acid and *p*-coumaric acid) have also been identified in wheats (Cooper et al., 1978). Phenolic aldehydes such as *p*-hydroxybenzaldehyde, vanillin, and syringaldehyde can be detected in wheat species of *Triticum dicoccoides* and *Aegilops geniculata*. In addition, veratraldehyde and protocatechualdehyde are found in wheat species of *Triticum dicoccoides* (Cooper et al., 1994).

Seitz (1989) identified sitostanyl, campestanyl and campesteryl ferulates in wheat grains (Figure 2.8); of these, campestanyl ferulate and sitostanyl ferulate are the main steryl ferulates. The total content of steryl ferulates in whole wheat grain ranges from 62 to 123 mg/kg (Hakala et al., 2002; Seitz, 1989).

Effect of Processing

Fresh wheat flour (Neepawa wheat) contains 71.4 mg/kg of phenolic acids. Ferulic acid is a major phenolic acid, constituting about 89.1% of total phenolic acids (Table 2.9). Over 84% of phenolic acids are present in the insoluble bound form. Storing wheat flour brings about the loss of free, soluble esters and insoluble bound phenolics due to destructive oxidation reactions (Sosulski et al., 1982).

Ferulic, syringic, caffeic, sinapic, *p*-coumaric, and vanillic acids and phenolic acid dimers have been identified in wheat gluten (Labat et al., 2000; Sosulski et al., 1982), with ferulic acid (49 to 177 mg/kg gluten of dry weight) predominant, followed by sinapic acid (14 to 35 mg/kg gluten of dry weight) (Labat et al., 2000). Moreover, Hilhorst et al. (1999) found that gluten prepared by hand washing Kolibri flour contains 80 mg of ferulic acid per kilogram of dry weight. The phenolics soluble in an acetone–methanol–water solvent system contain 64 to 85% of total sinapic acid and only 5 to 50% of total ferulic acid detected in wheat gluten (Labat et al., 2000). During gluten mixing, 21 to 66 and 22 to 94% of the initial soluble forms of ferulic and sinapic acids disappear, respectively (Jackson and Hoseney, 1986; Labat et al., 2000).

Abdel-Aal et al. (2001) investigated the distribution of phenolic acids in wheat milling fractions. About 73% of grain phenolic acids was found in the shorts and bran, but only 5% in first, second and third break milling fractions. Later, Hakala et al. (2002) studied the distribution of steryl ferulates in wheat milling products and detected only traces of steryl ferulates in low ash wheat flour; wheat brans contained 297 to 390 mg/kg of total steryl ferulates. Enrichment of low ash flour with aleurone layer increases the total ferulates level to 194 to 216 mg/kg.

NUTS

Nuts contain bioactive components; consumption of nuts has been linked with reducing the risk of coronary heart diseases (Hu et al., 1998; Sabate, 1999) and altering serum lipids (Kris-Etherton et al., 1999; Spiller et al., 1998).

Almonds

One of the world's most popular tree nuts, almonds (*Prunus amygdalus)* belong to the Rosaceae family (Menninger, 1977) and are used as snack food and as ingredients in many processed foods, including bakery and confectionary products. The U.S. is the largest producer of almonds in the world (Rosengarten, 1984). The almond fruit consists of the edible kernel, brown skin, shell and hull; blanched almonds have their shells and their skins removed. Removed shells, hulls and skins are usually discarded (Menninger, 1977; Rosengarten, 1984); however, they can be used as a source for dietary fiber, sweetener concentrate and antioxidants (Takeoka and Dao, 2003).

The antiradical activity of the almond and its components was recently reported by Siriwardhana and Shahidi (2002). Almond skin serves as a rich source of phenolic compounds (Lou et al., 2001), exhibiting an excellent scavenging activity for reactive oxygen species (Siriwardhana and Shahidi, 2002). Takeoka and Dao (2003) detected chlorogenic acid isomers, namely, chlorogenic acid (42.5 mg/100 g of fresh weight), cryptochlorogenic acid (7.9 mg/100 g of fresh weight) and neochlorogenic acid (3.0 mg/100 g of fresh weight) in almond (*Prunus dulcis* Mill.) hull. Furthermore, Sang et al. (2002a) identified a novel prenylated benzoic acid derivative, 3-prenyl-4-*O*-β-D-glucopyranosyloxy-4-hydroxybenzoic acid (Figure 2.17), as well as (+)-catechin, protocatechuic acid and ursolic acid (triterpenoid) in almond hulls. Later, Sang et al. (2002b) also isolated and identified additional phenolic compounds in almond skins:

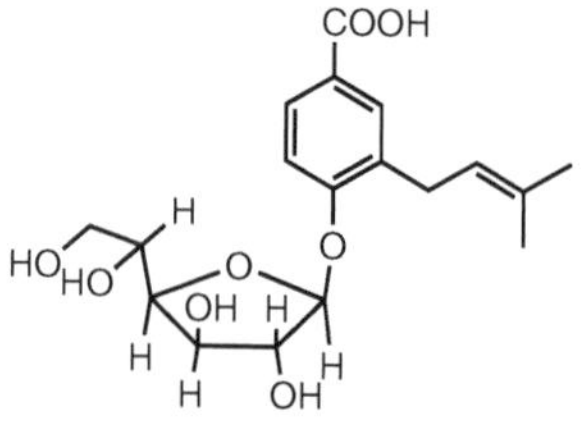

FIGURE 2.17 Chemical structures of prenylated benzoic acid derivative found in almond skins.

- Vanillic acid
- *p*-Hydroxybenzoic acid
- 3′-*O*-β-Methylquercetin 3-*O*-β-D-glucopyranoside
- 3′-*O*-β-Methylquercetin 3-*O*-β-D-galactopyranoside
- 3′-*O*-β-Methylquercetin 3-*O*-α-L-rhamnosylpyranoside-(1→6)-β-D-glucopyranoside
- Kaempferol 3-*O*-α-L-rhamnosylpyranoside-(1→6)-β-D-glucopyranoside
- Naringenin 7-*O*-β-D-glucopyranoside

All these compounds, except naringenin derivatives exhibit strong radical scavenging activity. Subsequently, Frison-Norrie and Sporns (2002) reported the presence of isorhamnetin 3-*O*-α-L-rhamnosylpyranoside-(1→6)-β-D-glucopyranoside, isorhamnetin 3-*O*-β-D-glucopyranoside, and kaempferol 3-*O*-β-D-glucopyranoside in almond skins. Later, Frison and Sporns (2002) also reported that isorhamnetin rutinoside was the most abundant flavonol glucoside in almond skins. Moreover, procyanidins with various degrees of polymerization have also been detected in the flesh and skin of almond fruit (Brieskorn and Betz, 1988; de Pascual-Teresa et al., 1998; Lou et al., 1999).

Betel Nuts

Betel nut or areca nut (*Areca catechu* L.) is used for human consumption in parts of Southeast Asia and the South Pacific. Used by approximately 10% of the world's population as an important component of their daily diet (Duke, 1985), betel nut also often serves as a breath-sweetening masticatory. In India, whole or broken betel nuts are usually processed by soaking or boiling with various plant extracts to decrease their astringency (Raghavan and Baruah, 1958). The most popular way to consume betel nuts is as "pan," also referred to as "betel quid." The major ingredients of betel quid are betel nut and slaked lime folded together in betel leaves, to which tobacco, catechu, clove, cardamon, anise seed, grated coconut, sugar and other spices or condiments are often added (Stich et al., 1982, 1983).

Betel nuts contain arecoline and other alkaloids responsible for stimulating effects on the nervous system and also have 26% tannins. The predominant class of tannins in betel nuts is gallotannic acid (Duke, 1985; Raghavan and Baruah, 1958). Presence of condensed tannins (proanthocyanidins) and structurally related monomeric and oligomeric flavan-3-ols has also been reported (Nonaka et al., 1981; Stitch et al., 1983). Butler and Pushpamma (1991) estimated the content of tannins in some betel nut varieties; the Chikni variety of betel nut contained 31 and 27% tannins as quantified by proanthocyanidin assay and vanillin assays, respectively. The content of tannins in betel nuts of Vakkalu variety is 8.7% (proanthocyanidin assay) and 20% (vanillin assay); however, a somewhat lower amount of tannin is found in betel quids due to the presence of higher amounts of tannin-free ingredients (Butler and Pushpamma, 1991). Later, Wang and Lee (1996) reported that the whole areca fruit contains condensed tannins (92 g/kg of dry weight), hydrolyzable tannins (69 g/kg of dry weight) and a nontannin phenol fraction (56 g/kg of dry weight). Catechin and epicatechin are the main phenolics detected in nontannin phenol fraction.

Betel products may be involved in carcinogenesis; possibly due to their tannin component (Giri et al., 1987; Gothoskar and Pai, 1986; Stich et al., 1982, 1983, 1984; Wenke et al., 1984), consumption of betel nuts has been related to the development of oral cancer deaths each year in Asia. In addition, betel tannins may bring about oral submucous fibrosis, a chronic disabling disease (Paymaster, 1962), and cause antinutritional effects such as diminishing palatibility or inhibiting the post-digestive metabolism (Salunkhe et al., 1990). Crude phenolic extracts of whole areca fruit exhibit marked antioxidative activity and antimutagenic effect toward *Salmonella typhimurium* TA98 and TA100 (Wang and Lee, 1996). Betel nut and betel quid tannins are less effective than sorghum tannins in precipitating proteins (Hagerman and Butler, 1978); however, their content is several times higher than that found in high-tannin sorghum varieties (Butler and Pushpamma, 1991). Therefore, consumption of betel nut products may result in serious nutritional consequences.

Cashew Nuts

Cashew nut is the kernel of *Anacardium occidentale* L. The testa of this nut is reddish-brown in color and serves as a good source of catechol-type tannin (Pillai et al., 1963). Main phenolic constituents of cashew kernel testa are (+)-catechin and (–)-epicatechin; accounting for 6 and 7.5% of the dried weight of cashew kernel testa, respectively, these constitute approximately 40% of the total phenolics in cashew nuts. The polymeric proanthocyanidins account for 40% of total testa phenolics; other components of testa phenolics include leucocyanidins and leucopelargonidins (Mathew and Parpia, 1970). Furthermore, nonylphenol has been identified

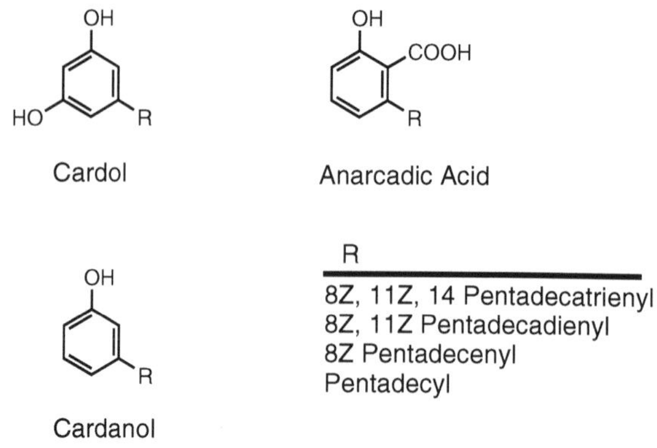

Cardol

Anarcadic Acid

Cardanol

R
8Z, 11Z, 14 Pentadecatrienyl
8Z, 11Z Pentadecadienyl
8Z Pentadecenyl
Pentadecyl

FIGURE 2.18 Chemical structures of anarcadic acids, cardols and cardanols.

in cashew apple headspace constituents and cinnamic acid in glycosidically bound components (Bicalho et al., 2000).

During moisture conditioning (a processing step prior to roasting) some nuts develop bluish-black patches on their surfaces —a discoloration that lowers the market value of such cashew nuts. There are a few indications that discoloration of cashew nut testa may be due to formation of complexes between iron and polyphenols. The darkened nuts have a higher level of iron than the good nuts (Farrer, 1935); Mathew and Parpia (1970) reported that polyphenols of cashew nut testa can form dark-colored complexes with iron. These dark pigments can easily be washed away with dilute acidic solutions.

Anacardic acids, cardols and cardanols are the major phenolic constituents of cashew nut shell liquid, a by-product obtained during the production of cashew nuts (Figure 2.18). The side chains of these phenolics comprise fully saturated pentadecyl groups and their unsaturated monoene, diene or triene analogues (Paramashivappa et al., 2001; Spencer et al., 1980). Cashew nut shell liquid and its phenolics are used as components of cement, specialty coatings and primers (Menon et al., 1985). The biological activities of anarcadic acids, cardols and cardanols such as antitumor (Kubo et al., 1993) and antiacne (Kubo et al., 1994a) properties, molluscicidical activity (Kubo et al., 1986), prostaglandin synthetase inhibition (Kubo et al., 1987), as well as their antibacterial activity against the Gram-negative bacterium *Helicobacter pylori* (Kubo et al., 1999) and Gram-positive bacteria (Himejima and Kubo, 1991), have been demonstrated.

These acids also exhibit inhibitory effects on urease (Kubo et al., 1999), α-glucosidase and aldose reductase (Toyomizu et al., 1993), tyrosinase (Kubo et al., 1994b), lipoxygenase (Grazzini et al., 1991; Shobba et al., 1994), and prostaglandin endoperoxide synthase (Grazzini et al., 1991). Paramashivappa et al. (2001) proposed a simple procedure for selective isolation of anacardic acids, cardanols and cardols from cashew nut shell liquid. This method involved precipitation of anacardic acid as calcium anacardate, treatment of the shell liquid with ammonia and extraction of cardanols into ethyl acetate/hexane. Subsequently, Kumar et al. (2002) described a method for isolation of cardanol from technical cashew nutshell oil (TCNSO), a mixture of cardanol, cardol and polymeric materials produced by roasting shell. This procedure involves dissolving TCNSO in a mixture of methanol and ammonia (8:5; v/v) and extracting cardanol by hexane. Cardol is then extracted from methanol–ammonia solution with a mixture of ethyl acetate and hexane.

Hazelnuts

The data on hazelnut (*Corylus* spp.) phenolics are still limited. Senter et al. (1983) reported that protocatechuic acid (0.36 mg/kg of hulls) is the predominant phenolic in hazelnut hulls. The levels of other phenolic acids (gallic acid, caffeic acid, vanillic acid and *p*-hydroxybenzoic acid) do not exceed 10 μg/kg of testa. Using carbon-13 cross-polarization magic-angle spinning NMR, Preston and Sayer (1992) subsequently detected only condensed tannins in the packing tissue inside shells of hazelnut. Later, Yurttas et al. (2000) isolated and tentatively identified six phenolic aglycones in Turkish and American hazelnuts: gallic acid, *p*-hydroxybenzoic acid, sinapic acid, quercetin, and caffeic acid and epicatechin. Furthermore, nonhydrolyzed extracts of hazelnut phenolics exhibit greater antioxidant activities than corresponding hydrolyzed extracts.

Peanuts

Nearly two thirds of the world's peanut production is pressed for oil; the peanut meal is used mostly as a feed for livestock. Defatted peanut flour may also serve as an alternative source of functional proteins with potential applications in food product formulations (McWatters and Cherry, 1975; McWatters and Holmes, 1979). However, peanut protein products contain phenolic compounds (Daigle et al., 1983; Maga and Lorenz, 1974) that have been implicated in flavor and color defects (Maga, 1978; Sosulski, 1979). Over half of the U.S. peanut crop is used for production of peanut butter (Potter, 1986).

Peanut protein products contain up to 2000 mg/kg of phenolic acids and up to 120 mg/kg of neutral phenolic compounds. Of these, protein isolates contain the least amount of total phenolic acids. Accounting for 40 to 68% of phenolic acids in all peanut products, *p*-coumaric acid (*trans* and *cis* forms combined) has been identified as the predominant phenolic acid in peanut protein products (Table 2.17) (Seo and Morr, 1985). Vanillic acid, *p*- and *o*-coumaric acids and ferulic acid may contribute significantly to the flavor characteristics of peanut flour because they are present at levels near or above their threshold values (Maga and Lorenz, 1974). Other studies have demonstrated that combinations of phenolic acids show a synergistic effect in lowering the taste threshold values (Maga and Lorenz, 1973).

Different parts of peanuts, including kernel, may contain *trans*-resveratrol, a major stilbene phytoalexin produced in response to exogenous stimuli such as fungal infection (Ingham, 1976). Growing immature peanut kernels contain higher levels of phytoalexins than mature ones (Dorner et al., 1989). The content of resveratrol in four raw shelled peanut samples available in the U.S. market ranges from 0.070 to 0.716 mg/kg in five boiled peanut samples from 1.779 to 7.092 mg/kg and in 15 samples of peanut butter is 0.148 to 0.504 mg/kg. Roasted peanuts contain the lowest

TABLE 2.17
Phenolic Acid Composition in Peanut Protein Products as Determined by HPLC[a]

Phenolic Acid	Laboratory Defatted Meal	Commercial Defatted Flour	Protein Isolate
Unknown	93.2	56.3	79.9
Unknown	104.8	64.4	13.3
Vanillic/Caffeic	69.6	96.5	35.4
Unknown	23.6	28.9	23.3
Syringic	14.2	12.4	13.3
p-Coumaric	1224.0	1193.0	537.0
Unknown	92.0	63.6	54.6
Ferulic	143.8	83.1	34.6
Sinapic	245.0	135.0	10.7
Isoferulic	23.5	22.6	8.2

[a] Milligrams per kilogram.

Source: Adapted from Seo, A. and Morr, C.V., 1985, *J. Food Sci.*, 50:262–263.

amount of resveratrol (Sobolev and Cole, 1999). Sanders et al. (2000) have reported that the content of resveratrol in 15 cultivars of market type peanuts, stored in the cold for up to 3 years, ranges from 0.02 to 1.79 mg/kg. Subsequently, Ibern-Gómez et al. (2000) detected the presence of *trans*-piceid (3-β-glucoside of *trans*-resveratrol) in peanuts.

The content of total stilbenes in peanut butter (natural and blended) is between 0.354 and 0.978 mg/kg. *Trans*-piceid comprises 11.7 to 35.8% of the total stilbenes. Health benefits related to resveratrol are associated with reduced risk of cancer (Jang et al., 1997) and cardiovascular disease (Goldberg, 1995). Some peanut products may also contain vanillin. Sobolev (2001) recently detected vanillin in boiled peanuts (1.61 to 1.71 mg/kg of kernel); peanuts are boiled in hulls high in lignins and vanillin is a product of oxidative hydrolysis of lignins (Fargues et al., 1996).

According to Sanders (1977), the bronze coloration of peanut hulls is due to the presence of tannins. Mature peanut hulls, seed coats, and fruits contain approximately 0.43, 6.04 and 0.04% tannins on a dry weight basis, respectively. Thus, most tannins are located in seed coats and hulls, while fruits (meat nut) are practically tannin free. The content of tannins changes with the maturation of peanuts. In hulls, tannin content increases markedly during the latter stages of maturation, while their content in skins increases rapidly in the early stages of maturation and then begins to decline (Sanders, 1977). The total content of phenolics in peanut hulls increases during the development of peanut from 33.4 mg/kg in hulls of peanuts harvested 74 days after planting (DAP) to 71.3 mg/kg of hulls at 144 DAP. Moreover, the concentration of luteolin in hulls at day 144 (28.6 g/kg) is 15 times that at day 74 (Yen et al., 1993).

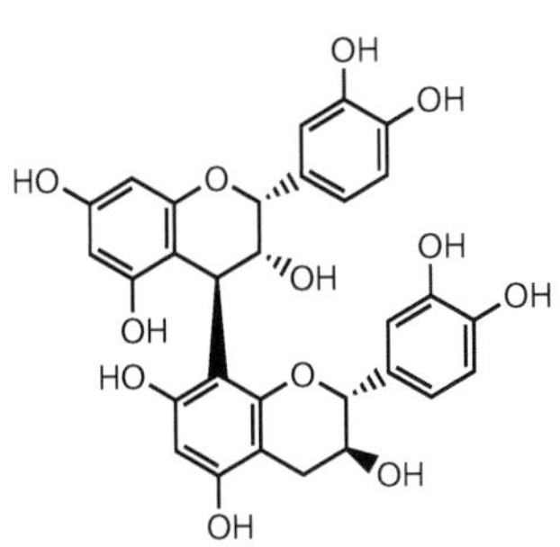

(1) Procyanidin B1

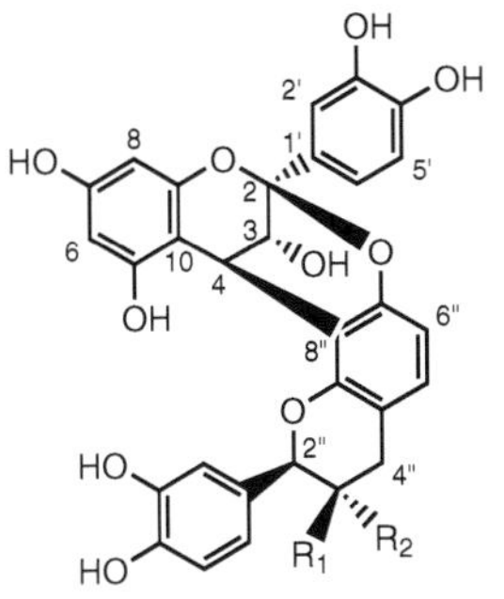

(2) R_1 = H, R_2 = OH ; EC-(4β→8; 2β→*O*→ 7)-C
(3) R_1 = OH, R_2 = H ; EC-(4β→8; 2β→*O*→ 7)-EC

FIGURE 2.19 Chemical stuctures of some dimeric procyanidins isolated from red peanut skin (EC, (–)-epicatechin; C, (+)-catechin).

Mature peanut skins contain 17% procyanidins by weight. Of these, 55% are low molecular weight oligomers of catechin and epicatechin that are soluble in ethyl acetate. Epicatechin-(4β→8; 2β→O-7)-catechin is found to be the predominant dimeric procyanidin in peanut skin, followed by epicatechin-(4β→8; 2β→O-7)-epicatechin and epicatechin-(4β→8)-catechin (procyanidin B1) (Figure 2.19). The higher molecular weight procyanidins soluble in ethyl acetate contain the (4β→8 or 4β→6) and (4β→8; 2β→O-7) interflavonoid bonds (Karchesy and Hemingway, 1986). However, the nature of flavonoids present in peanuts depends on their degree of maturity: the shells of immature peanuts contain predominantly eriodictyol while mature peanut shells are rich in luteolin (Daigle et al., 1988).

Pecans

Pecan kernels are semiperishable products that must be refrigerated in order to preserve their quality. They contain enough prooxidants and antioxidants to affect their storage stability. Of these, phenolic acids present in pecan kernels contribute to their enhanced storage stability. Senter et al. (1978) found that development of rancidity of pecan kernels closely parallels the content of total phenolics. On the other hand, leucocyanidins and leucodelphinidins participate in the development of red-brown discoloration in the testa of pecan (Senter et al., 1978).

The color of shelled pecan seed coats (testa) is a major criterion in their quality determination. Based on their color, pecans are graded into (1) light (golden color of testa), (2) light amber (light brown), (3) amber (medium brown) and (4) dark amber (dark brown). A light color is indicative of fully mature pecans that have been properly harvested, processed and stored. A darker color is caused by exposure of kernels to adverse conditions. The coloration of pecan kernels depends on growing conditions such as availability of moisture, location of tree within the orchard and horticultural practice (Woodroof, 1967), as well as the time of harvest (Heaton et al.,

TABLE 2.18
Changes in Total Phenols and Flavanols of Pecan Kernels Stored at 32°C, 50% RH[a]

Storage Time (weeks)	Pecan Variety			
	Stuart		Schley	
	Total Phenols	Total Flavanols	Total Phenols	Total Flavanols
0	10.65	4.43	8.30	3.63
6	14.20	4.15	9.30	3.65
12	11.15	3.88	9.10	3.70
15	11.55	3.58	7.80	2.78

Note: Total phenols determined by Folin-Ciocalteau assay; total flavanols by vanillin assay.

[a] Grams per kilogram.

Source: Adapted from Senter, S.D. et al., 1978, *J. Food Sci.*, 43:128–134.

1975). Storage at temperatures below –7°C protects the products against rancidity and discoloration for about 2.5 years.

The total content of phenolics in Stuart and Schley pecan varieties ranges from 7.80 to 14.20 mg/g of kernel (Table 2.18). However, Stuart varieties contain more phenolics than the Schley variety. Total flavanol content is between 2.78 and 4.43 mg/g of kernel (Senter et al., 1978). Thus, significant quantities of other phenolics are present in pecan kernels and may affect their quality.

Gallic, gentisic, vanillic, protocatechuic, coumaric, syringic, *p*-hydroxybenzoic and *p*-hydroxyphenylacetic acids have been identified in the methanolic extracts of pecan kernel (Senter et al., 1980). Of these, coumaric and syringic acids are present only in trace amounts; constituting approximately 78% of the total phenolic acids in the Stuart and Schley varieties, gallic acid has been found to be a predominant phenolic acid. Storage of pecan kernels decreases the content of their major phenolic acids, namely, gallic, gentisic, and vanillic acids by 39, 36, and 28%, respectively, after 12 weeks of storage in the dark at 21°C and 65% relative humidity (RH). These changes in the content of phenolic acids are closely related to the results of sensory evaluation of flavor and odor qualities of pecan kernels (Senter et al., 1980).

Senter et al. (1978) reported that redish-brown discoloration of pecan kernels may be brought about by the oxidation of endogenous leucocyanidin and leucodelphinidin to their respective phlobaphenes of varying degrees of polymerization and, to a lesser extent, by their subversion to cyanidin (Figure 2.20) and delphinidin. After 16 weeks of storage, pecan kernel at 32°C and 50% RH contain 10 times more phlobaphenes than anthocyanidins; over 50% decrease in the content of leucocyanidin has been noted. Phlobaphenes are reddish-colored, water-insoluble phenolic substances that are believed to be related to the co-occurrence of condensed tannins (Foo and Karchesy, 1988).

Tannins present in pecans may affect the color of nut meats as well as their flavor quality and acceptability. Some tannins present in shells may leach into the

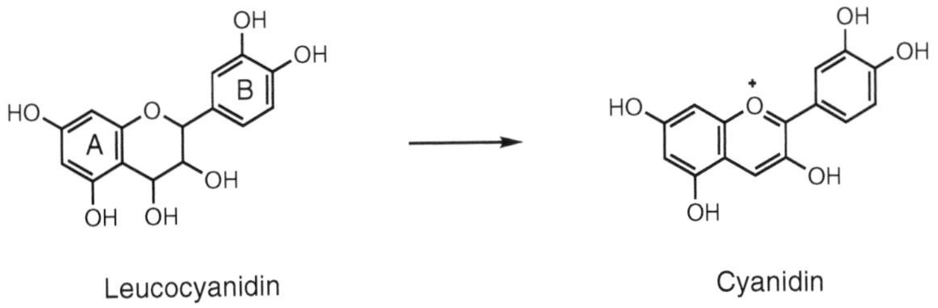

Leucocyanidin Cyanidin

FIGURE 2.20 Transformation of leucocyanidin to corresponding cyanidin.

meats during preconditioning (soaking) before pecans are cracked in the shelling process. Polles et al. (1981) determined the content of condensed tannins in 31 types of pecan nut kernels. Their content of tannins ranged from 0.70 to 1.71%, thus indicating a significant difference among different cultivars. According to Polles et al. (1981), the content of the tannins can be used as a trait for testing the desirability of pecan cultivars and examining which types are more suitable for processing with a minimum loss of quality. Using carbon-13 cross-polarization magic-angle spinning NMR, Preston and Sayer (1992) later detected only prodelphinidin polymer in the packing tissue inside pecan shells.

Walnuts

Walnuts are produced by tree nut cultivars of *Juglans regia* (family Junglandaceae). Production of this crop in the U.S. is mainly concentrated in California (Foreign Agricultural Service, 1998). Consumption of walnuts has a beneficial impact on human serum lipid profile by decreasing total and LDL cholesterol and increasing the level of HDL cholesterol (Chisholm et al., 1998; Lavendrine et al., 1999; Sabate et al., 1993; Zambon et al., 2000). Shelled walnuts contained 16 g/kg of total phenolics expressed as gallic acid equivalents (Anderson et al., 2001). Jurd (1956) identified ellagic acid, methyl gallate, gallic acid and a mixture ellagitannins in the skin of walnuts. Later, Jurd (1958) isolated a novel ellagitannin, isomeric with corilagin, from the walnut skin. Łuczak et al. (1989) identified 10 phenolic acids in the pericarp of walnut: *p*-hydroxybenzoic, vanillic, gentisic, protocatechuic, syringic, *p*-coumaric, gallic, ferulic, caffeic and sinapic acids. Bound phenolics (esterified + glycosylated) are the predominant form of phenolic acids and comprise 76% of total phenolic acids in walnuts. Syringic acid is the principal phenolic acid identified, accounting for 40.5 and 33.8% of total free and bound phenolic acids, respectively. Subsequently, Anderson et al. (2001) have reported that ellagitannins are the most abundant phenolics in walnuts. Of these, ellagic acid, valoneic acid dilactone, and pedunculagin have been identified in phenolic extracts from shelled walnuts (Figure 2.21).

Similar to other nuts, walnuts are prone to contamination with aflatoxins on infection with *Aspergillus flavus* and *Aspergillus parasiticus* (Henry et al., 1999); however, it has been shown that walnut husks are resistant to *A. flavus* growth (Mahoney et al., 1998). Walnuts contain a number of structurally related naphthoquinones, including 1,4-naphthoquinone, juglone (5-hydroxy-1,4-naphthoquinone),

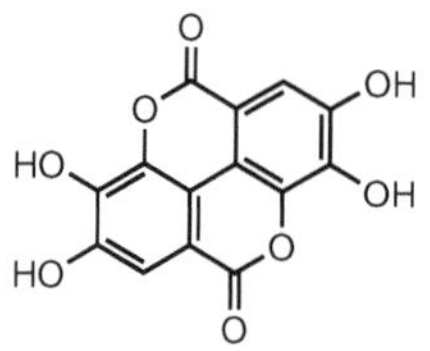

Ellagic acid

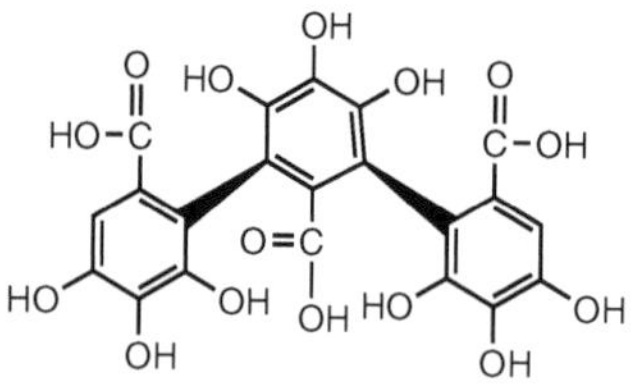

Valoneic Acid

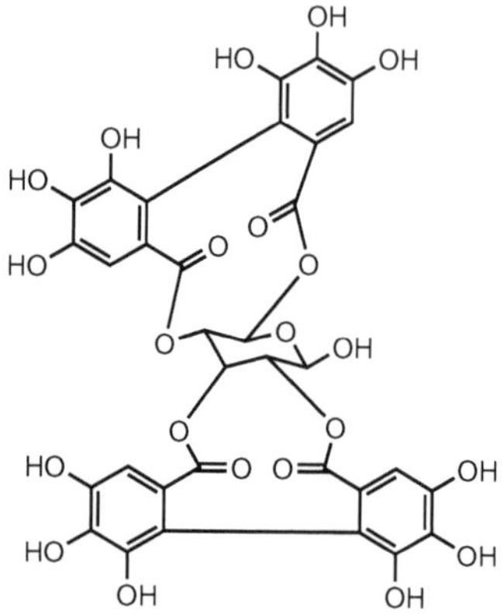

Pedunculagin

FIGURE 2.21 Chemical structures of some phenolic compounds found in walnuts.

1,4-Naphthoquinone

Juglone

2-Methyl-1,4-naphthoquinone

Plumbagin

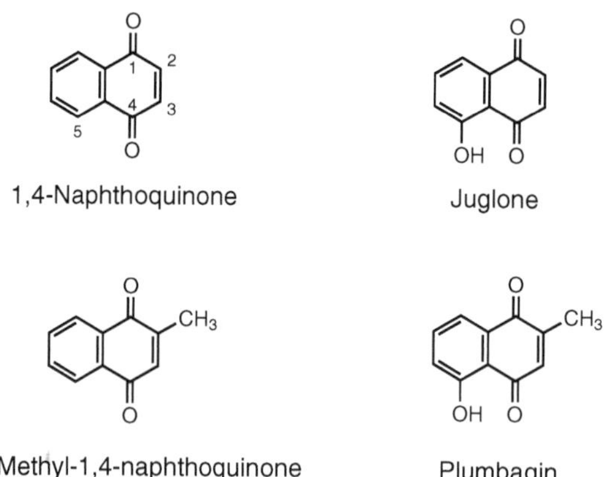

FIGURE 2.22 Chemical structures of some naphthoquinones found in walnuts.

2-methyl-1,4-naphthoquinone, and plumbagin (5-hydroxy-2-methyl-1,4-naphthoquinone) (Figure 2.22) (Leistner, 1981; Muller and Leistner, 1978a, b). Higher concentrations of these compounds have been found in husk tissues surrounding the nut (Binder et al., 1989). Recently, Mahoney et al. (2000) reported that naphthoquinones of walnut husk may suppress the growth of *A. flavus*. The growth of *A. flavus* is inhibited by 1,4-naphthoquinone at 170 μg/kg and juglone at 100 μg/kg; however, for plumbagin and 2-methyl-1,4-naphthoquinone, no growth of *A. flavus* has been observed at 50 μg/kg. These authors have also demonstrated that, at levels 10 μg/kg below that required for complete growth inhibition, juglone and plumbagin reduce aflatoxin production by 12 and 39% of the control, respectively.

REFERENCES

Abdel-Aal, E.-S.M., Hucl, P., Sosulski, F.W., Graf, R., Gillott, C., and Pietrzak, L. 2001. Screening spring wheat for midge resistance in relation to ferulic acid content. *J. Agric. Food Chem.*, 49:3559–3556.

Akihisa, E.M., Yasukawa, K., and Kasahara, Y. 1997. Triterpenoids from flowers of *Compositae* and their antiinflammatory effects. *Curr. Topics Phytochem.*, 1:137–144.

Akihisa, T., Yasukawa, K., Yamaura, M., Ukiya, M., Kimura, Y., Shimizu, N., and Arai, K. 2000. Triterpene alcohol and sterol ferulates from rice bran and their antiinflamatory effects. *J. Agric. Food Chem.*, 48:2313–2319.

Anderson, J.A. and Perkin, A.G. 1931. The yellow coloring matter of Khapli wheats, *J. Chem. Soc.*, 2624–2625.

Anderson, K.J., Teuber, S., Gobeille, A., Cremin, P., Waterhouse, A.L., and Steinberg, F.M. 2001. Walnut polyphenolics inhibit *in vitro* human plasma and LDL oxidation. *J. Nutr.*, 131:2837–2842.

Andreasen, M.F., Christensen, L.P., Meyer, A.S., and Hansen, A. 1999. Release of hydroxyxinnamic and hydroxybenzoic acids in rye by commercial plant cell wall degrading enzyme preparations. *J. Sci. Food Agric.*, 79:411–413.

Andreasen, M.F., Christensen, L.P., Meyer, A.S., and Hansen, A. 2000a. Ferulic acid dehydrodimers in rye (*Secale cereale* L.). *J. Cereal Sci.*, 31:301–307.

Andreasen, M.F., Christensen, L.P., Meyer, A.S., and Hansen, A. 2000b. Content of phenolic acids and ferulic acid dehydrodimers in 17 rye (*Secale cereale* L.) varieties. *J. Agric. Food Chem.*, 48:2837–2842.

Andreasen, M.F., Landbo, A.-K., Christensen, L.P., Lars, P., Hansen, A., and Meyer, A.S. 2001. Antioxidant effects of phenolic rye (*Secale cereale* L.) extracts, monomeric hydroxycinnamates, and ferulic acid dehydrodimers on human low-density lipoproteins. *J. Agric. Food Chem.*, 49:4090–4096.

Antony, U. and Chandra, T.S. 1998. Antinutrient reduction and enhancement in protein, starch, and mineral availability in fermented flour of finger millet (*Eleusine coracana).* *J. Agric. Food Chem.*, 46:2578–2582.

Arntfield, S.D., Ismond, M.A.H., and Murray, E.D. 1985. The fate of antinutritional factors during the preparation of a fababean protein isolate using a micellization technique. *Can. Inst. Food Sci. Technol. J.*, 18:137–143.

Asamarai, A., Addis, P.B., Epley, R.J., and Krick, T.P. 1996. Wild rice hull antioxidants. *J. Agric. Food Chem.*, 44:126–130.

Asquith, T.N., Izuno, C.C., and Butler, L.G. 1983. Characterization of the condensed tannin (proanthocyanidin) from a group II sorghum. *J. Agric. Food Chem.*, 31:1299–1303.

Bartolome, B., Faulds, C.B., and Williamson, G. 1997. Enzymic release of ferulic acid from barley spent grain. *J. Cereal Sci.*, 25:285–288.

Bate-Smith, E.C. and Rasper, V. 1969. Tannins of grain sorghum: luteoforol (leucoluteolidin) 3′,4,4′,5,7-pentahydroxyflavan. *J. Food Sci.*, 34:203–209.

Bendelow, V.M. and La Berge, D.E. 1979. Relationships among barley malt, and beer phenolics. *J. Am. Soc. Brew. Chem.*, 37:89–90.

Beninger, C.W., Hosfield, G.L., and Basset, M.J. 1999. Flavonoid composition of three genotypes of dry beans *(Phaseolus vulgaris)* differing in seedcoat coloring. *J. Am. Soc. Hort. Sci.*, 124:514–518.

Beninger, C.W. and Hosfield, G.L. 1999. Flavonol glycosides from Maltcalm dark red kidney bean: implications for the genetics of seed coat color in *Phaseolus vulgaris* L. *J. Agric. Food Chem.*, 47:4079–4082.

Beninger, C.W., Hosfield, G.L., and Nair, M.G. 1998. Flavonol glycosides from seed coat of a new Manteca-type dry beans (*Phaseolus vulgaris* L.). *J. Agric. Food Chem.*, 46:2906–2910.

Benson, G.O. and Pearce, R.B. 1987. Corn perspective and culture, in *Corn: Chemistry and Technology,* Watson, S.A. and Ramstad, P.E., Eds., American Association of Cereal Chemists, St. Paul, MN, 1–30.

Beta, T., Rooney, L.W., Marovatsanga, L.T., and Taylor, J.R.N. 2000. Effect of chemical treatments on polyphenols and malt quality in sorghum. *J. Cereal Sci.*, 31:295–302.

Bicalho, B., Pereira, A.S., Neto, R.A., Pinto, A.C., and Rezende, C.M. 2000. Application of high-temperature gas chromatography-mass spectrometry to the investigation of glycosidically bound components related to cashew apple (*Anacardium occidentale* L. Var. nanum) volatiles. *J. Agric. Food Chem.*, 48:1167–1174.

Binder, R.G., Benson, M.E., and Flath, R.A. 1989. 1,4-naphthoquinones from *Juglans. Phytochemistry,* 28:2799–2801.

Blessin, C.W., Van Etten, C.H., and Dimler, R.J. 1963. An examination of anthocyanogens in grain sorghums. *Cereal Chem.*, 40:241–250.

Bressani, R. 1975. Legumes in human diets and how they might be improved, in *Nutritional Improvement of Food Legumes by Breeding,* Milner, M., Ed., John Wiley & Sons, Inc. New York, 15–42.

Bressani, R. and Elias, L.G. 1980. The nutritional role of polyphenols in beans, in *Polyphenols in Cereal and Legumes*, Hulse, J.H., Ed., International Development Research Centre, Ottawa, Canada, 61–70.

Bressani, R., Elias, L.G., and Braham, J.E. 1982. Reduction of digestibility of legume proteins by tannins. *J. Plant Foods,* 4:43–55.

Brett, C.T., Wende, G., Smith, A.C., and Waldron, K.W. 1999. Biosynthesis of cell wall ferulate and diferulates. *J. Sci. Food Agric.*, 79:421–424.

Brieskorn, C.H. and Betz, R. 1988. Polymere Procyanidine, die praegenden Bestandteile der Mandel-Samenschale (Polymeric procyanidins: the characteristic constituents of the almond seedcoat). *Z. Lebensm. –Unters. Forsch.*, 187:347–353.

Briggs, D.E. 1978. *Barley,* Chapman & Hall, New York, NY, 128.

Bullard, R.G. and Elias, D.J., 1980. Sorghum polyphenols and bird resistance, in *Polyphenols in Cereal and Legumes*, Hulse, J.H., Ed., International Development Research Centre, Ottawa, Canada, 43–49.

Bushuk, W. 1976. History, world distribution, production and marketing, in *Rye: Production, Chemistry, and Technology,* Bushuk, W., Ed., American Association of Cereal Chemists, St. Paul, MN, 1–12.

Butler, L.G. 1992. Antinutritional effects of condensed and hydrolyzable tannins, in *Plant Polyphenols: Synthesis, Properties and Significance*, Hemigway, R.W. and Laks, P.E., Eds., Plenum Press, New York, 693–698.

Butler, L.G. and Pushpamma, P. 1991. Tannins in betel nut and its products consumed in India: comparison with high-tannin sorghum. *J. Agric. Food Chem.*, 39:322–326.

Butler, L.G. 1989. Sorghum polyphenols, in *Toxicants of Plant Origin,* vol. 4, Cheeke, P.R., Ed., CRC Press, Boca Raton, FL, 95–121.

Butler, L.G. 1982. Relative degree of polymerization of sorghum tannin during seed development and maturation. *J. Agric. Food Chem.*, 30:1090–1094.

Butler, L.G. and Price, M.L. 1977. Tannin biochemistry progress report, in Annual Report Inheritance and Improvement of Protein Quality and Content in Sorghu*m,* Rept. 13, Purdue University, West Lafayette, IN, 4.

Buttery, R.G. and Ling, L.C. 1998. Additional studies on flavor components of tortilla chips. *J. Agric. Food Chem.*, 46:2764–2769.

Campbell, A.D. 1992. Flour, flour mixtures, and cereal products, in *Food Theory and Applications*, 2nd ed., Bowers, J., Ed., MacMillan Publishing Company, New York, 259–358.

Cansfield, P.E., Marquardt, R.R., and Campbell, L.D. 1980. Condensed proanthocyanidins of faba beans. *J. Sci. Food Agric.*, 31:802–812.

Chavan, J.K., Kadam, S.S., Ghonsikar, C.P., and Salunkhe, D.K. 1979a. Removal of tannins and improvement of *in vitro* digestibility of sorghum seeds by soaking in alkali. *J. Food Sci.*, 44:1319–1321.

Chavan, J.K., Ghonsikar, C.P., Kadam, S.S., and Salunkhe, D.K. 1979b. Protein, tannin, and starch changes in developing seeds of low and high tannin cultivars of sorghum. *J. Food Biochem.*, 3:13–20.

Chen, K.T. and Geddes, W.F. 1945. Studies on wheat pigments, M.Sc. Thesis, University of Minnesota, St. Paul, MN.

Chibber, B.A.K., Mertz, E.T., and Axtell, J.D. 1978. Effect of dehulling on tannin content, protein distribution and quality of high and low tannin sorghum. *J. Agric. Food Chem.*, 26:679–682.

Chisholm, A., Mann, J., Skeaff, M., Frampton, C., Sutherland, W., Duncan, A., and Tiszavari, S. 1998. A diet rich in walnuts favourably influences plasma fatty acid profile in moderately hyperlipideamic subjects. *Eur. J. Clin. Nutr.*, 52:12–16.

Christensen, A.B., Gregersen, P.L., Olsen, C.E., and Collinge, D.B. 1998. A flavonoid 7-*O*-methyltransferase is expressed in leaves in response to pathogen attack. *Plant Mol. Biol.*, 36:219–227.

Collins, F.W. 1986. Oats phenolics: structure, occurence and function, in *Oats: Chemistry and Technology*, Webster, F.H., Ed., American Association of Cereal Chemists, St. Paul, MN, 227–295.

Collins, F.W. and Mullin, W.J. 1988. High performance liquid chromatographic determination of avenanthramides, N-aorylanthranilic acid alkaloids from oats. *J. Chromatogr.*, 445:363–370.

Collins, F.W. 1989. Oat phenolics: avenanthramides, novel substituted N-cinnamoylanthranilate alkaloids from oat groats and hulls. *J. Agric. Food Chem.*, 37:60–66.

Collins, F.W., McLachlan, D.C., and Blackwell, B.A. 1991. Oat phenolics: avenalumic acids, a novel group of bound phenolic acids from groats amd hulls. *Cereal Chem.*, 68:184–189.

Cooper, R., Gottlieb, H.E., and Lavie, D. 1978. Constituents of the Gramineae. II. Novel phenolic diglycerides from *Aegilops ovata. Phytochemistry,* 17:1673–1675.

Cooper, R., Lavie, D., Gutterman, Y., and Everani, M. 1994. The distribution of rare phenolic type compounds in wild and cultivated wheats. *J. Arid Environ.*, 27:331–336.

Cummings, D.P. and Axtell, J.D. 1973. Relationships of pigmented testa to nutritional quality of sorghum grain in inheritance and improvement of protein quality and content in sorghum, in *Res. Prog. Rep.*, Purdue University West Lafayette, IN, 112.

Daigle, D.J., Conkerton, E.J., Hammons, R.O., and Branch, W.D. 1983. A preliminary classification of selected white testa peanuts (*Arachis hypogaea* L.) by flavonoid analysis. *Peanut Sci.*, 10:40–43.

Daigle, D.J., Conkerton, E.J., Sanders, T.H., and Mixon, A.C. 1988. Peanut hull flavonoids: their relationship with peanut maturity. *J. Agric. Food Chem.*, 36:1179–1181.

Daniels, D.G.H. and Martin, H.F. 1961. Isolation of new antioxidant from oats. *Nature* (London), 191:1302.

Daniels, D.G.H. and Martin, H.F. 1965. Antioxidants in oats: diferulates of long chain diols. *Chem. Ind.* (London), 42:1763.

Daniels, D.G.H. and Martin, H.F. 1968. Antioxidants in oats: glyceryl esters of caffeic and ferulic acids. *J. Sci. Food Agric.*, 19:710–712.

Davis, A.B. and Hoseney, R.C. 1979. Grain sorghum condensed tannins. II Preharvest changes. *Cereal Chem.*, 56:314–316.

de Espana, M.E.F. 1977. Estudio sobre las posibles relaciones entre parametros fisicos, quimicos y nutricionales en *Phaseolus vulgaris*. B.S. Thesis CESNA/INCAP, cited in Bressani, R. and Elias, L.G. 1980. The nutritional role of polyphenols in beans, in *Polyphenols in Cereal and Legumes*, Hulse, J.H., Ed., The International Development Research Centre, Ottawa, Canada, 61–72.

Deguchi, T., Ohba, R., and Ueda, S. 1999. Effect of reactive oxygen and temperature on the formation of a purple pigment, hordeumin, from uncooked barley bran-fermented broth. *Biosci. Biotechnol. Biochem.*, 63:1151–1155.

Deguchi, T., Ohba, R., and Ueda, S. 2000a. Radical scavenging activity of a purple pigment, hordeumin, from uncooked barley bran-fermented broth. *J. Agric. Food Chem.*, 48:3198–3201.

Deguchi, T., Yoshimoto, M., Ohba, R., and Ueda, S. 2000b. Antimutagenicity of the purple pigment, hordeumin, from uncooked barley bran-fermented broth. *Biosci. Biotechnol. Biochem.*, 64:414–416.

Delcour, J.A. and Tuytens, G. 1984. Structure elucidation of three dimeric proanthocyanidins isolated from a commercial Belgium pilsner beer. *J. Inst. Brew.*, 90:153–161.

de Lumen, B.O. and Salamat, L.A. 1980. Trypsin inhibitor activity in winged beans (*Psophocarpus tetragonolobus* var. *Chimbu*) and the possible role of tannin. *J. Agric. Food Chem.*, 28:533–536.

Deokar, V.V. 1987. Investigation of nutritive value and polyphenol-protein interactions in pearl millet (*Pennisetum typhoides*), M.Sc. thesis, Mahatma Phule Agricultural Univ., Rahuri, India, cited in Salunkhe, D.K., Chavan, J.K., and Kadam, S.S. 1990. *Dietary Tannins: Consequences and Remedies*, CRC Press, Boca Raton, FL.

de Pascual-Teresa, S., Guitierrez-Fernandez, Y., Rivas-Gonzalo, J.C., and Santos-Buelga, C. 1998. Characterization of monomeric and oligomeric flavan-3-ols from unripe almond fruits. *Phytochem. Anal.*, 9:21–27.

Desai, B.B. and Zende, G.K. 1979. Role of bajra (*Pennisetum typhoides*) in human and animal nutrition. *Indian J. Nutr. Dietet.*, 16:390–396.

Deschner, E.E. 1992. Dietary quercetin and rutin: Inhibitors of experimental colonic neoplasia, in *Phenolic Compounds in Foods and Their Effects on Health. II Antioxidants and Cancer Prevention,* Huang, M.T., Ho, C-T., and Lee, C.Y., Eds., ACS Symposium Series 547, American Chemical Society, Washington, D.C., 265–268.

Deshpande, S.S., Sathe, S.K., Salunkhe, D.K., and Cornforth, D.P. 1982. Effects of dehulling on phytic acid, polyphenols, and enzyme inhbitors of dry beans (*Phaseolus vulgaris* L.). *J. Food Sci.*, 47:1846–1850.

Deshpande, S.S. and Cheryan, M. 1987. Determination of phenolic compounds of dry beans using vanillin, redox and precipitation assays. *J. Food Sci.*, 52:332–334.

Devlin, J.P.A. and Hargrave, K.D. 1985. Pulmonary and antiallergic drugs; design and synthesis. *Chem. Pharmacol. Drugs*, 4:191–312.

Diack, M. and Saska, M. 1994. Separation of vitamin E and γ-oryzanols from rice bran by normal-phase chromatography. *J. Am. Oil Chem. Soc.*, 71:1211–1217.

Dicko, M.H., Hilhorst, R., Gruppen, H., Traore, A.S., Laane, C., van Berkel, W.J.H., and Voragen, A.G.J. 2002. Comparison of content of phenolic compounds, polyphenol oxidase, and peroxidase in grains of fifty sorghum varieties from Burkina Faso. *J. Agric. Food Chem.*, 50:3780–3788.

Dietrych-Szostak, D. and Oleszek, W. 1999. Effect of processing on the flavonoid content in buckwheat (*Fagopyrum esculentum* Moench) grain. *J. Agric. Food Chem.*, 47:4384–4387.

Dimberg, L.H., Theander, O., and Lingnert, H. 1993. Avenanthramides — a group of phenolic antioxidants in oats. *Cereal Chem.*, 70:637–640.

Dimberg, L.H., Molteberg, E.L., Solheim, R., and Frolich, W. 1996. Variation in oat groats due to variety, storage, and heat treatment. I. Phenolic compounds. *J. Cereal Sci.*, 24:263–272.

Dini, I., Schettino, O., and Dini, A. 1998. Studies on the constituents of *Lupinus mutabilis* (Fabaceae). Isolation and characterization of two new isoflavonoid derivatives. *J. Agric. Food Chem.*, 46:5089–5092.

Doll, H., Holm, U., Bay, H., and Sogaard, B. 1994. Phenolic compounds in barley varieties with different degree of partial resistance against powdery mildew. *Acta Hort.* 381:576–582.

Dorner, J. W., Cole, R.J., Sanders, T.H., and Blankenship, P.D. 1989. Interrelationship of kernel water activity, soil temperature, maturity, and phytoalexin production in preharvest aflatoxin contamination of drout-stressed peanuts. *Mycopathologia*, 105:117–128.

Duh, P.-D., Yen, G.-C., Yen, W.-J., and Chang, L.-W. 2001. Antioxidant effects of water extracts from barley (*Hordeum vulgare* L.) prepared under different roasting temperatures. *J. Agric. Food Chem.*, 49:1455–1463.

Duke, J.A. 1981. Handbook of Legumes of World Economic Importance, Plenum Press, New York.

Duke, J.A. 1985. CRC Handbook of Medicinal Herbs, CRC Press, Boca Raton, FL, 57–58.

Durkee, A.B. 1977. Polyphenols of the bran-aleurone fraction of buckwheat seed (*Fagopyrum sagitatum*. Gilib). *J. Agric. Food Chem.*, 25:286–287.

Durkee, A.B. and Thivierge, P.A. 1977. Ferulic acid and other phenolics in oat seeds (*Avena sativa* L. var. Hinoat). *J. Food Sci.*, 42:551–552.

Ehmann, A. 1974. N-(*p*-coumaryl)-tryptamine and N-(ferulyl)-tryptamine in kernels of *Zea mays*. *Phytochemistry,* 13:1979–1983.

El-Basyouni, S. and Towers, G.H.N. 1964. The phenolic acids in wheat. 1. Changes during growth and development. *Can. J. Biochem.*, 42:203–210.

Elias, L.G., Fernandez, D.G., and Bressani, R. 1979. Possible effects of seed coat polyphenolics on the nutritional quality of bean protein. *J. Food Sci.*, 44:524–527.

Elkin, R.G., Freed, M.B., Hamaker, B.R., Zhang, Y., and Parsons, C.M. 1996. Condensed tannins are only partially responsible for variations in nutrient digestibilities of sorghum grain cultivars. *J. Agric. Food Chem.*, 44:848–853.

Emmons, C.L., Peterson, D.M., and Paul, G.L. 1999. Antioxidant capacity of oat (*Avena sativa* L.) extracts. 2. *In vitro* antioxidant activity and contents of phenolic and tecol antioxidants. *J. Agric. Food Chem.*, 47:4894–4898.

Emmons, C.L. and Peterson, D.M. 1999. Antioxidant activity and phenolic contents of oat groats and hulls. *Cereal Chem.*, 76:902–906.

Fargues, C., Mathias, A., and Rodrigues, A.E. 1996. Kinetics of vanillin production from Kraft lignin oxidation. *Ind. Eng. Chem. Res.*, 35:28–36.

Farrer, C.E. 1935. The determination of iron in biological materials. *J. Biol. Chem.*, 110:685–694.

Faulds, C.B. and Williamson, G. 1999. The role of hydroxycinnamates in the plant cell wall, review. *J. Sci. Food Agric.* 79:393–395.

Faurot, A., Saulnier, L., Bérot, S., Popineau, Y., Petit, M., Rouau, X., and Thibault, J.F. 1995. Large scale isolation of water-soluble pentosans from wheat flour. *Lebensm.-Wiss. Technol.*, 28:436–441.

Feenstra, W.J. 1960. Biochemical aspects of seedcoat colour inheritance in *Phaseolus vulgaris* L. Meded. *Landbouwhogesch. Wageningen*, 60:1–53.

Feng, Y., McDonald, C.E., and Vick, B.A. 1988. C-glycosylflavones from hard red spring wheat bran. *Cereal Chem.*, 65:452–456.

Feng, Y. and McDonald, C.E. 1989. Comparison of flavonoids in four classes of wheat. *Cereal Chem.*, 66:516–518.

Foo, L.Y. and Karchesy, J.J. 1988. Chemical nature of phlobaphenes, in *Chemistry and Significance of Condensed Tannins*, Hemingway, R.W. and Karchesy, J.J., Eds., Plenum Press, New York, 109–118.

Foreign Agricultural Service. 1998. FAS Online; U.S. Department of Agriculture, Washington, D.C; http://www.fas.usda.gov/htp/highlights/1998/98–03/uswalnut.html.

Fossen, T., Slimestad, R., and Andersen, O.M. 2001. Anthocyanins from maize (*Zea mays*) and red canarygrass (*Phalaris arundinacea*). *J. Agric. Food Chem.*, 49:2318–2321.

Frison, S. and Sporns, P. 2002. Variation in the flavonol glucoside composition of almond seed coats as determined by MALDI-TOF mass spectrometry. *J. Agric. Food Chem.*, 50:6818–6822.

Frison-Norrie, S. and Sporns, P. 2002. Identification and quantification of flavonol glycosides in almond seedcoats using MALDI-TOF-MS. *J. Agric. Food Chem.*, 50:2782–2787.

Fujimaki, M., Tsugita, T., and Kurata, T. 1977. Fractionation and identification of volatile acids and phenols in steam distillate of rice bran. *Agric. Biol. Chem.*, 41:1721–1725.

Fulcher, R.G. 1982. Fluorescence microscopy of cereals. *Food Microstruct.*, 1:167–175.

Fulcher, R.G., O'Brien, T.P., and Lee, J. 1972. Studies of the aleurone layer. I. Conventional and fluorescence microscopy of the cell wall with emphasis on phenol carbohydrate complexes in wheat. *Aust. J. Biol. Sci.*, 25:23–24.

Garcia, E., Filisetti, T.M.C.C., Udaeta, J.E.M., and Lajolo, F.M. 1998. Hard-to-cook beans (*Phaseolus vulgaris*): involvement of phenolic compounds and pectates. *J. Agric. Food Chem.*, 46:2110–2116.

Garcia-Conesa, M.T., Plumb, G.W., Waldron, K.W., Ralph, J., and Williamson, G. 1997. Ferulic acid dehydrodimers from wheat bran: isolation, purification and antioxidant properties of 8-*O*-4′-diferulic acid. *Redox. Rep.*, 3:319–323.

Geissmann, T. and Neukom, H. 1973. On the composition of the water soluble wheat flour pentosans and their oxidative gelation. *Lebensm.-Wiss. Technol.*, 6:59–62.

Giri, A.K., Banerjee, T.S., Talukder, G., and Sharma, A. 1987. Induction of sister chromatid exchange and dominant lethal mutation by "Katha" (catechu) in male mice. *Cancer Lett.*, 36:189–196.

Glennie, C.W., Kaluza, W.Z., and van Niekerk, P.J. 1981. High-performance liquid chromatography of procyanidins in developing sorghum grain. *J. Agric. Food Chem.*, 29:965–968.

Goldberg, D.M. 1995. Does wine work? *Clin. Chem.*, 41:14–16.

Goldstein, J.L. and Swain, T. 1963. Methods of determining the degree of polymerization of flavans. *Nature,* 198:587–578.

Gothoskar, S.V. and Pai, S.R. 1986. A study on the parotid gland tumor induced in Swiss mice by tannin containing fraction of betel nut. *Indian J. Exp. Bot.*, 24:229–231.

Goupy, P., Hugues, M., Boivin, P., and Amiot, M.J. 1999. Antioxidant composition and activity of barley (*Hordeum vulgarae)* and malt extracts and isolated phenolic compounds. *J. Sci. Food Agric.*, 79:1625–1634.

Grabber, J.H., Ralph, J., and Hatfield, R.D. 2000. Cross-linking of maize walls by ferulate dimerization and incorporation into lignin. *J. Agric. Food Chem.*, 48:6106–6113.

Gramshaw, J.W. 1968. Phenolic constituents of beer and beer materials III. Simple anthocyanogens from beer. *J. Inst. Brew.*, 74:20–38.

Gramshaw, J.W. 1969. Phenolic constituents of beer and brewing materials. IV. Anthocyanogens and catechins as haze precursors in beer. *J. Inst. Brew.*, 75:61–83.

Grazzini, R., Hesk, D., Heininger, E., Hildebrandt, G., Reddy, C.C., Cox-Foster, D., Medferd, J., Craig, R., and Mumma, R.O. 1991. Inhibition of lipoxygenase and prostaglandin endoperoxide synthase by anarcadic acids. *Biochem. Biophys. Res. Commun.*, 176:775–780.

Griffith, D.W. and Welch, R.W. 1982. Genotypic and environmental variation in the total phenol and flavanol contents of barley grain. *J. Sci. Food Agric.*, 33:521–527.

Gujer, R., Magnolato, D., and Self, R. 1986. Proanthocyanidins and polymeric flavonoids from sorghum. *Prog. Clin. Biol. Res.*, 213:159–161.

Gupta, R.K. and Haslam, E. 1978. Plant proanthocyanidins. Part 5. Sorghum polyphenols. *J. Chem. Soc. Perkin Trans.*1, 8:892–896.

Hagerman, A.E. and Butler, L.G. 1978. Protein precipitation method for quantitative determination of tannins. *J. Agric. Food Chem.*, 26:809–812.

Hahn, D.H., Faubion, J.M., and Rooney, L.W. 1983. Sorghum phenolic acids, their high performance liquid chromatography separation and their relation to fungal resistance. *Cereal Chem.*, 60:255–259.

Hahn, D.H., Rooney, L.W., and Earp, C.E. 1984. Tannins and phenols of sorghum. *Cereal Food World*, 29:757.

Hahn, D.H. and Rooney, L.W. 1986. Effect of genotype on tannins and phenols of sorghum. *Cereal Chem.*, 63:4–8.

Hakala, P., Lampi, A.-M., Ollilainen, V., Werner, U., Murkovic, M., Wähälä, K., Karkola, S., and Piironen, V. 2002. Steryl phenolic esters in cereals and their milling fractions. *J. Agric. Food Chem.*, 50:5300–5307.

Harborne, J.B. and Gavazzi, G. 1969. Effect of Pr and pr alleles on anthocyanin biosynthesis in *Zea mays*. *Phytochemistry,* 8:999–1001.

Harris, H.B. and Burns, R.E. 1970. Influence of tannin content on preharvest seed germination in sorghum. *Agron. J.,* 62:835–836.

Harris, P.J. and Hartley, R.D. 1976. Detection of bound ferulic acid in cell walls of the Gramineae by ultraviolet fluorescence microscopy. *Nature,* 259:508–510.

Hartley, R.D., Morrisson, W.H., Hilmmelsbach, D.S., and Borneman, W.S. 1990. Cross-linking of cell wall phenolic arabinoxylans in graminaceous plants. *Phytochemistry,* 29:3705–3709.

Hartley, R.D., Jones, E.C., and Wood, T.M. 1976. Lignin-carbohydrate linkages in plant cell walls. 3. Carbohydrates and carbohydrate esters or ferulic acid released from cell walls of *Lolium multiflorum* by treatment with cellulolytic enzymes. *Phytochemistry,* 15:305–307.

Haslam, E. 1977. Symmetry and promiscuity in procyanidin biochemistry. *Phytochemistry,* 16:1625–1640.

Hatfield, R.D., Ralph, J., and Grabber, J.H. 1999. Review. Cell cross-linking by ferulates and diferulates in grasses. *J. Sci. Food Agric.*, 79:403–407.

Heaton, E.K., Worthington, R.E., and Shewfelt, A.L. 1975. Pecan nut quality. Effect of time of harvest on composition, sensory and quality characteristics. *J. Food Sci.*, 40:1260–1263.

Hempel, J. and Bohm, H. 1996. Quality and quantity of prevailing flavonoid glycosides of yellow and green French beans (*Phaseolus vulgaris* L.). *J. Agric. Food Chem.*, 44:2114–2116.

Henry, S.H., Bosch, F.X., Troxell, T.C., and Bolger, P.M. 1999. Reducing liver cancer: global control of aflatoxin. *Science*, 286:2453–2454.

Hernanz, D., Nuñez, V., Sancho, A.I., Faulds, C.B., Williamson, G., Bartolome, B., and Gomez-Cordoves, C. 2001. Hydrocinnamic acids and ferulic acid dehydrodimers in barley and processed barley. *J. Agric. Food Chem.*, 49:4884–4888.

Herrmann, K. 1989. Occurrence and content of hydroxycinnamic and hydroxybenzoic acid compounds in foods. *CRC Crit. Rev. Food Sci. Nutr.*, 28:315–347.

Hertog, M.G.L., Hollman, P.C.H., and Katan, M.B. 1992. Content of potentially anticarcinogenic flavonoids of 28 vegetables and 9 fruits commonly consumed in the Netherlands. *J. Agric. Food Chem.*, 40:2379–2383.

Higuchi, T., Ito, Y., Shimada, M., and Kawamura, I. 1967. Chemical properties of milled wood lignin grasses. *Phytochemistry,* 6:1551–1556.

Hilhorst, R., Dunnewind, B., Orsel, R., Stegeman, P., van Vliet, T. Gruppen, H., and Schols, H.A. 1999. Baking performance, rheology, and chemical composition of wheat dough and gluten affected by xylanase and oxidative enzymes. *J. Food Sci.*, 64:453–457.

Himejima, M. and Kubo, I. 1991. Antibacterial agents from the cashew *Anacardium occidentale* (Anacardiacae) nut shell oil. *J. Agric Food Chem.*, 39:418–421.

Hincks, M.J. and Stanley, D.W. 1986. Multiple mechanisms of bean hardening, *J. Food Technol.*, 21:734–750.

Hinze, G. 1972. Millets in Colorado. Bull. 553S. Colorado State University Experiment Station, Fort Collins, 12 pp., cited in Lorenz, K. 1983. Tannins and phytate content in proso millets (*Panicum miliaceum*). *Cereal Chem.*, 60:424–429.

Hoseney, R.C., Variano-Marston, E., and Dendy, D.A.V. 1981. Sorghum and millet, in *Advances in Cereal Science and Technology*, vol. IV, Pomeranz, Y., Ed., American Association of Cereal Chemists, St Paul, MN, 71.

Hosny, M. and Rosazza, J.P. 1997. Structures of ferulic acid glycoside esters in corn hulls. *J. Nat. Prod.*, 60:219–222.

Hu, F.B., Stampfer, M.J., Manson, J.E., Rimm, E.B., Colditz, G.A., Rosner, B.A., Speizer, F.E., Hennekens, C.H., and Willett, W.C. 1998. Frequent nut consumption and risk of coronary heart disease in women: prospective cohort study. *Br. Med. J.*, 317:1341–1345.

Ibern-Gómez, M., Roig-Pérez, S., Lamuela-Raventós, and de la Torre-Boronat, M.C. 2000. Resveratrol and piceid levels in natural and blended peanut butters. *J. Agric. Food Chem.*, 48:6352–6354.

Iiyama, K., Lam, T-T., and Stone, B.A. 1994. Covalent cross links in the cell wall. *Plant Physiol.*, 104:315–320.

Iiyama, K., Lam, T.B., and Stone, B.A. 1990. Phenolic acid bridges between polysaccharides and lignin in wheat internodes. *Phytochemistry,* 29:733–737.

Ingham, J.L. 1976. 3,5,4′-Trihydroxystilbene as a phytoalexin from groundnuts (*Arachis hypogaea*). *Phytochemistry,* 15:1791–1793.

Ishii, T. and Hiroi, T. 1990. Linkage of phenolic acids to cell-wall polysaccharides of bamboo shoot. *Carbohydr. Res.*, 206:297–310.

Ishii, T. 1997. Structure and functions of feruolylated polysaccaharides, review. *Plant Sci.*, 127:111–127.

Iwata, K., Miwa, S., Inayama, T., Sasaki, H., Soeda, K., and Sugahara, T. 1990. Effects of kangra buckwheat on spontaneously hypertensive rats. *J. Kagawa Nutr. Coll.*, 21:55–61.

Izydorczyk, M.S., Biliarderis, C.G., and Bushuk, W. 1991. Physical properties of water-soluble pentosans from different wheat varieties. *Cereal Chem.*, 68:145–150.

Jackson, G.M. and Hoseney, R.C. 1986. Effect of endogenous phenolic acids on the mixing properties of wheat flours doughs. *J. Cereal Sci.*, 4:79–85.

Jambunathan, R. and Mertz, E.T. 1973. Relationship between tannin levels, rat growth, and distribution of protein in sorghum. *J. Agric. Food Chem.*, 21:692–696.

Jang, M., Cai, L., Udeani, G.O., Slowing, K.V., Thomas, C.F., Beecher, C.W.W., Fong, H.H.S., Farnsworth, N.R., Kinghorn, A.D., Mehta, R.G., Moon, R.C., and Pezzuto, J.M. 1997. Cancer chemopreventive activity of resveratrol, a natural product derived from grapes. *Science*, 275:218–220.

Jones, P.M.B. and Boulter, D. 1983a. The cause of reduced cooking rate in *Phaseolus vulgaris* following adverse storage conditions. *J. Food Sci.*, 48:623–626, 649.

Jones, P.M.B. and Boulter, D. 1983b. The analysis of development of hard-to-cook bean during storage of black beans (*Phaseolus vulgaris* L.). *Qual. Plant Plant Foods Hum. Nutr.*, 33:77–85.

Jurd, L. 1958. Plant polyphenols. III. The isolation of a new ellagitannin from the pellicle of the walnut. *J. Am. Chem. Soc.*, 80:2249–2252.

Jurd, L. 1956. Plant polyphenols. I. The polyphenolic constituents of the pellicle of the walnut (*Juglans regia*). *J. Am. Chem. Soc.*, 78:3445–3448.

Kadam, S.S., Kute, L.S., Lawande, K.M., and Salunke, D.K. 1982. Changes in chemical composition of winged bean (*Psophocarpus tetragonolobus* L.) during seed development. *J. Food Sci.*, 47:2051–2053, 2057.

Kambal, A.E. and Bate-Smith, E.C.A. 1976. Genetic and biochemical study on pericarp pigments in a cross between two cultivars of grain sorghum, *Sorghum bicolor. Heredity*, 37:413–416.

Kamisaka, S., Takeda, S., Takahashi, K., and Shibata, K. 1990. Diferulic and ferulic acid in the cell wall of *Avena coleoptiles:* their realtionship to mechanical properties of the cell wall. *Physiol. Plant*, 78:1–7.

Karchesy, J. and Hemingway, R.W. 1986. Condensed tannins: (4β→8; 2β→*O*→7)-linked procyanidins in *Arachis hypogea* L. *J. Agric. Food Chem.*, 34:966–970.

Kawashima, S., Hayashi, M., Takii, T., Kimura, H., Zhang, H.L., Nagatsu, A., Sakakibara, J., Murata, K., Oomoto, Y., and Onozaki, K. 1998. Serotonin derivative, N-(*p*-coumaroyl)serotonin, inhibits the production of TNF-α, IL-1α, IL-1β, and IL-6 by endotoxin-stimulated human blood monocytes. *J. Interferon Cytokine Res.*, 18:423–428.

Khetarpaul, N. 1988. Improvement of nutritional value of pearl millet by fermentation and utilisation of fermented product, Ph.D. thesis, Haryana Agricultural Univ. Hisar, India, cited in Khetarpaul, N. and Chauhan, B.M. 1990. Effects of germination and pure culture fermentation by yeasts and *Lactobacilli* on phytic acid and polyphenol content of pearl millet. *J. Food Sci.*, 55:1180; 1182.

Khetarpaul, N. and Chauhan, B.M. 1990. Effects of germination and pure culture fermentation by yeasts and *Lactobacilli* on phytic acid and polyphenol content of pearl millet. *J. Food Sci.*, 55:1180,1182.

King, H.G.C. 1962. Phenolic compounds of commercial wheat germ. *J. Food Sci.*, 27:446–454.

Kirby, L.T. and Styles E.D. 1970. Flavonoids associated with specific gene action in maize aleurone, and the role of light in substituting for the action of gene. *Can. J. Genet. Cytol.*, 12:934–940.

Kirchhoff, E. and Schieberle, P. 2001a. Determination of key aroma compounds in the crumb of a three-stage sourdough rye bread by stable isotope dilution assays and sensory studies. *J. Agric. Food Chem.*, 49:4304–4311.

Kitabayashi, H., Ujihara, A., Hirose, T., and Minami, M. 1995a. Varietal differences and heritability for rutin content in common buckwheat, *Fagopyrum esculentum* Moench. *Breed. Sci.*, 45:75–79.

Kitabayashi, H., Ujihara, A., Hirose, T., and Minami, M. 1995b. On the genetic differences for rutin content in buckwheat, *Fagopyrum tataricum* Gaertn. *Breed Sci.*, 45:189–194.

Kitta, K., Hagiwara, Y., and Shibamoto, T. 1992. Antioxidative activity of an isoflavonoid, 2"-*O*-glycosylisovitexin isolated from green barley leaves. *J. Agric. Food Chem.*, 40:1843–1845.

Kolattukudy, P.E., Espelie, K.E., and Soliday, C.L. 1981. Hydrophobic layers attached to cell walls and associated with waxes. Encycl. *Plant Physiol.* New Series 13B, p. 225.

Kreft, S., Knapp, M., and Kreft I. 1999. Extraction of rutin from buckwheat (*Fagopyrum esculentum* Moench) seeds and determination by capillary electrophoresis. *J. Agric. Food Chem.*, 47:4649–4652.

Kris-Etherton, P.M., Yu-Poth, S., Sabate, J., Ratcliff, H.E., Zhao, G., and Etherton, T.D. 1999. Nuts and their bioactive constituents: effects on serum lipids and other factors that affect disease risk. *Am. J. Clin. Nutr.*, 70:504S-511S.

Krueger, C.G., Vestling, M.M., and Reed, J.D. 2003. Matrix-assisted laser description/ionization time-of-flight mass spectrometry of heteropolyflavan-3-ols and glucosylated heteropolyflavans in sorghum [*Sorghum bicolor* (L.) Moench]. *J. Agric. Food Chem.*, 51:538–543.

Kubo, S., Komatsu, S., amd Ochi, M. 1986. Molluscicides from the cashew *Anarcadium occidentale* and their large scale isolation. *J. Agric Food Chem.*, 34:970–973.

Kubo, S., Kim, M., Naya, K., Komatsu, S., Yamagiwa, Y., Ohashi, K., Sakamoto, Y., Harakawa, S., and Kamikawa, T. 1987. Prostoglandin synthetase inhibitors from the African medicinal plant *Ozoroa mucronata. Chem. Lett.*, 1101–1104.

Kubo, I., Ochi, M., Vieira, P.C., and Komatsu, S. 1993. Antitumor agents from the cashew (*Anarcardium occidentale)* apple juice. *J. Agric. Food Chem.*, 41:1012–1015.

Kubo, I., Muroi, H., and Kubo, A. 1994a. Naturally occuring antiacne agents. *J. Nat. Prod.*, 57:59–62.

Kubo, I., Kinst-Hori, I., and Yokokawa, Y. 1994b. Tyrosinase inhibitors from *Anarcadium occidentale* fruits. *J. Nat. Prod.*, 57:545–551.

Kubo, J., Lee, J.R., and Kubo, I. 1999. Anti-*Helicobacter pyroli* agents from the cashew apple. *J. Agric. Food Chem.*, 47:533–537.

Kumar, P.P., Paramashivappa, R., Vithayathil, P.J., Rao, P.V.S., and Rao, A.S. 2002. Process for isolation of cardanol from technical cashew (*Anacardium occindentale* L.) nut shell liquid. *J. Agric. Food Chem.*, 50:4705–4708.

Labat, E., Morel, M.-H., and Rouau, X. 2000. Wheat gluten phenolic acids: occurrence and fate upon mixing. *J. Agric. Food Chem.*, 48:6280–6283.

Lam, T.B.T., Iiyama, K., and Stone, B.A. 1992. Cinnamic acid bridges between cell wall polysaccharides in wheat and phalaris internodes. *Phytochemistry*, 31:1179–1183.

Lasztity, R., Berndorfer-Kraszner, E., and Huszar, M. 1980. On the presence and distribution of some bioactive agents in oat varieties, in *Recent Progress in Cereal Chemistry*, Inglett, G.E. and Munck, L., Eds., Academic Press, New York, 429–445.

Lavendrine, F., Zmirou, D., Ravel, A., Balducci, F., and Alary, J. 1999. Blood cholesterol and walnut consumption: a cross-sectional survey in France. *Prev. Med.*, 28:333–339.

Leistner, E. 1981. Biosynthesis of plant quinones, in *Biochemistry of Plants,* Conn, E.E., Ed., Academic Press, New York, 403–423.

Lempereur, I., Rouau, X., and Abecassis, J. 1997. Genetic and agronomic variation in arabinoxylan and ferulic acid contents of durum wheat (*Triticum durum* L.) grain and its milling fractions. *J. Cereal Sci.*, 25:103–110.

Lempereur, I., Surget, A., and Rouau, X. 1998. Variability in dehydrodiferulic acid composition of durum wheat (*Triticum durum* Desf.) and distribution in milling fractions. *J. Cereal Sci.*, 28:251–258.

Lorenz, K. 1983. Tannins and phytate content in proso millets (*Panicum miliaceum*). *Cereal Chem.*, 60:424–429.

Lou, H., Yuan, H., Yamazaki, Y., Sasaki, T., and Oka, S. 2001. Alkaloids and flavonoids from peanut skin. *Planta Med.*, 67:345–349.

Lou, H., Yamazaki, Y., Sasaki, T., Uchida, M., Tanaka, H., and Oka, S. 1999. A-type proanthocyanidins from peanut skin. *Phytochemistry,* 51:297–308.

Łuczak, S., Świątek, L., and Zadernowski, R. 1989. Phenolic acids in leaves and pericarpium of walnut *Juglans regia,* L. *Acta Polon. Pharm.*, 46:494–498.

Luh, B.S., Ed., *Rice: Production and Utilization,* AVI Publishing Co., Westport, CT.

Luthar, Z. and Kreft, I. 1996. Composition of tannins in buckwheat (*Fagopyrum esculentum* Moench) seeds. *Res. Rep. Agric.*, 67:59–65.

Maga, J.A. and Lorenz, K. 1973. Taste thresholds values for phenolic acids that can influence flavor properties of certain flours, grains, and oilseeds. *Cereal Sci. Today*, 18:326–329.

Maga, J.A. and Lorenz, K. 1974. Gas-liquid chromatography separation of the free phenolic acid fractions in various oilseed protein sources. *J. Sci. Food Agric.*, 25:797–802.

Maga, J.A. 1978. Simple phenol and phenolic compounds in food flavor. *CRC Crit. Rev. Food Sci. Nutr.*, 11:323.

Mahajan, S. and Chauhan, B.M. 1987. Phytic acid and extractable phosphorus of pearl millet as affected by natural lactic acid fermentation. *J. Sci. Food Agric.*, 41:381–386.

Mahajan, S. and Chauhan, B.M. 1988. Effect of natural fermentation on the extractability of minerals from pearl millet flour. *J. Food Sci.*, 53:1576–1577.

Mahoney, N.E., Molyneux, R.J., and McGranahan, G. 1998. Effect of walnut tissues on *Apsergillus flavus* growth and aflatoxin production, in *Walnut Research Reports 1997,* Walnut Marketing Board, Sacramento, CA, 379–380.

Mahoney, N., Molyneux, R.J., and Campbell, B.C. 2000. Regulation of aflatoxin production by naftoquinones of walnut (*Juglans regia*). *J. Agric. Food Chem.*, 48:4418–4421.

Margna, U.V. and Margna, E.R. 1982. Differential nature of quantitative shifta of flavonoid accumulation in buckwheat seedlings of different ages. *Sov. Plant Physiol.*, 29:223–230.

Markwalder, H.U. and Neukom, H. 1976. Diferullic acid as possible crosslink in hemicelluloses from wheat germ. *Phytochemistry,* 15:836–837.

Marquardt, R.R., Ward, A.T., Campbell, L.D., and Cansfield, P.E. 1977. Purification, identification and characterization of a growth inhibitor in faba beans (*Vicia faba* L. var. minor). *J. Nutr.*, 107:1313–1324.

Marquardt, R.R., Ward, A.T., and Evans, L.E. 1978. Comparative cell wall constituent levels in tannin-free and tannin-containing cultivars of faba beans (*Vicia faba,* L.). *Can. J. Animal Sci.*, 58:775–781.

Martin-Tanguy, J., Pedrizet, E., Prevost J., and Martin, C. 1982. Hydroxycinnamic cid amides in fertile and cytoplasmic male sterile lines of maize. *Phytochemistry,* 21:1939–1945.

Martin-Tanguy, J., Cabanne, F., Perdrizet, E., and Martin, C. 1978. The distribution of hydroxycinnamic acid amides in flowering plants. *Phytochemistry,* 17:1927–1928.

Martin-Tanguy, J., Guillaume, J., and Kossa, A. 1977. Condensed tannins in horse bean seeds: chemical structure and apparent effect on poultry. *J. Sci. Food Agric.*, 28:757–764.

Mathew, A.G. and Parpia, H.A.B. 1970. Polyphenols of cashew kernel testa. *J. Food Sci.*, 35:140–145.

Maxson, E.D., Clark, L.E., Rooney, L.W., and Jonhson, J.W. 1972. Factors affecting the tannin content of sorghum grain as determined by two methods of tannin analysis. *Crop Sci.*, 12:233–235.

Maxson, E.D. and Rooney, L.W. 1972. Evaluation of methods for tannin analysis in sorghum grain. *Cereal Chem.*, 12:719–729.

Ma Yu and Bliss, F.A. 1978. Tannin content and inheritance in common beans. *Crop Sci.*, 18:201–204.

Mazza, G. 1993. Buckwheat, in *Encyclopedia of Food Science, Food Technology and Nutrition*, vol. 1, Macrae, R., Robinson, R.K., and Sadler, M.J., Eds., Academic Press, Toronto, Canada, 516–521.

McCallum, J.A. and Walker, J.R.L. 1991. Phenolic biosynthesis during grain development in wheat (*Trticum aestivum* L.). III. Changes in hydroxycinnamic acids during grain development. *J. Cereal Sci.*, 13:161–172.

McDougall, G.J. 1993. Phenolic cross-links in growth and development of plants, in *Polyphenolic Phenomena,* Scalbert, A., Ed., INRA Editions, Paris, 129–136.

McKeehan, J.D., Busch, R.H., and Fulcher, R.G. 1999. Evaluation of wheat (*Triticum aestivum* L.) phenolic acids during grain development and their contribution to *Fusarium* resistance. *J. Agric. Food Chem.*, 47:1476–1482.

McMillan, W.W., Wiseman, B.R., Burns, R.E., Harris, H.B., and Greene, G.L. 1972. Bird resistance in diverse germplasm of sorghum. *Agron. J.*, 64:821–822.

McMurrough, I., Loughrey, M.J., and Hennigan, G.P. 1983. Content of (+)-catechin and proanthocyanidins in barley and malt grain. *J. Sci. Food Agric.*, 34:62–72.

McMurrough, I., Madigan, D., and Smyth, M.R. 1996. Semipreparative chromatographic procedure for the isolation of dimeric and trimeric proanthocyanidins from barley. *J. Agric. Food Chem.*, 44:1731–1735.

McWatters, K.H. and Cherry, J.P. 1975. Functional properties of peanut paste as affected by moist heat treatment of full-fat peanuts. *J. Food Sci.*, 40:1257–1259.

McWatters, K.H. and Holmes, M.R. 1979. Influence of moist heat on solubility and emulsification properties of soy and peanut flours. *J. Food Sci.*, 44:474–479.

Mejia, E.G., 1979. Effect of various conditions on general aspects of bean hardening. Final Report UMU Fellow. Instituto de Nutricion de Centro America y Panama, Guatemala, cited in Sievwright, C.A. and Shipe, W.F. 1986. Effect of storage conditions and chemical treatments on firmness, *in vitro* protein digestibility, condensed tannins, phytic acid and divalent cations of cooked black beans (*Phaseolus vulgaris*). *J. Food Sci.*, 51:982–987.

Melake-Berhan, A., Butler, L.G., Ejeta, G., and Menkir, A. 1996. Grain mold resistance and polyphenol accumulation in sorghum. *J. Agric. Food Chem.*, 44:2428–2434.

Menninger, E.A. 1977. *Edible Nuts of the World,* Horticultural Books, Inc., Stuart, FL.

Menon, A.R., Pillai, C.K., Sudha, J.D., and Mathew, A.G. 1985. Cashew nut shell liquid — its polymeric and other industrial products. *J. Sci. Ind. Res.*, 44:324–338.

Molina, M.R., De la Fuente, G., and Bressani, R. 1975. Interrelationships between storage, soaking time, cooking time, nutritive value and other characteristics of the black beans (*Phaseolus vulgaris*). *J. Food Sci.*, 40:587–591.

Mueller-Harvey, I., Harley, R.D., Harris, P.J., and Curzon, E.H. 1986. Linkage of *p*-coumaryl and feruoyl groups to cell-wall polysaccharides of barley straw. *Carbohyd. Res.*, 148:71–85.

Muller, W.U. and Leistner, E. 1978a. Metabolic relation between naphtalene derivatives in *Juglans* (regia). *Phytochemistry,* 17:1735–1738.

Muller, W.U. and Leistner, E. 1978b. Aglycones and glycosides of oxygenated naphtalenes and glycosyltransferase from *Juglans* (spp.). *Phytochemistry,* 17:1739–1742.

Muneta, P. 1964. The cooking time of dry beans after extended storage. *Food Technol.*, 18:1240–1241.

Murray, E.D., Myers, C.D., and Barker, L.D. 1978. Protein product and process for preparing same. Canadian Patent, 1,028, 552.

Murray, E.D., Myers, C.D., Barker, L.D., and Maurice, T.J. 1981. Functional attributes of proteins — a noncovalent approach to processing and utilizing of proteins, in *Utilization of Protein Resources*, Stanley, D.W., Murray, E.D., and Lees, D.H., Eds., Food and Nutrition Press Inc., Westport, CT, 158.

Musehold, J. 1978. Dunnschitchromatographische Trennung von 5-alkyl-resorcinhomologen aus Getreidekornern (Thin-layer chromatographic separation of 5-alkyl-resorcinol homologs from cereals). *Z. Pflanzenzuecht*, 80:326–329.

Nakamura, Y. and Higuchi, T. 1978. Ester linkage of *p*-coumaric acid in bamboo lignin. III. Dehydrogenerative polymerization of coniferyl *p*-hydroxybenzoate and *p*-coumarate. *Cell. Chem. Technol.*, 12:209–221.

Nip, W. and Burns, E. 1969. Pigment characterisation in grain sorghum. I. Red varieties. *Cereal Chem.*, 46:490–495.

Nip, W. and Burns, E. 1971. Pigment characterisation in grain sorghum. I. White varieties. *Cereal Chem.*, 48:74–80.

Nonaka, G., Hsu, F., and Nishioka, I. 1981. Structures of dimeric, trimeric, and tetrameric procyanidins from Areca catechu L. *J. Chem. Soc. Chem. Commun.*, 781–783.

Norbaek, R., Brandt, K., and Kondo, T. 2000. Identification of flavone C-glycosides a new flavonoid chromophore from barley leaves (*Hordeum vulgare* L.) by improved NMR techniques. *J. Agric. Food Chem.*, 48:1703–1707.

Nordkvist, E., Salomonsson, A.-C., and Aman, P. 1984. Distribution of insoluble-bound phenolic acids in barley grain. *J. Sci. Food Agric.*, 35:657–661.

Norton, R.A. 1994. Isolation and identification of steryl cinnamic acid derivatives from corn bran. *Cereal Chem.*, 71:111–117.

Norton, R.A. 1995. Quantification of steryl ferulate and p-coumarate esters from corn and rice. *Lipids*, 30:269–274.

Nozzolillo, C. 1972. The site and chemical nature of red pigmentation in seedlings. *Can. J. Bot.*, 50:29–34.

Nozzilillo, C. 1973. A survey of anthocyanin pigments in seedling legumes. *Can. J. Bot.*, 51:911–915.

Nozzilillo, C. and MacNeill, J. 1985. Anthocyanin pigmentation in seedlings selected species of *Phaseolus* and *Vigna* (Fabaceae). *Can. J. Bot.*, 63:1066–1071.

Ohara, T.H., Ohinata, N., Maramatsu, N., and Matsuhashi, T. 1989. Determination of rutin in buckwheat by high performance liquid chromatography. *Nippon Skokuhin Kogyo Gakkaishi*, 36:114–120.

Ohba, R., Kainou, H., and Ueda, S. 1992. Production of purple pigment from green malt and dry malt grains by fermentation without steaming. *Appl. Microbiol. Biotechnol.*, 27:171–179.

Ohsawa, R. and Tsutsumi, T. 1995. Inter-varietal variations of rutin content in common buckwheat flour (*Fagopyrum esculentum* Moench.). *Euphytica*, 86:183–189.

Oomah, B.D., Campbell, C.G., and Mazza, G. 1996. Effects of cultivar and environment on phenolic acid in buckwheat. *Euphytica*, 90:73–77.

Oomah, B.D. and Mazza, G. 1996. Flavonoids and antioxidative activities in buckwheat. *J. Agric. Food Chem.*, 44:1746–1750.

Osawa, T., Katsuzaki, H., Hagiwara, Y., Hagiwara, H., and Shibamoto, T. 1992. A novel antioxidant isolated from young green barley leaves. *J. Agric. Food Chem.*, 40:1135–1138.

Osawa, T., Ramarathnam, N., Kawakishi, S., Namiki, M., and Tashiro, T. 1985. Antioxidative defense systems in rice hulls against damage by oxygen radicals. *Agric. Biol. Chem.*, 49:3085–3087.

Overland, M., Heintzman, K.B., Newman, C.W., Newman, R.K., and Ullrich, S.E. 1994. Chemical composition and physical characteristics of proanthocyanidin-free and normal barley isotypes. *J. Cereal Sci.*, 20:85–91.

Paramashivappa, R., Kumar, P.P., Vithayathil, P.J., and Rao, A.S. 2001. Novel method for isolation of major phenolic constituents from cashew (*Anacardium occidentale* L.) nut shell liquid. *J. Agric. Food Chem.*, 49:2548–2551.

Paymaster, J.C. 1962. Some observations on oral and pharyngeal carcinomas in the state of Bombay. *Cancer*, 15:578–583.

Peterson, D.M. and Qureshi, A.A. 1993. Genotype and environment effects on tocols of barley and oats. *Cereal Chem.*, 70:157–162.

Peterson, D.M. 2001. Oats antioxidants. *J. Cereal Sci.*, 33:115–129.

Peterson, D.M., Emmons, C.L., and Hibbs, A.H. 2001. Phenolic antioxidants and antioxidant activity in pearling fractions of oat groats. *J. Cereal Sci.*, 33:97–103.

Pillai, M.K.S., Kedlaya, K.J., and Selvarangam, R. 1963. Cashew skin as a tanning material. *Leather Sci.*, 10:317–318.

Pisha, E. and Pezzuto, J.M. 1994. Fruits and vegetables containing compounds that demonstrate pharmacological activity in humans, in *Economic and Medical Plant Research*, vol. 6, Wagner, H., Hikino, H., and Farnsworth, N.R., Eds., Academic Press, London, 189–223.

Polles, S.G., Hanny, B.W., and Harvey, A.J. 1981. Condensed tannins in kernels of thirty-one pecan (*Carya illinoensis (Wangenh) K. Koch)* cultivars. *J. Agric. Food Chem.*, 29:196–197.

Pollock, J.R.A., Pool, A.A., and Reynolds, T. 1960. Chemical aspects of malting. IX. Anthocyanogens in barley and other cereals and their fate during malting. *J. Inst. Brew.*, 66:389–394.

Porter, L.J. 1992. Structure and chemical properties of the condensed tannins, in *Plant Polyphenols*, Hemingway, R.W. and Laks, P.E., Eds., Plenum Press, New York, 245–258.

Potter, N.N. 1986. *Food Science*, 4th ed., AVI Book, Van Nostrand Reinhold, New York.

Preston, C.M. and Sayer, B.G. 1992. What's in a nutshell: and investigation of structure by carbon-13 cross-polarization magic-angle spinning nuclear magnetic resonance spectroscopy. *J. Agric. Food Chem.*, 40:206–210.

Price, M.L., Butler, L.G., Rogler, J.C., and Featherston, W.R. 1978. Detoxification of high tannin sorghum grain. *Nutr. Rep. Int.*, 17:229–236.

Price, M.L., Stromberg, A.M., and Butler, L.G. 1979a. Tannin content as a function of grain maturity and drying conditions in several varieties of *Sorghum bicolor* (L) Moench. *J. Agric. Food Chem.*, 27:1270–1274.

Price, M.L., Butler, L.G., Rogler, J.C., and Featherston, W.R. 1979b. Overcoming the nutritionally harmful effects of tannin in sorghum grain by treatment with inexpensive chemicals. *J. Agric. Food Chem.*, 27:441–445.

Price, M.L., Hagerman, A.E., and Butler, L.G. 1980. Tannin content of cowpeas, pigeon peas, and mung beans. *J. Agric. Food Chem.*, 28:459–461.

Pussayanawin, V., Wetzel, D.L., and Fulcher, R.G. 1988. Fluorescence detection and measurement of ferulic acid in wheat milling fractions by microscopy and HPLC. *J. Agric. Food Chem.*, 36:515–520.

Raghavan, V. and Baruah, H.K. Areca nut: India's popular masticatory-history, chemistry and utilization. *Econ. Bot.*, 12:315–345.

Ramachandra, G., Virupaksha, T.K., and Shadaksharaswamy, M. 1977. Relationship between tannin levels and *in vitro* protein digestibility in finger millet (*Eleusine coracana* Gaertn.). *J. Agric. Food Chem.*, 25:1101–1104.

Ramarathnam, N., Osawa, T., Namiki, M., and Kawakishi, S. 1989. Chemical studies on novel rice hull antioxidants. 2. Identification of isovitexin, a C-glycosyl flavonoid. *J. Agric. Food Chem.*, 37:316–319.

Ramarathnam, N., Osawa, T., Namiki, M., and Kawakishi, S. 1988. Chemical studies on novel rice hull antioxidants. 1. Isolation, fractionation and partial identification. *J. Agric. Food Chem.*, 36:732–737.

Reddy, G.M. 1964. Genetic control of leucoanthocyanidin formation in maize. *Genetics,* 50:485–489.

Reddy, N.R., Pierson, M.D., Sathe, S.K., and Salunkhe, D.K. 1985. Dry bean tannins: a review of nutritional implications. *J. Am. Oil Chem. Soc.*, 62:541–549.

Régnier, T. and Macheix, J.-J. 1996. Changes in wall-bound phenolic acid, phenyloalanine and tyrosine ammonia-lyases, and peroxidases in developing durum wheat grains (*Triticum turgidum* L. Var. Durum). *J. Agric. Food Chem.*, 44:1727–1730.

Reichert, R.D., Flemings, S.E., and Schwab, D.J. 1980a. Tannin deactivation and nutritional improvement of sorghum by anaerobic storage of H_2O or HCl or NaOH treated grain. *J. Agric. Food Chem.*, 28:824–829.

Reichert, R.D., Youngs, C.G., and Christensen, D.A. 1980b. Polyphenols in *Pennisetum* millet, in *Polyphenols in Cereals and Legumes*, Hulse, J.H., Ed., International Development Research Centre, Ottawa, Canada, 50–60.

Reichert, R.D. 1979. The pH-sensitive pigments in pearl millet. *Cereal Chem.*, 56:291–294.

Renger, A. and Steinhart, H. 2000. Ferulic acid dehydrodimers as structural elements in cereal dietary fibre. *Eur. Food Res. Technol.*, 211:422–428.

Ribereau-Gayon, P. 1972. *Plant Phenolics*, Oliver and Boyd, Edinburgh, U.K.

Riley, R.G. and Kolattukudy, P.E. 1975. Evidence for covalently attached *p*-coumaric and ferulic acids in cutins and suberins. *Plant Physiol.*, 56:650–654.

Rooney, L.W. and Serna-Saldivar, S.O. 1987. Food uses of whole corn and dry milled fractions, in *Corn, Chemistry and Technology*, Watson, S.A. and Ramstad, P.E., Eds., American Association of Cereal Chemists, St. Paul, MN, 399–430.

Rooney, L.W., Blakely, M.E., Miller, F.R., and Rosenow, D.T. 1980. Factors affecting the polyphenols of sorghum and their development and location in sorghum kernel, in *Polyphenols in Cereal and Legumes*, Hulse, J.H., Ed., International Development Research Centre, Ottawa, Canada, 25–35.

Rooney, L.W. and Miller, F.R. 1982. Variations in structure and kernel characteristics of sorghum, in *Proceedings Int. Symp. Sorghum Grain Quality,* Patancheru, A.P., Ed., ICRISAT, India, 143.

Rosengarten, F. *The Edible Nuts,* Walker and Company, New York.

Rostango, H.S., Featherston, W.R., and Rogler, J.C. 1973b. Studies on nutritive value of sorghum grains with varying tannin content for chicks. I. Growth studies. *Poult. Sci.*, 52:765–771.

Rostango, H.S., Rogler, J.C., and Featherston, W.R. 1973a. Studies on nutritive value of sorghum grains with varying tannin content for chicks. II. Amino acids digestibility studies. *Poult. Sci.*, 52:772–778.

Rozo, C. 1982. Effect of extended storage on the degree of thermal softening during cooking cell wall components and polyphenolic compounds of red kidney beans (*Phaseolus vulgaris*). *Diss. Abstr. Int. B.*, 42:4732.

Rybka, K., Sitarski, J., and Raczynska-Bojanowska, K. 1993. Ferulic acid in rye and wheat grain and grain dietary fibre. *Cereal Chem.*, 70:55–59.

Sabate, J. 1999. Nut consumption, vegetarian diets, ischemic heart disease risk, and all-cause mortality: evidence from epidemiological studies. *Am. J. Clin. Nutr.*, 70:500S-503S.

Sabate, J., Fraser, G.E., Burke, K., Knutsen, S.F., Bennett, H., and Lindsted, K.D. 1993. Effects of walnuts on serum lipid levels and blood pressure in normal men. *N. Engl. J. Med.*, 328:603–607.

Salomonsson, A.-C., Theander, O., and Aman, P. 1980. Composition of normal and high-lysine barleys. *Swed. J. Agric. Res.*, 10:11–16.

Salunkhe, D.K., Jadhav, S.J., Kadam, S.S., and Chavan, J.K. 1982. Chemical, biochemical, and biological significance of polyphenols in cereals and legumes. *CRC Crit. Rev. Food Sci. Nutr.*, 17:277–305.

Salunkhe, D.K., Chavan, J.K., and Kadam, S.S. 1990. *Dietary Tannins: Consequences and Remedies*, CRC Press, Boca Raton, FL, 29–76.

Sanders, T.H. 1977. Change in tannin-like compounds of peanut fruit parts during maturation. *Peanut Sci.*, 4:51–53.

Sanders, T.H., McMichael, R.W., and Hendrix, K.W. 2000. Occurrence of resveratrol in edible peanuts. *J. Agric. Food Chem.*, 48:1243–1246.

Sang, S., Lapsley, K., Rosen, R.T., and Ho, C.-T. 2002a. New prenylated benzoic acid and other constituents from almond hulls (*Prunus amygdalus* Batsch). *J. Agric. Food Chem.*, 50:607–609.

Sang, S., Lapsley, K., Jeong, W.-S., Lachance, P.A., Ho, C.-T., and Rosen, R.T. 2002b. Antioxidative phenolic compounds isolated from almond skins (*Prunus amygdalus* Batsch). *J. Agric. Food Chem.*, 50:2459–2463.

Sathe, S.K. and Salunkhe, D.K. 1981. Investigations on winged bean (*Psophocarpus tetragonolubus (*L.*)* DC*)* proteins and antinutritional factors. *J. Food Sci.*, 46:1389–1393.

Sato, H. and Sakamura, S. 1975. Isolation and identification of flavonoids in immature buckwheat seed (*Fagopyrum esculentum* Moench). *Agric. Chem. Soc. Jpn.*, 49:1204–1205 (in Japanese).

Schery, R.W. 1963. *Plants for Man,* Prentice-Hall, Inc., Englewood Cliffs, NJ.

Scalbert, A., Monties, B., Lallemand, J.Y., Guittet, E., and Rolando, C. 1985. Ether linkage between phenolic acids and lignin fractions from wheat straw. *Phytochemistry,* 24:1359–1362.

Schiller, H., Patz, B., and Schimmel, K.C. 1990. Klinische Studie mit einen Phytopharmakon zur Behandlung von Mikrozirkulationsstorungen. *Arztezeistschrift Naturheilverfarhren*, 31:819–826.

Seitz, L.M. 1989. Stanol and sterol esters of ferulic and *p*-coumaric acids in wheat, corn, rye, and triticale. *J. Agric. Food Chem.*, 37:662–667.

Senter, S.D., Forbus, Jr., W.R., and Smit, C.J.B. 1978. Leucoanthocyadin oxidation in pecan kernels: relation to discoloration and kernel quality. *J. Food Sci.*, 43:128–134.

Senter, S.D., Horvat, R.J., and Forbus, Jr., W.R. 1980. Relation between phenolic acid content and stability of pecan in accelerated storage. *J. Food Sci.*, 45:1380–1382, 1392.

Senter, S.D., Horvat, R.J., and Forbus, W.R. 1983. Comparative GLC-MS analysis of phenolic acids of selected tree nuts. *J. Food Sci.*, 48:788–789, 824.

Seo, A. and Morr, C.V. 1985. Activated carbon and ion exchange treatments for removing phenolics and phytate from peanut protein products. *J. Food Sci.*, 50:262–263.

Shellenberger, J.A. 1980. Advances in milling technology. *Adv. Cereal Sci. Technol.*, 3:227–270.

Shibuya, N. 1984. Phenolic acids and their carbohydrate esters in rice endosperm cell walls. *Phytochemistry,* 23:2233–2237.

Shin, T.-S. and Godberg, J.S. 1996. Changes of endogenous antioxidants and fatty acid composition in irradiated rice bran during storage. *J. Agric. Food Chem.*, 44:567–573.

Shobba, S.V., Ramadoss, C.S., and Ravindranath, B. 1994. Inhibition of soybean lipoxygenase-1 by anarcadic acids, cardols and cardanols. *J. Nat. Prod.*, 57:1755–1757.

Sievwright, C.A. and Shipe, W.F. 1986. Effect of storage conditions and chemical treatments on firmness, *in vitro* protein digestibility, condensed tannins, phytic acid and divalent cations of cooked black beans (*Phaseolus vulgaris*). *J. Food Sci.*, 51:982–987.

Singaravadivel, K. and Raj, S.A. 1979. Leaching of phenolic compounds during soaking of paddy. *J. Food Sci. Technol.*, 16:77–78.

Siriwardhana, S.S.K. and Shahidi, F. 2002. Antiradical activity of extracts of almonds and its by-products. *J. Am. Oil Chem. Soc.*, 79:903–908.

Slominski, B.A. 1980. Phenolic acids in the meal of developing and stored barley seeds. *J. Sci. Food Agric.*, 31:1007–1010.

Smith, M.M., Hartley, R.D., and Gaillard, T. 1981. Ferulic acid substitution of cell wall polysaccharides of wheat bran. *Biochem. Soc. Trans.*, 9:165.

Sobolev, V.S. and Cole, R.J. 1999. *trans*-Resveratrol content in commercial peanuts and peanut products. *J. Agric. Food Chem.*, 47:1435–1439.

Sobolev, V.S. 2001. Vanillin content in boiled peanuts. *J. Agric. Food Chem.*, 49:3725–3727.

Sosulski, F. 1979. Organoleptic and nutritional effects of phenolic compounds on oilseed protein products: a review. *J. Am. Oil Chem. Soc.*, 56:711–715.

Sosulski, F., Krygier, K., and Hogge, L. 1982. Free, esterified, and insoluble-bound phenolic acids. 3. Composition of phenolic acids in cereal and potato flours. *J. Agric. Food Chem.*, 30:337–340.

Sosulski, F. and Dabrowski, K. 1984. Composition of free and hydrolyzable phenolic acids in the flours and hulls of ten legume species. *J. Agric. Food Chem.*, 32:131–133.

Spencer, G.F., Tjarks, L.W., and Kleiman, R. 1980. Alkyl and phenyl-alkyl anarcadic acids from *Knema elegans* seed oil. *J. Nat. Prod.*, 43:724–730.

Spiller, G.A., Jenkins, D.A., Bosello, O., Gates, J.E., Cragen, L.N., and Bruce, B. 1998. Nuts and plasma lipids: an almond-based diet lowers LDL-C. *J. Am. Coll. Nutr.*, 17:285–290.

Srisuma, N., Hammerschmidt, R., Uebersax, M.A., Ruengsakulrach, S., Bennink, M.R., and Hosfield, G.L. 1989. Storage-induced changes of phenolic acids and the development of hard-to-cook in dry beans (*Phaseolus vulgaris,* var. Seafarer). *J. Food Sci.*, 54:311–314.

Stanton, W.R. and Francis, B.J. 1966. Ecological significance of anthocyanin in the seed coats of the *Phaseoleae. Nature,* 211:970–971.

Stewart, D., Robertson, G.W., and Morrison, I.M. 1994. Phenolic acid dimers in cell walls in barley. *Biol. Mass. Spectrom.*, 23:71–74.

Stich, H.F., Stich, W., Parida, B., and Parida, B.B. 1982. Elevated frequency of micronucleated cells in buccal mucosa of individuals at high risk for oral cancer: betel quid chewers. *Cancer Lett.*, 17:125–134.

Stich, H.F., Ohshima, H., Pignatelli, B., Michelson, J., and Bartsch, H. 1983. Inhibitory effect of betel nut extracts on endogenous nitrosation in humans. *J. Natl. Cancer Inst.*, 70:1047–1050.

Stich, H.F., Rosin, M.P., and Vallejera, M.O. 1984. Reduction with vitamin A and beta-carotene administration of proportion of micronucleated buccal mucosal cells in Asian betel nut and tobacco chewers. *Lancet*, 1204–1206.

Stoessl, A. 1965. The antifungal factors in barley. III. Isolation of *p*-coumarylagmatine. *Phytochemistry,* 4:973–976.

Strumeyer, D.H. and Malin, M.J. 1975. Condensed tannins in grain sorghum: isolation, fractionation, and characterizations. *J. Agric. Food Chem.*, 23:909–914.

Styles, E.D. and Ceska, O. 1975. Genetic control of 3-hydroxy- and 3-deoxy-flavonoids in *Zea mays. Phytochemistry,* 14:413–415.

Styles, E.D. and Ceska, O. 1977. The genetic control of flavonoid synthesis in maize. *Can. J. Genet. Cytol.*, 19:289–302.

Subba Rao, M.V.S.S.T. and Muralikrishna, G. 2002. Evaluation of antioxidant properties of free and bound phenolic acids from native and malted finger millet (Ragi, *Eleusine coracana*, Indaf-15). *J. Agric. Food Chem.*, 50:889–892.

Suman, C.N., Monteiro, P.V., Ramachandra, G., and Sudharshana, L. 1992. *In vitro* enzymic hydrolysis of the storage proteins of Japanese barnyard millet (*Echinochloa frumentacea*). *J. Sci. Food Agric.,* 58:505–509.

Swain, T. 1965. The tannins, in *Plant Biochemistry*, Bonner, J. and Varner, J.E., Eds., Academic Press, New York, 552–582.

Takeoka, G.R. and Dao, L.T. 2003. Antioxidant constituents of almond (*Prunus dulcis* (Mill.) D.A. Webb) hulls. *J. Agric. Food Chem.*, 51:496–501.

Takeoka, G.R., Dao, L.T., Full, G.H., Wong, R.Y., Harden, L.E., Edwards, R.H., and Berrios, J.J. 1997. Characterization of black beans (*Phaseolus vulgaris)* anthocyanins. *J. Agric. Food Chem.*, 45:3395–3400.

Thompson, R.S., Jacques, D., Haslam, E., and Tanner, R.J.N. 1972. Plant proanthocyanidins. Part I. Introduction: the isolation, structure and distribution in nature of plant procyanidins. *J. Chem. Soc. Perkin. Trans.*, I:1387–1399.

Toyomizu, M., Sugiyama, S., Jin, R.L., and Nakatsu, T. 1993. α–Glucosidase and aldose reductase inhibitors-constituents of cashew *Anarcadium occidentale,* nut shell liquids. *Phytother. Res.*, 7:252–254.

Tressl, R., Kossa, T., Renner, R., and Koppler, H. 1976. Gas chromatographic-mass spectroscopic investigations on the formation of phenolic and aromatic hydrocarbons in food. *Z. Lebensm. Unters. Forsch.*, 162:123–130.

Van Sumere, C.F., Hilderson, H., and Massart, L. 1958. Coumarins and phenolic acids of barley and malt husks. *Naturwissenschaften*, 45:292.

Van Sumere, C.F., Cottenie, J., De Greef, and Kint, J. 1972. Biochemical studies in relation to the possible germination regulatory role of naturally occuring coumarin and phenolics. *Recent Adv. Phytochem.*, 4:165–221.

Waldron, K.W., Parr, A.J., Ng, A., and Ralph, J. 1996. Cell wall esterified phenolic dimmers: identification and quantification by reversed phase high performance liquid chromatography and diode array detection. *Phytochem. Anal.*, 7:305–312.

Waldron, K.W., Ng, A., Parker, M.L., and Par, A.J. 1997. Ferulic acid dehydrodimers in cell walls of *Beta vulgaris* and their possible role in texture. *J. Sci. Food Agric.*, 74; 221–228.

Wang, X., Warkentin, T.D., Briggs, C.J., Oomah, B.D., Campbell, C.G., and Woods, S. 1998. Total phenolics and condensed tannins in field pea (*Pisum sativum* L.*)* and grass pea (*Lathyrus sativus* L.). *Euphytica*, 101:97–102.

Wang, C.-K. and Lee, W.-H. 1996. Separation, characteristics, and biological activities of phenolics in areca fruit. *J. Agric. Food Chem.*, 44:2014–2019.

Watanabe, M., Ohshita, Y., and Tsushida, T. 1997. Antioxidant compounds from buckwheat (*Fagopyrum esculentum* Moench) hulls. *J. Agric. Food Chem.*, 45:1039–1044.

Watanabe, M. 1998. Catechins as antioxidants from buckwheat (*Fagopyrum esculentum* Moench) groats. *J. Agric. Food Chem.*, 46:839–845.

Watanabe, M. 1999. Antioxidative phenolic compounds from Japanese barnyard millet (*Echinochloa utilis)* grains. *J. Agric. Food Chem.*, 47:4500–4505.

Watterson, J.J. and Butler, L.G. 1983. Occurence of an unusual leucoanthocyanidin and absence of proanthocyanidins in sorghum leaves. *J. Agric. Food Chem.*, 31:41–45.

Weidner, S., Amarowicz, R., Karamac, M., and Dabrowski, G. 1999. Phenolic acids of caryopses of two cultivars of wheat, rye and triticale that display different resistance to preharvest sprouting. *Eur. Food Res. Technol.*, 210:109–113.

Wenke, G., Brunnemann, K.D., Hoffmann, D., and Bhide, S.V.A. 1984. Study of betel quid carcinogenesis. *J. Cancer Res. Clin. Oncol.*, 108:110–113.

Wenkert, E., Loeser, E.M., Mahapatra, S.N., Schenker, F., and Wilson, E.M. 1964. Wheat bran phenols. *J. Org. Chem.*, 29:435–439.

Woodhead, S. 1981. Environmental and biotic factors affecting the phenolic concentration of different cultivars of sorghum bicolor. *J. Chem. Ecol.*, 7:1035–1047.

Woodroof, J.G. 1967. *Tree Nuts: Production, Processing, Products*, vol.2, AVI Publishing Co., Westport, CT.

Xing, Y.M. and White, P.J. 1997. Identification and function of antioxidants from oat groats and hulls. *J. Am. Oil Chem. Soc.*, 74:303–307.

Yasukawa, K. and Akihisa, T. 1997. Antiinfammatory and antitumor-promoting activities of sterols, triterpene alcohols and their derivatives. *Recent Res. Devel. Oil Chem.*, 1:115–125.

Yasukawa, K., Akihisa, T., Kimura, Y., Tamura, T., and Takido, M. 1998. Inhibitory effect of cycloartenol ferulate, a component of rice bran, on tumor promotion in two-stage carcinogenesis in mouse skin. *Biol Pharm. Bull.*, 21:1072–1076.

Yasumatsu, K., Nakayama, T.O.M., and Chichester, C.O. 1965. Flavonoids of sorghum. *J. Food Sci.*, 30:663–667.

Yen, G.-C., Duh, P.-D., and Tsai, C.L. 1993. Relationship between antioxidant activity and maturity of peanut hulls. *J. Agric. Food Chem.*, 41:67–70.

Yildzogle-Ari, N., Altan, V.M., Altinkurt, O., and Ozturk, Y. 1991. Pharmacological effects of rutin. *Phytother. Res.*, 5:19–23.

Yoshizawa, K., Komatsu, S., Takahashi, I., and Otsuka, K. 1970. Phenolic compounds in the fermented products. I. Origin of ferulic acid in sake. *Agric. Biol. Chem.*, 34:170–180.

Yu, J., Vasanthan, T., and Temelli, F. 2001. Analysis of phenolic acids in barley by high-performance liquid chromatography. *J. Agric. Food Chem.*, 49:4352–4358.

Yurttas, H.C., Schafer, H.W., and Warthesen, J.J. 2000. Antioxidant activity of nontocopherol hazelnut (*Corylus* spp.) phenolics. *J. Food Sci.*, 65:276–280.

Zadernowski, R., Nowak-Polakowska, H., and Rashed, A.A. 1999. The influence of heat treatment on the activity of lipo-and hydrophilic components of oat grain. *J. Food Proces. Preserv.*, 23:177–191.

Zadernowski, R., Pierzynowska-Korniak, G., Ciepielewska, D., and Fornal, Ł. 1992. Chemical characteristics and biological functions of phenolic acids of buckwheat and lentil seeds. *Fagopyrum*, 12:27–35.

Zambon, D., Sabate, J., Munoz, S., Campero, B., Casals, E., Merlos, M., Laguna, J.C., and Ros, E. 2000. Substituting walnuts for monosaturated fat improves the serum lipid profile of hypocholesterolemic men and women. A randomized crossover trial. *Ann. Intern. Med.*, 132:538–546.

Zupfer, J.M., Churchill, K.E., Rasmusson, D.C., and Fulcher, R.G. 1998. Variation in ferulic acid concentration among diverse barley cultivars by HPLC and microspectrophotometry. *J. Agric. Food Chem.*, 46:1350–1354.

3 Phenolic Compounds of Major Oilseeds and Plant Oils

INTRODUCTION

The predominant phenolic compounds of oilseed products belong to the phenolic acid, coumarin, flavonoid, tannin and lignin group of compounds (Ribereau-Gayon, 1972). The phenolic acids are hydroxylated derivatives of benzoic and cinnamic acids. Figure 3.1 shows the chemical structures of some phenolic acids and units of condensed tannins of some oilseeds; Table 3.1 gives the phenolic content in some oilseed flours.

RAPESEED AND CANOLA

INTRODUCTION

Rapeseed is conventionally processed to oil and a feed-grade meal. The use of rapeseed meal in human nutrition has been considered for many years; however, its present utilization is limited due to the presence of glucosinolates, phytates, phenolics and hulls. In spite of introduction of double-zero (low glucosinolate and low erucic acid) rapeseed varieties and patenting of a number of dehulling methods (Diosady et al., 1986; Greilsamer, 1983; Kozlowska et al., 1984; Sosulski and Zadernowski, 1981), the use of rapeseed meal as a source of food-grade protein is still limited by the presence of phytic acid and phenolic compounds, among others. Nonetheless, aqueous extraction and membrane processes have more recently been considered and are expected to lead to industrial developments. The phenolic content in rapeseed flour is much higher when compared to that found in flours obtained from other oleaginous seeds, exceeding about 28 times the level of phenolics in soybean flour (Table 3.1 and Table 3.2).

Phenolic compounds may contribute to the dark color, bitter taste and astringency of rapeseed meals (Naczk et al., 1998a, b). Phenolics or their degradation products can also form complexes with essential amino acids, enzymes and other proteins, thus lowering the nutritional value of rapeseed products. Therefore, phenolic compounds are important factors when considering rapeseed meal as a protein source in food formulations. However, the available information on undesirable effects of phenolics on the quality of rapeseed meals is still diverse and fragmentary (Kozlowska et al., 1975; Shahidi and Naczk, 1992; Sosulski, 1979).

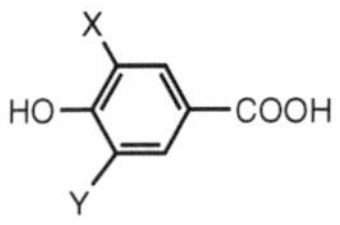

Benzoic Acid Derivatives	X	Y
p-Hydroxybenzoic acid	H	H
Vanillic acid	OCH_3	H
Syringic acid	OCH_3	OCH_3
Protocatechuic acid	H	H
Gallic acid	OH	OH

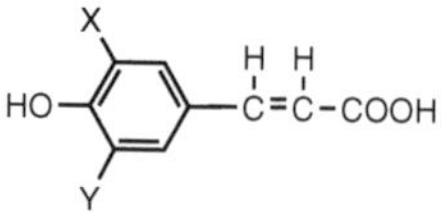

Cinnamic Acid Derivatives	X	Y
p-Coumaric acid	H	H
Caffeic acid	OH	H
Sinapic acid	OCH_3	OCH_3
Ferulic acid	OCH_3	H

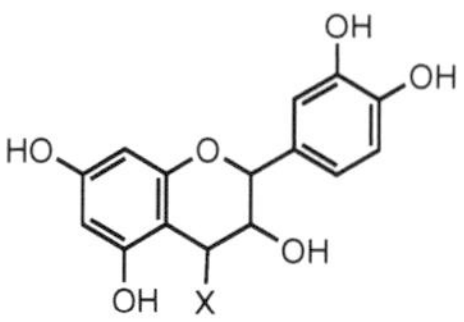

Catechin X = H

Leucocyanidin X = OH

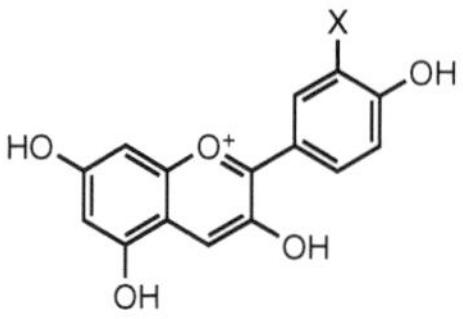

Pelargonidin X = H

Cyanidin X = OH

FIGURE 3.1 Chemical structures of basic phenolic acids and units of condensed tannins.

TABLE 3.1
Total Phenolic Acids in Some Oilseed Flours[a]

Phenolic Acid	Cottonseed Glanded	Sesame	Soybean	Flax	Peanut	Sunflower
p-Hydroxybenzoic	1.1	Trace	13.9	2.6	2.0	7.4
Vanillic	—	—	—	Trace	—	0.8
Syringic	—	—	28.9	0.4	—	3.6
trans-p-Coumaric	4.3	7.2	9.4	6.1	146.4	5.6
trans-Ferulic	4.5	5.7	15.7	37.6	16.2	7.2
trans-Caffeic	0.5	9.8	6.0	5.3	2.8	979.1
trans-Sinapic	—		—	29.1	8.1	—
Total	10.4	22.7	73.6	81.1	175.5	1003.7

[a] Milligram per 100 g of flour.

Source: Adapted from Dabrowski, K. and Sosulski, F., 1984, *J. Agric. Food Chem.*, 32:128–130.

TABLE 3.2
Total Content of Phenolic Acids in Oilseed Meals and Flours[a]

Product	Total Phenolics
Meals	
Evening Primrose	1040
Borage	720
Rapeseed	2570
Soybean	455
Flours	
Cottonseed	56.7
Peanut	63.6
Rapeseed	639.9
Soybean	23.4

[a] Milligrams per 100 g.

Source: Meals adapted from Zadernowski, R. et al., 1995, *Acta Acad. Agric. Techn. Olst.*, 27:107–118 (in Polish); soybean adapted from Naczk, M. et al., 1986, *Lebensm.-Wiss. u-. Technol.*, 19:13–16; and flours adapted from Kozlowska, H. and Zadernowski, R., 1988, paper presented at the 3rd North American Chemical Congress, Toronto, ON.

Phenolic Acids

Commonly cultivated rapeseed varieties, including canola (*Brassica napus* and *Brassica campestris*), have a similar phenolic content (Dabrowski and Sosulski 1984; Kozlowska et al., 1983; Krygier et al., 1982; Naczk and Shahidi, 1989, 1992; Naczk et al., 1986, 1998a, b; Shahidi et al., 1988). These are mainly located in seed cotyledons (Artz et al., 1986) and only trace amounts are found in seed coats. The total content of phenolic acids in various rapeseed protein products ranges from 1324.8 to 1837.0 mg/100 g of defatted meal and from 623.5 to 1280.9 mg/100 g of flour, on a dry weight basis (Table 3.3). On the other hand, rapeseed and canola hulls contain only 60 to 240 mg of sinapine per 100 g of sample, indicating the presence of embryo material in the hulls (Bell and Shires, 1982). Phenolic acids of rapeseed are in the free, esterified, glycosidic and insoluble-bound forms. Phenolic esters and free phenolic acids constitute as much as 80 and 16% of total phenolic compounds of rapeseed meals, respectively (Krygier et al., 1982; Naczk et al., 1986).

Tentative identification of principal phenolic acids of rapeseed has been reported (Durkee and Thivierge, 1975; Fenton et al., 1980; Kozlowska et al., 1975, 1983; Krygier et al., 1982). Sinapic acid is the predominant phenolic acid found in rapeseed (Clandinin, 1961), constituting over 73% of free phenolic acids and about 99% of the phenolic acids released from esters and glycosides (Dabrowski and Sosulski, 1984; Krygier et al., 1982) (Table 3.3). Minor phenolic acids are *p*-hydroxybenzoic, vanillic, gentisic, protocatechuic, syringic, *p*-coumaric, ferulic and caffeic acids

TABLE 3.3
Content of Free, Esterified and Insoluble Bound Phenolic Acids in Some Rapeseed Products

Product	Phenolic Acids (mg/100 g)			
	Free	Esterified	Insoluble-Bound	Total
Tower meal	244.0	1202.0	96.0	1542.0
Regent meal	262.0	1470.0	105.0	1837.0
Altex meal	248.0	1458.0	101.0	1807.0
Hu You 9 meal	119.1	1182.0	38.9	1340.0
Midas meal	143.5	1524.0	68.7	1736.2
Triton meal	61.5	1212.0	51.3	1324.8
Mustard meal	108.1	1538.0	22.4	1668.5
Tower flour	98.2	982.0	—	1066.5
Candle flour	84.5	1196.4	—	1280.9
Gorczanski flour	60.1	574.0	4.9	639.0
Start flour	71.8	700.0	4.7	776.5

Note: Expressed in sinapic acid equivalents.

Sources: Adapted from Krygier, K. et al., 1982, *J. Agric. Food Chem.*, 30:334–336; Kozlowska, H. et al., 1983, *J. Am. Oil Chem. Soc.*, 60:1119–1123; Naczk, M. and Shahidi, F., 1989, *Food Chem.*, 31:159–164; and Shahidi, F. and Naczk, M., 1990, in *Rapeseed/Canola: Production, Chemistry, Nutrition, and Processing Technology*, Van Nostrand Reinhold, New York, 291–306.

(Figure 3.1). In addition, chlorogenic acid has been found in trace amounts (Kozlowska et al., 1975, 1983; Krygier et al., 1982; Lo and Hill, 1972).

Free Phenolic Acids

Free phenolic acids present in rapeseed may contribute to the taste of rapeseed meals. They constitute 13 to 24% of total phenolic acids present in most rapeseed flours (Kozlowska and Zadernowski, 1988) and about 15% in canola meals (Table 3.3) (Naczk and Shahidi, 1989). However, flour obtained from Yellow Sarson variety contains only trace quantities of free phenolics (Krygier et al., 1982). No detectable amount of free phenolic acids has been found in hulls of Tower rapeseed (Durkee and Thivierge, 1975). Dabrowski and Sosulski (1984) have demonstrated that free phenolic acids are not present in 10 oilseed flours, including rapeseed (*B. campestris* L.).

Sinapic acid is a predominant phenolic acid; *p*-hydroxybenzoic, vanillic, gentisic, protocatechuic, syringic, *p*-coumaric, *cis*- and *trans*-ferulic, caffeic and chlorogenic acids have been found in small quantities (see Table 3.4 and Table 3.5). Both isomeric forms of sinapic acid, Z (*cis*) and E (*trans*), have been identified by a gas liquid chromatographic technique; these constitute up to 90% of the free phenolic acids present in rapeseed. However, an HPLC analysis did not confirm the presence of *cis*-sinapic acid (Kozlowska and Zadernowski, 1988).

Esterified Phenolic Acids

Esterified phenolic acids are the predominant form of phenolics in rapeseed, constituting up to 90% of phenolic acids present in rapeseed and canola (Table 3.3). The total content of phenolic acids liberated from soluble esters found in flours obtained from Polish rapeseed varieties (Start, Górczanski, and Bronowski), Candle variety of canola and white mustard ranges between 520 and 700 mg/100 g of sample (Kozlowska et al., 1983). However, Krygier et al. (1982) found that Tower and Candle flours contain much higher levels of esterified phenolics, ranging from 982 to 1196.4 mg/100 g of sample. They also reported that Tower hulls contain 110.0 mg of soluble esters per each 100 g of sample. On the other hand, rapeseed meals contained 1182 to 1524 mg of esterified phenolic acids per 100 g sample (Table 3.3) (Naczk and Shahidi, 1989, 1990).

Pokorny and Reblova (1995) have published an excellent review of sinapine, the choline ester of sinapic acid, and the predominant phenolic ester in rapeseed. According to Bouchereau et al. (1991), the composition of phenolic acids in sinapine is genetically controlled, but its content is affected by the cultivar and by growing conditions. The presence of sinapoylcholine (sinapine), feruloylcholine, isoferuloylcholine, coumaroylcholine, 4-hydroxybenzoylcholine and 3,4-dimethoxybenzoylcholine has been reported in rapeseed (Larsen et al., 1983) (Figure 3.2). Mueller et al. (1978) reported that *B. napus* cultivars of rapeseed contain significantly ($P < 0.01$) higher levels of sinapine (1.65 to 2.26%) than *B. campestris* cultivars (1.22 to 1.54% sinapine). Furthermore, Fenwick et al. (1984) demonstrated that *B. napus* meals contain 1.17 to 1.83% sinapine.

Similarly, a wide range of sinapine content (0.4 to 1.8%) has been observed in several cultivars of cruciferae crops, including the Brassica species (Kerber and Buchloh, 1980). However, Argentine and Polish rapeseed contain only 0.94 and 0.92% of sinapine, respectively (Austin and Wolff, 1968). Furthermore, Blair and Reichert (1984) found that the mean content of sinapine in defatted rapeseed and canola cotyledons is 2.67 and 2.85%, respectively. Later, Clausen et al. (1985) reported that the content of sinapine in *B. napus* cultivars ranges from 0.62 to 1.06% and that of *B. campestris* cultivars is 0.39 to 0.76%. Sinapic acid is the predominant phenolic acid and constitutes from 70.9 to 96.7% of the total content of esterified phenolic fractions (Table 3.4). Small quantities of other phenolic acids, such as *p*-hydroxybenzoic, vanillic, protocatechuic, synringic, *p*-coumaric, *cis*- and *trans*-ferulic, and caffeic acids, are also present in the hydrolyzate of the soluble esters extracted from Tower and Candle flours (Table 3.5) (Krygier et al., 1982). Of these, ferulic and caffeic acids are derived from choline and other ester bonds (Tantawy et al., 1983).

Sinapines are not the only phenolic esters found in rapeseed. Fenton et al. (1980) isolated sinapine and at least seven other compounds from rapeseed meals of Midas and Echo varieties that, upon hydrolysis, yielded sinapic acid. Zadernowski (1987) found that flours obtained from Polish rapeseed varieties contained from 0.16 to 0.65 mmol of phenolic acid glycosides per kilogram of flour. The compound 1-O-β-D glucopyranosyl sinapate is the predominant glycoside in rapeseed and canola varieties (Amarowicz and Shahidi, 1994; Amarowicz et al., 1995; Wanasundara et al., 1994).

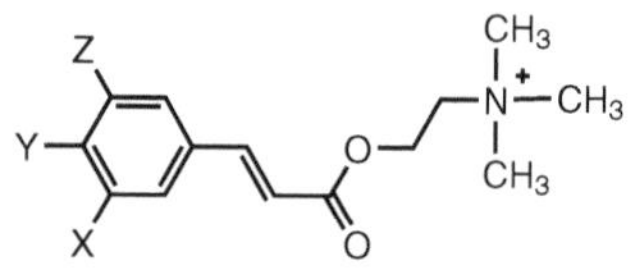

Sinapines	X	Y	Z
p-Coumaroylcholine	H	OH	H
Feruloylcholine	H	OH	OCH_3
Isoferuloylcholine	H	OCH_3	OH
Sinapine	OCH_3	OH	OCH_3
Sinapine glucoside	OCH_3	O-Glu	OCH_3

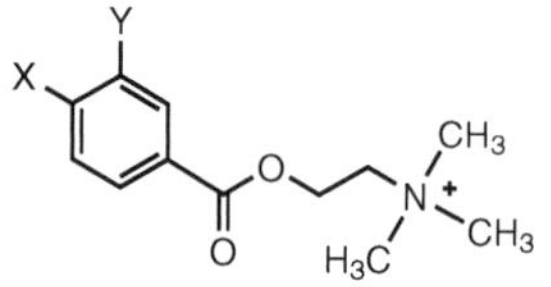

Sinapines	X	Y
4-Hydroxybenzoylcholine	OH	H
Hesperalin	OCH_3	OCH_3

FIGURE 3.2 Chemical structures of sinapines found in rapeseed.

TABLE 3.4
Contribution of Sinapic Acid in Different Fractions of Phenolic Constituents of Cruciferae Oilseeds

	Sinapic Acid Contribution (%) in Phenolic Acid Fraction			
Meal Sample	**Free**	**Esterified**	**Insoluble-Bound**	**Total**
Hu You 9	77.5	96.4	17.2	92.4
Midal	72.1	70.9	7.4	68.5
Triton	70.2	96.7	15.6	92.3
Mustard	85.4	72.6	32.1	72.8

Source: Adapted from Shahidi, F. and Naczk, M., 1990, in *Rapeseed/Canola: Production, Chemistry, Nutrition, and Processing Technology*, Van Nostrand Reinhold, New York, 291–306

Insoluble-Bound Phenolic Acids

Rapeseed meals contain 38.9 to 105.0 mg/100-g sample of insoluble phenolic acids expressed as sinapic acid equivalents (Naczk and Shahidi, 1989) (Table 3.3). Kozlowska et al. (1983) have shown that the total content of insoluble-bound

TABLE 3.5
Phenolic Acids Liberated from Soluble Esters Isolated from Rapeseed Flours[a]

Phenolic Acid	Yellow Sarson	Candle	Tower
p-Hydroxybenzoic	0.3	1.5	1.1
Vanillic	Trace	0.6	Trace
Protocatechuic	—	0.5	0.2
Syringic	—	Trace	Trace
p-Coumaric	—	Trace	2.2
cis-Ferulic	Trace	—	Trace
trans-Ferulic	5.9	5.5	8.7
Caffeic	—	—	1.1
cis-Sinapic	49.1	7.2	76.3
trans-Sinapic	712.3	1116.2	894.9

[a] Milligrams per 100 g.

Source: Adapted from Krygier, K. et al., 1982, *J. Agric. Food Chem.*, 30:334–336.

phenolics in rapeseed flours ranges from 3.2 in Yellow Sarson to 5.0 mg/100 g in Gorczanski cultivar of Polish rapeseed. Nine phenolic acids have been identified, among which sinapic acid is predominant, followed by *p*-coumaric and *trans*-ferulic acids. Krygier et al. (1982) demonstrated that 24.5 mg/100 g of insoluble-bound phenolic acids are present in Tower hulls and that protocatechuic acid is a predominant phenolic acid. However, they did not find any detectable quantities of bound phenolics in Yellow Sarson, Candle and Tower flours.

Condensed Tannins

Condensed tannins are complex phenolic compounds widely distributed in foods and feeds of plant origin in soluble and insoluble forms. The total amounts of condensed tannins in canola and rapeseed hulls, calculated as the sum of soluble and insoluble tannin contents, ranges from 1913 to 6213 mg/100 g of oil-free hulls (Naczk et al., 2000). Condensed tannins may form soluble and insoluble complexes with proteins, which could explain the antinutritional effects of tannin-containing ingredients in nonruminant (Martin-Tanguy et al., 1977) and ruminant (Kumar and Singh, 1984) feeds. Condensed tannins are also considered potent enzyme inhibitors due to their complexation with enzymes; for example, condensed tannins in rapeseed may cause tainting of eggs. Tannins appear to block the metabolism of trimethylamine (TMA) by inhibiting TMA oxidase, an enzyme that converts TMA to odorless, water-soluble TMA-oxide (Fenwick et al., 1984).

Soluble Condensed Tannins

Bate-Smith and Ribereau-Gayon (1959) first reported the presence of condensed tannins in rapeseed hulls. This finding was verified by Durkee (1971), who identified cyanidin, pelargonidin (Figure 3.1) and an artifact n-butyl derivative of cyanidin in the hydrolytic products of rapeseed hulls. Clandinin and Heard (1968) reported the presence of approximately 3% tannins in rapeseed meal as assayed by the tannin determination method in cloves and allspice (AOCS 1987). Fenwick and Hoggan (1976) showed that this value includes sinapine. Thus, the corrected concentration of tannins in rapeseed meals is 18 to 45% lower than that given by Clandinin and Heard. In addition, Fenwick et al. (1984) demonstrated that whole and dehulled Tower meals contain 2750 and 3910 mg of tannins per 100 g, respectively. Leung et al. (1979) reported that rapeseed hull contained only 100 mg acetone-extractable condensed tannins per 100 g sample and found that leucocyanidin is the basic unit of the isolated condensed tannins (Figure 3.1).

Mitaru et al. (1982) reported that hulls from rapeseed and canola seeds contain 20 to 220 mg of extractable tannins per 100 g sample. Blair and Reichert (1984) reported the presence of 90 to 390 mg of tannins per 100 g in the defatted rapeseed cotyledons and 230 to 540 mg of tannins per 100 g in the defatted canola cotyledons, as assayed by the modified vanillin method. They also found that sinapine does not interfere with the analysis of tannins. Later, Shahidi and Naczk (1988, 1989a, b) reported that canola varieties contains 682 to 772 mg of condensed tannins per 100 g of oil-free meal as assayed by the modified vanillin method. Only 556 and 426 mg of tannin per 100 g of oil-free meal are present in high glucosinolate Midas rapeseed and Hu You 9 Chinese cultivar of rapeseed, respectively. The latter results are somewhat higher than those reported by Blair and Reichert (1984), which may be due to existing differences in the solvent extraction systems employed for recovery of tannins.

The total content of soluble condensed tannins in hulls of some canola cultivars and Polish rapeseed varieties, as determined by the vanillin and proanthocyanidin assays, is summarized in Table 3.6; the content ranges from 23.4 to 2719 mg per 100 g of hulls. The least tannin content has been found in the sample of Westar hulls contaminated considerably with cotyledons (Naczk et al., 1994, 1998a, b, 2000, 2001a, b). Because seed coats of cereals and legumes are the primary locations of tannins in seeds (Salunkhe et al., 1990), the low tannin content in this sample may be explained as the effect of hull dilution with cotyledons. Thus, canola and rapeseed hulls contain up to 12 times more condensed tannins than those reported previously by Mitaru et al. (1982) and Leung et al. (1979). Up to 15-fold differences in condensed tannin values exist within the Westar, Cyclone and Leo varieties. According to Radhakrishnan and Sivaprasad (1980), the variation in tannin content of sorghum varieties ranges up to eight-fold. The content of tannins also depended on the stage of seed development, e.g., the maximum tannin content in sorghum is found in immature seeds (Price et al., 1979; Butler, 1982). Thus the differences in condensed tannin content within canola and rapeseed varieties may be due to the location and stage of seed development (Naczk et al., 1994, 1998a, b).

The biological and ecological role of tannins is attributed to their ability to bind and precipitate proteins (Bate-Smith, 1973; Hagerman and Butler, 1978; Salunkhe et al., 1990). The available information on the biological activity of canola and rapeseed condensed tannins is still diverse and fragmentary. Mitaru et al. (1982) reported that condensed tannins isolated from rapeseed hulls do not inhibit

TABLE 3.6
Content of Condensed Tannins in Canola Hulls as Determined by Chemical Assays[a]

Variety	Vanillin Assay[a]	Proanthocyanidin Assay[a]
Canola		
Cyclone		
Sample 1	1485 ± 63	1214 ± 125
Sample 2	1705 ± 12	1231 ± 35
Sample 3	929 ± 43	654 ± 77
Sample 4	414 ± 5	402 ± 42
Ebony	1335 ± 17	910 ± 110
Excel	83 ± 1	175 ± 16
Westar		
Sample 1	207 ± 1	286 ± 24
Sample 2	23 ± 6	92 ± 25
Vanguard	1296 ± 13	990 ± 60
Rapeseed		
Leo		
Sample 1	60 ± 7	69 ± 18
Sample 2	2719 ± 30	1539 ± 103
Mar	114 ± 2	157 ± 8
Marita	69 ± 3	115 ± 18
Idol	37 ± 17	76 ± 7

[a] Milligrams per 100 g of hulls.

Source: Adapted from Naczk, M. et al., 2000, *J. Agric. Food Chem.*, 48:1758–1762.

α-amylase activity *in vitro*, but Leung et al. (1979) found that tannins isolated from rapeseed hulls form a white precipitate after their addition to a 1% gelatin solution. The authors, however, did not attempt to quantify the biological activity of tannins. Later, Naczk et al. (1994) reported that condensed tannins, extracted from canola hull samples, precipitate 23.2 to 58.6 mg of dye-labeled BSA/g of hulls. The complete precipitation of dye-labeled BSA from the solution corresponded to 80 mg BSA/g of hulls.

The affinities of canola tannins for proteins have been characterized by the precipitation index (PI) calculated using the equation: PI = B/C, where B represents biological activity expressed as milligram of dye-labeled BSA precipitated per 1 g of hulls. The PI values for tannins isolated from samples of high-tannin canola hulls do not exceed 5.0 mg of BSA/mg of tannins but, for low-tannin canola hulls, PI ranges from 17.7 to 40.7 mg of BSA/mg of tannins. Thus, tannins isolated from low-tannin canola hulls show greater affinities for proteins than those from high-tannin canola hulls. This may arise from the existing differences in the molecular weight of tannins isolated from high- and low-tannin samples of canola hulls.

According to Porter and Woodruffe (1984), the ability of condensed tannins to precipitate proteins depends on their molecular weight. Subsequently, Naczk et al. (2001b) expressed the protein-precipitating potentials of condensed tannins in tannins from canola and rapeseed hulls as a slope of lines reflecting the amount of insoluble tannin–protein complex formed vs. the amount of tannins added to the reaction mixture. The slope values of lines obtained using the protein precipitation assay of Hagerman and Butler (1978) are 2.96 to 10.91 (absorbance unit at 510 nm/mg of added tannins), and those of lines obtained using the dye-labeled bovine serum albumin assay of Asquith and Butler (1985) range from 28 to 267 (percent of precipitated protein/mg of added tannins). For both assays a statistically significant ($P < 0.001$) correlation exists between the slopes and the condensed tannin contents in canola and rapeseed hulls.

Insoluble Condensed Tannins

The insoluble tannins may comprise 70.0 to 95.8% of the total condensed tannins in canola and rapeseed hulls. The differences in the content of insoluble condensed tannins among and within canola and rapeseed varieties are much lower than those found for soluble condensed tannins. The insolubility of condensed tannins may be the result of polymerization, as well as formation of insoluble complexes with the fiber and protein fractions of canola and rapeseed. The insoluble condensed tannins may be extracted into sodium dodecyl sulfate (SDS) solution. The amounts of SDS-extractable condensed tannins in canola and rapeseed hulls are comparable with those of soluble condensed tannins and comprise 4.7 to 14.7% of insoluble condensed tannins present in the hulls (Naczk et al., 2000).

Other Phenolic Compounds

Durkee and Harborne (1973) first observed the presence of flavonoid glycosides containing sinapic acid. These authors also identified a flavonol glycoside, kaempferol-7-glucoside-3-sophoroside, in the Brassica seed extracts. Later, Tantawy et al. (1983) reported the presence of two flavonoid glycosides in rapeseed meal, namely, 3-(*O*-sinapoyl sophoroside)-7-*O*-glucoside of kaempferol and 3-(*O*-sinapoyl glucoside)-7-*O*-sophoroside of kaempferol.

Effect of Processing

Phenolic Acids

Extraction

After oil extraction, meal contains a large amount of phenolic compounds, the content of which in rapeseed and other oilseed meals may be reduced by treatment with ammonia (gaseous or in alkanol solution). McGregor et al. (1983) found that gaseous ammoniation of *Brassica juncea* mustard meal removed up to 74% of sinapine. However, Fenwick et al. (1979, 1984) demonstrated that treatment of *B. napus* meal with ammonia or lime reduces sinapine content by approximately 90%. Similarly, Kirk et al. (1966) reported a drop of up to 90% in the sinapine of

TABLE 3.7
Removal of Phenolics and Tannins by Methanol–Ammonia–Water/Hexane Extraction System

Canola Variety	Removal (%)	
	Total Phenolic Acids	Tannins
Tower		
A	69.5 ± 2.8	70.1 ± 1.7
B	75.2 ± 2.8	78.0 ± 1.7
Regent		
A	78.1 ± 1.4	71.6 ± 2.0
B	74.6 ± 1.4	77.5 ± 2.0
Altex		
A	71.1 ± 1.8	72.6 ± 4.8
B	72.0 ± 1.8	84.9 ± 4.8

Note: A — meal extracted with 10% NH_3 in methanol and hexane; B — meal extracted with 10% NH_3 in 95% methanol and hexane.

Source: Adapted from Shahidi, F. and Naczk, M., 1988, in *Proceedings 14th Int. Conf. Group Polyphenols*, Bulletin de Liason, 14:89–92.

Crambe meal after its treatment with gaseous ammonia; however, aqueous ammonia is less effective. The exact mechanism of action of ammonia on sinapine is unknown. However, Austin and Wolff (1968) reported that, under alkaline conditions, sinapine is hydrolyzed to sinapic acid and choline.

Goh et al. (1982) used ethanol containing 0.2 M ammonia to extract Candle and Tower canola meals and removed up to 82 and 39% of their phenolic content, respectively. These authors did not offer any explanation for the difference in the efficiency of removal of phenolics from these meals. In another study, methanol–ammonia–water–hexane is used and results in the removal of 72.4% of the phenolics of canola meals in a laboratory scale process (Table 3.7) (Diosady et al., 1985; Naczk et al., 1986); its efficacy was improved to 80% in a semipilot extraction process (Diosady et al., 1987). The methanol–ammonia–water/hexane extraction system removed 82 and 50% of the esterified and free phenolic acids originally present in the seeds, respectively. However, the concentration of insoluble-bound phenolic acids was not affected (Naczk and Shahidi, 1989; Shahidi and Naczk, 1988). Dabrowski and Siemieniak (1987) found that about 90% of sinapine is removed after 30 min of batch extraction of rapeseed with methanol–ammonia.

Dabrowski and Sosulski (1983) reported that processing rapeseed to protein concentrates and isolates by typical industrial procedures reduces the content of phenolics by 60 to 83%. However, preparation of concentrates containing trace amounts of phenolics requires as much as seven extractions with 70% ethanol (Kozlowska and Zadernowski, 1988). Hydrothermal treatment of double-low rapeseed (low glucosinolates and low erucic acid content) of Jantar variety does not affect the phenolic acid content of the resultant flour; this treatment is an alternative to cooking during commercial processing of rapeseed. However, three additional extraction stages with 80% ethanol are required to reduce the content of phenolics to trace levels (Kozlowska and Zadernowski, 1988).

Heating

Cooking may affect the sinapine content in canola and rapeseed meals. Larsen et al. (1983) reported that 30-min toasting of dehulled and defatted Erglu rapeseed meal lowers its sinapine content by 17% — a decrease associated with the formation of oligomeric sinapine derivatives. Jensen et al. (1991) later demonstrated that heating reduces the content of sinapine, but increases the content of lignan-type products in rapeseed meal. Subsequently, by using model systems, Rubino et al. (1995, 1996) showed that sinapic acid readily converts to a lignan, thomasidioic acid (Figure 3.3) under alkaline conditions. These authors envisaged that additional studies would be required to establish whether a similar reaction occurs when canola meal is subjected to alkaline treatment.

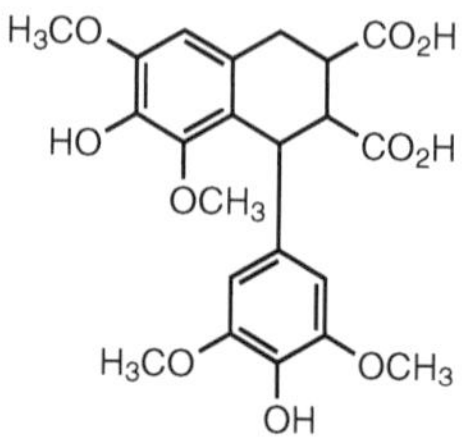

H_3CO CO_2H HO CO_2H OCH_3 H_3CO OCH_3 OH

FIGURE 3.3 Chemical structure of thomasidioic acid.

Condensed Tannins

Processing rapeseed with methanol alone extracts 16% of its tannins (Shahidi and Naczk, 1989a, b). Addition of 5% (v/v) water to methanol increases the efficiency of tannin extraction to 36%. Presence of ammonia in absolute or 95% methanol greatly enhances the extraction of tannins from rapeseed. Methanol–ammonia–water/hexane, however, is the most effective system for the removal of tannins (Table 3.7). The resultant meal contains from 4 to 33% of tannins originally present in the meals. This may be due to the extraction of tannins from the seed into the polar phase or decomposition of tannins to products insensitive to vanillin reagent. Ghandhi et al. (1975) found that ammonia depolymerized the tannins present in salseed meal so that the processed meal obtained is nontoxic and palatable. Moreover, upon alkali treatment, tannins may form phlobaphenes, which are chemically and nutritionally unreactive (Swain, 1979). On the other hand, Fenwick et al. (1984) reported that treatment of *B. napus* meals with ammonia or lime does not affect their tannin content appreciably.

Commercially, crushed rapeseed is generally cooked to 90 to 110°C for 20 to 25 min to facilitate oil release and to inactivate the enzyme myrosinase (Unger, 1990). Under experimental conditions (110°C, 30 min) simulating commercial cooking, heat treatment of canola hulls results in a decrease in the content of tannins as catechin equivalents in hulls from 2374 to 1944 mg/100-g sample. This treatment, however, increased the amount of precipitable phenolics by 18% (Naczk et al., 1996).

SOYBEANS

INTRODUCTION

There are three major species of soybean: *Glycine ussuriensis* (Regel and Maack wild), *Glycine max* (L.) and *Glycine gracilis* (Skvortzow intermediate). Of these, *G. max* (L.) Merr. (Leguminosae) is commonly cultivated throughout the world (Nwokolo and Smartt, 1966; Smith, 1978). Soybean seeds are used for the extraction of edible oil, human food and livestock feed. The highest human consumption of soybean is in Asia, where it has traditionally been processed into fermented (miso, natto, tempeh) and nonfermented (fresh soybean, soybean sprouts, soymilk, tofu) products (Wang and Murphy, 1996). Soybean grits, flours, flakes and protein concentrates and isolates are widely used as a food component in the U.S. and U.K. (Berk, 1992; Wiseman et al., 2002).

Fleury et al. (1991) reported the presence of anthocyanins, flavonols, flavones, isoflavones and chalcones and their derivatives with acetic, *p*-hydroxybenzoic, caffeic, coumaric, ferulic, gallic, malonic, hydroxycinnamic, oxalic, and sinapic acids. Of these, the isoflavones (Figure 3.4) are of much interest because of their estrogenic (Makela et al., 1995), cancer preventive, antimutagenic and antiproliferative (Hirano et al., 1994; Kuntz et al., 1999; Miyazawa et al., 1999, 2001; Wiseman, 1996), and antioxidant (Arora et al., 1998; Wei et al., 1995) activities. Table 3.8 gives the total content and distribution of isoflavone derivatives in some soy foods. Phenolic compounds have also been implicated in the formation of certain undesirable colors and flavors in soybean products (Matsuura et al., 1989; Sosulski, 1979; Wolf and Cowan, 1975).

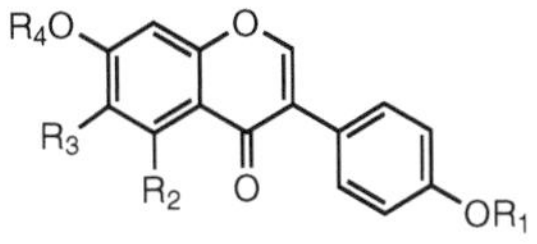

Compound	R_1	R_2	R_3	R_4
Daidzein	H	H	H	H
Formononetin	CH_3	H	H	H
Genistein	H	OH	H	H
Glycitein	H	H	OCH_3	H
Daidzin	H	H	H	glucose
Genistin	H	OH	H	glucose
Glycitin	H	H	OCH_3	glucose
6"- *O*-Acetyldaidzin	H	H	H	6"- *O*-acetylglucose
6"- *O*-Acetylgenistin	H	OH	H	6"- *O*-acetylglucose
6"- *O*-Acetylglycitin	H	H	OCH_3	6"- *O*-acetylglucose
6"- *O*-Malonyldaidzin	H	H	H	6"- *O*-malonylglucose
6"- *O*-Malonylgenistin	H	OH	H	6"- *O*-malonylglucose
6"- *O*-Malonylglycitin	H	H	OCH_3	6"- *O*-malonyldglucose

FIGURE 3.4 Chemical structures of isoflavones found in soybeans.

TABLE 3.8
Range of Isoflavone Content in Some Soy Foods[a]

Soy Food	Glucosides			Malonylglucoside			Acetylglucoside			Aglycon			Total
	DA	GE	GL	DA	GE	GL	DA	GE	GL	D	G	Gn	
Soymilk	20–94	25–130	4–16	6–42	14–45	0–5	0	4–10	0	1–3	1–4	0–1	80–165
Tofu	40–105	64–143	11–29	64–156	40–158	7–19	0–18	8–17	0	3–22	4–30	2–4	202–347
Soy sauce	0	0	0	0	0	0	0	0	0–3	1–6	0–4	0–3	1–14
Miso	62–66	89–92	10–13	14–32	17–37	0–8	8–14	20–23	0–9	25–38	29–43	12	228–235
Tempeh	93–105	206–226	14	64–66	111	8	35–49	46–50	0	77–85	89–103	8–9	531–538
Cooked burger	3–10	5–23	0–6	0–14	4–27	0–6	0–20	8–31	0	0–6	6–8	0–4	9–95

Note: DA — daidzin; GE — genistin; GL — glycitin; D — daidzein; G — genistein; Gn — glycitein.

[a] Milligram per kilogram of wet weight.

Source: Adapted from Murphy, P.A. et al., 1999, *J. Agric. Food Chem.*, 47:2697–2704.

Phenolic Acids

Maga and Lorenz (1974) reported that soybean flakes contain 25.6 mg of phenolic acid per 100-g sample, whereas Dabrowski and Sosulski (1984) demonstrated that soybean flours contain 73.6 mg of total phenolics per 100-g sample. Later, Naczk et al. (1986) reported the presence of 455 mg of total phenolics per 100 g of 48% commercial soybean flour.

Arai et al. (1966) identified at least seven phenolic acids in ethanolic extracts of defatted soy flour and reported that syringic acid is the predominant phenolic compound present. On the other hand, using an HPLC technique, How and Moor (1982) isolated 40 phenolic substances in the total phenolic fraction of defatted soy flakes. Of these, eight were tentatively identified as gentisic, vanillic, chlorogenic, syringic, *o*- and *p*-coumaric, ferulic, and salicylic acids. The compounds *o*- and *p*-coumaric and ferulic acids were predominant. Salicylic acid is the principal phenolic acid in Supra 620 soybean isolates (How and Moor, 1982). Traces of phenolic acids possessing a typical soy flavor have also been identified (Rackis et al., 1970). Maga and Lorenz (1974) reported that ferulic, syringic and vanillic acids are the major phenolic acids in soybean flours. Dabrowski and Sosulski (1984) also found *p*-hydroxybenzoic, *trans*-caffeic, *trans*-*p*-coumaric, *trans*-ferulic, and syringic acids (Figure 3.1) in soybean flours and demonstrated that syringic acid is the major phenolic acid present (Table 3.1).

Isoflavones

Several isoflavones, namely, daidzein (7,4′-dihydroxyisoflavone), glycitein (7,4′-dihydroxy-6-methylisoflavone) and genistein (6,7,4′-trihydroxyisoflavone) (Figure 3.4) have been isolated and identified in soybean protein products and found to possess an herb-like flavor, bitterness and astringency (Huang et al., 1979). These isoflavones occur in soybean and soy foods in the form of glucosides (daidzin, glycitin, genistin), malonylglucosides (6″-*O*-manoyldaidzin, 6″-*O*-manoylglycitin, 6″-*O*-manoylgenistin), acetylglucosides (6″-*O*-acetyldaidzin, 6″-*O*-acetylglycitin, 6″-*O*-acetylgenistin) and also in the free form (Barnes et al., 1994; Fleury et al., 1991; Hoeck et al., 2000; Kudou et al., 1991). Presence of formononetin (Murthy et al., 1986), isoformononetin (Ingham et al., 1981), and 6,7,4′-trihydroxyisoflavone (Gyorgy et al., 1964) has also been reported. Later, Hosny and Rosazza (1999) identified three novel isoflavones in soybean molasses:

- Glycitein 7-*O*-β-D-(2″,4″,6″-*O*-triacetyl)glucopyranoside
- 8-(γ-Hydroxy-γ,γ–dimethylpropyl)genistein 7-*O*-β–D-glucopyranoside
- 5-Hydroxy-8-(γ-hydroxy-γ,γ–dimethylpropyl)-3′-,4′-dimethoxyisoflavone-7-*O*-β–D-glucopyranoside

Recently, Hosny and Rosazza (2002) isolated and identified in soybeans two novel isoflavones, dihydrodaidzin and dihydrogenistin, and a new isoflavone, 2″,6″-diacetyloninin. Of these, glucosides were the predominant form of isoflavones in soybean products.

The content of isoflavones in soybean seeds is influenced by varietal and genetic differences, maturity stage, and climate and growing conditions (Dalais et al., 1997; Eldridge and Kwolek, 1983; Simonne et al., 2000; Wang and Murphy, 1994). The total content of isoflavones in soybeans ranges from 47.2 to 420.0 mg per 100 g (Eldridge and Kwolek, 1983; Simonne et al., 2000; Wang and Murphy, 1994, 1996) and, according to Wang and Murphy (1994), may vary from 117.6 to 174.9 mg/100 g among different locations within the same crop year. These authors also reported that Japanese soybean varieties have a higher level of 6″-*O*-malonylglycitin as well as higher ratios of 6″-*O*-malonyldaidzin to daidzin and 6″-*O*-malonylgenistin to daidzin than soybeans grown in the U.S. According to Tsukamoto et al. (1995), isoflavone content in soybean cultivars planted in April at Kyushu, Japan, is significantly lower than in those planted in June and July. Hutabarat et al. (2001) reported that the mean concentration of daidzein and genistein in soybeans grown in Indonesia is 50% higher ($P < 0.05$) than those grown in Australia.

Hoeck et al. (2000) observed varietal and growing location effects on isoflavone content in six soybean cultivars grown at eight locations for 2 years. These authors reported the existence of statistically significant differences ($P < 0.05$) in total and individual contents of isoflavones in soybeans grown at different locations. Furthermore, Fleury et al. (1991) noted that growing location induces a greater difference in isoflavone content (ratio of 4:1) than varietal type (ratio of 2:1). Depending on the genotype, immature soybean seeds harvested at 80% maturity stage may contain higher (Simonne et al., 2000) or lower (Kudou et al., 1991; Simonne et al., 2000) contents of isoflavones than mature samples (Table 3.9). This suggests that the regulation of isoflavone synthesis pathways may be influenced by soybean genotype (Simonne et al., 2000). In the seeds, isoflavones are mostly concentrated in the hypocotyl; cotyledons contain only 20% of the amount of isoflavones found in hypocotyl (Table 3.10).

TABLE 3.9
Content of Isoflavones at Immature[a] and Mature Stages of Soybean Development

Variety	Maturity Stage	Moisture (%)	Total (mg/100 g)
NTCPR93–40	Immature	65.1	720 ± 12.2
	Mature	13.0	225 ± 4.9
Honey brown	Immature	61.4	1283 ± 25.4
	Mature	13.0	644 ± 2.4
NTCPR93–286	Immature	60.5	810 ± 3.3
	Mature	13.0	377 ± 3.2
Hutcheson	Immature	51.1	2013 ± 53
	Mature	13.1	2545 ± 76.8

[a] Immature soybeans were harvested at 80% maturity.

Source: Adapted from Simonne, A.H. et al., 2000, *J. Agric. Food Chem.*, 48:6061–6069.

TABLE 3.10
Distribution of Soybean Isoflavones[a]

Variety	Hull	Hypocotyl	Cotyledon
Amsoy	10.6	1756.5	158.5
Tiger	20.0	1405.2	319.2

[a] Milligrams per 100 g.

Source: Adapted from Eldridge, A.C. and Kwolek, W.F., 1983, *J. Agric. Food Chem.*, 31:394–396.

Other Phenolic Compounds

Cyanidin 3-glucoside is the predominant anthocyanin in soybean seed coats (Kuroda and Wade, 1933). Pelargonidin 3-glucoside has also been identified in black-seeded soybean varieties (Yoshikura and Hamaguchi, 1969). Recently, Hosny and Rosazza (1999) identified two glycoside esters in the water-soluble portion of soybean molasses:

- 1-*O*-(E)-feruloyl[α-L-arabinofuranosyl-(1→3)[β-D-glucopyranosyl(1→6)]β–D-glucopyranose
- 1-*O*-(E)-3,4,5-trimethoxycinnamoyl[α-L-arabinofuranosyl(1→3)-*O*-β-D-glucopyranosyl(1→6)]β-D-glucopyranose

Soybeans may also contain isoflavonoid phytoalexins such as coumestrol (7,12-dihydroxy-coumestan), a coumestan isoflavone, and a pterocarpan phytoalexins such as glyceollins I, II and III (Figure 3.5). These phytoalexins are accumulated in soybean seeds after their exposure to microorganisms (Boue et al., 2000; Ebel et al., 1976). Knuckles et al. (1976) reported that the content of coumestrol in soybean seed is 0.12 mg/100 g on a dry basis (d.b.), while in soybean sprouts it is 7.11 mg/100 g d.b. Lookhart et al. (1979) found that soybean hulls contain the highest level of coumestrol; its content in seeds increases with increasing germination time. Recently, Hutabarat et al. (2001) reported that soymilk samples purchased in Indonesia contain 8.0 mg of coumestrol per 100 g of wet weight. Glyceollins have been shown to prevent the accumulation of aflatoxin B_1 in cultures of *Aspergillus flavus* (Song and Karr, 1993). Inoculation of living soybean plant with *Aspergillus* species increases the concentrations of coumestrol and glyceollins in the cotyledons (Boue et al., 2000). Recently, using matrix-assisted laser desorption/ionization time-of-flight mass spectrometry analysis, Takahata et al. (2001) detected procyanidins with up to 30 catechin/epicatechin units in brown soybean seed coats.

Effect of Processing

Soaking

Soaking is one of the steps in the production of soymilk, tempeh and tofu; it is used to aid in dehulling and grinding beans. During the soaking process, isoflavone

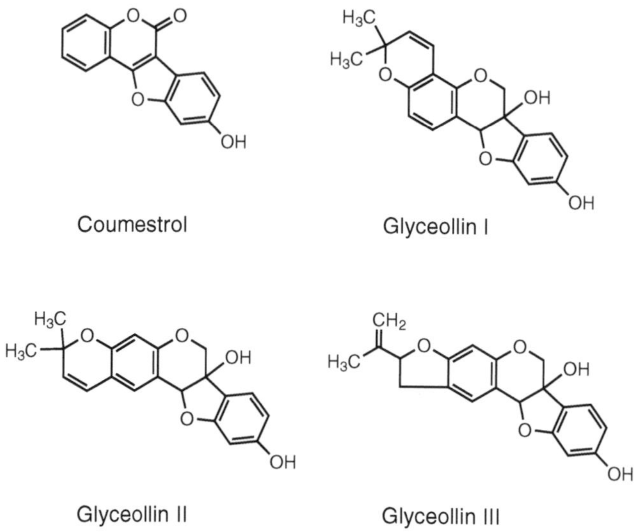

FIGURE 3.5 Chemical structures of soybean phytoalexins.

TABLE 3.11
Concentrations of Isoflavone Aglucones Daidzein and Genistein in Soaked Defatted Soybean Flakes[a]

Soaking Conditions	Daidzein	Genistein
6 h in 50°C distilled water	40.1 ± 9.9	35.5 ± 6.2
6 h in 50°C 0.25 $NaHO_3$	36.2 ± 5.2	31.3 ± 6.4
30 min in boiling 0.25 $NaHCO_3$	1.8 ± 0.01	2.3 ± 0.09

[a] Milligrams per 100 g, on a dry weight basis.

Source: Adapted from Ha, E.Y.A. et al., 1992, *J. Food Sci.*, 57:414–418.

glucosides may be hydrolyzed by β–glucosidases to their aglycones, daidzein and genistein (Figure 3.4). The resulting isoflavone aglycones have more objectionable flavors than their glucoside precursors (Matsuura et al., 1989). The role of β-glucosidases in this reaction was confirmed by Matsuura and Obata (1993) by the addition of glucono-δ-lactone, β–glucosidase competitive inhibitor to the soaking water. The concentration of isoflavone aglycones depends on the conditions of soaking process. Soaking soybeans for 6 h at 50°C in distilled water or in a 0.25% sodium bicarbonate solution results in the formation of substantial amounts of isoflavone aglycones in soybean products (Table 3.11) (Ha et al., 1992). Furthermore, soaking soybeans for 10 to 12 h at room temperature in tap water results in loss of about 10% of isoflavones originally present (Wang and Murphy, 1996). According to these authors, soaking water contained 34% of daidzein and 18% of genistein originally present in the beans.

Heating

Heat treatment of soybean is used to inactivate trypsin inhibitors, kill bacteria, reduce raw and beany flavors, and improve nutritional value of products. Cooking beans is one of the steps employed in the production of tempeh and tofu (Shurtleff and Aoyagi, 1979) because cooking increases leaching of isoflavones to the cooking water. Simmone et al. (2000) reported that blanching immature beans for 1 min in boiling water or boiling the beans for 35 min in water results in similar losses (54 to 56%) of isoflavones originally present in the beans. Furthermore, during production of tempeh most of the isoflavones are leached out into the cooking water (Wang and Murphy, 1996). In addition, it was found that cooking does not affect the retention of isoflavones in tofu because the cooking water is incorporated into the final product.

Heat treatment alters the distribution of individual isoflavones in cooked soy foods. Cooking water slurries of soybean or dehulled soybeans brings about a drastic reduction in the concentration of malonylated isoflavone glucosides, but increases the content of isoflavone glucosides, acetyl isoflavone glucosides and aglycones (Kudou et al., 1991; Simonne et al., 2000; Wang and Murphy, 1996). Kudou et al. (1991) suggested that malonyl glucosides are transformed into acetyl isoflavone glucoside derivatives during heating.

Unfermented Soy Products

Soy Protein Isolates

Approximately 53% of isoflavones originally present in raw materials are lost during the processing of soybean into soy protein isolate (Wang and Murphy, 1996). Major losses of isoflavones occur during alkaline extraction of proteins; the alkaline step also modifies the distribution of isoflavone forms. The fraction soluble in alkaline solution contains elevated levels of aglycones compared to those found in raw soybean. Only 25.8% retention of isoflavones is achieved during processing of soybean to soy protein isolate (Wang et al., 1998). The loss of isoflavones is distributed among the wash water (21.6%), solid waste (19%) and whey (14.3%); however, approximately 19.4% of isoflavones is not accounted for in the mass balance study conducted by Wang et al. (1998).

Soymilk

Soymilk may be produced from soy isolates and whole soybeans. Wang and Murphy (1996) did not observe any loss of isoflavones during soymilk production. The total content of isoflavones in U.S. retail samples of soymilk ranges from 8.0 to 16.5 mg/100 g of wet weight. These isoflavones mostly comprise their glucoside and malonylglucoside derivatives. The contents of daidzin and genistin are between 2.0 to 9.4 and 2.5 to 13.0 mg per 100 g of wet weight, respectively (Murphy et al., 1999). Recently, Hutabarat et al. (2001) demonstrated that starting material used for production of soymilk does not affect the contents of daidzein and genistein on a dry weight basis. However, up to four-fold differences on a dry weight basis have been observed in the content of daidzein and genistein between samples of soymilk brands purchased in Australia and Indonesia. Variations in these levels may be due to different cultivars of soybean used and techniques employed for soymilk production (Hutabarat et al., 2001).

Tofu

The production of tofu involves soaking soybeans in water, washing them, and then grinding them in water. The slurry is subsequently cooked and filtered to remove insoluble matter known as okara from the soymilk. Following this, calcium ion is added to soymilk and proteins are coagulated by heat to form a curd (tofu). Traditional tofu, called momem or cotton tofu, contains only 33% (based on dry matter) of isoflavones originally present in soybean seeds (i.e., 21.7 mg/100 g of seeds). Approximately 43% of the total mass of isoflavones is lost during coagulation of curd, but the distribution of isoflavone forms in tofu is only slightly altered when compared to that of seeds. Isoflavones are not destroyed during tofu production, but rather distributed into soaking water, okara, whey and tofu (Wang and Murphy, 1996). The total content of isoflavones in tofu marketed in the U.S. ranges from 20.2 to 34.7 mg/100 g of wet weight (Murphy et al., 1999). On the other hand, samples of raw tofu purchased in Singapore contained between 25.6 and 31.9 mg of isoflavones per 100 g of wet weight, while those of cooked tofu varied from 23.1 to 28.7 mg/100 g of wet weight (Franke et al., 1999).

Fermented Soy Products

Fermented Soybean Curd

Fermented soybean curd is a popular Chinese soy food. It is a soft, cheese-like fermented product made of tofu classified into white and red fermented soybean curds (Liu et al., 1988). Chung (2000) identified eight phenolic compounds in volatile flavor components of red fermented soybean curds in concentrations of 0.0033 to 0.22 mg/100-g sample: eugenol, coumarin, phenol, 4-ethyl-2-methoxyphenol, 4-methylphenol, 4-ethylphenol, ethyl cinnamate, and 2-methoxy-4-vinylphenol.

Soy Pasta

Soybean pasta is an important Oriental soy food known as miso in Japan, taucho in Indonesia, jang in Korea, taotsi in the Philippines and jiang or chiang in China (Hutabarat et al., 2001; Lin, 1991). The production of soy pasta involves the preparation of koji (see the next section), which may then be mixed with cereals such as rice or barley (Chiou and Cheng, 2001; Zarkadas et al., 1997). The mash is then cooked after addition of whole soybean mash and salt and inoculated with a mixture of microorganisms and fermented. The total content of isoflavones in miso sold in the U.S. ranges from 22.8 to 23.5 mg/100 g of wet weight. The distribution of isoflavone forms in miso is summarized in Table 3.8 (Murphy et al., 1999). On the other hand, samples of soy pasta purchased in Indonesia have contained 12.8 to 27.6 mg of daidzein per 100 g of wet weight and 10.1 to 23.4 mg of genistein per 100 g of wet weight (Hutabarat et al., 2001).

Soy Sauce

In the production of soy sauce, part of the raw material is first used to prepare a cultured material called koji. During koji production (2 to 3 days), proteins and starches are digested by enzymes produced by *Aspergillus* species and a part of the ferulic acid is converted to vanillin and vanillic acid, while *p*-hydroxycinnamic acid is formed and partially converted to *p*-hydroxybenzoic acid (Yokosuka, 1986). In addition, isoflavone glycosides daidzin and genestin are transformed into daidzein and genistein (Chiou and Cheng, 2001). Koji is then mixed with the remaining raw material and brine to produce a wet mash or moromi; the mixture is finally fermented to afford soy sauce. During the fermentation process, ferulic and *p*-hydroxycinnamic acids are metabolized into 4-ethylguaiacol and *p*-ethylphenol, respectively. These phenolics are considered important components of soy sauce flavor (Yokosuka, 1986) and are mainly derived from wheat components. Because isoflavone glucosides, malonylglucosides, and acetylglucosides are degraded by the microorganisms, soy sauce is considered a poor source of isoflavones (Murphy et al., 1999). On the other hand, novel tartaric acid isoflavone derivatives may be formed. Kinoshita et al. (1997) reported that Japanese soy sauces contain conjugated ethers of tartaric acid with daidzein, genistein, and 8-hydroxygenistein. These authors developed a chemometric technique for differentiating the Japanese soy sauces based on HPLC patterns for these novel isoflavone derivatives.

Tempeh

Tempeh is a fermented product indigenous to Indonesia. Fermentation brought about by genus *Rhizopus* significantly alters distribution of isoflavones in the final product. Tempeh contains 6.5 times higher levels of aglycones and 57% lower levels of glucosides than those in the original raw material (Wang and Murphy, 1994). This increase in aglycones may be due to the hydrolytic action of β-glucosidase from fungi (Murakami et al., 1984). Tempeh purchased in Indonesia and Australia has contained 74.5 and 17.6 mg/100 g of daidzein, respectively, and 85.4 and 33.3 mg/100 g of genistein, respectively, on a dry weight basis. These differences in the isoflavone levels may originate from the soybean cultivar used and processing steps employed (Hutabarat et al., 2001).

OTHER OILSEEDS

Borage

The seeds of borage (*Borago officinalis* L.) plant serve as a good source of an oil containing up to 25% of γ-linolenic acid (18:3ω6) (Whipkey et al., 1988). Available information on the phenolic compounds of borage seeds is still diverse and fragmentary. Recently, Wettasinghe et al. (2001) fractionated ethanolic extract of borage meal into four fractions using a Sephadex LH-20 column chromatography. Using TLC methodology, these authors located one and two strong antioxidative spots in fractions I and IV, respectively. These compounds were then positively identified as rosmarinic, syringic and sinapic acids.

Recently, Zadernowski et al. (2002) reported that defatted borage meal contains 18.5 mg of total phenolic acids per 100 g of meal. Free phenolics were the predominant form of phenolic acids and comprised 69.3% of the total phenolic acids in borage meal. Ferulic acid was the principal free phenolic acid identified, accounting for 50.2% of total free phenolic acids of borage. Protocatechuic, *p*-hydroxybenzoic, and *p*-hydroxyphenyllactic acids contributed 40.5% to the total free phenolic acids. The total concentration of 2-hydroxy-4-methoxybenzoic, gallic, *p*-coumaric and cinnamic acids did not exceed 1.4 mg/100 g. On the other hand, phenolic acids liberated from soluble esters contributed 40.4% to the total phenolic acids; hydroxycaffeic acid was the major phenolic acid in this fraction. Small quantities (0.1 to 0.3 mg/100 g) of 2-hydroxy-4-methoxybenzoic, caffeic and sinapic acids were also present in borage seeds. In addition, up to eight minor esterified phenolic acids (below 0.1 mg/100 g) were detected in the meal.

Cottonseed

Cotton is a plant grown in the tropical and subtropical regions of the world for food and fiber use; its hulls and linters comprise about 45% and its kernels 55% of cottonseed weight. Gossypol is the major phenolic component in pigment glands of cottonseed kernels at 1400 to 13,000 mg/kg sample (Heywood et al., 1986; Lawhon et al., 1977; McCarty et al., 1996; Smith, 1970). Other gossypol-like pigments include gossyfulvin, gossypurpurin and gossycaerulin (Chamkasem, 1988). Gossypol [1,1′-,6,6′-,7,7′-hexahydroxy-5,5′-diisopropyl-3,3′-dimethyl-(2,2′-binaphtalene)-8,8′-dicarboxaldehyde] is a reactive triterpenoid aldehyde (Figure 3.6) that exists as (+)- and (–)-enantiomers because of restricted rotation around the binaphtyl ring (Bailey et al., 2000).

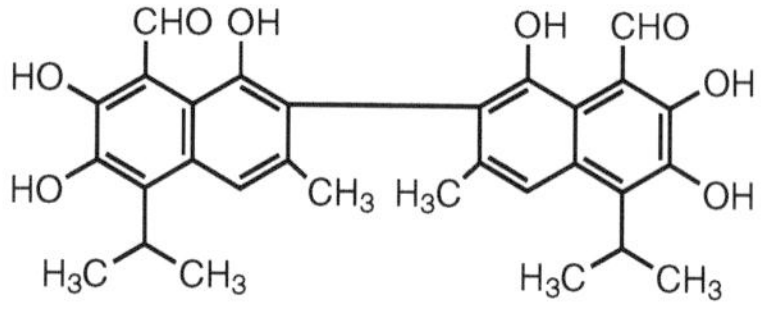

FIGURE 3.6 Chemical structure of gossypol.

Gossypol is present in cottonseed kernels in the free and bound forms. Gossypol and gossypol derivatives that can be extracted into 70% aqueous acetone are defined as free gossypol (AOCS Ba7–58, 1987). On the other hand, total gossypol (free and bound) is defined as gossypol that can be hydrolyzed and complexed with 3-amino-1-propanol in a dimethylformamide solution to form a diaminopropanol complex (AOCS Ba 8–78, 1987). Wang and Plhak (2000) have suggested that some gossypol may also be present in cottonseed in oxidized or degraded forms. Oxidation may result in the loss of aldehydic group, but degradation may increase the level of gossypol derivatives.

Gossypol interacts with the ε-amino group of lysine and thus may make it unavailable during digestion. In addition, gossypol is toxic to monogastric animals, but does not affect the ruminants (Berardi and Goldblatt, 1980). Recently, Bailey (2000) demonstrated that broilers fed on crushed cottonseed with a high (+)- to (–)-gossypol enantiomer ratio have significantly higher feed-to-weight gain ratios than those fed on crushed cottonseed with a lower (+)- to (–)-gossypol enantiomer ratio. Current U.S. FDA regulations limit the content of free gossypol in edible cottonseed flour to 450 mg/kg (Salunkhe et al., 1992). However, the Protein Advisory Group of the United Nations Food and Agriculture and World Health organizations (FAO/WHO) has limited the content of free gossypol in edible cottonseed flour to 600 mg/kg and total gossypol to 12,000 mg per kilogram (Lusas and Jividen, 1987). Furthermore, *p*-hydroxybenzoic, *trans-p*-coumaric, *trans*-ferulic and *trans*-caffeic acids have been detected in glanded and glandless cottonseed meals (Dabrowski and Sosulski, 1984).

Evening Primrose

The seeds of evening primrose *(Oenothera biennis* L.) serve as good sources of the essential fatty acid γ-linolenic acid (18:3ω6) (Hudson, 1984). This fatty acid is prone to oxidation, but despite this, the oil in the seeds resists oxidation. This suggests that these oleaginous seeds must contain some potent antioxidants (Amarowicz et al., 1999). Lu and Foo (1995), Shahidi et al. (1997a) and Birch et al. (2001) demonstrated that this antioxidative activity in evening primrose may be due to the presence of phenolic compounds.

The total content of phenolic acids extracted from evening primrose cake into water and 70% aqueous acetone is 580 and 180 mg/g of crude extract, respectively. Lu and Foo (1995) reported the presence of procyanidine gallate oligomer evening primrose seeds; later, Shahidi et al. (1997a) demonstrated that they contain catechins as well as dimers and trimers of proanthocyanidins. Furthermore, Balasinska and Troszynska (1998) reported the presence of a phenolic acid in acetone–water extract of evening primrose seeds, but did not attempt to identify phenolic acids present. Subsequently, using nuclear magnetic resonance spectroscopy and electron impact mass spectrometry, Wettasinghe et al. (2001, 2002) identified (+)-catechin, (–)-epicatechin and gallic acid in acetone extracts from evening primrose seeds. Recently, Hamburger et al. (2002) discovered three triterpenoid caffeates in cold-pressed, nonraffinated evening primrose oil: 3-*O*-*trans*-caffeoyl derivatives of betulinic, morolic, and oleanolic acid (Figure 3.7). These authors also demonstrated that these lipophilic caffeates display a strong radical scavenging activity against DPPH (2,2-diphenyl-1-picrylhydrazyl radical) and are potent inhibitors of cyclooxygenase-1 and -2, as well as neutrophil elastase *in vitro.*

Evening primrose seeds contain 12.4 mg of total phenolic acids per 100 g of defatted meal (Zadernowski et al., 2002). Free phenolic acids comprise 69% of total phenolic acid content. Protocatechuic acid is the predominant phenolic acid, constituting 58.5% of the total free phenolic acid present. Salicylic, *p*-hydroxybenzoic, 2-hydroxy-4-methoxybenzoic, vanillic, *m*- and *p*-coumaric, gallic, ferulic and caffeic acids are found in smaller quantities (above 0.1 mg/100 g). Furthermore, gallic and protocatechuic acids have been identified as the principal phenolic acids in the fractions of phenolics liberated from esters and glycosides (Zadernowski et al., 2002).

Flaxseed

Flaxseed is grown mostly for industrial purposes; only in certain cultures has it served as a source of edible oil as well as medicine (Wanasundara et al., 1997).

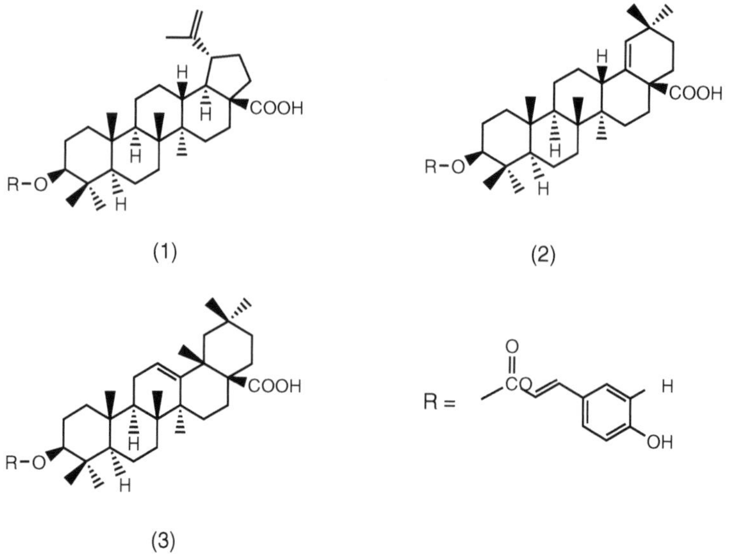

(1) (2) (3)

R =

FIGURE 3.7 Chemical structures of 3-*O-trans*caffeoyl derivatives of (1) betulinic, (2) morolic and (3) oleanolic acids found in evening primrose oil.

Nutritional and health benefits from the consumption of flaxseed and its components have recently been documented (Cunnane and Thompson, 1995; Haumann, 1993; McCord and Rao, 1997; Oomah and Mazza, 1998; Thompson, 1995; Thompson et al., 1996a, b). Flaxseed is gaining popularity as a health food and as an ingredient in muffins, breads and breakfast cereals (Jenkins, 1995; McCord and Rao, 1997).

Flaxseed is a rich source of phenolic acids, containing 800 to 1000 mg of total phenolic acids per 100 g of seeds. The content of esterified and etherified phenolic acids is up to 500 mg/100 g and 300 to 500 mg/100 g, respectively. Esterified phenolic acids comprise 48 to 66% and are not dependent on the cultivar (Oomah et al., 1995). Furthermore, Varga and Diosady (1994) have found that soluble and insoluble phenolic acids constitute 54 and 26 to 29% of total phenolic acids in flaxseed flour, respectively. *Trans*–ferulic and *trans*-sinapic acids are the major phenolice acids and *trans*-caffeic, *p*-coumaric and *p*-hydroxybenzoic are the minor phenolic acids found in dehulled, defatted flaxseed meal (Figure 3.1) (Dabrowski and Sosulski, 1984). However, Harris and Haggerty (1993) reported that the content of ferulic and chlorogenic acids accounted for 84% of total phenolic acids in methanolic extracts of defatted flaxseed meal. In addition, *p*-coumaric acid glucoside (linocinnamarin) and ferulic acid glucoside have been identified in flaxseed (Johnsson et al., 2002; Westcott and Muir, 1996a; Westcott et al., 2000).

Flaxseed also contains flavonoids, coumarins, and lignans (Amarowicz et al., 1994; Bambagiotti-Alberti et al., 1994a, b; Caragay, 1992; Johnsson et al., 2000, 2002; Meagher et al., 1999; Obermeyer et al., 1995; Thompson et al., 1997; Westcott and Muir, 1996b, 2000). The total content of flavonoids in the seeds ranges from 35 to 71 mg/100 g (Oomah et al., 1996) and flavone C- and O-glycosides are the

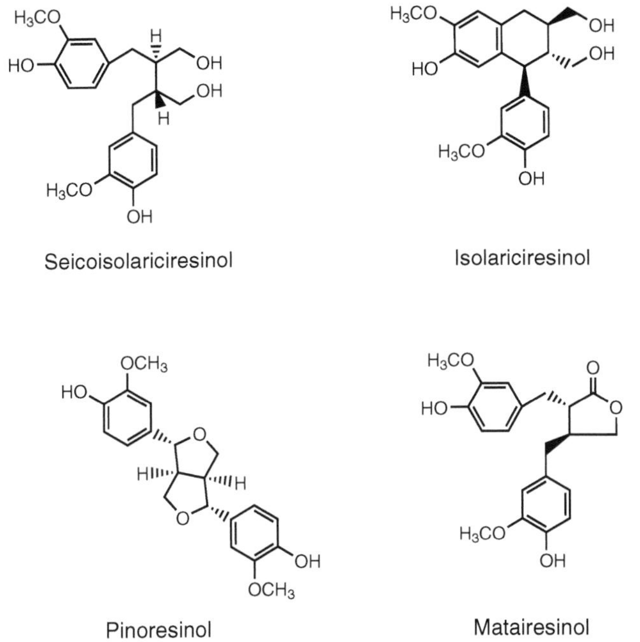

Seicoisolariciresinol

Isolariciresinol

Pinoresinol

Matairesinol

FIGURE 3.8 Chemical structures of lignans found in flaxseed.

major flavonoids present in flaxseed cotyledons (Ibraham and Shaw, 1970). Secoisolariciresinol diglucoside (SDG) has been identified as a major lignan of flaxseed (Bakke and Klosterman, 1956; Mazur et al., 1996; Meagher et al., 1999); isolariciresinol, pinoresinol, and matairesinol (Figure 3.8) have been identified as minor lignan components (Meagher et al., 1999). The content of SDG in flaxseed is significantly affected by the cultivar, year of harvest and, to lesser extent, growing location (Westcott and Muir, 1996). Mazur et al. (1996) reported that the content of secoisolariciresinerol and matairesinol in flaxseed is 370 and 1.087 mg/100 g, respectively. Secoisolariciresinol diglucoside and matairesinol may be converted into two mammalian lignans, enterodiol and enterolactone, by intestinal microflora (Axelson et al., 1982; Bambagiotti-Alberti et al., 1994a,b; Borriello et al., 1985; Heinonen et al., 2001; Setchell et al., 1983; Thompson et al., 1991; Wanasundara et al., 1997). Flaxseed meal is used as an ingredient in a number of baked goods in which the SDG content may range from 8.9 to 280 mg per bread loaf and from 121 to 1360 mg/kg (dry basis) of flax cookies, bagels, and muffins (Muir and Westcott, 2000)

Oomah and Mazza (1997) reported that abrasive dehulling of flaxseed concentrated phenolic acids in the dehulled flaxseed and observed that changes in phenolics are highly cultivar dependent ($P \leq 0.0001$). Upon dehulling, the total content of phenolic acids of McGregor seeds changes from 700 to 1980 mg/100 g, but only marginal changes are observed for cultivars NorMan, Omega and Vimy. Furthermore, Varga and Diosady (1994) showed that extraction of flaxseed meal with methanol–ammonia lowers the content of soluble esterified phenolic acids and insoluble bound phenolic acids by 20 and 29%, respectively.

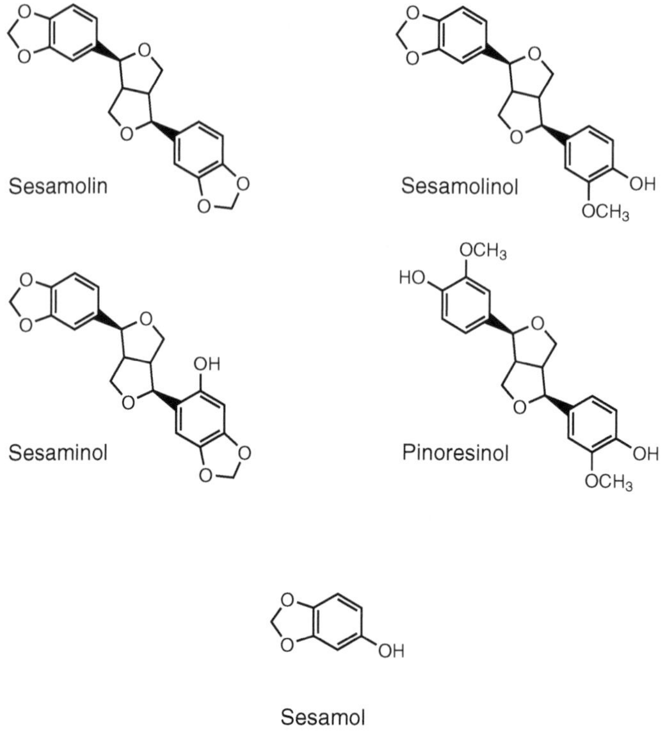

FIGURE 3.9 Chemical structures of some phenolic compounds found in sesame seeds.

SESAME SEEDS

Sesame seeds have been grown for their oil and proteins; defatted sesame seed meals contain up to 50% protein (Sen and Bhattacharyya, 2001). Sesame seed is a popular snack and confectionary ingredient in such products as halva; sesame oil is obtained from unroasted and roasted seeds, usually by expeller pressing (Wanasundara et al., 1997).

Sesame seeds contain lignophenols and carboxyphenols (Budowski, 1950, 1964; Budowski et al., 1950; Fukuda et al., 1981, 1986a, b, c, 1994; Shahidi et al., 1997b; Sirato-Yasumoto et al., 2000). The major lignans of sesame seed are sesamin (200 to 500 mg/100 g) and sesamolin (200 to 300 mg/100 g). In addition, sesamolinol and sesaminol are found in seeds and oil (Figure 3.9) (Nagata et al., 1987; Osawa et al., 1985). Fresh oils obtained from coated Egyptian and Sudanese sesame seed varieties contain 649.1 and 579.1 mg of sesamin per 100 g of oil and 183.3 and 349.2 mg sesamolin per 100 g of oil, respectively (Shahidi et al., 1997b). A number of pinoresinol glucosides have also been identified in sesame seed, namely (Katsuzaki et al., 1994).

- Pinoresinol 4′-*O*-β-D-glucopyranosyl (1→6)-β-D-glucopyranoside
- Pinoresinol 4′-*O*-β-D-glucopyranosyl (1→2)-β-D-glucopyranoside
- Pinoresinol 4′-*O*-β-D-glucopyranosyl (1→2)-β-D-glucopyranosyl (1→2)-β-D-glucopyranoside

Sesame seed lignans have shown beneficial physiological effects in experimental animals and humans (Akimoto et al., 1993; Hirata et al., 1996; Hirose et al., 1991; Ogawa et al., 1995; Sirato-Yasumoto et al., 2001).

Purified unroasted sesame oil contains from 50 to 100 mg of sesaminol per 100 g; the refining process brings about an increase in sesaminol level in sesame oil (Wanasundara et al., 1997). According to Fukuda et al. (1986a, b, 1994), sesaminol may be formed from sesamolin through intermolecular group transformation involving formation of oxonium ion and then the carbon–carbon bond. On the other hand, thermal decomposition of sesamolin leads to the formation of sesamol, while its hydrolysis affords sesamol and sesamin (Fukuda et al., 1994). Exposure of sesame lignans to elevated temperatures brings about their degradation. Only 40.5% of sesaminol added to corn oil is detected after 6 h of heating at 180°C, while all added sesamol is decomposed after 4 h heating at 180°C (Fukuda et al., 1994). Shahidi et al. (1997b) also reported that processing sesame oil at 65°C for 35 days has a more drastic effect on sesomolin than on sesamin.

Sesame flour also contains *trans*-caffeic, *trans*-*p*-coumaric and *trans*-ferulic acids and traces of *p*-hydroxybenzoic acid (Table 3.1) (Dabrowski and Sosulski, 1984). Moreover, the presence of guaiacol and 2-methoxy-5-(1-propenyl)phenol has been detected in the volatile components of roasted sesame seeds (Shimoda et al., 1997). Two potent antioxidants have also been isolated from suspension of sesame callus (Mimura et al., 1994):

- Acteoside (3,4-dihydroxy-β-phenethyl-*O*-α-rhamnopyranosyl (1→3)-4-*O*-caffeoyl-β-glucopyranoside)
- 3,4-Dihydroxy-β-phenyl-*O*-ethylcarboxyl-*O*-α-rhamnopyranosyl (1→3)-4-*O*-caffeoyl-β-glucopyranoside (SI-1)

Sunflower

The presence of phenolics in sunflower has restricted its widespread use for human consumption and industrial food applications (Delic et al., 1975). The major phenolics in sunflower seeds are chlorogenic, quinic and caffeic acids (Figure 3.10) (Sosulski, 1979), which occur in the defatted sunflower meal at about 2700, 400 and 200 mg/100 g, respectively (Carter et al., 1972). The minor phenolic acids in sunflower seeds are *trans*-ferulic, *p*-hydroxybenzoic, *trans*-*p*-coumaric, syringic and vanillic acids (Figure 3.1 and Table 3.1) (Dabrowski and Sosulski, 1984).

The occurrence and the content of chlorogenic acid (Figure 3.10) are responsible for dark discoloration of food products and may also decrease protein availability (Carter et al., 1972; Dryden and Satterlee, 1978; Robertson, 1975). Interaction of phenolics with proteins is a well-known phenomenon (Carter et al., 1972; Sosulski

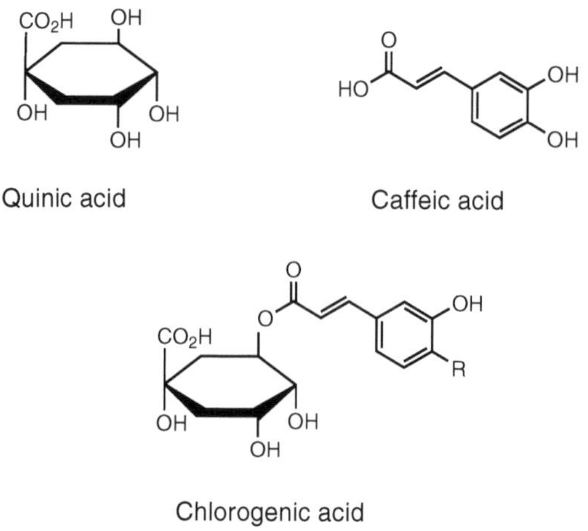

FIGURE 3.10 Chemical structures of chlorogenic, quinic and caffeic acids.

et al., 1973); digestive enzymes such as trypsin, lipase and amylase are affected (Salunkhe et al., 1992). Futhermore, interaction of chlorogenic acid with low molecular weight albumins (<5000 Da) of sunflower is well known (Prasad, 1988; Sabir et al., 1974). These interactions involve sulfhydryl or amino groups of proteins that decrease the nutritional quality of proteins because such condensation products cannot be metabolized by humans (Desphande et al., 1984).

Chlorogenic acid may be removed from sunflower meal by solvent extraction procedures involving water and alcohol (Rutkowski, 1972) or acidic butanol (Sodini and Canella, 1977). Partial removal of chlorogenic acid with water, 40% (v/v) aqueous methanol, 50% (v/v) aqueous isopropanol, ethanol or acetone has also been reported (Gheyasuddin et al., 1970; Sosulski and McCleary, 1972; Sripad and Narasinga Rao, 1987). In addition, dehulling seeds and soaking in citric acid or sodium bisulfite significantly reduces the level of phenolic compounds in sunflower (Bau et al., 1983).

PLANT OILS

Olive Oil

The production of virgin olive oil involves mechanical pressing of mesocarp of drupes of olive trees (*Olea europea* L.), washing, decanting, centrifuging and selective filtering — virgin olive oil is consumed without refining (Caponio et al., 2001; Wanasundara et al., 1997). The major olive oil producers are Spain, Italy, and Greece (Brenes et al., 1999). Phenolic compounds affect the sensory characteristics and autoxidative stability of virgin olive oil (Baldioli et al., 1996; Briante et al., 2002; Gutierrez Gonzales-Quijano et al., 1977; Le Tutor and Guedon, 1992; Visioli et al., 1998, 1999). Olive oil phenolics also possess pharmacological (Le Tutor and Guedon, 1992; Lipworth et al., 1997; Manna et al., 1999; Medina et al., 1999;

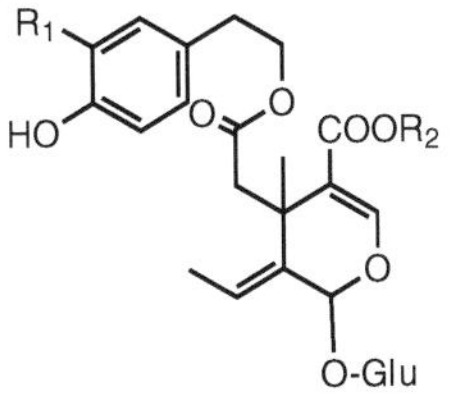

Compound	R_1	R_2
Oleuropein	OH	CH_3
Demethyloleuropein	OH	H
Ligstroside	H	CH_3

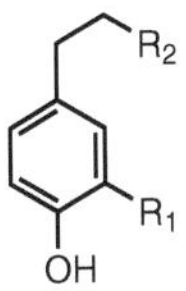

Compound	R_1	R_2
Tyrosol	H	OH
Hydroxytyrosol	OH	OH
Verbascoside	OH	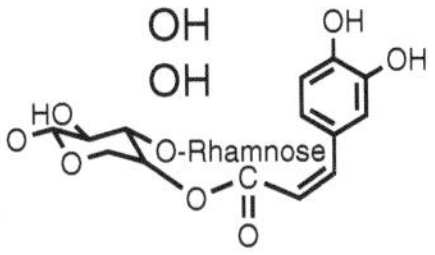

FIGURE 3.11 Chemical structures of some phenolics found in olive oil.

Trichopouolu, 1995; Visioli et al., 2000; Visioli and Galli, 2002) and antimicrobial (Brenes et al., 1992, 1995; Fleming et al., 1973) properties.

Olive fruits may contain up to 80 mg of polyphenols per 100-g sample that are responsible for the unique flavor of virgin olive oil (Visioli et al., 2000). The total phenolic content and the distribution of phenolic components are affected by the cultivar, growing location, and the degree of ripeness (Amiot et al., 1986; Capiono et al., 2001). Oleuropein content decreases as the olive fruits ripen (Amiot et al., 1989; Caponio et al., 2001; Esti et al., 1998), while the content of demethyloleuropein and 2,4-dihydroxyphenylethanol increases (Brenes et al., 1995; Esti et al., 1998).

Secoiridoids, oleuropein, demethyloleuropein, and ligstroside are the main phenolic glucosides and verbascoside (caffeoylrhamnosylglucoside of hydroxytyrosol) is the main hydroxycinnamic acid derivative of olive fruit (Figure 3.11) (Amiot et al., 1986; Angerosa et al., 1995, 1996; Borzillo et al., 2000; Gariboldi et al., 1986; Kuwajima et al., 1988; Maestro-Duran et al., 1994; Ryan and Robards, 1998; Saija and Uccella, 2001; Soler-Rivas et al., 2000; Uccella, 2001). The aglycone of oleuropein is the ester of elenolic acid with 2-(3,4-dihydroxyphenyl)ethanol (hydroxytyrosol) (3,4-DHPEA-EA) and the aglycone of ligstroside is the ester of elenolic acid with 2-(4-hydroxyphenyl)ethanol (tyrosol) (Figure 3.12) (Tsimidou, 1998).

Oleuropein is the major phenolic compound responsible for the development of bitterness in olive fruits (Panizzi et al., 1960). Moreover, phenyl alcohols such as hydroxytyrosol (3,4-DHPEA) and tyrosol (p-HEPA), as well as verbascoside (Figure 3.11) and phenolic acids (including hydroxycinnamic, hydroxybenzoic, hydroxycaffeic, and hydroxyphenylacetic acids), have been reported in olive fruits (Bastoni et al., 2001; Bianco and Uccella, 2000; Servilli et al., 1999a, b; Vierhuis et al., 2001). In addition, flavonoids such as quercetin, rutin, luteolin 7-glucoside and apigenin glucosides (Amiot et al., 1986; Mazza and Miniati, 1993; Rovellini et al., 1997; Vlahov, 1992) and hydroxyisochromans, namely, 1-phenyl-6,7-dihydroxyisochroman and 1-(3′-methoxy-4′-hydroxy)phenyl-6,7-dihydroxyisochroman

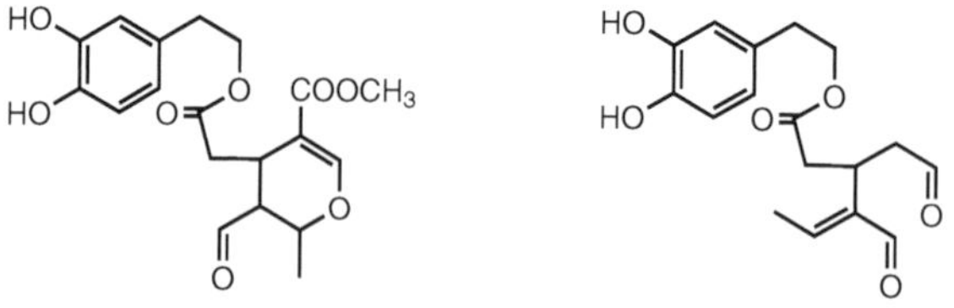

FIGURE 3.12 Chemical structures of aldehydic and dialdehydic forms of oleuropein aglycone (3,4-DHPEA-EA = aldehydic form of oleuropein aglycone; 3,4-DHPEA-EDA = dialdehydic form of decarboxymethyloleuropein aglycone; 3,4-DHPEA = 2-(3,4-dihydroxyphenyl)ethanol; EA = elenolic acid; and EDA = dialdehydic form of elenoic acid).

(Bianco et al., 2001), have been identified in olive fruits. Unprocessed olive fruits also contains 3,4-dihydroxyphenylglycol (Bianchi and Pozzi, 1994) and anthocyanins (Macheix et al., 1990). Lactic acid fermentation of table olives brings about complete hydrolysis of oleuropein and luteolin 7-glucoside to hydroxytyrosol, tyrosol and luteolin (Blekas et al., 2002); this is catalyzed by β-glucosidase and esterase from *Lactobacillus plantarum* strains (Ciafardini et al., 1994; Marsilio et al., 1996).

The content of hydroxytyrosol and hydroxytyrosol derivatives (oleuropein, oleuropein aglycone, demethyloleuropein and hydroxytyrosol glucosides) in Greek, Portuguese, Italian and Spanish table olive fruits ranges from 100 to 430 mg/kg and from 3670 to 5610 mg/kg, respectively (Bastoni et al., 2001). Recently, Blekas et al. (2002) detected a high level of hydroxytyrosol (250 to 760 mg/kg) in Kalamata and Spanish-style green olives. The content of verbascoside in olive fruits from Italian cultivars is 160 to 3200 mg/kg (Romani et al., 1999), while the concentrations of rutin and luteolin 7-glucoside, two main flavonoids in olive fruits (Amiot et al., 1986; Vlahov, 1992), are 110 to 660 and 5 to 600 mg/kg, respectively (Esti et al., 1998; Panizzi, 1960). In addition, cyanidin 3-*O*-glucoside (50 to 880 mg/kg) and cyanidin-3-*O*-rutinoside (250 to 3300 mg/kg) are the most abundant anthocyanins in olive fruits (Romani et al., 1999; Vázquez-Roncero and Maestro-Duran, 1970).

The composition of phenolics in some virgin olive oil samples is shown in Table 3.12. The composition of phenolics is affected by the cultivar, degree of ripeness, climate (Amiot et al., 1986; Caponio et al., 2001), type of crushing machine and the conditions used during malaxation (stirring olive fruit paste at elevated temperature) (Catalano and Caponio, 1996; Vierhuis et al., 2001). According to Medina et al. (1999), the total content of phenolics in extra virgin oil is 86 mg of gallic acid equivalents (GAE) per 100 g of oil, whereas refined olive oil contains only 0.8 mg GAE per 100 g of oil.

Simple phenolic compounds of virgin olive oils include hydroxytyrosol, tyrosol, vanillic acid, *p*-coumaric acid and ferulic acid (Montedoro et al., 1992, 1993). Recently, Mateos et al. (2001) detected cinnamic acid and 2-(4-hydroxyphenyl)ethyl acetate in olive oil. In addition, the presence of vanillin and 4-(acetoxyethyl)-1,2-dihydroxybenzene (Brenes et al., 1999) and flavonoids such as luteolin and apigenin has been reported (Brenes et al., 1999; Rovellini et al., 1997). Luteolin in olive oil

TABLE 3.12
Phenolic Compounds in Some Extra Virgin Oils[a]

	Cultivar			
	Picual	**Picudo**	**Anbequina**	**Empeltre**
Hydroxytyrosol	17.8	7.5	11.4	1.1
Tyrosol	10.5	10.0	8.4	3.2
Vanillic acid	0.1	0.2	0.1	0.1
Vanillin	0.6	0.2	0.3	0.3
3,4-DHPEA-AC	12.8	2.4	7.4	2.8
p-Coumaric acid	0.3	0.1	0.1	0.1
3,4-DHPEA-EDA	82.0	14.0	28.6	14.3
p-HPEA-EDA	31.0	7.3	7.5	7.0
1-Acetoxypinoresinol	4.9	6.8	36.4	31.5
Pinoresinol	29.5	31.2	34.0	19.0
3,4-DHPEA-EA	359.7	40.9	29.0	15.9
p-HPEA-EA	48.7	7.0	4.4	1.5

Note: 3,4-DHPEA-AC: 4-(acetoxyethyl)-1,2-dihydroxybenzene; 3,4-DHPEA-EDA: dialdehydic form of elenolic acid linked to hydroxytyrosol; *p*-HPEA-EDA: dialdehydic form of elenolic acid linked to tyrosol; 3,4-DHPEA-EA: oleuropein aglycone; *p*-HPEA-EA: ligtroside aglycone.

[a] Milligrams per kilogram.

Source: Adapted from Brenes, M. et al., 2000, *J. Am. Oil Chem. Soc.*, 77:715–720.

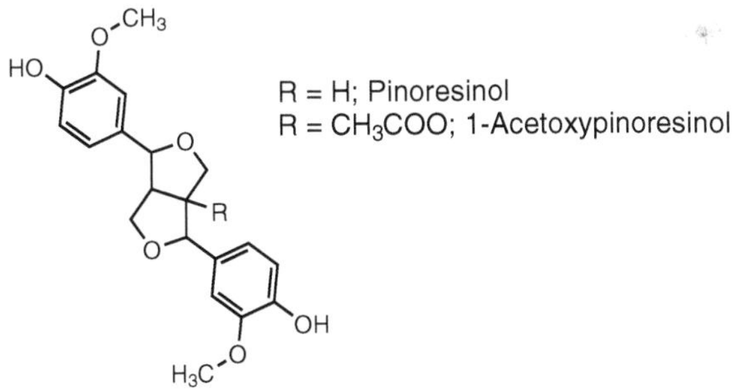

FIGURE 3.13 Chemical structures of lignans found in olive oil.

originates from rutin and from luteolin 7-glucoside while apigenin arises from apigenin glucosides (Brenes et al., 1999). Other phenolics identified in olive oils are oleuropein and ligstroside aglycons and the dialdehydic form of elenolic acid linked to hydroxytyrosyl and tyrosol (Figure 3.12) (Angerosa et al., 1995, 1996; Brenes et al., 1999; Pirisi et al., 2000). Moreover, Brenes et al. (2000) have identified two new phenolic compounds in olive oil, namely, pinoresinol and 1-acetoxypinoresinol (Figure 3.13). The contents of each of these lignans is 5 to 67 mg/kg of oil.

During crushing and malaxing processes, oleuropein and demethyloleuropein are hydrolyzed by endogenous glycosidases to dialdehydic form of decarboxymethyloleuropein aglycone (3,4-DHPEA-EDA) and aldehydic form of oleuropein aglycone (3,4-DHPEA-EA) (Figure 3.12). The aglycons become soluble in the oil phase while the glycosides remain in the wastewater phase (Cortesi et al., 1995). During these processes some isoforms of oleuropein are also generated by keto-enolic tautomeric equilibrium, which involves opening the secoiridoid ring (Angerosa et al., 1995; Gariboldi et al., 1986). Furthermore, during the malaxation process some losses in secoiridoids and phenyl alcohol contents have been observed (Servilli et al., 1999a, b). These phenolic losses may be due to enzymatic oxidation by endogenous oxireductases (Sciancalepore, 1985) and nonenzymatic oxidation (Servilli et al., 1998), as well as their complexation with polysaccharides (McManus et al., 1985). Nitrogen flushing as well as addition of cell wall-degrading enzymes during malaxation increases the concentration of phenolics in virgin olive oil (Ranalli and De Mattia, 1997; Servilli et al., 1992; Vierhuis et al., 2001).

Cinquanta et al. (1997) reported increased levels of hydroxytyrosol and tyrosol during olive oil storage and linked this increase to the hydrolysis of complex phenols (secoiridoids). Subsequently, Brenes et al. (2001) noted a rise in the concentration of hydroxytyrosol and tyrosol in virgin olive oil stored in darkness at 30°C and attributed this to the hydrolysis of secoiridoid aglycones to hydroxytyrosol and tyrosol. Moreover, these authors reported that the hydrolysis of secoiridoid aglycones is influenced by the acidity of the oil and the filtration step. Later, Okogeri and Tasioula-Margari (2002) reported a significant decrease in the content of complex phenols (secoiridoids) in virgin olive oil stored under diffused light or in the dark, but did not detect any increase in the levels of hydroxytyrosol and tyrosol.

Hydroxytyrosol derivatives are most sensitive to thermal oxidation of oil. No hydroxytyrosol derivatives have been detected at peroxide values of 20 to 25 meq/kg of oil. However, tyrosol remains stable even at peroxide value of 70 meq/kg oil (Nissiotis and Tasioula-Margari, 2001). Heating olive oil at 180°C brings about a rapid decrease in polyphenolic compounds due to their thermal destruction or oxidative degradation. On the other hand, boiling olive oil–water mixtures results in the hydrolysis of secoiridoid aglycones to simple phenolics and migration of hydrophilic phenols from oil into the water phase (Brenes et al., 2002; Sacchi et al., 2002).

Palm Oil

Palm oil, an important source of edible oil in many tropical countries, is obtained from the mesocarp of the oil palm fruit (Wanasundara et al., 1997). Because most published studies focus on tocols in palm oil, the available information on palm oil phenolics is still fragmentary. Palm oil is a good source of tocols such as α-tocopherol, α-tocoenol, α-tocotrienol, β-tocotrienol, γ-tocotrienol, and δ-tocotrienol (Choo, 1994; Choo et al., 1996; Goh et al., 1985; Strohschein et al., 1999). The content of total tocols in palm oil ranges from 800 to1000 mg/kg (Choo, 1994; Choo et al., 1996). Tocotrienols are the predominant fraction of palm oil tocols, comprising up to 78% of total tocols (Ab Gapor, 1990; Choo, 1994; Choo et al., 1996).

Rice Bran Oil

Rice bran, a by-product of rice milling, contains from 15 to 20% of an oil considered a valuable domestic oil in many countries (Salunkhe et al., 1992). The content of unsaponifiable matter in rice bran oil is about 2% (Salunkhe et al., 1992). The unsaponifiables of rice bran oil comprise sterols, higher alcohols and γ-oryzanol (Nicolosi et al., 1994). γ-Oryzanol is a mixture of ferulic acid esters of triterpenoid alcohol and has been shown to lower blood cholesterol (Nicolosi et al., 1994). Xu and Godber (1999) and Xu et al. (2001) identified 10 components of γ-oryzanols (Figure 3.14):

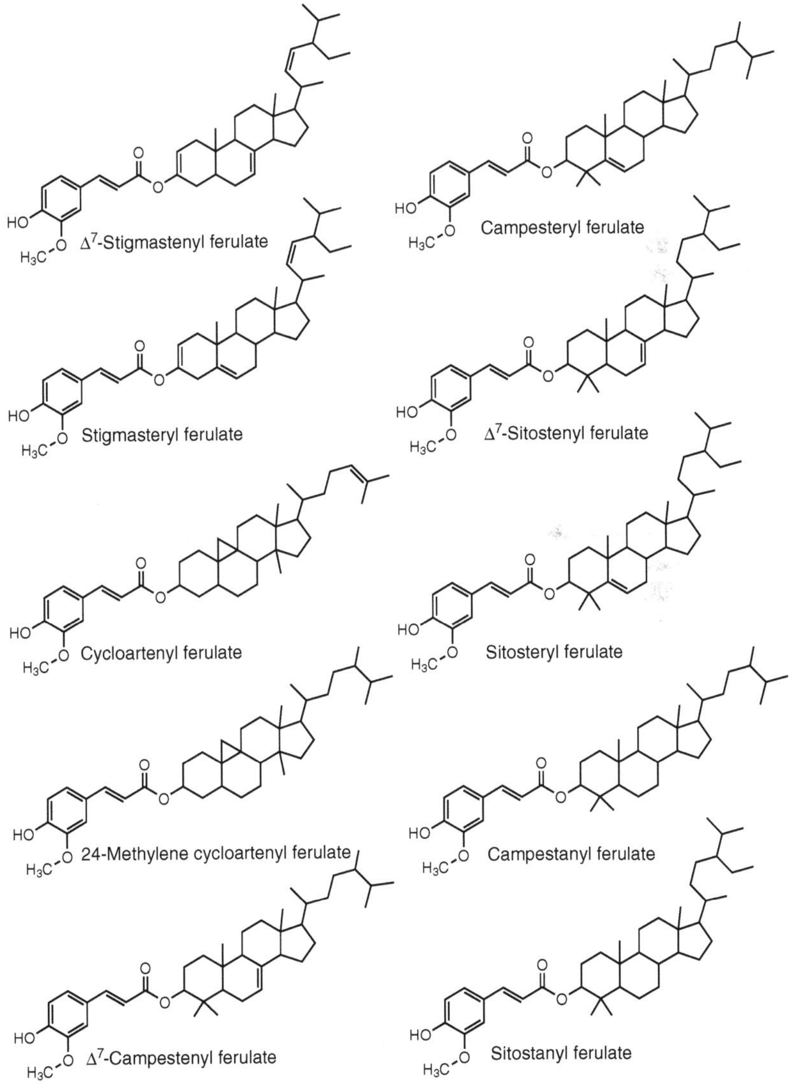

FIGURE 3.14 Chemical structures of γ-oryzanol components.

- Δ^7-Stigmastenyl ferulate
- Stigmasteryl ferulate
- Cycloartenyl ferulate
- 24-Methylenecycloartanyl ferulate
- Δ^7-Campestenyl ferulate
- Campesteryl ferulate
- Δ^7-Sitostenyl ferulate
- Sitosteryl ferulate
- Compestanyl ferulate
- Sitostanyl ferulate

Of these, 24-methylenecycloartanyl ferulate, cycloartenyl ferulate and campesteryl ferulate are the major constituents of γ-oryzanol.

REFERENCES

Ab Gapor, M.T. 1990. Content of vitamin E in palm oil and its antioxidant activity. *Palm Oil Dev.,* 12:25–27.

Akimoto, K., Kigawa, Y., Akamatsu, T., Hirose, N., Sugano, M., Shimuzu, S., and Yamada, H. 1993. Suppressive effect of sesamin against liver damage caused by alcohol or carbotetrachloride in rodents. *Ann. Nutr. Metabol.*, 37:218–224.

Amarowicz, R. and Shahidi, F. 1994. Chromatographic separation of glucopyranosyl sinapate from canola meal. *J. Am. Oil Chem. Soc.*, 71:551–552.

Amarowicz, R., Wanasundara, P.K.J.P.D., and Shahidi, F. 1994. Chromatographic separation of flaxseed phenolics. *Nahrung*, 1994:520–526.

Amarowicz, R., Karamac, M., Rudnicka, B., and Ciska, E. 1995. TLC separation of glucopyranosyl sinapate and other phenolic compounds. *Fat Sci. Technol.*, 97:330–333.

Amarowicz, R., Raab, B., and Karamac, M. 1999. Antioxidative activity of an ethanolic extract of evening primrose. *Nahrung*, 43:216–217.

American Oil Chemist's Society (AOCS). 1987. *Official Methods and Recommended Practices of the American Oil Chemist's Society*, Firestone, D., Ed., AOCS Press, Champaign, IL.

Amiot, M.J., Fleuriet, A., and Macheix, J.J. 1986. Importance and evolution of phenolic compounds in olive during growth and maturation. *J. Agric. Food Chem.*, 34:823–826.

Amiot, M.J., Fleuriet, A., and Macheix, J.J. 1989. Accumulation of oleuropein derivatives during olive maturation. *Phytochemistry*, 28:67–69.

Angerosa, F., d'Alessandro, N., Corana, F., and Mallerio, G. 1996. Characterization of phenolic and secoiridoid aglycons present in virgin oil by gas chromatography-chemical ionization mass spectrophotometry. *J. Chromatog.*, 736:195–203.

Angerosa, F., d'Alessandro, N., Konstantinou, P., and Di Giacinto, L. 1995. GC-MS evaluation of phenolic compounds in virgin olive oil. *J. Agric. Food Chem.*, 43:1802–1807.

Arai, S., Suzuki, H., Fujimaki, M., and Sakurai, Y. 1966. Studies on flavor components of soybeans. Part 2. Phenolic acids in defatted soybean flours. *Agric. Biol. Chem.*, 30:364–369.

Arora, A., Nair, M.G., and Strasburg, G.M. 1998. Antioxidant activities of isoflavones and their biological metabolites in liposomal system. *Arch. Biochem. Biophys.*, 356:133–141.

Artz, W.E., Swanson, B.G., Sendzicki, B.J., Rasyid, A., and Birch, R.E.W. 1986. Protein–procyanidin interaction and nutritional quality of dry beans, in *Plant Proteins: Applications, Biological Effects and Chemistry,* Ory, R.L., Ed., ACS Symposium Series 312, American Chemical Society, Washington, D.C., 126–137.

Asquith, T.N. and Butler, L.G. 1985. Use of dye-labeled protein as spectrophotometric assay for protein precipitants such as tannins. *J. Chem. Ecol.*, 11:1535–1544.

Austin, F.L. and Wolff, I.A. 1968. Sinapine and related esters in seed meal of *Crambe abyssinica. J. Agric. Food Chem.*, 16:132–135.

Axelson, M., Sjovall, B., Gustafsson, B., and Setchell, K.D.R. 1982. Origin of lignans in mammals and identification of a precursor from plants. *Nature*, 298:659–670.

Bailey, C.A., Stipanovic, R.D., Ziehr, M.S., Haq, A.U., Sattar, M., Kubena, L.F., Kim, H.L., and de Vieira, R.M. 2000. Cottonseed with a high (+) – to (–)- gossypol enantiomer ratio favorable to broiler production. *J. Agric. Food Chem.*, 48:5692–5695.

Bakke, J.E. and Klosterman H.J. 1956. A new diglucoside from flaxseed. *Proc. N. Dakota Acad. Sci.*, 10:227–235.

Balasinska, B. and. Troszynska, A. 1998. Total antioxidative activity of evening primrose (*Oenothera paradoxa*) cake extract measured *in vitro* by liposome model and murine L1210 cells. *J. Agric. Food Chem.*, 46:3558–3563.

Baldioli, M., Servili, M., Perretti, G., and Montedoro, G.F. 1996. Antioxidant activity of tocopherols and phenolic compounds of virgin olive oil. *J. Am. Oil Chem. Soc.*, 73:1589–1593.

Bambagiotti-Alberti, M., Coran, S.A., Ghiara, C., Giannellini, V., and Raffaelli, A. 1994a. Revealing the mammalian lignan precursor secoiriciresinol diglucoside in flaxseed by ionspray mass spectrometry. *Rapid Commun. Mass Spectrom.*, 8:595–598.

Bambagiotti-Alberti, M., Coran, S.A., Ghiara, C., Giannellini, V., and Raffaelli, A. 1994b. Investigation of mammalian lignan precursors in flaxseed: first evidence of secoisolariciresinol diglucoside in two isomeric forms by liquid chromatography/mass spectrometry. *Rapid Commun. Mass Spectrom.*, 8:929–932.

Barnes, S., Kirk, M., and Coward, L. 1994. Isoflavones and their conjugates in soy foods: extraction conditions and analysis by HPLC-mass spectrophotometry. *J. Agric. Food Chem.*, 42:2466–2474.

Bastoni, L., Bianco, A., Piccioni, F., and Uccella, N. 2001. Biophenolic profile in olives by nuclear magnetic resonance. *Food Chem.*, 73:145–151.

Bate-Smith, E.D. and Ribereau-Gayon, P. 1959. Leucoanthocyanins in seeds. *Qual. Plant Mater. Veg.*, 5:189–198.

Bau, H.M., Mohtadi-Nia, D.J., Mejean, L., and Debry, G. 1983. Preparation of colorless sunflower protein products: effect of processing on physicochemical and nutritional properties. *J. Am. Oil Chem. Soc.*, 60:1141–1148.

Bell, J.M. and Shires, A. 1982. Composition and digestibility by pigs of hull fractions from rapeseed cultivars with yellow or brown seed coats. *Can. J. Animal Sci.*, 62:557–565.

Berardi, L.C. and Goldblatt, L.A. 1980. Gossypol, in *Toxic Constituents of Plant Foodstuff*; Liener, I.E., Ed., Academic Press, New York, 183.

Berk, Z. 1992. Technology of production of edible flours and protein products from soybeans. Food and Agriculture Organization of the United Nations, Rome, Italy.

Bianchi, G. and Pozzi, N. 1994. 3,4-Dihydroxyphenylglycol, a major C_6-C_2 phenolic in *Olea europaea* fruits. *Phytochemistry,* 35:1335–1337.

Bianco, A., Coccioli, F., Guiso, M., and Marra, C. 2001. The occurrence in olive oil of a new class of phenolic compounds: hydroxy-isochromans. *Food Chem.*, 77: 405–411.

Bianco, A. and Uccella, N. 2000. Biophenolic compounds of olives. *Food Res. Int.*, 33:475–485.

Birch, A.E., Fenner, G.P., Watkins, R., and Boyd, L.C. 2001. Antioxidant properties of evening primrose seed extracts. *J. Agric. Food Chem.*, 49:4502–4507.

Blair, R. and Reichert, R.D. 1984. Carbohydrate and phenolic constituents in a comprehensive range of rapeseed and canola fractions: nutritional significance for animals. *J. Sci. Food Agric.*, 35:29–35.

Blekas, G., Vassilakis, C., Harizanis, C., Tsimidou, M., and Boskou, D.G. 2002. Biophenols in table olives. *J. Agric. Food Chem.*, 50:3688–3692.

Borriello, S.P., Setchell, K.D.R., Axelson, M., and Lawson, A.M. 1985. Production and metabolism of lignans by human faecal flora. *J. Appl. Bacteriol.*, 58:37–43.

Borzillo, A., Iannota, N., and Uccella, N. 2000. Oinotria table olives: quality evaluation during ripening and processing by biomolecular components. *Eur. Food Res. Technol.*, 212:113–121.

Bouchereau, A., Hamelin, J., Lamour, I., Renard, M., and Larher, F. 1991. Distribution of sinapine and related compounds in seeds of Brassica an allied genera. *Phytochemistry*, 30:1873–1881.

Boue, S.E., Carter, C.H., Ehrlich, K.C., and Cleveland, T.E. 2000. Induction of the soybean phytoalexins coumestrol and glyceollin by *Aspergillus*. *J. Agric. Food Chem.*, 48:2167–2172.

Brenes, M., Garcia, A., Dobarganes, M.C., Velasco, J., and Romero, C. 2002. Influence of thermal treatments simulating cooking processes on the polyphenol content in virgin oil. *J. Agric. Food Chem.*, 50:5962–5967.

Brenes, M., Garcia, A., Garcia, P., and Garrido, A. 2001. Acid hydrolysis of secoiridoid aglycons during storage of virgin olive oil. *J. Agric. Food Chem.*, 49:5609–5614.

Brenes, M., Hidalgo, F.J., Garcia, A., Rios, J.J., Garcia, P., Zamora, R., and Garrido, A. 2000. Pinoresinol and 1-acetoxypinoresinol, two new phenolic compounds identified in olive oil. *J. Am. Oil Chem. Soc.*, 77:715–720.

Brenes, M., Garcia, A., Garcia, P., Rios, J.J., and Garrido, A. 1999. Phenolic compounds of Spanish olive oils. *J. Agric. Food Chem.*, 47:3535–3540.

Brenes, M., Rejano, L., Garcia, P., Sanchez, A.H., and Garrido, A. 1995. Biochemical changes in phenolic compounds during Spanish-style green olive processing. *J. Agric. Food Chem.*, 43:2702–2706.

Brenes, M., Garcia, P., and Garrido, A. 1992. Phenolic compounds related to the black color formed during the processing of ripe olives. *J. Agric. Food Chem.*, 40:1192–1196.

Briante, R., Patumi, M., Terenziani, S., Bismuto, E., Febbraio, F., and Nucci, R. 2002. *Olea europea* L. leaf extract and derivatives: antioxidant properties. *J. Agric. Food Chem.*, 50: 4934–4940.

Budowski, P. 1964. Recent research on sesamin, sesamolin, and related compounds. *J. Am. Oil Chem. Soc.*, 41:280–285.

Budowski, P. 1950. Sesame oil. III. Antioxidant properties of sesamol. *J. Am. Oil Chem. Soc.*, 27:264–267.

Budowski, P., O'Connor, R.T., and Field, E.T. 1950. Sesame oil. IV. Determination of free and bound sesamol. *J. Am. Oil Chem. Soc.*, 27:307–310.

Butler, L.G. 1982. Relative degree of polymerization of sorghum tannin during seed development and maturation. *J. Agric. Food Chem.*, 30:1090–1092.

Caponio, F., Gomes, T., and Pasqualone, A. 2001. Phenolic compounds in virgin olive oils: influence of the degree of olive ripeness on organoleptic characteristics and shelf-life. *Eur. Food Res. Technol.*, 212:329–333.

Caragay, A. 1992. Cancer-preventive foods and ingredients. *Food Technol.*, 46:65–68.

Carter, C.M., Gheyasuddin, S., and Matill, K.F. 1972. The effect of chlorogenic, caffeic and quinic acids on the solubility and color of protein isolates, especially from sunflower seed. *Cereal Chem.*, 49:508–514.

Catalano, P. and Caponio, F. 1996. Machines for olive paste preparation producing quality virgin olive oil. *Fett. Lipid*, 98:408–412.

Chamkasem, K. 1988.Gossypol analysis in cottonseed oil by HPLC. *J. Am. Oil Chem. Soc.*, 65:1601–1605.

Chiou, R.Y.-Y. and Cheng, S.-L. 2001. Isoflavone transformation during soybean koji preparation and subsequent miso fermentation supplemented with ethanol and NaCl. *J. Agric. Food Chem.*, 49:3656–3660.

Choo, Y.M. 1994. Palm oil carotenoids. *Nutr. Bull.*,15:130–137.

Choo, Y.M., Yap, S.C., Ooi, A.N., Ma, A.X., Goh, S.H., and Ong, A.S.H. 1996. Recovered oil from palm-pressed fibre: a good source of natural carotenoids, vitamin E, and sterols. *J. Am. Oil Chem. Soc.*, 73:599–602.

Chung, H.Y. 2000. Volatile flavor components in red fermented soybean (*Glycine max*) curds. *J. Agric. Food Chem.*, 48:1803–1809.

Ciafardini, G., Marsilio, V., Lanza, B., and Pozzi, N. 1994. Hydrolysis of oleuropein by *Lactobacillus plantarum* strains associated with olive fermentation. *Appl. Environ. Microbiol.*, 60:4142–4147.

Cinquanta, L., Esti, M., and La Notte, E. 1997. Evolution of phenolic compounds in virgin olive oil during storage. *J. Am.Oil Chem. Soc.*, 74:1259–1264.

Clandinin, D.R. 1961. Effect of sinapine, the bitter substance in rapeseed oil meal, on the growth of chickens. *Poultry Sci.*, 40:484–487.

Clandinin, D.R. and Heard, J. 1968. Tannins in prepress-solvent and solvent processed rapeseed meal. *Poultry Sci.*, 47:688–689.

Clausen, S., Larsen, L., Ploeger, A., and Sorensen, H. 1985. Aromatic choline esters in rapeseed. *World Crops*, 11:61–72.

Cortesi, N., Azzolini, M., Rovellini, P., and Fedeli, E. 1995. Minor polar components of virgin olive oils: hypothetical structure by LC-MS. *Riv. Ital. Sost. Grasse*, LXXII, 241–251.

Cunnane, S.C. and Thompson, L.U. 1995. *Flaxseed in Human Nutrition*, AOCS Press, Champaign, IL.

Dabrowski, K. and Siemieniak, B. 1987. Removal of sinapine from rapeseed using various solvents and extraction conditions, in *Proceedings, 7th International Rapeseed Congress,* vol. VI, Poznan, Poland, 1476–1481.

Dabrowski, K. and Sosulski, F. 1984. Composition of free and hydrolyzable phenolic acids in defatted flours of ten oilseeds. *J. Agric. Food Chem.*, 32:128–130.

Dabrowski, K. and Sosulski, F. 1983. Extraction of phenolic compounds from canola during protein concentration and isolation, in *Proceedings of 6th International Rapeseed Congress,* vol. II, Paris, 1138–1142.

Dalais, F.M., Wahlquist, M.L., and Rice, G.E. 1997. Variation in isoflavonoid phytoestrogen content of soybean grown in Australia. *Proc. Nutr. Soc. Aust.*, 21:161.

Delic, I., Vucurevic, N., and Stojanovic, S. 1975. Investigation of inactivation of chlorogenic acid from sunflower meal under *in vitro* conditions in mice. *Acta Vet.* (Belgrade), 25:115–119.

Deshpande, S.S., Sathe, S.K., and Salunkhe, D.K. 1984. Chemistry and safety of plant polyphenols, in *Nutritional and Toxicological Aspects of Food Safety*, Friedman, M., Ed., Plenum, New York, 457–495.

Diosady, L.L., Naczk, M., and Rubin, L.J. 1985. The effect of ammonia concentration on the properties of canola meals produced by ammonia–methanol–water–hexane system. *Food Chem.*, 18:121–130.

Diosady, L.L., Rubin, L.J., Tar, C.G., and Etkin, B. 1986. Air classification of rapeseed meal using Tervel Separator. *Can. J. Chem. Eng.*, 64:768–774.

Diosady, L.L., Tar, C.G., Rubin, L.J., and Naczk, M. 1987. Scale-up of the production of glucosinolate-free canola meal. *Acta Aliment.*, 16:167–179.

Dryden, M.J. and Saterlee, L.D. 1978. Effect of free and bound chlorogenic acid on the *in vitro* protein digestibility and *Tetrahymena* based PER an a casein model system. *J. Food Sci.*, 43:650–651.

Durkee, A.B. 1971. The nature of tannins in rapeseed (*Brassica campestris*). *Phytochemistry*, 10:1583–1585.

Durkee, A.B. and Harborne, J.B. 1973. Flavanol glycosides in *Brassica* and *Sinapis*. *Phytochemistry*, 12:1085–1089.

Durkee, A.B. and Thivierge, P.A. 1975. Bound phenolic acids in *Brassica* and *Sinapis* oilseeds. *J. Food Sci.*, 40:820–822.

Ebel, J., Ayers, A.R., and Albersheim, P. 1976. Host-pathogen interactions. XII. Response of suspension–cultured soybean cells to elicitor isolated from *Phytophtora megasperma* var. *sojae,* a fungal pathogen of soybeans. *Plant Physiol.*, 57:775–779.

Eldridge, A.C. and Kwolek, W.F. 1983. Soybean isoflavones: effect of environment and variety on composition. *J. Agric. Food Chem.*, 31:394–396.

Esti, M., Cinquanta, L., and La Notte, E. 1998. Phenolic compounds in different olive varieties. *J. Agric. Food Chem.*, 46:32–35.

Fenton, T.W., Leung, J., and Clandinin, D.R. 1980. Phenolic components of rapeseed meal. *J. Food Sci.*, 45:1702–1705.

Fenwick, R.G. and Hoggan, S.A. 1976. The tannin content of rapeseed meals. *Br. Poultry Sci.*, 17:59–62.

Fenwick, G.R., Hobson-Frohock, A., Land, D.G., and Curtis, R.F. 1979. Rapeseed meal and egg taint: treatment of rapeseed meal to reduce tainting potential. *Br. Poultry Sci,.* 20:323–329.

Fenwick, G.R., Curl, C.L., Pearson, A.W., and Butler, E.J. 1984. The treatment of rapeseed meal and its effect on the chemical composition and egg tainting potential. *J. Sci. Food Agric.*, 35:757–761.

Fleming, H.P., Walter, W.M., and Etchells, J.L. 1973. Antimicrobial properties of oleuropein and products of its hydrolysis from green olives. *Appl. Microbiol.*, 26:777–782.

Fleury, Y., Welti, D.H., Philippossian, G., and Magnolato, D. 1991. Soybean (manoyl) isoflavones. Characterization and antioxidant properties, in *Phenolic Compounds in Food and Their Effects on Health II. Antioxidants and Cancer Prevention,* Huang, M.-T., Ho, C.-T., and Lee, C.Y., Eds., ACS Symposium Series 507, American Chemical Society, Washington, D.C., 98–113.

Franke, A.A., Hankin, J.H., Yu, M.C., Maskarinec, G., Low, S.-H., and Custer, L.J. 1999. Isoflavone levels in soy foods consumed by multiethnic populations in Singapore and Hawaii. *J. Agric. Food Chem.*, 47:977–986.

Fukuda, Y., Osawa, T., and Namiki, M. 1981. Antioxidants in sesame seed. *Nippon Shokuhin Kogyo Gakkaishi*, 28:461–464.

Fukuda, Y., Nagata, M., Osawa, T., and Namiki, M. 1986a. Contribution of lignan analogues to antioxidative activity of refined unroasted sesame seed oil. *J. Am. Oil Chem. Soc.*, 63:1027–1031.

Fukuda, Y., Isobe, M., Nagata, M., Osawa, Y., and Namiki, M. 1986b. Acidic transformation of sesamolin, the sesame-oil constituent, into antioxidant. *Heterocycles*, 24:923–926.

Fukuda, Y., Nagata, M., Osawa, T., and Namiki, M. 1986c. Chemical aspects of the antioxidative activity of roasted sesame oil, and the effect of using oil for frying. *Agric. Biol. Chem.*, 50:857–862.

Fukuda, Y., Osawa, T., Kawakishi, S., and Namiki, M. 1994. Chemistry of lignan antioxidants in sesame seed and oil, in *Food Phytochemicals for Cancer Prevention II. Teas, Spices and Herbs,* Ho, C.-T., Osawa, T., Huang, M.-T., and Rosen, R.T., Eds., ACS Symposium Series 547, American Chemical Society, Washington, D.C., 264–274.

Gandhi, V.M., Cheryan, K.K., Mulky, M.J., and Menon, K.K.G. 1975. Utilization of nontraditional indigenous oilseed meal. 2. Studies on the detoxified salseed meal. *J. Oil Technol. Assoc. India* (Bombay), 7:44–50.

Gariboldi, P., Jommi, G., and Verotta, L. 1986. Secoiridoids from *Olea europea. Phytochemistry*, 25:865–869.

Goh, Y.K., Shires, A.R., Robblee, A.R., and Clandinin, D.R. 1982. The effect of ammoniation on the sinapine content of canola meal. *Br. Poultry Sci.*, 23:121–128.

Goh, S.H., Choo, Y.M., and Ong, A.S.H. 1985. Minor constituents of palm oil. *J. Am. Oil Chem. Soc.*, 62:237–240.

Greilsamer, B. 1983. Depiculage-tirage des graines de colza (Dehulling and extraction of rapeseed), in *Proceedings of the Sixth International Rapeseed Congress*, vol. 2, Paris, 1496–1500.

Gutierrez Gonzales-Quijano, R., Janer del Valle, C., Janer del Valle, M.L., Gutierrez Rosales, F., and Vasquez Roncero, A. 1977. Relationship between polyphenol content and the quality and stability of virgin olive oil. *Grasas Aceites*, 28:101–106.

Gyorgy, P., Murata, K., and Ikehata, H. 1964. Antioxidants isolated from fermented soybeans (tempeh). *Nature*, 203:870–872.

Ha, E.Y.A., Mor, C.V., and Seo, A. 1992. Isoflavone aglucones and volatile organic compounds in soybeans: effect of soaking. *J. Food Sci.*, 57:414–418.

Hagerman, A.E. and Butler, L.G. 1978. Protein precipitation method for the quantification of tannins. *J. Agric. Food Chem.*, 26:809–812.

Hamburger M., Riese, U., Graf, H., Melzig, M.F., Ciesielski, S., Baumann, D., Dittman, K., and Wegner, C. 2002. Constituents in evening primrose oil with radical scavenging, cyglooxygenase, and neutrophil elastase inhibitory activities. *J. Agric. Food Chem.*, 50:5533–5538.

Harris, R.K. and Haggerty, W.J. 1993. Assays for potentially carcinogenic phytochemicals in flaxseed. *Cereal Foods World*, 38:147–151.

Haumann, B.F. 1993. Designing foods. *Int. News Fats Oils Rel. Mater.*, 4:345–373.

Heinonen, S., Liukkonen, K., Poutanen, K., Wahala, K., Deyama, T., Nishibe, S., and Adlercreutz, H. 2001. *In vitro* metabolism of plant lignans: new precursors of mammalian lignans enterolactone and enterodiol. *J. Agric. Food Chem.*, 49:3178–3186.

Heywood, R., Lloyd, G.K., Maged, S.K., and Gopinath, C. 1986. The toxicity of gossypol to male rat. *Toxicity*, 40:279–284.

Hirano, T., Gotoh, M., and Oka, K. 1994. Natural flavonoids and lignans are potent cytostatic agents against human leukimic HL-60 cells. *Life Sci.*, 146:294–306.

Hirata, F., Fujita, K., Ishikura, Y., Hosoda, K., Ishikawa, T., and Nakamura, H. 1996. Hypocholesterolemic effect of sesame lignan in human. *Atherosclerosis*, 122:135–136.

Hirose, Y., Inoue, T., Nishhara, K., Sugano, M., Akimoto, K., Shimizu, S., and Yamada, H. 1991. Inhibition of cholesterol absorption and synthesis in rats by sesamin. *J. Lipid Res.*, 32:629–638.

Hoeck, J.A., Fehr, W.R., Murphy, P.A., and Welke, G.A. 2000. Influence of genotype and environment on isoflavone contents of soybean. *Crop Sci.*, 40:48–51.

Hosny, M. and Rosazza, J.P.N. 1999. Novel isoflavone, cinnamic acid, and triterpenoid glycosides in soybean molasses. *J. Nat. Prod.*, 62:853–858.

Hosny, M. and Rosazza, J.P.N. 2002. New isoflavone and triterpene flycosides from soybeans. *J. Nat. Prod.*, 65:805–813.

How, J.S.L. and Morr, C.V. 1982. Removal of phenolic compounds from soy protein extracts using activated carbon. *J. Food Sci.*, 47:933–940.

Huang, A.S., Hseih, O., and Chang, S.S. 1979. Characterization of nonvolatile constituents responsible for bitter and astringent taste in soybean protein. Presented at the 39th Annual Meeting of the Institute of Food Technologists, St. Louis, MO.

Hudson, B.J.F. 1984. Evening primrose (*Oenothera spp.*) oil and seed. *J. Am. Oil Chem. Soc.*, 61:540–543.

Hutabarat, L.S., Greenfield, H., and Mulholland, M. 2001. Isoflavones and coumestrol in soybeans and soybean products from Australia and Indonesia. *J. Food Comp. Anal.*, 14:43–58.

Ibraham, R.K. and Shaw, M. 1970. Phenolic constituents of the oil flax (*Linum usitatissimum*). *Phytochemistry*, 9:1855–1858.

Ingham, J.L., Keen, N.T., Mulheirn, L.J., and Lyne, R.L. 1981. Inducibly formed isoflavonoids from leaves of soybean. *Phytochemistry*, 20:795–798.

Jenkins, D.J.A. 1995. Incorporation of flaxseed or flaxseed components into cereal foods, in *Flaxseed in Human Nutrition,* Cunnane, S. and Thompson, L.U., Eds., AOCS Press, Champaign, IL, 281–294.

Jensen, S., Olsen, H., and Sorensen, H. 1991. Aqueous enzymatic processing of rapeseed for production of high quality products, in *Canola and Rapeseed: Production, Chemistry, Nutrition, and Processing Technology,* Shahidi, F., Ed., AVI Books, New York, 331–343.

Johnsson, P., Peerlkamp, N., Kamal-Eldin, A., Andersson, R.E., Andersson, R., Lundgren, L.N., and Åman, P. 2002. Polumeric fractions containing phenol glucosides in flaxseed. *Food Chem.*, 76:207–212.

Johnsson, P., Kamal-Eldin, A., Lundgren, L.N., and Åman, P. 2000. HPLC method for analysis of secoisolariciresinol diglucoside in flaxseeds. *J. Agric. Food Chem.*, 48:5216–5219.

Katsuzaki, H., Osawa, T., and Kawakishi, S. 1994. Chemistry and antiosidative activity of lignan glucosides in sesame seed, in *Food Phytochemicals for Cancer Prevention II. Teas, Spices and Herbs,* Ho, C.-T., Osawa, T., Huang, M.-T., and Rosen, R.T., Eds., ACS Symposium Series 547, American Chemical Society, Washington, D.C., 275–280.

Kerber, E. and Buchloh, G. 1980. The sinapine content of crucifer seeds. *Agnew. Bot.*, 54:47–54.

Kinoshita, E., Ozawa, Y., and Aishima, T. 1997. Novel tartaric acid isoflavone derivative that play key roles in differentiating Japanese soy sauces. *J. Agric. Food Chem.*, 45:3753–3759.

Kirk, L.D., Mustakas, G.C., and Griffin, E.L., Jr. 1966. Crambe seed processing improved feed meal by ammoniation. *J. Am. Oil Chem. Soc.*, 43:550–555.

Knuckles, B.E., de Fremery D., and Kohler, G.O. 1976. Coumestrol content of fractions obtained by wet processing of alfalfa. *J. Agric. Food Chem.*, 24:1177–1180.

Kozlowska, H., Sabir, M.A., Sosulski, F.W., and Coxworth, E. 1975. Phenolic constituents of rapeseed flour. *Can. Inst. Food Sci. Technol.* J., 8:160–163.

Kozlowska, H., Rotkiewicz, D.A., Zadernowski, R., and Sosulski, F.W. 1983. Phenolic acids in rapeseed and mustard. *J. Am. Oil Chem. Soc.*, 60:1119–1123.

Kozlowska, H., Zadernowski, R., and Lossow, B. 1984. Equipment for dehulling of *Brassica* seed, especially rapeseed. Polish Patent No. 37490.

Kozlowska, H. and Zadernowski, B. 1988. Phenolic compounds of rapeseed as factors limiting the utilization of protein in nutrition. Paper presented at the 3rd North American Chemical Congress, Toronto, ON.

Krygier, K., Sosulski, F.W., and Hogge, L. 1982. Free, esterified and insoluble phenolic acids. 2. Composition of phenolic acids in rapeseed flour and hulls. *J. Agric. Food Chem.*, 30:334–336.

Kudou, S., Fluery, Y., Welti, D., Magnolato, D., Uchida, T., Kitamura, K., and Okubo, M. 1991. Malonyl isoflavone glucosides in soybean seeds (*Glycine max Merrill*). *Agric. Biol. Chem.*, 55:2227–2233.

Kumar, R. and Singh, M. 1984. Tannins: their adverse role in ruminant nutrition. *J. Agric. Food Chem.*, 32:447–453.

Kuntz, S., Wenzel, U., and Daniel, H. 1999. Comparative analysis of the effects of flavonoids on profileration, cytotoxicity, and apoptosis in human colon cancer cells. *Eur. J. Nutr.*, 38:133–142.

Kuroda, C. and Wade, M. 1933. The coloring matter of kuro-mame. *Proc. Imp. Acad.* (Tokyo), 9:17–18, in *Chemical Abstracts*, CA 27:2448.

Kuwajima, H., Uemura, T., Takaishi, K., Inoue, K., and Inouye, H. 1988. A secoroid glucoside from *Olea europea. Phytochemistry*, 27:1757–1759.

Larsen, L., Olsen, O., Plaeger, A., and Sorensen, H. 1983. Phenolic choline esters in rapeseed: possible factors affecting nutritive value and quality, in *Proceedings of the Sixth International Rapeseed Congress*, vol. 2, Paris, 1577–1582.

Lawhon, J.T., Cater, C.M., and Mattil, K.F. 1977. Evaluation of the food use potential of 16 varieties of cottonseed. *J. Am. Oil Chem. Soc.*, 54:75–80.

Le Tutor, B. and Guedon, D. 1992. Antioxidative activities of *Olea europea* leaves and related phenolic compounds. *Phytochemistry*, 31:1173–1178.

Leung, J., Fenton, T.W., Mueller, M.M., and Clandinin, D.R. 1979. Condensed tannins of rapeseed meal. *J. Food Sci.*, 44:1313–1316.

Lin, S. 1991. Fermented soya foods, in *Developments in Food Proteins*, Hudson, B.J.F., Ed., Elsevier, London, 167–193.

Lipworth, L., Martinez, M.E., Angell, J., Hsien, C.C., and Trichopouolu, A. 1997. Olive oil and human cancer assessment of evidence. *Prev. Med.*, 26:181–190.

Liu, C.-L., Kubota, K., and Kabayashi, A. 1988. Aroma components of tofu. *Nippon Nigeikagaku Kaishi*, 62:1201–1205.

Lo, M.T. and Hill, D.C. 1972. Composition of the aqueous extracts of rapeseed meals. *J. Sci. Food Agric.*, 23:823–830.

Lookhart, G.L., Finney, P.L., and Finney, K.F. 1979. Note on coumestrol in soybeans and fractions at various germination times. *Cereal Chem.*, 56:495–496.

Lu, F. and Foo, L.Y. 1995. Phenolic antioxidant components of evening primrose, in *Nutrition, Lipids, Health and Disease,* Niki, E. and Packer, L., Eds., AOCS Press, Champaign IL, 86–95.

Lusas, E.W. and Jividen, G.M. 1987. Glandless cottonseed: a review of the first 25 years of processing and utilization of research. *J. Am. Oil Chem. Soc.*, 64:839–854.

Macheix, J.-J., Fleuriet, A., and Billot, J. 1990. *Fruit Phenolics*, CRC Press, Boca Raton, FL, 111–112.

Maestro-Duran, R., Leon Cabello, R., and Ruiz Gutierrez, V. 1994. Phenolic compounds from olive (*Olea europa). Grasas Aceites*, 45:265–269.

Maga, J.A. and Lorenz, K. 1974. Gas-liquid chromatography separation of free phenolic acid fractions in various oilseed protein sources. *J. Sci. Food Agric.*, 25:797–802.

Makela, S.I., Pylkkanen, L.H., Santti, R.S.S., and Adlecreutz, H. 1995. Dietary soybean may be antiestrogenic in male mice. *J. Nutr.*, 125:437–445.

Manna, C., Galletti, P., Cucilla, V., Montedoro, G., and Zappia, V. 1999. Olive oil hydroxyl protects human erythrocytes against oxidative damages. *J. Nutr. Biochem.*, 10:159–165.

Marsilio, V., Lanza, B., and Pozzi, N. 1996. Progress in table olive debittering: degradation *in vitro* of oleuropein and its derivatives by *Lactobacillus plantarum. J. Am. Oil Chem. Soc.*, 73:593–597.

Martin-Tanguy, J., Guillaume, J., and Kossa A. 1977. Condensed tannins of horse bean seeds: chemical structure and apparent effects on poultry. *J. Sci. Food Agric.*, 28:757–764.

Mateos, R., Espartero, J.L., Trujillo, M., Rios, J.J., Leon-Camacho, M., Alcudia, F., and Cert, A. 2001. Determination of phenols, flavones, and lignans in virgin olive oils by solid-phase extraction and high-performance liquid chromatography with diode array ultraviolet detection. *J. Agric. Food Chem.*, 49:2185–2192.

Matsuura, M. Obata, A., and Fukushima, D. 1989. Objectionable flavor of soy milk developed during the soaking of soybeans and its control. *J. Food Sci.*, 56:602–605.

Matsuura, M. and Obata, A. 1993. β–Glucosidases from soybeans hydrolyze daidzin and genistin. *J. Food Sci.*, 58:144–147.

Mazur, W.M., Fotsis, T., Oajala, S., Salakka, A., and Adlecreutz, H. 1996. Isotope dilution gas chromatographic–mass spectrometric method for determination of isoflavonoids, coumestrol and lignans in food samples. *Anal. Biochem.*, 233:169–180.

Mazza, G. and Miniati, E. 1993. *Anthocyanins in Fruits and Vegetables*, CRC Press, Boca Raton, FL., 64–67.

McCarty, J.C., Hedin, P.A., and Stipanovic, R.D. 1996. Cotton *Gossypium* spp. plant gossypol contents of selected GL_2 and GL_3 alleles. *J. Agric. Food Chem.*, 44:613–616.

McCord, H. and Rao, L. 1997. Top seed: with its healing power, flax is the next nutritional star. *Prevention Mag.*, (4):81–85.

McGregor, D., Blake, J., and Pickard, M. 1983. Detoxification of Brassica juncea with ammonia, in *Proceedings of the Sixth International Rapeseed Congress*, vol. 2, Paris, 1426–1430.

McManus, J.P., Davis, K.G., Beart, D.J.E., Gaffney, S.H., Lilley, T.H., and Haslam, E. 1985. Polyphenol interactions. Part I. Introduction: some observations on the reversible complexation with proteins and polysaccharides. *J. Chem. Soc. Perkin Trans.*, 2, 1429–1438.

Meagher, L.P., Beecher, G.R., Flanagan, V.P., and Li, B.W. 1999. Isolation and characterization of the lignans, isolariciresinol and pinoresinol, in flaxseed meal. *J. Agric. Food Chem.*, 47:3173–3180.

Medina, I., Satue-Gracia, M.T., German, J.B., and Frankel, E.N. 1999. Comparison of natural polyphenol antioxidants from extra virgin oil with synthetic antioxidants in tuna lipids during thermal oxidation. *J. Agric. Food Chem.*, 47:4873–4879.

Mimura, A., Takebayashi, K., Niwamo, M., Takahara, Y., Osawa, T., and Tokuda, H. 1994. Antioxidative and anticancer components produced by cell sesame cultures, in *Food Phytochemicals for Cancer Prevention II. Teas, Spices and Herbs,* Ho, C.-T., Osawa, T., Huang, M.-T., and Rosen, R.T., Eds., ACS Symposium Series 547, American Chemical Society, Washington, D.C., 281–294.

Mitaru, B.N., Blair, R., Bell, J.M., and Reichert, R.D. 1982. Tannin and fiber contents of rapeseed and canola hulls. Notes. *Can. J. Animal Sci.*, 62:661–663.

Miyazawa, M., Sakano, K., Nakamura, S.-I., and Kosaka, H. 2001. Antimutagenic activity of isoflavone from *Pueraria lobata. J. Agric. Food Chem.* 49:336–341.

Miyazawa, M., Sakano, K., Nakamura, S.-I., and Kosaka, H. 1999. Antimutagenic activity of isoflavones from soybean seeds (*Glycine max* Merrill). *J. Agric. Food Chem.*, 47:1346–1349.

Montedoro, G.F., Servilli, M., Baldioli, M., and Miniati, E. 1992. Simple and hydrolyzable phenolic compounds in virgin oil. 1. Their extraction, separation, and quantitative and semiquantitative evaluation by HPLC. *J. Agric. Food Chem.*, 40:1571–1576.

Montedoro, G.F., Servilli, M., Baldioli, M., Selvaggini, R., Miniati, E., and Macchioni, A. 1993. Simple and hydrolyzable compounds in virgin olive oil. 3. Spectroscopic characterization of the secoiridoid derivatives. *J. Agric. Food Chem.*, 41:2228–2234.

Mueller, M.M., Ryl, E., Fenton, T.W., and Clandinin, D.R. 1978. Cultivar and growing location differences on the sinapine content of rapeseed. *Can. J. Animal Sci.*, 58:579–583.

Muir, A.D. and Westcott, N.D. 2000. Quantification of the lignan secoisolaricisresinerol diglucoside in baked goods containing flaxseed and flax meal. *J. Agric. Food Chem.*, 48:4048–4052.

Murakami, H., Asakawa, T., Terao, J., and Matsushita, S. 1984. Antioxidative stability of tempeh and liberation of isoflavones by fermentation. *Agric. Biol. Chem.*, 48:2971–2975.

Murphy, P.A., Song, T., Buseman, G., Barua, K., Beecher, G.R., Trainer, D., and Holden, J. 1999. Isoflavones in retail and institutional soy foods. *J. Agric. Food Chem.*, 47:2697–2704.

Murthy, M.S.R., Venkata Rao, E., and Ward, R.S. 1986. Carbon-13 nuclear magnetic resonance spectra of isoflavones. *Magn. Reson. Chem.*, 24:225–230.

Naczk, M., Diosady, L.L., and Rubin, L.J. 1986. The phytate and complex phenol content of meals produced by alkanol-ammonia/hexane extraction of canola. *Lebensm.-Wiss. u-. Technol.*, 19:13–16.

Naczk, M. and Shahidi, F. 1989. The effect of methanol-ammonia-water treatment on the content of phenolic acids of canola. *Food Chem.*, 31:159–164.

Naczk, M. and Shahidi, F. 1992. Phenolic constituents of rapeseed, in *Plant Polyphenols: Synthesis, Properties, Significance,* Hemingway, R.W. and Laks, P.E., Eds., Plenum Press, New York, 895–910.

Naczk, M., Nichols, T., Pink, D., and Sosulski, F. 1994. Condensed tannins in canola hulls. *J. Agric. Food Chem.*, 42:2196–2200.

Naczk, M., Oickle, D., Pink, D., and Shahidi, F. 1996. Protein precipitating capacity of crude canola tannins: effect of pH, tannin and protein concentrations. *J. Agric. Food Chem.*, 44:1444–1448.

Naczk, M., Amarowicz, R., Sullivan, A., and Shahidi, F. 1998a. Current research developments on polyphenolics of rapeseed/canola: a review. *Food Chem.,* 62:489- 502.

Naczk, M., Amarowicz, R., and Shahidi, F. 1998b. Role of phenolics in flavor of rapeseed protein products, in *Food Flavors: Formation, Analysis, and Packaging Influences,* Contis, E.T., Ho, C.-T., Mussinan, C.J., Parliament, C.J., Shahidi, F., and Spanier, A.M., Eds., Elsevier Science B.V., Amsterdam, 597–609.

Naczk, M., Amarowicz, R., Pink, D., and Shahidi, F. 2000. Insoluble condensed tannins of canola/rapeseed. *J. Agric. Food Chem.*, 48:1758–1762.

Naczk, M., Amarowicz, R., and Shahidi, F. 2001a. Canola/rapeseed hull phenolics as potential free radical scavengers, in *Food Flavors and Chemistry: Advance of the New Millenium,* Spanier, A., Shahidi, F., Parliament, T.H., Mussinan, C., Ho, C.-T., and Tatras Contis, E., Eds., The Royal Society of Chemistry, Thomas Graham House, Cambridge, U.K., 583–591.

Naczk, M., Amarowicz, R., Zadernowski, R., and Shahidi, F. 2001b. Protein-precipitating capacity of crude condensed tannins of canola and rapeseed hulls. *J. Am. Oil Chem. Soc.*, 78:1173–1178.

Nagata, M., Osawa, T., Namiki, M., and Fukuda, Y. 1987. Stereochemical structure of antioxidative bisepoxylignans, sesaminol and its isomers, transformed from sesamolin. *Agric. Food Chem.*, 51:1285–1289.

Nicolosi, R.J., Rogers, E.J., Ausman, L.M., and Orthoefer, F.T. 1994. Rice bran oil and its health benefits,. In *Rice Science and Technology,* Marshall, W.E. and Wadsworth, J.I., Eds., Marcel Dekker, New York, 421.

Nissiotis, M.E. and Tasioula-Margari, M. 2001. Changes in the antioxidant concentration of virgin olive oil during thermal oxidation, in *Food Flavors and Chemistry: Advance of the New Millenium,* Spanier, A., Shahidi, F., Parliament, T.H., Mussinan, C., Ho, C.-T., and Tatras Contis, E., Eds., The Royal Society of Chemistry, Thomas Graham House, Cambridge, U.K., 578–582.

Nwokolo, E. and Smartt, J. 1996. *Food and Feed from Legumes and Oilseeds,* Chapman & Hall, London, 90–101.

Obermeyer, W.R., Musser, S.M., Betz, J.M., Casey, R.E., Pohland, A.E., and Page, S.W. 1995. Chemical studies of phytoestrogens and related compounds in dietary supplements. *Proc. Soc. Exp. Biol. Med.*, 208:6–12.

Ogawa, H., Sasagawa, S., Murakami, T., and Yoshizumi, H. 1995. Sesame lignans modulate cholesterol metabolism in stroke-prone spontaneously hypersensitive rat. *Clin. Exp. Pharmacol. Physiol.*, 22:S310-S312.

Okogeri, O. and Tasioula-Margari, M. 2002. Changes occuring in phenolic compounds and α–tocopherol of virgin olive oil during storage. *J. Agric. Food Chem.*, 50:1077–1080.

Oomah, B.D., Kenaschuk, E.O., and Mazza, G. 1995. Phenolic acids in flaxseed. *J. Agric. Food Chem.*, 43:2016–2019.

Oomah, B.D., Mazza, G., and Kenaschuk, E.O. 1996. Flavonoid content of flaxseed. Influence of cultivar and environment. *Euphytica*, 90:163–167.

Oomah, B.D. and Mazza, G. 1997. Effect of dehulling on chemical composition and physical properties of flaxseed. *Lebensm.-Wiss.u.-Technol.*, 30:135–140.

Oomah, B.D. and Mazza, G. 1998. Flaxseed products for disease prevention, in *Functional Foods: Biochemical and Processing Aspects,* Mazza, G., Ed., Technomic Publishing Co., Lancaster, PA, 91–138.

Osawa, T., Nagata, M., Namiki, M., and Fukuda, Y. 1985. Sesaminol, a novel antioxidant isolated from sesame seeds. *Agric. Biol. Chem.*, 49:3351–3352.

Panizzi, L., Scarpati, M.L., and Oriente, E.G., 1960. Structure of oleuropein, bitter glycoside with hypotensive action of olive oil. Note II. *Gazz. Chim. Ital.*, 90:1449–1485.

Pirisi, F.M., Cabras, P., Cao, C.F., Migliorini, M., and Muggelli, M. 2000. Phenolic compounds in virgin olive oil. 2. Reappraisal of the extraction, HPLC separation, and quantification procedures. *J. Agric. Food Chem.*, 48:1191–1196.

Pokorny, J. and Reblova, Z. 1995. Sinapines and other phenolics of Brasicaceae seeds. *Potrav. Vedy*, 13:155–168.

Porter, L.J. and Woodruffe, J. 1984. Haemalysis: the relative astringency of proanthocyanidin polymers. *Phytochemistry*, 23:1255–1256.

Price, M., Stromberg, A.M., and Butler, L.G. 1979. Tannin content as a function of grain maturity and drying conditions in several varieties of sorghum bicolor (L.) Moench. *J. Agric. Food Chem.*, 27:1270–1274.

Prasad, D.T. 1988. Studies on interactions of sunflower albumins with chlorogenic acid. *J. Agric. Food Chem.*, 36:450–452.

Rackis, J.J., Sessa, D.J., and Honig, D.H. 1970. Flavor problems of vegetable food proteins. *J. Am. Oil Chem. Soc.*, 56:262–271.

Radhakrishnan, M.R. and Sivaprasad, J. 1980. Tannin content of sorghum varieties and their role in bioavailability. *J. Agric. Food Chem.*, 28:55–57.

Ranalli, A. and De Mattia, G. 1997. Characterization of olive oil production with new enzyme processing aid. *J. Am. Oil Chem. Soc.*, 74:1105–1113.

Ribereau-Geyon, P. 1972. *Plant Phenolics,* Oliver and Boyd, Edinburg, U.K.

Robertson, J.A. 1975. Use of sunflower seed in food products. *CRC Crit. Rev. Food Sci. Nutr.*, 6:201–240.

Romani, A., Mulinacci, N., Pinalli, P., Vincieri, F., and Cimato, A. 1999. Polyphenolic content in five Tuscany cultivars of *Olea europaea* L. *J. Agric. Food Chem.*, 47:964–967.

Rovellini, P., Cortesi, N., and Fedeli, E. 1997. Analysis of flavonoids from *Olea europea* by HPLC-UV and HPLC-electrospray-MS. *Riv. Ital. Sostanze Grasse*, 74:273–279.

Rubino, M., Arntfield, S., and Charlton, J. 1995. Conversion of phenolics to lignans: sinapic acid to thomasidioic acid. *J. Am. Oil Chem. Soc.*, 72:1465–1470.

Rubino, M., Arntfield, S., and Charlton, J. 1996. Evaluation of alkaline conversion of sinapic acid to thomasidioic acid. *J. Agric. Food Chem.*, 44:1399–1402.

Rutkowski, S. 1972. Oilseed proteins and their characteristics. *Riv. Ital. Sost. Grasse*, 49:416–427.

Ryan, D. and Robards, K. 1998. Phenolic compounds in olives. *Analyst*, 123:31R-44R.

Sabir, M.A., Sosulski, F.W., and Finlayson, A.J. 1974. Chlorogenic acid-protein interaction in sunflower. *J. Agric. Food Chem.*, 22:575–578.

Sacchi, R., Paduano, A., Fiore, F., Medaglia, D.D., Ambrosiano, M.L., and Medina, I. 2002. Partitioning behavior of virgin oil phenolic compounds in oil-brine mixtures during thermal processing of fish canning. *J. Agric. Food Chem.*, 50:2830–2835.

Saija, A. and Uccella, N. 2001. Olive biophenols: functional effects on human wellbeing. *Trends Food Sci. Technol.*, 11:357–363.

Salunkhe, D.K., Chavan, J.K., and Kadam, S.S. 1990. *Dietary Tannins: Consequences and Remedies*, CRC Press, Boca Raton, FL, 29; 122.

Salunkhe, D.K., Chavan, J.K., Adsule, R.N., and Kadam, S.S. 1992. *World Oilseeds,* Van Nostrand Reinhold, New York.

Sciancalepore, V. 1985. Enzymatic browning in five olive varieties. *J. Food Sci.*, 50:1194–1195.

Sen, M. and Bhattacharyya, D.K. 2001. Nutritional quality of sesame seed protein fraction extracted with isopropanol. *J. Agric. Food Chem.*, 49:2641–2646.

Servilli, M., Begliomini, A.L., Montedoro, G.F., Petrucioli, M., and Federici, F. 1992. Utilization of yeast pectinase in olive oil extraction and red wine making processes. *J. Sci. Food Agric.*, 58:355–362.

Servilli, M., Baldiolli, M., Selvaggini, R., Mariotti, F., Federici, E., and Montedoro, G.F. 1998. Effect of malaxation under N_2 flush on phenolic and volatile compounds of virgin olive oil. *Adv. Plant Lipid Res.*, 307–310.

Servilli, M., Baldiolli, M., Selvaggini, R., Miniati, E., and Montedoro, G.F. 1999a. HPLC evaluation of phenols in olive fruits, virgin olive oil, vegetation water and pomace and 1D-and 2-D-NMR characterization. *J. Am. Oil Chem. Soc.*, 76:873–882.

Servilli, M., Baldiolli, M., Mariotti, F., and Montedoro, G.F. 1999b. Phenolic composition of olive fruit and virgin olive oil: distribution in the constitutive parts of fruit and evolution during mechanical extraction process. *Acta Hortic.*, 474:609–619.

Setchell, K.D.R., Lawson, A.M., McLaughlin, L.M., Patel, S., Kirk, D.N., and Axelson, M. 1983. Measurement of eneterolactone and eneterodiol, the first mammalian lignans, using stable isotope dilution and gas chromatography-mass spectrometry. *Biomed. Mass Spectrom.*, 10:227–235.

Shahidi, F. and Naczk, M. 1988. Effect of processing on the phenolic constituents of canola, in *Proceedings 14th Int. Conf. Group Polyphenols*, Bulletin de Liason, 14:89–92.

Shahidi, F., Naczk, M., Rubin, L.J., and Diosady, L.L. 1988. A novel approach for rapeseed and mustard seed-removal of undesirable constituents by methanol-ammonia. *J. Food Protec.*, 51:743–749.

Shahidi, F. and Naczk, M. 1989a. Effect of processing on the content of condensed tannins in rapeseed meals. A research note. *J. Food Sci.*, 54:1082–1083.

Shahidi, F. and Naczk, M. 1989b. Solvent extraction of tannins from canola. Poster at 50th Annual IFT Meeting. June 25–29, Chicago, IL.

Shahidi, F. and Naczk, M. 1990. Removal of glucosinolates and other antinutrients from canola and rapeseed by methanol/ammonia processing, in *Rapeseed/Canola: Production, Chemistry, Nutrition, and Processing Technology,* Shahidi, F., Ed., Van Nostrand Reinhold, New York, 291–306.

Shahidi, F. and Naczk, M. 1992. An overview of the phenolics of canola and rapeseed: chemical, sensory, and nutritional significance. *J. Am. Oil Chem. Soc.*, 69: 917–924.

Shahidi, F., Amarowicz, R., He, Y., and M. Wetasinghe, M. 1997a. Antioxidant activity of phenolic extracts of evening primrose *(Oenothera biennis):* a preliminary study. *J. Food Lipids*, 4:75–86.

Shahidi, F., Amarowicz, R., Abu-Gharbia, H.-A., and Shehata, A.A.J. 1997b. Endogenous antioxidants and stability of sesame oil as affected by processing and storage. *J. Am. Oil Chem. Soc.*, 74:143–148

Shimoda, M., Nakada, Y., Nakashima, M., and Osajima, Y. 1997. Quantitative comparison of volatile flavor compounds in deep-roasted and light-roasted sesame seed oil. *J. Agric. Food Chem.*, 45:3193–3196.

Shurtleff, W. and Aoyagi, A. 1979. *The Book of Tempeh*, Harper and Row, New York, 173–198.

Simonne, A.H., Smith, M., Weaver, D.B., Vail, T., Barnes, S., and Wei, C.I. 2000. Retention and changes of soy isoflavones and carotenoids in immature soybean seeds (Edamame) during processing. *J. Agric. Food Chem.*, 48:6061–6069.

Sirato-Yasumoto, S., Katsuta, M., Okuyama, Y., Takahashi, Y., and Ide, T. 2001. Effect of sesame seeds rich in sesamin and sesamolin on fatty acid oxidation in rat liver. *J. Agric. Food Chem.*, 49:2647–2651.

Sirato-Yasumoto, S., Ide, T., Takahashi, Y., Okuyama, Y., and Katsuta, M. 2000. New sesame line having high lignan content in seed and its functional activity. *Breed. Res.*, 2 (Suppl. 2):184.

Smith, K.J. 1970. Practical significance of gossypol in feed formulation. *J. Am. Oil Chem. Soc.*, 47:448–450.

Smith, A.K. 1978. *Soybeans: Chemistry and Technology,* vol. 1, AVI Books, New York, 1–26.

Sodini, G. and Canella, M. 1977. Acid butanol removal of color forming phenols from sunflower meal. *J. Agric. Food Chem.*, 25:822–825.

Soler-Rivas, C., Espin, J.C., and Wichers, H.J. 2000. Oleuropein and related compounds. *J. Sci. Food Agric.*, 80:1013–1023.

Song, D.K. and Karr, A.L. 1993. Soybean phytoalexin, glyceollin, prevents accumulation of aflatoxin B_1 in cultures of *Aspergillus flavus. J. Chem. Ecol.*, 19:1183–1194.

Sosulski, F. and McCleary, C.W. 1972. Diffusion extraction of chlorogenic acid from sunflower kernels. *J. Food Sci.*, 37:253–256.

Sosulski, F.W. 1979. Organoleptic and nutritional effects of phenolic. A review. *J. Am. Oil Chem. Soc.*, 56:711–715.

Sosulski, F.W. and Zadernowski, R. 1981. Fractionation of rapeseed meal into flour and hull component. *J. Am. Oil Chem. Soc.*, 58:96–98.

Sripad, G. and Narasinga Rao, M.S. 1987. Effects of methods to remove polyphenols from sunflower meal on physico-chemical properties of proteins. *J. Agric. Food Chem.*, 35:962–967.

Strohschein, S., Rentel, C., Lacker, T., Bayer, E., and Albert, K. 1999. Separation and identification of tocotrienol isomers by HPLC and HPLC-NMR coupling. *Anal. Chem.*, 71:1780–1785.

Takahata, Y., Ohnishi-Kameyama, M., Furuta, S., Takahashi, M., and Suda, I. 2001. Highly polymerized procyanidins in brown seed coat with high radical-scavenging activity. *J. Agric. Food Chem.*, 49:5843–5847.

Tantawy, B., Robin, J.P., and Tollter, M.Th. 1983. Characterisation de deux glycosides acyles du kaempferol dans les extraits ethanolique du tourteau du colza 'o-thio,' in *Proc. 6th Int. Rapeseed Congr.,* vol. II, Paris, 1313–1320.

Thompson, L.U., Robb, P., Serraino, M., and Cheung, F. 1991. Mammalian lignan production from various foods. *Nutr. Cancer*, 16:43–52.

Thompson, L.U. 1995. Flaxseed, lignans and cancer, in *Flaxseed in Human Nutrition,* Cunnane, S. and Thompson, U.L., Eds., AOCS Press, Champaigne, IL, 219–236.

Thompson, L.U., Rickard, S.E., Orcheson, L.J., and Seidl, M.M. 1996a. Flaxseed and its lignan and oil components reduce mammary growth at a late stage of carcinogenesis. *Carcinogenesis*, 17:1373–1376.

Thompson, L.U., Seidl, M.M., Rickard, S.E., Orcheson, L.J., and Fong, H.H. 1996b. Antitumorigenic effect of mammalian lignan precursors from flaxseed. *Nutr. Cancer*, 26:159–165.

Thompson, L.U., Rickard, S.E., Cheung, F., Kenaschuk, E., and Obermeyer, W.R. 1997. Variability in anticancer lignan levels in flaxseed. *Nutr. Cancer*, 27:26–30.

Trichopouolu, A. 1995. Olive oil and breast cancer. *Cancer Causes Control.*, 6:475–476.

Tsimidou, M. 1998. Polyphenols and quality of virgin olive oil in retrospect. *Ital. J. Food Sci.*, 10:99–116.

Tsukamoto, C., Shimada, K., Igita, S., Kudou, M., Kokubun, K., Okubo, K., and Kitamura, K. 1995. Factors affecting isoflavone content in soybeans seeds: changes in isoflavones, saponines and composition of fatty acids at different temperatures during seed development. *J. Agric. Food Chem.*, 43:1184–1192.

Uccella, N. 2001. Olive biophenols: novel ethnic and technological approach. *Trends Food Sci. Technol.*, 11:357–363.

Unger, E.H. 1990. Commercial processing of canola and rapeseed: crushing and oil extraction, in *Rapeseed/Canola: Production, Chemistry, Nutrition and Processing Technology*, Shahidi, F., Ed., Van Nostrand Reinhold, New York, pp. 235–250.

Varga, T.K. and Diosady, L. 1994. Simultaneous extraction of oil and antinutritional components from flaxseed. *J. Am. Oil Chem. Soc.*, 71:603–607.

Vázquez-Roncero, A. and Maestro-Duran, R. 1970. Los colorantes anthocianicos de la aceituna madura. I. Estudio cualitativo. *Grasas Aceites*, 21:208–214.

Vierhuis, E., Servili, M., Baldioli, M., Schols, H.E., Voragen, A.G.J., and Montedoro, G.F. 2001. Effect of enzyme treatment during mechanical extraction of olive oil on phenolic compounds and polysaccharides. *J. Agric. Food Chem.*, 49:1218–1223.

Visioli, F., Bellomo, G., and Galli, C. 1998. Free radical-scavenging properties of olive oil polyphenols. *Biochem. Biophys. Res. Commun.*, 247:60–64.

Visioli, F., Romani, A., Mulinacci, N., Zarini, S., Conte, D., Vincieri, F.F., and Galli, C. 1999. Antioxidant and other biological activities of olive oil mill waste waters. *J. Agric. Food Chem.*, 47:3397–3401.

Visioli, F., Caruso, D., Galli, C., Viapiani, S., Galli, G., and Sala, A. 2000. Olive oils rich in natural catecholic phenols decrease isoprostane excretion in humans. *Biochem. Biophys. Res. Commun.*, 278:797–799.

Visioli, F. and Galli, C. 2002. Biological properties of olive oil phytochemicals. *CRC Crit. Rev. Food Sci. Nutr.*, 42:209–221.

Vlahov, G. 1992. Flavonoids in three olive (*Olea europea*) fruit varieties during maturation. *J. Food Sci,*. 58:157–159.

Wanasundara, J., Amarowicz, R., and Shahidi, F. 1994. Isolation and identification of antioxidative component of canola meal. *Food Chem.*, 42:1285–1290.

Wanasundara, J., Shahidi, F., and Shukla, V. 1997. Endogenous antioxidants from oilseeds and edible oils. *Food Rev. Int.*, 13:225–292.

Wang, C. and Murphy, P. 1994. Isoflavone composition of American and Japanese soybean in Iowa: effects of variety, crop year, and location. *J. Agric. Food Chem.*, 42:1674–1677.

Wang, H.-J. and Murphy, P. 1996. Mass balance study of isoflavones during soybean production. *J. Agric. Food Chem.*, 44:2377–2383.

Wang, C., Ma, Q., Pagadala, S., Sherrard, M.S., and Krishnan, P.G. 1998. Changes in isoflavones during processing of soy protein isolates. *J. Am. Oil Chem. Soc.*, 75:337–341.

Wang, C., Sherrard, M., Pagadala, S., Wixon, R., and Scott, R.A. 2000. Isoflavone content among the group 0 and II soybeans. *J. Am. Oil Chem. Soc.*, 77:483–487.

Wang, X. and Plhak, L.C. 2000. Production, characterization, and application of antigossypol polyclonal antibodies. *J. Agric. Food Chem.*, 48:5109–5116.

Wei, H., Bowen, R., Cai, Q., Barnes, S., and Wang, Y. 1995. Antioxidant and antipromotional effects of the soybean isoflavone genistein. *Proc. Soc. Exp. Biol. Med.*, 208:124–130.

Westcott, N.D. and Muir, A.D. 1996a. Process for extracting and purifying lignans and cinnamic acid derivatives from flaxseed. PCT patent No. WO9630468A2.

Westcott, N.D. and Muir, A.D. 1996b. Variation in the concentration of the flaxseed lignan concentration with variety, location and year, in *Proceedings of the 56th Flax Institute of the United States,* Flax Institute of the United States, Fargo, ND, 77–80.

Westcott, N.D. and Muir, A.D. 2000. Overview of flaxseed lignans. *INFORM,* 11:118–121.

Westcott, N.D., Hall, T.W., and Muir, A.D. 2000. Evidence for the occurrence of the ferulic acid derivative in flaxseed meal, in *Proceedings of the 58th Flax Institute of the United States,* The Flax Institute of the United States, Fargo, ND, 49–52.

Wettasinghe, M., Shahidi, F., Amarowicz, R., and Aboud-Zaid, M.M. 2001. Phenolic acids in defatted seeds of borage (*Borago officinalis* L.). *Food Chem.*, 75:49–56.

Wettasinghe, M., Shahidi, F., and Amarowicz, R. 2002. Identification and quantification of low molecular weight phenolic antioxidants in seeds of evening primrose (*Oenethera biennis* L.). *J. Agric. Food Chem.*, 50:1267–1271.

Whipkey, A., Simon, J.E., and Janick, J. 1988. *In vivo* and *in vitro* lipid accumulation in *Borago officinalis* L. *J. Am. Oil Chem. Soc.*, 65:979–984.

Wiseman, H., Casey, K., Clarke, D.B., Barnes, K.A., and Bowey, E. 2002. Isoflavone aglycon and glucoconjugate content of high- and low-soy UK Foods used in nutritional studies. *J. Agric. Food Chem.*, 50:1404–1410.

Wiseman, H. 1996. Role of dietary phyto-oestrogens in the protection against cancer and heart disease. *Biochem. Comp. Food*, 24:795–800.

Wolf, W.J. and Cowan, J.C. 1975. *Soybeans as a Food Source.* CRC Press, Boca Raton, FL.

Xu, Z. and Godber, J.S. 1999. Purification and identification of components of γ-oryzanol in rice bran oil. *J. Agric. Food Chem.*, 47:2724–2728.

Xu, Z., Hua, N., and Godber, J.S. 2001. Antioxidant activity of tocopherols, tocotrienols, and γ-oryzanol from rice bran against cholesterol oxidation accelerated by 2,2′-azobis(2-methylpropionamidine) hydrochloride. *J. Agric. Food Chem.*, 49:2077–2081.

Yokosuka, T. 1986. Soy sauce biochemistry. *Adv. Food Res.*, 30:196–331.

Yoshikura, K. and Hamaguchi, Y. 1969. Anthocyanins of the black soybean. *Eiyo To Shokuryo*, 22:367–370, in *Chemical Abstracts*, CA 72, 63,600.

Zadernowski, R. 1987. Studies on phenolic compounds of rapeseed flours. *Acta Acad. Agric. Technol. Olst.*, 21F:1–55 (in Polish).

Zadernowski, R., Nowak-Polakowska, H., and Lossow, B. 1995. Natural antioxidants in selected plant seeds. *Acta Acad. Agric.Techn. Olst.*, 27:107–118 (in Polish).

Zadernowski, R., Naczk, M., and Nowak-Polakowska, H. 2002. Phenolic acids of borage (*Borago officinalis* L.) and evening primrose (*Oenothera biennis* L.). *J. Am. Oil Chem. Soc.*, 79:335–338.

Zarkadas,C.G., Voldeng, H.D., Yu, Z.R., and Choi, V.K. 1997. Determination of the protein quality of three new northern adapted cultivars of common and miso type soybeans by amino acids analysis. *J. Agric. Food Chem.*, 45:1161–1168.

4 Phenolic Compounds in Fruits and Vegetables

INTRODUCTION

Because they exhibit health-promoting effects such as reducing blood pressure and lowering incidences of cancer and cardiovascular diseases, fruits and vegetables are excellent sources of phenolics. (Ames et al., 1993; Ascherio et al., 1992; Block et al., 1992; Cook and Samman, 1996; Facino et al., 1999; Hertog et al., 1993, 1997; Hollman, 2001; Hollman et al., 1996; Huang et al., 1992; Leake, 1997; Middleton and Kandaswami, 1994; Morazzoni and Bombardelli, 1996; Ness and Powless, 1997; Steinmetz and Potter, 1991a, b, 1996; Swanson, 1998; Willet, 1994). Moreover, phenolics and related enzymes affect the quality of fruits and vegetables (Amiot et al., 1997; Brouillard et al., 1997; Clifford, 1997; Crouzet et al., 1997; Lea, 1992; Macheix and Fleuriet, 1993; Nicolas et al., 1993; Parr and Bowell, 2000; Tomás-Barberán and Espin, 2001; Tomás-Barberán and Robins, 1997; Zobel, 1997). Some phenolics are unique for certain fruits: for example, cinnamic esters of tartaric acid in grapes (de Simon et al., 1992), phlorizin in apples (Sanoner et al., 1999) and flavanone glycosides in citrus fruits (Louche et al., 1998; Marini and Balestrieri, 1995; Mouly et al., 1994; Ooghe and Detavernier, 1997). However, published data on the content of phenolics in fruits and vegetables are still incomplete and often restricted to a few cultivars.

FRUITS

BERRIES

Small berries constitute an important source of potential health-promoting phytochemicals; these include fruits of the *Vaccinium, Ribes, Rubus* and *Fragaria* genera (Fukumoto and Mazza, 2000; Kähkonen et al., 1999; Wang et al., 1996). Lowbush and highbush blueberry, bilberry, cranberry and lingonberry are examples of *Vaccinium* genus and blackberries, red and black raspberries are examples of *Rubus* genus; gooseberries, jostaberries and currants belong to the *Ribes* genus and strawberry to the *Fragaria* genus. These fruits are rich sources of flavonoids and other phenolics that display potential health-promoting effects (Block et al., 1992; Bomser et al., 1996; Feldman, 2001; Landbo and Meyer, 2001; Saito et al., 1998). For example, over 180 *Vaccinum*-based pharmaceuticals have been introduced to the market (Kalt and Dufour, 1997). Grapes are one of the world's largest berry crops. Approximately 80% of the total crop is utilized for wine making; 13% is consumed as table grapes and 7% processed into juice and raisins. Grapes belonging to the species *Vitis vinifera* L. are predominantly cultivated in Europe, while those belonging to species *Vitis labrusca* and *Vitis rotundifolia* are grown in North America (Mazza, 1995).

Benzoic Acid Derivatives	X	Y
p-Hydroxybenzoic Acid	H	H
Vanillic Acid	OCH_3	H
Syringic Acid	OCH_3	OCH_3
Protocatechuic Acid	H	H
Gallic Acid	OH	OH

Cinnamic Acid Derivatives	X	Y
p-Coumaric Acid	H	H
Caffeic Acid	OH	H
Sinapic Acid	OCH_3	OCH_3
Ferulic Acid	OCH_3	H

FIGURE 4.1 Chemical structures of phenolic acids.

Bilberries

Bilberry (*Vaccinium myrtillus*) is a rich source of phenolic compounds, namely, phenolic acids and flavonoids; bilberry juice has been used in medicine for its pharmacological properties (Azar et al., 1987). The total content of phenolics in bilberry fruit ranges from 33,000 to 38,200 mg/kg of dry weight as gallic acid equivalents (Kähkönen et al., 2001). The presence of caffeic and chlorogenic acids in bilberry fruit was first reported by Friedrich and Schönert (1973). Later, Brenneisen and Steinegger (1981a, b) found four more phenolic acids in bilberry fruit, namely, *p*-coumaric, ferulic and syringic acids, and a derivative of *p*-hydroxybenzoic acid. Subsequently, Azar et al. (1987) confirmed the presence of previously reported phenolic acids, but also found other phenolic acids such as *o*- and *m*-coumaric, gallic, *m*- and *p*-hydroxybenzoic, protocatechuic and vanillic acids in bilberry juice (Figure 4.1). The total content of hydroxycinnamic and hydroxybenzoic acids is 1130 to 2310 mg of chlorogenic acid equivalent/kg of dry weight and 33 to 58 mg gallic acid equivalent/kg of dry weight, respectively (Kähkönen et al., 2001).

Four flavonol glycosides, namely, quercetrin (quercetin 3-rhamnoside), isoquercetrin (quercetin 3-glucoside), hyperin (quercetin 3-galactoside), and astragalin (kaempferol 3-glucoside), have also been found in bilberry fruit and juice (Figure 4.2) (Azar et al., 1987; Brenneisen and Steinegger, 1981a, b; Friedrich and Schönert, 1973). The total content of flavonols in berries is between 540 and 1300 mg of rutin equivalent/kg of dry weight (Kähkönen et al., 2001). In addition, several flavan-3-ols, including (+)catechin and (–)epicatechin (Figure 4.3) (Brenneisen and Steinegger, 1981a, b; Friedrich and Schönert, 1973) and procyanidins B1, B2, B3, and B4 (Figure 4.4) (Brenneisen and Steinegger, 1981a, b), have also been detected. The reported values on the total anthocyanin content range from 3700 to 6980 mg/kg of fresh weight (Kalt et al., 1999a; Nyman and Kumpulainen, 2001; Senchuk and Borukh, 1976) and from 22,980 to 30,900 mg/kg of dry weight (Kähkönen et al., 2001). Ten anthocyanins are found in bilberry fruit powder (Figure 4.5) (Chandra et al., 2001):

- Cyanidin-3-galactoside
- Cyanidin-3-glucoside
- Delphinidin-3-galactoside

Compound	R_1	R_2	R_3	R_4
Flavones				
Apigenin	H	H	H	H
Diosmin	rutinosyl	OH	CH_3	H
Isohoifolin	rutinosyl	H	H	H
Luteolin	H	OH	H	H
Neodiosmin	neohesperidosyl	OH	CH_3	H
Rhoifolin	neohesperidosyl	H	H	H
Flavonols				
Kaempferol	H	H	H	OH
Rutin	H	OH	H	*O*-rutinoside
Quercetin	H	OH	H	OH
Quercitrin	H	OH	H	*O*-rhamnoside
Isoquercitrin	H	OH	H	*O*-glucoside
Hyperin	H	OH	H	*O*-galactoside
Astragalin	H	H	H	*O*-glucoside
Reynoutrin	H	OH	H	*O*-xyloside
Avicularin	H	OH	H	*O*-arabinoside

FIGURE 4.2 Chemical structures of some flavones and flavonols and their glycosylated derivatives.

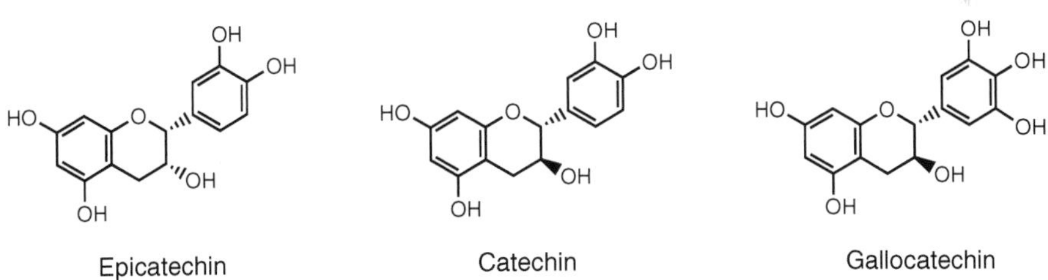

FIGURE 4.3 Chemical structures of catechins.

- Malvidin-3-arabinoside
- Malvidin-3-glucoside
- Malvidin-3-galactoside
- Peonidin-3-galactoside
- Petunidin-3-galactoside

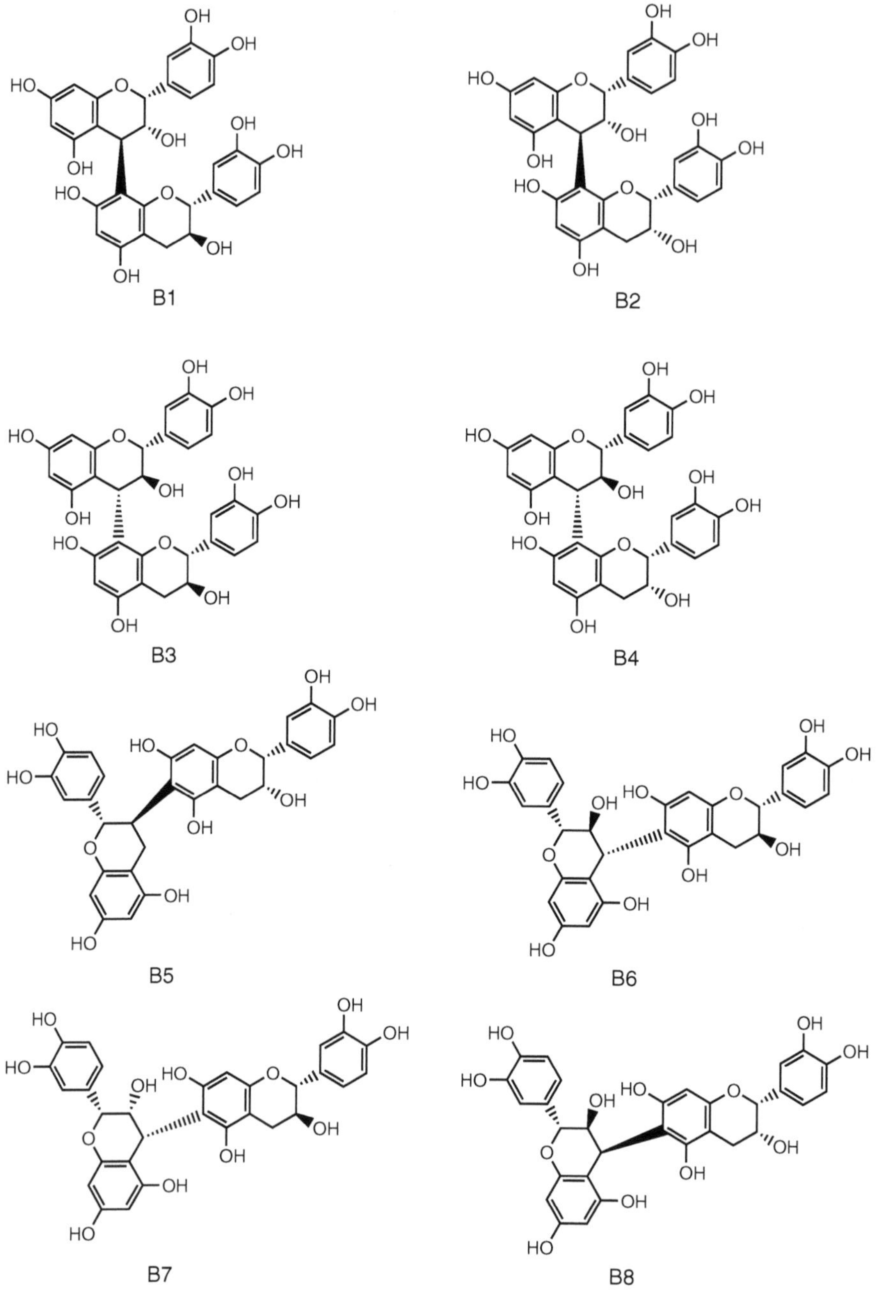

FIGURE 4.4 Chemical structures of procyanidin dimers type B.

- Petunidin-3-glucoside
- Petunidin-3-arabinoside

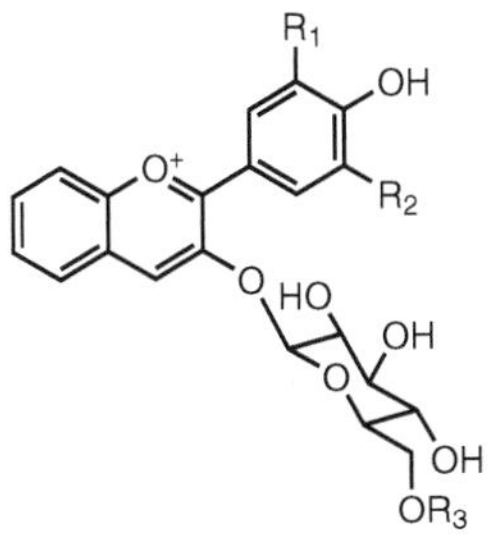

Compound	R_1	R_2	R_3
Cyanidin-3-*O*-glucoside	OH	H	H
Cyanidin-3-*O*-rutinoside	OH	H	rhamnosyl
Delphinidin-3-*O*-glucoside	OH	OH	H
Delphinidin-3-*O*-rutinoside	OH	OH	rhamnosyl
Pelargonidin-3-*O*-glucoside	H	H	H
Pelargonidin-3-*O*-rutinoside	H	H	rhamnosyl
Petunidin-3-*O*-glucoside	OCH_3	OH	H
Petunidin-3-*O*-rutinoside	OCH_3	OH	rhamnosyl
Malvidin-3-*O*-glucoside	OCH_3	OCH_3	H
Malvidin-3-*O*-rutinoside	OCH_3	OCH_3	rhamnosyl

FIGURE 4.5 Chemical structures of some anthocyanins.

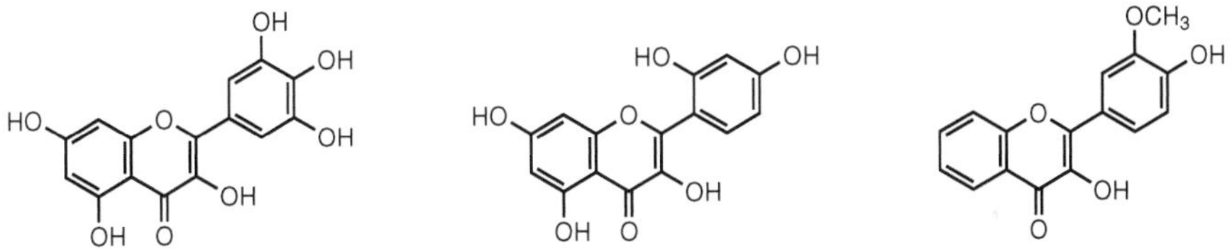

FIGURE 4.6 Chemical structures of myricetin, morin and isorhamnetin.

Blackberries

The total content of phenolics in Georgia-grown blackberries is 4865.3 mg/kg of fresh weight expressed as gallic acid equivalents (Sellappan et al., 2002). Several phenolic acids have been detected in blackberries, namely, gallic, caffeic, ferulic, *p*-coumaric and ellagic acids (Figure 4.1). Of these, ellagic acid is the major phenolic (Sellappan et al., 2002). Ripe fruits of thornless blackberry (*Rubus* sp.) cultivars contain 5 to 35 mg/kg of quercetin and 1 to 3 mg/kg of fresh weight of kaempferol (Bilyk and Sapers, 1986); however, Sellappan et al. (2002) detected only catechin (2657.5 to 3128.6 mg/kg) and myricetin (99.9 mg/kg) (Figure 4.6) in Georgia-grown blackberries.

The total content of anthocyanins in blackberries ranges from 1165.9 to 1528 mg/kg of fresh weight expressed as cyanidin-3-glucoside equivalents (Sellappan et al., 2002; Wang and Lin, 2000). Several anthocyanins have been identified in blackberry extracts, namely:

- Cyanidin-3-galactoside
- Cyanidin-3-glucoside
- Cyanidin-3-arabinoside
- Cyanidin-3-xyloside
- Malvidin-3-arabinoside
- Pelargonidin-3-glucoside (Dugo et al., 2001)
- Cyanidin-3-rutinoside (Mazza and Miniati, 1993; Nybom, 1968)
- Cyanidin-3-sophoroside
- Cyandin-3-glucosylrutinoside
- Cyanidin-3-rutinoside (Figure 4.5) (Hong and Wrolstad, 1990; Mazza and Miniati, 1993)

In addition, Sapers et al. (1986) detected two acylated anthocyanins in unripe blackberries and reported that their levels rapidly decrease during the ripening process. Later, Fang-Chiang (2000) detected cyanidin-3-glucoside acylated with malonic acid in blackberry while Stintzing et al. (2002) isolated cyanidin dioxalyl-glucoside-a novel zwitterionic anthocyanin from evergreen blackberry (*Rubus lacitianus* Wild.). Of these, cyanidin-3-glucoside was the major anthocyanin in blackberry (Garcia-Viguera et al., 1997; Goiffon et al., 1991; Hong and Wrolstad, 1990; Mazza and Miniati, 1993; Sapers et al., 1986; Versari et al., 1997).

Blackcurrants

Blackcurrants are commercially grown for production of juices and jams; these fruits are also a component of fruit-based yogurts. Blackcurrant berries serve as a rich source of phenolic compounds such as anthocyanins, flavonoids, phenolic acids and proanthocyanidins (Constantino et al., 1992, 1993; Demina, 1974; Foo and Porter, 1981; Häkkinen and Auriola, 1998; Koeppen and Herrmann, 1977; Le Lous et al., 1975; Stöhr and Herrmann, 1975). The total content of phenolics in blackcurrant berries is between 6940 and 38,200 mg/kg expressed as gallic acid equivalent (Kähkönen et al., 2001; Moyer et al., 2002). Blackcurrant anthocyanins are located in the skin of berries (Iversen, 1999); their total content ranges from 1560 to 10,640 mg/kg expressed as cyanidin 3-glucoside equivalents (Iversen 1999; Kähkönen et al., 2001; Koeppen and Herrmann, 1977; Moyer et al., 2002). The predominant anthocyanins of blackcurrants are delphinidin-3-rutinoside, cyanidin-3-rutinoside, delphinidin-3-glucoside and cyanidin-3-glucoside (Figure 4.5) (Iversen, 1999; Matsumoto et al., 2001; Mazza and Miniati, 1993; Millet et al., 1984; Renault et al., 1997). These four main anthocyanins make up >97% of total anthocyanins in blackcurrants (Slimstad and Solheim, 2002).

Other anthocyanins detected in blackcurrants include

- Pelargonidin-3-rutinoside
- Cyanidin-3-sophoroside

- Delphinidin-3-sophoroside (Le Lous et al., 1975)
- Peonidin-3-glucoside
- Peonidin-3-rutinoside
- Malvidin-3-glucoside
- Malvidin-3-rutinoside
- Cyanidin-3-arabinoside
- Cyanidin-3-(6″-coumarylglucoside)
- Delphinidin-3-(6″-coumarylglucoside) (Slimstad and Solheim, 2002)

Blackcurrants also serve as an important source of dietary flavonols (Häkkinen and Auriola, 1998; Häkkinen et al., 1999a; Starke and Herrmann, 1976). Flavonols in blackcurrants are present predominantly in the conjugated form and free flavonols contribute only 1.13 to 5.12% to total flavonol contents (Amakura et al., 2000). Recently, Mikkonen et al. (2001) reported great variability in the flavonol content among different blackcurrant cultivars, but did not find any consistent differences within the same cultivar, grown conventionally, and those grown on organic farms. The total content of flavonols in blackcurrant cultivars varies from 157 to 870 mg/kg of fresh weight (Kähkönen et al., 2001; Mikkonen et al., 2001). Myricetin (89 to 245 mg/kg of fresh weight) (Figure 4.6) is the most abundant flavonol in blackcurrants, followed by quercetin (52 to 122 mg/kg of fresh weight) and kaempferol (9 to 23 mg/kg of fresh weight) (Figure 4.2) (Häkkinen et al., 2000a; Mikkonen et al., 2001).

Myricetin is the least stable flavonol in blackcurrant. Postharvest storage temperature has a major affect on the content of flavonols; considerably lower myricetin levels (about 20%) are observed for samples stored 1 day at 22°C in comparison to those stored at 5°C. Myricetin content is also significantly lowered in blackcurrants stored for 6 months at –20°C (Häkkinen et al., 2000a). The content of flavonols in juice is significantly affected by the method of juice extraction. Blackcurrant juice obtained by cold-pressing berries contains twice as many flavonols as that extracted from berries by steam (Häkkinen et al., 2000a). Furthermore, approximately 62% of total anthocyanins are extracted into the raw juice after treatment of blackcurrants with PectinexTM Ultra SP enzyme, but only 24% of anthocyanins remain after pasteurization of juice (Iversen, 1999).

Blackcurrant seeds are a rich source of potent antioxidant because seed oil containing high levels of polyunsaturated fatty acids, especially γ-linolenic acid (Traitler et al., 1984; Zhao et al., 1994), is stable in the intact seed (Lu et al., 2002; Lu and Foo, 2003). Two novel noncyanogenic nitrile-containing compounds were recently isolated by Lu et al. (2002) from seed residue and identified as nigrumin-5-*p*-coumarate and nigrumin-5-ferulate (Figure 4.7). Later, Lu and Foo (2003) identified a number of phenolic acids and their derivatives in blackcurrant seeds, namely, caffeic acid, ferulic acid, *p*-coumaric acid, protocatechuic acid, gallic acid, *p*-hydroxybenzoic acid, 1-cinnamoyl-β-D-glucoside, 1-*p*-coumaroyl-β-D-glucoside. In addition, these authors detected the presence of 3-glucosides and 3-rutinosides of delphinidin and cyanidin and a number of flavonoids such as taxifolin (dihydroquercetin), aureusidin, 3-glucosides of quercetin, kaempferol and myricetin, as well as 3-rutinosides of quercetin and myricetin (Figure 4.2, Figure 4.6, and Figure 4.7).

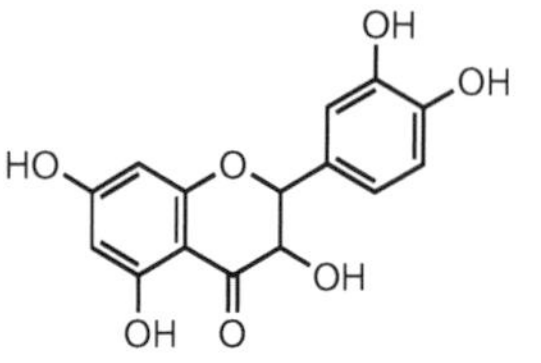

Dihydroquercetin

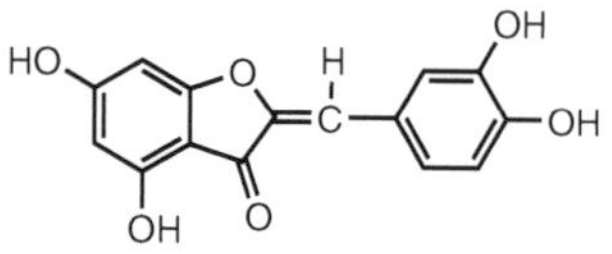

Aureusidin

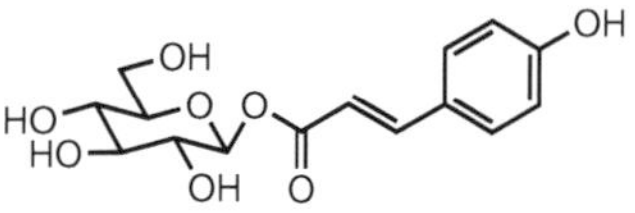

1-*p*-Coumaroyl-β-D-glucoside

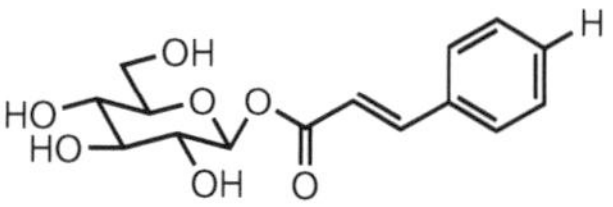

1-*p*-Cinnamoyl-β-D-glucoside

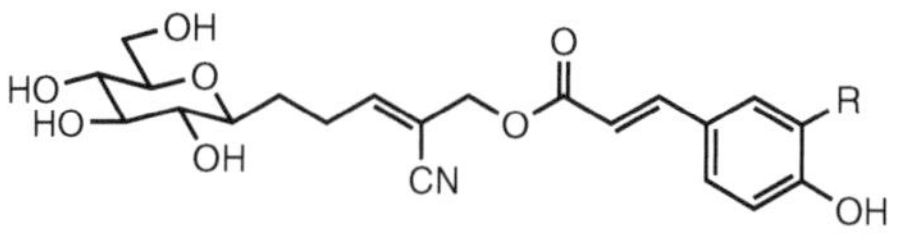

R=H; Nigrumin-5-*p*-coumarate R=OCH_3; Nigrumin-5-ferulate

FIGURE 4.7 Chemical structures of some phenolic compounds found in blackcurrant seeds.

Blueberries

Lowbush (wild) and highbush (cultivated) blueberries have been used for many years for production of berries on a commercial scale (Kalt et al., 2001). Blueberries are a rich source of phenolic acids, catechins, flavonols, anthocyanins and proanthocyanidins (de Pascual-Teresa et al., 2000; Gao and Mazza, 1994; Kader et al., 1996; Kalt and Dufour, 1997; Kalt and McDonald, 1996; Prior et al., 1998; Sellappan et al., 2002; Smith et al., 2000; Stöhr and Herrmann, 1976). Blueberry phenolics display inhibitory effects against chemically induced carcinogenesis (Bomser et al., 1996). The content of phenolics in blueberry is affected by the degree of maturity at harvest, genetic differences (cultivar), preharvest environmental conditions, postharvest storage conditions and processing.

Ehlenfeldt and Prior (2001) surveyed the total content of phenolics and anthocyanins in 87 highbush blueberry (*Vaccinium corymbosum* L.) and species-introgressed highbush blueberry cultivars. The total content of phenolics in these berries ranged from 430 to 1990 (950 on average) mg/kg of fresh weight expressed as gallic acid equivalents, while the total anthocyanin content was between 890 and 3310 (1790 on average) mg/kg of fresh weight expressed as cyanidin-3-glucoside equivalents. Kalt et al. (2000) also surveyed the total phenolic and anthocyanin contents in 100 highbush blueberry varieties and 155 lowbush blueberry clones. The mean total anthocyanin content in highbush and lowbush blueberries was 1180 and 1630 mg of cyanidin-3-glucoside equivalents/kg of fresh weight, respectively. On the other hand, the mean total phenolic content in highbush and lowbush blueberries is 1910 and 3760 mg gallic acid equivalents/kg of fresh weight. Thus, lowbush blueberries contain almost 50% more anthocyanins and over 90% more total phenolics than highbush blueberries. Recently, Sellappan et al. (2002) reported that the total content of anthocyanins and phenolics in Georgia-grown rabbiteye blueberries (*Vaccinium ashei* Reade) is between 127 and 1973.4 mg of cyanidin-3-glucoside equivalents/kg of fresh weight and from 2700.2 to 6690.1 mg of gallic acid equivalents/kg of fresh weight, respectively. Subsequently, Moyer et al. (2002) found that the total content of anthocyanins and total phenolics in 29 *Vaccinium* species ranges from 340 to 515 mg of cyanidin 3-glucoside equivalents/kg of fresh weight (on average 2300 mg/kg) and from 2110 to 9610 mg of gallic acid equivalents/kg of fresh weight (on average 5210 mg/kg), respectively.

Several phenolic acids have been identified in rabbiteye blueberries (*V. ashei* Reade) and southern highbush blueberries (*V. corymbosum* L.), namely (Figure 4.1):

Gallic (0 to 2589 mg/kg of fresh weight)
Caffeic (0 to 63.2 mg/kg of fresh weight)
p-Coumaric (24 to 157 mg/kg of fresh weight)
Ferulic (30.2 to 169.7 mg/kg of fresh weight)
Ellagic acids(2.2 to 66.5 mg/kg of fresh weight)

Rabbiteye blueberries also contain a higher level of catechin (145.3 to 3874.8 mg/kg of fresh weight) than southern highbush blueberries (98.7 to 292.8 mg/kg of fresh weight). The contents of myricetin (Figure 4.6), quercetin, and kaempferol (Figure 4.2) are 67.2 to 99.9, 58.2 to 146, and 25.1 to 37.2 mg/kg of fresh weight, respectively (Sellappan et al., 2002). On the other hand, Bilyk and Sapers (1986) detected only 24 to 28.5 mg quercetin/kg of fresh weight in highbush blueberries.

A number of anthocyanins have been isolated and identified in lowbush (wild), highbush, rabbiteye and Tifblue blueberries, namely, 3-galactosides and 3-arabinosides of cyanidin, delphinidin, peonidin, petunidin and malvidin, and 3-glucosides of cyanidin, delphinidin, peonidin, petunidin and malvidin (Figure 4.5) (Ballington et al., 1987; Francis et al., 1966; Mazza and Miniati, 1993; Prior et al., 2001). Small quantities of 6-acetyl-glucosides of cyanidin, delphinidin, malvidin, peonidin, petunidin, and 6-acetyl-galactoside of malvidin have also been detected in lowbush blueberries (Prior et al., 2001). Of these, malvidin-3-glucoside, malvidin-3-galactoside, cyanidin-3-arabinoside, delphinidin-3-arabinoside, petunidin-3-galactoside, petunidin-3-galactoside and delphinidin-3-galactoside were the major anthocyanins in lowbush and Tifblue blueberries (Ballington et al., 1987; Francis et al., 1966; Mazza and Miniati, 1993; Prior et al., 2001).

Prior et al. (2001) identified oligomeric B-type procyanidins (from monomeric to octamers) with (+)catechin and/or (–)epicatechin units linked through (4→8 or 4→6 bonds) in highbush and lowbush blueberries (Figure 4.8). Later, Gu et al. (2002) reported that lowbush blueberry contains 19,990 mg/kg of dry weight of procyanidins.

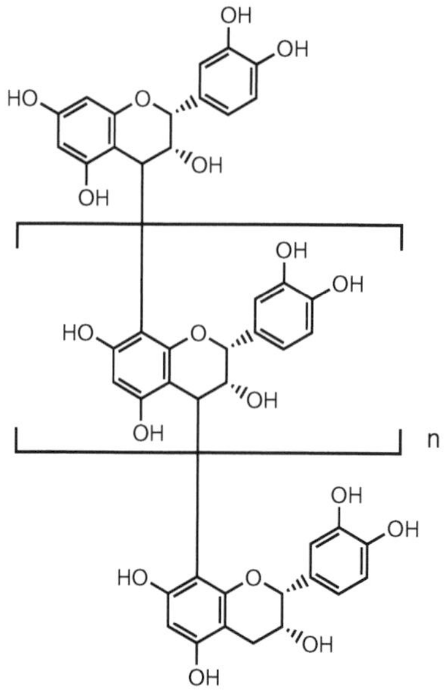

FIGURE 4.8 Chemical structures of oligomeric (n = 1 to 8) and polymeric (n > 8) procyanidins.

The polymeric procyanidins comprise over 76% of total procyanidins and are a mixture of polymers with a degree of polymerization ranging from 14.4 to 114.1 as determined by thiolysis (Gu et al., 2002).

Prior et al. (1998) demonstrated that late harvest berries of *V. ashei* Reade cultivars Tifblue and Brightwell exhibit greater antioxidant activities than those from early harvest berries (Prior et al., 1998). On the other hand, Kalt et al. (1999b) and Kalt and McDonald (1996) showed that postharvest storage conditions also affect antioxidant potential of blueberries as determined by ORAC (oxygen radical absorbance capacity) measurements. Almost a 20% increase in anthocyanins was observed in lowbush blueberries (*Vaccinium angustifolium* Ait cultivars) stored for 2 weeks at 1°C (Kalt and McDonald, 1996). However, a similar increase in anthocyanins was noted in highbush blueberries (*V. corybosum* L. cultivar Bluecrop) stored for 8 days at 20°C, but not in berries stored for the same period at 0, 10 and 30°C (Kalt et al., 1999b).

Recently, Connor et al. (2002) evaluated the effect of extended storage (up to 5 weeks at 5°C) on the antioxidant potential of berries harvested from nine blueberry cultivars. These authors demonstrated that extended storage of blueberries does not significantly decrease their antioxidant potential. The antioxidant activity of blueberry juice is affected by the extraction temperature, pH and addition of oxygen to juice (Kalt et al., 2000). Processing of blueberries to juice has a detrimental effect on their phenolics; only 32, 35, 43 and 53% of anthocyanins, flavonols, procyanidins and chlorogenic acids present remain in their corresponding juices, respectively. These losses are believed to be due to the activity of native polyphenol oxidase during milling and depectinization of berries (Skrede et al., 2000).

Cranberries

The fruits of American cranberries, *Vaccinium macrocarpon,* and European cranberries, *Vaccinium oxycoccus,* possess a distinctive flavor and a bright red color; they are sold fresh or processed into sauce, concentrates and juice. Cranberry fruits serve as an excellent source of anthocyanins (Hong and Wrolstad, 1986; Mazza and Miniati, 1993; Prior et al., 2001; Zapsalis and Francis, 1965), flavonol glycosides (Kandil et al., 2002; Puski and Francis, 1967), proanthocyanidins (Foo and Porter, 1980; Hale et al., 1986; Kandil et al., 2002; Prior et al., 2001) and phenolic acids (Zuo et al., 2002). Cranberry proanthocyanidins display a potent bacterial antiadhesion activity (Howell, 2002), inhibit the sialyllactose specific adhesion of *Helicobacter pyroli* to human gastric mucus (Burger et al., 2002), and lower the rate of cardiovascular diseases (Reed, 2002).

Cranberries contain about 1g/kg of fresh weight of phenolic acids (Zuo et al., 2002), predominantly as glycosides and esters (Chen et al., 2001). Only small quantities of phenolic acids are detected in the free form (Zuo et al., 2002). Twelve phenolic acids have been identified in cranberry (Figure 4.1) (Marwan and Nagel, 1982; Zuo et al., 2002):

- *o*-Hydroxybenzoic
- *m*-Hydroxybenzoic
- *p*-Hydroxybenzoic
- *p*-Hydroxyphenyl acetic
- 2,3-Dihydroxybenzoic
- 2,4-Dihydroxybenzoic
- Vanillic
- *o*-Hydroxycinnamic
- Caffeic
- *p*-Coumaric
- Ferulic
- Sinapic

In addition, bound gallic, chlorogenic, *p*-coumaric and *p*-hydroxybenzoic acids have been detected in cranberry pomace by Zheng and Shetty (2000). Sinapic, caffeic and *p*-coumaric acids are the most abundant bound phenolic acids and coumaric, 2,4-dihydroxybenzoic and vanillic acids the predominant free phenolic acids found in cranberry (Zuo et al., 2002).

Resveratrol has also been detected in cranberry fruit (Figure 4.9). The concentration of resveratrol in cranberry juice (0.25 mg/kg) is similar to that found in Concord grape juice (0.36 mg/kg). Raw cranberry juice contains predominantly *trans*-resveratrol, while processing increases the level of *cis*-resveratrol (Wang et al., 2002a).

Several flavonols have been identified in cranberry fruits, namely quercetin, quercetin 3-*O*-galactoside (hyperin), quercetin 3-*O*-arabinoside, quercetin 3-*O*-rhamnoside (quercitrin) (Figure 4.2), myricetin 3-*O*-arabinoside and 3-*O*-digalactoside (Kandil et al., 2002; Puski and Francis, 1967). The total content of quercetin

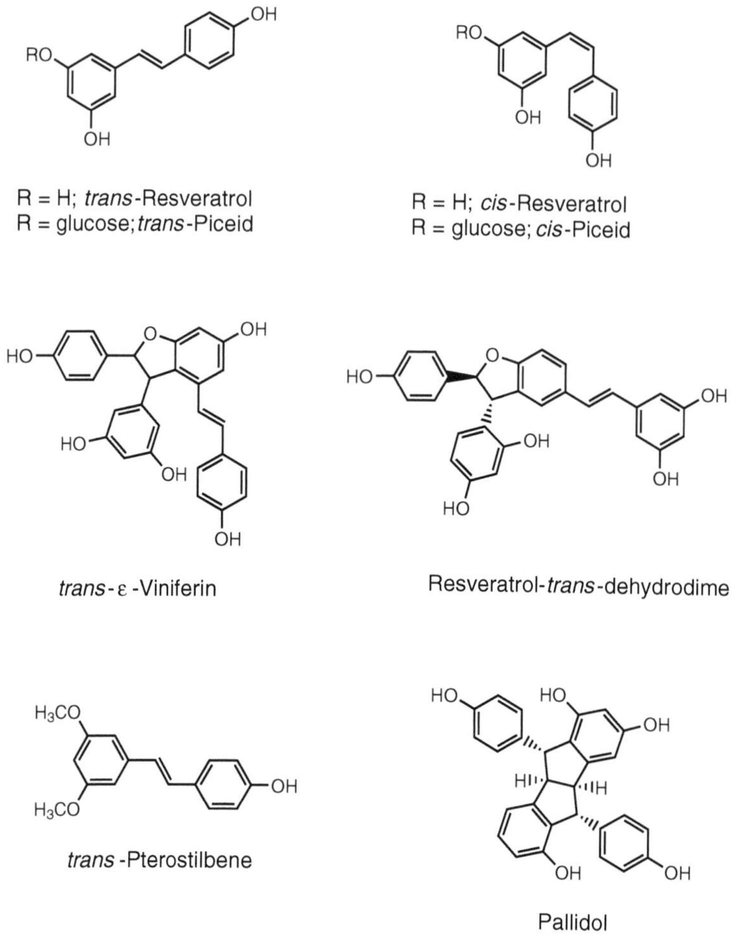

R = H; *trans*-Resveratrol
R = glucose; *trans*-Piceid

R = H; *cis*-Resveratrol
R = glucose; *cis*-Piceid

trans-ε-Viniferin

Resveratrol-*trans*-dehydrodime

trans-Pterostilbene

Pallidol

FIGURE 4.9 Chemical structures of stilbens.

and myricetin in cranberry fruit ranged from 73 to 250 and from 4.0 to 26.7 mg/kg fresh of weight, respectively. Small quantities of kaempferol (0.6 to 2.7 mg/kg fresh weight) were also detected in some cranberry varieties (Amakura et al., 2000; Bilyk and Sapers, 1986).

The predominant anthocyanins in American cranberries are 3-*O*-galactosides and 3-*O*-arabinosides of cyanidin and peonidin, while European cranberries contain 3-*O*-glucosides of cyanidin and peonidin (Figure 4.5) (Mazza and Miniati, 1993; Sapers and Hargrave, 1987; Zapsalis and Francis, 1965). Recently, Prior et al. (2001) reported the presence of small quantities of petunidin 3-*O*-galactoside and cyanidin 3-*O*-glucoside in American cranberry fruits. The total content of anthocyanins in cranberry fruits ranges from 180 to 656 mg/kg of fresh weight (Bilyk and Sapers, 1986; Wang and Stretch, 2001); these are located under the fruit skin (Francis, 1957; Sapers et al., 1983; Vorsa and Welker, 1985). Cyanidins comprise approximately 55% of total anthocyanins in cranberry (Prior et al., 2001). The content of anthocyanins

TABLE 4.1
Total Anthocyanin Content of Fruit of Cranberry Cultivars Stored for 3 Months at Different Storage Temperatures[a]

Cultivar	Initial	Storage Temperature, °C		
		0	5	15
Ben Lear	250 ± 13	460 ±25	546 ±29	766 ±35
Cropper	198 ±13	348 ±16	492 ±30	604 ±21
Crowley	656 ±20	654 ±34	808 ±37	1092 ±29
Early Black	634 ±15	724 ±52	886 ±41	1172 ±32
Franklin	541 ±36	604 ±18	696 ±32	936 ±14
Howes	235 ±11	330 ±14	494 ±16	710 ±22
Pilgrim	207 ±11	308 ±15	386 ±10	600 ±23
Stevens	228 ±13	316 ±12	474 ±25	656 ±18
Wilcox	243 ±17	548 ±27	646 ±28	724 ±17

[a] Milligrams of cyanidin 3-*O*-galactoside per kilogram of fresh weight.

Source: Adapted from Wang, S.Y. and Stretch, A.L., 2001, *J. Agric. Food Chem.*, 49:969–974.

is affected by cultivar, fruit size and pre- and postharvest conditions (Cracker, 1971; Galletta, 1975; Hall and Stark, 1972; Sapers et al., 1983; Vorsa and Welker, 1985; Wang and Stretch, 2001). Smaller berries contain higher levels of anthocyanins than those of larger ones due to the location of anthocyanins in the fruits (Vorsa and Welker, 1985; Wang and Stretch, 2001). Storing cranberries for 3 months at temperatures of up to 15°C significantly increases their total anthocyanin content (Table 4.1); however, this increase is somewhat lower when cranberry fruits are stored at the recommended storage temperature of 2 to 4°C (Hardenburg et al., 1986; Wang and Stretch, 2001).

Whole cranberries contain approximately 17 mg/kg of total proanthocyanidin, while 2.16 to 2.23 mg/L of total procyanidins are found in cranberry juices (Prior et al., 2001). The polymeric proanthocyanidins comprise 63% of total proanthocyanidins in cranberries (Gu et al., 2002). Foo and Porter (1980) reported that the ratio of procyanidins (with units containing two hydroxyl groups on B-ring) to prodelphinidins (containing three hydroxyl groups on B-ring) is 78:22 in European cranberries. Later, Foo et al. (2000) isolated and identified three proanthocyanidin trimers with A-type doubly linked interflavonoid linkages in ripe American cranberry fruits, namely:

- Epicatechin-(4β→6)-epicatechin-(4β→8, 2β→*O*→7)-epicatechin
- Epicatechin-(4β→8, 2β→*O*→7)-epicatechin-(4β→6)-epicatechin
- Epicatechin-(4β→8)-epicatechin-(4β→8, 2β→*O*→7)-epicatechin

Subsequently, Prior et al. (2001) reported that (–)epicatechin and its dimers and A-type trimers (Figure 4.10) are the predominant proanthocyanidins in cranberries.

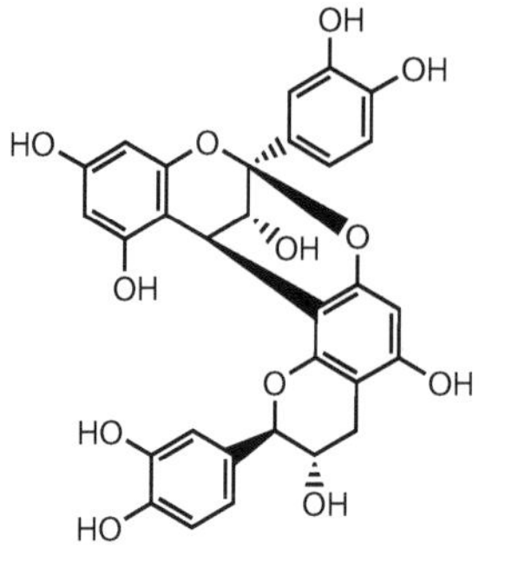

A1

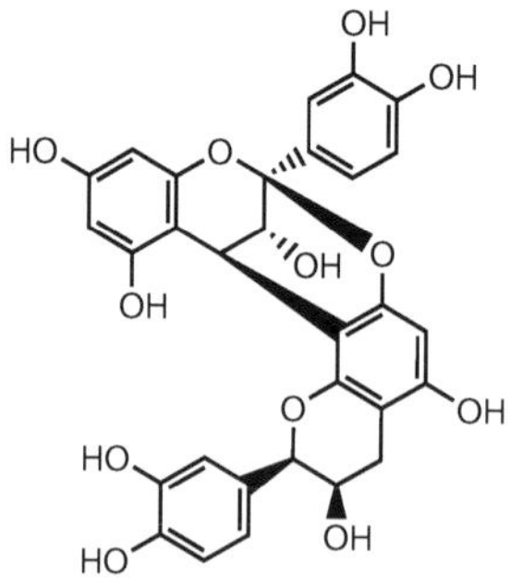

A2

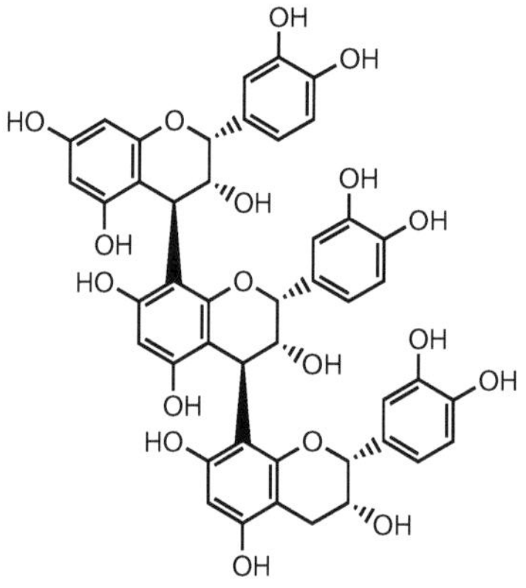

C1

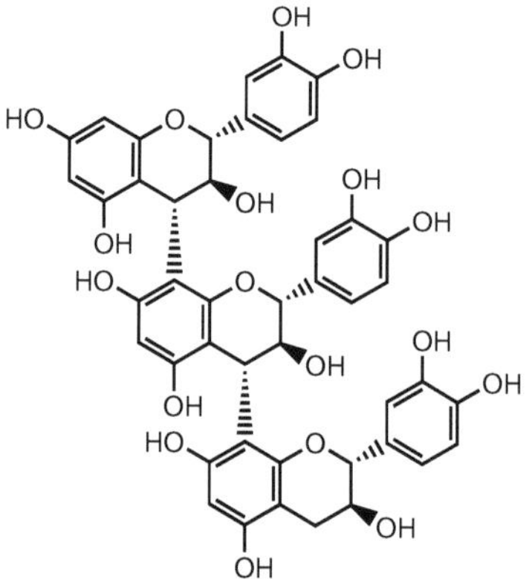

C2

FIGURE 4.10 Chemical structures of some procyanidin dimers type A and trimers type C.

These authors also detected trace amounts of B-type (single linked interflavonoid linkages) and A-type tetramers. Recently, Kandil et al. (2002) detected (–)epicatechin, (+)catechin, gallocatechin and epigallocatechin (Figure 4.3) as well as higher molecular weight proanthocyanidins in the American cranberry.

Grapes

The chemistry of grape phenolics is discussed in Chapter 5 and Mazza (1995) also provides an excellent review on the types of anthocyanins and their distribution in grapes; therefore, only some aspects of the chemistry of grape phenolics are discussed in this section. The main phenolics identified in *Vitis vinifera* berries are listed in Table 4.2 and the total content of phenolics in some table grape varieties is summarized in Table 4.3. Anthocyanins are the predominant phenolics of red table grape varieties, while flavan-3-ols are the main phenolics in white table grape varieties (Cantos et al., 2002; de Simón et al., 1993; Ingalsbe et al., 1963; Mazza, 1995; Spanos and Wrolstad, 1992). These phenolics contribute to the sensory quality of grape products (Jaworski and Lee, 1987).

TABLE 4.2
Main Phenolics Identified in *Vitis Vinera* Berries

Phenolic acids
p-hydroxybenzoic; o-hydroxybenzoic; salicylic; gallic; cinnamic; p-coumaroylartaric (= coutaric); caffeoyltartaric (= caftaric); feruloylartaric (= fertaric); p-coumaroyl glucose; feruloylglucose; glucose ester of coutaric acid

Anthocyanins
Cyanidin 3-glucoside; cy 3-acetylglucoside; cy 3-p-coumaryl-glucoside; peonidin 3-glucoside; pn 3-acetylglucoside; pn 3-p-coumarylglucoside; pn 3-caffeylglucoside (?); delphinidin 3-glucoside; dp 3-acetylglucoside; dp 3-p-coumarylglycoside; petunidin 3-glucoside; pt 3-p-coumarylglucoside; malvidin 3-glucoside; mv 3-acetylglucoside; mv 3-p-coumaryglucoside; mv 3-caffeylglucoside

Flavonols
Kaempferol 3-glucoside; k 3-glucuronide; k 3-glucosylarabinoside (?); k 3-galactoside; quercetin 3-glucoside; q 3-glucoronide; q 3-rutinoside; q 3-glucosylgalactoside (?); q 3-glucosylxyloside (?); iso-rhamnetic 3-glucoside

Flavan-3-ols and tannins
(+)catechin; (–)epicatechin; (+)gallocatechin; (–)epigallocatechin; epicatechin-3-O-gallate; procyanidins B1, B2, B3, B4, C1, C2, polymeric forms of condensed tannins

Flavanonols
Dihydroquercetin 3-rhamnoside (= astilbin); dihydrokaempferol 3-rhamnoside (= engeletin)

Source: Adapted from Macheix, J.J. et al., 1990, *Fruit Phenolics*, CRC Press, Boca Raton, FL.

Pomace (skin and seeds), a by-product from processing grape to juice and wine, comprises about 13% of the amount of processed berries (Torres and Bobet, 2001) and may also contain stems when wines are made from nondestemmed crop. Grape seeds, stems and skins are a rich source of health-promoting flavonoids such as proanthocyanidins, flavonols and flavan-3-ols (Cheynier and Rigaud, 1986; Diplock et al., 1998; Roggero et al., 1988; Ruf, 1999; Souquet et al., 1996, 2000; Su and Singleton, 1969; Yang et al., 1997). Of these, proanthocyanidins are the major polyphenols in grape seeds, stems and skins. Procyanidins are the predominant proanthocyanidins in grape seeds, while procyanidins and prodelphinidins are dominant in grape skins and stems (Escribano-Bailón et al., 1992a, b, 1995; Souquet et al., 1996, 2000; Thorngate and Singleton, 1994). The procyanidin dimers identified in grape seeds and their contents are listed in Table 4.4, Table 4.5 and Figure 4.4. In addition, 11 trimers and one tetramer have also been detected in grape seeds (Boukharta, 1988; da Silva et al., 1991; Escribano-Bailón et al., 1992b; Romeyer et al., 1986; Santos-Buelga et al., 1995). The total contents of dimers, dimer gallates and trimers in seeds from grape cultivars grown in Ontario are 160 to 3750 mg of B2 equivalents, traces of 1080 mg of B2–3′-*O*-gallate quivalents, and traces of 840 mg of B2 equivalents/kg of seeds (Fuleki and da Silva, 1997).

Approximately 55% of grape seed procyanidins are of polymeric type (degree of polymerization, DP ≥ 5) (Figure 4.8) (Prieur et al., 1994), while the ratios of polymeric procyanidins (DP ≥ 4) to monomeric (catechin + epicatechin) are 5.2 to 8.9 (Peng et al., 2001). Grape seeds contain polymeric procyanidins from 33,200 to 50,700 mg/kg in seeds or from 1680 to 3190 mg/kg in berries (Labarbe et al., 1999,

TABLE 4.3
Total Contents of Phenolics in Some White and Red Table Grape Varieties[a]

Total Content of	Red Globe (R)	Flame (R)	Crimson (R)	Napoleon (R)	Superior (W)	Dominga (W)	Moscatel Italica (W)
Phenols[b]	225.4	361.2	131.9	135.9	135.7	114.9	145.1
Hydroxycinnamates[c]	8.4	47.6	9.5	9.5	9.0	25.0	16.3
Flavonols[d]	61.3	53.8	12.8	32.4	64.0	32.7	47.7
Flavan-3-ols[e]	40.4	109.1	41.1	18.3	62.7	57.2	81.1
Anthocyanins[f]	115.3	150.7	69.5	75.7			

Note: Abbreviations: R = red; W = white.

[a] Milligrams per kilogram of fresh weight.
[b] Total phenols = total hydroxycinnamates + total flavonols + total flavan-3-ols + total anthocyanins.
[c] Total hydroxycinnamates expressed as chlorogenic acid equivalents.
[d] Total flavonols expressed as quercetin 3-rutinoside equivalents.
[e] Total flavan-3-ols expressed as catechin equivalents.
[f] Total anthocyanins expressed as cyanidin 3-rutinoside equivalents.

Source: Adapted from Cantos, E. et al., 2002, *J. Agric. Food Chem.*, 50:5691–5696.

TABLE 4.4
Oligomeric Procyanidin Dimers Identified in Grape Seeds

Procyanidins	Compound	Reference
B1	Ec-(4β→8)-Cat	Lea, A.G.H. et al. (1979)
B2	Ec-(4β→8)-Ec	Lea, A.G.H. et al. (1979)
B3	Cat-(4α→8)-Cat	Lea, A.G.H. et al. (1979)
B4	Cat-(4α→8)-Ec	Lea, A.G.H. et al. (1979)
B5	Ec-(4β→6)-Ec	da Silva, J.M. et al. (1991)
B6	Cat-(4β→6)-Cat	da Silva, J.M. et al. (1991)
B7	Ec-(4β→6)-Cat	Boukharta, M. (1998)
B8	Cat-(4β→6)-Ec	da Silva, J.M. et al. (1991)
B1–3-O-gallate	Ec-3-O-gallate-(4β→8)-Cat	da Silva, J.M. et al. (1991)
B2–3-O-gallate	Ec-3-O-gallate-(4β→8)-Ec	da Silva, J.M. et al. (1991)
B2–3′-O-gallate	Ec-(4β→8)-Ec-3-O-gallate	Boukharta, M. (1988)
B4–3′-O-gallate	Cat-(4β→6)-Ec-3-O-gallate	da Silva, J.M. et al. (1991)
B7–3-O-gallate	Ec-3-O-gallate-(4β→6)-Cat	Santos de Buelga, C. et al. (1995)
B2–3,3′-di-O-gallate	Ec-3-O-gallate-(4β→8)-Ec-3-O-gallate	da Silva, J.M. et al. (1991)

Note: Ec = (–)epicatechin; Cat = (+)catechin.

TABLE 4.5
Content of Catechins and Procyanidin Dimers in Seeds of Different *Vitis Vinifera* Cultivars Grown in Ontario in 1993[a]

	Catechins		Procyanidins			
Cultivar	Catechin	Epicatechin	B1	B2	B3	B4
Cabernet Franc	960	1360	260	790	210	430
Gamay	1140	1140	620	930	710	1490
Merlot	640	790	200	480	80	210
Pinot Noir	2440	1930	560	900	300	590
Riesling	250	240	110	290	20	120

[a] Milligrams per kilogram.

Source: Adapted from Fuleki, T. and da Silva, J.M., 1997, *J. Agric. Food Chem.*, 45:1156–1160.

Souquet et al., 2000). The mean degree of polymerization for proanthocyanidins isolated from the seeds of grapes (cv. Cabernet franc) ranges from 4.7 to 17.4. For those isolated from grape skin, it is between 9.3 and 73.8 (Labarbe et al., 1999; Souquet et al., 2000) and for those extracted from grape stems between 4.9 and 27.6 (Souquet et al., 2000).

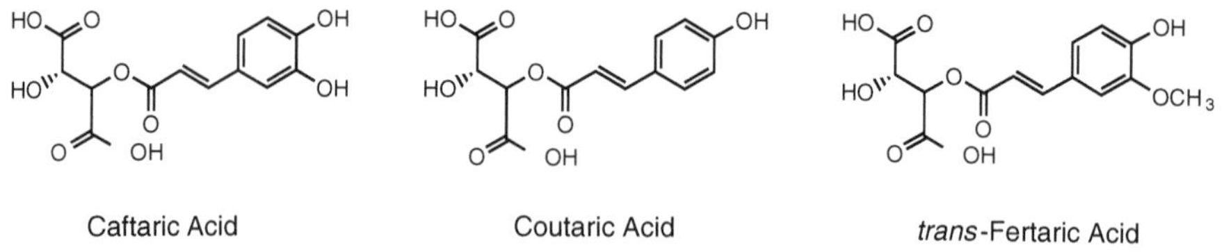

Caftaric Acid Coutaric Acid *trans*-Fertaric Acid

FIGURE 4.11 Chemical structures of caftaric, coutaric and *trans*-fertaric acids.

Using matrix-assisted laser desorption/ionization time-of-flight mass spectrometry (MALDI-TOF MS), Yang and Chien (2000) later observed the presence of procyanidin oligomers of up to nanomers in grape seed extracts. Similar results were obtained by Saucier et al. (2001), who reported that the mean degree of polymerization of proanthocyanidins isolated from seeds of cv. Merlot is 3.1 to 12.2. Furthermore, the galloylation extent of proanthocyanidins isolated from grape seed was much higher (20.4%) than that for proanthocyadins from grape stems (15.6%) and grape skins (2.7%) (Labarbe et al., 1999; Souquet et al., 2000). The composition of grape skin proanthocyanidins is influenced by the stage of berry ripening. The change in the color of berry from green to red effects an increase in the mean degree of polymerization in the (–)epigallocatechin extension units and in the level of anthocyanins associated with the proanthocyanidin fraction (Kennedy et al., 2001).

Other phenolics detected in whole grape berries, grape skins and stems include phenolic acids:

- Caftaric acid (*trans*-caffeoyltartaric acid)
- Coutaric acid (*p*-coumaryltartaric acid)
- *trans*-Fertaric acid (Figure 4.11) (Cantos et al., 2002; Singleton et al., 1978; Souquet et al., 2000; Vrhovšek, 1998)

flavonols:

- Quercetin 3-glucuronide
- Quercetin 3-glucoside (Figure 4.2)
- Myricetin 3-glucuronide
- Myricetin 3-glucuronide (Cheynier and Rigaud, 1986; Moskowitz and Hrazdina, 1981; Souquet et al., 2000)

and flavanonols:

- Astilbin (dihydroquercetin 3-rhamnoside)
- Engeletin (dihydrokaempferol 3-rhamnoside) (Figure 4.12) (Lu and Foo, 1999; Souquet et al., 2000; Trousdale and Singleton, 1983).

Grape berries contain from 269 to 467 mg/kg of total hydroxycinnamoyltartaric acids (HCAs). Caftaric acid (117.5 to 369.5 mg/kg) and *trans*-coutaric acid (55.3 to 93 mg/kg) are the predominant HCAs in berries, while *cis*-coutaric acid (11.8 to

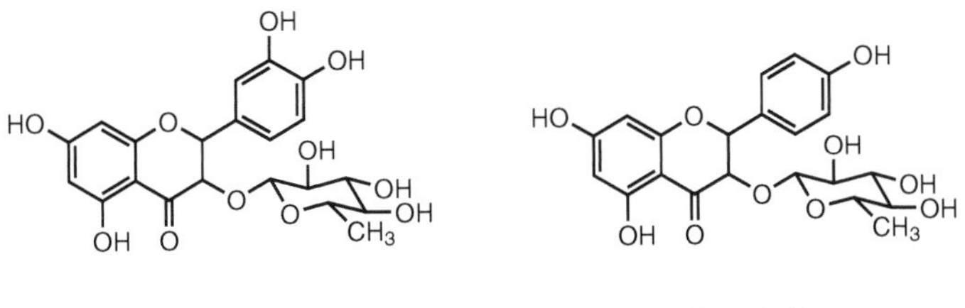

FIGURE 4.12 Chemical structures of some flavanonols.

21.0) and fertaric acids (1.7 to 16.8 mg/kg) are minor HCAs present (Vrhovšek 1998).

Stilbenes such as *trans*- and *cis*-resveratrols (3,5,4′-trihydroxystilbene), *trans*- and *cis*-piceids (3-*O*-β-D-glucosides of resveratrol), *trans*- and *cis*-astringins (3-*O*-β-D-glucosides of 3′-hydroxyresveratrol), *trans*- and *cis*-resveratrolosides (4′-*O*-β-D-glucosides of resveratrol) pterostilbene (a dimethylated derivative of stilbene) are grapevine phytoalexins found in grape leaves and berries (Figure 4.9). In berries stilbenes are mostly located in the grape skin (Jaendet et al., 2002; Langcake and Pryce, 1977; Langcake et al., 1979; Versari et al., 2001; Wang et al., 2002a; Waterhouse and Lamuela-Raventós, 1994). Of these, *cis*-piceid is the major stilbene found in berry skins during fruit ripening (39.5 mg/kg of fresh weight at 60 days after véraison), while resveratrol is the predominant stilbene in wilting berries (28 mg/kg at day 74) (Versari et al., 2001). Moreover, *trans*-piceid (3.38 mg/L) is the major resveratrol derivative in red grape juices, while *cis*-piceid (0.26 mg/L) occurs in white grape juices (Romero-Pérez et al., 1999). In grape berries, *p*-coumaroyl-CoA and malonyl-CoA serve as substrates for the synthesis of stilbene catalyzed by stilbene synthase (Fritzemeier and Kindl, 1981). The synthesis of stilbenes in grapevine is stimulated by stress factors such as fungal infection (*Botris cinerea*), injury, UV radiation, and wilting as well as such factors as grape cultivar, berry development stage, and soil practices (Bavaresco et al., 1997; Cantos et al., 2000, 2001b; Douillet-Breuil et al., 1999; Jaendet et al., 1991; Langcake and Pryce, 1976; Siemann and Creasy, 1992; Versari et al., 2001).

Jaendet et al. (1991) reported that a correlation exists between the accumulation of resveratrol in grape berries at different stages of development and that of grape leaves. Douillet-Breuil et al. (1999) noticed an increase in production of resveratrol and ε-viniferin (resveratrol (*E*)-dehydrodimer) in grape berry skins subjected to short UV-C radiation, but not piceid and pterostilbene. Subsequently, Cantos et al. (2000) reported that postharvest treatment of grape berries with UV-B or UB-C radiation increases the level of resveratrol in berries by two- or three-fold, respectively. Meanwhile, Adrian et al. (2000) noticed that UV-C radiation has a more pronounced effect on the production of stilbenes in healthy berries than in berries stressed by fungus. Later, Cantos et al. (2001b) reported an 11-fold increase in accumulation of resveratrol in berries treated with UV radiation pulses (total radiation power of 510W, radiation time of 30 sec, lamp distance of 40 cm from the berries) and then stored for 3 days at 20°C.

On the other hand, Versari et al. (2001) demonstrated that UV radiation of berries has a more pronounced effect on the synthesis of resveratrol during the early stages of fruit development than during ripening. Extensive fungal infection of grapes brings about biotransformation of stilbenes catalyzed by stilbene oxidase or grapevine peroxidases into ε-viniferin (resveratrol (*E*)dehydrodimer), α-viniferin, pallidol (Figure 4.9), leachinol F and restrytisols A to C (Breuil et al., 1998; Cichewicz et al., 2000; Morales et al., 1997; Versari et al., 2001). Later, Waffo-Teguo et al. (2001) isolated and identified two more products of stilbene oxidative degradation, namely, resveratrol (*E*)-dehydrodimer 11-*O*-β-D-glucopyranoside and resveratrol (*E*)-dehydrodimer 11′-*O*-β-D-glucopyranoside. In addition, they reported that resveratrol (*E*)-dehydrodimer 11-*O*-β-D-glucopyranoside and resveratrol (*E*)-dehydrodimer exhibit a nonspecific inhibitory activity against cyclooxygenase-1 and -2. Furthermore, *trans*- and *cis*-resveratrols display inhibitory effects against tyrosinase kinase (Jayatilake et al., 1993) and platelet aggregation (Chung et al., 1992), while piceids are effective against platelet aggregation (Chung et al., 1992; Orsini et al., 1997) and human LDL oxidation (Frankel et al., 1995).

Chloroanisoles are responsible for a musty off-flavor in stored raisins. These compounds may originate in raisins from chemical reactions and microbial and airborne contaminations (Maarse et al., 1988; Reineccius, 1991). The formation of chloroanisoles involves chlorination of phenols derived from the Shikimic pathway by fungal or host chloroperoxidases followed by their methylation (Liardon et al., 1991). Aung et al. (1996) demonstrated the formation of chloroanisoles in Thompson seedless, dry-on-the-vine raisins under sterile, nonsterile and $a_w < 0.8$ conditions. The raisins used in this study were sterilized with propylene oxide or hydrogen peroxide.

Torres and Bobet (2001) generated a series of novel flavanol antioxidant from grape by-products by subjecting the extracts of polymeric proanthocyanidins to depolymerization in the presence of cystamine. The terminal flavan-3-ols units were released as such, while the extension units were cleaved as thio-derivatives. Three major aminoethylthio-flavan-3-ols conjugates were isolated and identified in the reaction mixture: 4β-(2-aminoethylthio)epicatechin, 4β-(2-aminoethylthio)epicatechin 3-*O*-gallate, and 4β-(2-aminoethylthio)catechin (Figure 4.13). The antioxidant activities of these novel flavan-3-ol conjugates were similar to those displayed by their nonderivatized counterparts.

Raisins are important processed grape products. Italy, France and the U.S. are the world's largest producers of raisins (Pollack and Perez, 1997). Karadeniz et al. (2000) evaluated the effect on the composition of phenolic in raisins of sun-drying grapes (sun-dried raisins), dipping grapes into hot water (87 to 93°C) for 15 to 20 s before drying at 71°C for 20 to 24 h (dipped raisins), and dipping grapes into hot water followed by 5- to 8-h treatment with sulfur dioxide and then drying at 63°C (golden raisins). Oxidized hydroxycinnamic acids, formed upon the action of polyphenoloxidases, were only found in sun-dried and dipped raisins. According to Aguilera et al. (1987), at least a 2-min dip of berries into water at 93°C is required to inactivate polyphenoloxidases in Sultana grapes. The loss of hydroxycinnamic acids and flavonols during processing of grapes to raisins is in the order of 90 and 62%, respectively; procyanidins are degraded completely (Karadeniz et al., 2000).

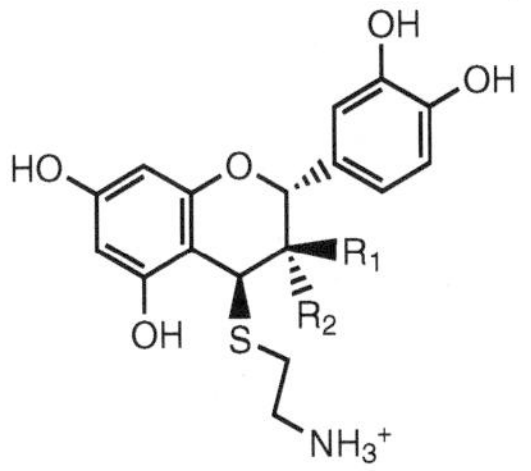

Compound	R_1	R_2
4β-(2-Aminoethylthio)-epicatechin	H	OH
4β-(2-Aminoethylthio)-catechin	OH	H
4β-(2-Aminoethylthio)-epicatechin-3-O-gallate	H	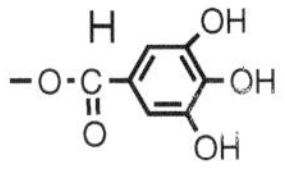 $-O-C(=O)-C_6H_2(OH)_3$

FIGURE 4.13 Chemical structures of 4-β-(2-aminoethylthio)-flavan-3-ols.

Cheynier and da Silva (1991) demonstrated that caftaric acid *o*-quinones may be involved in the oxidation of procyanidins through a coupled oxidation mechanism, leading to the reduction of caftaric acid quinones back to caftaric acid. Drying grape pomace may be an essential step in the utilization of this by-product for the production of pharmaceuticals. Larrauri et al. (1997) reported that drying red grape pomace at 60°C does not have a detrimental effect on the phenolics and their antioxidant activity. On the other hand, a significant reduction in the extractability of total phenolics and proanthocyanidins and their antioxidant activity was noted when grape pomace was dried at 100 and 140°C.

Raspberries

Raspberries (*Rubus idaeus* L.) are a rich source of phenolic compounds. The content of phenolics in raspberries may be influenced by cultivar, maturity, processing and geographic area of origin (Rommel et al., 1992; Wang and Lin, 2000). The total phenolics in ripe raspberries range from 1137 to 29,900 mg/kg of fresh weight expressed as gallic acid equivalents (de Ancos et al., 2000; Kähkönen et al., 2001; Liu et al., 2002; Moyer et al., 2002; Mullen et al., 2002b; Wang and Lin, 2000). Anthocyanins (Boyles and Wrolstad, 1993), flavonols and their conjugates (Henning, 1981; Rommel and Wrolstad, 1993a; Rommel et al., 1992), and ellagic acid and its derivatives (Rommel and Wrolstad, 1993b) are the major phenolics present in raspberries.

The total content of ellagic acid in fresh raspberry fruits ranges from 172.9 to 244.4 mg/kg of fresh weight (Amakura et al., 2000; de Ancos et al., 2000), while in raspberry juices ellagic acid is present at 5.7 to 80.4 μg/kg (Rommel and Wrolstad, 1993b). According to Häkkinen et al. (1999b), ellagic acid comprises 88% of the total phenolics; de Ancos et al. (2000) have reported that ellagic acid represents 13.8 to 19% of the total phenolic acids. Ellagic acid in raspberries is present in the free form as ellagic acid glycosides and as ellagitannins. Kähkönen et al. (2001) reported that raspberry fruits contain from 16,920 to 17,540 mg of ellagitannins/kg of fresh weight expressed as ellagic acid equivalents. Rommel et al. (1992) and Rommel and Wrolstad (1993b) detected 16 derivatives of ellagic acid in raspberry juices, but did not characterize these compounds any further. These authors suggested that a mixture of ellagic acid with varying methylation, methoxylation and glycolysation patterns, as well as ellagitannins, is present. Later, Zafrilla et al. (2001) and Mullen et al. (2002a) isolated five ellagic acid derivatives, but identified only three derivatives of ellagic acid, namely, 4-arabinoside, 4′(4″-acetyl) arabinoside and 4′-(4″-acetyl) xyloside.

In addition, several ellagitannins have been characterized in raspberries, namely, potentillin and casuarictin and their C-O oxidatively coupled dimeric forms as well as β–1,2,3-tri-*O*-galloyl-4,6-hexahydroxydiphenoyl-D-glucose, pedunculagin, and 2,3-di-*O*-galloyl-4,6-(hexahydroxydiphenoyl)-D-glucose (Figure 4.14) (Haddock et al., 1982; Haslam, 1989; Haslam and Lilley, 1985; Haslam et al., 1982). Subsequently, Mullen et al. (2002a, b) identified and quantified two ellagitannins in Glen Ample raspberries, namely, lambertianin C (18.7 μg/kg of gallic acid equivalent) and sanguiin H-6 (69.2 μg/kg of gallic acid equivalent), as major contributors to the antioxidant capacity of raspberries.

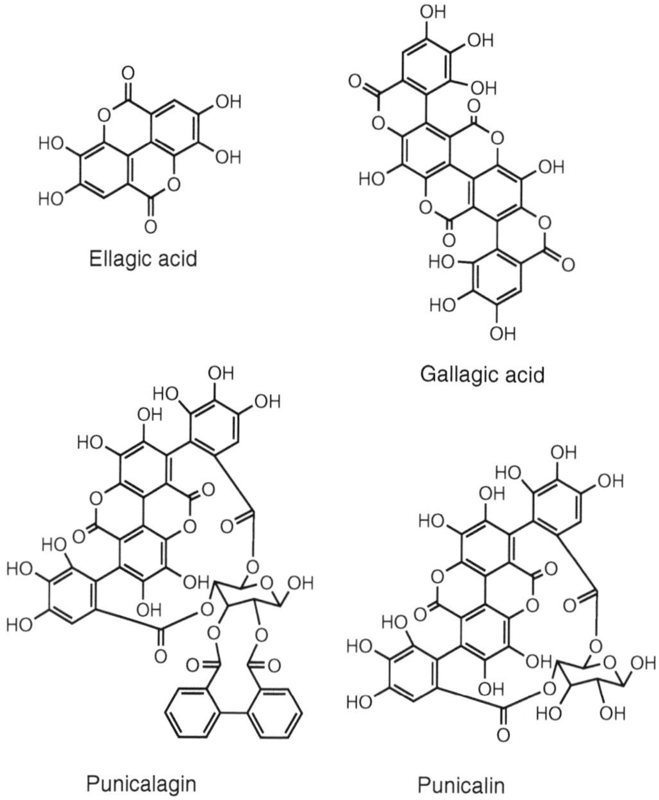

Ellagic acid

Gallagic acid

Punicalagin

Punicalin

FIGURE 4.14 Chemical structures of some phenolic compounds found in raspberries and pomegranate fruits.

The total flavonoid content in raspberry fruits is between 842 and 1034 mg/kg of fresh weight expressed as (+)catechin equivalents (Liu et al., 2002). Flavonols and their conjugates are the major flavonoids in raspberry. According to Häkkinen et al. (2000a), red raspberries contain 9.5 mg/kg of fresh weight of flavonols expressed as aglycones. Zafrilla et al. (2001) and Kähkönen et al. (2001) have reported values of 70 to 300 mg/kg of fresh weight of quercetin 3-glucoside. On the other hand, Rommel and Wrolstad (1993b) found that total content of flavonols in 55 samples of experimental and commercial red raspberry juices ranges from 15.7 to 286 μg/kg. Several flavonol glycosides have been identified by Ryan and Coffin (1971), Henning (1981) and Mullen et al. (2002a) in fresh raspberry fruits: quercetin 3-glucuronide, quercetin 3-glucoside, quercetin 3-xylosylglucuronide, quercetin 3-galactoside, methylquercetin-pentoside (probably xyloside) and kaempferol 3-glucoside. Later, Rommel and Wrolstad (1993a) detected more quercetin and kaempferol derivatives, but did not attempt to characterize them. Of these, quercetin 3-glucuronide is the predominant flavonol glycoside in red raspberry juices (Rommel and Wrolstad, 1993a)

Total anthocyanin levels in fruits are between 520 (red raspberry) and 6270 (black raspberry) mg/kg and in juices range from 1.7 (in yellow raspberry) to 1972 mg/kg (black raspberry) (Table 4.6) (Boyles and Wrolstad, 1993; Kähkönen et al., 2001; Liu et al., 2002; Moyer et al., 2002; Rommel et al., 1990; Wang and Lin, 2000; Wrolstad et al., 1993). Rommel et al. (1990) and Wrolstad et al. (1993) evaluated the anthocyanin profiles in selected commercial and laboratory-prepared red raspberry juices. Several anthocyanins were identified, namely, cyanidin-3-sophoroside, cyanidin-3-glucorutinoside, cyanidin-3-glucoside, pelargonidin-3-sophoroside, cyanidin-3-rutinoside and pelargonidin-3-glucorutinoside. In addition, de Ancos et al. (2000) detected malvidin-3-glucoside and delphinidin-3-glucoside in Spanish raspberry cultivars and Mullen et al. (2002a) reported the presence of cyanidin-3,5-diglucoside, cyanidin-3-sambubioside, cyanidin-3-xylosylrutinoside, pelargonidin-3-glucoside and pelargonidin-3-rutinoside in Glen Ample raspberries. Of these, cyanidin-3-sophoroside and cyanidin-3-glucoside are the predominant anthocyanins in raspberry fruits and juices.

Processing red raspberry into jam reduces the content of quercetin 3-glucoside by 6% and kaempferol 3-glucoside by 20%, while storage of jam accelerates these losses. After 6 months of storage, 40 and 50% of quercetin and kaempferol 3-glucosides are lost in the jam, respectively (Zafrilla et al., 2000). Short-term storage (3 days at 4°C or 24 h at 18°C), mimicking the possible routes the fresh fruits may take after harvest before reaching the consumer's table, brings about only an increase in the level of ellagitannins in the fruit (ellagic acid — 528%, lambertianin C — 145% and sanguiin H-6 — 118% of the values for fresh fruit) (Mullen et al., 2000b). Long-term frozen storage (365 days at –20°C) of raspberry fruits decreases the content of ellagic acid in the fruits by 14 to 21% (de Ancos et al., 2000). de Ancos et al. (2000) also noted that stability of anthocyanins during frozen storage is dependent on the harvesting time and observed a slight increase (5 to 17%) in the total anthocyanin content for early raspberry cultivars, but a slight decrease (4 to 17.5%) in total pigments for late raspberry cultivars. In addition, de Ancos et al. (2000) observed that cyanidin 3-glucoside was the least stable anthocyanin during long-term frozen storage of raspberry fruits. Cyanidin 3-glucoside is the most unstable

TABLE 4.6
Total Anthocyanin and Phenolics Content of Juices from Various Cultivars of Raspberry Fruits at Different Maturity Stages[a]

Cultivar	Maturity Stage	Total Anthocyanins[b]	Total Phenolics[c]
Jewel	Green	17 ± 6	3380 ± 71
(black raspberry)	Pink	228 ± 54	1900 ± 35
	Ripe	1972 ± 35	2670 ± 43
Autumn Bliss	Green	16 ± 5	1740 ± 37
(red raspberry)	Pink	64 ± 12	1040 ± 15
	Ripe	750 ± 38	2450 ± 61
Summit	Green	22 ± 6	2150 ± 41
(red raspberry)	Pink	159 ± 12	1220 ± 21
	Ripe	995 ± 38	2580 ± 56

[a] Milligrams per kilogram of fresh weight.
[b] Expressed as cyanidin 3-glucoside equivalents.
[c] Expressed as gallic acid equivalents.

Source: Adapted from Wang, S.Y. and Lin, H.-S., 2000, *J. Agric. Food Chem.*, 48:140–146.

and cyanidin 3-sophoroside the most stable pigment during fermentation of raspberry pulp; over 50% of total anthocyanins is lost after 2 months' storage of raspberry wine at 20°C (Rommel et al., 1990).

Strawberries

The total content of anthocyanins, the total flavonols and the total phenolics in strawberries grown in Finland ranges from 1840 to 2320 mg as cyanidin 3-glucoside equivalents/kg, 63 to 200 mg as rutin equivalents/kg and 16,000 to 24,100 mg as gallic acid equivalents/kg of fresh weight. On the other hand, the total ellagitaninns, total hydroxybenzoic acid and total hydroxycinnamic acid contents in these strawberries are 810 to 1840 mg of ellagic acid equivalents/kg, 110 to 500 mg of gallic acid equivalents/kg, and 470 to 630 mg of chlorogenic acid equivalents/kg of fresh weight, respectively (Kähkönen et al., 2001).

p-Coumarylglucose, quercetin 3-glucoside, quercetin 3-glucuronide, kaempferol 3-glucoside, kaempferol 3-glucuronide (Macheix et al., 1989), dihydroflavonol (Wang et al., 2002b) and ellagic acid (Maas et al., 1990) have been identified in strawberry fruits (Figure 4.2). Ellagic acid, quercetin and kaempferol derivatives are predominantly located in the external tissues, while a similar concentration of *p*-coumarylglucose has been found in the external and internal tissues of strawberry (Gil et al., 1997) (Table 4.7).

Several anthocyanins have been identified in strawberry fruits, namely, pelargonidin-3-glucoside, cyanidin-3-glucoside, and pelargonidin-3-rutinoside (Gil et al., 1997; Mazza and Miniati, 1993). In addition, Wang et al. (2002b) have detected the presence of pelargonidin-3-glucoside-succinate and cyanidin-3-glucoside-succinate.

TABLE 4.7
The Average Content of Selected Phenolics in External and Internal Tissues of cv. Selva Strawberry Fruit[a]

Phenolic	External Tissue	Internal Tissue
Quercetin derivatives	63.8 ± 5.6	3.3 ± 2.1
Kaempferol derivatives	21.8 ± 2.3	3.3 ± 0.6
Ellagic acid	33.3 ± 3.6	8.4 ± 1.1
p-Coumarylglucose	15.0 ± 5.2	17.2 ± 5.1
Total anthocyanins[b]	195.3 ± 11.3	55.1 ± 6.1

[a] Milligrams per kilogram of fresh weight.
[b] Expressed as cyanidin 3-glucoside equivalents.

Source: Adapted from Gil, M.I. et al., 1997, *J. Agric. Food Chem.*, 45:1662–1667.

Of these, pelargonidin-3-glucoside is the predominant anthocyanin in strawberry, comprising approximately 77% of total anthocyanin content in fruit (Gil et al., 1997; Mazza and Miniati, 1993; Wang et al., 2002b). However, pelargonidin-3-glucoside and cyanidin-3-glucoside contribute to the red color of strawberries (Timberlake and Bridle, 1982). Anthocyanins are mainly located in the external tissues of strawberry (Table 4.7) and cyanidin-3-glucoside is found only in the external berry tissues (Gil et al., 1997). Ripe strawberries contain about six times more anthocyanins than berries at their pink color stage (Wang and Lin, 2000).

Processing strawberries by cooking with sugar reduces the content of ellagic acid by 20% (Häkkinen et al., 2000b), and flavonols by 15 to 20% (Häkkinen et al., 2000a). On the other hand, postharvest storage of strawberry fruits for 24 h at 5°C increases their quercetin and kaempferol levels by 10 to 20% in comparison with berries stored for 24 h at 22°C. Prolonged frozen storage (9 months at –20°C) markedly increases the content of quercetin (by 35%), but has a detrimental effect on kaempferol, of which only traces are detected after 6 months of frozen storage. Similar changes in flavonols have been detected in strawberry jams stored up to 9 months at 5 and –20°C. The increase in quercetin level in frozen berries and strawberry jams has been associated with the degradation of cell wall structures (Häkkinen et al., 2000a). Storing strawberries in air increases the content of anthocyanins and flavonols in the external and internal tissues, while storage in CO_2-enriched atmospheres decreases the content of anthocyanins and flavonols, mostly in the internal berry tissues (Gil et al., 1997).

Other Berries

Berries of black chokeberry (*Aronia melanocarpa*) are a rich source of anthocyanins. Methanolic (MeOH +3% formic acid) extract from aronia berries contains 623 g/kg of dry extract of total anthocyanins. Cyanidin 3-galactoside (57%), cyanidin 3-arabinoside (29%), cyanidin 3-xyloside (7%), cyanidin 3-glycoside (4%) and cyanidin 3-glucoside (3%) are the major anthocyanins found in aronia extract. Small

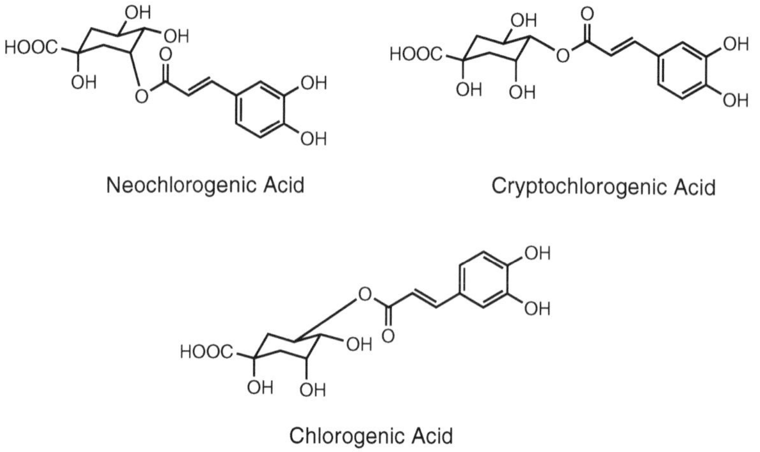

FIGURE 4.15 Chemical structures of chlorogenic acid isomers.

quantities of caffeic acid derivatives (3.22 g/kg) and quercetin derivatives (0.82 g/kg) have also been detected (Espin et al., 2000).

Drupes

Cherries

Sweet cherries (*Prunus avium* L.) and tart cherries (*Prunus cerasus* L.) are rich sources of flavonoids. Sweet cherries are grown commercially as table or maraschino cherries as well as ingredients of fruit cocktail (Mazza and Miniati, 1993). Schaller and Von Elbe (1970) identified six isomers of caffeoylquinic acid (Figure 4.15), four isomers of *p*-coumarylquinic acid and two free phenolic acids, in tart cherries cv. Montmorency. Möller and Herrmann (1983) reported that neochlorogenic acid and 3′-*p*-coumarylquinic acid are the major hydroxycinnamoylquinic acids in sweet cherries. Later, Gao and Mazza (1995) confirmed this finding and reported that the content of neochlorogenic acid in 11 cultivars and hybrids of sweet cherries ranges from 240 to 1280 mg/kg, while *p*-coumarylquinic acid ranges from 230 to 1310 mg/kg of pitted cherries. Subsequently, Wang et al. (1999a) identified three new phenolic compounds: 2-hydroxy-3-(*o*-hydroxyphenyl) propanoic acid, 1-(3′,4′-dihydroxycinnamoyl)-cyclopenta-2,5-diol and 1-(3′,4′-dihydroxycinnamoyl)-cyclopenta-2,3-diol (Figure 4.16), as well as chlorogenic acid methyl ester in ethyl acetate extracts of cv. Balaton cherries. The antioxidant activities of these phenolics are similar to those exhibited by commercial antioxidants such as TBHQ and BHT (Wang et al., 1999a).

The total content of anthocyanins in pitted dark cherries ranges from 82 to 298 mg/kg and in light-colored pitted cherries from a few to 41 mg/kg (Gao and Mazza, 1995). Tart cherries cv. Balaton (235.9 mg/kg) contain three times more anthocyanin than does tart cherry cv. Montmorency (75.3 mg/kg) (Wang et al., 1997). The presence of cyanidin 3-rutinoside and cyanidin 3-glucoside in sweet cherries

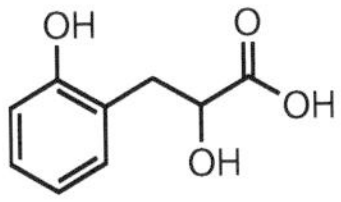

2-Hydroxy-3-(*o*-hydroxyphenyl)-propanoic Acid

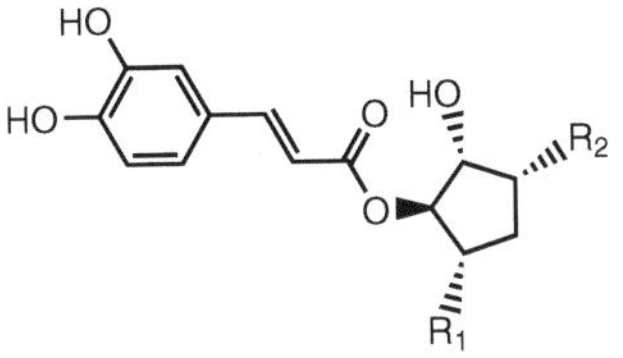

Compound	R_1	R_2
1-(3',4'-Dihydroxycinnamoyl)-cyclopenta-2,5-diol	OH	H
1-(3',4'-Dihydroxycinnamoyl)-cyclopenta-2,3-diol	H	OH

FIGURE 4.16 Chemical structures of some phenolic compounds isolated from sweet cherries.

was first detected by Willstästter and Zollinger (1916) and then confirmed by Robinson and Robinson (1931). Harborne and Hall (1964) detected cyanidin 3-glucosylrutinoside in six out of seven tart cherry varieties. Later, peonidin and its two glycosidic derivatives were also found in cv. Bing cherries (Lynn and Luh, 1964), while Okombi (1979) identified peonidin 3-rutinoside as the dominant anthocyanin in Bigarreau Napolèon. Furthermore, peonidin 3-rutinoside, peonidin, cyanidin, cyanidin 3-sophoroside, cyanidin 3-rutinoside and cyanidin 3-glucoside were found in tart cherry cv. Montmorency (Figure 4.5) (Chandra et al., 1992; Dekazos, 1970; Schaller and Von Elbe, 1968). Subsequently, Gao and Mazza (1995) reported that cyanidin 3-glucoside and cyanidin 3-rutinoside are the main anthocyanins, while peonidin 3-glucoside, peonidin 3-rutinoside and pelargonidin 3-rutinoside are the minor anthocyanins in dark sweet cherries. In addition, tart cherries contain cyanidin 3-glucosylrutinoside, cyanidin 3-rutinoside, and cyanidin 3-glucoside (Chandra et al., 2001; Wang et al., 1997).

The presence of kaempferol 3-glucoside and kaempferol 3-rutinoside in Montmorency cherries was reported by Schaller and Von Elbe (1970). Later, Shrikhande and Francis (1973) identified several more quercetin and kaempferol derivatives, namely, kaempferol-3-rhamnosylglucoside, quercetin-3-rhamnosylglucoside, quercetin-3-glucoside, quercetin-4′-glucoside, and tentatively detected kaempferol-3-rhamnoside-4′-glucoside, and kaempferol-4′-glucoside in sour cherries (*P. cerasus* L. cv. Montmorency). A number of flavonoids have also been detected in the bark of cherry trees (*P. cerasus*) (Geibel, 1995; Geibel and Feucht, 1991; Geibel et al., 1990):

- Pinostrobin
- Naringenin
- Prunin
- Sakuranetin
- Chrysin
- Sakuranin
- Naringenin
- Prunetin
- Genistein
- Tectochrysin
- Tectochrysin 5-glucoside
- Genistein 5-glucoside
- Prunetin 5-glucoside
- Dihydrowogonin 7-glucoside

Recently, Wang et al. (1999b) identified seven flavonoid compounds in Montmorency and Balaton tart cherries, namely, naringenin, genistein, quercetin 3-rhamnoside, rhamnasin, genistein 7-glucoside, kaempferol 6″-*O*-L-rhamnopyranosyl-β-D-glucopyranoside, and 6.7-dimethoxy-5,8,4′-trihydroxyflavone. Of these, 6.7-dimethoxy-5,8,4′-trihydroxyflavone and quercetin 3-rhamnoside are the most active antioxidants at 10-μM concentrations.

Nectarines and Peaches

Hydroxycinnamic acid derivatives, anthocyanins, flavonols and flavan-3-ols are the most prevalent phenolics in peaches and nectarines (Bengoechea et al., 1997; Cheng and Crisosto, 1995; Henning and Herrmann, 1980b; Hernandez, 1997; Luh et al., 1967; Mosel and Herrmann, 1974b; Risch and Herrmann, 1988b; Senter and Callaghan, 1990; Senter et al., 1989; Talcott et al., 2000a; Tomás-Barberán et al., 2001). Nectarines and peaches exhibit similar phenolic profiles (Tomás-Barberán et al., 2001). The content of total phenolics in whole peaches ranges from 213 to 1800 mg/kg of fresh weight (Carbonaro et al., 2002; Chang et al., 2000; Imeh and Khokhar, 2002; Van Buren, 1970). In addition, the total phenolics content in clingstone peach flesh is between 432.9 and 768 mg of gallic acid equivalents/kg, but varies in peach peels from 910.9 to 1922.9 mg of gallic acid equivalents/kg (Chang et al., 2000).

Chlorogenic acids, caffeic acid, catechin and procyanidin B3 (catechin-(4β→8)-catechin) are the major phenolics in peaches (Figure 4.1, Figure 4.3, Figure 4.4, and Figure 4.15) (Lee et al., 1990). More recently, Tomás-Barberán et al. (2001) found that chlorogenic and neochlorogenic acids are the main hydroxycinnamic acid derivatives present, while procyanidin B1 (epicatechin-(4β→8)-catechin), catechin and epicatechin are the predominant flavan-3-ols in peaches and nectarines. Higher levels of hydroxycinnamic acids, flavan-3-ols, and flavonols have been found in peach peel compared to peach flesh (Chang et al., 2000; Tomás-Barberán et al., 2001) (Table 4.8).

Anthocyanins are predominantly located in the skin of peaches and nectarines (Chang et al., 2000; Tomás-Barberán et al., 2001) (Table 4.8). Small quantities of pigments may also be found in tissues near the stone. Cyanidin 3-glucoside and

TABLE 4.8
Ranges of Phenolic Contents in the Peel and Flesh of Five Cultivars of Ripe White and Yellow Flesh Nectarines and Peaches[a]

Phenolic Compounds	White Flesh Cultivars				Yellow Flesh Cultivars			
	Nectarines		Peaches		Nectarines		Peaches	
	Peel	Flesh	Peel	Flesh	Peel	Flesh	Peel	Flesh
Hydroxycinnamic acids[b]	77.7–727.1	50.5–460.1	76.5–498.5	76.5–321.8	137.7–429.6	66.1–186.8	153.5–347.4	59.2–170.8
Flavan-3-ols[c]	133.6–1106	23.0–443.7	296.5–1166	117.1–695.8	128.0–621.2	30.2–214.6	225.4–663.0	93.2–357.0
Flavonols[d]	0–10.2	32.0–73.6	6.1–47.7	0–11.4	49.3–91.1	6.9–21.1	34.8–74.1	13.8–19.3
Anthocyanins[e]	99.0–172.7	0–6.7	54.4–142.6	0–17.6	33.5–260.9	0–31.4	85.6–273.6	0–8.7

[a] Milligrams per kilogram of fresh weight.
[b] Expressed as chlorogenic acid equivalents.
[c] Expressed as catechin equivalents.
[d] Expressed as quercetin 3-rutinoside.
[e] Expressed as cyanidin 3-rutinoside.

Source: Adapted from Tomás-Barberán, F.A. et al., 2001, *J. Agric. Food Chem.*, 49:4748–4760.

cyanidin 3-rutinoside are the main pigments in nectarines and peaches (Figure 4.5) (Hsia et al., 1965; Mazza and Miniati, 1993; Tomás-Barberán et al., 2001); some cultivars may also contain cyanidin 3-acetylglucoside and cyanidin 3-galactoside (Tomás-Barberán et al., 2001). Quercetin 3-glucoside and quercetin 3-rutinoside are the major flavonols in nectarines and peaches and are mainly located in the skin (Table 4.19) (Tomás-Barberán et al., 2001).

Manufacturing fruit juices consists of processing fruits to more stable products such as concentrates and purees and reconstituting to juices by adding water. Production of purees and concentrates may include removing stones, chemical peeling, enzymatic treatment, and pressing. Thus, juices made from peeled fruits may differ in their phenolic composition from those made from whole fruits. Chlorogenic acid, caffeic acid, protocatechuic acid, catechin, and procyanidin B2 (epicatechin-(4β→8)-epicatechin), B3 (catechin-(4β→8)-catechin) and B4 (catechin-(4β→8)-epicatechin) have been found in peach puree (Bengoechea et al., 1997).

Plums and Prunes

Plums include fruits of *Prunus domestica, Prunus salicina, Prunus spinosa, Prunus subcordiata,* and *Prunus insititia*; prune is the dried fruit of some cultivars of *P. domestica* (Pijipers et al., 1986). Phenolic compounds detected in plums include phenolic acid derivatives, coumarins, flavan-3-ols and anthocyanins (Henning and Herrmann, 1980a; Herrmann, 1989; Macheix et al., 1989; Stöhr et al., 1975). Of these, HCA derivatives are the major phenolics, comprising 84 to 90% of total phenolics (Donovan et al., 1998; Herrmann, 1989). Recently, Fang et al. (2002) identified over 30 HCAs in prunes. Of these, six are novel HCA isomers and contain two caffeic acids and quinic acid in their structures; however, the molecular structures of these compounds are not fully elucidated.

Chlorogenic acids (Figure 4.15) are the major phenolic acid derivatives in plums and prunes. Of these, neochlorogenic acid (3-*O*-caffeoylquinic acid, 3-CQA) is the major chlorogenic acid in plums and prunes (Donavan et al., 1998; Fang et al., 2002; Herrmann, 1989; Möller and Herrmann, 1983; Nakatani et al., 2000; Raynal et al., 1989; Tomás-Barberán et al., 2001). Möller and Herrmann (1983) detected 88 to 731 mg/kg of 3-CQA, 15 to 129 mg/kg of chlorogenic acid (5-*O*-caffeoylquinic acid, 5-CQA) and 56 mg/kg of cryptochlorogenic acid (4-*O*-caffeoylquinic acid, 4-CQA) in fresh plums. Following this, Raynal et al. (1989) reported that 3-CQA comprises approximately 50% of total phenolics in exocarp and pulp of fresh plums. Later, Donovan et al. (1998) found 807and 1306 mg/kg of 3-CQA and 144 and 436 mg/kg of 5-CQA in fresh prune-making plums and pitted prunes, respectively. Subsequently, Nakatani et al. (2000) detected 3-CQA (1228 to 1485 mg/kg), 4-CQA (288 to 351 mg/kg) and 5-CQA (53 to 77 mg/kg) in prunes. The contents of neochlorogenic and chlorogenic acids are 2.4 and 12.4 times higher in exocarp than in the pulp (Raynal et al., 1989)

Recently, Tomás-Barberán et al. (2001) have also reported that hydroxycinnamic acid derivatives are mainly located in the plum peel (Table 4.9). The total HCA content in the peel and the flesh of plums ranges from 115 to 375 mg/kg and from 16.3 to 194 mg/kg of sample, respectively. In addition, a number of minor phenolic

TABLE 4.9
Content of Hydroxycinnamic Acid Derivatives and Flavan-3-ols in the Peel and Flesh of Some Plum Cultivars[a]

Cultivar	Part	3-CQA[b]	Total HCA[c]	A-type	Procyanidin dimers[d] B1	B2	B4	Total Flavan-3-ols
Angeleno	Peel	40 ± 11	115 ± 15	477 ± 66	81 ± 15	340 ± 70	52 ± 29	1151 ± 208
	Flesh	17 ± 4	17 ± 4	ND	31 ± 15	120 ± 42	154 ± 55	377 ± 126
Black Beauty	Peel	225 ± 0.3	282 ± 2	257 ± 10	292 ± 4	137 ± 3	668 ± 24	1650 ± 20
	Flesh	103 ± 3	113 ± 3	103 ± 9	136 ± 24	88 ± 5	137 ± 5	566 ± 51
Red Beauty	Peel	326 ± 20	375 ± 25	75 ± 10	268 ± 62	72 ± 19	253 ± 52	847 ± 171
	Flesh	184 ± 6	194 ± 6	37 ± 3	41 ± 11	41 ± 3	42 ± 2	182 ± 15
Wickson	Peel	135 ± 7	181 ± 11	78 ± 6	540 ± 13	nd	485 ± 28	1534 ± 54
	Flesh	87 ± 4	90 ± 4	ND	110 ± 17	nd	46 ± 6	156 ± 23

Note: ND = not detected.

[a] Milligrams per kilogram of fresh weight.
[b] Neochlorogenic acid expressed as chlorogenic acid.
[c] Hydroxycinnamic acid derivatives expressed as chlorogenic acid.
[d] Expressed as catechin.

Source: Adapted from Tomás-Barberán, F.A. et al., 2001, *J. Agric. Food Chem.*, 49:4748–4760.

acids and their glycosides have been identified in prunes: gallic, protocatechuic, *p*-hydroxybenzoic, vanillic, methoxybenzoic, caffeic, syringic, methoxycinnamic, *p*-coumaric and ferulic acids (Figure 4.1) (Fang et al., 2002; Kayano et al., 2002). Drying plums brings about degradation of HCA linked to the polyphenoloxidase activity in the fruit because more HCAs are degraded when plums are dried at a lower drying temperature. Furthermore, HCAs present in the exocarp disappear more rapidly than those found in the flesh of the fruit and neochlorogenic acid degrades more rapidly during drying than other chlorogenic acid isomers (Raynal et al., 1989).

The anthocyanins in the ripe fruit of *P. domestica* L. are mainly located in the skin and the outer cells of the flesh (Mazza and Miniati, 1993). The total content of anthocyanins in the peel and the flesh of ripe plums is 129 to 1614 and 0 to 28.4 mg/kg of fresh weight, respectively (Tomás-Barberán et al., 2001). Presence of four anthocyanins, namely, 3-glucosides and 3-rutinosides of cyanidin and peonidin, has been detected in Victoria plums and in sloe (*P. spinosa)* (Figure 4.5) (Van Buren, 1970). Presence of the same anthocyanins in plums was reported by Macheix et al. (1989) and Raynal et al. (1989). Later, Hong and Wrolstad (1990) detected seven anthocyanins in a plum juice concentrate, but fully characterized only two of them: cyanidin 3-glucoside and cyanidin 3-rutinoside. Subsequently, Tomás-Barberán et al. (2001) detected 3-glucoside, 3-rutinoside, 3-galactoside and 3-acetylglucoside of cyanidin, but failed to find peonidin derivatives in plums. Of these, cyanidin 3-glucoside and cyanidin 3-rutinoside are the major anthocyanins in plums.

Anthocyanins degrade rapidly during drying of plums. For example, Raynal et al. (1989) detected only 15% cyanidin 3-rutinoside and 17% peonidin 3-rutinoside in plums dried at 95°C for 1 h. The rate of degradation of anthocyanins increased as the drying temperature increases. No anthocyanins have been found in prunes and prune juices (Fang et al., 2002; Raynal and Moutounet, 1989; Van Gorsel et al., 1992). It has been suggested that loss of anthocyanins involves enzymatic degradation attributed to the presence of quinones and thermal degradation linked with the increase in chalcones (Raynal and Moutounet, 1989).

The total content of flavonol glycosides (Figure 4.3) in a ripe plum ranges from 186 to 352 mg of rutin equivalents/kg of fresh weight (Tomás-Barberán et al., 2001). The fruit of *P. domestica* and *P. salicina* contains 3-arabinoside-7-rhamnosides, 3-rutinosides, 3-glucosides, and 3-galactosides of quercetin and kaempferol, as well as quercetin 3-xyloside. On the other hand, quercetin glycosides have only been detected in the fruit of *P. spinosa* (Macheix et al., 1989). Rutin (quercetin 3-rutinoside) is the predominant flavonol glycoside in plums and prunes (Fang et al., 2002; Henning and Herrmann, 1980a, b; Raynal et al., 1989; Starke and Herrmann, 1976) and is only found in the exocarp (Raynal et al., 1989).

According to Mosel and Herrmann (1974b), plums contain only 20 to 40 mg of flavan-3-ols/kg of fresh weight. Recently, Tomás-Barberán et al. (2001) have reported that the total content of flavan-3-ols in plums ranges from 662 to1837 and from 138 to 618 mg of catechin equivalents/kg of fresh weight of peel and fruit flesh, respectively. A similar content of flavan-3-ols has been reported by de Pascual-Teresa et al. (2000). No detectable levels of flavan-3-ols have been found in prunes, however (Fang et al., 2002). The discrepancies in the reported values may be due to the differences in the methodologies applied for their quantitation. The major flavan-3-ols identified in plums are procyanidins B1, B2, B4 as well as A-type dimers (Figure 4.4 and Figure 4.10) (Tomás-Barberán et al., 2001).

POMES

Apples

The content of phenolics in apples is affected by variety, maturity, harvesting season, processing, cultural conditions, crop load, development of infection, fruit position within the canopy, and geographic origin (Awad et al., 2000a, b, 2001; Burda et al., 1990; Escarpa and González, 1998; Ingle and Hyde, 1968; Lattanzio et al., 2001; Lea and Beech, 1978; Lea and Timberlake, 1978; Macheix et al., 1989; Mangas et al., 1999; Mayr et al., 1995; McRae et al., 1990; Mosel and Herrmann, 1974a; Pinicelli et al., 1997; Spanos et al., 1990; Stopar et al., 2002; van der Sluis et al., 2001). Hydroxycinnamic acid derivatives, flavan-3-ols (monomeric and oligomeric), flavonols and their conjugates, and dihydrochalcones are the major phenolics in apples (Table 4.10) (Dick et al., 1987; Duggan, 1969c; Foo and Lu, 1999; Guyot et al., 1997, 1998, 2001; Lea, 1978, 1984, 1990; Lea and Arnold, 1978; Lea and Timberlake, 1974; Lister et al., 1994; Lu and Foo, 1997; McRae et al., 1990; Nicolas et al., 1994; Ohnishi-Kameyama et al., 1997; Oleszek et al., 1988; Pérez-Ilzarbe et al., 1991, 1992; Schmidt and Neukom, 1969; Spanos and Wrolstad, 1992; Tomás-Barberán et al., 1993; Van Buren, 1970; Van Buren et al., 1976; Vallés et al., 1994; Whiting and Coggins, 1975; Wilson, 1981). In addition, anthocyanins are found in the skin of some red apple varieties (Alonso-Salces et al., 2001; Lancaster, 1992; Soji et al., 1999; Sun and Francis, 1967; Timberlake and Bridle, 1971; van der Sluis et al., 2001), but not in the vacuoles of epidermal and subepidermal cells (Strack and Wray, 1994). Of these, only procyanidins are able to bind to cell wall matrix (Renard et al., 2001).

Phenolics are involved in the defense mechanism in apple against fungal pathogens such as *Venturia* sp., *Gloeosporium* sp., *Sclerotinia fructigena*, and *Botrytis cinerea.* Infection of apple tissue brings about an increase in polyphenol oxidase activity leading to acceleration of polyphenol oxidation. Thus, the oxidation products of polyphenols play an important role in apple tissues' resistance to pathogens (Feucht et al., 1992; Ju et al., 1996; Lattanzio et al., 1994, 2001; Nicholson and Hammerschmidt, 1992; Noveroske et al., 1964a, b; Oszmianski and Lee, 1991; Piano et al., 1997; Prusky, 1996; Raa and Overeem, 1968). Recently, Whitaker et al. (2001) identified two novel fatty acid esters of phenolic alcohols in epiticular wax of apple fruit as *E* and *Z* isomers of *p*-coumaryl alcohol.

The total content of phenolics in epidermis, parenchyma, core, and seeds of apple is above 5000 mg/kg (Burda et al., 1990; Golding et al., 2001; Guyot et al., 1998), but most phenolics found in the entire apple are located in the parenchyma (64.9%) (Table 4.11) (Guyot et al., 1998). The total content of phenolics in most apple varieties is between 1000 and 6000 mg/kg of fresh weight, but in some apple cultivars may be 10,000+ mg/kg of fresh weight (Table 4.10) (Gorinstein et al., 2001; Herrmann, 1973; Podsędek et al., 2000; Sanoner et al., 1999; Van Buren, 1970); in the apple peel total phenolics range from 1275 (Granny Smith) to 4036

TABLE 4.10
Content of Phenolic Compounds in Cortex of Some Apple Varieties[a]

Variety	Total Phenolics[b]	Total Catechins[b]	Procyanidins[b]	Chlorogenic Acid[c]	p-Coumarylquinic Acid[c]	Total Dihydrochalcones[d]
Golden Delicious	1280	102	761	132	20	26
Judor	1100	106	515	338	59	26
Guillevic	1740	Traces	1066	465	134	38
Petit Jaune	2220	177	1372	415	29	28
Binet Rouge	2500	241	1254	601	176	51
Juliana	2250	154	1551	522	72	122
Clozette	2400	351	1144	832	84	40
Avrolles	2560	Traces	1528	154	104	80
Dous Moen	3130	418	1902	967	100	39
Antoinette	3350	466	1796	641	36	55
Bedan	3210	627	2417	649	147	42
Jeanne Renard	6000	1464	4731	666	50	82

[a] Milligrams per kilogram of fresh weight.
[b] Catechin + epicatechin; expressed as (–)epicatechin equivalents.
[c] Assayed according to their own response factors.
[d] Phlorizidin + phloretin xyloglucoside; expressed as phlorizidin equivalents.

Source: Adapted from Sanoner, P. et al., 1999, *J. Agric. Food Chem.*, 47:4847–4853.

TABLE 4.11
Distribution of Each Class Phenolic Compounds among Various Tissues of the Kermerrien Cider Apple Fruit[a]

	Epidermis Tissue (%)	Parenchyma Tissue (%)	Core (%)	Seeds (%)	Total in Apple Fruit (mg/kg fresh weight)
Procyanidins	27.6	65.0	7.1	0.3	3236
Catechins[b]	17.8	72.1	9.9	0.01	739
Flavonols[c]	100.0	ND	ND	ND	23
Dihydrochalcones[d]	14.2	27.9	39.6	17.8	169
Hydroxycinnamic acids[e]	14.7	67.5	17.4	0.4	837
Total phenolics	23.8	64.9	10.3	1.0	5004

Note: ND = not detected.

[a] Percent of total phenols in apple fruit.
[b] Expressed as catechin equivalents.
[c] Expressed as hyperoside equivalents.
[d] Expressed as phloridzin equivalents.
[e] Expressed as chlorogenic acid equivalents.

Source: Adapted from Guyot, S. et al., 1998, *J. Agric. Food Chem.*, 46:1698–1705.

(Lady Williams) mg/kg of fresh weight (Golding et al., 2001). Total content of phenolics is also influenced by crop load of the apple tree. Decrease in crop load for var. Jonagold (*Malus x domestica* Borkh.) from 157 to 30 fruits per crown increased the total content phenolics from 1300 to 1680 mg/kg of fresh weight (cortex + skin) (Stopar et al., 2002). On the other hand, the total content of phenolics in 46 samples of juices made of Spanish cider apple juice cultivars varied from 750 to 2420 (harvest 1995) and from 570 to 20,600 mg/L (harvest 1994) of tannic acid equivalents (Mangas et al., 1999). Furthermore, the level of phenolic compounds varies among individual apples of the same variety. A 10 to 30% variation in the content of chlorogenic acid, phlorizidin, epicatechin and quercetin glycosides between individual Jonagold apples has been detected by van der Sluis et al. (2001).

Hydroxycinnamic acids (HCAs; Figure 4.1) are important substrates for polyphenol oxidase. Their oxidation and condensation products not only contribute to the formation of yellow-brown pigments in apple fruits but may also enhance the oxidation of other phenolics such as flavan-3-ols by coupled mechanism (Amiot et al., 1992, 1993; Brugirard and Tavernier, 1952; Nicolas et al., 1994; Oszmianski and Lee, 1990). On the other hand, chlorogenic acid (Figure 4.15) in combination with phlorizidin (Figure 4.17) and polyphenol oxidase may contribute to quiescence (limited infections with no lesion produced) (Prusky, 1996) of the *Phlyctaena vagabunda* while the fruit is on the tree (Lattanzio et al., 2001). In addition, microbial decomposition of HCAs by bacteria of genus *Lactobacillus* and yeasts of genera *Saccharomyces* and *Brattanomyces* may lead to the formation of volatile phenols,

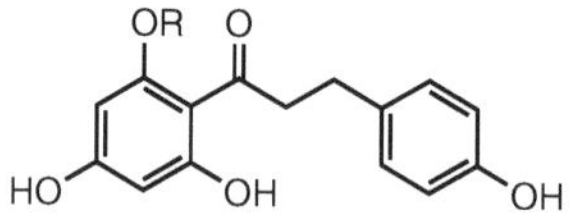

OR O

HO OH OH

R = H; Phloretin
R = glucose; Phloridzin

FIGURE 4.17 Chemical structures of dihydrochalcones found in apple fruits.

accumulation of which above their flavor thresholds is responsible for the development of off-flavor in cider juice (Beech and Carr, 1977; Chatonnet et al., 1992, 1993).

HCAs are mainly located in the parenchyma zone of the fruit (Table 4.11). Chlorogenic acid is the major HCA found in the apple fruit and accounts for 79, 76, 79, and 87% of total HCA in epidermis, parenchyma, core, and seeds, respectively (Guyot et al., 1998). Other isomers of chlorogenic acid and *p*-coumarylquinic acid isomers, as well as *p*-coumaric, ferulic, and caffeic acids, have also been detected in apples (Bengoechea et al., 1997; Durkee and Poapst, 1965; Nicolas et al., 1994). Podsędek et al. (2000) recently reported that the content of chlorogenic acid (CA) in 10 Polish apple varieties ranged from 30 to 430 mg/kg of fresh weight in eight apple varieties grown in the U.S. and varied from 1.5 to 228 mg/kg of fresh weight (Coseteng and Lee, 1987; Lee and Wrolstad, 1988); in 14 French apple varieties (12 cider and 2 juice) it was between 132 and 967 mg/kg of fresh weight (Table 4.10) (Sanoner et al., 1999). Apple purées and concentrates contain 25.8 to 68.2 and 38.9 to 81.3 mg/L of CA (Bengoechea et al., 1997), respectively, while the concentration of CA in juices pressed from 46 apple juice varieties ranges from 21.2 to 350.5 (harvest 1995) and 25.1 to 377.1 (harvest 1994) mg/L (Mangas et al., 1999).

The presence of several anthocyanins, namely, cyandin 3-galactoside (ideain), cyanidin-3-rutinoside, malvidin-3-glucoside, and malvidin-3,5-diglucoside, has been confirmed in Starking Delicious apple juice. Of these, cyanidin-3-galactoside (ideain) is the predominant anthocyanin in the apple juice and peel (Alonso-Salces et al., 2001; Shoji et al., 1999). Fermentation of Starking Delicious apple juice markedly decreases the concentration of ideain, but induces the formation of anthocyanin polymers. Further, a decrease in cyanidin-3-galactoside content has been noted during maturation of wine (Shoji et al., 1999).

Flavonol glycosides are quantitatively considered to be a minor class of phenolics and are found predominantly in the epidermis tissue of apple fruits (Table 4.11) (Guyot et al., 1998; Lister et al., 1994; Pérez-Ilzarbe et al., 1991; Price et al., 1999; van der Sluis et al., 2001). These phenolics may be involved in the regulation of apple texture as β-galactosidase inhibitors (Dick and Smith, 1990; Dick et al., 1985; Lidster and McRae, 1985, Lidster et al., 1986). Several quercetin glycosides have been identified in apple fruit: rutin, hyperin (quercetin-3-β-D-galactoside), isoquercitrin (quercetin-3-β-D-glucoside), reynoutrin (quercetin-3-β-D-xyloside), avicularin (quercetin-3-α-L-arabinofuranoside), and quercitrin (quercetin-3-α-L-rhamnoside) (Figure 4.2) (Burda et al., 1990; Coseteng and Lee, 1987; Dick et al., 1987; Golding et al., 2001; Guyot et al., 1998; Lommen et al., 2000; Oleszek et al., 1988; van der Sluis et al., 2001). In addition, Lommen et al. (2000) have detected one quercetin glycoside in apple peel and tentatively identified it as quercetin 3-β-L-arabinopyranoside or 3-β-L-apiofuranoside. More precise identification of this flavonol glycoside is still needed.

Guyot et al. (1998) detected 129 mg of total flavonols/kg of peel, but only 23 mg/kg of fresh weight of the entire Kermerrien cider apple. Subsequently, Lommen et al. (2000) reported that the content of individual flavonol glycosides in apple peel ranges from 200 to 500 mg/kg of peel. Later, van der Sluis et al. (2001) found 60 to 98 mg of total flavonol glycosides/kg of apple (fresh weight), while the level of individual flavonol glycosides in Jonagold apple peel was between 20 (isoquercitrin) to 159 (avicularin) mg/kg of peel. Moreover, Alonso-Salces et al. (2001) reported that flavonol glycosides are one of the main classes of phenolics in the peel of 14 Basque cider apple varieties, comprising 30 to 40% of total polyphenols in the peel. Significant levels of flavonol glycosides have also been found in apple juices produced by diffusion extraction technology (Spanos et al., 1990).

Phloridzin (phloretin 2′-β-D-glucoside) and phloretin 2′-β-D-xylosyl-(1→6)- β-D-glucoside are the major dihydrochalcones found in apple fruits (Figure 4.17) (Durkee and Poapst, 1965; Lommen et al., 2000). Alonso-Salces et al. (2001) have reported the presence of two additional dihydrochalcones in the apple peel; however, they have not attempted to identify the compounds involved. The orange oxidation products of dihydrochalcones contribute to the color of apple and cider juices (Lea, 1984; Oszmianski and Lee, 1991). Dihydrochalcones are distributed in the entire fruit (Table 4.11); however, in the seeds (3416 mg/kg of seed) they comprise about 66% of total phenols, while in the parenchyma and epidermis tissues they comprise less than 3% of total phenols in each tissue (Guyot et al., 1998).

On the other hand, Alonso-Salces et al. (2001) have reported that dihydrochalcones constitute 17 to 57% and 4 to 13% of total phenolics in the peel and pulp of 15 Basque apple varieties, respectively. The total content of dihydrochalcones in apple cortex ranges from 26 to 122 mg of phloridzin equivalents/kg of fresh weight (Table 4.10) (Guyot et al., 1998; McRae et al., 1990; van der Sluis et al., 2001); in the peel it is between 60 and 500 mg/kg (Burda et al., 1990; Guyot et al., 1998; Lommen et al., 2000). In addition, industrially manufactured apple purées contain 10.4 to 29.0 mg/L of phloridzin, 3 to 7.4 mg/L of phloretin 2′-xyloglucose, and 0.8 to 2.5 mg/L of unknown phloretin derivatives (Bengoechea et al., 1997).

The only catechins detected in apples are (+)catechin and (–)epicatechin (Figure 4.3). Catechins are prone to enzymatic oxidation and together with chlorogenic acid contribute to the development of brown pigments in damaged apples (Goupy et al., 1995, Macheix et al., 1990; Nicolas et al.1994). The content of catechins is high during early stages of apple fruit development and then drops drastically as the fruit size increases (Mosel and Herrmann, 1974a). Catechins account for 6 to 21% of total phenolics in the cortex of ripened fruits (Table 4.10) (Sanoner et al., 1999). The content of catechin in the cortex of fruit apple ranges from 0 to 154 mg/kg of fresh weight, while epicatechin is present between 0 and 1410 mg/kg of fresh weight (Coseteng and Lee, 1987; Mangas et al., 1999; Podsędek et al., 2000; Sanoner et al., 1999; van der Sluis et al., 2001). Moreover, depending on variety, the apple skin may contain up to10 times more epicatechins than the flesh (Burda et al., 1990).

Procyanidins are found in the entire apple fruit, but their concentration gradually increases from 1232 mg/kg in the seeds to 4964 mg/kg in the epidermal tissue (Guyot et al., 1998). The occurrence of high levels of procyanidins in the epidermis tissue may be due to their protective role against fungi (Feucht et al., 1992). Procyanidins detected in the epidermis and parenchyma tissues account for 65 and 74.5% of the total phenolics in each tissue, respectively. However, globally, the procyanidins found in the parenchyma tissue comprise 65% of the total procyanidins in the entire fruit (Table 4.11) (Guyot et al., 1998). Apple procyanidins are a mixture of oligomers and polymers made of (–)epicatechin and (+)catechin as constitutive units (Figure 4.8) (Lea, 1990). These phenolic compounds are responsible for the bitterness and astringency of cider juice, and partly for the formation of haze in apple juice (Beveridge and Tait, 1992; Beveridge et al., 1997; Lea, 1990; Lea and Arnold, 1978). Lea and Arnold (1978) related the bitterness and astringency of English ciders to the degree of procyanidin polymerization. Low molecular weight procyanidins (up to tetramers) contribute to the bitterness of cider apples; high molecular weight procyanidins affect astringency. In addition, the balance between astringency and bitterness defines the "mouthfeel" and body of cider apples (Lea, 1990).

Lea and Timberlake (1974) isolated three dimeric procyanidins, B1, B2 and B5 (Figure 4.4), and two unknown trimers in English ciders. Following this, Lea (1978) identified one of the procyanidin trimers as C1 (Figure 4.10) and also detected the presence of more polymerized procyanidins with at least seven-fold degree of polymerization. Subsequently, Vallés et al. (1994) detected two dimeric procyanidins, B1 and B2, one trimer, C1, one tetramer and one unknown procyanidin in juices made from Spanish apple varieties. Bengoechea et al. (1997) identified only two dimeric procyanidins, B1 and B2, in apple purées, but not in apple concentrates.

Later, Guyot et al. (1998) reported that high molecular weight procyanidins represent over 26% of total procyanidins in the apple fruit and those with the highest average degree of polymerization (11.2) are located in the seeds. On the other hand, the average degree of polymerization in apple tissues ranges from 4.4 for procyanidins in core to 6.0 for procyanidins in the epidermis. Subsequently, Sanoner et al. (1999) detected more polymerized procyanidins in the cortex of French apple varieties with the average degree of polymerization varying from 4.5 (var. Judor) to 50.3 (var. Avrolles). Later, Guyot et al. (2001) reported the presence of procyanidins with an average degree of polymerization of up to 190 in the cortex of two cider apple varieties (*Malus domestica;* Kermerrien and Avrolles). These authors reported that only oligomeric procyanidin fractions with average degrees of polymerization of 2 to 8 are eluted using normal-phase HPLC, while more polymeric procyanidin fractions with average degree of polymerization ranging from 7 to190 are isolated on reversed-phase HPLC.

Pears

The phenolic constituents of pears are hydroxycinnamic acids, arbutin (4-hydroxyphenyl-β-D-glucopyranoside), flavonols, catechins and procyanidins. The average concentration of phenolic compounds in Portuguese pear is 3.7 g/kg of fresh pulp (Ferreira et al., 2002). Presence of chlorogenic acid isomers, *p*-coumarylquinic and dicaffeoylquinic acids in pears has been reported by several researchers (Amiot et al., 1992; Billot et al., 1978; Cartwright et al., 1955; Challice and Williams, 1972; Hulme, 1958; Mosel and Herrmann, 1974a; Sioud and Luh, 1966; Spanos and Wrolstad, 1992;

Wald et al., 1989). Oleszek et al. (1994) have identified *p*-coumarylmalic acid while Andrade et al. (1998) found *p*-hydroxybenzoic acid in pears. Of these, 3-*O*-caffeoylquinic and 5-*O*-caffeoylquinic acids are the major phenolic acids of pear puree (Andrade et al., 1998). The content of caffeoylquinic acid in pears ranged from 40 to 140 mg/kg of fresh weight (Amiot et al., 1995; Blankenship and Richardson, 1985; Ferreira et al., 2002).

Pears also contain a mixture of quercetin, isoquercetin, kaempferol and isorhamnetin (3′-methylether of quercetin) glycosides (Figure 4.2) (Andrade et al., 1998; Duggan, 1967, 1969a, b; Herrmann, 1976; Nortje and Koeppen, 1965; Oleszek et al., 1994; Sioud and Luh, 1966; Spanos and Wrolstad, 1992). Flavonols are mostly located in the pear skins; Herrmann (1976) reported that the skin and peel of Williams Christ pears contain 28 mg/kg of quercetin and 12 mg/kg of kaempferol, while the total content of flavonols in remaining pear tissue did not exceed 0.1 mg/kg. Nortje and Koeppen (1965) detected five flavonols in Bon Chretien pears and positively identified four of them: isoquercetin, isorhamnetin 3-*O*-glucoside, isorhamnetin 3-*O*-rhamnoglucoside, and isorhamnetin 3-*O*-rhamnogalactoside. Later, Sioud and Luh (1966) detected only two flavonols in Barlett pears, identifying one as quercetin 3-*O*-glucoside.

Subsequently, Duggan (1969b) confirmed the presence of quercetin 3-*O*-glucoside in Packingham pears and detected seven additional flavonols. Of these, four were positively identified as isorhamnetin 3-*O*-glucoside, isorhamnetin 3-*O*-rutinoside, isorhamnetin 3-*O*-galactoside and quercetin 7-xyloside. Oleszek et al. (1994) detected three quercetin and five isorhamnetin glycosides in pears:

- Quercetin 3-*O*-rutinoside
- Quercetin 3-*O*-glucoside
- Quercetin 3-*O*-malonylglucoside
- Isorhamnetin 3-*O*-rutinoside
- Isorhamnetin 3-*O*-galactorhamnoside
- Isorhamnetin 3-*O*-glucoside
- Isorhamnetin 3-*O*-malonylgalactoside
- Isorhamnetin 3-*O*-malonylglucoside

In addition, Andrade et al. (1998) reported the presence of quercetin 3-*O*-galactoside and quercetin 3-*O*-rhamnoside in Rocha pear puree.

Other phenolic compounds found in pears include (+)catechin and (−)epicatechin (Figure 4.3) (Andrade et al., 1998; Mosel and Herrmann, 1974a; Ranadive and Haard, 1971; Sioud and Luh, 1966). The content of monomeric flavan-3-ols in pears is 13 to 92 mg/kg of fresh flesh; (−)epicatechin is the most abundant flavan-3-ol present (Amiot et al., 1995; Ferreira et al., 2002; Mosel and Herrmann, 1974a). Procyanidins B1 and B2 (Figure 4.4) have been detected in Barlett pears (Spanos and Wrolstad, 1992) and procyanidins with a mean degree of polymerization of 13 to 44 detected in Portuguese pears, comprising about 96% of total phenolics (Ferreira et al., 2002). Two anthocyanins have also been found in pears, namely, cyanidin 3-galactoside and cyanidin 3-arabinoside (Francis, 1970; Timberlake and Bridle, 1971).

Quince Fruits

Raw quince fruit (*Cydonia oblonga*) is generally consumed after cooking or processing into jam and jelly. Andrade et al. (1998) detected a number of phenolic compounds in quince fruit puree and identified six of them: *cis*- and *trans*-3-*O*-caffeoylquinic acid, *cis*- and *trans*-5-*O*-caffeoylquinic acid, quercetin 3-rhamnoside, quercetin 3-xyloside, quercetin 3-galactoside and rutin. Later, Silva et al. (2000) identified two additional phenolics, 4-*O*-caffeoylquinic and procyanidin B_3, as well as four unknown procyanidin polymers in quince jam. Subsequently, three more phenolic compounds have been identified — 3,5-dicaffeoylquinic acid in the pulp and peel, and kaempferol 3-glucoside and kaempferol 3-rutinoside only in the peel (Silva et al., 2002). These authors also reported that the total content of phenolics in the pulp ranges from 11.7 to 268.3 mg/kg and between 243 and 1738 mg/kg in the peel. Furthermore, rutin and 5-*O*-caffeoylquinic acid are the major phenolics in the peel, whereas 3-*O*- and 5-*O*-caffeoylquinic acids are dominant in the pulp (Silva et al., 2002). Two anthocyanins, cyanidin 3-glucoside and cyanidin 3,5-diglucoside, have also been identified in quince fruits with red pigmented skin (Markh and Kozenko, 1965).

Tropical Fruits

Banana

Bananas (*Musa* sp.) are popular edible fruits. The ripening stages of banana are scored based on the color of the peel, from 1 for all green to 8 for duffel. Bananas are harvested at ripening stages 1 to 3 (all green to half green), and usually sold at ripening stages 6 to 7 (yellow to star) (Kanazawa and Sakakibara, 2000). The major phenolics of banana include catecholamines, naringin and rutin (Drell, 1970; Griffiths, 1959; Kanazawa and Sakakibara, 2000; Undenfriend et al., 1959). Dopamine and norepinephrine (α-(aminomethyl)-3,4-hydroxybenzyl alcohol) are the predominant catecholamines in banana peel and pulp; phenolics accumulate mainly in the peel (Table 4.12) (Kanazawa and Sakakibara, 2000). Dopamine displays anti-inflammatory activity and protects against intestinal injury by acting as modulator of eicosanoid synthesis (Alanko, et al., 1992; MacNaughton and Wallace, 1989).

During the late stages of ripening, dopamine is converted to salsolinol (1-methyl-6,7-dihydroxy-1,2,3,4-tetrahydroxyisoquinoline) (Figure 4.18) (Riggin et al., 1976) via nonenzymatic Pictet-Spengler condensation (Whaley and Govindachari, 1951). The biosynthesis of salsolinol involves complexation of dopamine with acetaldehyde to form an imine, which then cyclizes to salsolinol (Whaley and Govindachari, 1951). The synthesis of salsolinol is accelerated during ripening of banana due to increased formation of acetaldehyde from ethanol produced during the postclimacteric phase (Riggin et al., 1976). Salsolinol has been implicated in the black appearance of overripe bananas (Riggin et al., 1976). Recently, Sojo et al. (2000) demonstrated that salsolinol is oxidized to corresponding quinone by banana pulp polyphenol oxidase.

Tressl and Drawart (1973) identified eugenol methyl ether, eugenol, elimicin, 3,4-dimethoxytoluene and 5-methoxyeugenol in banana volatiles. These compounds are formed in banana from phenylalanine via Shikimic acid pathway. Later, Pérez

TABLE 4.12
Levels of Phenolics in Unripened and Ripened Cavendish Banana[a]

Phenolic	Unripe Banana[b]		Ripe Banana[c]	
	Peel	Pulp	Peel	Pulp
Dopamine	8650–19,400	47–100	800–560	25.0–100
Norepinephrine	550–1180	8–17	0–240	8.2–16
Naringin	1200–2600	0–650	280–950	0.0–33
Rutin	160–230	0–48	110–160	0

[a] Milligrams per kilogram of fresh weight.
[b] Ripening stages 1 to 3 (all green to half green).
[c] Ripening stages 6 to 7 (full yellow to star).

Source: Adapted from Kanazawa, K. and Sakakibara, H., 2000, *J. Agric. Food Chem.*, 48:844–848.

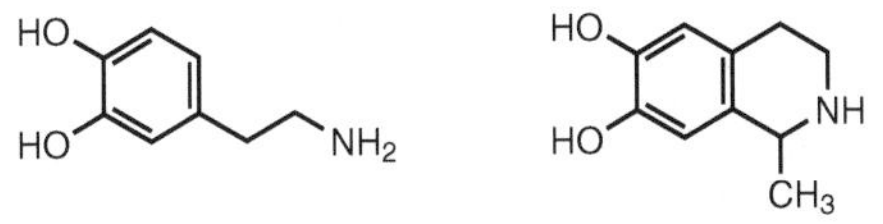

Dopamine   Salsolinol

FIGURE 4.18 Chemical structures of some phenolic compounds found in banana fruit.

et al. (1997) reported that these phenolics are present in glycosidically bound form in ripe banana. Glucosides of pelargonidin and cyanidin have been found in *Musa coccinea*, while glucoside of delphinidin and cyanidin is present in *Musa laterita, Musa Balbisiana* and *Musa velutina* (Simmonds, 1954a). In addition, methylated anthocyanidin glucosides of peonidin, petunidin, and malvidin have been detected in *Musa acuminata* (Figure 4.5) (Simmonds, 1954a, b).

Citrus Fruits

Cinnamic acid derivatives, coumarins and flavonoids (flavonones, flavones and flavonols) are the major groups of phenolic compounds occurring in citrus fruits in the free form and/or as glycosides (Horowitz and Gentili, 1977; Maier and Matzler, 1967; Manthey and Grohmann, 2001; Rouseff and Ting, 1979; Wilfried et al., 1994a, b). Flavanone glycosides, phlorin and cinnamic acid derivatives have been proposed for species differentiation and detection of adulterations (Louche et al., 1998; Marini and Balestrieri, 1995; Mouly et al., 1993, 1994, 1995, 1997; Ooghe and Detavernier, 1997; Rouseff, 1988; Rouseff et al., 1987). The peel is the primary waste product from processing citrus fruits. For example, orange peels comprise approximately 45% of the fruit mass.

The various co-products obtained from citrus peel include citrus peel molasses and juices, cold-pressed citrus peel oil, pectin, limonene and dried citrus pulp

(Braddock, 1995; Kesterson and Braddock, 1976). For example, peel juice is obtained by liming and then pressing the peel, while molasses is obtained by evaporation of the peel juice to a liquor of at least 35.5°Brix (Grohmann et al., 1999; Manthey and Grohmann, 2001). The highest concentration of flavanones and flavanone glycosides occurs in the peel (Bocco et al., 1998; Horowitz, 1961; Kanes et al., 1993; McIntosh and Mansell, 1997), e.g., orange and grapefruit peels contain between 13.5 and 22.3 g of flavonoids/kg of dry weight (Bocco et al., 1998; Sinclair, 1972). Other groups of phenolic compounds such as hydroxycinnamates, psolarens and polymethoxylated flavones have also been detected in citrus peels (Kanes et al., 1993; Manthey and Grohmann, 2001; Peleg et al., 1991; Risch and Herrmann, 1988a; Stanley and Vannier, 1957; Tatum and Berry, 1979).

Phenolic acids are predominantly located in the flavedo of citrus fruit (Manthey and Grohmann, 2001; Peleg et al., 1991). Caffeic, *p*-coumaric, ferulic and sinapic acids (Figure 4.1) have been detected in citrus fruits and their co-products in the form of esters, amides and glycosides (Bocco et al., 1998; Fallico et al., 1996; Manthey and Grohmann, 2001; Mouly et al., 1997; Peleg et al., 1991; Rapisarda et al., 1998; Risch and Herrmann, 1988a). Only very small amounts of free phenolic acids are present (Manthey and Grohmann, 2001; Peleg et al., 1991). Ferulic acid has been found to be the major phenolic acid in the Shamuti orange fruit (Peleg et al., 1991) and citrus peel and their co-products (Bocco et al., 1998; Manthey and Grohmann, 2001), while coumaric and ferulic acids are present in Italian blood orange juices (Rapisarda et al., 1998).

The total hydroxycinnamic acid content in juices of different Italian blood orange varieties ranges from 53.1 to 158.8 mg/L (Rapisarda et al., 1998), while in citrus peels and seeds it is 144 to 2956 mg/kg of dry matter (Bocco et al., 1998). HCAs have been associated with the formation of off-flavors in citrus fruits and their products. Ferulic and *p*-coumaric acid esters may be enzymatically hydrolyzed during thermal processing and storage and then decarboxylate to unpleasant compounds such as *p*-vinylguaiacol (PVG) and *p*-vinylphenol (Fallico et al., 1996; Klaren-De Wit et al., 1971; Lee and Nagy, 1990; Naim et al., 1988, 1993). Accumulation of PVG can be controlled by storing orange juice under nitrogen, as well as in the presence of cupric ions and L-cysteine (Peleg et al., 1992, Naim et al., 1993).

Mouly et al. (1997) isolated and identified cinnamoyl-β-D-glucopyranoside in blood oranges grown in Florida. Two other glycosides, namely, phlorin (3,5-dihydroxyphenyl β-D-glucopyranoside) (Figure 4.19) and coniferin (coniferyl alcohol 4-β-D-glucopyranoside), have been detected in citrus fruit juices, citrus peels and/or their by-products (Braddock and Bryan, 2001; Hammond et al., 1996; Horowitz and Gentili, 1961; Johnson et al., 1995; Louche et al., 1998; Manthey and Grohmann, 2001). According to Louche et al. (1998), phlorin mainly accumulates in the albedo of the peel (0.8 to 1.1 mg/kg). Based on the level of this phenolic

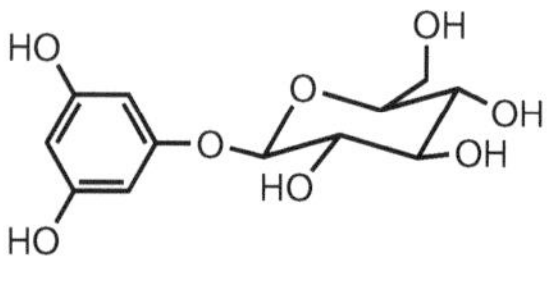

FIGURE 4.19 Chemical structure of phlorin.

compound (1.1 to 25.5 mg/L), Mouly et al. (1997) suggested it as a natural marker for authentication of Florida blood orange juice. On the other hand, Rapisarda et al. (1998) proposed to use hydroxycinnamate profiles as markers for Italian blood orange juices. Subsequently, Louche et al. (1998) suggested phlorin as peel marker in orange juices because a high level of phlorin in juice would indicate adulteration with peels when pulp wash is added to the juice.

Accumulation of certain coumarins, namely, xanthyletin, seselin and scoparone (6,7-dimethoxycoumarin), has been associated with the infection of citrus fruits by pathogens (Figure 4.20) (Afek and Sztejnberg, 1988, 1993; Ben-Yehoshua et al., 1988; Khan et al., 1985; Kim et al., 1991; Stange et al., 1993; Vernenghi et al., 1987). Treating tangelo Nova fruits with 0.5% Bromatox, a phytoregulator, significantly enhances the resistance of fruits to fungi. The treated wounded fruits accumulate more scoparone than the untreated ones (Ortuno et al., 1997). Also, treating citrus fruits with UV radiation stimulates the biosynthesis of coumarins (Ben-Yehoshua et al., 1992; Riov et al., 1971, 1972). Increased accumulation of scoparone has also been noted in uninfected, wounded citrus fruits. For example, 274 mg scoparone/kg of fresh weight has been detected in wounded fruits after 7 days following wounding (Ortuno et al., 1997). Furthermore, auraptene (7-genaryloxycoumarin) (Figure 4.20) has been found in a number of citrus species (Gray and Waterman, 1978; Masuda et al., 1992).

Ogawa et al. (2000) evaluated the auraptene content in 77 citrus species. These authors reported that the peels contain up to 16.6 g of auroptene/kg of dry weight; in juice sacs, up to 10.32 g of auroptene/kg of dry weight is present. The largest quantities of auroptene were detected in *Citrus-Poncirus trifoliata* hybrids. Later, Takahashi et al. (2002) reported that *Citrus natsudaidai* and *Citrus hassaku* peels contained over 0.2 g of auraptene/kg of fresh peel. In addition, several coumarins, including 8-geranyloxypsolaren, 5-geranyloxypsolaren (bergamottin) and 5-geranyloxy-7-methoxycoumarin, have been identified in lemon peels; these are abundantly present in the flavedo of the peel (116, 138 and 83.2 mg/kg of fresh weight, respectively) as well as in a cold-pressed lemon oil (1850, 1700 and 1700 mg/kg, respectively) (Miyake et al., 1999). Recently, Takahashi et al. (2002) developed a rapid method for preparing an auraptene-enriched product from orange peel oils by using SP-70 adsorbent (styrene-divinylbenzene synthetic adsorbent). Auraptene exerts inhibitory effects toward chemical cancerogenesis in some rodent models (Murakami et al., 1997; Tanaka et al., 1997, 1998a, b).

Polymethoxylated flavones (Figure 4.21), a group of phenolic compounds unique for *Citrus* species, are mostly accumulated in the peel (Ortuno et al., 2002) and their profile in citrus fruit is characteristic of each species (Gaydou et al., 1987; Mizuno et al., 1991). Nobiletin (5,6,7,8,3′, 4′-hexamethoxyflavone) and sinensetin (5,6,7,3′,4′-pentamethoxyflavone) have been identified in orange peel, while tangeretin (5,6,7,8,4′-pentamethoxyflavone) (Kefford and Chandler, 1970; Ting and Attaway, 1971), tetramethylscutellarein (5,6,7,4′-tetramethoxyflavone), 3,5,6,7,8,3′,4′-heptamethoxyflavone, nobiletin and sinensetin have been found in tangerine oil (Gaydou et al., 1987). Tangeretin, 3,5,6,7,8,3′,4′-heptamethoxyflavone, 5,7,8,4′-tetramethoxyflavone, and 5,7,8,3′,4′-pentamethoxyflavone have been found in grapefruits (Harborne, 1967; Venkataraman, 1975). Later, sinensetin, isosinensetin

(a)

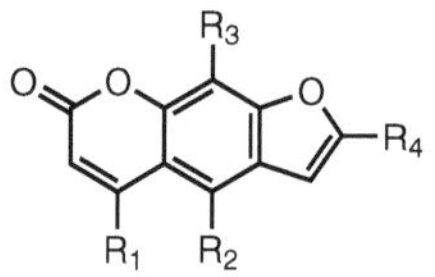

Compound	R_1	R_2	R_3	R_4
Psolaren	H	H	H	H
Bergapten	H	OCH_3	H	H
Xanthotoxin	H	H	OCH_3	H
Isopimpenellin	H	OCH_3	OCH_3	H
4,5',8-Trimethylpsolaren	CH_3	H	CH_3	CH_3

(b)

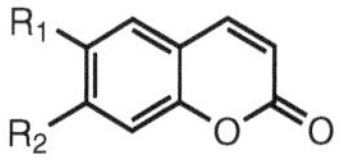

Compound	R_1	R_2
Esculatin	OH	OH
Scopoletin	OCH_3	OH
Scopanone	OCH_3	OCH_3
Umbelliferone	H	OH

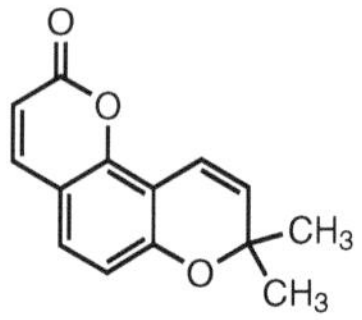

Seselin

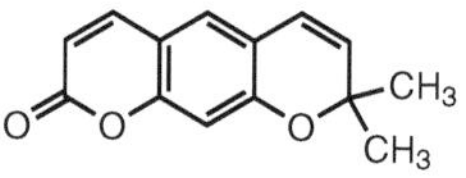

Xanthyletin

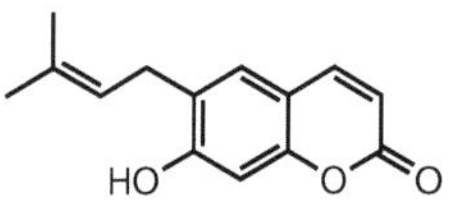

Suberosin

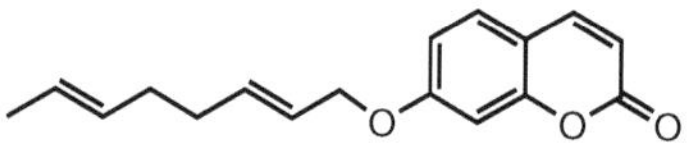

Auraptene

FIGURE 4.20 Chemical structures of furanocoumarins (A) and coumarins (B).

(5,7,8,3′,4′-pentamethoxyflavone), hexa-*O*-methylquercetagetin (3,5,6,7,3′,4′-hexamethoxyflavone), nobiletin, 3,5,6,7,8,3′,4′-heptamethoxyflavone, tetramethylscutellarein, and 5-desmethylnobiletin (5-hydroxy-6,7,8,3′,4′-pentamethoxyflavone) were detected in the peel molasses from orange, tangerine and grapefruit (Manthey and Grohmann, 2001). Subsequently, Kawaii et al. (1999c, 2001) detected nobiletin (7 to 173 mg/kg of dry weight), 3,5,6,7,8,3′,4′-heptamethoxyflavone (0 to 87 mg/kg of dry weight), natsudaidain (5,6,7,8,3′,4′-hexametoxymethoxyflavone) (0 to 69 mg/kg of dry weight) and tangeretin 98 to 62 (mg/kg of dry weight) in edible parts of *Citrus* fruits.

The concentration of individual polymethoxylated flavones is affected by the stage of *Citrus* fruit development. In tangelo Nova fruits, the highest concentration of nobiletin, sinensetin and tangeretin is found in immature fruits, while 3,5,6,7,8,3′,4′-heptamethoxyflavone is dominant in fruits reaching maturity (Ortuno

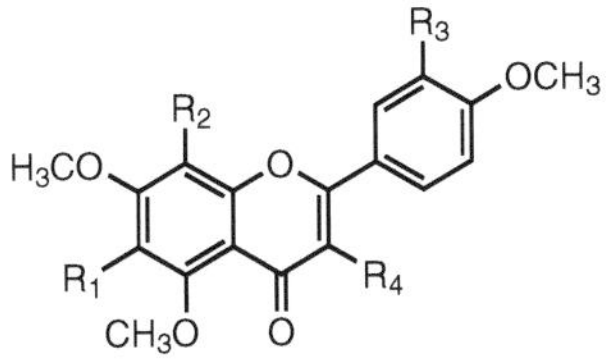

Compound	R_1	R_2	R_3	R_4
Sinensetin	OCH_3	H	OCH_3	H
Tetramethylscutellarein	OCH_3	H	H	H
Isosinensetin	H	OCH_3	OCH_3	H
Nobiletin	OCH_3	OCH_3	OCH_3	H
Tangeretin	OCH_3	OCH_3	H	H
Heptamethoxyflavone	OCH_3	OCH_3	OCH_3	OCH_3
Natsudaidain	OCH_3	OCH_3	OCH_3	OH

FIGURE 4.21 Chemical structures of some polymethoxylated flavones found in *Citrus* fruits.

TABLE 4.13
Content of Polymethoxylated Flavones in Peel Oils of Various Citrus Fruit Samples[a]

Citrus Fruit	Tangeritin	Nobiletin	Sinsensetin	Total
Orange	0.70	0.50	0.30	1.88
Common Mandarin	2.80	2.00	0.07	6.49
King Mandarin	2.50	0.60	0.20	6.00
Clementine	1.10	0.40	0.07	3.60
Tangeritin	1.50	1.50	0.07	3.64

[a] Grams per liter.

Source: Adapted from Gaydou, E.M. et al., 1987, *J. Agric. Food Chem.*, 35:525–529.

et al., 1999). The accumulation of polymethoxylated flavones in fruits can be modulated by the application of growth regulators during fruit development. It has been shown that 7 days after the treatment of tangelo Nova fruits with 100 mg/kg of benzylaminopurine, a growth cytokinin regulator, the level of polymethoxylated flavones in treated fruits increases by 12% for tangeretin, 20% for nobiletin and sinensetin and 29% for heptamethoxyflavone (Ortuno et al., 2002). The total content of polymethoxylated flavones in peel oil of citrus fruits ranges from 1.88 g/L for orange (*Citrus sinensis*) to 6.49 g/L for common mandarin (Table 4.13) (Gaydou et al., 1987); in peel molasses from orange, tangerine and grapefruit it varies from 0.08 to 1.13% (Manthey and Grohmann, 2001). Polymethoxylated flavones exert potent anticarcinogenic, anti-inflammatory and cardioprotective activities (Benavente-Garcia et al., 1997; Kawaii et al., 1999a, b; Manthey et al., 1999, 2001; Miyazawa et al., 1999).

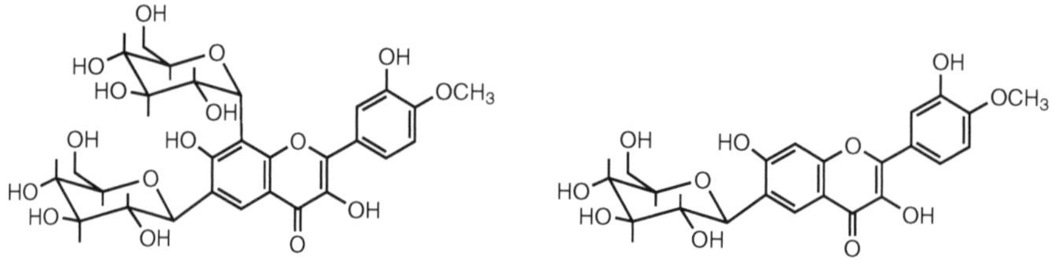

FIGURE 4.22 Chemical structures of C-glycosyldiosmins found in lemon peel.

Several glycosylated flavones have been detected in citrus fruits. Of these, diosmin (4′methoxy-5,7,3′-trihydroxyflavone-7-rutinoside) and neodiosmin (4′methoxy-5,7,3′-trihydroxyflavone-7-neohepseridoside) (Figure 4.2) have also been found in immature citrus fruits (Benavente-Garcia et al., 1993; Del Rio et al., 1991; Marin and Del Rio, 2001). The maximum accumulation of these two flavones occurs during the first stages of fruit development (lineal phase) and then their levels gradually decrease (Benavente-Garcia et al., 1993). The concentration of diosmin in immature citrus fruits ranges from 0 for *Citrus aurantium* cultivars to over 30 g/kg of dry weight for *Citrus medica* cv. Buda's fingers and *Citrus limon* cv. Meyer, while in mature fruits it does not exceed 4 g/kg of dry weight.

On the other hand, the content of neodiosmin, found in *C. aurantium* cultivars, is 4.8 to 9 g/kg of dry weight in immature fruits and 0.45 to 1.1 g/kg of dry weight in mature fruits (Marin and Del Rio, 2001). Diosmin and neodiosmin have been found in greater concentrations in flavedo than in albedo of the peel; only small quantities of these flavones are present in the pulp. Application of hormonal treatment with 6-benzylaminopurine and 2,4-dichlorophenoxyacetic acid during the early stages of fruit development greatly increases the content of diosmin and neodiosmin in immature fruits (Marin and Del Rio, 2001). In addition, citrus peel contains small quantities of C-glucosylflavones such as 8-di-*C*-β-glucosyldiosmin (216 mg/kg) and 6-*C*-β-glucosyldiosmin (34.8 mg/kg) (Figure 4.22) (Miyake et al., 1997). Diosmin is used in the treatment of circulatory system diseases (Galley and Thiollet, 1993; Tsouderos 1991), in the relief of the symptoms of severe hemorrhoids (Godeberg, 1994; Meyer, 1994) and in inflammatory disorders (Jean and Bodinier, 1994).

Presence of four flavanones, namely, naringenin (5,7,4′-trihydroxyflavanone), eriodictyol (5,7,3′,4′-tetrahydroxyflavanone), hesperetin (4′-methoxy-5,7,3′-trihydroxyflavanone), and isosakuranetin (4′-methoxy-5,7-dihydroxyflavanone), in citrus fruits has been reported (Figure 4.23). These aglycones, as well as narirutin 4′-*O*-glucoside, are glycosylated in the 7-position with disaccharides rutinose or neohesperidose. Both disaccharides are made of rhamnose and glucose and differ only in their linkage type; 1→6 for rutinose and 1→2 for neohesperidose.

According to Kanes et al. (1992), flavanone glycosides such as naringin, neoeriocitrin and hesperidin comprise 50 to 80% of total flavonoids in *Citrus* fruits. The predominant glycosides in grapefruits are naringin, naringenin 7-neohesperidoside and narirutin, naringenin 7-rutinoside. On the other hand, dominant flavanone

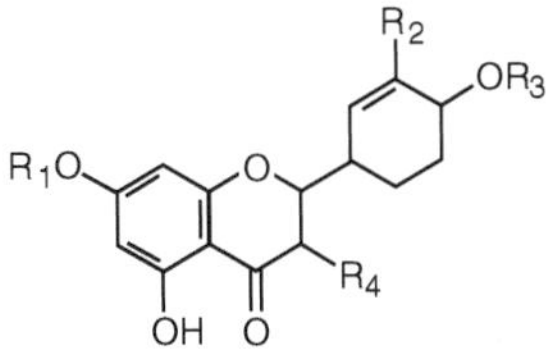

Compound	R_1	R_2	R_3	R_4
Eriocitrin	rutinosyl	OH	H	H
Hesperidin	rutinosyl	OH	CH_3	H
Neoeriocitrin	neohesperidosyl	OH	H	H
Naringenin	H	H	H	H
Naringin	neohesperidosyl	H	H	H
Neohesperidin	neohesperidosyl	OH	CH_3	H
Neoponcirin	rutinosyl	H	CH_3	H
Narirutin	rutinosyl	H	H	H
Poncirin	neohesperidosyl	H	CH_3	H
Taxifolin	H	OH	H	OH

FIGURE 4.23 Chemical structures of flavanones and their glycosylated derivatives.

TABLE 4.14
Content of Flavanone Glycosides in Citrus Juices[a]

Fruit	Narirutin	Naringin	Hesperedin	Neohesperedin
Sour orange	<3	133–262	<3	97–209
Sweet orange	18–65	—	122–254	—
Grapefruit	23–124	73–419	4–16	4–10
Pumello	—	40–144	<3	—

[a] Milligrams per liter.

Source: Adapted from Rouseff, R.I. et al., 1987, *J. Agric. Food Chem.*, 35:1027–1030.

glycosides in sweet oranges are 7-rutinosides, namely, narirutin, hesperedin, and hesperetin 7-rutinoside. Sour oranges contain 7-neohesperidosides, namely, naringin and neohesperidin as well as hesperetin 7-neohesperidoside (Table 4.14). Hesperedin, narirutin and didymin (isosakuranetin 7-rutinoside) are the major flavanone glycosides in Navel oranges (Gil-Izquierdo et al., 2002) as well as blood oranges (Mouly et al., 1997). Lemon peel is a rich source of glycosylated flavonones such as neoeriocitrin, eriodictyol 7-neohesperidoside (6.1 g/kg), naringin (6.1 g/kg) and neosheperidin (4.4 g/kg) (Bocco et al., 1998). In addition, hesperedin is the predominant glycoside in orange peel and orange peel juice, while narirutin dominates in orange peel molasses (Table 4.15) (Manthey and Grohmann, 1996). Flavanone glycosides contribute to the taste and bitterness of citrus juice; their fungistatic and fungitoxic properties are well established (Kefford and Chandler, 1970). However,

TABLE 4.15
Content of Flavanone Glycosides in Valencia Orange Peel and Peel By-Products[a]

Compound	Peel	Peel Juice	Peel Molasses
Eriocitrin	1010	145	787
Narirutin	1998	116	2050
Narirutin 4′-O-glucoside	835	60	933
Hesperedin	19,170	408	961
Isosakuranetin rutinoside	1858	27	256

[a] Milligrams per kilogram.

Source: Adapted from Manthey, J.A. and Grohmann, K., 1996, *J. Agric. Food Chem.*, 44:811–814.

flavanone glycosides exert only a weak antiproliferative activity toward cancer cell lines (Kawaii et al., 1999a).

A number of anthocyanins have been identified in blood oranges (Lee, 2002; Maccarone et al., 1983, 1985, 1998; Trifiro et al., 1995). The content of anthocyanins in the blood orange fruit is affected by the variety, stage of fruit ripeness and storage conditions (Rapisarda and Giulffrida, 1992; Rapisarda et al., 2000, 2001). Cyanidin 3-glucoside, cyanidin-3-(6″-malonyl)-glucoside and cyanidin 3-rhamnoside are the major anthocyanins in Italian blood oranges (Maccarone et al., 1983, 1985, 1998; Trifiro et al., 1995), while cyanidin 3-glucoside and cyanidin-3-(6″-malonyl)-glucoside are present in high amounts in Budd blood oranges (Lee, 2002). Cyanidin 3-glucoside comprises 40 to 60% of total anthocyanins in blood oranges grown in Italy (Maccarone et al., 1985) and malonylated anthocyanins account for over 50% of the total anthocyanins in Budd blood oranges (Lee, 2002).

Mango

One of the most popular edible fruits for its tasty, fleshy mesocarp, mango (*Magnifera indica* L.) is widely grown in many tropical and subtropical regions (Arogba, 2000). A number of phenolics have been detected in mango fruits: gallic acid, *m*-digallic acid, β-glucogallin (β-1-*O*-galloyl-D-glucose), gallotannins, quercetin, isoquercitrin, mangiferin, ellagic acid (Bathia et al., 1967; El Ansari et al., 1971; Saleh and El Ansari, 1975), and peonidin 3-galactoside (Figure 4.24) (Proctor and Creasy, 1969). Volatile components of mango fruits contain a number of phenolic constituents. Adedeji et al. (1992) reported the presence of eugenol, ferulic acid and thymol in glycosidically bound aroma compounds in African mango (*Mangifera indica* L.). Later, Sakho et al. (1997) positively identified two phenolic glucosides in African mango aroma: eugenyl and vanillyl β-D-glucopyranosides.

Phenolic compounds are also involved in allergic contact dermatitis (similar to poison ivy) among workers harvesting green mango fruits from trees. Bandyopadhyay et al. (1985) isolated and identified the mango fruit allegern as 5-[2(Z)-heptadecenyl] resorcinol. Aqueous decoction of mango stem bark has been used for treatment of

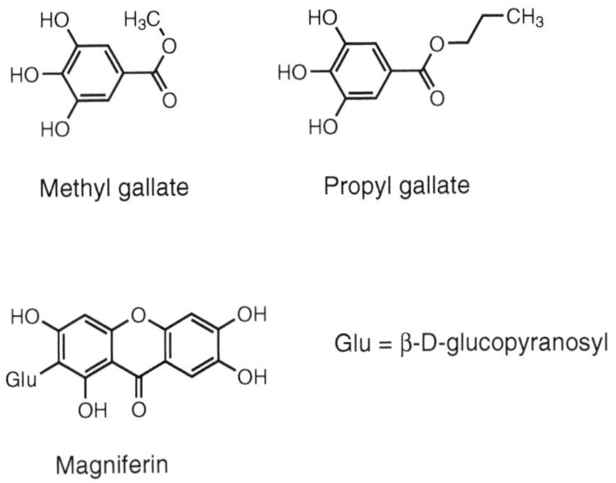

Methyl gallate

Propyl gallate

Glu = β-D-glucopyranosyl

Magniferin

FIGURE 4.24 Chemical structures of some phenolic compounds found in mango fruit.

diarrhea, menorrhagia, scabies, syphilis, anemia, cutaneous infections and diabetes (Guha et al., 1996; Miura et al., 2001a, b; Napralert Data Base, 2002; Sato et al., 1992; Yoshimi et al., 2001). In Cuba this decoction is carried out on an industrial scale for use as a nutritional supplement and a component of cosmetics and in phytomedicine (Núnéz-Selles et al., 2000). The decoction contains 94 g/kg of dry weight of total phenolics expressed as gallic acid equivalents. Seven phenolic constituents have been found in this decoction: gallic acid, 3,4-dihydroxybenzoic acid, methyl gallate, propyl gallate, magniferin, (+)catechin, and (–)epicatechin. Of these, propyl gallate, magniferin and (+)catechin are the major phenolics of mango stem bark decoction (Núnéz-Selles et al., 2002). Dry mango kernel contains 64 g/kg of dry weight of total phenolics, which are a mixture of tannins, gallic acid and epicatechin. Tannins are the predominant phenolic constituents of mango kernel; hydrolyzable tannins comprise 75% of total tannins present (Arogba, 2000).

Passion Fruits

The purple passion fruit *(Passiflora edulis)* and the yellow passion fruit *(Passiflora edulis* f. *flavicarpa)* are two edible passion fruits processed into juice. Several phenolics have been identified in the volatiles of passion fruit. Winter and Kloti (1972) identified eugenol in the aroma compounds of yellow passion fruits. Later, Chassagne et al. (1995) found eugenol and methyl salicylate in the extracts of passion fruits; eugenol was abundant in juice (920 mg/kg) and peels (1720 mg/kg), while methyl salicylate was present only in the juice (Chassagne et al., 1997). Several phenolic glycosides have been characterized in purple and yellow passion fruit volatiles: β-D-glucopyranoside of eugenol and methyl salicylate, as well as 6-*O*-α-L-rhamnopyranosyl-β-D-glucopyranoside of methyl salicylate (Chassagne et al., 1997, 1998). Moreover, several anthocyanins have been detected in the outer cortex of purple passion fruit, namely, pelargonidin 3-diglucoside (Pruthi et al., 1961) delphinidin 3-glucoside (Harborne, 1967), but not in passion fruit juice (Pruthi, 1963).

Pomegranate

Pomegranate (*Punica granatum)* is one of the oldest known edible fruits and is grown in Afghanistan, China, India, Iran, Japan, Mediterranean countries, Russia and the U.S.; it is sold raw or processed to juice and sauce. The total phenolic content in pomegranate juice ranges from 1808 to 2566 mg/L, similar to that found in Cabernet Sauvignon red wine (Gil et al., 2000). Pomegranate is a rich source of hydrolyzable tannins and anthocyanins. The total content of anthocyanins and hydrolyzable tannins in pomegranate juice ranges from 161.9 to 387.4 mg/L and from 417.3 to 556.6 mg/g, respectively (Gil et al., 2000). Pomegranate juice displays antiatherogenic properties *in vivo* (Aviram et al., 2000); tannins from the bark show antitumoral activity (Kashiwada et al., 1992) and antiviral activity against genital herpes (Zhang et al., 1995).

According to Harborne (1967), delphinidin 3,5-diglucoside is the predominant anthocyanin in pomegranate juice. Later, Du et al. (1975) identified 3-glucosides and 3,5-diglucosides of pelargonidin, delphidin and cyanidin in the seed coat of pomegranate fruit. Recently, Gil et al. (2000) reported that cyanidin 3-glucoside (59.5 to 128.3 mg/L) is the major anthocyanin in pomegranate juice. Other identified anthocyanins include delphinidin 3-glucoside (23.6 to 95.2 mg/L), cyanidin 3,5-diglucoside (31.4 to 71.4 mg/L), delphinidin 3,5-diglucoside (21.1 to 61.1 mg/L) and pelargonidin 3-glucoside (3.9 to 8.5 mg/L).

A number of hydrolyzable tannins have been isolated and identified in pomegranate leaves, bark and fruits (Gil et al., 2000; Mayer et al., 1977; Nawwar et al., 1994; Tanaka et al., 1985, 1986 a, b). Hydrolyzable tannins of pomegranate include gallotannins, ellagic acid tannins and gallagyl tannins such as punicalagin and punicalin (Figure 4.14). Of these, gallagyl tannins are the major hydrolyzable tannins in commercial pomegranate juices because the juice contains 1500 to 1900 mg/L of punicalagin (Gil et al., 2000). In addition, small quantities of several flavan-3-ols have been detected in pomegranate: catechin (4.0 mg/kg), epicatechin (0.8 mg/kg), gallocatechin (1.7 mg/kg), and procyanidins B1 (1.3 mg/kg) and B3 (1.6 mg/kg of fresh weight) (de Pascual-Teresa et al., 2000).

Star Apple

Star apple (*Chrysophyllum cainato)* bears a pear-shaped fruit that is red-purple or pale-green in color. When cut in half, the fruit looks like a star-shaped array of eight segments. The sweet and aromatic flesh of this fruit is edible, but the skin and rind are considered inedible (Morton, 1987). In folk medicine, star apple fruit or its decoction is used for the treatment of laryngitis, pneumonia, diabetes mellitus (Morton, 1987), diarrhea, fever and venereal disease (Coee and Anderson, 1996). Recently, Luo et al. (2002) identified nine phenolic compounds in apple star fruit: (+)catechin, (–)epicatechin, (+)gallocatechin, (–)epigallocatechin, quercetin, quercitrin, isoquercitrin, myricitrin and gallic acid. The content of these phenolic compounds ranges from 0.5 (quercetin) to 7.3 ((–)epicatechin) mg/kg of fresh weight.

VEGETABLES

Bulbs, Roots, and Tubers

Carrots

The total content of soluble phenolics in carrots ranges from 5088 to 7699 mg/kg of dry weight (Talcott and Howard, 1999a). Phenolic acids and isocoumarins, predominant phenolics in carrots, contribute to plant tissue's defense mechanism against infections or injuries (Babic et al., 1993). Phenolic acid content in carrot peels may increase up to seven-fold when carrots are subjected to abiotic stress during postharvest handling and storage (Sarker and Phan, 1979). Phenolics have also been associated with imparting bitter and sour tastes to stored and processed carrots (Cantwell et al., 1989; Sarker and Phan, 1979) and may result in color loss in processed strained carrots (Talcott and Howard, 1999c; Talcott et al., 2000b).

The total content of phenolic acids in fresh carrots ranges from 77.2 mg/kg of fresh weight for yellow varieties to 746.4 mg/kg of fresh weight for purple varieties (Alasalvar et al., 2002); in processed strained carrots the amount is between 24.9 and 156.2 mg/kg of dry weight (Talcott and Howard, 1999a). Major phenolic acids in carrots include various hydroxycinnamic acid derivatives, *p*-hydroxybenzoic, and syringic acids (Figure 4.1) (Babic et al., 1993; Talcott and Howard, 1999a). However, Babic et al. (1993) did not find *p*-hydroxybenzoic acid and its esters in freshly shredded carrots, but observed an increase in their contents during storage at 4°C. Moreover, Sarker and Phan (1979) identified only caffeic, isochlorogenic and chlorogenic acids in orange carrot roots (Figure 4.15). Recently, Alasalvar et al. (2002) reported that orange, purple, yellow and white carrot varieties contain mainly hydroxycinnamic acid derivatives:

- 3′-Caffeoylquinic acid (neochlorogenic acid)
- 5′-Caffeoylquinic acid (chlorogenic acid)
- 3′-, 4′- and 5′-Feruoylquinic acids
- 3′- and 5′-*p*-Coumaroylquinic acids
- 3′,4′- and 3′,5′-Dicaffeoylquinic acids
- 3′,4′- and 3′,5′-Diferuoylquinic acids

The total content of phenolic acids esterified to cell wall material of carrots is between 324.8 mg/kg of cell wall carbohydrate in mature, and 661.1 mg/kg in stored, carrots (Parr et al., 1997; Ng et al., 1998a). Ten phenolic acids have been identified in alkaline hydrolyzates from carrot cell wall material. Of these, *p*-hydroxybenzoic acid is the predominant phenolic acid in cell wall material. Over 30% of total ferulic acid exists in the dehydrodimer form as 5,8′-(benzofuran form), 8,8′-(aryltetralin form) and 8-*O*-4′-diferulic (β–[4–(2–carboxyvinyl)-2-methoxyphenoxy]-4-hydroxy-3-methoxycinnamic) acids; 8-*O*-4′-diferulic acid is the major dehydrodimer in mature carrots (Figure 4.25) (Hartley and Harris, 1981; Ng et al., 1998a; Parr et al., 1997). Cell wall material also contains small amounts of vanillic, *p*-coumaric, *trans*- and *cis*-ferulic acids as well as *p*-hydroxybenzaldehyde and vanillin (Massiot et al., 1988; Ng et al., 1998a; Parr et al., 1997).

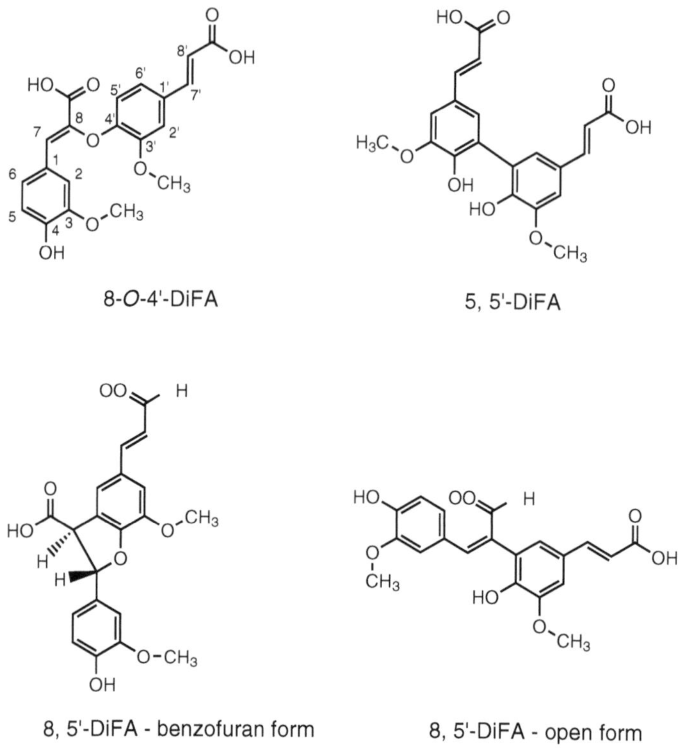

FIGURE 4.25 Chemical structures of ferulic acid dehydrodimers.

Two isocoumarins, namely, 3-methyl-6-methoxy-8-hydroxy-3,4-dihydroisocoumarin (6-methoxymellein) and 3-methyl-6,8-dihydroxy-3,4-dihydroisocoumarin (6-hydroxymellein), have been detected in carrots (Figure 4.26) (Coxon et al., 1973; Harding and Heale, 1980; Marinelli et al., 1990; Sondheimer, 1957). The presence of furanocoumarins has also been reported in diseased carrot tissues (Ceska et al., 1986; Harborne, 1985). The reported levels of 6-methoxymellein in whole carrots range from 0 to 400 mg/kg of dry weight (Talcott and Howard, 1999a). Isocoumarins are predominantly accumulated in the periderm tissue of carrot root and their concentrations incrementally decrease from the peel to vascular tissues (Talcott and Howard, 1999b). Both 6-hydroxymellein and 6-methoxymellein exhibit toxic effects on animal cells, microorganisms and plant cells. The toxic effect exerted by 6-methoxymellein on Chinese hamster cells (EC_{50} = 0.46 mM) (Superchi et al., 1993) are much lower than that exerted on microorganisms and plant cells (EC_{50} = 0.04 to 0.05 mM) (Kurosaki et al., 1984; Harding and Heale, 1980; Hoffman and Heale, 1987; Marinelli et al., 1989).

The accumulation of 6-methoxymellein has been associated with the development of bitterness in carrots (Lafuente et al., 1989; Sarker and Phan, 1979; Sondheimer, 1957; Talcott and Howard, 1999a). The formation of 6-methoxymellein in carrots is induced by their exposure to ethylene (Lafuente et al., 1989; Sarker and Phan, 1979; Sondheimer, 1957; Talcott and Howard, 1999b) and to UV light (Mercier

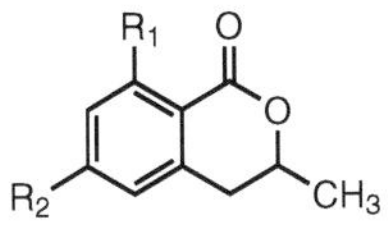

Compound	R_1	R_2
6-Methoxymellein	OH	OCH_3
6-Hydroxymellein	OH	OH
6,8-Dimethoxymellein	OCH_3	OCH_3

FIGURE 4.26 Chemical structures of some dihydroisocoumarins.

et al., 1994), by microbial infection (Kurosaki and Nishi, 1983; Hoffman et al., 1988), by wounding (Talcott and Howard, 1999b), by elevated storage temperatures (Lafuente et al., 1989), and by pectinolytic enzymes (Marinelli et al., 1990, 1994; Movahedi and Heale, 1990). Whole carrots exposed to ethylene contain 15 to 21% more 6-methoxymellein in the root tip than in the other root sections. Moreover, removal of the tip also elevates the content of 6-methoxymellein in the tip section of carrot root, while the removal of crown significantly increases the concentration of 6-methoxymellein in the root section next to the crown (Talcott and Howard, 1999b).

Onions

One of the most widely consumed vegetables, onions are classified based on their color into yellow, red, and white; based on their taste, they are divided into sweet and nonsweet products. Onions are rich in flavonoids (Bilyk et al., 1984; Crozier et al., 1997; Price and Rhodes, 1997; Rhodes and Price, 1996) and serve as one of the major sources of flavonols such as quercetin, isorhamnetin, myricetin and kaempferol conjugates in the diet (Leighton et al., 1992; Tsushida and Suzuki, 1995). Of these, quercetin is the major flavonol in onions (Rhodes and Price, 1996; Sellappan and Akoh, 2002). Levels of quercetin glucosides are much higher in onions than those in other vegetables (Table 4.16) (Crozier et al., 1997; Hertog et al., 1992a; Mizuno et al., 1992). The content of quercetin conjugates in bulbs of red onion cultivars ranges from 110 to 295 mg of quercetin equivalents/kg and between 119 and 286 mg of quercetin equivalents/kg in bulbs of yellow onion cultivars (Bilyk et al., 1984; Patil and Pike, 1995; Patil et al., 1995; Tsushida and Suzuki, 1996), and varies from 185 to 634 mg of quercetin equivalents/kg in bulbs of white onion cultivars (Crozier et al., 1997). The total content of myricetin and kaempferol in Vidalia onions ranges from 27.7 to 41.3 and from 15.4 to 19.8 mg/kg of fresh weight, respectively (Sellappan and Akoh, 2002).

Leighton et al. (1992) identified four flavonoids in onions: quercetin diglucoside (quercetin 3,4′-*O*-diglucoside), quercetin monoglucoside (quercetin 4′-*O*-glucoside), isorhamnetin monoglucoside and quercetin. Later, Park and Lee (1996) detected the presence of rutin, isorhamnetin and kaempferol (Figure 4.2) and identified isorhamnetin 4′-*O*-glucoside. Subsequently, Price and Rhodes (1997) reported that quercetin 3,4′-*O*-diglucoside and quercetin 4′-*O*-glucoside are the major flavonols in edible portions of onions. In addition, 20 minor quercetin derivatives have been detected (Rhodes and Price, 1996).

TABLE 4.16
Total Content of Quercetin Conjugates in Lettuce, White Onions and Tomatoes[a]

Vegetable	Content
"Round" lettuce var. Cortina	11
"Green Salad Bowl" lettuce	147
White onions	185–634
Spanish tomatoes var. Assun	3.5–4.4
Spanish tomatoes var. Daniella	2–8.7
Scottish tomatoes var. Spectra	4.6–11.2
Dutch beef tomatoes var. Trust	2.2–6.8
Spanish cherry tomatoes var. Paloma	23–203
English cherry tomatoes var. Favorita	17–77

[a] Milligrams per kilogram of fresh weight.

Source: Adapted from Crozier, A. et al., 1997, *J. Agric. Food Chem.*, 45:590–595.

The flavonols are mostly concentrated in the skin, which contains from 5.3 to 34.15 g of quercetin equivalents/kg of fresh weight; of these, 23 to 53% of total quercetin is present as aglycone. In addition, the skin contains 6 to 677 mg of kaempferol/kg of fresh weight (Bilyk et al., 1984). The tissue and spatial distribution of quercetin and its two major glucosides, quercetin 3,4′-diglucoside (Q_{DG}) and quercetin 4′-monoglucoside (Q_{MG}), have been evaluated by Hirota et al. (1998). The highest level of flavonols is detected in abaxial epidermis of scales while the lowest amount is found in the mesophyl. Moreover, approximately 50% of flavonols accumulate in the top quarter part of the scales. The mole ratios of Q_{DG}/Q_{MG} in the outer onion scales are lower than those in the inner onion scales. The outer scales are more aged than those of the inner scales (Barden et al., 1987); thus, the aging of scales may promote more rapid synthesis of Q_{MG} (Hirota et al., 1998) or induce enzymatic hydrolysis of Q_{DG} to Q_{MG} (Tsushida and Suzuki, 1996).

Onions also contain small quantities of phenolic acids bound to cell walls. Of these, protocatechuic acid has been detected only in the papery scales, while *p*-hydroxybenzaldehyde and vanillin have been identified only in the fleshy scales. In addition, ferulic, *p*-hydroxybenzoic, vanillic and coumaric acids have been found in the papery and fleshy scales (Ng et al., 2000). Moreover, *p*-hydroxybenzoic acid-*O*-β-D-glucoside (9 mg/kg), vanillic acid-*O*-β-D-glucoside (3 mg/kg) and protocatechuic acid-4-*O*-β-D-glucoside (91 mg/kg) have been detected in onion peels (Herrmann, 1989).

Red onions contain a number of anthocyanins that are mostly concentrated in the skin and the outer fleshy layer (Table 4.17) (Gennaro et al., 2002). Fuleki (1971) and Herrmann (1976) identified cyanidin 3-glucoside, peonidin 3-glucoside and cyanidin 3-laminariobioside in the outer layer of onions. Later, Ferreres et al. (1996) detected cyanidin 3-glucoside and cyanidin 3-arabinoside and their malonylated

TABLE 4.17
Distribution of Main Anthocyanins in Different Parts of Tropea Red Onions[a]

Anthocyanin	Dry Skin	Outer Fleshy Layer	Edible Portion
Delphinidin 3-glucosylglucoside	7831 ± 75	582 ± 29	815 ± 37
Cyanidin 3-(6″-malonylglucoside)	7340 ± 154	1250 ± 51	188 ± 8
Cyanidin 3-(6″-glucosylglucoside)	4565 ± 84	625 ± 17	125 ± 4

[a] Milligrams per kilogram of dry matter.

Source: Adapted from Gennaro, L. et al., 2002, *J. Agric. Food Chem.*, 50:1904–1910.

TABLE 4.18
Effect of Storing and Processing on Total Quercetin Conjugate Losses in Onions

Onion	Method of Processing	Loss of Quercetin (%)	Reference
White-skinned	Frying, 3 min	21	Crozier, A. et al. (1997)
	Boiling, 15 min	75	
	Microwaving, 1.3 min	64	
Brown-skinned	Storage, 24 weeks	16	Price, K.R. et al. (1997)
	Boiling, 20 min	21.9	
	Frying, 15 min	29	
Red-skinned	Storage, 24 weeks	35.4	
	Boiling, 20 min	14.3	
	Frying, 15 min	23	
Orange-skinned	Chopping	1	Makris, D. and Rossiter, J.T. (2001)
	Boiling, 60 min	20.6	
Red-skinned	Peeling	21	Gennaro, L. et al. (2002)

derivatives in red onions. Subsequently, Donner et al. (1997) found peonidin 3-malonylglucoside; Gennaro et al. (2002) recently found delphinidin and petunidin derivatives in red onions. According to Fossen et al. (1996) and Donner et al. (1997), cyanidin 3-(6″-malonylglucoside), cyanidin 3-(6″-malonylglucoside-3″-glucosylglucoside) and cyanidin 3-glucoside comprise over 95% of total anthocyanins in whole red onions. On the other hand, Gennaro et al. (2002) have demonstrated that cyanidin derivatives constitute over 50% of total anthocyanins and delphinidin derivatives comprise about 30% of total anthocyanins in whole red onions.

Common domestic storing and processing, such as peeling, chopping, boiling, microwaving, and frying, reduce the total content of quercetin conjugates from 1% in the case of chopping onions to up to 75% for boiling them in water (Table 4.18). Price et al. (1997) reported an overall loss of 25% in quercetin glycosides in onions boiled for 20 min; however, no preferential leaching of quercetin 3,4′-*O*-diglucoside (Q_{DG}), quercetin, 4′-*O*-glucoside (Q_{MG}) and quercetin (Q) into boiling water was noted. Subsequently, Hirota et al. (1998) found over 50% decrease in the content of Q_{DG} and Q_{MG} after 25 and 15 min of boiling, respectively. Later, Makris and Rossiter (2001) reported an overall loss of 20.5% of total flavonols after boiling bulbs in water for 60 min. These authors demonstrated that 8.4, 37.6 and 43.2% of Q_{DG}, Q_{MG}, and Q, respectively, leached into the cooking water. Blanching may also yield higher loss of flavonols; according to Ewald et al. (1999), blanching onions decreases the content of quercetin and kaempferol by 39 and 64%, respectively.

Parsnip

Five furanocoumarins, angelicin, psolaren, bergapten (5-methoxypsolaren), xanthotoxin (8-methoxypsolaren) and isopimpinellin (5,8-dimethoxypsolaren), have been detected in parsnip roots (Figure 4.20) (Alley, 1987; Ostertag et al., 2002). Of these, angelicin and xanthotoxin are the predominant furanocoumarins in parsnips. Ivie et al. (1981) reported that raw parsnip roots contain up to 40 mg/kg of total furanocoumarins. Later, Ostertag et al. (2002) reported that the total content of furanocoumarins in freshly harvested parsnips did not exceed 2.5 mg/kg of fresh weight; however, in parsnips purchased from the market the amount reached 49 mg/kg of fresh weight. Peeling parsnip roots removes only 30% of the total furanocoumarins present, while boiling and microwaving do not reduce the level of furanocoumarins in the roots to any great extent (Ivie et al., 1981). Storing whole parsnips at +4°C for 7 days effects an increase in furanocoumarins from 1 mg/kg of fresh weight to 33 mg/kg of fresh weight; however, storing them at –18°C for up to 50 days does not markedly affect the content of furanocoumarins (Ostertag et al., 2002).

Potato

Friedman (1997) has published an excellent review on the chemistry, biochemistry, and dietary role of potato polyphenols; some aspects of the chemistry of potato phenolics is discussed here. Chlorogenic acid is the predominant phenolic in potato tuber, constituting up to 90% of total phenolics (Friedman, 1997; Malmberg and Theander, 1985). Chlorogenic acids in potato tuber contribute to the defense mechanism of the plant against insects and microbial pathogens (Friedman, 1997). In addition, chlorogenic acids are linked with after-cooking blackening of potato tubers. A colorless iron-chlorogenic chelate, formed during cooking of susceptible potato cultivars, oxidizes to produce a bluish-gray ferric complex upon subsequent exposure of cooked potato to air (Griffiths et al., 1995). The after-cooking blackening of potato tubers is influenced by fertilization, soil, and climatic conditions as well as storage period and exposure to light (Griffiths and Bain, 1997; Silva et al., 1991). Approximately 50% of chlorogenic acid isomers are found in potato skin and adjoining tissues, while in the tuber cortex the level of chlorogenic acids gradually decreases from outside toward the center of potato tuber (Friedman, 1997; Hasegawa et al., 1966; Reeve et al., 1969).

The total content of monomeric anthocyanins in different red-fleshed potato cultivars (*Solanum tuberosum* and *Solanum stenotum*) ranges from 24 to 403 mg/kg of tuber (Rodriguez-Saona et al., 1998). Higher levels of anthocyanins have been detected in the skin of purple potato cv. Urenika (5078 mg/kg) than in the flesh (1836 mg/kg) (Lewis, 1996). On the other hand, a somewhat higher concentration of anthocyanin has been detected in the flesh than in the skin of red-fleshed potatoes. Pelargonidin 3-rutinoside-5-glucoside acylated with *p*-coumaric acid is the predominant anthocyanin (70% of total anthocyanins) (Rodriguez-Saona et al., 1998).

Wounding and/or challenge of potato with phytopathogens brings about a rapid modification of cell wall in periderm, wound periderm, and endo- and exodermal cells known as suberization. In potato, suberized cells become an effective defensive barrier against dehydration and pathogen penetration (Kolattukudy, 1987; Lulai and Orr, 1994). The suberin is composed of polyphenolic and aliphatic domains. The polyphenolic domain is linked to the cell wall; the aliphatic domain is covalently linked to the polyphenolic domain (Kolattukudy, 1987; Lulai and Corsini, 1998; Razem and Bernards, 2002), which makes up 50 to 75% of suberized cell wall (Dean and Kolattukudy, 1977). The polyphenolic domains contain not only lignins but also small amounts of monolignols (coniferyl and sinapyl alcohols) as well as significant amounts of nonlignin precursors such as hydroxycinnamic acids and their derivatives (Figure 4.27) (Bernards et al., 1995; Negrel et al., 1996; Razem and Bernards, 2002). Kolattukudy (1980) has suggested that polymerization of phenolics during the suberization process occurs via a peroxidases–hydrogen peroxide-mediated free radical coupling. Recently, Razem and Bernards (2002) demonstrated the involvement of hydrogen peroxide in the development of polyphenolic domain in suberized cell wall.

Fresh-cutting and subsequent cold storage of potatoes at 4°C induces the biosynthesis of chlorogenic acids and flavonols. Three flavonols, namely, quercetin 3-rutinoside, quercetin diglucoside and querectin 3-glucosylrutinoside, have been detected in potato after a lag period of 3 days of cold storage. A similar lag period has also been observed for chlorogenic acid. Depending on the cultivar, the total contents of flavonols and chlorogenic acids in fresh-cut potatoes after 6 days of cold storage are 60 to 140 and 70 to 300 mg/kg of fresh weight, respectively. Cold storage of fresh-cut potatoes under light enhances the accumulation of flavonols (Tudela et al., 2002); prolonged cold storage of potato tubers at 5°C decreases the level of chlorogenic acids (Percival and Baird, 2000).

The light-induced accumulation of chlorogenic acid is also affected by the cold-storage period in the dark. The rate of chlorogenic acids accumulation is three to four times greater in potato tubers exposed to light after 2 weeks of cold storage in the dark than in tubers exposed to light after 3 months of dark cold storage (Percival and Baird, 2000). Boiling, steam-cooking, frying and microwaving significantly reduce the level of phenolics in potato tubers. Cooked potato strips retain about 50% of flavonols, while the retention of chlorogenic acids is influenced by the method of cooking (Tudela et al., 2002). About 55% of the initial chlorogenic acid is detected in microwaved potato strips while only 24% is retained in fried potato strips (Tudela et al., 2002).

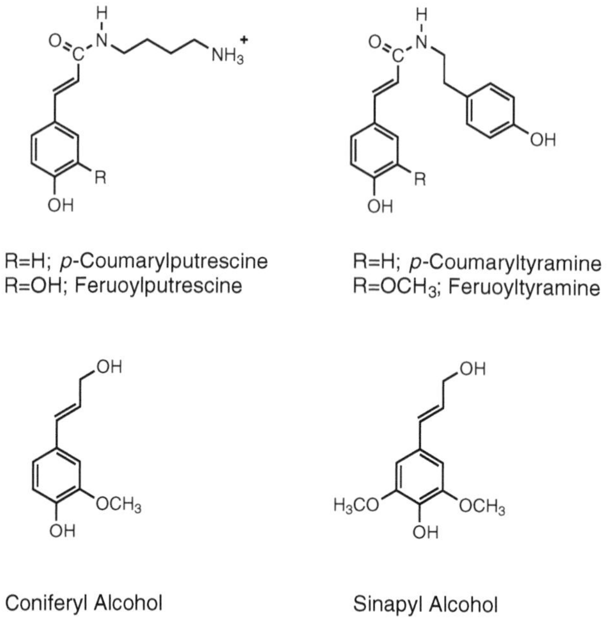

R=H; *p*-Coumarylputrescine
R=OH; Feruoylputrescine

R=H; *p*-Coumaryltyramine
R=OCH_3; Feruoyltyramine

Coniferyl Alcohol

Sinapyl Alcohol

FIGURE 4.27 Chemical structures of some phenolic compounds detected in suberizing potato tuber.

Red Beetroot

The content of phenolics in various red beetroot (*Beta vulgaris)* parts decreases in the order of peel, crown and flesh. Betanin is the predominant phenolic compound in red beetroot (Figure 4.28) and β-D-fructofuranosyl-α-D-(6-*O*-(*E*)-feruloylglucopyranoside (FAE) is the main known ferulic acid ester found in the peel. Beet root cell wall material also contains high levels of dehydrodimers of ferulic acids such as 5,5′-, 8-*O*-4′-, and 5,8′-diferulic acids (Ng et al., 1998b; Steglich and Strack, 1991; Waldron et al., 1997). A two-fold increase in the content of dehydrodimers of ferulic acid has been noted in samples of beetroot treated with hydrogen peroxide (Ng et al., 1998b). In addition, *p*-hydroxybenzoic, *cis*- and *trans*-ferulic, *trans*-coumaric, and vanillic acids, as well as *p*-hydroxybenzaldehyde and vanillin, have been detected in beetroot. Of these, ferulic acids (156.6 mg/kg of fresh weight) are found to be the predominant phenolic acids in red beetroot (Ng et al., 1998b). Cold storage of red beetroot for 9 months decreases the content of betanin from 38.7 to 30.7 g/kg of dry weight, but has a minor effect on the total content of phenolics. On the other hand, cold storage has resulted in an increase in the content of FAE in peels from trace to 357.7 mg/kg of dry weight (Kujala et al., 2000). Furthermore, pasteurization of red beetroot nectar reduces the content of betanin by 50% (Patkai et al., 1997).

Sweet Potato

The total content of phenolics in seven sweet potato tubers (*Ipomoea batatas*) ranges from 117 to 467 mg of chlorogenic acid equivalents/kg of fresh weight (Walter and

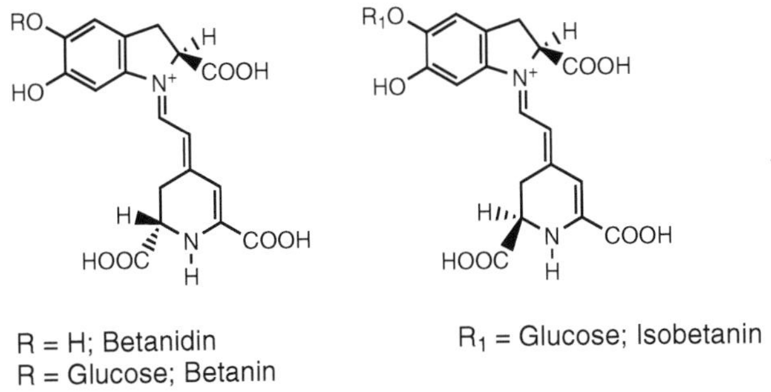

R = H; Betanidin
R = Glucose; Betanin

R_1 = Glucose; Isobetanin

FIGURE 4.28 Chemical structures of some betanins.

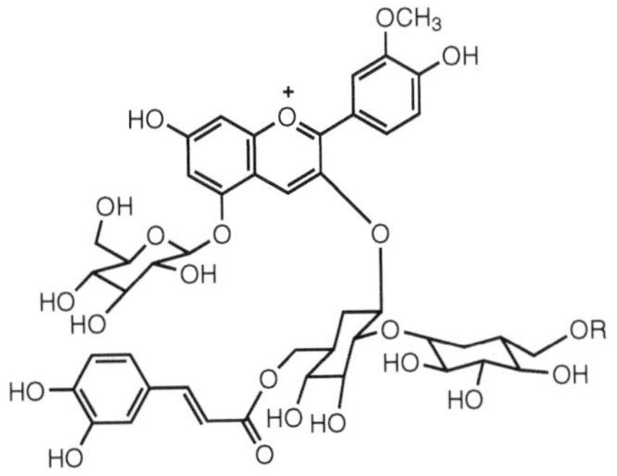

Anthocyanin	R
Peonidin-3-(dicaffeyl)-sophoroside-5-glucoside	caffeic acid
Peonidin-3-(*p*-hydroxybenzylcaffeyl)-sophoroside-5-glucoside	*p*-hydroxybenzoic acid
Peonidin-3-caffeylsophoroside-5-glucoside	H
Peonidin-3-(ferulylcaffeyl)-sophoroside-5-glucoside	ferulic acid

FIGURE 4.29 Chemical structures of some anthocyanins isolated from purple-fleshed sweet potato.

Purcell, 1979). Anthocyanins and chlorogenic acids are the major phenolics found in sweet potato tubers (Suda et al., 2002; Walter and Purcell, 1979).

The purple-fleshed sweet potato tuber is a good source of anthocyanins; the major anthocyanin pigments identified in sweet potato include mono- and diacetylated forms of cyanidin and peonidin (Goda et al., 1997; Odake et al., 1992; Suda et al., 2002; Terahara et al., 1999). Figure 4.29 shows the chemical structures of some anthocyanins detected in purple-fleshed sweet potato. The total content of anthocyanins in sweet potato is 590 mg of peonidin 3-caffeoylsophoroside-5-glucoside equivalent/kg (Furuta et al., 1998). Acylated peonidin and acylated cyanidin constitute 74 and 19% of anthocyanin pigments in sweet potato cv. Ayamurasaki, respectively (Tsukui et al., 1999).

The greens of sweet potato (*Ipomoea batatas* (L) Lam) cultivars are used in many parts of the world as fresh vegetables (Nwinyi, 1992; Villareal et al., 1982;

Woolfe, 1992). These greens are a rich source of antioxidants, proteins, vitamin B and minerals such as iron, calcium and zinc (Islam et al., 2002a; Woolfe, 1992). The total phenolic content in sweet potato greens ranges from 14.2 to 171 g of chlorogenic acid equivalents/kg of dry matter. The leaves of sweet potato are the richest source of phenolics (61.9 g/kg of dry matter), followed by petioles (29.7 g/kg of dry matter) and stems (18.8 g/kg of dry matter) (Islam et al., 2002b). Six phenolic acids have been identified in sweet potato leaves: 3,5-di-*O*-caffeoylquinic acid, 4,5-di-*O*-caffeoylquinic acid, chlorogenic acid, 3,4-di-*O*-caffeoylquinic acid, 3,4,5-tri-*O*-caffeoylquinic acid and caffeic acid. Of these, 3,4,5-tri-*O*-caffeoylquinic acid and 4,5-di-*O*-caffeoylquinic acid are the predominant phenolics in sweet potato greens (Islam et al., 2002b).

LEAVES AND STEMS

Asparagus

Ferulic and *p*-coumaric acids have been detected only in the cell wall material of white asparagus and are predominantly concentrated in the basal region of the spear. No phenolic acid dimers are present, suggesting that cross linking polysaccharides via phenolic acid dimers is not an important component of cell wall structure. A slight increase in the phenolic acid levels is detected after 21-day storage of asparagus at 4°C (Rodriguez et al., 1999). Later, Rodriguez-Arcos et al. (2002) detected the presence of ferulic acid dehydrodimers, namely, 5,5′-, 8,5′- and 8,8′-diferulic acid, 8-*O*-4′diferulic acid, 8,5′-diferulic acid benzofuran form and 8,8′-diferulic acid aryltetralyn form, in green asparagus (Figure 4.25). The ferulic acid dehydrodimers comprise >60% of total ferulic acid and >70% after 3 days of storage at 21°C. These authors also noticed a three-fold increase in the total content of ferulic acids during storage, particularly in the cell walls located in the basal section of the stem.

Eleven phenolic compounds have been detected in the aroma components of cooked asparagus: phenol, *o*-, *m*- and *p*-cresols, guaiacol, acetovanillone, 4-hydroxybenzaldehyde, vanillin, veratrol, 4-vinylphenol and 4-vinylguaiacol. These phenols are products of thermal fragmentation of free *p*-coumaric acid and ferulic acid released from cell wall materials by enzymes (Tressl et al., 1977).

Rutin is the major flavonoid found in asparagus spears, but eight more flavonoid (quercetin and kaempferol) glycosides have also been detected (Figure 4.2) (Billau et al., 1985; Dame et al., 1957; Makris and Rossiter, 2001; Stevenson, 1950; Woldecke and Herrmann, 1974c). The content of rutin in green asparagus ranges from 250 to 1000 mg/kg of fresh weight (Makris and Rossiter, 2001; Stevenson, 1950), while in white asparagus it is between 1 and 20 μg/kg of fresh weight (Billau et al., 1985; Woldecke and Herrmann, 1974c). Chopping and maceration of asparagus spears decreases rutin content by 18.5%; boiling for 60 min in water results in a 43.9% decrease (Makris and Rossiter, 2001).

Rutin is also involved in the development of objectionable black discoloration on the surface of canned green asparagus within a few minutes to 3 h after the can is opened (Lueck, 1970) and in the formation of highly visible white and grayish deposits on the surface of pickled green asparagus after 6 months of storage (Fuleki,

1999). The black discoloration is caused by the formation of a rutin–ferric ion complex, while crystalline deposits on the surface of pickled green asparagus are the result of lower solubility of rutin in acidic solutions.

Celery

The major phenolic components of celery include linear furanocoumarins (psolarens), psolaren, bergapten (5-methoxypsolaren), xanthotoxin (8-methoxypsolaren) and isopimpinellin (5,8-dimethoxypsolaren) as shown in Figure 4.20 (Beier et al., 1983; Diawara et al., 1995). Their presence was first reported by Scheel et al. (1963), who isolated and identified two psolarens, 4,5′,8-trimethylpsolaren and 8-methoxypsolaren, in diseased celery. Subsequently, Wu et al. (1972) reported the presence of xanthotoxin and bergapten in diseased celery. Later, Tang et al. (1990) identified psolaren, 9-methoxypsolaren, 4-methoxypsolaren and 4,9-dimethoxypsolaren in volatiles, while Trumble et al. (1992) and Nigg et al. (1997) detected several psolarens in stems, leaves and petioles of healthy fresh celery (*Apium graveolens* L.). Synthesis of psolarens in plants may be induced by a number of environmental stresses such as pathogens, cold temperature (Beier and Oertli, 1983), acidic atmospheric pollution (Dercks et al., 1990), and even fungicide treatment (Nigg et al., 1997).

Except for isopimpinellin (Ashwood-Smith et al., 1992; Hudson et al., 1987), these phenolic compounds have been linked to acute (Austad and Kavli, 1983) and chronic (Seligman et al., 1987) dermatitis in celery handlers. The total content of furanocoumarins at levels of 7 to 9 mg/kg and ≥18 mg/kg of fresh weight may cause chronic (Seligman et al., 1987) and acute (Austad and Kavli, 1983) dermatitis responses, respectively. Furthermore, ingestion of celeriac may cause a serious photoxic burn in humans (Ljunggren, 1990). Psolarens have also been implicated in photocarcinogenicity in mice and humans (Young, 1990), as well as initiation of skin cancer (Berenbaum, 1991; Musajo and Rodighiero, 1962; Stern et al., 1979;). The World Health Organization (International Agency for Research on Skin Cancer, 1983) has listed psolarens as causative agents of skin cancer.

The total content of furanocoumarins in six Florida celery cultivars processed similar to kitchen preparation ranged between 11.7 and 49.6 mg/kg. Of these, the Florida 296 cultivar contained the highest level of total furanocoumarins; the contents of furanocoumarins in remaining cultivars did not differ from one another (Nigg et al., 1997). Similar levels of furanocoumarins have also been reported by others (Dercks et al., 1990; Diawara et al., 1993; Trumble et al., 1992). Celery leaves contain 7 to 15 times more furanocoumarins than stalks (Diawara et al., 1993; Trumble et al.1992). The total content of furanocoumarins in the outer leaves and inner mature leaves of celery ranges between 40 and 49.8 mg/kg of fresh weight and between 8 and 12 mg/kg of fresh weight, respectively; however, in the heart leaf and different parts of petioles (outer, inner and heart), the amounts do not exceed 3.8 mg/kg of fresh weight (Table 4.19) (Diawara et al., 1995). In addition, seasonal concentrations of bergapten in celery leaves and petioles are much higher than those of xanthotoxin and psolaren. However, the concentration of bergapten in celery petioles declines by 55 to 75% as the plant matures (Diawara et al., 1995).

TABLE 4.19
Distribution of Main Psolarens[a] in Different Parts of Celery Variety Tall Utah 5270R (*A. graveolens* L. var. dulce)

Plant Part	Bergapten	Psolaren	Xanthotoxin
Outer leaf	28.00 ± 4.55	3.94 ± 1.21	17.90 ± 3.66
Inner leaf	4.80 ± 0.57	0.17 ± 0.13	3.00 ± 0.55
Heart leaf	2.42 ± 0.43	0.02 ± 0.02	1.29 ± 0.24
Petiole	2.69 ± 0.18	0.01 ± 0.01	1.06 ± 0.15
Roots	0.57 ± 0.12	0.04 ± 0.01	0.49 ± 0.12

[a] Milligrams per kilogram of fresh weight.

Source: Adapted from Diawara, M.M., et al., 1995, *J. Agric. Food Chem.*, 43:723–727.

Beier and Oertli (1983) reported a significant increase in the contents of psolaren, bergapten and xanthotoxin in celery treated with copper sulfate and stored for 3 days at 4°C. Later, Trumble et al. (1992) reported that treatment of celery with pesticides (prometryn, naled, methomyl) did not significantly affect the concentration of coumarins. Subsequently, Nigg et al. (1997) demonstrated that treatment of celery with fungicides such as Bravo 500, Manzate-D or Kocide 101 efects a two- to four-fold increase in the content of bergapten in leaves and stalk, two- to three-fold increase of xanthotoxin in stalk, and two- to three-fold increase of isopimpinellin in leaves. Acidic pollution may also increase the content of furanocoumarins in leaves and petioles of celery up to 540% (to 135 mg/kg of fresh weight) and 440% (to 55.56 mg/kg of fresh weight), respectively. Moreover, a 10-fold increase in the level of furanocoumarins has been noted in celery infected with *Erwinia caratovora* cv. Caratovora (Surico et al., 1987), *Fusarium oxysporum* (Heath-Pagliuso et al.1992) and *Sclerotinia sclerotiorum* (Chaudhary et al., 1985). However, growth of celery under elevated atmospheric carbon dioxide conditions does not significantly increase the content of furanocoumarins in celery (Reitz et al., 1997).

Apigenin and luteolin have been identified in acidic extracts of celery (Figure 4.2). Celery consumed in the Netherlands contains 22 and 108 mg/kg of fresh weight of luteolin and apigenin, respectively (Hertog et al., 1992a,b). Similar levels of luteolin and apigenin have been found in celery purchased in Glasgow, U.K., (Table 4.20) (Crozier et al., 1997) and in Malaysia (Miean and Muhamed, 2001).

Endive

The total content of flavonoids in various cultivars of endive (*Cichorium endivia*) may range from <2 to 246 mg/kg of fresh sample weight (DuPont et al., 2000; Hertog et al., 1992a). Flavonoids found in endive are a mixture of kaempferol and quercetin conjugates, namely, kaempferol 3-glucuronide, kaempferol 3-*O*-(6-*O*-malonyl)-glucoside, kaempferol 3-*O*-glucoside (DuPont et al., 2000; Herrmann, 1976), quercetin 3-rhamnoside and quercetin 3-galactoside (Goupy et al., 1990; Hertog et al., 1992a;

TABLE 4.20
Content of Luteolin and Apigenin in Commercial Celery[a]

Celery	Apigenin	Luteolin
White stalks	0–104	0–40
White hearts	17	6.6
Green hearts	191	35

[a] Milligrams per kilogram of fresh weight.

Source: Adapted from Crozier, A. et al., 1997, *J. Agric. Food Chem.*, 45:590–595.

Rees and Harborne, 1984). Of these, kaempferol 3-glucuronide comprises over 70% of total flavonoids (DuPont et al., 2000). Small quantities of hydroxycinnamoyl derivatives such as dicaffeoyl tartaric esters and caffeoyl esters (DuPont et al., 2000; Goupy et al., 1990; Winter and Herrmann, 1986; Woldecke and Herrmann, 1974a) have also been detected in endive extracts. Foliar parenchyma contain higher levels of phenolics than veins; phenolics are located in the foliar parenchyma in the vacuoles, preferentially in the green over the etiolated foliar parenchyma (Goupy et al., 1990). Exposure of shredded endives, sealed in polyethylene bags, to light at 22°C for 48 h reduces the total content of flavonoids and also alters their composition (DuPont et al., 2000). Presence of phenol and 2-ethylphenol in neutral volatiles from blended endive has also been confirmed (Gotz-Schmidt and Schreier, 1986).

Lettuce

Caffeoyltartaric, chlorogenic, dicaffeoyltartaric and 3′,5′-dicaffeoylquinic acids have been identified in red lettuce (cv. Lollo Rosso) (Ferreres et al., 1997a), as well as in iceberg and romaine lettuce (Cantos et al., 2001a; Hyodo et al., 1978; Ke and Saltveit, 1989; Winter and Herrmann, 1986). A number of quercetin conjugates have also been detected in red pigmented and green leaf lettuce. These include quercetin 3-(6-malonylglucoside), quercetin 3-glucoside, quercetin 3-glucuronide, and quercetin 3-rhamnoside (DuPont et al., 2000; Ferreres et al., 1997a). In addition, green leaf lettuce contains quercetin 3-galactoside (DuPont et al., 2000), while quercetin 3-(6-malonylglucoside)-7-glucoside, and cyanidin 3-*O*-glucoside cyanidin 3-*O*-[(6-*O*-malonyl)glucoside] are found in red pigmented lettuce (DuPont et al., 2000; Ferreres et al., 1997a;Yamaguchi et al., 1996).

The total content of phenolic acids in whole red pigmented lettuce ranges from 65 to 270 mg/kg of fresh weight (Winter and Herrmann, 1986), while romaine, iceberg and butter leaf lettuce contain between 2.83 and 45 mg of phenolic acids/kg of fresh weight (Cantos et al., 2001a; Tomás-Barberán et al., 1997). Red lettuce is a good source of flavonoids (Table 4.21) (Crozier et al., 1997; DuPont et al., 2000; Gil et al., 1998a). According to Bilyk and Sapers (1985), green leaf and head lettuce varieties contain 2 to 54 and 1 to 28 mg quercetin/kg of fresh weight, respectively. The outer and inner leaves of Lollo Rosso red lettuce contain 911 and 450 mg of quercetin/kg of fresh weight, respectively (Crozier et al., 1997); its red tissues

TABLE 4.21
Total Content of Flavonoids in Whole Lettuce[a]

Variety	Total Flavonoids	Reference
Round	11 ± 0.5	Crozier, A. et al. (1997)
Green salad bowl	147 ± 5.2	
Lollo biondo	94 ± 4.6	
Iceberg	0.3 ± 0.05	DuPont, M.S. et al. (2000)
Green salad bowl	19.9 ± 2.0	
Lollo biondo	95.7 ± 4.2	
Lollo rosso	207.0 ± 13.0	
Red oak leaf	76.2 ± 4.0	

[a] Milligrams per kilogram of fresh weight.

TABLE 4.22
Total Phenolic Acids and Flavonoids Content in Various Tissues of Lollo Rosso Lettuce[a]

Tissue	Phenolic Acids	Flavonoids
White	213 ± 60	43 ± 9
Green	570 ± 54	244 ± 20
Red	1696 ± 166	1384 ± 197

[a]Milligrams per kilogram of fresh weight.

Source: Adapted from DuPont, M.S. et al., 2000, *J. Agric. Food Chem.*, 48:3957–3964.

contain three to eight times more phenolics and 5 to 30 times more flavonoids than its white and green tissues (Table 4.22) (DuPont et al., 2000).

Exposure of shredded green lettuce leaves to fluorescent light for 48 h at 22°C affects the total flavonoid content and the distribution of quercetin conjugates. The changes in quercetin conjugates are brought about by their demalonation (DuPont et al., 2000; Gil et al., 1998a). Gil et al. (1998a) have suggested that demalonation leads first to glucosides and then to formation of acetyl derivatives. However, according to Horowitz and Asen (1989), changes in the flavonoid malonylglycosides may be initiated by their decarboxylation. Wounding lettuce leaves increases the content of phenolic acids two-fold in the stems of red pigmented lettuce leaf (Ferreres et al., 1997a); in stems of iceberg lettuce leaves the increase is 10-fold (Cantos et al., 2001b; Tomás-Barberán et al., 1997). Chlorogenic acid is the predominant phenolic acid accumulated in processed midribs of lettuce leaves (Cantos et al., 2001a; Ferreres et al., 1997a). In addition, a marked increase in the activity of phenylalanine ammonia-lyase (PAL) in wounded lettuce leaves has been observed (Cantos et al.,

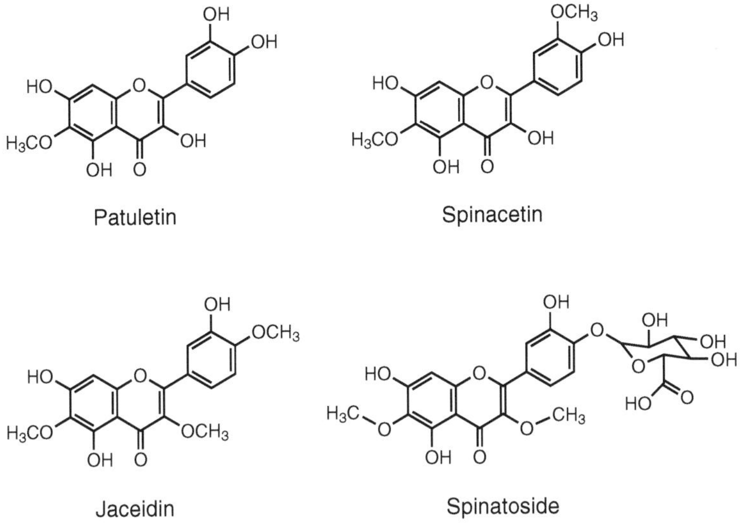

FIGURE 4.30 Chemical structures of some phenolic compounds found in spinach.

2001a; Tomás-Barberán et al., 1997). However, no correlation exists between browning of lettuce leaves and the increase in phenolic acid and PAL activities (Cantos et al., 2001a).

Spinach

Total phenol content in fresh spinach (*Spinacia oleracea*) leaves is 1629 to 4835 mg of chlorogenic acid equivalents/kg of fresh weight. Moreover, spring-grown spinach contains a higher level of total phenols than that of spinach grown in the fall (Howard et al., 2002). Spinach leaves are a rich source of flavonoids; fresh-cut spinach contains from 807 to 2241 mg of flavonoids/kg of fresh weight (Gil et al., 1999; Howard et al., 2002). Spinach flavonoids exhibit antioxidant (Gil et al., 1999), antiinflammatory (Lomnitski et al., 2000a, b), antimutagenic (Edenharder et al., 2001) and anticarcinogenic (Nyska et al., 2001) properties. The spinach flavonoids include

- Patuletin (quercetagetin 6-methyl ether)
- Spinacetin (quercetagetin 6,3′-dimethyl ether) (Zane and Wender, 1961)
- Spinatoside (3,6-dimethylquercetagetin 4′-*O*-glucuronide) (Figure 4.30) (Wagner et al., 1977)
- Jaceidin 4′-glucuronide
- Patuletin 3-gentobioside
- Patuletin 3-glycosyl-(1→6)-[apiosyl)(1→2)]-glucoside
- Spinacetin 3-glycosyl(1→6)-[apiosyl(1→2)]-glucoside
- Patuletin 3-(2″-feruloylglycosyl)(1→6)-[apiosyl(1→2)]-glucoside
- Spinacetin 3-(2″-feruloylglycosyl)(1→6)-[apiosyl(1→2)]-glucoside

- Spinacetin 3-(2″-*p*-coumaroylglycosyl) (1→6)-[apiosyl(1→2)]-glucoside
- Spinacetin 3-gentiobioside (Aritomi et al., 1984, 1986; Ferreres et al., 1997b; Howard et al., 2002)

In addition, Edenharder et al. (2001) and Howard et al. (2002) have identified the 4′-glucuronides of

- 5,7,4′-Trihydroxy-3,6,3′-trimethoxyflavone
- 5,3′,4′-Trihydroxy-3-methoxy-6,7-methylenedioxyflavone
- 5,4′-Dihydroxy-3,3′-dimethoxy-6,7-methylenedioxyflavone (Aritomi and Kawasaki, 1984)
- 5,3′-Dihydroxy-4′-methoxy-6,7-methylenedioxyflavonol 3-*O*-β-glucuronide
- 5,2′,3′-Trihydroxy-4′-methoxy-6,7-methylenedioxyflavonol 3-*O*-β-glucuronide
- 5-Hydroxy-3′,4′-dimethoxy-6,7-methylenedioxyflavonol 3-*O*-β-glucuronide
- 5,6,3′-Trihydroxy-7,4′-dimethoxyflavonol 3-*O*-β-glucuronide
- 5,6-Dihydroxy-7,3′,4′-trimethoxyflavonol 3-*O*-β-glucuronide
- 5,6,4′-Trihydroxy-7,3′-dimethoxyflavonol 3-*O*-disaccharide
- 5,6,3′4′-Tetrahydroxy-7-methoxyflavonol 3-*O*-disaccharide
- 5,8,4′-Trihydroxyflavanone
- 7,8,4′-Trihydroxyflavanone

Patuletin derivatives, spinacetin derivatives, jaceidin and flavones account for 21.8, 32.4, 6.3, and 39.5% of the total flavonoids in spinach leaves, respectively (Howard et al., 2002). Storage of spinach up to 7 days under modified atmosphere at 10°C does not affect the total flavonoid content. After 10 min of cooking in water (spinach-to-water ratio of 1:4) at 90°C, the total content of flavonoids in spinach is reduced by 50%. Flavonoid glucuronides have been extracted into cooking water more than other flavonoid compounds (Gil et al., 1999).

Swiss Chard

The total content of flavonoids in leaves of green and yellow Swiss chard cultivars ranges from 2.4 to 3.0 g and from 2.1 to 2.3 g of flavonoid/kg of fresh weight, respectively. Five flavonoids isolated from green cultivars are identified as 2″-xyloside vitexin, 6″-malonyl-2″-xyloside vitexin, 3-gentiobioside kaempferol, 3-gentiobioside isorhamnetin and 3-vicianoside isorhamnetin. Only 2″-xyloside vitexin and its 6″-malonylated derivative have been detected in the yellow cultivar. Of these flavonoids, 2″-xyloside vitexin and 6″-malonyl-2″-xyloside vitexin comprise over 60% of total flavonoids in green and yellow cultivars (Gil et al., 1998b). Boiling the leaves of Swiss chard in water for 10 min removes about 30 and 50% of flavonoids initially present in green and yellow cultivars, respectively (Gil et al., 1998b). On the other hand, storing them under modified atmosphere (7% O_2 and 10% CO_2) does not affect the total flavonoid content to any great extent (Gil et al., 1998b).

Other Vegetables

Avocado

Golan et al. (1977) reported that mesocarp (edible portion of avocado) of Fuerte cultivar of avocado contains 287 mg of phenolics/kg of fresh weight expressed as chlorogenic acid equivalents; they identified four phenolic acids, namely, *p*-coumaric, caffeic, ferulic and protocatechuic acids. Later, Torres et al. (1987) found that the total content of phenolics in mesocarps of four avocado cultivars ranged from 1100 to 1800 mg/kg of fresh weight expressed as gallic acid equivalents. These authors identified 16 phenolic acids in avocado mesocarp: *p*-hydroxybenzoic, *o*-pyrocatechuic, protocatechuic, α–,β– and γ–resorcylic, caffeic, *o*-, *m*- and *p*-coumaric, and syringic, ferulic, sinapic, isovanillic and vanillic acids. Two anthocyanins, cyanidin 3-galactoside and cyanidin 3,5-diglucoside acylated with *p*-coumaric acid, have also been detected (Prabha et al., 1980).

Pepper

Consumed as fresh or pickled fruits, as well as spices and seasonings, over 200 *Capsicum* cultivars are cultivated around the world (Contreras-Padilla and Yahia, 1998). The quality of peppers is affected by their color, aroma and pungency. Capsaicinoids are responsible for the development of pungency in pepper (Kirschbaum-Titze et al., 2002a, b).

Total content of soluble phenolics in pepper is between 1180 and 3849 mg of chlorogenic acid equivalents/kg of fresh weight (Lee et al., 1995; Racchi et al., 2002). Flavonoids and capsaicinoids are the predominant phenolics found in pepper. Small amounts of phenolic acids such as protocatechuic, chlorogenic, coumaric and ferulic acids are also detected. Of these, protocatechuic, chlorogenic and coumaric acids are the major phenolic acids in the fruit of pepper (Figure 4.1 and Figure 4.15) (Contreras-Padilla and Yahia, 1998; Estrada et al., 2000). Various phenolic acid glycosides have been identified; these include *cis-p*-coumaric acid 4-*O*-β-D-glucoside, *trans*-sinapoyl β-D-glucoside and vanilloyl β-D-glucoside (Iorizzi et al., 2001). The total content of phenolic acids declines during fruit ripening from 215 μg/kg (14 days after flowering) to 73 μg/kg (42 days after flowering) of dry weight (Estrada et al., 2000). This decrease in phenolic acids has been linked to increase in capsaicinoid content (Contreras-Padilla and Yahia, 1998).

Quercetin and luteolin (Figure 4.2), two major flavonoids found in pepper fruits (*Capsicum* species), are present in fruits in the form of glycosidic conjugates (Iorizzi et al., 2001; Sukrasno and Yeoman, 1993). The total content of flavonoids in pepper cultivars ranges from 1.75 mg/kg of fresh weight in *Capsicum chinese* cv. Red Savina to 851.53 mg/kg of fresh weight in yellow wax (*Capsicum annuum* L.) cv. Hungarian Yellow (Howard et al., 2000; Lee et al., 1995). Recently, Iorizzi et al. (2001) reported the occurrence of lignin glycoside icaraside E_5, a potential antioxidant containing a vanillyl moiety, in *Capsicum* species.

Genus *Capsicum* contains a unique group of phenolic compounds, known as capsaicinoids, responsible for the development of pungency in the pepper fruit (Figure 4.31) (Bennett and Kirby, 1968; Leete and Louden, 1968). These phenolics

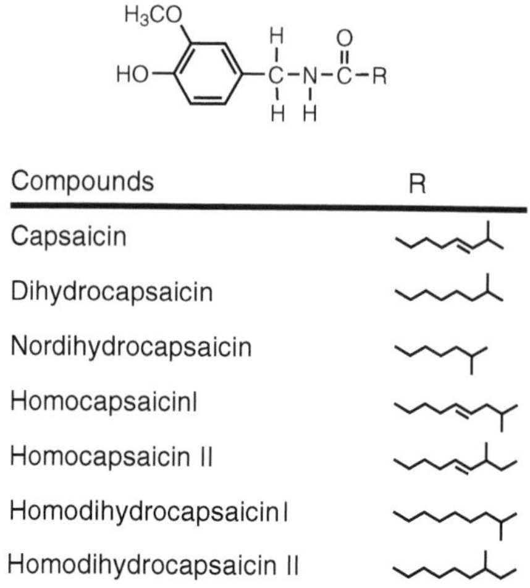

Compounds	R
Capsaicin	
Dihydrocapsaicin	
Nordihydrocapsaicin	
Homocapsaicinl	
Homocapsaicin II	
Homodihydrocapsaicin I	
Homodihydrocapsaicin II	

FIGURE 4.31 Chemical structures of some capsaicinoids.

are amides of vanillylamine and C_8 and C_{13} branched fatty acids and their synthesis occurs on the membrane of vacuoles in the placenta of the fruit (Fujikawe et al., 1982; Gassner, 1973; Suzuki and Iwai, 1984) by condensation of vanillylamine with fatty acids. The fatty acid moiety of the capsaicinoid is derived from L-valine, L-leucine, and L-isoleucine; the vanillylamine is formed from L-phenylalanine via phenylpropanoid pathway (Bennett and Kirby, 1968; Curry et al., 1999; Estrada et al., 1999a; Fujikawe et al., 1982; Leete and Louden, 1968; Sukrasno and Yeoman, 1993; Suzuki and Iwai, 1984). Furthermore, Sukrasno and Yeoman (1993) suggested that, in *Capsicum frutescens*, glycosylated phenolics may act as a source of intermediates for capsaicinoids synthesis. Later, Sakamoto et al. (1994) found no correlation between the accumulation of free phenolics and capsaicinoids in *Capsicum annuum.* Furthermore, individual capsaicinoids are not transformed into one another in the pepper fruit (Kopp and Jurenitsch, 1982), but their levels are controlled by the pool of CoA derivatives of fatty acids (Fujikawe et al., 1980; Suzuki and Iwai, 1984).

Capsaicinoids are accumulated predominantly in the epidermal tissue of the placenta (Ishikawa et al., 1998; Iwai et al., 1979; Rowland et al., 1983; Suzuki et al., 1980). Over 15 capsaicinoids have been isolated and identified (Bennett and Kirby, 1968; Constant and Cordell, 1996; Suzuki and Iwai, 1984). Of these, capsaicin (8-methyl-N-vanillyl-6-nonenamide) and dihydrocapsaicin contribute about 90% to the total pungency. Nordihydrocapsaicin is considered a third major pungent principle in the fruit of pepper (Iwai et al., 1979; Kosuge and Furata, 1970). The fruits of nonpungent cultivars of *Capsicum,* on the other hand, contain nonpungent capsaicinoid-like substances (CLS) that are esters of vanillyl alcohol and fatty acids (Yazawa et al., 1989). Kobata et al. (1999) proposed to name this group of compounds "capsinoids" (Figure 4.32). Capsilate (4-hydroxy-3-methoxybenzyl (*E*)-8-methyl-6-nonenoate), dihydrocapsiate (4-hydroxy-3-methoxybenzyl-8-methyl-6-nonanoate) and nordihydrocapsiate (4-hydroxy-3-methoxybenzyl-7-methyloctananoate) are the predominant capsinoids in nonpungent pepper fruits,

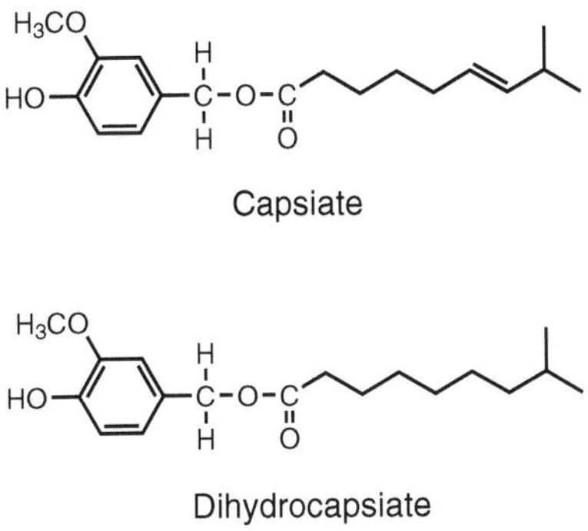

FIGURE 4.32 Chemical structures of some capsaicinoid-like substances.

but several minor CLS compounds have also been detected (Kobata et al., 1998, 1999).

The total content of capsaicinoids in the fruits of pepper ranges from 189 to 778 mg/kg of fresh weight (Contreras-Padilla and Yahia, 1998; Kirschbaum-Titze et al., 2002a, b; Rowland et al., 1983) and is affected by the species, growing season, ripening, and environmental factors such as light, temperature, fertilizer supplementation, as well as processing after harvesting (Contreras-Padilla and Yahia, 1998; Estrada et al., 1997, 1998, 1999a, b, 2000; Govindarajan, 1986; Iwai et al., 1979; Jurenitsch et al., 1979). *C. frutescens* varieties produce more pungent fruits compared to those of *C. annuum* varieties (Jurenitsch et al., 1979). The first signs of capsaicinoid accumulation in fruits are detected 15 to 20 days after flowering.

As *Capsicum* fruits begin developing, the content of capsaicinoids levels off between 25 and 50 days after flowering (Estrada et al., 1997, 1998, 1999b, 2000; Iwai et al., 1979; Suzuki et al., 1981) and then declines (Contreras-Padilla and Yahia, 1998). Oxidative degradation of capsaicinoids and/or phenolic precursors of capsaicin biosynthesis by peroxidase isoenzyme B_6, located in the placental epidermal cells (Bernal et al., 1993a, b, c, 1994, 1995; Contreras-Padilla and Yahia, 1998), have been implicated in the loss of pungency in later stages of fruit development. Bernal et al. (1995) demonstrated involvement of hot pepper peroxidase in oxidation of capsaicin precursors, caffeic and ferulic acids, leading to insolubilization of phenylpropanoid precursors. Later, Bernal and Ros Barcelo (1996) also identified 5,5′-dicapsaicin, 4′-*O*-5-dicapsaicin ether and dehydrogenation polymers with high molecular weights as products of capsaicin oxidation by peroxidase.

The fruits of pepper harvested between June and September contain a higher level of capsaicinoids compared to those harvested in October (Estrada et al., 1999b). Variations in capsaiciniod levels have also been found among fruits harvested at the same time from the same plant as well as plants subjected to identical growing conditions (Kirschbaum-Titze et al., 2002b). The content of capsaicinoids does not correlate with color (Kirschbaum-Titze et al., 2002a), but is affected by the location of fruit on the plant. The apical fruits contain higher levels of capsaicinoids (1090 mg/kg of dry weight) compared to those harvested from the middle and basal (780 mg/kg of dry weight) segments of the plant (Estrada et al., 2002).

Upon heating, capsaicin decomposes to three major products: vanillin, 8-methyl-6-nonenamide and 8-methyl-6-nonenoic acid. Other minor products of thermal decomposition include heptane, 2-methoxyphenol, 2-methoxy-4-methylphenol, methylvanillin, n-vanillyl-di(8-methyl-6-nonen)imide, iso- and nonanamide, pentamide and nonanic acid (Henderson and Henderson, 1992). Storing halved fruits at 4°C for 15 days does not have any detrimental effect on the capsaicinoid content in fruits, but storing minced fruits at 4°C results in the loss of 90% of the original amount of capsaicinoids after several days. Much lower losses of capsaicinoids are observed when minced fruits are stored under nitrogen (Kirschbaum-Titze et al., 2002a). Enzymatic metabolism has been found responsible for the reduction of capsaicinoids in minced fruits; this was affected by temperature and water activity (Titze and Petz, 1999).

Tomato

The total flavonol content of tomatoes grown in different countries ranges from 1.3 to 36.4 mg of quercetin/kg of fresh weight (Dewanto et al., 2002; Stewart et al., 2000) but, according to Hertog et al. (1992a) and Crozier et al. (1997), the total quercetin content is between 1.3 and 203 mg quercetin/kg of fresh weight. Flavonols are located mostly in the tomato skin and only small quantities are found in the flesh and seeds (Table 4.23) (Stewart et al., 2000).

The content of flavonols in tomatoes is affected by the cultivar, harvesting season (Crozier et al., 1997; Stewart et al., 2000) and environmental stresses such as the amount of daylight and temperature, nutritional status of soil, application of plant growth regulators, and invasion with pathogens (Bongue-Bartelsman and Phillips, 1995; Christie et al., 1994; Dixon and Paiva, 1995; Flors et al., 2001; Landry et al., 1995). Cherry tomatoes contain a much higher level of flavonols (17 to 203 mg/kg of fresh weight) than larger size tomato cultivars (2.2 and 11.2 mg/kg of fresh weight) (Crozier et al., 1997).

TABLE 4.23
Distribution of Flavonols in Spanish Cherry Tomatoes cv. Paloma

	(%) Total Flavonols			
Sample	Free	Conjugated Quercetin	Conjugated Kaempferol	Total Flavonols (mg/kg f.w.[a])
Whole	2.8	92.5	4.7	25.3 ± 1.3
Skin	0.8	96.2	3.0	143.3 ± 5.8
Flesh	8.3	75.0	17.7	1.2 ± 0.1
Seed	20.0	66.7	13.3	1.5 ± 0.1

[a] f.w. = fresh weight.

Source: Adapted from Stewart, A.J. et al., 2000, *J. Agric. Food Chem.*, 48:2663–2669.

Furthermore, marked seasonal variations in the flavonol content have been observed. The content of flavonols in Spanish cherry tomatoes var. Paloma, harvested in February 1997 and February 1998, varied between 10.2 and 27.8 mg of quercetin/kg of fresh weight (Stewart et al., 2000), while that of English cherry tomatoes var. Favorita, harvested between February 1995 and August 1995, ranged from 17 to 77 mg of quercetin/kg of fresh weight (Crozier et al., 1997). Exogenous application of a plant growth regulator (mixture of 10 mM adipic acid monoethyl ester, 7.5 mM 1,2,3,4-tetra-O-acetyl-β–D-glucopyranose, and 2.5 mM furfuralamine) during the vegetative development cycle of the tomato plant (at a 0.2% dose) effects an increase in the content of soluble phenolics and flavonoids by over 100% (Flors et al., 2001).

Quercetin conjugates are the predominant form of flavonols found in tomatoes, but smaller quantities of kaempferol conjugates and traces of free aglycons have also been detected (Crozier et al., 1997; Stewart et al., 2000). Flavonols of tomatoes are a mixture of quercetin 3-rhamnosylglucoside (rutin), quercetin 3-rhamnosyldiglucoside, kaempferol 3-rhamnosylglucoside and kaempferol 3-rhamnosyldiglucoside (Woldecke and Herrmann, 1974b). Presence of *p*-coumaric acid conjugate of rutin has also been reported (Buta and Spaulding, 1997). Of these, rutin is the major flavonol in tomatoes (Stewart et al., 2000; Woldecke and Herrmann, 1974b).

Immature green tomato fruits contain a high level of chlorogenic acid in the pericarp and pulp. The level of chlorogenic acid rapidly declines as the color of the fruit changes from green to pink and then to red. The content of rutin also reaches a maximum in green tomato and then decreases during successive stages of fruit development (Buta and Spaulding, 1997; Fleuriet and Macheix, 1981). On the other hand, *p*-coumaric acid glucoside has been found only in the pulp and reaches a maximum level at the breaker stage (i.e., when the fruit color changes from green to pink). However, no rapid decline in *p*-coumaric acid glucoside content has been noted at successive stages of fruit development (Buta and Spaulding, 1997). Changes in chlorogenic acid and rutin are similar to those observed for auxin (indole-3-acetic acid) metabolism. Therefore, rutin (Figure 4.20) and chlorogenic acid (Figure 4.15) have been suggested as regulators of auxin metabolism during ripening of tomato fruit (Buta and Spaulding, 1994; Hocher et al., 1992; Jacobs and Rubery, 1988). Cohen (1996) demonstrated that supplying tomato fruits with exogenous indole-3-acetic acid delayed their ripening.

Frying, boiling or microwaving removes 35 to 78% of quercetin conjugates originally present in tomatoes. These losses may be due to the degradation or extraction of flavonols from tomato by water (Crozier et al., 1997). Tomato juice and puree are a rich source of flavonols. Processing tomatoes to juice and puree increases the content of free quercetin by up to 30%, an increase that may be brought about by enzymatic hydrolysis of quercetin conjugates. Tomato juice and puree contain 15.2 to 16.9 mg/L and 16.6 to 72.2 mg/kg of fresh weight of flavonols, respectively. On the other hand, canned tomatoes are a poor source of flavonols (Stewart et al., 2000).

REFERENCES

Adedeji, J., Hartman, T.G., Lech, J., and Ho, C.-T. 1992. Characterization of glycosidically bound aroma compounds in the African mango (*Magnifera indica* L.). *J. Agric. Food Chem.*, 40:659–661.

Adrian, M., Jeandet, P., Douillet-Breuil, A.C., Tesson, L., and Bessis, R. 2000. Stilbene content of mature *Vitis vinifera* berries in response to UV-C elicitation. *J. Agric. Food Chem.*, 48:6103–6105.

Afek, U. and Sztejnberg, A. 1993. Temperature and gamma irradiation effects on scoparone, a phytoalexin associated with resistance of citrus to *Phytophtora citriphtora. Phytopatology*, 83:753–758.

Afek, U. and Sztejnberg, A. 1988. Accumulation of scoparone, a phytoalexin associated with resistance of citrus to *Phytophtora citriphtora. Phytopatology*, 78:1678–1682.

Aguilera, J.M., Oppermann, K., and Sanchez, F. 1987. Kinetics of browning of Sultana grapes. *J. Food Sci.*, 52:990–993.

Alanko, J., Riutta, A., and Vappatalo, H. 1992. Effects of catecholamines on eicosanoid synthesis with special reference to prostanoid/leukotriene ratio. *Free Rad. Biol. Med.*, 13:677–688.

Alasalvar, C., Grigor, J.M., Zhang, D., Quantick, P.C., and Shahidi, F. 2002. Comparison of volatiles, phenolics, sugars, antioxidant vitamins, and sensory quality of different colored carrot varieties. *J. Agric. Food Chem.*, 50:2039–2041.

Alley, A. 1987. Parsnips and furanocoumarins. *Food Chem. Toxicol.*, 25:634–635.

Alonso-Salces, R.M., Korta, E., Barranco, A., Berrueta, L.A., Gallo, B., and Vicente, F. 2001. Determination of polyphenolic profiles of Basque cider apple varieties using accelerated solvent extraction. *J. Agric. Food Chem.*, 49:3761–3767.

Amakura, Y., Umino, Y., Tsuji, S., and Tonogai, Y. 2000. Influence of jam processing on the radical scavenging activity and phenolic content in berries. *J. Agric. Food Chem.*, 48:6292–6297.

Ames, B.M., Shigena, M.K., and Hagen, T.M. 1993. Oxidants, antioxidants and the degenerative diseases of aging. *Proc. Natl. Acad. Sci. U.S.A.*, 90:7915–7922.

Amiot, M.J., Tacchini, M., Aubert, S., and Nicolas, J. 1992. Phenolic composition and browning susceptibility of various apple cultivars at maturity. *J. Food Sci.*, 57:958–962.

Amiot, M.J., Aubert, S., and Nicolas, J. 1993. Phenolic composition and browning susceptibility of various apple and pear cultivars at maturity. *Acta Hortic.*, 343:67–69.

Amiot, M.J., Tacchini, M., Aubert, S.Y., and Oleszek, W. 1995. Influence of cultivar, maturity stage, and storage conditions on phenolic composition and enzymatic browning of pear fruits. *J. Agric. Food Chem.*, 43:1132–1137.

Amiot, M.J., Fleuriet, A., Cheynier, V., and Nicolas, J. 1997. Phenolic compounds and oxidative mechanisms in fruit and vegetables, in *Phytochemistry of Fruit and Vegetables, Proceedings of the Phytochemical Society of Europe 41*, Tomás-Barberán, F.A. and Robins, R.J., Eds., Clarendon Press, Oxford, U.K., 51–86.

Andrade, P.B., Carvalho, A.R.F., Seabra, R.M., and Ferreira, M.A. 1998. A previous study of phenolic profiles of quince, pear, and apple purees by HPLC diode array detection for the evaluation of quince puree genuineness. *J. Agric. Food Chem.*, 46:968–972.

Aritomi, M. and Kawasaki, T. 1984. Three highly oxygenated flavone glucuronides in leaves of *Spinacia oleracea. Phytochemistry*, 23:2043–2047.

Aritomi, M., Komori, T., and Kawasaki, T. 1986. Flavonol glycosides in leaves of *Spinacia oleracea. Phytochemistry*, 25:231–234.

Arogba, S.S. 2000. Mango (*Magnifers indica*) kernel: chromatographic analysis of tannin, and stability study of the associated polyphenol oxidase activity. J. *Food Comp. Anal.*, 13:149–156.

Ascherio, A., Rimm, E.B., Giovannucci, E.L., Colditz, G.A., Rosner, B., Willet, W.C., Sacks, F., and Stampfer, M.J. 1992. A prospective study of nutritional factors and hypertension among U.S. men. *Circulation*, 86:1475–1484.

Ashwood-Smith, M.J., Ceska, O., and Warrington, P.J. 1992. Isopimpinellin is photobiologically inactive. *Photochem. Photobiol.*, 38:113–118.

Aung, L.H., Smilanick, J.L., Vail, P.V., Hartsell, P.L., and Gomez, E. 1996. Investigation into origin of chloroanisoles causing musty off-flavor in raisins. *J. Agric. Food Chem.*, 44:3294–3296.

Austad, J. and Kavli, G. 1983. Phototoxic dermatitis by celery infected by *Sclerotina sclerotiorum*. *Contact Dermatitis*, 9:448–451.

Aviram, M., Dornfeld, L., Rosenblat, M., Volkova, N., Kaplan, M., Coleman, R., Hayek, T., Presser, D., and Furhman, B. 2000. Pomegranate juice consumption reduces oxidative stress, atherogenic modification and platelet aggregation: studies in humans and atherosclerotic apolipoprotein E-deficient mice. *Am. J. Clin. Nutr.*, 71:1062–1076.

Awad, M.A., Wagenmakers, P.S., and de Jager, A. 2001. Effect of light environment on flavonoid and chlorogenic acid levels in the skin of 'Jonagold'apples. *Sci. Hortic.*, 88:289–298.

Awad, M.A., de Jager, A., Dekker, M., Jongen, W., van der Krol, A., and van der Plas, L. 2000a. Flavonoids and chlorogenic acid levels in apple skin as affected by mineral nutrition and crop load, in *Proc. 20th Int. Conf. Polyphenols*, Freising-Weihenstephen, Germany, 2:657–658.

Awad, M.A., de Jager, A., and van Westing, L.M. 2000b. Flavonoid and chlorogenic acid levels in apple fruit: characterization of variation. *Sci. Hortic.*, 83:249–263.

Azar, M., Verette, E., and Brun, S. 1987. Identification of some phenolic compounds in bilberry juice *Vaccinium myrtillus*. *J. Food Sci.*, 52:1255–1257.

Babic, J., Amiot, M.J., Nguyen-The, C., and Aubert, S. 1993. Changes in phenolic content in fresh ready-to-use shredded carrots during storage. *J. Food Sci.*, 58:351–355.

Ballington, J.R., Ballinger, W.E., and Maness, E.P. 1987. Interspecific differences in the percentage of anthocyanins, aglycones, and aglycone-sugars in the fruit of seven species of blueberries. *J. Am. Soc. Hortic. Sci.*, 112:859–864.

Bandyopadhyay, C., Gholap, A.S., and Mamdapur, V.S. 1985. Characterization of alkenylresorcinol in Mango (*Magnifera indica* L.) latex. *J. Agric. Food Chem.*, 33:377–379.

Barden, J.A., Halfacre, R.G., and Parrish, D.J. 1987. *Plant Science*, McGraw-Hill, New York.

Bathia, V.K., Ramanathan, J.D., and Seshadri, T.R. 1967. Constitution of magniferin. *Tetrahedron,* 23:1363–1368.

Bavaresco, L., Pettegolli, D., Cantù, E., Fregoni, M., Chiusa, G., and Trevisan, M. 1997. Elicitation and accumulation of stilbene phytoalexin in grapevine berries infected by *Botris cinerea*. *Vitis*, 36:77–83.

Beech, F.W. and Carr, J.G. 1977. Cider and perry, in *Alcoholic Beverages*, Rose, A.H., Ed., Academic Press, London, 139–313.

Beier, R.C. and Oertli, E.H. 1983. Psolaren and other linear furanocoumarins as photoalexins in celery. *Phytochemistry*, 22:2595–2597.

Beier, R.C., Ivie, G.W., Oertli, E.H., and Holt, D.L. 1983. HPLC analysis of linear furanocoumarins (psolarens) in healthy celery (*Apium* graveolens). *Food Chem. Toxicol.*, 21:163–165.

Ben-Yehoshua, S., Rodov, V., Kim, J., and Carmeli, S. 1992. Performed and induced antifungal materials of citrus fruits' relation to the enhancements of decay resistance by heat and UV treatments. *J. Agric. Food Chem.*, 40:1217–1221.

Ben-Yehoshua, S., Shapiro, B., Kim, J.J., Sharoni, J., Carmeli, S., and Kashman, Y., 1988. Resistance of citrus fruit to pathogens and its enhancement by curing, in *Proceedings of the 6th International Citrus Congress*, Goren R. and Mendel, K., Eds., Balaban Publishing, Rehovot, Israel, 1371–1374.

Benavente-Garcia, O., Castillo, J., Marin, F.R., Ortuno, A., and Del Rio, J.A. 1997. Uses and properties of citrus flavonoids. *J. Agric. Food Chem.*, 45:4505–4515.

Benavente-Garcia, O., Castillo, J., and Del Rio Conesa, J.A. 1993. Changes in neodiosmin levels during the development of *Citrus aurantium* leaves and fruits. Postulation of a neodiosmin biosynthetic pathway. *J. Agric. Food Chem.*, 41:1916–1919.

Bengoechea, M.L., Sancho, A.I., Bartollomé, B., Estrella, I., Gómez-Cordovés, and Hernandez, M.T. 1997. Phenolic comoposition of industrial manufactured purées and concentrates from peach and apple fruits. *J. Agric. Food Chem.*, 45:4071–4075.

Bennett, D.J. and Kirby, G.W. 1968. Constitution and biosynthesis of capsaicin. *J. Chem. Soc.*, C:442–446.

Berenbaum, M.R. 1991. Coumarins, in *Herbivores, Their Interactions with Secondary Plant Metabolites*, vol. I., Rosenthal, G.A. and Barenbaum, M.R., Eds., Academic Press, New York, 468.

Bernal, M.A. and Ros Barcelo, A. 1996. 5,5′-Dicapsaicin, 4′-O-5dicapsaicin ether, and dehydrogenation polymers with high molecular weights are main products of the oxidation of capsaicin by peroxidase from hot pepper. *J. Agric. Food Chem.*, 44:3085–3089.

Bernal, M.A., Calderon, A.A., Pedreno, M.A., Ferrer, M.A., Merino de Caceres, F., and Ros Barcelo, A. 1995. Oxidation of capsaicin and capsaicin phenolic precursors by basic peroxidase isoenzyme B6 from hot pepper. *J. Agric. Food Chem.*, 43:352–355.

Bernal, M.A., Merino de Caceres, F., and Ros Barcelo, A. 1994. Histochemical localization of peroxidase in *Capsicum* fruits. *Lebensm. Wiss.u.-Technol.*, 27:197–198.

Bernal, M.A., Pedreno, M.A., Calderon, A.A., Munoz, R., Ros Barcelo, A., and Merino de Caceres, F. 1993a. The subcellar localization of isoperoxidases in *Capsicum annuum* leaves and their different expression in vegetative and flowered plants. *Ann. Bot.*, 72:415–421.

Bernal, M.A., Calderon, A.A., Pedreno, M.A., Munoz, R., Ros Barcelo, A., and Merino de Caceres, F. 1993b. Dihydrocapsaicin oxidation by *Capsicum annuum* (var. annuum) peroxidase. *J. Food Sci.*, 58:611–613.

Bernal, M.A., Calderon, A.A., Pedreno, M.A., Munoz, R., Ros Barcelo, A., and Merino de Caceres, F. 1993c. Capsaicin oxidation by peroxidase from *Capsicum annuum* (var. annuum) fruits. *J. Agric. Food Chem.*, 41:1041–1044.

Bernards, M.A., Lopez, M.L., Zajicek, J., and Lewis, N.G. 1995. Hydroxycinnamic acid–derived polymers constitute the polyaromatic domain of suberin. *J. Biol. Chem.*, 270:7382–7386.

Beveridge, T., Harrison, J.E., and Weintraub, S.E. 1997. Procyanidin contribution to haze formation in anaerobically produced apple juice. *Lebesm.-Wiss.u.-Technol.*, 30:594–601.

Beveridge, T. and Tait, V. 1993. Structure and composition of apple juice haze. *Food Struct.*, 12:195–198.

Billau, W., Buchloh, G., Geiger, H., and Hartman, H.D. 1985. HPLC-analysis of lignan presursors in shoots of white asparagus (*Asparagus officinalis* L.) grown under three temperature regimes and two different types of soils. *Asparagus Res. Newslett.*, 3(1):1–2.

Billot, J., Hartmann, C., Macheix, J.J., and Rateau, J. 1978. Phenolic compounds during Passe-Crassane pear growth. *Physiol. Veg.*, 16:693–714.

Bilyk, A., Cooper, P.L., and Sapers, G.M. 1984. Varietal differences in distribution of quercetin and kaempferol in onion (*Allium cepa* L.) tissue. *J. Agric. Food Chem.*, 32:274–276.

Bilyk, A. and Sapers, G.M. 1985. Distribution of quercetin and kaempferol in lettuce, kale, chive, garlic chive, leek, horseradish, and red cabbage tissue. *J. Agric. Food Chem.*, 33:226–228.

Bilyk, A. and Sapers, G.M. 1986. Varietal differences in quercetin, kaempferol, and myricetin contents of highbush blueberry, cranberry, and thornless blackberry fruits. *J. Agric. Food Chem.*, 34:585–588.

Blankenship, S.M. and Richardson, D.G. 1985. Changes in phenolic acids and internal ethylene during long-term cold storage of pears. *J. Am. Soc. Hortic. Sci.*, 110:336–339.

Block, G., Patterson, B., and Subar, A. 1992. Fruits, vegetables and cancer prevention: a review of the epidemiological evidence. *Nutr. Cancer*, 18:1–29.

Bocco, A., Cuvelier, M.-E., Richard, H., and Berset, C. 1998. Antioxidant activity and phenolic composition of peel and seed extracts. *J. Agric. Food Chem.*, 46:2123–2129.

Bomser, J., Madhavi, D.L., Singletary, K., and Smith, M.L. 1996. *In vitro* anticancer activity of fruit extracts from *Vaccinum* species. *Planta Med.*, 62:212–216.

Bongue-Bartelsman, M. and Phillips, D.A. 1995. Nitrogen stress regulates gene expression of enzymes in the flavonoid biosynthetic pathway of tomato. *Plant Physiol. Biochem.*, 33:539–546.

Boukharta, M. 1988. Study of *Vitis vinifera* flavonoids: structures of procyanidins of grape seeds, canes and leaves. Ph.D. Thesis, Institut Polytechnique de Lorraine, Nancy, France; cited in Fuleki, T. and da Silva, R.J.M. 1997. Catechin and procyanidin composition of seeds from grape cultivars grown in Ontario. *J. Agric. Food Chem.*, 45:1156–1160.

Boyles, M.J. and Wrolstad, R.E. 1993. Anthocyanin composition of red raspberry juice: Influences of cultivar, processing, and environmental factors. *J. Food Sci.*, 58:1135–1141.

Braddock, R.J. and Bryan, C.R. 2001. Extraction parameters and capillary electrophoresis analysis of limonin glucoside and phlorin in citrus by-products. *J. Agric. Food Chem.*, 49:5982–5988.

Braddock, R.J. 1995. By-products of citrus fruits. *Food Technol.*, 49:74–77.

Brenneisen, V.R. and Steingger, E. 1981a. Zur analytik der polyphenole der früchte von *Vaccinium myrtillus* L. (Ericaceae) (Analysis of polyphenols from *Vaccinium myrtillus* L. (Ericaceae) fruit. *Pharm. Acta Helv.*, 56:180–185.

Brenneisen, V.R. and Steinegger, E. 1981b. Quantitativer Vergeleich der polyphenole in früchten von *Vaccinium myrtillus* L. Unterschiedlichen reifegrades (Quantitative comparison of polyphenols from *Vaccinium myrtillus* L. (Ericaceae) fruit. *Pharm. Acta Helv.*, 56:341–345.

Breuil, A.C., Adrian, M., Pirio, N., Meunier, P., Bessis, R., and Jeandet, P. 1998. Metabolism of stilbene phytoalexins by *Botris cinerea*. 1. Characterization of a resveratrol dehydrodimer. *Tetrahedron Lett.*, 39:537–540.

Brouillard, R., Figueiredo, P., Elhabiri, M., and Dangles, O. 1997. Molecular interactions of phenolic compounds in relation to colour of fruit and vegetables, in *Phytochemistry of Fruit and Vegetables, Proceedings of the Phytochemical Society of Europe 41*, Tomás-Barberán, F.A. and Robins, R.J., Eds., Clarendon Press, Oxford, U.K., 29–50.

Brugirard, A. and Tavernier, J. 1952. Les matières tannoïdes dans les cidres et les poirés. *Ann. Technol. Agric.*, 3:311–343.

Burda, S., Oleszek, W., and Lee, C.Y. 1990. Phenolic compounds and their changes in apples during maturation and storage. *J. Agric. Food Chem.*, 38:945–948.

Burger, O., Weiss, E., Sharon, N., Tabak, M., Neeman, I., and Ofek, I. 2002. Inhibition of *Helicobacter pyroli* adhesion to human gastric mucus by a high-molecular-weight constituent of cranberry juice. *CRC Crit. Rev. Food Sci. Nutr.*, 42(Suppl.):279–284.

Buta, J.G. and Spaulding, D.W. 1997. Endogenous levels of phenolics in tomato fruit during growth and maturation. *J. Plant Growth Regul.*, 16:43–46.

Buta, J.G. and Spaulding, D.W. 1994. Changes in indole-3-acetic acid and abscisic acid levels during tomato (*Lycospersicon esculentum*) fruit development and ripening. *J. Plant Growth Regul.*, 13:163–166.

Cantos, E., Espin, J.C., and Tomás-Barberán, F. 2002. Varietal differences among polyphenols profiles of seven table grape cultivars studied by LC-DAD-MS-MS. *J. Agric. Food Chem.*, 50:5691–5696.

Cantos, E., Espin, J.C., and Tomás-Barberán, F.A. 2001a. Effect of wounding on phenolic enzymes in six minimally processed lettuce cultivars upon storage. *J. Agric. Food Chem.*, 49:322–330.

Cantos, E., Espin, J.C., and Tomás-Barberán, F.A. 2001b. Postharvest induction modelling method using UV irradiation pulses for obtaining resveratrol-enriched table grapes: a new "functional" fruit? *J. Agric. Food Chem.*, 49:5052–5058.

Cantos, E., Garcia-Viguera, C., de Pascual-Teresa, S., and Tomás-Barberán, F.A. 2000. Effect of postharvest ultraviolet irradiation on resveratrol and other phenolics of cv. Napoleon table grapes. *J. Agric. Food Chem.*, 48:4606–4612.

Cantwell, M., Yang, S.F., Rubatzky, V., and Lafuente, M.T. 1989. Isocoumarin content of carrots as influenced by ethylene concentration, storage temperature and stress conditions. *Acta Hortic.*, 258:523–534.

Carbonaro, M., Mattera, M., Nicoli, S., Bergamo, P., and Cappelloni, M. 2002. Modulation of antioxidant compounds in organic vs. conventional fruit (peach, *Prunus persica* L., and pear, *Pyrus communis*, L). *J. Agric. Food Chem.*, 50:5458–5462.

Cartwright, R.A., Roberts, E.A., Flood, E.A., and Williams, A.H. 1955. The suspected presence of *p*-coumarylquinic acids in tea, apple, and pear. *Chem. Ind.*, 1955:1062–1063.

Ceska, O., Chaudhary, S.K., Warrington, P.J., and Ashwood-Smith, M.J. 1986. Furanocoumarins in the cultivated carrot, *Daucus carota*. *Phytochemistry*, 25:81–83.

Challice, J.C. and Williams, A.H. 1972. Phenolic compounds of genus *Pyrus*. *Phytochemistry*, 11:37–44.

Chandra, A., Rana, J., and Li, Y. 2001. Separation, identification, quanatitation, and method validation of anthocyanins in botanical supplement raw materials by HPLC and HPLC-MS. *J. Agric. Food Chem.*, 49:3515–3521.

Chandra, A., Nair, M.G., and Iezzoni, A. 1992. Evaluation and characterization of the anthocyanin pigment in tart cherries (*Prunus cerasus* L.). *J. Agric. Food Chem.*, 40:967–969.

Chang, S., Tan, C., Frankel, N., and Barrett, D.M. 2000. Low-density lipoprotein activity of phenolic compounds and polyphenol oxidase activity in selected clingstone peach cultivars. *J. Agric. Food Chem.*, 48:147–151.

Chassagne, D., Crouzet, J., Bayonove, C.L., and Baumes, R.L. 1998. Identification of passion fruit glycosides by gas chromatography/mass spectrophotometry. *J. Agric. Food Chem.*, 46:4552–4557.

Chassagne, D., Crouzet, J., Bayonove, C.L., and Baumes, R.L. 1997. Glycosidically bound eugenol and methyl salicylate in the fruit of edible Passiflora species. *J. Agric. Food Chem.*, 45:2685–2689.

Chassagne, D., Bayonove, C., Crouzet, J., and Baumes, R.L. 1995. Formation of aroma by enzymic hydrolysis of glycosidically bound components of passion fruit, in *Bioflavour 95, Analysis, Precursor Studies, Biotechnology*, Etievant, P. and Schreier, P., Eds., INRA, Paris, 217–222.

Chatonnet, P., Dubourdieu, D., Boidron, J.-N., and Pons, M. 1992. The origin of ethylphenols in wines. *J. Sci. Food Agric.*, 60:165–178.

Chatonnet, P., Dubourdieu, D., Boidron, J.-N., and Lavigne, V. 1993. Synthesis of volatile phenols by Saccharomyces cerevisiae in wines. *J. Sci. Food Agric.*, 62:191–202.

Chaudhary, S.K., Ceska, O., Warrington, P.J., and Ashwood-Smith, M.J. 1985. Increased furocoumarin content in celery during storage. *J. Agric. Food Chem.*, 33:1153–1157.

Chen, H., Zuo, Y., and Deng, Y. 2001. Separation and determination of flavonoids and other phenolic compounds in cranberry juice by high-performance liquid chromatography. *J. Chromatogr.*, 913:387–395.

Cheng, G.W. and Crisosto, C.H. 1995. Browning potential, phenolic composition and polyphenoloxidase activity of buffer extracts of peach and nectarine skin tissue. *J. Am. Soc. Hortic. Sci.*, 120:835–838.

Cheynier, V. and da Silva, R. 1991. Oxidation of grape procyanidins model solutions containing trans-caffeoryltartaric acid and polyphenoloxidase. *J. Agric. Food Chem.*, 39:1047–1049.

Cheynier, V. and Rigaud, J. 1986. HPLC separation and characterization of flavonols in the skin of *Vitis vinifera* var. Cinsault. *Am. J. Enol. Vitic.*, 37:248–252.

Christie, P.J., Alfenito, M.R., and Walbot, V. 1994. Impact of low-temperature stress on general phenylpropanoid and anthocyanin pathways: enhancement of transcript abundance and anthocyanin pigmentation in maize seedlings. *Planta*, 194:541–549.

Chung, M.I., Teng, C.M., Cheng, K.L., Ko, F.N., and Lin, C.N. 1992. An antiplatelet principle of *Veratrum formosanum*. *Planta Med.*, 58:274–276.

Cichewicz, R.H., Kouzi, S.A., and Hamann, M.T. 2000. Dimerization of resveratrol by the grapevine pathogen *Botrytis cinerea*. *J. Nat. Prod.*, 63:29–33.

Clifford, M.N. 1997. Astringency, in *Phytochemistry of Fruit and Vegetables, Proceedings of the Phytochemical Society of Europe 41*, Tomás-Barberán, F.A. and Robins, R.J., Eds., Clarendon Press, Oxford, U.K., 87–108.

Coee, F.G. and Anderson, G.J. 1996. Ethnobotany of the Garifuna of Eastern Nicaragua. *Econ. Bot.*, 50:71–107.

Cohen, J.D. 1996. *In vitro* tomato fruit cultures demonstrate a role for indole-3-acetic acid in regulating fruit ripening. *J. Am. Soc. Hort. Sci.*, 121:520–524.

Connor, A.M., Luby, J.J., Hancock, J.F., Berkheimer, S., and Hanson, E.J. 2002. Changes in fruit antioxidant activity among blueberry cultivars during cold-temperature storage. *J. Agric. Food Chem.*, 50:893–898.

Constant, H. and Cordell, G.A. 1996. Nonivamide, a constituent of *Capsicum oleoresin*. *J. Nat. Prod.*, 59:425–426.

Constantino, L., Albasini, A., Rastelli, G., and Benvenuti, S. 1992. Activity of polyphenolic extracts as scavengers of superoxide radicals and inhibitors of xanthine oxidase. *Planta Med.*, 58:342–344.

Constantino, L., Rastelli, G., Rossi, T., Bertoldi, M., and Albasini, A. 1993. Antilipoperoxidant activity of polyphenolic extracts of *Ribes nigrum* L. *Plant. Med. Phytother.*, 26:207–214.

Contreras-Padilla, M. and Yahia, E.M. 1998. Changes in capsicinoids during development, maturation, and senscence of Chile peppers and relation with peroxidase activity. *J. Agric. Food Chem.*, 46:2075–2079.

Cook, N.C. and Samman, S. 1996. Flavonoids-chemistry, metabolism, cardioprotective effects, and dietary sources. *Nutr. Biochem.*, 7:66–76.

Coseteng, M.Y. and Lee, C.Y. 1987. Changes in apple polyphenoloxidase and polyphenol concentrations in relation to degree of browning. *J. Food Sci.*, 52:985–989.

Coxon, D.T., Curtis, R.F., Price, R.K., and Levett, G. 1973. Abnornal metabolites produced in *Daucus carota* roots stored under conditions of stress. *Phytochemistry,* 12:1881–1885.

Cracker, L.E. 1971. Postharvest color promotion in cranberry with ethylene. *HortScience*, 6:865–869.

Crouzet, J., Sakho, M., and Chassagne, D. 1997. Fruit aroma precursors with special reference to phenolics, in *Phytochemistry of Fruit and Vegetables, Proceedings of the Phytochemical Society of Europe 41*, Tomás-Barberán, F.A. and Robins, R.J., Eds., Clarendon Press, Oxford, U.K., 109–124.

Crozier, A., Lean, M.E.J., McDonald, M.S., and Black, C. 1997. Quantitative analysis of the flavonoid content of commercial tomatoes, onions, lettuce, and celery. *J. Agric. Food Chem.*, 45:590–595.

Curry, J., Aluru, M., Mendoza, M., Navarez, J., Melendrez, M., and Connell, M.A. 1999. Transcripts for possible capsaicinoid biosynthetic genes are differentially accumulated in pungent and nonpungent *Capsicum* spp. *Plant Sci.*, 148:47–57.

Dame, C., Chichester, C.O., and Marsh, G.L. 1957. Studies of processed all-green aparagus. I. Quantitative analysis of soluble compounds in respect to strain and harvest variables, and their distribution within the asparagus spear. *Food Res.*, 22:658–672.

da Silva, J.M., Rigaud, J., Cheynier, V., Cheminat, A., and Moutounet, M. 1991. Procyanidin dimers and trimers from grape seeds. *Phytochemistry*, 30:1250–1264.

Dean, B.B. and Kolattukudy, P.E. 1977. Biochemistry of suberization. Incorporation of [1–14C]oleic acid and [1–14C]acetate into aliphatic components of suberin in potato tuber disk (*Solanum tuberosum*). *Plant Physiol.*, 59:48–54.

de Ancos, B., González, E.M., and Cano, M.P. 2000. Ellagic acid, vitamin C, and total phenolic contents and radical scavenging capacity affected by freezing and frozen storage in raspberry fruit. *J. Agric. Food Chem.*, 48:4565–4570.

Dekazos, E.D. 1970. Anthocyanin pigments in red tart cherries. *J. Food Sci.*, 35:237–241.

Demina, T.G. 1974. Anthocyanins of several varieties of blackcurrant. Biol. *Aktiv. Soedn. Rast. Sib. Flory,* 23–26, in *Chemical Abstracts,* CA 82, 40688e.

Del Rio, J.A., Benavente, O., Castillo, J., and Borrego, F. 1991. Neodiosmin a flavone glycoside from *Citrus aurantium. Phytochemistry,* 31:1148–1153.

de Pascual-Teresa, Santos-Buelga, C., and Rivas-Gonzalo, J. 2000. Quantitative analysis of flavan-3-ols in Spanish foodstuffs and beverages. *J. Agric. Food Chem.*, 48:5331–5337.

Dercks, W., Trumble, J., and Winter, C. 1990. Impact of athmospheric pollution on linear furanocoumarin content in celery. *J. Chem. Ecol.*, 16:443–454.

de Sim6n, F.B., Hernández, T., and Estrella, I. 1993. Phenolic composition of white grapes (Var. Airen). Changes during ripening. *Food Chem.*, 47:52.

de Sim6n, F.B., Peréz-Ilzarbe, J., Hernández, T., G6mez-Cordovés, C., and Estrella, I. 1992. Importance of phenolic compounds for the characterization of fruit juices. *J. Agric. Food Chem.*, 40:1531–1535.

Dewanto, V., Wu, X., Adom, K.K., and Liu, R.H. 2002. Thermal processing enhances the nutritional value of tomatoes by increasing total antioxidant activity. *J. Agric. Food Chem.*, 50:3010–3014.

Diawara, M.M., Trumble, J., and Quiros, C.F. 1993. Linear furanocoumarins of three celery breeding lines: implications for intergrated pest mangement. *J. Agric. Food Chem.*, 41:819–824.

Diawara, M.M., Trumble, J.T., Quiros, C.F., and Hansen, R. 1995. Implications of distribution of linear furanocoumarins within celery. *J. Agric. Food Chem.*, 43:723–727.

Dick, A.J. and Smith, K.C. 1990. Quercetin glucosides and galactosides: substrates and inhibitors of apple β-galactosidase. *J. Agric. Food Chem.*, 38:923–926.

Dick, A.J., Redden, P.R., DeMacro, A.C., Lidster, P.D., and Grindley, T.B. 1987. Flavonoid glycosides of Spartan apple peel. *J. Agric. Food Chem.*, 35:529–531.

Dick, A.J., Williams, R., Bearne, S.L., and Lidster, P.D. 1985. Quercetin glycosides and chlorogenic acid: inhibitors of apple β-galactosidase and of apple softening. *J. Agric. Food Chem.*, 33:798–800.

Diplock, A.T., Charleux, J.L., Crozier-Willi, G., Kok, F.J., Rice-Evans, C., Roberfroid, M., Stahl, W., and Viña-Ribes, J. 1998. Functional foods and defense against reactive oxidative species. *Br. J. Nutr.*, 80(Suppl. 1):S77-S112.

Dixon, R.A. and Paiva, N.L. 1995. Stress-induced phenylpropanoid metabolism. *Plant Cell*, 7:1085–1097.

Donner, H., Gao, L., and Mazza, G. 1997. Separation and characterization of simple and malonylated anthocyanins in red onions, *Allium cepa* L. *Food Res. Int.*, 30:637–643.

Donovan, J.L., Meyer, A.S., and Waterhouse, A.L. 1998. Phenolic composition and antioxidant activity of prunes and prune juice (*Prunus domestica*). *J. Agric. Food Chem.*, 46:1247–1252.

Douillet-Breuil, A.-C., Jeandet, P., Adrian, M., and Bessis, R. 1999.Changes in phytoalexin content of various *Vitis* Spp.in response to ultraviolet C elicitation. *J. Agric. Food Chem.*, 47:4456–4461.

Drell, W. 1970. Separation of catecholamines from catechol acids by alumina. *Anal. Biochem.*, 34:142–151.

Du, C.T., Wang, P.L., and Francis, F.J. 1970. Anthocyanins of pomegranate, *Punica granatum*. *J. Food Sci.*, 40:417–418.

Duggan, M.B. 1967. Identification of the plant material by its phenolic content. *J. Assoc. Off. Anal. Chem.*, 50:727–734.

Duggan, M.B. 1969a. Methods of examination of flavonoids in fruits: application to flavonol glycosides and aglycones in apples, pears and strawberries. *J. Assoc. Off. Anal. Chem.*, 52:1038–1043.

Duggan, M.B. 1969b. Identity and occurrence of certain flavonol glycosides in four varieties of pears. *J. Agric. Food Chem.*, 17:1098–1101.

Duggan, M.B. 1969c. Methods for examination of flavonoids in fruits: application to flavonol glycosides and aglycones of apples, pears, and strawberries. *Anal. Chem.*, 52:1038–1043.

Dugo, P., Mondello, L., Errante, G., Zappia, G., and Dugo, G. 2001. Identification of anthocyanins in berries by narrow-bore high-performance liquid chromatography with electrospray ionization detection. *J. Agric. Food Chem.*, 49:3987–3992.

DuPont, M.S., Mondin, Z., Williamson, G., and Price K.R. 2000. Effect of variety, processing and storage on the flavonoid glycoside content and composition of lettuce and endive. *J. Agric. Food Chem.*, 48:3957–3964.

Durkee, A.B. and Poapst, P.A. 1965. Phenolic constituents in core tissues and ripe seed of McIntosh apples. *J. Agric. Food Chem.*, 13:137–139.

Edenharder, R., Keller, G., Platt, K.L., and Unger, K.K. 2001. Isolation and characterization of structurally novel antimutagenic flavonoids from spinach (*Spinacia oleracea*). *J. Agric. Food Chem.*, 49:2767–2673.

Ehlenfeldt, M.K. and Prior, R.L. 2001. Oxygen radical absorbance (ORAC) and phenolic and anthocyanin concentrations in fruits and leaf tissues of highbush blueberry. *J. Agric. Food Chem.*, 49:2222–2227.

El Ansari, M.A., Reddy, K.K., Sastry, K.N.S., and Nayudamma, Y. 1971. Polyphenols of *Magnifera indica. Phytochemistry*, 10:2239–2241.

Escarpa, A. and González, M.C. 1999. High-performance liquid chromatography with diode-array detection for the determination of phenolic compounds in peel and pulp from different apple varieties. *J. Liq. Chromatogr.*, 15:637–646.

Escribano-Bailón, T., Gutiérrez-Fernandez, Y., Rivaz-Gonzalo, J.C., and Santos-Buelga, C. 1992a. Characterization of procyanidins of *Vitis vinifera* variety Tinta del Pais grape seeds. *J. Agric. Food Chem.*, 40:1794–1799.

Escribano-Bailón, T., Gutiérrez-Fernandez, Y., Rivaz-Gonzalo, J.C., and Santos-Buelga, C. 1992b. Analysis of flavan-3-ol from *Vitis vinifera* variety Tempranillo grape seeds. *Proc. Groupe Polyphenols*, 16:129–132.

Escribano-Bailón, T., Guerra, M.T., Rivas-Gonzalo, J.C., and Santos-Buelga, C. 1995. Proanthocyanidins in skins from different grape varieties. *Z. Lebensm. Unters. Forsch.*, 200:221–224.

Espin, J.C., Soler-Rivas, C., Wichers, H.J., and Garcia-Viguera, C. 2000. Anthocyanin-based natural colorants: a new source of antiradical activity in foodstuff. *J. Agric. Food Chem.*, 48:1588–1592.

Estrada, B., Bernal, M.A., Diaz, J., Pomar, F., and Merino, F. 2002. Capsaicinoids in vegetative organs of *Capsicum annuum* L. in relation to fruiting. *J. Agric. Food Chem.*, 50:1188–1191.

Estrada, B., Bernal, M.A., Diaz, J., Pomar, F., and Merino, F. 2000. Fruit development in *Capsicum annuum*. Changes in capsaicin, lignin, free phenolics, and peroxidase patterns. *J. Agric. Food Chem.*, 48:6234–6239.

Estrada, B., Pomar, F., Diaz, J., Merino, F., and Bernal, A. 1999a. Pungency level in fruits of the Padron pepper with different water supply. *Sci. Hort.* Amsterdam, 81:385–396.

Estrada, B., Diaz, J., Merino, F., and Bernal, M.A. 1999b. The effect of seasonal changes on the pungency level of Padron pepper fruits. *Capsicum Eggplant Newslett.*, 18:28–31.

Estrada, B., Pomar, F., Diaz, Z., Merino, F., and Bernal, A. 1998. Effects of mineral feritilizer supplementation on fruit development and pungency level of Padron pepper fruits. *Capsicum Eggplant Newslett.*, 73:493–497.

Estrada, B., Pomar, F., Diaz, Z., Merino, F., and Bernal, A. 1997. Evolution of capsaicinoids in *Capsicum annuum* L. var. annuum cv. Padron at different growth stages after flowering. *Capsicum Eggplant Newslett.*, 16:60–64.

Ewald, K., Fjelkner-Modig, S., Johansson, K., Sjöholm, I., and Akesson, B. 1999. Effect of processing on major flavonoids in processed onions, green beans, and peas. *Food Chem.*, 64:231–235.

Facino, R.M., Carini, M., Aldini, G., Berti, F., Rossoni, G., Bombardelli, E., and Morazzoni, P. 1999. Diet enriched with procyanidins enhances antioxidant activity and reduces myocardial postischemic damage to rats. *Life Sci.*, 64:627–642.

Fallico, B., Lanza, M.C., Maccarone, E., Asmundo, C.N., and Rapisarda, P. 1996. Role of hydroxycinnamic acids and vinylphenols in the flavor alteration of blood orange juices. *J. Agric. Food Chem.*, 44:2654–2657.

Fang, N., Yu, S., and Prior, R.L. 2002. LC/MS/MS characterization of phenolic constituents in dried plums. *J. Agric. Food Chem.*, 50:3579–3585.

Fang-Chiang, H.J. 2000. Anthocyanin pigments, nonvolatile acid and sugar composition of blackberries. M.S. thesis, Oregon State University, cited in Stintzing, F.C., Stintzing, A.S., Carle, R., and Wrolstad, R.E. 2002. A novel zwitterionic anthocyanin from evergreen blackberry (*Rubus laciniatus* Wild.). *J. Agric. Food Chem.*, 50:396–399.

Feldman, E.B. 2001. Fruits and vegetable and the risk of stroke. *Nutr. Rev.*, 59:24–27.

Ferreira, D., Guyot, S., Marnet, N., Delgadillo, I., Renard, C.M.G.C., and Coimbra, M.A. 2002. Composition of phenolic compounds in a Portuguese pear (*Pyrus communis* L. var. S. Bartolomeu) and changes after sun drying. *J. Agric. Food Chem.*, 50:4537–4544.

Ferreres, F., Gil, M.I., and Tomás-Barberán, F.A. 1996. Anthocyanins and flavonoids from shredded red onion and change during storage in perforated films. *Food Res. Int.*, 29:389–395.

Ferreres, F., Gil, M.I., Castaner, M., and Tomás-Barberán, F.A. 1997a. Phenolic metabolites in red pigmented lettuce (*Lactuca sativa*). Changes with minimal processing and cold storage. *J. Agric. Food Chem.*, 45:4249–4254.

Ferreres, F., Castaner, M., and Tomas-Barberan, F.A. 1997b. Acylated flavonol glycosides from spinach leaves (*Spinacia oleracea*). *Phytochemistry*, 45:1701–1705.

Feucht, W., Treutter, D., and Christ, E. 1992. The precise localization of catechins and proanthocyanidins in the protective layers around fungal infection. *Z. Pflanzenkrankh. Pflanzenschutz*, 99:404–413.

Fleuriet, A. and Macheix, J.-J. 1981. Quinyl esters and glucose derivatives of hydroxycinnamic acids during growth and ripening of tomato fruit. *Phytochemistry*, 20:667–671.

Flors, V., Miralles, C., Cerezo, M., Gonzalez-Bosch, C., and Garcia-Agustin, P. 2001. Effect of a novel chemical mixture on senescence processes and plant–fungus interaction in Solanasceae plants. *J. Agric. Food Chem.*, 49:2569–2575.

Foo, L.Y. and Porter, L.J. 1980. The phytochemistry of proanthocyanidin polymers. *Phytochemistry*, 19:1747–1754.

Foo, L.Y. and Porter, L.J. 1981. The structure of tannins of some edible fruits. *J. Sci. Food Agric.*, 32:711–716.

Foo, L.Y. and Lu, Y. 1999. Isolation and identification of procyanidins in apple pomace. *Food Chem.*, 64:511–518.

Foo, L.Y., Lu, Y., Howell, A.B., and Vorsa, N. 2000. A-type proanthocyanidin trimers from cranberry that inhibit adherence of uropathogenic P-fimbriated *Escherichia coli. J. Nat. Prod.*, 63:1225–1228.

Fossen, T., Andersen, O.M., Ovstedal, D.O., Pedersen, A.T., and Raknèss, A. 1996. Characteristic anthocyanin pattern from onion and other *Allium* spp. *J. Food Sci.*, 61:703–706.

Francis, F.J. 1957. Color and pigment measurement in fresh cranberries. *Proc. Am. Soc. Hortic. Sci.*, 69:296–301.

Francis, F.J., Harborne, J.B., and Barker, W.G. 1966. Anthocyanins in the lowbush blueberry, *Vaccinium angustifolium. J. Food Sci.*, 31:583–587.

Francis, F.J. 1970. Anthocyanins in pears. *HortScience*, 5:42.

Frankel, E.N., Waterhouse, A.L., and Teissedre, P.L. 1995. Principal phenolic phytochemicals in selected California wines and their antioxidant activity in inhibiting oxidation of human low-density lipoproteins. *J. Agric. Food Chem.*, 43:890–894.

Friedman, M. 1997. Chemistry, biochemistry, and dietary role of potato polyphenols. A review. *J. Agric. Food Chem.*, 45:1523–1540.

Friedrich, V.H. and Schönert, J. 1973. Untersuchungen über einige inhaltsstoffe der blätter und früchte von *Vaccinium myrtillus. Planta Med.*, 24:90–100.

Fritzemeier, K.H. and Kindl, H. 1981. Coordinate induction by UV light of stilbene synthase phenylalanine ammonialyase and cinnamate 4-hydroxylyase in leaves of Vitaceae. *Planta*, 151:48–52.

Fujikawe, H., Suzuki, T., and Iwai, K. 1982. Intracellular distribution of enzymes and intermediates involved in the biosynthesis of capsaicin and its analogs in *Capsicum* fruits. *Agric. Biol.* Chem., 46:2685–2689.

Fujikawe, H., Suzuki, T., Oka, S., and Iwai, K. 1980. Enzymatic formation of capsaicinoids from vanillylamine and iso-type fatty acids by cell extracts of *Capsicum annuum* var. annuum cv. Karayatsubusa. *Agric. Biol. Chem.* (Tokyo), 44:2907–2912.

Fukumoto, L.R. and Mazza, G. Assessing antioxidant and prooxidant activities of phenolic compounds. *J. Agric. Food Chem.*, 48:3597–3604.

Fuleki, T. 1971. Anthocyanins in red onion, *Allium cepa. J. Food Sci.*, 36:101–104.

Fuleki, T. 1999. Rutin, the main component of surface deposits on pickled green asparagus. *J. Food Sci.*, 64:252–254.

Fuleki, T. and da Silva, R.J.M. 1997. Catechin and procyanidin composition of seeds from grape cultivars grown in Ontario. *J. Agric. Food Chem.*, 45:1156–1160.

Furuta, S., Suda, I., Nishiba, Y., and Yamakawa, O. 1998. High tert-butylperoxyl radical scavenging activities of sweet potato cultivars with purple flesh. *Food Sci. Technol. Int. Tokyo*, 4:33–35.

Galletta, G.J. 1975. Blueberries and cranberries, in *Advances in Fruit Breeding*, Janick, J. and Moore, J.N., Eds., Purdue University Press, West Lafayette, IN, 154–196.

Galley, P. and Thiollet, M.A. 1993. A double-blind placebo-controlled trial of a new veno-active flavonoid fraction in the treatment of symptomatic capillary fragility. *Int. Angiol.*, 12:69–72.

Gao, L. and Mazza, G. 1995. Characterization, quantitation, and distribution of anthocyanins and colorless phenolics in sweet cherries. *J. Agric. Food Chem.*, 43:343–346.

Gao, L. and Mazza, G. 1994. Quantification and distribution of simple and acylated anthocyanins and other phenolics in blueberries. *J. Food Sci.*, 59:1057–1059.

Garcia-Viguera, C., Zafrilla, P., and Tomàs-Barberàn, F.A. 1997. Determination of authenticity of fruit jams by HPLC of anthocyanins. *J. Sci. Food Agric.*, 73:207–213.

Gassner, G. 1973. *Mikroskopische Untersuchung pflanzlicher Lebesmittel*, 4th ed., G. Fischer, Stuttgart, Germany, 254–260.

Gaydou, E.M., Bianchini, J.P., and Ramananarivo, R.P. 1987. Orange and mandarin oils differentation using polymetoxylated flavone composition. *J. Agric. Food Chem.*, 35:525–529.

Geibel, M. 1995. Sensitivity of the fungus *Cytospora personii* to the flavonoids of *Prunus cerasus. Phytochemistry*, 38:599–601.

Geibel, M. and Feucht, W. 1991. Flavonoid 5-glucosides from *Prunus cerasus* bark and their characteristic weak glucosidic bonding. *Phytochemistry*, 30:1519–1521.

Geibel, M., Geiger, H., and Treuter, D. 1990. Tectochrysin 5-and genistein 5-glucosides from the bark of *Prunus cerasus. Phytochemistry*, 29:1351–1353.

Gennaro, L., Leonardi, C., Esposito, F., Salucci, M., Maiani, G., Quaglia, G., and Fogliano, V. 2002. Flavonoid and carbohydrate contents in Tropea red onions: effects of home-like peeling and storage. *J. Agric. Food Chem.*, 50:1904–1910.

Gil, M.I., Holcroft, D.M., and Kader, A.A. 1997. Changes in strawberry anthocyanins and other phenolics in response to carbon dioxide treatments. *J. Agric. Food Chem.*, 45:1662–1667.

Gil, M.I., Castaner, M., Ferreres, F., Artes, F., and Tomás-Barberán, F. 1998a. Modified-atmosphere packaging of minimally processed "Lollo Rosso" (*Latuca sativa*). *Z. Lebensm. Unters. Forsch.*, 206:350–354.

Gil, M.I., Ferreres, F., and Tomás-Barberán, F.A. 1998b. Effect of modified atmosphere packaging on the flavonoids and vitamin C content in minimally processed Swiss chard (*Beta vulgaris* subspecies cycla). *J. Agric. Food Chem.*, 46:2007–2012.

Gil, M.I., Ferreres, F., and Tomás-Barberán, F.A. 1999. Effect of postharvest storage and processing on the antioxidant constituents (flavonoids and vitamin C) of fresh-cut spinach. *J. Agric. Food Chem.*, 47:2213–2217.

Gil, M.I., Tomás-Barberán, F.A., Hess-Pierce, B., Holcrof, D.M., and Kader, A.A. 2000. Antioxidant activity of pomegranate juice and its relationship with phenolic composition and processing. *J. Agric. Food Chem.*, 48:4581–4589.

Gil-Izquierdo, A., Gil, M.I., and Ferreres, F. 2002. Effect of processing techniques at industrial scale on orange juice antioxidant and beneficial health compounds. *J. Agric. Food Chem.*, 50:5107–5114.

Goda, Y., Shimizu, T., Kato, Y., Nakamura, M., Maitani, T., Yamada, T., Terahara, N., and Yamaguchi, M. 1997. Two acylated anthocyanins from purple sweet potato. *Phytochemistry*, 44:183–186.

Godeberg, P. 1994. Daflon 500 mg in the treatment of hemorroidal disease: a demonstrated efficacy in comparison with placebo. *Angiology*, 45:574–578.

Goiffon, J.-P., Brun, M., and Bourrier, M.J. 1991. HPLC of red fruit anthocyanins. *J. Chromatogr. A*, 537:101–121.

Golan, A., Kahn, V., and Sadovski, A.Y. 1977. Relationship between polyphenols and browning in avocado mesocarp. Comparison between the Fuerte and Lerman cultivars. *J. Agric. Food Chem.*, 25:1253–1260.

Golding, J.B., McGlasson, W.B., Wyllie, S.G., and Leach, D.N. 2001. Fate of apple peel phenolics during cold storage. *J. Agric. Food Chem.*, 49:2283–2289.

Gorinstein, S., Zachwieja, Z., Folta, M., Barton, H., Piotrowicz, J., Zemser, M., Weisz, M., Trakhtenberg, S., and Màrtin-Belloso, O. 2001. Comparative contents of dietary fibers, total phenolics, and minerals in persimmons and apples. J. *Agric. Food Chem.*, 49:952–957.

Gotz-Schmidt, E.-M. and Schreier, P. 1986. Neutral volatiles from blended endive (*Cichorium endivia* L.). *J. Agric. Food Chem.*, 34:212–215.

Goupy, P.A., Amiot, M.J., Richard-Forget, F., Duprat, F., Aubert, S., and Nicolas, J. 1995. Enzymatic browning of model solutions and apple phenolic extracts by apple polyphenoloxidase. *J. Food Sci.*, 60:497–501.

Goupy, P.A., Varoquaux, P.J.A., Nicolas, J.J., and Macheix, J.J. 1990. Identification and localization of hydroxycinnamoyl and flavonol derivatives from endive (*Cichorium endivia* L. cv. Geante Maraichere) leaves. *J. Agric. Food Chem.*, 38:2116–2121.

Govindarajan, V.S. 1986. *Capsicum* production, technology, chemistry, and quality. III. Chemistry of the color, aroma, and pungency stimuli. *CRC Crit. Rev. Food Sci. Nutr.*, 25:245–355.

Gray, A.I. and Waterman, P.G. 1978. Coumarins in Rutaceae. *Phytochemistry*, 17:845–864.

Griffiths, D.W. and Bain, H. 1997. Photo-induced changes in the concentration of individual chlorogenic isomers in potato (*Solanum tuberosum*) tubers an their complexation with ferric ions. *Potato Res.*, 40:307–315.

Griffiths, D.W., Dale, M.F., and Bain, H. 1995. Photo-induced changes in the total chlorogenic acid content of potato (*Solanum tuberosum*) tubers. *J. Sci. Food Agric.*, 68:105–110.

Griffiths, L.A. 1959. Detection and identification of the polyphenol substrate of the banana. *Nature*, 184:58–59.

Grohmann, K., Manthey, J.A., Cameron, R.G., and Buslig, B.S. 1999. Purification of citrus peel juice and molasses. *J. Agric. Food Chem.*, 47:4859–4867.

Gu, L., Kelm, M., Hammerstone, J.F., Beecher, G., Cunningham, D., Vannozzi, S., and Prior, R.L. 2002. Fractionation of polymeric procyanidins from lowbush blueberry and quantification of procyanidins in selected foods with an optimized normal-phase HPLC-MS fluorescent detection method. *J. Agric. Food Chem.*, 50:4852–4860.

Guha, S., Ghosal, S., and Chattopadhyav, V. 1996. Antitumor, immunomodulatory and anti-HIV effect of magniferin, a naturally occuring glucosylxanthone. *Chemotheraphy* (Basel), 42:443–451.

Guyot, S., Marnet, N., and J.-F., Drilleau, J.F. 2001. Thiolysis-HPLC characterization of apple procyanidins covering large range of polymerization states. *J. Agric. Food Chem.*, 49:14–20.

Guyot, S., Marnet, N., Laraba, D., Sanoner, P., and Drilleau, J.-F. 1998. Reversed-phase HPLC and characterization of the four main classes of phenolic compounds in different tissue zones of French cider apple variety (*Malus domestica* var. Kermerrien). *J. Agric. Food Chem.*, 46:1698–1705.

Guyot, S., Doco, T., Souquet, J.M., Moutounet, M., and Drilleau, J.F. 1997. Characterization of highly polymerized procyanidins in cider apple (*Malus domestica* var. Kermerrien) skin and pulp. *Phytochemistry*, 44:351–357.

Haddock, E.A., Gupta, R.K., Al-Shafi, S.M.K., Layden, K., Haslam, E., and Magnolato, D. 1982. The metabolism of gallic acid in plants: biogenetic and molecular taxonomic considerations. *Phytochemistry*, 21:1049–1062.

Häkkinen, S., Kärenlampi, S.O., Mykkänen, H.M., and Törrönen, A.R. 2000a. Influence of domestic processing and storage on flavonol content in berries. *J. Agric. Food Chem.*, 48:2960–2965.

Häkkinen, S., Kärenlampi, S.O., Mykkänen, H.M., Heinonen, I.M., and Törrönen, A.R. 2000b. Ellagic acid content in berries: influence of domestic processing and storage. *Eur. Food Res. Technol.*, 212:75–80.

Häkkinen, S., Heinonen, I.M., Kärenlampi, S.O., Mykkänen, H.M., and Törrönen, A.R. 1999a. Content of flavonols quercetin, myrecitin, and kaempferol in 25 edible berries. *J. Agric. Food Chem.*, 47:2274–2279.

Häkkinen, S., Kärenlampi, S.O., Heinonen, I.M., Mykkänen, H.M., Ruuskanen, J., and Törrönen, A.R. 1999b. Screening of selected flavonoids and phenolic acids in 19 berries, *Food Res. Int.*, 32:345–353.

Häkkinen, S. and Auriola, S. 1998. High-performance liquid chromatography with electrospray ionization mass spectrometry and diode array ultraviolet detection in the identification of flavonol aglycones and glycosides in berries. *J. Chromatogr. A*, 829:91–100.

Hale, M.L., Francis, F.J., and Fagerson, I.S. 1986. Detection of enocyanin in cranberry juice cocktail by HPLC anthocyanin profile. *Food Chem.*, 51:1511–1513.

Hall, I.V. and Stark, R. 1972. Anthocyanin production in cranberry leaves and fruit, related to cool temperatures at low light intensity. *Hort. Res.*, 12:183–186.

Hammond, D.A., Lea, A.G.H., and Rousseau, L. 1996. Detection of the addition of pulwash to orange juice using two new HPLC procedures. *Int. Fruchtsaft-Union, Wiss.-Technol. Komm.*, 24:233–237.

Harborne, J.B. 1985. Phenolics and plant defense, in *Annual Proceedings of the Phytochemical Society of Europe 25*, Van Sumere, C.F. and Lea, P.J., Eds., Clarendon Press, Oxford, U.K., 25:257.

Harborne, J.B. 1967. *Comparative Biochemistry of the Flavonoids*, Academic Press, London.

Harborne, J.B. and Hall, E. 1964. Plant polyphenols. XIII. The systematic distribution and origin of anthocyanins containing branched trisaccharides. *Phytochemistry*, 3:453–463.

Hardenburg, R.E., Watada, A.E., and Wang, C.Y. 1986. The commercial storage of fruits, vegetables, and florist and nursery stocks. U.S. Department of Agriculture Handbook No. 66, U.S. Government Printing Office, Washington, D.C., 37–38,

Harding, V.K. and Heale, J.B. 1980. Isolation and identification of the antifungal compounds accumulating in the induced response of carrot slices to *Botrytis cinerea*. *Physiol. Plant Pathol.*, 17:277–289.

Hartley, R.D. and Harris, P.J. 1981. Phenolic constituents of the cell walls of dicotyledons. *Biochem. Syst. Ecol.*, 9:189–203.

Hasegawa, D., Johnson, R.M., and Gould, W.A. 1966. Effect of cold storage on chlorogenic acid content of potatoes. *J. Agric. Food Chem.*, 14:165–169.

Haslam, E. 1989. *Plant Polyphenols. Vegetable Tannins Revisited.* Cambridge University Press, Cambridge, U.K., 90–146.

Haslam, E. and Lilley, T.H. 1985. New polyphenols for old tannins. The biochemistry of plant phenolics, in *Annual Proceedings of the Phytochemical Society of Europe 25*, Van Sumere, C.F. and Lea, P.J., Eds., Clarendon Press, Oxford, U.K., 25:237–256.

Haslam, E., Haddock, E.A., Gupta, R.K., Al-Shafi, S.M.K., and Layden, K. 1982. The metabolism of gallic acid and hexahydroxydiphenic acid in plants. Part 1. Naturally occurring galloyl esters. *J. Chem. Soc. Perkin Trans.*, 1, 2515–2524.

Heath-Pagliuso, S., Matlin, S.A., Fang, N., Thompson, R.H., and Rappaport, L. 1992. Stimulation of furanocoumarin accumulation in celery and celeriac tissues by *Fusarium oxysporum* f. sp. apii. *Phytochemistry*, 31:2683–2688.

Henderson, D.E. and Henderson, S.K. 1992. Thermal decomposition of capsaicin. 1. Interaction with oleic acid at high temperatures. *J. Agric. Food Chem.*, 40:2263–2268.

Henning, W. 1981. Flavonol glycosides of strawberries (*Fragaria x ananassa* Duch.), raspberries (*Rubus idaeus* L.) and blackberries (*Rubus fructicosus* L.) 14. Phenolics of fruits. *Z. Lebensm. Unters. Forsch.*, 173:180–187.

Henning, W. and Herrmann, K. 1980a. Flavonol glycoside of plums of the species *Prunus domestica* L. and *Prunus salicina* Lindley. 12. Phenolics of fruits. *Z. Lebensm. Unters. Forsch.*, 171:111–118.

Henning, W. and Herrmann, K. 1980b. Flavonol glycosides of apricots (*Prunus armeniaca* L.) and peaches (*Prunus persica* Batsch). *Z. Lebensm. Unters. Forsch.*, 171:183–188.

Hernández, M.T. 1997. Phenolic composition of industrially manufactured purées and concentrates from peach and apple fruits. *J. Agric. Food Chem.*, 45:4071–4075.

Herrmann, K. 1989. Occurrence and content of hydroxycinnamic and hydroxybenzoic acid compounds in foods. *CRC Crit. Rev. Food Sci. Nutr.*, 28:315–347.

Herrmann, K. 1976. Flavonols and flavones in food plants: a review. *J. Food Technol.*, 11:433–448.

Herrmann, K. 1973. Die phenolischen inhaltsstoffe des obstes. I. Bischerige kenntnisse uber vorkmmen, gehalte sowie veranderungen wahrend des fruchtwachstums. *Lebensm. Unters. Forsch.*, 151:41–51.

Hertog, M.G.L., van Poppel, G., and Verhoeven, D. 1997. Potentially anticarcinogenic secondary metabolites from fruit and vegetables, in *Phytochemistry of Fruit and Vegetables, Proceedings of the Phytochemical Society of Europe 41*, Tomás-Barberán, F.A. and Robins, R.J., Eds., Clarendon Press, Oxford, U.K., 313–330.

Hertog, M.G.L., Freskens, E.J.M., Hollman, P.C.H., Katan, M.B., and Kromhout, D. 1993. Dietary antioxidant flavonoids and risk of coronary heart disease: the Zutphen Elderly Study. *Lancet*, 342:1007–1011.

Hertog, M.G.L., Hollman, P.C.H., and Katan, M.B. 1992a. Content of potentially anticancerogenic flavonoids of 28 vegetables and 9 fruits commonly consumed in The Netherlands. *J. Agric. Food Chem.*, 40:2379–2383.

Hertog, M.G.L., Hollman, P.C.H., and Venena, D.P. 1992b. Optimization of a quantitative HPLC determination of potentially anticarcinogenic flavonoids in vegetables and fruits. *J. Agric. Food Chem.*, 40:1591–1596.

Hirota, S., Shimoda, T., and Takahama, U. 1998. Tissue and spatial distribution of flavonol and peroxidase in onion bulbs and stability of flavonol glucosides during boiling of the scales. *J. Agric. Food Chem.*, 46:3497–3502.

Hocher, V., Sotta, B., Maldiney, R., Bonnet, M., and Migniac, E. 1992. Changes in indole-3-acetic acid levels during tomato (*Lycopersicon esculentum*) seed development. *Plant Cell Rep.*, 11:253–256.

Hoffman, R. and Heale, J.B. 1987. Cell death, 6-methoxymellein accumulation, and induced resistance to *Botrytis cinerea* in carrot root slices. *Mol. Plant. Pathol.*, 30:67–75.

Hoffman, R., Roebroeck, E., and Heale, J. 1988. Effects of ethylene biosynthesis in carrots slices on 6-methoxymellein accumulation and resistance to *Botris cinerea. Physiol. Plant.*, 73:71–76.

Hollman, P.C.H. 2001. Evidence for health benefits of plant phenols: local or systemic. *J. Sci. Food Agric.*, 81:842–852.

Hollman, P.C.H., Hertog, M.G.L., and Katan, M.B. 1996. Analysis and health benefits of flavonoids. *Food Chem.*, 57:43–46.

Hong, V. and Wrolstad, R. 1990. Characterization of anthocyanin-containing colorants and fruit juices by HPLC/photodiode array detection. *J. Agric. Food Chem.*, 38:698–708.

Hong, V. and Wrolstad, R. 1986. Cranberry juice composition. *J. Assoc. Off. Anal. Chem.*, 69:199–207.

Horowitz, R.M. and Asen, S. 1989. Decarboxylation and exchange reactions in flavonoid glycoside malonates. *Phytochemistry,* 28:2531–2532.

Horowitz, R.M. and Gentili, B. 1977. Flavonoids constituents of citrus, in *Citrus Science and Technology,* Nagy S., Shaw, P.E., and Veldhuis, M.K., Eds., AVI Publishing, Westport, CT, vol. 1, 397–426.

Horowitz, R.M. and Gentili, B. 1961. Phenolic glycosides of grapefruit: a relation between bitterness and structure. *Arch. Biochem. Biophys.*, 92:191–192.

Horowitz, R.M. 1961. The citrus flavonoids, in *The Orange. Its Biochemistry and Physiology,* Sinclair, W.B., Ed., University of California, Division of Agricultural Sciences, Los Angeles, 334–372.

Howard, L.R., Talcott, S.T., Brenes, C.H., and Villalon, B. 2000. Changes in phytochemical and antioxidant activity of selected pepper cultivars (*Capsicum* species) as influenced by maturity. *J. Agric. Food Chem.*, 48:1713–1720.

Howard, L.R., Pandjaitan, N., Morelock, T., and Gil, M.I. 2002. Antioxidant capacity and phenolic content of spinach as affected by genetics and growing season. *J. Agric. Food Chem.*, 50:5891–5896.

Howell, A.B. 2002. Cranberry proanthocyanidins and the maintaince of urinary tract health. *CRC Crit. Rev. Food Sci. Nutr.*, 42(Suppl.):273–278.

Hsia, C.L., Luh, B.C., and Chichester, C.O. 1965. Anthocyanins in freestone peaches. *J. Food Sci.*, 30:5–12.

Huang, M.-T., Osawa, T., Ho, C.-T., and Rosen, R.T. 1992., Eds., *Food Phytochemicals for Cancer Prevention I. Fruits and Vegetables.* ACS Symposium Series 546, American Chemical Society, Washington, D.C.

Hudson, J.B., Miki, N., and Towers, G.H.N. 1987. Isopimpinellin is not phototoxic to viruses and cells. *Planta Med.*, 53:306–307.

Hulme, A.C. 1958. Some aspects of the biochemistry of apple and pear fruits. *Adv. Food Res.*, 8:351–355.

Hyodo, H., Kuroda, H., and Yang, S.F. 1978. Induction of phenylalanine ammonia-lyase and increase in phenolics in lettuce leaves in relation to the development of russet spotting caused by ethylene. *Plant Physiol.*, 62:31–35.

Imeh, U. and Khokhar, S. 2002. Distribution of conjugated and free phenols in fruits: antioxidant activity and cultivar variations. *J. Agric. Food Chem.*, 50:6301–6306.

Ingalsbe, D.W., Neubert, A.M., and Carter, G.H. 1963. Concord grape pigments. *J. Agric. Food Chem.*, 11:263–268.

Ingle, M. and Hyde, J.F. 1968. Effect of bruising on discoloration and concentration of phenolic compounds in apple tissue. *Proc. Am. Soc. Hortic. Sci.*, 93:738–745.

International Agency for Research on Skin Cancer. 1983. Evaluation of the carcinogenic risk of chemicals to humans. Supplement 4: Chemicals, industrial process and industries associated with cancer in humans; International Agency for Research on Cancer. Lyon, France.

Iorizzi, M., Lanzotti, V., De Marino, S., Zollo, F., Blanco-Molina, M., Macho, A., and Munoz, E. 2001. New glycosides from *Capsicum annuum* L. var. acuminatum. Isolation, structure determination, and biological activity. *J. Agric. Food Chem.*, 49:2022–2029.

Ishikawa, K., Janos, T., Sakamoto, S., and Nunomura, O. 1998. The contents of capsaicinoids and their phenolic intermediates in various tissues of the plants of *Capsicum annuum* L. *Capsicum Eggplant Newslett.*, 17:222–225.

Islam, M.S., Yoshimoto, M., Yamakawa, O., Ishiguro, K., and Yoshinaga, M. 2002a. Antioxidative compounds in the leaves of sweet potato. *Sweet Potato Res. Front*, 13:4.

Islam, M.S., Yoshimoto, M., Yahara, S., Okuno, S., Ishiguro, K., and Yamakawa, O. 2002b. Identification and characterization of foliar polyphenolic composition in sweet potato (*Ipomoea batatas* L.) genotypes. *J. Agric. Food Chem.*, 50:3718–3722.

Iversen, C.K. 1999. Black currant nectar. Effect of processing and storage on anthocyanin and ascorbic acid content. *J. Food Sci.*, 64:37–41.

Ivie, G.W., Holt, D.L., and Ivey, M.C. 1981. Natural toxicants in human foods: psolarens in raw and cooked parsnip root. *Science,* 213:909–910.

Iwai, K., Suzuki, T., and Fujiwake, H. 1979. Formation and accumulation of pungent principle of hot pepper fruits, capsaicin and its analogues in *Capsicum annuum* var. annuum cv. Karayatsubusa at different stages after flowering. *Agric. Food Chem.*, 43:2493–2498.

Jacobs, M. and Rubery, P.H. 1988. Naturally occurring auxin transport regulators. *Science*, 241:346–349.

Jaendet, P., Douillet-Breuil, A.-C., Bessis, R., Debord, S., Sbaghi, M., and Adrian, M. 2002. Phytoalexins from the Vitaceae: biosynthesis, phytoalexin gene expression in transgenic plants, antifungal activity, and metabolism. *J. Agric. Food Chem.*, 50:2731–2741.

Jaendet, P., Bessis, R., and Gautheron, B. 1991. The production of resveratrol (3,5,4′-trihydrostilbene) by grape berries in different developmental stages. *Am. J. Enol. Vitic.*, 42:41–46.

Jaworski, A.W. and Lee, C.Y. 1987. Fractionation and HPLC determination of grape phenolics. *J. Agric. Food Chem.*, 35:257–259.

Jayatilake, G.S., Jayasuriya, H., Lee, E.S., Koonchanok, N.M., Geahlen, R.L., Ashendel, C.L., McLaughlin, J.L., and Chang, C.L. 1993. Kinase inhibitors from *Polygonum cuspidatum*. *J. Nat. Prod.*, 56:1805–1810.

Jean, T. and Bodinier, M.C. 1994. Mediators involved in inflammation: effects of Daflon 500 mg on their release. *Angiology,* 45:554–559.

Johnson, R.L., Htoon, A.K., and Shaw, K.J. 1995. Detection of orange peel extract in orange juice. *Food Aust.*, 47:426–433.

Ju, Z., Yuan, Y., Liu, C., Zhan, S., and Wang, M. 1996. Relationship among simple phenol, flavonoid, and anthocyanin in apple fruit at harvest and scald suceptibility. *Postharvest Biol. Technol.*, 8:83–93.

Jurenitsch, J., Kubelka, W., and Jentzsch, K. 1979. Identification of cultivated taxa of *Capsicum*. Taxonomy, anatomy, and composition of pungent principle. *Planta Med.*, 35:174–183.

Kader, F., Rovel, B., Girardin, M., and Metche, M. 1996. Fractionation and identification of the phenolic compounds of highbush blueberries (*Vaccinium corymbosum* L.). *Food Chem.*, 55:35–40.

Kähkönen, M.P., Hopia, A.I., and Heinonen, M. 2001. Berry phenolics and their antioxidant activity. *J. Agric. Food Chem.*, 49:4076–4082.

Kähkönen, M.P., Hopia, A.I., Vuorela, H.L., Rauha, J., Pihlaja, K., Kujala, T.S., and Heinonen, M. 1999. Antioxidant activity of plant extracts containing phenolic compounds. *J. Agric. Food Chem.*, 47:3954–3962.

Kalt, W., Ryan, D.A.J., Duy, J.C., Prior, R.L., Ehlenfeldt, M.K., and Kloet, S.P.V. 2001. Interspecific variation in anthocyanins, phenolics, and antioxidant capacity among genotypes of highbush and lowbush blueberries (*Vaccinium section cyancoccus* spp.). *J. Agric. Food Chem.*, 49:4761–4767.

Kalt, W., McDonald, J.E., and Donner, H. 2000. Anthocyanins, phenolics and antioxidant capacity of processed lowbush blueberry products. *J. Food Sci.*, 65:390–393.

Kalt, W., McDonald, J.E., Ricker, R.D., and Lu, X. 1999a. Anthocyanin content and profile within and among blueberry species. *Can. J. Plant Sci.*, 79:617–623.

Kalt, W., Forney, C.F., Martin, A., and Prior, R.L. 1999b Antioxidant capacity, vitamin C, phenolics and anthocyanins after fresh storage of small fruits. *J. Agric. Food Chem.*, 47:4638–4644.

Kalt, W. and Doufour, D. 1997. Health functionality of bluberries. *Hortic. Technol.*, 7:216221.

Kalt, W. and McDonald, J.E. 1996. Chemical composition of lowbush blueberry cultivars. *J. Am. Soc. Hort. Sci.*, 121:142–146.

Kanazawa, K. and Sakakibara, H. 2000. High content of dopamine, a strong antioxidant, in Cavendish banana. *J. Agric. Food Chem.*, 48:844–848.

Kandil, F.E., Smith, M.A.L., Rogers, R.B., Pepin, M.-F., Song, L.L., Pezzuto, J.M., and Seigler, D.S. 2002. Composition of a chemopreventive proanthocyanidin-rich fraction from cranberry fruits responsible for the inhibition of 12-O-tetradecanoyl phorbol-13-acetate (TPA)-induced ornithine dacarboxylase (ODC) activity. *J. Agric. Food Chem.*, 50:1063–1069.

Kanes, K., Tisserat, B., Berhow, M., and Vandercook, C. 1992. Phenolic composition of various tissues in Rutaceae species. *Phytochemistry,* 31:967–974.

Kanes, K., Tisserat, B., Berhow, M., and Vandercook, C. 1993. Phenolic composition of various tissues in Rutaceae species. *Phytochemistry,* 32:1529–1536.

Karadeniz, F., Durst, R.W., and Wrolstad, R.E. 2000. Polyphenolic composition of raisins. *J. Agric. Food Chem.*, 48:5343–5350.

Kashiwada, Y., Nonaka, G.I., Nishioka, I., Chang, J.J., and Lee, K.H. 1992. Antitumor agents. 129. Tannins and related compounds as selective cytotoxic agents. *J. Nat. Prod.*, 55:1033–1043.

Kawaii, S., Tomono, Y., Katase, E., Ogawa, K., Nonomura-Nakano, M., Nesumi, H., Yoshida, T., Sugiura, M., and Yano, M. 2001. Quantitative study of fruit flavonoids in citrus hybrids of King (*C. nobilis*) and Mukaku Kishu (*C. kinokuni*). *J. Agric. Food Chem.*, 49:3982–3986.

Kawaii, S., Tomono, Y., Katase, E., Ogawa, K., and Yano, M. 1999a. Antiprofilerative activity of flavonoids on several cancer lines. *Biosci. Biotechnol. Biochem.*, 63:896–899.

Kawaii, S., Tomono, Y., Katase, E., Ogawa, K., and Yano, M. 1999b. Antiproliferative effects of the readily extractable fractions prepared from various citrus juices on several cancer lines. *J. Agric. Food. Chem.*, 47:2509–2512.

Kawaii, S., Tomono, Y., Katase, E., Ogawa, K., and Yano, M. 1999c. Quantitation of flavonoid constituents in citrus fruits. *J. Agric. Food Chem.*, 47:3565–3571.

Kayano, S.-I., Kikuzaki, H., Fukutsuka, N., Mitani, T., and Nakatani, N. 2002. Antioxidant activity of prune (*Prunus domestica* L.) constituents and a new synergist. *J. Agric. Food Chem.*, 50:3708–3712.

Ke, D. and Saltveit, M.E. 1989. Wound-induced ethylene production, phenolic metabolism and susceptibility to russet spotting in iceberg lettuce. *Physiol. Plant.*, 76:412–418.

Kefford, J.B. and Chandler, B.V. 1970. *The Chemical Constituents of Citrus Fruits*, Academic Press, New York, 113.

Kennedy, J.A., Hayasaka, Y., Vidal, S., Waters, E., and Jones, G.P. 2001. Composition of grape skin proanthocyanidins at different stages of berry developments. *J. Agric. Food Chem.*, 49:5348–5355.

Kesterson, J.W. and Braddock, R.J. 1976. By-products and speciality products of Florida citrus. *Univ. Fla. Agric. Exp. Stn. Tech. Bull.*, No. 784:8.

Khan, A., Kunesch, G., Chuilon, S., and Ravise, A. 1985. Structure and biological activity of xanthyletin, a new phytoalexin of citrus. *Fruits*, 40:807–811.

Kim, J.J., Ben-Yehoshua, S., Shapiro, B., Henis, Y., and Carmeli, S. 1991. Accumulation of scoparone in heat-treated lemon fruit inoculated with *Penicillium digitatum* Sacc. *Plant Physiol.*, 97:880–885.

Kirschbaum-Titze, P., Hiepler, C., Mueller-Seitz, E., and Petz, M. 2002a. Pungency in paprika (*Capsicum annuum*). 1. Decrease of capsaicinoid content following cellular disruption. *J. Agric. Food Chem.*, 50:1260–1263.

Kirschbaum-Titze, P., Mueller-Seitz, E., and Petz, M. 2002b. Pungency in paprika (*Capsicum annuum*). 2. Heterogenity of capsaicinoid content of individual fruit from one plant. *J. Agric. Food Chem.*, 50:1264–1266.

Klaren-De Wit, M., Frost, D.J., and Ward, J.P. 1971. Formation of *p*-vinylguaiacol oligomers in the thermal decarboxylation of ferulic acid. *Recl. Trav. Chim. Pays-Bas.*, 90:906–911.

Kobata, K., Sutoh, K., Todo, T., Yazawa, S., Iwai, K., and Watanabe, T, 1999. Nordihydrocapsiate, a new capsinoid from the fruits of nonpungent pepper, *Capsicum annuum*. *J. Nat. Prod.*, 62:335–336.

Kobata, K., Todo, T., Yazawa, S., Iwai, K., and Watanabe, T. 1998. Novel capsaicinoid-like substances, capsiate and dihydrocapsiate, from fruits of a nonpungent cultivar, CH-19 sweet, of pepper (*Capsicum annuum*, L.). *J. Agric. Food Chem.*, 46:1695–1697.

Koeppen, B.H. and Herrmann, K. 1977. Flavonoid glycosides and hydroxycinnamic acid esters of blackcurrants. *Z. Lebensm. Unters. Forsch.*, 164:263–268.

Kolattukudy, P.E. 1987. Lipid-derived defensive polymers and waxes and their role in plant microbe interactions, in *The Biocehemistry of Plants, a Comprehensive Treatise*, Stumph, P.K., Ed., Academic Press, Orlando, FL, 291–314.

Kolattukudy, P.E. 1980. Biopolyester membranes of plants: cutin and suberin. *Science*, 208:990–1000.

Kopp, B. and Jurenitsch, J. 1982. Biosynthesis of capsaicinoids in *Capsicum annuum* L. var. annuum. III. Problem of the formation of capsaicin and dihydrocapsaicin. *Sci. Pharm.*, Vienna, 50:150–157.

Kosuge, S. and Furata, M. 1970. Studies on pungent principle of *Capsicum*. Part XIV: chemical constitution of the pungent principle. *Agric. Food Chem.*, 34:248–256.

Kujala, T.S., Loponen, K.D., Klika, K.D., and Pihlaja, K. 2000. Phenolics and betacyanins in red beetroot (*Beta vulgaris*) root: distribution and effect of cold storage on the content of total phenolics and three individual compounds. *J. Agric. Food Chem.*, 48:5338–5342.

Kurosaki, F., Matsui, K., and Nishi, A. 1984. Production and metabolism of 6-methoxymellein in cultured carrot cells. *Physiol. Plant Pathol.*, 25:313–322.

Kurosaki, F. and Nishi, A. 1983. Isolation and antimicrobial activity of the phytoalexin 6H methoxymellein from cultured carrot cells. *Phytochemistry*, 22:669–672.

Labarbe, B., Cheynier, V., Brossaud, F., Souquet, J.-M., and Moutounet, M. 1999. Quantitative fractionation of grape proanthocyanidins according to their degree of polymerization. *J. Agric. Food Chem.*, 47:2719–2723.

Lafuente, M.T., Cantwell, M., Yang, S.F., and Rubatzky, V. 1989. Isocoumarin content of carrots as influenced by ethylene concentration, storage temperature and stress conditions. *Acta Hort.*, 258:523–534.

Lancaster, J.E. 1992. Regulation of skin color in apples. *CRC Crit. Rev. Plant Sci.*, 10:487–502.

Langcake, P., Cornford, C.A., and Pryce, R.J. 1979. Identification of pterostilbene as a phytoalexin from *Vitis vinifera* leaves. *Phytochemistry*, 18:1025–1027.

Langcake, P. and Pryce, R.J. 1977. A new class of phytoalexins from grapevine. *Experentia*, 33:151–152.

Langcake, P. and Pryce, R.J. 1976. The production of resveratrol by *Vitis vinifera* and other members of Viticeae as a response to infection or injury. *Physiol. Plant Pathol.*, 9:77–86.

Landbo, A.K. and Meyer, A.S. 2001. Enzyme-assisted extraction of antioxidative phenols from black currant juice press residues (*Ribes nigrum*). *J. Agric. Food Chem.*, 49:3169–3177.

Landry, L.G., Chapple, C.C.S., and Last, R.L. 1995. Arabidopsis mutants lacking phenolic sunscreens exhibit enhanced ultraviolet-B injury and oxidative damage. *Plant Physiol.*, 109:1159–1166.

Larrauri, J.A., Rupérez, P., and Saura-Calixto, F. 1997. Effect of drying temperature on the stability of polyphenols and antioxidant activity of red grape pomace peels *J. Agric. Food Chem.*, 45:1390–1393.

Lattanzio, V., Di Venere, D., Linsalata, V., Bertolini, P., Ippolito, A., and Salerno, M. 2001. Low temperature metabolism of apple phenolics and quiescence of *Phlyctaena vagabunda*. *J. Agric. Food Chem.*, 49:5817–5821.

Lattanzio, V., Cardinali, A., and Palmieri, S. 1994. The role of phenolics in postharvest physiology of fruits and vegetables: browning reactions and fungal diseases. A review. *Ital. J. Food Sci.*, 6:3–22.

Lea, A.G.H. 1992. Flavor, color, and stability of fruit products: the effect of polyphenols, in *Plant Polyphenols: Synthesis, Properties, Significance*, Hemingway, R.W., Laks, P.E., and Branham, S.J., Eds., Plenum Press, New York, 523–538.

Lea, A.G.H. 1990. Bitterness and astringency: the procyanidins of fermented apple ciders, in *Bitterness in Food and Beverages*, Roussef, R.L., Ed., Elsevier, Oxford, U.K., 123–143.

Lea, A.G.H. 1984. Farb-und Gerbstoffe in englischen mostäpfeln (Tannin and colour in English cider apples). *Flüssiges Obst.*, 8:356–361.

Lea, A.G.H., Bridle, P., Timberlake, C.F., and Singleton, V. 1979. The procyanidins of white grapes and wines. *Am. J. Enol. Vitic.*, 30:289–300.

Lea, A.G.H. 1978. The phenolics of ciders: oligomeric and polymeric procyanidins. *J. Sci. Food Agric.*, 29:471–478.

Lea, A.G.H. and Beech, F.W. 1978. The phenolics of ciders: effect of cultural conditions. *J. Sci. Food Agric.*, 29:493–496.

Lea, A.G.H. and Arnold, G.M. 1978. The phenolics of ciders: bitterness and astringency. *J. Sci. Food Agric.*, 29:478–483.

Lea, A.G.H. and Timberlake, C.F. 1978. The phenolics of ciders: effect of processing conditions. *J. Sci. Food Agric.*, 29:482–492.

Lea, A.G.H. and Timberlake, C.F. 1974. The phenolics of ciders. 1. Procyanidins. *J. Sci. Food Agric.*, 25:1537–1545.

Leake, D.S. 1997. The possible role of antioxidants in fruit and vegetables in protecting against coronary heart disease, in *Phytochemistry of Fruit and Vegetables, Proceedings of the Phytochemical Society of Europe 41*, Tomás-Barberán, F.A. and Robins, R.J., Eds., Clarendon Press, Oxford, U.K., 287–312.

Lee, H.S. 2002. Characterization of major anthocyanins and the color of red-fleshed Budd blood orange (*Citrus sinensis*). *J. Agric. Food Chem.*, 50:1243–1246.

Lee, Y., Howard, L.R., and Villalon, B. 1995. Flavonoids and antioxidant activity of fresh pepper (*Capsicum annuum*) cultivars. *J. Food Sci.*, 60:473–476.

Lee, C.Y., Kagan, V., Jaworski, A.W., and Brown, S.K. 1990. Enzymatic browning in relation to phenolic compounds and polyphenoloxidase activity among various peach cultivars. *J. Agric. Food Chem.*, 38:99–101.

Lee, H.S. and Nagy, S. 1990. Formation of 4-vinylguaiacol in adversely stored orange juice measured by improved HPLC method. *J. Food Sci.*, 55:162–163,166.

Lee, H.S. and Wrolstad, R.E. 1988. Apple juice composition: sugar: nonvolatile acids, and phenolic profile. *J. Assoc. Off. Anal. Chem.*, 71:789–794.

Leete, E. and Louden, M.C.L. 1968. Biosynthesis of capsaicin and dihydrocapsaicin in *Capsicum frutescens. J. Am. Chem. Soc.*, 90:6837–6841.

Leighton, T., Ginther, C., Fluss, L., Harter, W.K., Cansado, J., and Notario, V. 1992. Molecular characterisation of quercetin and quercetin glycosides in Allium vegetables: their effects on malignant cell transformation, in *Phenolic Compounds in Food and Their Effects on Health. II Antioxidants and Cancer Prevention*, Huang, M.-T., Ho, C.-T., and Lee, C.Y., Eds., ACS Symposium Series 507, American Chemical Society, Washington, D.C., 220–238.

Le Lous, J., Majoie, B., Moriniére, J.L., and Wulfert, E. 1975. Étude des flavonoides de *Ribes nigrum. Ann. Pharm. Fr.*, 33:393–399, in *Chemical Abstracts* CA 84, 176692a.

Lewis, C.A. 1996. Biochemistry and regulation of anthocyanin synthesis in potato and other tuber-bearing Solanum species. Ph.D. thesis, University of Canterbury, Christchurch, N.Z., cited in Rodriguez-Saona, L.E., Giusti, M.M., and Wrolstad, R.E. 1998. Anthocyanin pigment composition of red-fleshed potatoes. *J. Food Sci.*, 63:458–465.

Liardon, R., Braendlin, N., and Spadone, J.C. 1991. Biogenesis of Rio flavor impact compounds: 2,4,6-trichloanisole, in *14th International Colloquium on Coffee*, San Francisco, CA, cited in Aung, L.H., Smilanick, J.L., Vail, P.V., Hartsell, P.L.,and Gomez, E. 1996. Investigation into origin of chloroanisoles causing musty off-flavor in raisins. *J. Agric. Food Chem.*, 44:3294–3296.

Lidster, P.D., Dick, A.J., DeMarco, A., and McRae, K.B. 1986. Application of flavonoid glycosides and phenolic acids to suppress firmness loss in apples. *J. Am. Soc. Hortic. Sci.*, 111:892–896.

Lidster, P.D. and McRae, K.B. 1985. Effects of vacuum infusion of a partially purified β-galactosidase inhibitor on apple quality. *Hort. Sci.*, 20:80–82.

Lister, C.E., Lancaster, J.E., and Sutton, K.H. 1994. Developmental changes in the concentration and composition of flavonoids in skin of a red and green apple cultivar. *J. Sci. Food Agric.*, 64:155–161.

Liu, M., Li, X.Q., Weber, C., Lee, C.Y., Brown, J., and Liu, R.H. 2002. Antioxidant and antiproliferative activities of raspberries. *J. Agric. Food Chem.*, 50:2926–2930.

Ljunggren, B. 1990. Severe phototoxic burn following celery ingestion. *Arch. Dermatol.*, 126:1334–1336.

Lommen, A., Godejohann, M., Venema, D.P., Hollman, P.C.H., and Spraul, M. 2000. Application of directly coupled HPLC-NMR-MS to the identification and confirmation of quercetin glycosides and phloretin glycosides in apple peel. *Anal. Chem.*, 72:1793–1797.

Lomnitski, L., Nyska, A., Ben-Shaul, V., Maronpot, R.R., Haseman, J.K., Levin-Harrus, T., Bergman, M., and Grossman, S. 2000a. Effects of antioxidants apocynin and natural water-soluble antioxidant on cellular damage induced by lipopolysaccharide in the rat. *Toxicol. Pathol.*, 28:580–587.

Lomnitski, L., Foley, J., Grossman, S., Ben-Shaul, V., Maronpot, R.R., Moomaw, C., Carbonatto, M., and Nyska, A. 2000b. Effects of apocynin and natural antioxidant from spinach on inducible nitric oxide synthase and cyclooxygenase-2 induction in lipopolysaccharide-induced hepatic injury in the rat. *Pharmacol. Toxicol.*, 87:18–25.

Louche, L.M.-M., Gaydou, E.M., and Lesage, J.-C. 1998. Determination of phlorin as peel marker in orange (*Citrus sinensis*) fruits and juices. *J. Agric. Food Chem.*, 46:4191–4197.

Lu, Y. and Foo, L.Y. 2003. Polyphenolic constituents of blackcurrant seed residue. *Food Chem.*, 80:71–76.

Lu, Y., Foo, L.Y., and Wong, G. 2002. Nigrumin-5-p-coumarate and nigrumin-5-ferulate, two unusual nitrile-containing metabolites from black currant (*Ribes nigrum*) seed. *Phytochemistry*, 59:465–468.

Lu, Y. and Foo, L.Y. 1999. The polyphenol constituents of grape pomace. *Food Chem.*, 65:1–8.

Lu, Y. and Foo, L.Y. 1997. Identification and quantification of major polyphenols of apple pomace. *Food Chem.*, 59:187–194.

Lueck, R.H. 1970. Black discoloration in canned asparagus: interrelations of iron, tin, oxygen, and rutin. *J. Agric. Food Chem.*, 18:607–612.

Luh, B.S., Hsu, E.T. and Stachowicz, K. 1967. Polyphenolic compounds in canned Cling peaches. *J. Food Sci.*, 32:251–258.

Lulai, E.C. and Corsini, D.L. 1998. Differential deposition of suberin phenolic and aliphatic domains and their roles in resistance to infection during potato tuber (*Solanum tuberosum* L.) wound healing. *Physiol. Mol. Plant Pathol.*, 53:209–222.

Lulai, E.C. and Orr, P. 1994. Techniques for detecting and measuring developmental and maturational changes in tuber native periderm. *Am. Potato J.*, 71:489–505.

Luo, X.-D., Basile, M.J., and Kenelly, E.J. 2002. Polyphenolic antioxidants from fruits of *Chrysophyllum cainito* L. (Star apple). *J. Agric. Food Chem.*, 50:1374–1382.

Lynn, D.Y.C. and Luh, B.S. 1964. Anthocyanin pigments in Bing cherries. *J. Food Sci.*, 29:735–743.

Maarse, H., Nijssen, L.M., and Angelino, S.A.G.F. 1988. Halogenated phenols and chloroanisols: occurrence, formation and prevention, in *Characterization, Production and Application of Food Flavours*, Rothe, M., Ed., Akademic Verlag, Berlin, 43–63.

Maas, J.L., Galetta, G.J., and Stoner, G.D. 1990. Ellagic acid, anticarcinogen in fruit, especially in strawberries: a review. *Hort. Sci.*, 26:10–14.

Maccarone, E., Rapisarda, P., Fanella, F., Arena, E., and Mondello, L. 1998. Cyanidin-3-(6″-malonyl)-glucoside. An important anthocyanin of blood orange juice. *Ital. Food Sci.*, 10:367–372.

Maccarone, E., Maccarone, A., and Rapisarda, P. 1985. Acylated anthocyanins from oranges. *Ann. Chim.*, 75:79–86.

Maccarone, E., Maccarone, A., Perrini, G., and Rapisarda, P. 1983. Anthocyanins from the Moro orange juice. *Ann. Chim.*, 73:533–539.

Macheix, J.J., Fleuriet, A., and Billot, J. 1990. *Fruit Phenolics*. CRC Press, Boca Raton, FL.

Macheix, J.J. and Fleuriet, A. 1993. Phenolics in fruits and fruit products: progress and prospects, in *Polyphenolic Phenomena*, Scalbert, A., Ed., INRA, Paris, 157–164.

MacNaughton, W.K. and Wallace, J.L. 1989. A role of dopamine as endogenous protective factor in the rat stomach. *Gastroenterology*, 96:972–980.

Maier, V.P. and Matzler, D.M. 1967. Grapefruit phenolics. II. Principal aglycons of endocarp and peel and their possible biosynthetic relationship. *Phytochemistry*, 6:1127–1135.

Makris, D. and Rossiter, J.T. 2001. Domestic processing of onion bulbs (*Allium cepa*) and asparagus spears (*Asparagus officinalis*): effect on flavonol content and antioxidant status. *J. Agric. Food Chem.*, 49:3216–3222.

Malmberg, A.G. and Theander, O. 1985. Determination of chlorogenic acid in potato tubers. *J. Agric. Food Chem.*, 33:549–551.

Mangas, J.J., Rodriguez, R., Suárez, B., Picinelli, A., and Dapena, E. 1999. Study of the phenolic profile of cider apple cultivars at maturity by multivariate techniques. *J. Agric. Food Chem.*, 47:4046–4052.

Manthey, J.A., Grohmann, K., Mantanari, A., Ash, K., and Manthey, C.L. 1999. Polymethoxylated flavones derived from citrus suppress tumor necrosis factor-α expression by human monocytes. *J. Nat. Prod.*, 62:441–444.

Manthey, J.A. and Grohmann, K. 2001. Phenols in citrus peel by-products. Concentrations of hydroxycinnamates and polymethoxylated flavones in citrus peel molasses. *J. Agric. Food Chem.*, 49:3268–3273.

Manthey, J.A., Guthrie, N., and Grohmann, K. 2001. Biological properties of citrus flavonoids pertaining to cancer and inflammation. *Curr. Med. Chem.*, 8:135–153.

Manthey, J.A. and Grohmann, K. 1996. Concentrations of hesperidin and other orange peel flavonoids in citrus processing by-products. *J. Agric. Food Chem.*, 44:811–814.

Marin, F.R. and Del Rio, J.A. 2001. Selection of hybrids and edible citrus species with high content in the diosmin functional compound. Modulating effect of plant growth regulators on contents. *J. Agric. Food Chem.*, 49:3356–3362.

Marinelli, V.K., Zanelli, U., Nuti Ronchi, V., Pini, D., and Salvadori, P. 1989. Toxicity of 6-methoxymellin to carrot cells suspension cultures. *Votr. Plfanzenz.*, 15:23–26.

Marinelli, V.K., Nuti Ronchi, V., Pini, D., and Salvadori, P. 1990. Induction of 6-methoxymellein and 6-hydroxymellein production in carrot cells. *Phytochemistry*, 29:849–851.

Marinelli, F., Ronchi, V.N., and Salvadori, P. 1994. Elicitor induction of enzyme activities and 6-methoxymellein production in carrot cell suspension culture. *Phytochemistry*, 35:1457–1460.

Marini, D. and Balestrieri, F. 1995. Multivariate analysis of flavanone glycosides in citrus juices. *Ital. J. Food Sci.*, 3:255–264.

Markh, A.T. and Kozenko, S.I. 1964. Biochemical characteristics of quince fruits. *Prikl. Biokhim. Mikrobiol.*, 380–386, in *Chemical Abstracts*, CA 64, 1014d.

Marwan, A.G. and Nagel, C.W. 1982. Identification of the hydroxycinnamic acid derivatives in cranberries. *J. Food Sci.*, 47:774–782.

Massiot, P., Rouau, X., and Thibault, J.-F. 1988. Isolation and characterization of cell-wall fibres of carrot. *Carbohydr. Res.*, 172:217–227.

Masuda, T., Muroya, Y., and Nakatani, N. 1992. Coumarin constituents of the juice oil from citrus hassaku and their sposmolytic activity. *Biosci. Biotechnol. Biochem.*, 56:1257–1260.

Matsumoto, H., Hanamura, S., Kawakami, T., Sato, Y., and Hirayama, M. 2001. Preparative-scale isolation of four anthocyanin components of black currant (*Ribes nigrum*) fruits. *J. Agric. Food Chem.*, 49:1541–1545.

Mayer, W., Görner, A., and Andrä, K. 1977. Punicalagin and punicalin, zei gerbstoffe aus den schalen der granatäpfel Liebigs. *Ann. Chem.*, 1976–1986.

Mayr, U., Treutter, D., Santos-Buelga, C., Bauer, H., and Feucht, W. 1995. Developmental changes in the phenol concentrations of 'Golden Delicious' apple fruits and leaves. *Phytochemistry*, 38:1151–1155.

Mazza, G. 1995. Anthocyanins in grapes and grape products. *CRC Crit. Rev. Food Sci. Nutr.*, 35:341–371.

Mazza, G. and Miniati, E. 1993. *Anthocyanins in Fruits, Vegetables, and Grains*, CRC Press, Boca Raton, FL.

McIntosh, C.A. and Mansell, R.L. 1997. Three-dimensional distribution of limonin, limonoate A-ring monolactone, and naringin in the fruit tissues of three varieties of *Citrus paradisi*. *J. Agric. Food Chem.*, 45:2876–2883.

McRae, K.B., Lidster, P.D., De Marco, A.C., and Dick, A.J. 1990. Comparison of the polyphenol profiles of apple fruit cultivars by correspondence analysis. *J. Sci. Food Agric.*, 50:329–342.

Mercier, J., Aurl, J., and Julien, C. 1994. Research note. Effects of food preparation on isocoumarin, 6-methoxymellein, content of UV-treated carrots. *Food Res. Int.*, 27:401–404.

Meyer, O. 1994. Safety and security of Daflon 500 mg in venous insufficiency and hemorroidal disease. *Angiology*, 45:579–584.

Middleton, E., Jr. and Kandaswami, C. 1994. The impact of plant flavonoids on mammalian biology: implications for immunity, inflamation and cancer, in *The Flavonoids, Advances in Research Since 1986*, Harborne, J.B., Ed., Chapman & Hall, London, 619–652.

Miean, K.H. and Mohamed, S. 2001. Flavonoid (myricetin, kaempferol, luteolin, and apigenin) content of edible tropical plants. *J. Agric. Food Chem.*, 49:3106–3112.

Mikkonen, T.P., Määttä, K.R., Hukkanen, A.T., Kokko, H.I.,Törrönen, A.R., Kärenlampi, S.O., and Karjalainen, R.O. 2001. Flavonol content varies among black currant cultivars. *J. Agric. Food Chem.*, 49:3274–3277.

Millet, J., Chicaud, C., Legendre, C., and Fontaine, M. 1984. Improvement of blood filtrability in cynomolgus monkeys given a fat diet and treated with a purified extract of blackcurrant anthocyanosides. *J. Pharmacol.* (Paris), 15:439–445.

Miura, T., Fukuta, C., Ichiki, H., Iwamoto, N., Kato, M., Kubo, M., Komastu, Y., Ishida, T., Okada, M., and Taginawa, K. 2001a. Hypolipidemic activity of magniferin in cholestorol-fed mice. *Wakan Iyakugaku Zasshi*, 18:118–121.

Miura, T., Fukuta, C., Hashimoto, I., Iwamoto, N., Kato, M., Kubo, M., Ishihara, E., Komastu, Y., Okada, M., Ishida, T., and Taginawa, K. 2001b. Antidiabetic activity of a xanthone compound, magniferin. *Phytomedicine,* 8:85–87.

Miyazawa, M., Okuno, Y., Fukuyama, M., Nakamura, S., and Kosaka, H. 1999. Antimutagenic activity of polymetoxylated flavonoids from *Citrus aurantium. J. Agric. Food Chem.*, 47:5239–5244.

Miyake, Y., Murakami, A., Sugiyama, Y., Isobe, M., Koshimizu, K., and Ohigashi, H. 1999. Identification of coumarins from lemon fruit (*Citrus limon*) as inhibitors of *in vitro* tumor promotion and superoxide and nitric oxide generation. *J. Agric. Food Chem.*, 47:3151–3157.

Miyake, Y., Yamamoto, K., Morimitsu, Y., and Osawa, T. 1997. Isolation of C-glucosylflavone from lemon peel and antioxidative activity of flavonoid compounds in lemon fruit. *J. Agric. Food Chem.*, 45:4619–4623.

Mizuno, M., Tsushida, H., Kozukue, N., and Mizuno, S. 1992. Rapid quantitative analysis and distribution of free quercetin in vegetables and fruits. *Nippon Shokuhin Kogyo Gakkaishi*, 39: 88–92.

Mizuno, M., Iinuma, M., Ohara, M., Tanaka, T., and Iwasama, M. 1991. Chematoxomy on the genus Citrus based on polymethoxyflavones. *Chem. Pharm. Bull.*, 39: 945–949.

Möller, B. and Herrmann, K. 1983. Quinic acid esters of hydroxycinnamic acids in stone and pome fruit. *Phytochemistry*, 22:477–481.

Morales, M., Alcántara, J., and Barceló, A.R. 1997. Oxidation of *trans*-resveratrol by a hypodermal peroxidase isoenzyme from Gamay rouge grape (*Vitis vinifera*) berries. *J. Enol. Vitic.*, 48:33–38.

Morazzoni, P. and Bombardelli, E. 1996. *Vaccinum myrtillis* L. *Fitoterapia*, 57:3–29.

Morton, J.F. 1987. *Fruits of Warm Climates*. Morton, Miami, FL, 408–410.

Mosel, H.D. and Herrmann, K. 1974a. Changes in catechins and hydroxycinnamic acid derivatives during development of apples and pears. *J. Sci. Food Agric.*, 25:251–256.

Mosel, H.D. and Herrmann, K. 1974b. The phenolics of fruits. III. The contents of catechins and hydroxycinnamic acids in pomes and stone fruits. *Z. Lebensm. Unters. Forsch.*, 154:6–11.

Moskowitz, A.H. and Hrazdina, G. Vacuolar contents of fruit subepidermal cells from *Vitis* species. *Plant Physiol.*, 68:686–692.

Mouly, P.P., Gaydou, E.M., Faure, R., and Estienne, J.M. 1997. Blood orange juice authentication using cinnamic acid derivatives. Variety differentiations with flavanone glycoside content. *J. Agric. Food Chem.*, 45:373–377.

Mouly, P.P., Arzouyan, C.R., Gaydou, E.M., and Estienne, J.M. 1995. Chromatographie des flavanonosides des jus de differentes varietes de pamplemouses. Differentiation par analyses of statistiques multidimensionnelles. *Analusis*, 23:336–341.

Mouly, P.P., Arzouyan, C.R., Gaydou, E.M., and Estienne, J.M. 1994. Differentiation of citrus juices by factorial discriminant analysis using liquid chromatography of flavanone glycosides. *J. Agric. Food Chem.*, 42:70–79.

Mouly, P.P., Gaydou, E.M., and Estienne, J. 1993. Column liquid chromatography determination of flavanone glycosides in citrus. Application to grapefruit and sour orange juice adulterations. *J. Chromatogr.*, 634:129–134.

Movahedi, S. and Heale, J.B. 1990. The role of aspartic proteinase and endo-pectin lyase enzymes in the primary stages of infection and pathogenesis of various host tissues by different isolates of *Botrytis cinerea* Pers ex. Pers. *Physiol. Mol. Plant Pathol.*, 36:303–324.

Moyer, R.A., Hummer, K.E., Finn, C.E., Frei, B., and Wrolstad, R.E. 2002. Anthocyanins, phenolics, and antioxidant capacity of diverse small fruits: *Vaccinum, Rubus*, and *Ribes*. *J. Agric. Food Chem.*, 50:519–525.

Mullen, W., McGinn, J., Lean, M.E.J., MacLean, M.R., Gardner, P., Duthie, G.G., Yokota, T., and Crozier, A. 2002a. Ellagitannins, flavonoids, and other phenolics in red raspberries and their contribution to antioxidant capacity and vasorelaxation properties. *J. Agric. Food Chem.*, 50:5191–5196.

Mullen, W., Stewart, A.J., Lean, M.E.J., Gardner, P., Duthie, G.G., and Crozier, A. 2002b. Effect of freezing and storage on the phenolics, ellagitannins, flavonoids, and antioxidant capacity of red raspberries. *J. Agric. Food Chem.*, 50:5197–5201.

Murakami, A., Kuki, W., Takahashi, Takahashi, Y., Yonei, H., Nakamura, Y., Ohto, Y., Ohigashi, H., and Koshmizu, K. 1997. Auraptene, a citrus coumarin, inhibits 12-O-tetradecanoylphorbol-13-acetate-induced tumor promotion in ICR mouse skin, possibly through suppression of superoxide generation in leukocytes. *Jpn. J. Cancer Res.*, 88:4443–4452.

Musajo, L. and Rodighiero, G. 1962. The skin-photosensitizing furanocoumarins. *Experentia*, 18:152–200.

Naim, M., Striem, B.J., Kanner, J., and Peleg, H. 1988. Potential of ferulic acid as precursor to off-flavors in stored orange juice. *J. Food Sci.*, 53:500–503, 512.

Naim, M., Zuker, I., Zehavi, U., and Rouseff, R.L. 1993. Inhibition by thiol compounds of off-flavor formation in stored orange juice. II. Effect of L-cysteine and N-acetylcysteine on *p*-vinylguaiacol formation. *J. Agric. Food Chem.*, 41:1359–1361.

Nakatani, N., Kayano, S.-I., Kikuzaki, H., Sumino, K., Katagiri, K., and Mitani, T. 2000. Identification, quantitative determination, and antioxidative activities of chlorogenic acid isomers in prune (*Prunus domestica* L.). *J. Agric. Food Chem.*, 48:5512–5516.

Napralert Database 2002; University of Illinois, Chicago. http://www.ag.uiuc.edu./ffh/napra.html.

Nawwar, M.A.M., Husein, S.A.M., and Merfort, I. 1994. Spectral analysis of polyphenols from *Punica granatum*. *Phytochemistry*, 36:793–798.

Negrel, J., Pollet, B., and Lapierre, C. 1996. Ether-linked ferulic acid amides in natural and wound periderms of potato tubers. *Phytochemistry*, 43:1195–1199.

Ness, A.R. and Powless, J.W. 1997. Fruit and vegetables, and cardiovascular disease: a review. *Int. J. Epidemiol.*, 26:1–13.

Ng, A., Parr, J., Ingham, L.M., Rigby, N.M., and Waldron, K.W. 1998a. Cell wall chemistry of carrots (*Daucus carota* cv. Amstrong) during maturation and storage. *J. Agric. Food Chem.*, 46:2933–2939.

Ng, A., Harvey, A.J., Parker, M.L., Smith, A.C., and Walderon, K.W. 1998b. Effect of oxidative coupling on the thermal stability of texture and cell wall chemistry of beet root (*Beta vulgaris*). *J. Agric. Food Chem.*, 46:3365–3370.

Ng, A., Parker, M.L., Parr, A.J., Saunders, P.K., Smith, A.C., and Waldron, K.W. 2000. Physicochemical characteristics of onion (*Allium cepa* L.) tissues. *J. Agric. Food Chem.*, 48:5612–5617.

Nicholson, R.E. and Hammerschmidt, R. 1992. Phenolic compounds and their role in disease resistance. *Annu. Rev. Phytopathol.*, 30:369–389.

Nicolas, J.J., Richard-Forget, F.C., Goupy, P.M., Amiot, M., and Aubert, S.Y. 1994. Enzymatic browning reaction in apple and apple products. *CRC Crit. Rev. Food Sci. Nutr.*, 34:109–157.

Nicolas, J.J., Cheynier, V., Fleuriet, A., and Rouet-Mayer, M.-A. 1993. Polyphenols and enzymatic browning, in *Polyphenols Phenomena*, Scalbert, A., Ed., INRA, Paris, 165–176.

Nigg, H.N., Strandberg, J.O., Beier, R.C., Petersen, H.D., and Harrison, J.M. 1997. Furanocoumarins in Florida celery varieties increased by fungicide treatment. *J. Agric. Food Agric.*, 45:1430–1436.

Nortje, B.K. and Koeppen, B.H. 1965. The flavonol glycosides in the fruit of *Pyrus communis* L. cultivar Bon Chretien. *Biochem. J.*, 97:209–211.

Noveroske, R.L., Kuæ, J., and Williams, E.B. 1964a. Oxidation of phlorizidin and phloretin related to resistance of *Malus* to *Venturia inaequalis*. *Phytopatology*, 54:92–97.

Noveroske, R.L., Williams, E.B, and Kuæ, J. 1964b. β-Glucosidase and phenoloxidase in apple leaves and their possible resistance to *Venturia inaequalis*. *Phytopathology*, 54:98–103.

Núñez-Selles, A.J., Vélez-Castro, H.T., Agüero-Agüero, J., González-González, J., Naddeo, F., De Simone, F., and Rastrelli, L. 2002. Isolation and quantitative analysis of phenolic antioxidants, free sugars, and polyols from mango (*Magnifera indica* L.) stem bark aqueous decoction used in Cuba as a nutritional supplement. *J. Agric. Food Chem.*, 50:762–766.

Núñez-Selles, A.J., Capote, H.R., Agüero-Agüero, J., Garrido-Garrido, G., Delgado-Hernandéz, R., Martinez-Sánchez, G., León-Fernandéz, O., and Morales-Segura, M. 1999. New antioxidant product derived from *Magnifera indica* L., in *Abstracts of Papers of the 218th National Meeting of the American Chemical Society*, New Orleans, LA, American Chemical Society, Washington, D.C.

Nwinyi, S.C.O. 1992. Effect of age at shoot removal on tuber and shoot yields at harvest of five sweet potato (*Ipomoea batatas* (L.) Lam) cultivars. *Field Crop Res.*, 29:47–54.

Nybom, N., 1968. Cellulose thin layers for analysis of anthocyanins, with special reference to the anthocyanins of black raspberries. *J. Chromatogr.*, 38:382–387

Nyman, A.N. and Kumpulainen, J.T. 2001. Determination of anthocyanidins in berries and red wine by high-performance liquid chromatography. *J. Agric. Food Chem.*, 49:4183–4187.

Nyska, A., Lomnitski, L., Spalding, J., Dunson, D.B., Goldsworthy, T.L., Grossman, S., Bergman, M., and Boorman, G. 2001. Topical and oral administration of water-soluble antioxidant from spinach reduces the multiplicity of papillomas in the Tg.AC mouse model. *Toxicol Lett.*, 122:33–44.

Odake, K., Terahara, N., Saito, N., Toki, K., and Honda, T. 1992. Chemical structure of two anthocyanins from purple sweet potato *Ipomoea batatas*. *Phytochemistry*, 31:2127–2130.

Ogawa, K., Kawasaki, A., Yoshida, T., Nesumi, H., Nakano, M., Ikoma, Y., and Yano, M. 2000. Evaluation of auraptene content in citrus fruits and their products. *J. Agric. Food. Chem.*, 48:1763–1769.

Ohnishi-Kameyama, M., Yanagida, A., Kanda, T., and Nagata, T. 1997. Identification of catechin oligomers from apple (*Malus pumila* cv. Fuji) in matrix-assisted laser desorption/ionization time-of-flight mass spectrometry and fast-atom bombardment mass spectrometry. *Mass. Spectrom.*, 11:31–36.

Okombi, G. 1979. The pigments of cherries, *Prunus avium* (L.), variety Bigarreau Napolèon: influence of growth, maturity, and preservation. Ph.D. thesis, Universitè d'Orlèans, France; cited in Gao, L. and Mazza, G. 1995. Characterization, quantitation, and distribution of anthocyanins and colorless phenolics in sweet cherries. *J. Agric. Food Chem.*, 43:343–346.

Oleszek, W., Amiot, M.J., and Aubert, S.Y. 1994. Identification of some phenolics in pear fruit. *J. Agric. Food Chem.*, 42:1261–1265.

Oleszek, W., Lee, C.Y., Jaworski, A.W., and Price, K.R. 1988. Identification of some phenolic compounds in apples. *J. Agric. Food Chem.*, 36:430–432.

Ooghe, W.C. and Detavernier, C.M. 1997. Detection of the addition of *Citrus reticulata* and hybrids of *Citrus sinensis* by flavonoids. *J. Agric. Food Chem.*, 45:1633–1637.

Orsini, F., Pelizzoni, F., Verotta, L., and Aburjai, T. 1997. Isolation, synthesis, and antiplatelet aggregation activity of resveratrol 3-O-D-glucopyranoside and related compounds. *J. Nat. Prod.*, 60:1082–1087.

Ortuno, A.M., Arcas, M.C., Botia, J.M., Fuster, M.D., and Del Rio, J.A. 2002. Increasing resistance against *Phytophora citrophthora* in tangelo Nova fruits by modulating polymethoxyflavones levels. *J. Agric. Food Chem.*, 50:2836–2839.

Ortuno, A.M., Arcas, M.C., Benavente-Garcia, O., and Del Rio, J.A. 1999. Evolution of polymethoxy flavones during development of *Tangelo Nova* fruits. *Food Chem.*, 66:217–220.

Ortuno, A., Botia, J.M., Fuster, M.D., Porras, I., Garcia-Lidon, A., and Del Rio, J.A. 1997. Effect of scoparone (6,7-dimethoxycoumarin) biosynthesis on the resistance of tangelo Nova, *Citrus paradisi*, and *Citrus aurantium* fruits against *Phytophthora parasitica*. *J. Agric. Food Chem.*, 45:2740–2743.

Ostertag, E., Becker, T., Ammon, J., Bauer-Aymanns, H., and Schrenk, D. 2002. Effects of storage conditions on furocoumarin level in intact, chopped, or homogenized parsnips. *J. Agric. Food Chem.*, 50:2565–2570.

Oszmianski, J. and Lee, C.Y. 1991. Enzymatic oxidation of phloretin glucoside in model system. *J. Agric. Food Chem.*, 39:1050–1052.

Oszmianski, J. and Lee, C.Y. 1990. Enzymatic oxidative reaction of catechin and chlorogenic acid in a model system. *J. Agric. Food Chem.*, 38:1202–1204.

Park, Y.-K. and Lee, C.Y. 1996. Identification of isorhamnetin 4′-glucoside in onions. *J. Agric. Food Chem.*, 44:34–36.

Parr, A.J. and Bolwell, G.P. 2000. Phenols in plant and men. The potential for possible nutritional enhancement of the diet by modifying the phenol content or profile. *J. Sci. Food Agric.*, 80:985–1012.

Parr, A.J., Ng, A., and Waldron, K.W. 1997. Ester linked phenolic components of carrot cell walls. *J. Agric. Food Chem.*, 45:2468–2471.

Patil, B.S. and Pike, L.M. 1995. Distribution of quercetin content in different rings of various coloured onion (*Allium cepa* L.) cultivars. *J. Hortic. Sci.*, 70:643–650.

Patil, B.S., Pike, L.M., and Sun Yoo K. 1995. Variation in the quercetin content in different colored onions (*Allium cepa* L.). *J. Am. Hortic. Soc. Sci.*, 120:909–913.

Patkai, G., Barta, J., and Varsanyi, I. 1997. Decomposition of anticarcinogen factors of the beetroot during juice and nectar production. *Cancer Lett.*, 114:105–106.

Peleg, H., Naim, M., Rouseff, R.L., and Zehavi, U. 1991. Distribution of bound and free phenolic acids in oranges (*Citrus sinensis*) and grapefruit (*Citrus paradis*). *J. Sci. Food Agric.*, 60:417–426.

Peleg, H., Naim, M., Zehavi, U., Rouseff, R.L., and Nagy, S. 1992. Pathways of 4-vinylguaiacol formation from ferulic acid in model solution of orange juice. *J. Agric. Food Chem.*, 40:764–767.

Peng, Z., Hayasaka, Y., Iland, P.G., Sefton, M., Høj, P., and Waters, E.J. 2001. Quantitative analysis of polymeric procyanidins (tannins) from grape (*Vitis vinifera*) seeds by reverse phase high-performance liquid chromatography. *J. Agric. Food Chem.*, 49:26–31.

Percival, G.L. and Baird, L. 2000. Influence of storage upon light-induced chlorogenic acid accumulation in potato tubers (*Solanum tuberosum* L.). *J. Agric. Food Chem.*, 48:2476–2482.

Pérez, A.G., Cert, A., Rios, J.J., and Olias, J.M. 1997. Free and glycosidically bound volatile compounds from two banana cultivars: Valery and Pequéna Enana. *J. Agric. Food Chem.*, 45:4393–4397.

Pérez-Ilzarbe, J., Hernández, T., and Estrella, I. 1991. Phenolic compounds in apples: varietal differences. *Z. Lebensm. Unters. Forsch.*, 192:551–554.

Pérez-Ilzarbe, J., Martínez, V., Hernández, T., and Estrella, I. 1992. Liquid chromatographic determination of apple pulp procyanidins. *J. Liq. Chromatogr.*, 15:637–646.

Piano, S., Neyrotti, V., Migheli, Q., and Gullino, M.L. 1997. Biocontrol of *Metschnikowia pulcherrima* against Botrytis postharvest rot in apple. *Postharvest Biol Technol.*, 11:131–140.

Pijipers, D., Constant, J. G., and Jansen, K. 1986. *The Complete Book of Fruit.* Multimedia Publications Ltd., London, 98–103.

Pinicelli, A., Suárez, B., and Mangas, J.J. 1997. Analysis of polyphenols in apple products. *Z. Lebensm. Unters. Forsch.*, 204:48–51.

Podsędek, A., Wilska-Jeszka, J., Anders, B., and Markowski, J. 2000. Compositional characterisation of some apple varieties. *Eur. Food Res. Technol.*, 210:268–272.

Pollack, S. and Perez, A. 1997. Prospects favorable for U.S. grape industry. *Agricultural Outlook*, U.S. Department of Agriculture, Washington, D.C., 6:7–10.

Prabha, T.N., Ravindranath, B., and Patwardhan, M.V. 1980. Anthocyanins of avocado (*Persea americana*). *J. Food Sci. Technol.*, 17:241–242.

Price, K.R., Prosser, T., Richetin, A.M.F., and Rhodes, M.J.C. 1999. A comparison of the flavonol content in dessert, cooking and cider-making apples: distribution within the fruit and effect of juicing. *Food Chem.*, 66:489–494.

Price, K.R., Bacon, J.R., and Rhodes, M.J.C. 1997. Effect of storage and domestic processing on the content of flavonol glucosides in onion (*Allium cepa*). *J. Agric. Food Chem.*, 45:938–942.

Price, K.R. and Rhodes, M.J.C. 1997. Analysis of the major flavonol glycosides present in four varieties of onions (*Allium cepa*) and changes in composition resulting from autolysis. *J. Sci. Food Agric.*, 74:331–339.

Prieur, C., Rigaud, J., Cheynier, V., and Moutounet, M. 1994. Oligomeric and polymeric procyanidins from grape seeds. *Phytochemistry*, 36:781–784.

Prior, R.L., Lazarus, S.A., Cao, G., Muccitelli, H., and Hammerstone, J. F. 2001. Identification of procyanidins and anthocyanins in bluberries and cranberries (*Vaccinium* spp.) using high-performance liquid chromatography/mass spectrometry. *J. Agric. Food Chem.*, 49:1270–1276.

Prior, R., Cao, G., Martin, A., Sofic, E., McEven, J., O'Brien, C., Lischner, N., Ehlenfeldt, M., Kalt, W., Krewer, G., and Mainland, M. 1998. Antioxidant capacity as influenced by total phenolic and anthocyanin content, maturity, and variety of *Vaccinium* species. *J. Agric. Food Chem.*, 46:2686–2693.

Proctor, T.A. and Creasy, L.L. 1969. The anthocyanins of mango fruit. *Phytochemistry*, 8:2108.

Prusky, D. 1996. Pathogen quiescence in postharvest diseases. *Annu. Rev. Phytopathol.*, 34:413–434.

Pruthi, J.S. 1963. Physiology, chemistry and technology of passion fruit. *Adv. Food Res.*, 12:203–282.

Pruthi, J.S., Susheela, R., and Lal, G. 1961. Anthocyanin pigments in passion fruit skin, *J. Food Sci.*, 26:385–388.

Puski, G. and Francis, F.J. 1967. Flavonol glycosides in cranberries. *J. Food Sci.*, 35:527–530.

Raa, J. and Overeem, J.C. 1968. Transformation reactions of phlorizidin in the presence of apple leaf enzymes. *Phytochemistry*, 7:721–731.

Racchi, M., Daglia, M., Lanni, C., Papetti, A., Govoni, S., and Gazzani, G. 2002. Antiradical activity of water soluble components in common diet vegetables. *J. Agric. Food Chem.*, 50:1272–1277.

Ranadive, A.S. and Haard, N.F. 1971. Changes in polyphenolics on ripening of selected pear varieties. *J. Sci. Food Agric.*, 22:86–89.

Rapisarda, P., Bellomo, S.E., and Intelisano, S. 2001. Storage temperature effects on blood orange fruit quality. *J. Agric. Food Chem.*, 49:3230–3235.

Rapisarda, P., Fanella, F., and Maccarone, E. 2000. Reliability of analytical methods for determining anthocyanins in blood orange juices. *J. Agric. Food Chem.*, 48:2249–2252.

Rapisarda, P., Carollo, G., Fallico, B., Tomaselli, F., and Maccarone, E. 1998. Hydroxycinnamic acids as markers of Italian blood orange juices. *J. Agric. Food Chem.*, 46:464–470.

Rapisarda, P. and Giuffrida, A. 1992. Anthocyanin level in Italian blood oranges. *Proc. Int. Soc. Citriculture*, 3:1130–1133.

Raynal, J., Moutounet, M., and Souquet, J.-M. 1989. Intervention of phenolic compounds in plum technology. 1. Changes during drying. *J. Agric. Food Chem.*, 37:1046–1050.

Raynal, J. and Moutounet, M. 1989. Intervention of phenolic compounds in plum technology. 2. Mechanisms of anthocyanins degradation. *J. Agric. Food Chem.*, 37:1051–1053.

Razem, F.A. and Bernards, M.A. 2002. Hydrogen peroxide is required for poly(phenolic) domain formation during wound-induced suberization. *J. Agric. Food Chem.*, 50:1009–1015.

Reed, J. 2002. Cranberry flavonoids, atherosclerosis and cardiovascular health. *CRC Crit. Rev. Food Sci. Nutr.*, 42(Suppl.):301–316.

Rees, S.B. and Harborne, J.B. 1984. Flavonoids and other phenolics of *Cichorium* and related members of Lactuceae (Compositae*). Bot. J. Linn. Soc.*, 89:313–319.

Reeve, R.M., Hautala, E., and Weaver, M.L. 1969. Anatomy and compositional variation within the potatoes. 2. Phenolics, enzymes and other minor components. *Am. Potato J.*, 46:374–386.

Reineccius, G. Off-flavors in foods. *CRC Crit. Rev. Food Sci. Nutr.*, 29:381–402.

Reitz, S.R., Karowe, D.N., Diawara, M.M., and Trumble, J.T. 1997. Effects of elevated atmospheric carbon dioxide on the growth and linear furanocoumarin content of celery. *J. Agric. Food Chem.*, 45:3642–3646.

Renard, C.M.G.C., Baron, A., Guyot, S., and Drilleau, J.F. 2001. Interactions between apple cell walls and native apple polyphenols: quantification and some consequences. *Int. J. Biol. Macromol.*, 29:115–125.

Renault, J.H., Thepenier, P., Zeches, H.M., Le, M.-O., Durand, A., Foucault, A., and Margraft, R. 1997. Preparative separation of anthocyanins by gradient elution centrifugal partition chromatography. *J. Chromatogr. A.*, 763:345–352.

Rhodes, M.J.C. and Price, K.R. 1996. Analytical problems in the study of flavonoid compounds in onions. *Food Chem.*, 57:113–117.

Riggin, R.M., McCarthy, M.J., and Kissinger, P.T. 1976. Identification of salsolinol as a major dopamine metabolite in the banana. *J. Agric. Food Chem.*, 24:189–191.

Riov, J., Monselise, S.P., and Goren, R. 1972. Stimulation of phenolic biosynthesis in citrus fruit peel by gamma radiation. *Radiat. Res. Rev.*, 3:417–427.

Riov, J., Goren, R.., Monselise, S.P., and Kahan, R.S. 1971. Effect of gamma radiation on the synthesis of scopoletin and scopolin in grapefruit peel in relation to radiation damage. *Radiat. Res.*, 45:326–334.

Risch, B. and Herrmann, K. 1988a. Hydroxycinnamic acid derivatives in citrus fruits. *Z. Lebensm. Unters. Forsch.*, 186:530–534.

Risch, B. and Herrmann, K. 1988b. Content of hydroxycinnamic acid derivatives and catechins in pome and stone fruit. *Z. Lebensm. Unters. Forsch.*, 186:225–230.

Robinson, G.M. and Robinson, R.A. 1931. A survey of anthocyanins. *Biochem. J.*, 25:1687; cited in Gao, L. and Mazza, G. 1995. Characterization, quantitation, and distribution of anthocyanins and colorless phenolics in sweet cherries. *J. Agric. Food Chem.*, 43:343–346.

Rodriguez-Arcos, R., Smith, A.C., and Waldron, K.W. 2002. Effect of storage on wall-bound phenolics in green asparagus. *J. Agric. Food Chem.*, 50:3197–3203.

Rodriguez, R., Jimenez, A., Guillen, R., Heredia, A., and Fernandez-Bolanos, J. 1999. Postharvest changes in white asparagus cell wall during refrigerated storage. *J. Agric. Food Chem.*, 47:3551–3557.

Rodriguez-Saona, L.E., Giusti, M.M., and Wrolsdat, R.E. 1998. Anthocyanin pigment composition of red-fleshed potatoes. *J. Food Sci.*, 63:458–465.

Roggero, J.P., Larice, J.L., Rocheville-Divorne, C., Archier, P., and Coen, S. 1988. Composition anthocyanique des cépages. I: Essai de classification par analyse en composantes principales et par analyse factorielle discriminante. *Rev. Fr. d'Enol.*, 112:41–48.

Romero-Pérez, A.I., Ibern-Goméz, M., Lamuela-Raventós, R.M., and de la Torre-Boronat, M.C. 1999. Piceid, the major resveratrol derivative in grape juices. *J. Agric. Food Chem.*, 47:1533–1536.

Romeyer, F.M., Macheix, J.J., and Sapis, J.C. 1986. Changes and importance of oligomeric procyanidins during maturation of grape seeds. *Phytochemistry*, 25:219–221.

Rommel, A. and Wrolstad, R.E. 1993a. Composition of flavonols in red raspberry juice as influenced by cultivar, processing, and environmental factors. *J. Agric. Food Chem.*, 41:1941–1950.

Rommel, A. and Wrolstad, R.E. 1993b. Ellagic acid content of red raspberry juice as influenced by cultivars, processing, and environmental factors. *J. Agric. Food Chem.*, 41:1951–1960.

Rommel, A., Wrolstad, R.E., and Durst, R.W. 1992. Red raspberry phenolics: influences of processing, variety, and environmental factors, in *Phenolic Compounds in Foods and Their Effects on Health I*, Ho, C.-T., Lee, C.Y., and Huang, M.-T., Eds., ACS Symposium Series 506, American Chemical Society, Washington, D.C., 259–286.

Rommel.A., Heatherbell, D.A., and Wrolstad, R.E. 1990. Red raspberry juice and wine: effect of processing and storage on anthocyanin pigment composition, color and appearance. *J. Food Sci.*, 55:1011–1017.

Rouseff, R.I. 1988. Differentiating citrus juices using flavanone glycoside concentration profiles, in *Aldurations of Fruit Juice Beverages*, Nagy, S., Attaway, J.A., and Rhodes, M.E., Eds., Marcel Dekker, New York, 49–64.

Rouseff, R.I., Martin, S.F., and Youtsy, C.O. 1987. Quantitative survey of narirutin, naringin, hesperidin and neohesperidin in citrus. *J. Agric. Food Chem.*, 35:1027–1030.

Rouseff, R.L. and Ting, S.V. 1979. Quantification of polymethoxylated flavones in orange juices by high-performance liquid chromatography. *J. Chromatogr.*, 176:75–87.

Rowland, B.J., Villalon, B., and Burns, E.E. 1983. Capsaicin production in sweet bell and pungent Jalapeno peppers. *J. Agric. Food Chem.*, 31:484–487.

Ruf, J.C. 1999. Wine and polyphenols related to platelet aggregation and atherothrombosis. *Drug Exp. Clin. Res.*, 25:125–131.

Ryan, J.J. and Coffin, D.E. 1971. Flavonol glucuronides from red raspberry, *Rubus idaeus* (Rosaceae). *Phytochemistry*, 10:1675–1677.

Saito, M., Hosoyama, Ariga, T., Kataoka, S., and Yamaji, N. 1998. Antiulcer activity of grape seed extract and procyanidins. *J. Agric. Food Chem.*, 46:1460–1464.

Sakamoto, S., Goda, Y., Maitani, T., Yamada, T., Nunomura, O., and Ishikawa, K. 1994. High-performance liquid chromatographic analyses of capsaicinoids and their phenolic intermediates in *Capsicum annuum* to characterize their biosynthetic status. *Biosci. Biotechnol. Biochem.*, 58:1141–1142.

Sakho, M., Chassagne, D., and Crouzet, J. 1997. African Mango glycosidically boound volatile compounds. *J. Agric. Food Chem.*, 45:883–888.

Saleh, N.A.M. and El Ansari, M.A. 1975. Polyphenols in twenty local varieties of *Magnifera indica. Planta Med.*, 28:124–130.

Sanoner, P., Guyot, S., Marnet, N., Molle, D., and Drilleau, J.-F. 1999. Polyphenol profiles of French cider apple varieties (*Malus domestica* sp.). *J. Agric. Food Chem.*, 47:4847–4853.

Santos de Buelga, C., Francia-Aricha, E.M., and Escribano-Bail6n, M.T. Comparative flavan-3-ol composition of seeds from different grape varieties. *Food Chem.*, 53:197–201.

Sapers, G.M. and Hargrave, D.L. 1987. Proportions of individual anthocyanins in fruits of cranberry cultivars. *J. Am. Soc. Hortic. Sci.*, 112:100–104.

Sapers, G.M., Hicks, K.B., Burgher, A.M., Hargrave, D.L., Sondey, S.M., and Bilyk, A. 1986. Anthocyanin patterns in ripening thornless balckberries. *J. Am. Soc. Hort. Sci.*, 111:945–950.

Sapers, G.M., Phillips, J.G., Rudolf, H.M., and DiVito, A.M. 1983. Cranberry quality: selection for breeding programs. *J. Am. Soc. Hortic. Sci.*, 108:241–246.

Sarker, S.K. and Phan, C.T. 1979. Naturally occuring and ethylene-induced phenolic compounds in the carrot root. *J. Food Protec.*, 42:526–534.

Sato, T., Kawamoto, A., Tamura, S., Tatsum, Y., and Fujii, T. 1992. Mechanism of antioxidant action of Puararia glycoside (PG)-1 (an isoflavonoid) and magniferin (a xanthonid). *Chem. Pharm. Bull.*, 40:721–724.

Saucier, C., Mirabel, M., Daviaud, F., Longieras, A., and Glories, Y. 2001. Rapid fractionation of grape seed proanthocyanidins. *J. Agric. Food Chem.*, 49:5732–5735.

Schaller, D.R. and Von Elbe, J.H. 1970. Polyphenols in Montmorency cherries. *J. Food Sci.*, 35:762–765.

Schaller, D.R. and Von Elbe, J.H. 1968. The minor pigment component of Montmorency cherries. *J. Food Sci.*, 3:442–443.

Scheel, L.D., Perone, V.B., Larkin, R.L., and Kupel, R.E. 1963. The isolation and characterization of two phototoxic furanocoumarins (psolarens) from diseased celery. *Biochemistry*, 2:1127–1131.

Schmidt, H.W.H. and Neukom, H. 1969. Identification of the main oligomeric proanthocyanidin occurring in apples. *J. Agric. Food Chem.*, 17:344–346.

Seligman, P.J., Mathias, C.G.T., O'Malley, M.A., Beier, R., Fehrs, L.J., Serrill, W.S., and Halperin, W.E. 1987. Photodermatitis from celery among grocery store workers. *Arch. Dermatol.*, 123:1478–1482.

Sellappan, S. and Akoh, C.C. 2002. Flavonoids and antioxidant capacity of Georgia-grown Vidalia onions. *J. Agric. Food Chem.*, 50:5338–5442.

Sellappan, S., Akoh, C.C., and Krewer, G. 2002. Phenolic compounds and antioxidant capacity of Georgia-grown bluberries and blackberries. *J. Agric. Food Chem.*, 50:2431–2438.

Senchuk, G.V. and Borukh, I.F. 1976. Wild berries of Belorussia. *Rastit. Resur.*, 12:113–117.

Senter, S.D. and Callahan, A. 1990. Variability in the quantities of condensed tannins and other major phenols in peach fruit during maturation. *J. Food Sci.*, 55:1585–1587; 1602.

Senter, S.D., Robertson, J.A., and Meredith, F.I. 1989. Phenolic compounds in mesocarp of Cresthaven peaches during storage and ripening. *J. Food Sci.*, 54:1259–1260, 1268.

Shoji, T., Yanagida, A., and Kanda, T. 1999. Gel permeation chromatography of anthocyanin pigments from rosé cider and red wine. *J. Agric. Food Chem.*, 47:2885–2890.

Shrikhande, A.J. and Francis, F.J. 1973. Flavone glycosides of sour cherries. *J. Food Sci.*, 38:1035–1037.

Siemann, E.H. and Creasy, L.L. 1992. Concentration of the phytoalexin resveratrol in wine. *Am. J. Enol. Vitic.*, 43:49–52.

Silva, B.M., Andrade, P.B., Ferreres, F., Dominigues, A.L., Seabra, R.M., and Ferreira, M.A. 2002. Phenolic profile of quince fruit (*Cyclonia oblonga* Miller) (pulp and peel). *J. Agric. Food Chem.*, 50:4615–4618.

Silva, B.M., Andrade, P.B., Mendes, G.C., Valentao, P., Seabra, R.M., and Ferreira, M.A. 2000. Analysis of phenolic compounds in the evaluation of commercial quince jam authenticity. J. *Agric. Food Chem.*, 48:2853–2857.

Silva, G.H., Chase, R.W., Hammerschmidt, R., and Cash, J.N. 1991. After-cooking darkening of Spartan Pearl potatoes as influenced by location, phenolic acids, and citric acid. *J. Agric. Food Chem.*, 39:871–873.

Simmonds, N.W. 1954a. Anthocyanins in bananas. *Ann. Bot.*, (London) 28:471–482.

Simmonds, N.W. 1954b. Anthocyanins in bananas. *Nature,* (London) 173:402.

Sinclair, W.B. 1972. *The Grapefruit. Its Composition, Physiology, and Products*, University of California, Division of Agricultural Sciences, Los Angeles, 268–274.

Singleton, V.L., Timberlake, C.F., and Lea, A.G.F. 1978. The phenolic cinnamates of white grapes and wine. *J. Sci. Food Agric.*, 29:403–410.

Sioud, F.B. and Luh, B.S. 1966. Polyphenolic compounds in pear puree. *Food Technol.*, 20:535–538.

Skrede, G., Wrolstad, R.E., and Durst, R.W. 2000. Changes in anthocyanins and polyphenolics during juice processing of highbush blueberries (*Vaccinium corymbosum* L.). *J. Food Sci.*, 65:357–364.

Slimstad, R. and Solheim, H. 2002. Anthocyanins from black currants (*Ribes nigrum* L.). *J. Agric. Food Chem.*, 50:3228–3231.

Smith, M.A.L., Marley, K.A., Seigler, D., Singletary, K.W., and Meline, B. 2000. Bioactive properties of wild blueberry fruits. *J. Food Sci.*, 65:352–356.

Sojo, M.M., Nuñez-Delicado, E., Sánchez-Ferrer, A., and Garcia-Carmona, F. 2000. Oxidation of salsolinol by banana pulp polyphenol oxidase and its kinetic synergism with dopamine. *J. Agric. Food Chem.*, 48:5543–5547.

Sondheimer, E. 1957. The isolation and identification of 3-methyl-6-methoxy-8-hydroxy-3,4-dihydroisocoumarin from carrots. *J. Am. Chem. Soc.*, 79:5036–5039.

Souquet, J.M., Labarbe, B., Le Guernevé, C., Cheynier, V., and Moutounet, M. 2000. Phenolic composition of grape stems. *J. Agric. Food Chem.*, 48:1076–1080.

Souquet, J.M., Cheynier, V., Brossaud, F., and Moutounet, M. 1996. Polymeric proanthocyanidins from grape skins. *Phytochemistry*, 43:509–512.

Spanos, G.A. and Wrolstad, R.E. 1992. Phenolics of apple, pear, and white grape juices and their changes with processing and storage — a review. *J. Agric. Food Chem.*, 40:1478–1487.

Spanos, G.A., Wrolstad, R.E., and Heatherbell, D.A. 1990. Influence of processing and storage on the phenolic composition of apple juice. *J. Agric. Food Chem.*, 38:1572–1579.

Stange, Jr., R.R., Midland, S.L., Eckert, J.W., and Sims, J.J. 1993. An antifungal compound produced by grapefruit and Valencia orange after wounding of the peel. *J. Nat. Prod.*, 56:1627–1627.

Stanley, W.L. and Vannier, S.H. 1957. Chemical composition of lemon oil. I. Isolation of a series of substituted coumarins. *J. Am. Chem. Soc.*, 79:3488–3491.

Starke, H. and Herrmann, K. 1976. The phenolic of fruits. VIII. Changes in flavonol concentration during fruit development. *Z. Lebensm. Unters. Forsch.*, 161:131–135.

Steglich, W. and Strack, D. 1991. Betalains, in *The Alkaloids*, Brossi, A., Ed., Academic Press, Orlando, FL, 1–62.

Steinmetz, K.A. and Potter, J.D. 1996. Vegetables, fruits, and cancer prevention: a review. *J. Am. Diet. Assoc.*, 96:1027–1039.

Steinmetz, K.A. and Potter, J.D. 1991a. Vegetables, fruits, and cancer. I. Epidemiology. *Cancer Causes Control*, 2:325–357.

Steinmetz, K.A. and Potter, J.D. 1991b. Vegetables, fruits, and cancer. II. Mechanisms. *Cancer Causes Control*, 2:427–442.

Stern, R.S., Thibodeau, L.A., Kleinerman, R.A., Parrish, J., and Fitzpatrick, T.B. 1979. Risk of cutaneous carcinoma in patients treated with oral methoxalen phytochemotherapy for psiorasis. *N. Engl. J. Med.*, 300:809–813.

Stevenson, A.E. 1950. Rutin content of asparagus. *Food Res.*, 15:150–154.

Stewart, A.J., Bozonnet, S., Mullen, W., Jenkins, G.I., Lean, M.E.J., and Crozier, A. 2000. Occurrence of flavonols in tomatoes and tomato-based products. *J. Agric. Food Chem.*, 48:2663–2669.

Stintzing, F.C., Stintzing, A.S., Carle, R., and Wrolstad, R.E. 2002. A novel zwitterionic anthocyanin from evergreen blackberry (*Rubus laciniatus* Willd.). *J. Agric. Food Chem.*, 50:396–399.

Stöhr, H. and Herrmann, K. 1975. Phenolics of fruits. VI. Phenolics of currants, gooseberries, and blueberries and the changes in phenolic acids and catechins during the development of blackcurrants. *Z. Lebensm. Unters. Forsch.*, 159:31–37.

Stöhr, H., Mosel, H.D., and Herrmann, K. 1975. The phenolics of fruits. VII. The phenolics of cherries and plums and the changes in catechins and hydroxycinnamic acid derivatives during the development of fruits. *Z. Lebensm. Unters. Forsch.*, 159:85–91.

Stopar, M., Bolcina, U., Vanzo, A., and Vrhovsek, U. 2002. Lower crop load for cv. Jonagold (*Malus x domestica* Borkh.) increases polyphenol content and fruit quality. *J. Agric. Food Chem.*, 50:1643–1646.

Strack, D. and Wray, V. 1994. The anthocyanins, in *The Flavonoids: Advances in Research Since 1986*, Harborne, J.B., Ed., Chapman & Hall, London, 1063–1072.

Su, C.T. and Singleton, V.L. 1969. Identification of three flavan-3-ols from grape. *Phytochemistry*, 8:1553–1558.

Suda, I., Oki, T., Masuda, M., Nishiba, Y., Furuta, S., Matsugano, K., Sugita, K., and Terahara, N. 2002. Direct absorption of acylated anthocyanin in purple-fleshed sweet potato into rats. *J. Agric. Food Chem.*. 50:1672–1676.

Sukrasno, N. and Yeoman, M.M. 1993. Phenylpropanoid metabolism during growth and development of *Capsicum frutescens* fruits. *Phytochemistry*, 32:839–844.

Sun, B.H. and Francis, F.J. 1967. Apple anthocyanins: identification of cyanidin-7-arabinoside. *J. Food Sci.*, 32:647–649.

Superchi, S., Pini, D., Salvadori, P., Marinelli, F., Rainaldi, G., Zanelli, U., and Nuti Ronchi, V. 1993. Synthesis and toxicity to mammalian cells of the carrot dihydroxycoumarins. *Chem. Res. Toxicol.*, 6:46–49.

Surico, G., Varvaro, L., and Solfrizzo, M. 1987. Linear furanocoumarin accumulation in celery plants with *Erwinia caratovora* cv. Caratovora. *J. Agric. Food Chem.*, 35:406–409.

Suzuki, T. and Iwai, K. 1984. Constituents of red pepper species: chemistry, biochemistry, pharmacology, and food science of the pungent principle of *Capsicum* species, in *The Alkaloids*, Brossi, A., Ed., Academic Press, Orlando, FL., vol. 23, 227–299.

Suzuki, T., Kawada, T., and Iwai, K. 1981. The precursors affecting the composition of capsaicin and its analogues in the fruits of *Capsicum annuum* cv. Karayatsubusa. *Agric. Biol. Chem.*, 45:535–537.

Suzuki, T., Fujikawe, H., and Iwai, K. 1980. Intracellular localization of capsaicin and its analogues in *Capsicum* fruit. I. Microscopic investigation of the structure of the placenta of *Capsicum annuum* var. annuum cv. Karayatsubusa. *Plant Cell Physiol.*, 21:839–853.

Swanson, C.A. 1998. Vegetables, fruits, and cancer risk: the role of phytochemicals, in *Phytochemicals: A New Paradigm*, Bidlack, W.R., Omaye, S.T., Meskin, M.S., and Jahner, D., Eds., Technomic Publishing Co. Inc., Lancaster, PA, 1–10.

Takahashi, Y., Inaba, N., Kuwahara, S., Kuki, W., Yamane, K., and Murakami, A. 2002. Rapid and convenient method for preparing auraptene-enriched product from hassaku peel oil: implication for cancer-preventive food additives. *J. Agric. Food Chem.*, 50:3193–3196.

Talcott, S.T., Howard, L.R., and Brenes, C.H. 2000a. Contribution of periderm material and blanching time to the quality of pasteurized peach puree. *J. Agric. Food Chem.*, 48:4590–4596.

Talcott, S.T., Howard, L.R., and Brenes, C.H. 2000b. Antioxidant changes and sensory properties of carrot puree processed with and without periderm tissue. *J. Agric. Food Chem.*, 48:1315–1321.

Talcott, S.T. and Howard, L.R. 1999a. Chemical and sensory quality of processed carrot puree as influnced by stress-induced phenolic compounds. *J. Agric. Food Chem.*, 47:1362–1366.

Talcott, S.T. and Howard, L.R. 1999b. Determination and distribution of 6-methoxymellein in fresh and processed carrot puree by rapid spectrophotometric assay. *J. Agric. Food Chem.*, 47:3237–3242.

Talcott, S.T. and Howard, L.R. 1999c. Phenolic autoxidation is responsible for color degradation in processed carrot puree. *J. Agric. Food Chem.*, 47:2109–2115.

Tanaka, T., Kawabata, K., Kakumota, M., Makita, H., Hara, A., Murakami, A., Kuki, W., Takahashi, Y., Yonei, H., Koshimizu, K., and Ohigashi, H. 1997. Citrus auraptene inhibits chemically induced colonic aberrant crypt foci in male F344 rats. *Carcinogenesis,* 18:2151–2161.

Tanaka, T., Kawabata, K., Kakumota, M., Hara, A., Murakami, A., Kuki, W., Takahashi, Y., Yonei, H., Maeda, M., Ota, T., Odashima, S., Yamane, T., Koshimizu, K., and Ohigashi, H. 1998a. Citrus auraptene exerts dose-dependent chemopreventive activity in large bowel tumorgenesis: the inhibition correlates with suppression of cell profileration and lipid peroxidation and with induction of phase II drug-metabolizing enzymes. *Cancer Res.*, 58:2550–2556.

Tanaka, T., Kawabata, K., Kakumota, M., Matsunaga, K., Mori, H., Murakami, A., Kuki, W., Takahashi, Y., Yonei, H., Satoh, K., Hara, A., Maeda, M., Ota, T., Odashima, S., Yamane, T., Koshimizu, K., and Ohigashi, H. 1998b. Chemoprevention of 4-nitroquinoline-1-oxide-induced oral carcinogenesis by citrus auraptene in rats. *Carcinogenesis*, 19:425–431.

Tanaka, T., Nonaka, G., and Nishioka, I. 1986a. Tannins and related compounds. XLI. Isolation and characterization of ellagitannins, puicacorteins A, B, C and D and punigluconin from the bark of *Punicum granatum* L. *Chem. Pharm. Bull.*, 34:656–663.

Tanaka, T., Nonaka, G., and Nishioka, I. 1986b. Tannins and related compounds. XL. Revision of the structures of punicalin and punicalagin, and isolation and characterization of 2-O-galoylpunicalin from the bark of *Punica granatum* L. *Chem. Pharm. Bull.*, 34:650–655.

Tanaka, T., Nonaka, G., and Nishioka, I. 1985. Punicafolin and ellagitannin from the leaves of *Punica granatum. Phytochemistry*, 24:2075–2078.

Tang, J., Zhang, Y., Hartman, T.G., Rosen, R.T., and Ho, C.-T. 1990. Free and glycosidically bound volatile compounds in fresh celery (*Apium graveolens* L.). *J. Agric. Food Chem.*, 38:1937–1940.

Tatum, J.H. and Berry, R.H. 1979. Coumarins and psolarens in grapefruit peel oil. *Phytochemistry*, 18:500–502.

Terahara, N., Shimizu, T., Kato, Y., Nakamura, M., Maitani, T., Yamaguchi, M., and Goda, Y. 1999. Six diacetylated anthocyanins from the storage roots of purple sweet potato, *Ipomoea batatas. Biosci. Biotechnol. Biochem.*, 63:1420–1424.

Thorngate, J.H. and Singleton, V.L. 1994. Localization of procyanidins in grape seeds. *Am. J. Vitic.*, 45:259–262.

Timberlake, C.F. and Bridle, P. 1982. Distribution of anthocyanins in food plant, in *Anthocyanins as Food Color*, Markakis, P., Ed., Academic Press, New York, 137.

Timberlake, C.F. and Bridle, P. 1971. The anthocyanins of apples and pears. The occurrence of acyl derivatives. *J. Sci. Food Agric.*, 22:509–513.

Ting, S.V. and Attaway, J.A. 1971. Citrus fruits, in *The Biochemistry of Fruits and Their Products*, Hulme, A.C., Ed., Academic Press, London, vol. 2, 107.

Titze, P. and Petz, U. 1999. Untersuchungen zum Capsaicinoidgehalt in frishen Gewurzpaprika (Studies on the capsaicinoid content in fresh paprika). *Lebensmittelchemie*, 53:78.

Tomás-Barberán, F.A., Gil, M.I., Cremin, P., Waterhouse, A.L., Hess-Pierce, B., and Kader, A.A. 2001. HPLC-DAD-ESIMS analysis of phenolic compounds in nectarines, peaches, and plums. *J. Agric. Food Chem.*, 49:4748–4760.

Tomás-Barberán, F.A. and Espin, J.C. 2001. Phenolic compounds and related enzymes as determinants of quality in fruits and vegetables. *J. Sci. Food Agric.*, 81:853–876.

Tomás-Barberán, F.A., Gil, M.I., Castaner, M., Artes, F., and Saltveit, M.E. 1997. Effect of selected browning inhibitors on phenolic metabolism in stem tissue of harvested lettuce. *J. Agric. Food Chem.*, 45:583–589.

Tomás-Barberán, F.A. and Robins, R.J. 1997. Introduction, in *Phytochemistry of Fruit and Vegetables, Proceedings of the Phytochemical Society of Europe 41*, Tomás-Barberán, F.A. and Robins, R.J., Eds., Clarendon Press, Oxford, U.K., 1–10.

Tomás-Barberán, F.A., García-Viguera, C., Nieto, J.L., Ferreres, F., and Tomás-Lorente, F. 1993. Dihydrochalcones from apple juices and jams. *Food Chem.*, 46:33–36.

Torres, J.L. and Bobet, R. 2001. New flavanol derivatives from grape (*Vitis vinifera*) by-products. Antioxidant aminoethylthio-flavan-3-ol conjugates from a polymeric waste fraction used as a source of flavanols. *J. Agric. Food Chem.*, 49:4627–4634.

Torres, A.M., Mau-Lastovicka, T., and Rezaaiyan, R. 1987. Total phenolics and high-performance liquid chromatography of phenolic acids of avocado. *J. Agric. Food Chem.*, 35:921–925.

Traitler, H., Winter, H., Richli, U., and Ingenbleek, Y. 1984. Characterization of γ-linolenic acid in *Ribes* seed. *Lipids*, 19:923–928.

Tressl, R., Bahri, D., Holzer, M., and Kossa, T. 1977. Formation of flavor components in asparagus. 2. Formation of flavor components in cooked asparagus. *J. Agric. Food Chem.*, 25:459–463.

Tressl, R. and Drawert, F. 1973. Biogenesis of banana volatiles. *J. Agric. Food Chem.*, 21:560–565.

Trifiro, A., Gherardi, S., and Calza, M. 1995. Effects of storage time and temperature on the quality of fresh juice from pigmented oranges. *Ind. Conserve*, 70:243–251,

Trousdale, E.K. and Singleton, V.L. 1983. Astilbin and engeletin in grapes and wine. *Phytochemistry*, 22:619–620.

Trumble, J.T., Millar, J.G., Ott, D.E., and Carson, W.C. 1992. Seasonal patterns and pesticidal effects on the phototoxic linear furanocoumarins in celery, *Apium graveolens* L. *J. Agric. Food Chem.*, 40:1501–1506.

Tsouderos, Y. 1991. Venous tone: are the phlebotonic properties of a theurapeutic benefit? A comprehensive view of our experience with Daflon 500 mg. *Z. Kardiol.*, 80:S95–S101.

Tsukui, A., Sukuki, A., Komaki, K., Terahara, N., Yamakawa, O., and Hayashi, K. 1999. Stability and composition ratio of anthocyanin pigments from *Ipomoea batatas* Poir. *Jpn. Soc. Food Sci. Technol.*, 46:148–154, cited in Suda, I., Oki, T., Masuda, M., Nishiba, Y., Furuta, S., Matsugano, K., Sugita, K., and Terahara, N. 2002. Direct absorption of acylated anthocyanin in purple-fleshed sweet potato into rats. *J. Agric. Food Chem.*, 50:1672–1676.

Tsushida, T. and Suzuki, M. 1995. Isolation of flavonoid glycosides in onion and identification by chemical synthesis of the glycosides. *Nipp. Shok. Kaga Kaishi*, 42:100–108.

Tsushida, T. and Suzuki, M. 1996. Content of flavonol glucosides and some properties of enzymes metabolising the glucosides in onion. *Nipp. Shok. Kaga Kaishi*, 43:642–649.

Tudela, J.A., Cantos, E., Espín, J.A., Tomás-Barberán, F.A., and Gil, M.I. 2002. Induction of antioxidant flavonol biosynthesis in fresh-cut potato. Effect of domestic cooking. *J. Agric. Food Chem.*, 50:5925–5931.

Undenfriend, S., Lovenberg, W., and Sjodedsma, A. 1959. Physiologically active amines in common fruits and vegetables. *Arch. Biochem. Biophys.*, 85:487–490.

Vallés, B.S., Victorero, J.S., Alonso, J.J.M., and Gomis, D.B. 1994. High-performance liquid chromatography of the neutral phenolic compounds of low molecular weight in apple juice. J. *Agric. Food Chem.*, 42:2732–2736.

Van Buren, J., de Vos, L., and Pilnik, W. 1976. Polyphenols in Golden Delicious apple juice in relation to method of preparation. *J. Agric. Food Chem.*, 24:448–451.

Van Buren, J. 1970. Fruit phenolics, in *Biochemistry of Fruits and Their Products*, Hulme, A.C., Ed., Academic Press, London, 269–304.

van der Sluis, A.A., Dekker, M., de Jager, A., and Jongen, W.M.F. 2001. Activity and concentration of polyphenolic antioxidants in apple: effect of cultivar, harvest year, and storage conditions. *J. Agric. Food Chem.*, 49:3606–3613.

Van Gorsel, H., Li, C., Kerbel, E.L., Smiths, M., and Kadar, A.A. 1992. Compositional characterization of prune juice. *J. Agric. Food Chem.*, 40:784–789.

Venkataraman, K. 1975. Flavones, in *The Flavonoids*, Harborne, J.B., Mabry, T.J., and Mabry, H., Eds., Chapman & Hall, London, 267.

Vernenghi, A., Ramiandrasoa, F., Chuilon, S., and Ravise, A. 1987. Phytoalexines des citrus: seseline proprietes inhibitrices et modulation de synthese. *Fruits,* 42:103–111.

Versari, D., Barbanti, S., Biesenbruch, S., and Farnell, P.J. 1997. Analysis of anthocyanins in red fruits by use of HPLC/spectral array detection. *Ital. J. Food Sci.*, 9:141–148.

Versari, A., Parpinello, G.P., Tornielli, G.B., Ferrarini, R., and Giulivo, C. 2001. Stilbene compounds and stilbene synthase expression during ripening, wilting and UV treatment in grape cv. Corvina. *J. Agric. Food Chem.*, 49:5531–5536.

Villareal, R.L., Tsou, S.C., Lo, H.F., and Chiu, S.C. 1982. Sweet potato tips as vegetables, in *Sweet Potato: Proceedings of the First International Symposium*, Villareal, R.L. and Griggs, T.D., Eds., AVRDC, Shauna, Taiwan, 313–320.

Vorsa, N. and Welker, W.V. 1985. Relationship between fruit size and extractable anthocyanin content in cranberry. *HortScience*, 20:402–403.

Vrhovŝek, U. 1998. Extraction of hydroxycinnamoyltartaric acids from berries of different grape varieties. *J. Agric. Food Chem.*, 46:4203–4208.

Waffo-Teguo, P., Lee, D., Cuendet, M., Merillon, J.-M., Pezzuto, J.M., and Kinghorn, A.D. Two new stilbene dimer glucosides from grape (*Vitis vinifera*) cell cultures. *J. Nat. Prod.*, 64:136–138.

Wagner, H., Maurer, I., Farkas, L., and Strelisky, J. 1977. Synthese von Polyhydroxy-Flavonol methyläthern mit potentieller cytotoxischer Wirksamkeit I. Synthese von Quercetagetin- und Gossypetin-dimethyläthern zum Strukturbeweis neuer Flavonole aus *Parthenium-Chrysosplenium-*, *Larrea-* und *Spinacia*-Arten. *Tetrahedron*, 33:1405–1409.

Wald, B., Wray, V., Galensa, R., and Herrmann, K. 1989. Malonated flavonol glycosides and 3,5-dicaffeoylquinic acid from pears. *Phytochemistry*, 28:663–664.

Waldron, K.W., Ng, A., Parker, M.L., and Parr, A.J. 1997. Ferulic acid dehydrodimers in the cell walls of *Beta vulgaris* and their possible role in texture. *J. Sci. Food Agric.*, 74:221–228.

Walter, W.M. and Purcell, A.E. 1979, Evaluation of several methods for analysis of sweet potato phenolics. *J. Agric. Food Chem.*, 27:942–946.

Wang, Y., Catana, F., Yang, Y., Roderick, R., and van Breemen, R.B. 2002a. An LC-MS method for analyzing total resveratrol in grape juice, cranberry juice, and in wine. *J. Agric. Food Chem.*, 50:431–435.

Wang, S.Y., Zheng, W., and Galleta, G.J. 2002b. Cultural systems affect fruit quality and antioxidant capacity in strawberries. *J. Agric. Food Chem.*, 50:6534–6542.

Wang, S.Y. and Stretch, A.L. 2001 Antioxidant capacity in cranberry is influenced by cultivar and storage temperature. *J. Agric. Food Chem.*, 49:969–974.

Wang, S.Y. and Lin, H.-S. 2000. Antioxidant activity in fruits and leaves of blackberry, raspberry, and strawberry varies with cultivar and development stage. *J. Agric. Food Chem.*, 48:140–146.

Wang, H., Nair, M.G., Strasburg, G.M., Booren, A.M., and Gray, J.I. 1999a. Novel antioxidant compounds from tart cherries (*Prunus cereus*). *J. Nat. Prod.*, 62:86–88.

Wang, H., Nair, M.G., Strasburg, G.M., Booren, A.M., and Gray, J.I. 1999b. Antioxidant polyphenols from tart cherries (*Prunus cerasus*) *J. Agric. Food Chem.*, 47:840–844.

Wang, H., Nair, M.G., Iessoni, A.F., Strasburg, G.M., Booren, A.M., and Gray, J.I. 1997. Quantification and characterization of anthocyanins in Balaton tart cherries. *J. Agric. Food Chem.*, 45:2556–2560.

Wang, H., Cao, G., and Prior, R.L. 1996. Total antioxidant capacity of fruits. *J. Agric. Food Chem.*, 44:701–705.

Waterhouse, A.L. and Lamuela-Raventós, R.M. 1994. The occurrence of piceid, a stilbene glucoside, in grape berries. *Phytochemistry*, 27:571–573.

Whaley, W.M. and Govindachari, T.R. 1951. The Pictet-Spengler synthesis of tetrahydroisoquinolines and related compounds. *Org. React.*, 6:151–190.

Whitaker, B.D., Schmidt, W.F., Kirk, M.C., and Barnes, S. 2001. Novel fatty acid esters of *p*-coumaryl alcohol in epiticular wax of apple fruit. *J. Agric. Food Chem.*, 49:3783–3792.

Whiting, G.C., and Coggins, R.A. 1975. Estimation of monomeric phenolics of ciders. *J. Sci. Food Agric.*, 26:1833–1838.

Wilfried, C.O., Sigrid, J.O., Detavernier, C.M., and Huyghebaert, A. 1994a. Characterization of orange juice (*Citrus sinensis*) by flavanone glycosides. *J. Agric. Food Chem.*, 42:2183–2190.

Wilfried, C.O., Sigrid, J.O., Detavernier, C.M., and Huyghebaert, A. 1994b. Characterization of orange juice (*Citrus sinensis*) by polymethoxylated flavanone glycosides. *J. Agric. Food Chem.*, 42:2191–2195.

Willet, C.W. 1994. Micronutrients and cancer risk. *Am. J. Clin. Nutr.*, 59:162S–165S.

Willstästter, R. and Zollinger, E.H. 1916. Investigation of anthocyanin. XIV. Pigments of cherries and wild plums. *Justus Liebigs Ann. Chem.*, 412:1654, cited in Gao, L. and Mazza, G. 1995. Characterization, quantitation, and distribution of anthocyanins and colorless phenolics in sweet cherries. *J. Agric. Food Chem.*, 43:343–346.

Wilson, E.L. High-performance liquid chromatography of apple juice phenolic compounds. *J. Sci. Food Agric.*, 32:257–264.

Winter, M. and Herrmann, K. 1986. Esters and glucosides of hydroxycinnamic acids in vegetables. *J. Agric. Food Chem.*, 34:616–620.

Winter, M. and Kloti, R. 1972. Uber das Aroma der Gelben Passionsfrucht. *Helv. Chim. Acta*, 55:1916–1921.

Woldecke, M. and Herrmann, K. 1974a. Isolation and identification of the flano(ol)glycosides of the endive (*Cichorium endivia* L.) and the lettuce (*Lactuca sativa* L.) (in German). *Z. Naturforsch.*, 29C:355–361.

Woldecke, M. and Herrmann, K. 1974b. Flavanole und flavone der Gemusearten. III. Flavanole und flavone der tomaten und des gemusepaprikas. *Z. Lebensm. Unters. Forsch.*, 155:216–219.

Woldecke, M. and Herrmann, K. 1974c. Flavonole und der Flavone der Gemusearten und Flavonole des Spargels. *Z. Lebensm. Unters. Forsch.*, 155:151–154.

Woolfe, J.A. 1992. *Sweet Potato. An Untapped Food Resource*, Cambridge University Press, Cambridge, U.K., 118–187.

Wrolstad, R.E., McDaniel, M.R., Durst, R.W., Michaels, N., Lampi, K.A., and Beaudry, E.G., 1993. Composition and sensory characterization of red raspberry juice concentrated by direct-osmosis or evaporation. *J. Food Sci.*, 58:633–637.

Wu, C.M., Koehler, P.E., and Ayres, J.C. 1972. Isolation and identification of xanthotoxin (8-methoxypsolaren) and bergapten (5-methoxypsolaren) from celery infected with *Sclerotinia sclerotiorum*. *Appl. Microbiol.*, 23:237–247.

Yamaguchi, M.-A., Kawanobu, S., Maki, T., and Ino, I. 1996. Cyanidin 3-malonylglucoside and malonyl-coenzyme A: anthocyanidin malonyltransferase in *Lactuca sativa* leaves. *Phytochemistry*, 42:661–663.

Yang, Y. and Chien, M. 2000. Chracterization of grape procyanidins using high-performance liquid chromatography/mass spectrometry and matrix-assisted laser desorption/ionization time-of-flight mass spectrometry. *J. Agric. Food Chem.*, 48:3990–3996.

Yang, C.S., Lee, M.-J., Chen, L., and Yang, G.-Y. 1997. Polyphenols as inhibitors of carcinogenesis. *Environ. Health Perspect.*, 105 (Suppl. 4):971–976.

Yazawa, S., Suetome, N., Okamoto, K., and Namiki, T. 1989. Content of capsaicinoids and capsaicinoid-like substances in fruit of pepper (*Capsicum annuum* L.) hybrids made with "CH-19 sweet" as a parent. *J. Jpn. Soc. Hortic. Sci.*, 58:601–607.

Yoshimi, N., Matsunaga, K., Katayama, M., Yamada, Y., Kuno, T., Qiao, Z., Hara, A., Yamahara, T., and Mori, H. 2001. The inhibitory effects of magniferin, a naturally occurring glucosylxanthone, in bowel carcinogenesis of male F344 rats. *Cancer Lett.*, 163:163–170.

Young, A.R. 1990. Photocarcinogenicity of psolarens used in PUVA treatment-present status in mouse and man. *J. Photochem. Photobiol. B*, 6:237–247.

Zafrilla, P., Ferreres, F., and Tomás-Barberán, F.A. 2001. Effect of processing and storage on the antioxidant ellagic acid derivatives and flavonoids of red raspberry (*Rubus idaeus*) jams. *J. Agric. Food Chem.*, 49:3651–3655.

Zane, E. and Wender, S.H. 1961. Flavonols in spinach leaves. *J. Org. Chem.*, 26:4718–4719.

Zapsalis, C. and Francis, F.J. 1965. Cranberry anthocyanins. *J. Food Sci.*, 30:396–399.

Zhang, J., Zhan, B., Yao, X., and Song, J. 1995. Antiviral activity of Punica granatum L. against genital herpes virus *in vitro*. *Zhogguo Zhongyao Zazhi*, 20:556–558.

Zhao, S., Fu, L., Yu, Y., and Liu, M. 1994. Nutrients in the seed of six kinds of *Ribes* plants. *Yingyang Xuebao*, 16:232–235, in *Chemical Abstracts*, CA 122, 289363q.

Zheng, Z. and Shetty, K. 2000. Solid-state bioconversion of phenolics from cranberry pomace and role of *Letinus edodes* β-glucosidase. *J. Agric. Food Chem.*, 48:895–900.

Zobel, A.M. 1997. Coumarins in fruit and vegetables, in *Phytochemistry of Fruit and Vegetables, Proceedings of the Phytochemical Society of Europe 41*, Tomás-Barberán, F.A. and Robins, R.J., Eds., Clarendon Press, Oxford, U.K., 173–204.

Zuo, Y., Wang, C., and Zhan, J. 2002. Separation, characterization, and quantification of benzoic and phenolic antioxidants in American cranberry fruit by GC-MS. *J. Agric. Food Chem.*, 50:3789–3794.

5 Phenolic Compounds of Beverages

INTRODUCTION

Popular beverages in the world include tea, coffee, cocoa, beer, wine and fruit juices; all of these beverages contain phenolic compounds. The content of phenolics in beverages depends on species, degree of maturity and processing and climatic factors of the starting materials.

TEA

Tea has been cultivated and consumed in China and Asia for over 2000 years. Tea beverage is prepared by brewing the leaves of the plant *Camellia sinensis* (Balentine et al., 1997; Harbowy and Balentine, 1997; Ukers, 1994; Wilson and Clifford, 1992). Approximately 20% of world tea production is in the form of green tea — a product obtained by rapid inactivation (dry heating or steaming) of the phenol oxidase (phenolase) present in tea leaves. Lack of enzymatic oxidation of phenolics gives the green tea its light color and characteristic astringency. Red and yellow teas are semifermented products with some of the characteristics of green tea and black tea. Phenolics of oolong and Pounchong teas are 70 and 30% oxidized, respectively (Balentine et al., 1997).

Black tea is one of the most popular beverages in the world and is traditionally made from flush that comprises the apical bud with two or three leaves. Production of black tea involves six stages:

1. Flush is plucked at different intervals, depending on the season, to obtain the best possible chemical composition (Sanderson, 1964). Plucking may be done by hand or using harvesting machines.
2. Flush is dried out in a process called withering. During this period a number of chemical changes takes place (Graham, 1983).
3. Dried tea leaves are rolled to disrupt the tissue, which brings about a cellular decompartmentation and provides conditions required for oxidation of polyphenols in the cell vacuole by the tea oxidative enzymes.
4. Broken tea leaves are fermented at room temperature in the presence of oxygen and high humidity. During this process polyphenols in tea undergo nonenzymatic and enzymatic oxidation, leading to the formation of theaflavins and thearubigins (Robertson, 1992).

TABLE 5.1
Content of Polyphenols in Various Types of Green and Black Tea[a]

Tea	Content
Green Tea	
Finest grade (Japan)	132
Popular grade (Japan)	229
Popular grade (China)	258
Black Tea	
High grown (Sri Lanka)	280
Low grown (Sri Lanka)	302

[a] Grams per kilogram of dry weight.

Source: Adapted from Wickremasinghe, R.L. 1978. Tea, *Adv. Food Res.*, 24:229–286.

5. The fermentation process is terminated by firing with hot air, during which time and temperature are manipulated to obtain a final moisture content of about 3 to 5%.
6. Fermented tea leaves are graded (Graham, 1983).

Composition of Polyphenolics

Polyphenolic compounds constitute up to 35% of dry weight of tea (Table 5.1). The major constituents of tea polyphenolics are flavanols, such as (–) epicatechin gallate, (–) epigallocatechin, (–) epigallocatechin gallate, (–) epicatechin, (+) catechin (Figure 5.1), and their derivatives (Table 5.2), flavonols (quercetin, kaempferol and their glycosides), flavones (vitexin, isovitexin), phenolic acids, and depsides (gallic acid, chlorogenic acid, theogallin) (Balentine et al., 1997; Finger et al., 1991; Harbowy and Balentine, 1997; Wickremasinghe, 1978). The composition of unprocessed and black tea polyphenols is summarized in Table 5.3. According to Haslam (1989), tea flush (the immature vegetative portions of the tea plant) may contain up to 30% polyphenols, of which the major components are flavan-3-ols.

This finding was later confirmed by Yamanishi et al. (1995), who reported that the total content of polyphenols in tea flush is between 20 and 35% and that flavanols comprise up to 90% of total polyphenols. Soluble polyphenols constitute about 15% (w/w) of black tea. However, their composition may vary depending on the variety of tea, its geographical origin, and environmental conditions, as well as agronomic situation. According to Chen et al. (2001), the total contents of epicatechin derivatives in Chinese black teas is between 2.4 and 5.1 g/kg; in Chinese oolong tea it ranges from 41.4 and 46.3 g/kg and in Chinese green tea between 80 and 144.4 g/kg. Epigallocatechin gallate is the most abundant epicatechin derivative in teas.

Recently, Sano et al. (1999) isolated two novel catechin derivatives with a potent antiallergic activity from Taiwanese oolong tea and elucidated their structures as

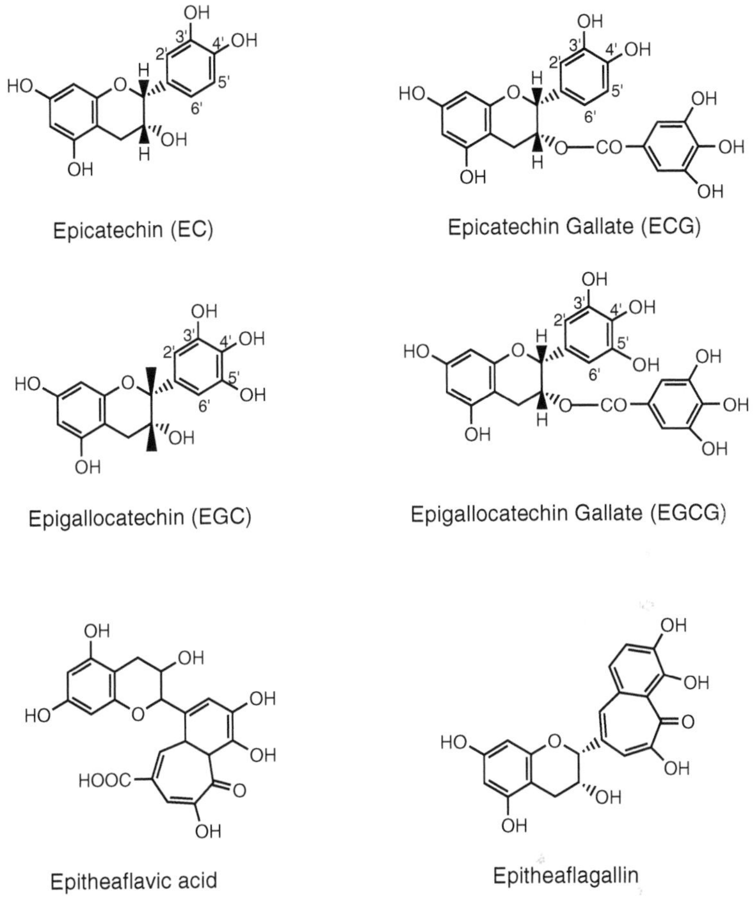

Epicatechin (EC)

Epicatechin Gallate (ECG)

Epigallocatechin (EGC)

Epigallocatechin Gallate (EGCG)

Epitheaflavic acid

Epitheaflagallin

FIGURE 5.1 Structures of catechins, epitheaflavic acid and epitheaflagallin found in tea.

(–)-epigallocatechin 3-*O*-(3-*O*-methyl)gallate (C-1) and (–)-epigallocatechin 3-*O*-(4-*O*-methyl)gallate (C-2). Of the green tea leaves tested, 13 of 15 contain from 0.2 to 8.2 g of C-1/kg; C-2 is detected only in one cultivar. On the other hand, oolong tea cultivar, Tong ting, contains 3.6 and 2.0 g/kg of C-1 and C-2, respectively. These authors demonstrated that oral doses between 5 and 50 mg/kg of C-1 and C-2 show greater inhibition of type I allergic (anaphylactic) reaction in mice, sensitized with ovalbumin than the major tea catechin (–)-epigallocatechin 3-*O*-gallate. Recently, Lakenbrink et al. (1999) isolated two novel proanthocyanidins from Chinese green tea (Lung Ching from Zhejiang) and elucidated their structures as epiafzelechingallate-(4β→8)-epicatechingallate and epiafzelechingallate-(4β→6)-epicatechin-gallate.

Fermentation and Polyphenolics

A critical process for production of good quality tea, fermentation leads to the formation of orange-yellow to red-brown pigments and volatile flavor compounds

TABLE 5.2
Content of Some Flavanols in Fresh Tea Flush[a]

Compound	Content[b]
(-) Epicatechin	10–30
(-) Epicatechin gallate	30–60
(-) Epigallocatechin	30–60
(-) Epigallocatechin gallate	90–130
(+) Catechin	10–20
(+) Gallocatechin	30–40

[a] Grams per kilogram of dry weight.

Source: Adapted from Wickremasinghe, R.L. 1978. Tea, *Adv. Food Res.*, 24:229–286.

TABLE 5.3
Polyphenolic Composition of Tea (Assam Variety)

Phenolic Compounds	Content[a]
Unprocessed tea	
Flavanols (catechins + gallocatechins)	170–300
Flavonols + flavonol glycosides	30–40
Flavandiols (leucoanthocyanidins)	20–30
Phenolic acids + depsides	50
Black tea	
Dialysable thearubigins + bisflavanols	20–40
Other soluble thearubigins	15
Theaflavins	10–20
Phenolic acids and depsides	40
Unchanged flavanols	10–30
Flavonols + flavonol glycosides	20–30

[a] Grams per kilogram of dry weight.

Source: Adapted from Stagg, G.V. and Millin, D.J., 1975, *J. Sci. Food Agric.*, 26:1439–1459.

(Balentine et al., 1997; Harbowy and Balentine, 1997). During this process polyphenolases catalyze the oxidation of polyphenolic compounds present in tea leaves to their corresponding *o*-quinones (Figure 5.2). Oxidation of flavan-3-ols, leading to the formation of theaflavins and therubigins responsible for the formation of the characteristic color and flavor of fermented tea, is catalyzed by catechol oxidase (Robertson, 1992). Therefore, control of the fermentation process has important effects on the flavor and color of tea, which depend on the degree of oxidation of

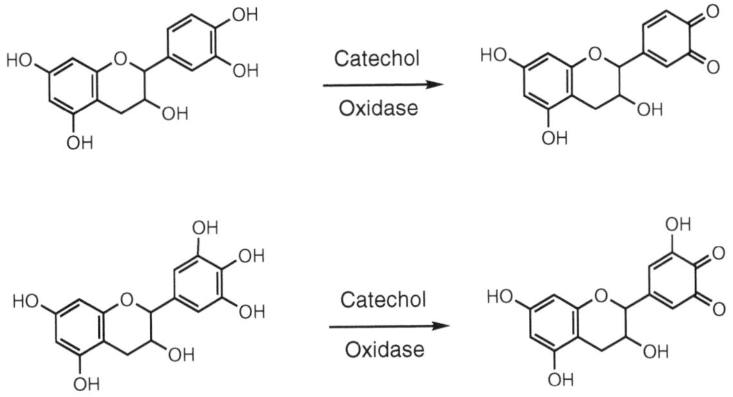

FIGURE 5.2 Oxidation of catechin (top) and epigallocatechin (bottom) to o-quinones.

tea phenolics. The rate and degree of polyphenolic oxidation depends on the composition, distribution, and content of flavanols in fresh tea shoot, activity of oxidizing enzymes, degree of tissue damage and cellular disruption, and temperature and oxygen content of the fermenting tea leaves. The polyphenolase is a copper-containing enzyme with a molecular weight of 140,000 Da located in the leaf microsomes (Balentine, 1992; Graham, 1983). A correlation exists between the flavor and activity of tea polyphenolases as reported by Roberts (1952).

Flavanols such as catechins and their derivatives, as well as flavonols such as quercetin, myricetin and kaempferol, constitute 15 to 25% of dry weight of tea. The content of some flavanols found in fresh tea flush is given in Table 5.2 and Table 5.3. Epigallocatechin gallate and epigallocatechin constitute the major catechins in green tea leaves. During fermentation, oxidation of flavanols to quinones is followed by a Michael addition of gallocatechin quinone to catechin quinone, carbonyl addition across the ring and decarboxylation (Balentine, 1992). This leads to the formation of theaflavins and finally to polymeric compounds referred to as thearubigins (Figure 5.3). It has also been suggested that thearubigins are formed from theaflavins via coupled oxidation with other polyphenolic compounds (Roberts, 1962). In addition, during the initial stages of fermentation, thearubigins may also be formed directly from flavanols by oxidative transformation (Brown et al., 1969). Therefore, reactions leading to the formation of theaflavins and thearubigins are competitive.

As the fermentation progresses, fewer theaflavins and more thearubigins are found in fermented tea leaves. Small quantities of other oxidation products have also been identified in black tea, namely, theaflavic acids and theaflagallins (Figure 5.1). Theaflavic acids are products of the reaction of gallic acid and catechin quinones, while theaflagallins are products of the reaction of gallic acid with gallocatechin quinones (Robertson, 1992). The composition of phenolic compounds in black tea is given in Table 5.4.

The values provided in Table 5.3 indicate that fermentation significantly reduces the total content of flavanols in processed tea; however, other polyphenolic substances in their monomeric form (flavonols and flavonol glycosides) remain unchanged throughout the fermentation process and thus coexist with oxidized

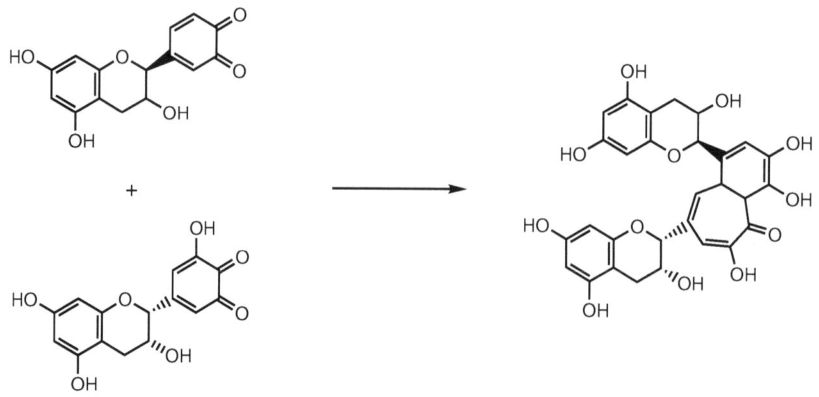

FIGURE 5.3 Formation of theaflavin from oxidized flavan-3-ols.

TABLE 5.4
Effect of Fermentation on Distribution of Catechin Derivatives in Taiwanese Teas[a]

Tea Variety	EGCG	EGC	ECG	EC	C	GC	C-1	Total
Benihomare								
Green tea	97.4	45.8	22.7	11.1	1.8	2.4	5.8	187.0
Oolong tea	92.0	35.2	20.2	9.0	2.1	2.2	4.7	165.4
Black tea	3.4	1.3	0.5	0.8	0.2	0.7	—	6.9
Izumi								
Green tea	131.1	47.6	29.1	10.6	3.6	3.6	0.4	226.0
Oolong tea	109.9	38.3	25.1	8.6	3.3	3.9	0.2	189.3
Black tea	4.8	0.9	1.0	1.0	0.5	0.7	0.2	9.1

Note: EGCG = epigallocatechin gallate; EGC = epigallocatechin; ECG = epicatechin gallate; EC = epicatechin; C = catechin; GC = gallocatechin; C-1 = epigallocatechin 3-*O*-(3-*O*-methyl)gallate.

[a] Grams per kilogram of dry weight.

Source: Adapted from Sano, M. et al., 1999, *J. Agric. Food Chem.*, 47:1906–1910.

polyphenols in black tea. Changes brought about by fermentation in the distribution of catechins in some Chinese tea cultivars are shown in Table 5.4. Theaflavins and thearubigins are considered primary and secondary products of polyphenol oxidation, respectively (Haslam, 1989). The relative content of theaflavins and thearubigins is determined by the fermentation temperature, pH and plucking intervals.

Low-temperature fermentation enhances the production of theaflavins during the initial stages and reduces the chance of conversion of theaflavins to thearubigins during the latter stages. On the other hand, tea produced at higher fermentation temperatures contains more thearubigin pigments, including a nondialyzable polymeric fraction (Cloughley, 1980). This nondialyzable material is composed largely of an ill-defined polyphenols–protein complex. It contains a polyphenolic component formed by excessive polymerization during the fermentation process (Millin et al., 1969; Sanderson et al., 1972). Subramanian et al. (1999) showed that lowering the pH of macerated tea leaves from 5.5 (the normal pH of tea leaves) to 4.5 increases the content of theaflavins up to 40%. These researchers also demonstrated, *in vitro*, that hydrogen peroxide is generated during oxidation of catechins by polyphenol oxidase. This hydrogen peroxide is utilized for oxidation of theaflavins to thearubigins. The content of theaflavins also depends on plucking intervals. A shorter plucking interval produces tea with a higher content of theaflavins (Owuor, 1990).

Sanderson et al. (1972) studied the biochemistry of tea fermentation in order to determine the role of various flavanols as precursors of polyphenolic compounds found in black tea. A number of flavanols isolated from fresh tea leaves were subjected to oxidation catalyzed by tea enzymes. Results of this study indicate that theaflavins are formed according to the following oxidative condensation reactions:

- (–)Epicatechin + (–)epigallocatechin + 3/2 $O_2 \rightarrow$ theaflavin + CO_2
- (–)Epicatechin + (–)epigallocatechin-3-gallate + 3/2 $O_2 \rightarrow$ theaflavin gallate A + CO_2
- (–)Epicatechin-3-gallate + (–)epigallocatechin + 3/2 $O_2 \rightarrow$ theaflavin gallate B + CO_2
- (–)Epicatechin-gallate + (–)epigallocatechin-3-gallate + 3/2 $O_2 \rightarrow$ theaflavin digallate + CO_2

During the fermentation of flavanols some bisflavanols (colorless, reactive substances also known as theasinensins) are also formed (Figure 5.4). Their rearrangement leads to the formation of undefined black tea flavonoids (Hashimoto et al., 1988; Nonaka et al., 1983). Reactions leading to the formation of bisflavanols are given below:

- 2(–)epigallocatechin-3-gallate + 1/2 $O_2 \rightarrow$ bisflavanol A + H_2O
- (–)epigallocatechin-3-gallate + (–)epigallocatechin + 1/2 $O_2 \rightarrow$ bisflavanol B + H_2O
- 2(–)epigallocatechin + 1/2 $O_2 \rightarrow$ bisflavanol C + H_2O

Theaflavins are yellow-orange in color and contribute to the astringency as well as another flavor characteristic of black tea known as “briskness” (Mcdowell et al., 1995; Sanderson and Graham, 1973). Hilton and Ellis (1972) reported a correlation between the content of theaflavins and taste sensory scores of tea from Central Africa. Oxidized flavanols may also react with amino acids and carotenoids. In addition, Strecker degradation of amino acids may lead to the formation of various aldehydes, and ionones and their derivatives may be formed from carotenoids. These reaction products contribute to the aroma of fermented tea (Sanderson et al., 1971).

Thearubigins, the polymeric form of oxidized flavanols, are dominant phenolics of black tea (Table 5.5). These heterogenous oxidation products may be derived not

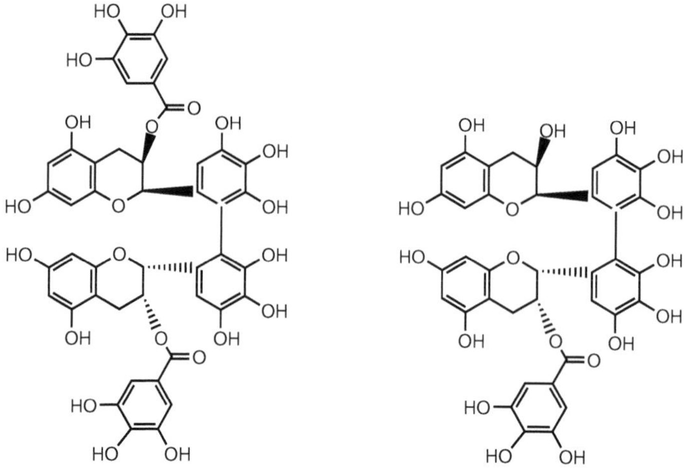

TheasinensinA
(Bisflavanol A)

TheasinensinB
(Bisflavanol B)

FIGURE 5.4 Chemical structure of bisflavanol A, also known as theasinensin A.

only from theaflavins but also from oxidative condensation of all tea flavanols (Sanderson et al., 1972). Recently, Tanaka et al. (2001) demonstrated two types of oxidative dimerization of theaflavins to thearubigins. One reaction leads to the formation of bistheaflavin A via self-association of theaflavin into antiparallel dimer at the benzotropolone rings, followed by their oxidative C–C coupling. Another reaction leads to the formation of bistheaflavin B via autoxidation of theaflavins. According to these authors, bistheaflavin B is probably formed by intermolecular cyclization between dehydrotheaflavin and dehydrotheanaphthoquinone.

Thearubigins are black-brown in color and contribute to the color, acidity, body and astringency of tea. Thearubigins found in hot water extracts of tea are mixtures of polymers with a wide molecular weight range. While some thearubigins contain up to 100 flavonoid subunits, on average, four to five subunits of flavonoid thearubigins are found in hot water extracts of tea. These compounds are heterogenous polymers of catechin and flavan-3-ol gallates linked at C4, C6, C8, C2′, C5′ and C6′ positions in flavon-3-ol units (Ozawa et al., 1996). Two fractions have been isolated from black tea thearubigins, namely, theafulvin and oolongtheanin (Bailey et al., 1992; Hashimoto et al., 1988). Acid hydrolysis of thearubigins results in the formation of cyanidin and delphinidin (Balentine, 1992; Sanderson, 1972).

Brewing

Tea beverage is commonly prepared by brewing tea in hot water in a proportion of 1g of tea leaf to 100 mL of water. The concentration of solids in tea infusion after 3 min of brewing ranges from 2.5 to 3.5 g/L, but black and green tea infusions differ in their composition of phenolics as affected by the manufacturing process (Balentine et al., 1997).

TABLE 5.5
Composition of Phenolics in Black Tea Leaves[a]

Polyphenols	Content
Catechin	2.3
Epicatechin	4.1
Epicatechin gallate	8.0
Epigallocatechin	10.5
Epigallocatechin gallate	16.5
Flavonol glycosides	0.5
Theaflavin	2.5
Theaflavin gallate	6.6
Thearubigins	59.5

[a] Grams per kilogram of dry weight.

Source: Adapted from Balentine, D., 1992, in *Phenolic Compounds in Food and Their Effects on Health,* Ho, C.-T., Lee, C.Y., and Huang, M.T., Eds., American Chemical Society, Washington, D.C., 102–117.

Green tea infusions have a clear, greenish-yellow color and a balanced taste of bitterness, astringency, and sweet aftertaste (Sanderson et al., 1976; Wickremasinghe, 1978). The distinct sensory quality of green tea extract is affected by the addition of sugar, milk, lemon and heat processing as well as storage (Wang et al., 2000). Therefore, production of a high-quality canned green tea beverage is a dificult task. Recently, Wang et al. (2000) evaluated the effect of 1-min heat treatment at 121°C and accelerated 12-day storage at 50°C (storage conditions equivalent to 70-day storage at 20°C) on flavanols and sensory qualities of green tea extracts. Green tea infusions used in this study were prepared by extraction of green tea with hot water (80°C) at a 1:160 ratio (tea leaves to water, w/w). Epigallocatechin gallate (EGCG) and epigallocatechin (EGC) accounted for 80% of total epicatechins of green tea infusion. After processing and accelerated storage, 86% of EGCG and 79% of EGC were not accounted for in the tea infusion. Therefore, according to these authors, changes in EGCG and EGC may be responsible for changes in color and taste of green tea brought about by processing and storage. Chen et al. (2001) observed epimerization of EGCG to gallocatechin gallate when pure EGCG is subjected to autoclaving at 120°C for 20 min.

The stability of green tea catechins is pH dependent. Green tea catechins are very unstable at pH values of >7.0 and degrade in a few minutes, but very stable at pH < 4.0 for at least 18 h. Moreover, boiling green tea catechins in water (pH = 4.9) for 7 h reduces the total content of catechins by only 15% (Zhu et al., 1997). The mechanism of catechin breakdown under alkaline conditions is still unknown. However, it has been demonstrated that at pH >7.0, catechins tend to form semi-quinone free radicals (Guo et al., 1996; Yoshioka et al., 1991). Later, Chen et al. (1998) observed that addition of ascorbic acid at a level of 0.2g/L significantly

TABLE 5.6
Content of Polyphenols in Consumer Tea Leaves and Brews

Tea Source	Polyphenols	Tea Leaves[a]	Tea Brew[b]	
			40–60	120
U.K.	Total phenolics	160.0	591–694	843
U.S.		162.9	533	967
U.K.	Catechins	13.4	39.1–48.1	56.2
U.S.		36.5	18.5	37.1
U.K.	Theaflavins	15.4	55.3–70.2	77.4
U.S.		11.7	24.4	39.4
U.K.	Flavonols	8.6	50.2–54.0	63.0
U.S.		8.8	37.8	67.2

[a] Grams per kilogram.
[b] Milligram per liter.

Source: Adapted from Lakenbrink, C. et al., 2000, *J. Agric. Food Agric. Chem.*, 48:2848–2852.

improves the stability of green tea catechins at pH > 7.0. These authors suggested that ascorbic acid might protect catechins by recycling the catechin free radical forms or by slowing down their oxidation by removing oxygen from the solution.

Liebert et al. (1999) evaluated the effect of different brewing times and conditions on the extraction of total phenolics from black tea. All tea infusions were prepared at tea leaves-to-water ratio of 1:77. The total content of phenolics in tea infusion ranges from 338 mg/L after 0.5 min to 684 mg/L after 10 min of brewing time. Stirring black tea during infusion increased the total phenolic content in tea extract up to 40%. The content of flavonol glycosides in tea infusions is also affected by brewing conditions and varies from 36.5 to 88.3 mg/L (Price et al., 1998). Arts et al. (2000) later reported that the maximum catechin concentration in tea infusion is nearly reached after 5 min of brewing. Lakenbrink et al. (2000) subsequently reported that only 35 to 55% of total phenolics, total flavonoids, catechins and theaflavins originally present in tea leaves are extracted at brew times of up to 2 min (Table 5.6).

COFFEE

Coffee beans are seeds of red cherries of evergreen shrubs belonging to the *Rubiaceae* subfamily *Cinchonoideae* tribe *Coffeae*. Two genera containing numerous species and varieties of coffee, namely, *Coffea* and *Psilanthus*, belong to this tribe. An example is *Coffea* subgenera *Coffea*, which contains 85 species of coffee. Of these, only two species, known as *Coffea arabica* and *Coffea canephora* (commonly referred to as arabicas and robustas, respectively), are of commercial importance (Flament, 1995).

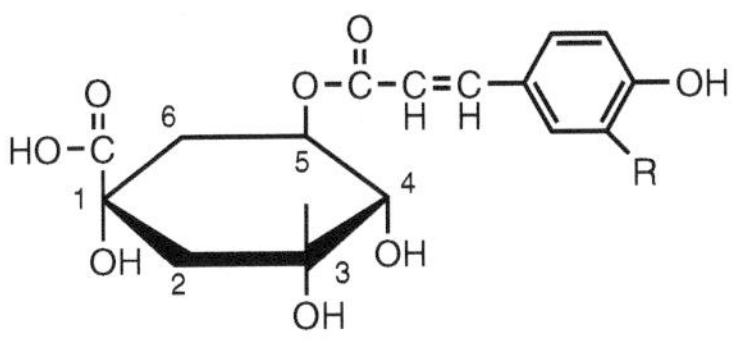

R = H	5-*p*-Coumaroylquinic Acid
R = OH	5-Caffeoylquinic Acid
R = OCH_3	5-Feruloylquinic Acid

FIGURE 5.5 Structure of chlorogenic acid and its derivatives.

The cherry usually has one or two seeds (beans) that account for 55% of the weight of fruit dry matter. The pericarp pulp, a by-product obtained during wet-processing of fruit, accounts for 29% of the cherry. Most of the pulp solids are currently discarded. Because this may bring about serious pollution problems, attempts have been made to utilize coffee pulp as feed for cattle, swine and poultry. However, the presence of antinutritional factors such as caffeine and polyphenols in the pulp used in animal feed impairs the growth rate of experimental animals when used in excess of 10% of rations. Elucidation of the exact chemical nature of compounds responsible for these effects requires further investigation (Bressani and Graham, 1980; Clifford and Ramirez-Martinez, 1991).

Green Coffee Beans

Green coffee beans may contain at least five major groups of chlorogenic acid isomers: caffeoylquinic acids (CQA), dicaffeoylquinic acids (diCQA), feruoylquinic acids (FQA), *p*-coumaroylquinic acids (CoQA) and caffeoylferuloylquinic acids (CFQA). Figure 5.5 depicts the general structure of chlorogenic acid. The contents of total chlorogenic, caffeoylquinic, and feruloylquinic acids for four varieties of green coffee beans are summarized in Table 5.7 and Table 5.8. Data presented in these tables indicate that Robusta coffee has a substantially higher content of chlorogenic acids. The total content of chlorogenic acids in *C. arabica* and *C. canephora* is 5 to 8% and 7 to 10% on a dry weight basis, respectively (Clifford, 1985), and varies with maturity (Clifford and Kazi, 1987). The content of chlorogenic acids in *C. cenophora* accessions grown in Colombia is even higher and ranges from 12 to 13.5% (dry basis) (Guerrero et al., 2001). Vanillin, 4-ethylguaiacol and 4-vinylguaiacol have been indentified among the 16 potent odorants in raw Arabica coffee beans. However, only the concentration of 4-vinylguaiacol exceeds the odor threshold (Czerny and Grosch, 2000).

Many attempts have been made to find a correlation between the level of total, or any specific isomer of, chlorogenic acid and any quality characteristic of beverages such as astringency. However, data available in the literature to indicate such

TABLE 5.7
Content of Chlorogenic Acid Isomers in Some Green Coffee Beans

Isomer	Samples[a]				
	B-1	B-2	B-3	B-4	B-5
3-CQA	3.0	2.6	2.5	2.3	2.4
4-CQA	4.9	4.8	5.4	4.8	5.2
5-CQA	45.6	39.7	47.2	39.2	40.0
Total CQA	53.3	47.1	55.1	46.3	47.7
5-FQA	7.9	3.3	3.5	3.3	3.4
3,4-diCQA	4.4	0.9	0.8	1.0	0.9
3,5-diCQA	5.6	2.6	2.2	3.0	2.5
4,5-diCQA	0.5	2.3	1.6	2.6	2.1
Total diCQA	10.5	5.8	4.5	6.6	5.6
Total chlorogenic acids	71.7	56.2	63.1	56.2	56.7

Note: Results are average of duplicate determinations for dry matter; CQA = caffeoylquinic acids; FQA = feruloylquinic acids; diCQA = dicaffeoylquinic acids; B-1 = *Coffea canephora*; B-2 = Timor hybrid (*C. arabica* x *C. robusta*); B-3 = catimor (Timor hybrid x *C. arabica* var. caturra); B-4 = *Coffea arabica* var. caturra vermelho; B-5 = *Coffea arabica* var. bourbon vermelho.

[a] Grams per kilogram of dry weight.

Source: Adapted from Clifford, M.N. and Ramirez-Martinez, J.R., 1991, *Food Chem.*, 40:35–42.

TABLE 5.8
Content of Feruloyiquinic Acids, Caffeoylquinic Acids, and Total Chlorogenic Acids in Green Coffee Beans[a]

Phenolic Acids	Arabica		Robusta	
	Santos	Sao Paulo	Ghana	Uganda
Feruoylquinic acids				
Range	2.3–3.3	0–2.1	11.6–12.0	5.4–6.8
Mean	2.8	0.6	11.8	6.1
Caffeoylquinic acids				
Range	60.8–62.6	56.2–58.2	79.2–84.3	77.1–80.9
Mean	61.7	57.2	81.8	79.1
Total chlorogenic acids				
Range	64.2–64.8	56.5–59.1	92.6–94.7	83.9–86.6
Mean	64.5	57.8	93.6	85.2

[a] Grams per kilogram of dry weight.

Source: Adapted from Clifford, M.N. and Wright, J., 1976, *J. Sci. Food Agric.*, 27:73–84.

correlations are scarce. Nonetheless, it is commonly recognized that Arabica coffee with a lower content of chlorogenic acid is superior to Robusta coffee. Recently, Guerrero et al. (2001) developed a multivariate model for prediction of coffee genotype that correlates HPLC chromatographic profiles of 80% aqueous ethanolic extracts of coffee with chlorogenic acid contents.

Roasted Coffee Beans

Phenols play an important role in the formation of coffee flavor. From over 800 volatile compounds found in roasted coffee aroma, only 42 have been identified as phenols (Flament, 1989, 1995; Nijssen et al., 1996). Phenolic compounds are indogenous or are produced from thermal degradation of chlorogenic acid and lignins. Time and temperature of roasting influence the phenols produced and their composition. Roasting results in a significant decline in the content of chlorogenic acid in green coffee beans; the content of chlorogenic acid decreases rapidly as temperature increases throughout the roasting cycle (Merrit and Proctor, 1959a). A longer roasting time enhances the decomposition of chlorogenic acids in coffee. During a 7-min roasting, 77 and 60.9% of chlorogenic acids present in Robusta green beans of Tanzania and Arabica green beans of Guatemala are decomposed, respectively (Table 5.9). Light roasting produces a higher level of caffeoylquinic, dicaffeoylquinic, and feruloylquinic acids in Robusta coffee than in Arabica coffee.

The rates of degradation of chlorogenic acids in these two types of coffee may change significantly as the roasting process continues. Darker Arabica coffee contains more caffeoylquinic acids than Robusta coffee; however, the total content of dicaffeoylquinic and feruloylquinic acids is higher in darker Robusta coffee (Table 5.10). Thus, the generally accepted inferior quality of Robusta as compared with Arabica coffee may be due to the presence of a higher concentration of dicaffeoylquinic and feruloylquinic acids (Trugo and Macrae, 1984a). The dicaffeoylquinic acid isomers have a "peculiar lingering metallic taste" that may play a role in the acceptability of coffee (Clifford and Ohiokpehai, 1983). Figure 5.6 outlines the degradation of quinic and caffeic acids to phenolics found in coffee aroma; the content of some phenols found in coffee aroma is provided in Table 5.11. Catechol is the predominant volatile phenolic compound found in coffee aroma, followed by 4-ethylguaiacol, 4-ethylcatechol, pyrogallol, quinol, and 4-vinylcatechol. A higher guaiacol content in the roasted Robusta compared to Arabica may be associated with greater destruction of feruloylquinic acid during the roasting process. According to Czerny et al. (1999), 4-vinylquaiacol is one of the character impact odorants of roasted Arabica coffee. The compounds 4-vinylguaiacol, 4-ethylguaiacol and vanillin are the predominant volatiles in roasted Arabica beans (Czerny and Grosch, 2000).

The ratio of 5-caffeoylquinic acid to caffeine has been proposed as an indicator for monitoring the roasting process of coffee. During roasting, the content of chlorogenic acid declines while caffeine remains virtually unchanged. Because of this, caffeine may be used as an internal standard, thus allowing measurement of the level of chlorogenic acid in roasted coffee bean independent of weight losses due to formation of volatiles during roasting. This ratio may be useful for measuring the compositional changes of roasted coffee, especially those produced by fast roasting

TABLE 5.9
The Decline in Chlorogenic Acid Contents during Roasting of Coffee Beans[a]

Sample	Roasting		Chlorogenic Acid Isomers			
	Time	Loss	CQA	diCQA	FQA	Total
Arabicas						
Colombia	0	0	42.8	9.1	3.3	55.2
	not specified		16.9	2.5	0.5	19.9
Guatemala	0	0	57.6	8.7	2.5	68.8
	420	—	23.8	2.2	0.9	26.9
	600	—	19.8	1.5	0.8	22.2
	780	—	7.1	0.3	0.3	7.7
	1140	—	2.2	0.1	0.1	2.4
Tanzania	0	0	32.6	12.0	2.9	51.9
	—	7	21.1	5.9	1.8	31.0
	—	9	13.7	4.8	1.8	22.1
	—	10	11.9	4.6	1.8	20.0
	—	11	8.8	Nd	0.3	9.1
Robusta						
Uganda	0	—	68.3	13.8	6.0	88.1
	300	—	30.2	2.9	2.4	35.4
	420	—	17.8	1.4	1.5	20.7
	840	—	5.2	0.5	0.5	6.2
	960	—	1.4	0.2	0.1	1.8

Note: CQA = caffeoylquinic acids; diCQA = dicaffeoylquinic acids; FQA = feruoylquinic acids.

[a] Grams per kilogram of dry weight.

Source: Adapted from Smith, A.W., 1985, in *Coffee, Vol. 1. Chemistry,* Clark, R.J. and Macrae, R., Eds., Elsevier Applied Science, London, 1–41.

(3 to 5 min of roaster residence time). Shorter roasting times promote the so-called "case roasting," i.e., uneven roasting from the surface to the center of the individual bean (Purdon and McCamey, 1987).

Instant Coffee

The content of chlorogenic acids in instant coffee depends on the nature of roasting and extraction conditions; Table 5.12 shows the content of chlorogenic acid isomers in five selected commercial instant coffees. The major isomer found in all the samples is 5-caffeoylquinic acid, which represents about 30% of the total, whereas the sum of the caffeoylquinic acid isomers accounts for approximately 70% of the phenolic compounds. The feruloylquinic and dicaffeoylquinic acids accounted for about 20 and 10% of the instant coffee phenolics, respectively. The decaffeination process brings about considerable loss in the chlorogenic-acid content of coffee. Decaffeinated coffee may contain approximately three times less chlorogenic acid (3.6%) than that found in "mild" coffee (10.7%) (Trugo and Macrae, 1984b).

TABLE 5.10
Effect of Roasting of Coffee Beans on Content of Chlorogenic Acids[a]

Coffee	Chlorogenic Acid	Green Coffee	Type of Roasting			
			Light	Medium	Dark	Very Dark
Arabica	Total CQA	57.61	23.78	19.84	7.10	2.22
(Guatemala)	5-FQA	2.49	0.86	0.84	0.30	0.08
	Total di CQA	8.67	2.24	1.53	0.31	0.12
	Total	68.77	26.88	22.21	7.71	2.42
	Total CQA	68.23	30.20	17.82	5.17	1.41
Robusta	5-FQA	6.04	2.39	1.50	0.46	0.11
(Uganda)	Total diCQA	13.77	2.85	1.42	0.52	0.24
	Total	88.04	35.44	20.74	6.15	1.76

Notes: CQA = caffeoylquinic acids; diCQA = dicaffeoylquinic acids; 5-FQA = 5-feruloylquinic acid. Roasting times at 205°C: light = 7 min; dark = 13 min; very dark = 19 min.

[a] Grams per kilogram of dry weight.

Source: Adapted from Trugo, L.Z. and Macrae, R., 1984b., *Analyst*, 109:263–266.

Brewing Method

The extraction of chlorogenic acids into the coffee beverage depends on a number of factors, including the proportion of ground coffee to water, freshness of roasted coffee, grind, method of brewing, temperature of water and length of time that water is in contact with the grounds.

Higher brewing temperatures result in greater extraction of chlorogenic acid from the coffee grounds into the brew. The extraction rate increases over the 10 min of brewing generally employed, although this increase is less pronounced after the first 2 min of brewing. About 69% of the chlorogenic acid is extracted into the coffee brew under conditions normally encountered during coffee brewing, i.e., 2-min contact time at 93°C (Table 5.13) (Merrit and Proctor, 1959). Holding the coffee brew at elevated temperatures results in loss of chlorogenic acid, which depends on temperature and time. About 15% of chlorogenic acid is lost after 24 h of holding at 83°C. The decrease in chlorogenic acid may play an important role in flavor changes in the coffee brew (Segall and Proctor, 1959).

Table 5.14 summarizes the amount of chlorogenic acid extracted by pouring 150 mL of boiling water over 6, 9 and 12 g of coffee. The brew obtained using 6 g of coffee contains 1.34 to 1.67 g of chlorogenic acid per liter. When 12 g of coffee is used, the amount of chlorogenic acid ranges between 2.62 and 3.65 g per liter. The pH of coffee is not significantly affected by the content of chlorogenic acid (Hamboyan and Pink, 1990).

Feathering of Coffee Cream

Polyphenolic compounds present in coffee, notably chlorogenic and caffeic acids, are implicated in curdling of unstabilized coffee cream added to hot coffee, also

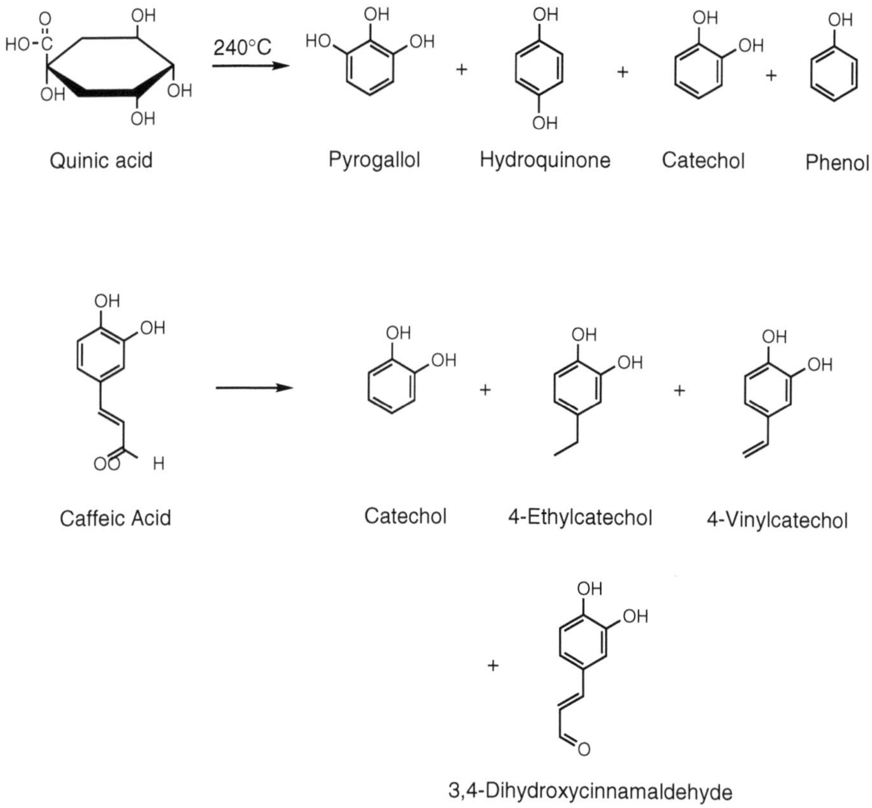

Quinic acid → (240°C) Pyrogallol + Hydroquinone + Catechol + Phenol

Caffeic Acid → Catechol + 4-Ethylcatechol + 4-Vinylcatechol + 3,4-Dihydroxycinnamaldehyde

FIGURE 5.6 Degradation of quinic acid and caffeic acid during roasting.

known as feathering (Charley, 1982). The extent of curdling, however, is not uniquely correlated with the content of chlorogenic acid in the coffee. The concentration of chlorogenic acid together with a parameter describing the shape of UV spectrum in the 230- to 340-nm region may serve as a reliable criterion for predicting the occurrence of feathering in hot coffee (Hamboyan et al., 1989).

COCOA

Cocoa beans are seeds found in woody plants (shrubs or small trees) belonging to the genus *Theobroma*. Only two species of this genus, *Criollo*, bearing warty fruits containing white or faintly purple seeds, and *Forastero*, bearing smoother fruits containing seeds of a deeper purple shade, are recognized from a commercial standpoint. Numerous hybrids containing genes of both species have recently been developed (Flament, 1989).

Manufacturing cocoa involves fermentation, drying, cleaning, roasting, and milling steps. The fermentation process reduces the astringency, acidity and bitterness of cocoa (Bonvehi and Coll, 1997a, b; Luna et al., 2002). The precursors of cocoa aroma are formed during the fermentation and drying/roasting steps. Drying/roasting

TABLE 5.11
Content of Phenols in Coffee Aroma[a]

Compound	Arabica	Arabusta	Robusta
Phenol	13.0	9.5	17.0
2-Methylphenol	1.2	0.7	1.1
4-Methylphenol	1.3	0.3	1.0
3-Methylphenol	0.7	1.0	1.2
3-Ethylphenol	—	0.4	—
4-Vinylphenol	0.2	0.2	0.2
Guaiacol	2.7	3.9	8.4
4-Ethylguaiacol	0.3	1.2	5.6
4-Vinylguaiacol	9.5	18.4	19.5
Vanillin	5.2	4.4	5.0
Catechol	80.0	95.0	120.0
4-Methylcatechol	16.0	10.0	13.0
Quinol	40.0	25.0	30.0
4-Ethylcatechol	37.0	20.0	80.0
4-Vinylcatechol	25.0	15.0	25.0
Pyrogallol	45.0	25.0	35.0
1,2,4-Trihydroxybenzene	20.0	6.0	13.0
3,4-Dihydroxycinnamaldehyde	10.0	5.0	12.0
3,4-Dihydroxybenzaldehyde	20.0	8.0	9.0

[a] Milligrams per kilogram of dry weight.

Source: Adapted from Smith, A.W., 1985, in *Coffee,* Vol. 1. *Chemistry,* Clark, R.J. and Macrae, R., Eds., Elsevier Applied Science, London, 1–41.

contributes significantly to changes in the composition of cocoa beans and is responsible for the development of the specific cocoa aroma due to partial degradation of cocoa components and their subsequent interactions (Cros et al., 1999; Rohan, 1969; Ziegleder and Biehl, 1988). After fermentation cocoa beans contain about 50 g of water per 100 g; therefore, drying is also necessary to preserve beans (Hor et al., 1984). Cocoa beans that contain more than 6 to 7 g of water per 100-g sample are prone to microbial (fungi) and enzymatic (lipoxygenase, lipase, peroxidase and polyphenol oxidase) degradation (Villeneuve et al., 1985). Cocoa liquor, a component of all chocolate products, is prepared by finely grinding the nib of the cocoa bean, while cocoa powder is prepared by removing part of the cocoa butter from the liquor (Apgar and Tarka, 1998; Vinson et al., 1999).

Composition of Polyphenolics

Polyphenolic compounds comprise approximately 2% of fresh unfermented cocoa beans (*Theobroma cacao*) (Porter et al., 1991). Epicatechin, catechin, epigallocatechin and epicatechin-based procyanidins have been identified as the predominant phenolics in the flesh of fresh unfermented cocoa seeds (Porter et al., 1991;

TABLE 5.12
Chlorogenic Acid Isomers Contents in Some Commercial Instant Coffees

Isomer	Samples[a]				
	C-1	C-3	C-5	C-7	C-9
3-CQA	13.8	9.0	18.9	7.5	10.0
4-CQA	16.5	10.4	22.5	8.7	11.6
5-CQA	22.5	13.8	35.0	10.2	16.9
Total CQA	52.8	33.2	76.4	26.4	38.5
3-FQA	2.5	1.7	3.5	1.5	2.1
4-FQA	5.9	4.4	7.6	3.2	4.9
5-FQA	5.8	3.2	8.2	2.7	4.1
Total FQA	14.2	9.3	19.3	7.4	11.1
3,4-diCQA	2.1	1.0	3.9	0.9	1.2
3,5-diCQA	1.6	0.7	2.9	0.7	0.8
4,5-diCQA	0.3	1.4	4.8	0.7	1.0
Total diCQA	4.0	3.8	11.6	2.3	3.0
Total chlorogenic acids	71.0	45.6	107.3	36.1	52.6

Notes: Results are average of duplicate determination for dry matter; CQA = caffeoylquinic acids; FQA = feruloylquinic acids; diCQA = dicaffeoylquinic acids. C-1 to C-9 samples are commercial instant coffees.

[a] Grams per kilogram of dry weight.

Source: Adapted from Trugo, L.Z. and Macrae, R., 1984, *Analyst*, 109:263–266.

TABLE 5.13
Effect of Brewing Conditions (Temperature and Time) on Extraction of Chlorogenic Acid from Coffee[a]

Brewing Temperature	Brewing Time (min)				
	0.5	1.0	2.0	5.0	10.0
100	15.8	16.9	22.6	26.8	29.8
120	20.8	22.2	26.3	29.8	34.9
140	22.3	23.0	26.7	33.9	37.3
160	23.6	25.1	29.6	37.2	38.4
180	35.8	28.2	33.3	39.2	41.6
200	26.9	30.0	35.4	41.5	43.7

Note: Roasted ground coffee contained 51.3 (g/kg) chlorogenic acid.

[a] Grams per kilogram of dry weight.

Source: Adapted from Merrit, M.C. and Proctor, B.E., 1959b, *Food Res.*, 24:735–743.

TABLE 5.14
The Content of Total Chlorogenic Acid and pH of Coffee Beverage as Affected by Portion of Coffee Grinds Used

Coffee Sample	Portion of Grinds Used (g)	Chlorogenic Acid (g/L)	pH
CD	12	3.65	4.99
	6	1.79	5.01
CS	12	2.67	5.58
	9	1.92	5.70
	6	1.34	5.79
IG	12	2.86	5.32
	9	1.95	5.31
	6	1.37	5.32
TC	12	2.62	5.23
	9	2.17	5.00
NS	12	2.80	5.07
	9	2.39	5.31
	6	1.67	5.16

Source: Adapted from Hamboyan, L. and Pink, D., 1990, *J. Dairy Sci.*, 57:227–232.

Thompson et al., 1972; Villeneuve et al., 1989). The following have been identified as the major flavanoids in fresh cocoa beans:

- Epicatechin-(4β→8)-catechin (procyanidin B-1)
- Epicatechin-(4β→8)-epicatechin (procyanidin B-2)
- [Epicatechin-(4β→8)]$_2$-epicatechin (procyanidin C-1) (Thompson et al., 1972)
- Epicatechin-(2β→5, 4β→6)-epicatechin
- 3T-*O*-β-D-galactopyranosyl-*ent*-epicatechin-(2α→7, 4α→8)-epicatechin
- 3T-*O*-β-L-arabinopyranosyl-(2α→7, 4α→8)-epicatechin (Porter et al., 1991)

Presence of higher molecular weight proanthocyanidin oligomers with up to 9 monomer units (nonamers) in raw cocoa and up to 10 units (decamers) in chocolate have been reported (Hammerstone et al., 1999; Porter et al., 1991). In addition, two quercetin glycosides, the 3-*O*-glucoside and 3-*O*-galactoside, have been identified in cotyledons of fresh cocoa beans (Jalal and Collin, 1977).

Polyphenolic compounds comprise 12 to 18% of the weight of whole dry bean and are implicated in the formation of the characteristic flavor and color of fermented cocoa (Bracco et al., 1969). Approximately 35% of polyphenol content of unfermented *Forastero* cocoa beans is (–)-epicatechin (Forsyth, 1955, 1963). On the other hand, the content of (–)epicatechin in freshly harvested *Catongo* and *Forastero* cocoa beans of verified genetic origin ranges from 34.65 to 43.27 mg/g of defatted sample as determined by high-performance liquid chromatography (Table 5.15 and

TABLE 5.15
Concentration of (–) Epicatechin in Defatted Cocoa Beans Unaffected by Postharvest Variables[a]

Clone	Variety	Content
NA-22—23	Nacional	37.63 ± 0.22
N-22-A3	Nacional	34.65 ± 0.20
UP-667	Trinitario	41.33 ± 0.54
UF-11	Trinitario	39.33 ± 0.71
UF-296	Trinitario	36.33 ± 1.30
EEG-29	Amazon Forastero	35.33 ± 1.18
SIC-250	Amazon Forastero	43.27 ± 0.44

Notes: Pods opened, testa removed, and cotyledons freeze-dehydrated.

[a] Grams per kilogram of dry weight.

Source: Adapted from Kim, H. and Keeney, P.G., 1984, *J. Food Sci.*, 49:1090–1092.

TABLE 5.16
Concentration of (–)Epicatechin in Fermented Defatted Cocoa Beans from Shipments Representing Several Countries of Production[a]

Bean Source	Content
Ivory Coast	6.22
Maracaibo (Venezuela)	3.62
Samoa	10.64
Trinidad	4.68
Bahia (Brazil)	8.23
Ghana	4.52
Lagos (Nigeria)	4.68
Costa Rica	16.52
Arriba (Ecuador)	8.08
Jamaica	2.66

Note: Results are mean of duplicate injections of a single extract.

[a] Grams per kilogram of dry weight.

Source: Adapted from Kim, H. and Keeney, P.G., 1984, *J. Food Sci.*, 49:1090–1092.

Table 5.16) (Kim and Keeney, 1984). Moreover, clovamide (N-(3′,4′-dihydroxy-*trans*-cinnamoyl)-3-(3,4-dihydroxyphenyl)-L-alanine) and deoxyclovamide (N-(4′-hydroxy-*trans*-cinnamoyl)-3-(3-hydroxyphenyl)-L-alanine), as well as quercetin 3-*O*-β-D-glucopyranoside and quercetin 3-*O*-α-D-arabinopyranoside, have been isolated and identified in cocoa liquor (Sanbongi et al., 1998).

The purple color of unfermented cocoa beans is due to the presence of anthocyanins, which constitute about 0.5% of the weight of the defatted beans. Two main components of anthocyanin fraction of cocoa beans are 3-β-D-galactosidyl- and 3-α-L-arabinosyl-cyanidins (Forsyth and Quesnel, 1957). Polyphenols present in cocoa and chocolate are thought to be anticariogenic, but some may also have toxic and carcinogenic effects (Singleton and Kratzer, 1973; Singleton, 1981).

According to Bonvehi and Coll (1998), the total content of low molecular weight phenolics in cocoa powder, such as phenol, 2-methoxyphenol, 3-methylphenol, 4-methylphenol, 2,3-dimethylphenol, 3-ethylphenol, 4-ethylphenol, 3,4-dimethylphenol and 3,5-dimethylphenol, should not exceed 9.6 mg/kg. Increased levels of these phenolics indicate contamination with smoke and contribute to the development of smoky taste in cocoa powder. Poor drying and storage conditions are responsible for the contamination of cocoa powder with smoke. A sensorially acceptable cocoa powder should contain no more, per kilogram of sample, than 2 mg of phenol, 0.9 mg of 3-methylphenol, 0.55 mg of 2,3-dimethylphenol, 0.9 mg of 3-ethylphenol, and 0.7 mg of 4-ethylphenol.

Effect of Processing

Fermentation is a very complex process involving external microbiological processes and internal structural changes as well as enzymic reactions. Lactic acid- and acetic acid-producing bacteria and yeast are predominant microorganisms found in the fermented mass. Microbial processes such as formation of acetic acid bring about an increase in the temperature of fermented mass that often exceeds 45°C. The combination of increased temperature and ingress of acids inhibits germination of seeds and also contributes to structural changes in fermented beans, resulting in removal of the compartmentation of enzymes and substrates (Biehl and Adomako, 1983). Accumulation of acetic acid during fermentation lowers the pH of cocoa beans. This modifies biochemical processes leading to the development of cocoa flavor (Jinap, 1994), such as formation of amino acid and peptides (Biehl et al., 1993), reduction of sugars (Rohan and Stewart, 1966a, b) and loss of solubility by phenolics (Villeneuve et al., 1989).

Commercially, the degree of bean fermentation is determined by cutting 100 beans and recording the color of their cotyledons. Presence of brown color indicates that beans are fully fermented. During fermentation and subsequent drying, flavor precursors responsible for the development of the characteristic flavor of cocoa are formed (Rohan, 1963, 1964, 1967; Rohan and Connell, 1964; Rohan and Stewart, 1967a, b). These precursors include flavonoids (catechin, epicatechin and gallocatechin) (Kharlamova, 1964), amino acids and sugars; however, unfermented cocoa beans do not contain these aroma precursors (Knapp, 1937). Recently, Luna et al.

(2002) demonstrated the existence of a positive correlation between polyphenol levels and bitterness and astringency of cocoa liquors and also showed that polyphenols are essential contributors to the overall sensory characteristics of Ecuadorian cocoa liquors.

Fermentation and drying bring about complex chemical changes in cocoa polyphenolic compounds. Kim and Keeney (1984) reported a 90% drop in the concentration of epicatechin. Later, Porter et al. (1991) reported a similar drop in the concentration of tannins, but a large increase in the relative content of procyanidin B-1. Structural changes in cocoa phenolics generally occur after 24 to 48 h of bean fermentation. At this point, (–)epicatechin molecules diffuse from their storage cells, undergo oxidation in the presence of polyphenol oxidase, and polymerize with one another or with anthocyanidins to form complex phenolics (tannins) (Forsyth, 1963; Wong et al., 1990). The content of (–)epicatechin decreases sharply after 2 or 3 days of fermentation (Figure 5.7), indicating the diffusion of (–)epicatechin from storage cells and the subsequent oxidation. The enzymic oxidation may continue during the drying stage, although some published data demonstrate a decline in the activity of cocoa polyphenol oxidase as the fermentation process progresses (Villeneuve et al., 1985). Thus, an increase in the oxidation rate may be due to greater penetration of oxygen into the cocoa bean mass. The enzymic oxidation of (–)epicatechins and leucocyanidins results in the formation of a brown color characteristic of chocolate caused by the production of melanin and melanoproteins (Griffiths, 1957). Eventually, when the moisture content in beans is insufficient, the enzymic oxidation reaction is inhibited. Another pathway for the loss of (–)epicatechin is its spontaneous nonenzymic oxidation.

During the fermentation process, anthocyanins are hydrolyzed to anthocyanidins that polymerize along with simple catechins to form complex tannins (Forsyth, 1963). Anthocyanins usually disappear rapidly during the fermentation process (Forsyth, 1952). After 4 days of fermentation, the content of anthocyanins decreases by up to 7% of its initial value (Figure 5.8). Underfermented beans are purple or slaty due to the presence of anthocyanin pigments, whereas fully fermented beans are quite brown. Thus, the content of anthocyanin may be considered a good index for determining the degree of cocoa bean fermentation. Further reduction in anthocyanin content is brought about by the drying process. Depending on the length of fermentation time, drying reduces the content of anthocyanins of beans by 13 to 44%. However, although drying unfermented beans results in loss of 79% of their anthocyanin, they still contain seven times more anthocyanin pigments than those of fermented products. A 97% loss of phenolics results from 4-day fermentation of beans followed by drying (Pettipher, 1986).

The fermentation and drying processes alter the content and composition of polyphenolic compounds. Fermented cocoa beans contain a lower amount of low molecular weight phenolics and an enhanced content of condensed phenolics. The latter compounds are involved in protein–phenol interactions and may contribute to the low digestibility and poor biological value of cocoa proteins (Chatt, 1953). Formation of the protein–phenol complex also reduces the bitterness and astringency associated with the presence of polyphenolics and reduces the unpleasant flavors

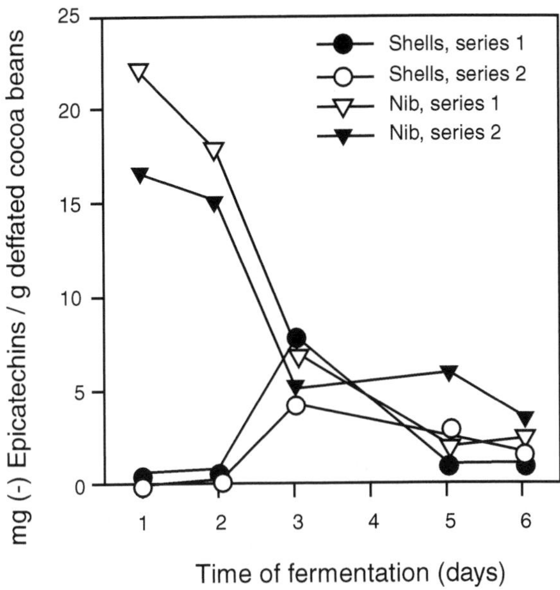

FIGURE 5.7 Changes in (–)epicatechin of Trinidad-Jamaica hybrid of cocoa beans during fermentation. (Adapted from Kim, H. and Keeney, P.G., 1984, *J. Food Sci.*, 49:1090–1092.)

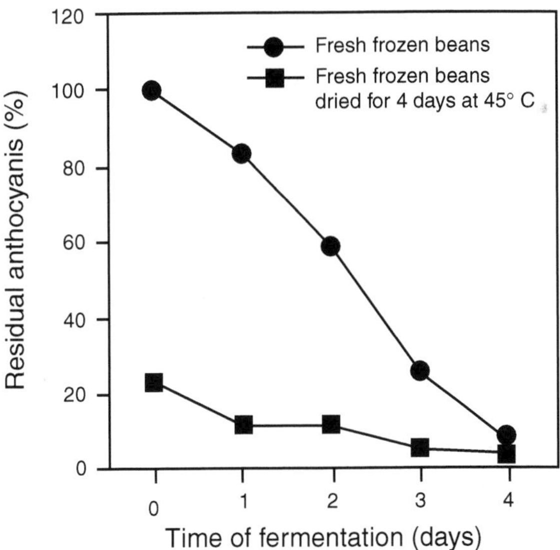

FIGURE 5.8 The effect of fermentation with or without drying on the anthocyanin content of cocoa beans. (Adapted from Pettipher, G.L., 1986, *J. Sci. Food Agric.*, 37:289–296.)

and odors in roasted beans (Griffiths, 1957). On the other hand, the low molecular weight polyphenols still present in chocolate may be responsible for its astringent and bitter taste.

BEER

The characteristic flavor of beer arises from malt, the predominant source of carbohydrates, and hops, a source of bitter compounds. The production of beer involves malting, mashing, fermentation and storage. The malting and mashing produce a wort, an aqueous extract of malted barley and hops. Yeasts are added to this wort and utilize nutrients present in it to produce ethanol, carbon dioxide, and other by-products. The final product after removal of yeast is beer, which contains some malt and hops constituents as well as substances produced by metabolism of yeast.

Composition of Polyphenolics

Brewing material such as barley malt and other raw or malted grains (rice, corn, wheat) used as adjuncts, hops and other aromatic materials contribute phenolic and polyphenolic compounds to the early stages of the brewing process. Further polymerization of phenolics and formation of polyphenols can occur during wort boiling, and possibly during fermentation and storage of beer. Beer polyphenolics furnish color, impart astringent taste, serve as a reservoir for oxygen reduction and substrate browning, and participate in precipitation of poorly coagulable beer proteins.

Phenolic compounds of beer contribute to its flavor quality and are also implicated in nonbiological haze formation (Charalambous, 1974; Dadic, 1974a, b; Gramshaw, 1970). Phenolic substances in beer are present in various forms; their volatility depends on their molecular weight (Table 5.17). Beer contains 283 mg/L of nontannin and nonflavonoid phenolic compounds (Fantozzi et al., 1998). Phenolic compounds of beer are classified as beneficial, harmful, or neutral in so far as their influence on beer stability and sensory properties is concerned (Dadic, 1971). Some beer phenolics may also act as antioxidants or may contribute to the formation of carbonyls in beer. It has been found that ferulic acid markedly increases the formation of carbonyls in beer aging under high and low air conditions. Moreover, ferulic acid and quercetin have been found to be active promoters of diacetyl formation in aging beer (Dadic, 1974b). Some 67 different phenolic compounds have been identified in beer. Simple phenolics, aromatic carboxylic- and phenol carboxylic acids, hydroxycoumarins, catechins, leucoanthocyanidins, anthocyanidins, flavonols, flavonones, flavones, prenylated flavonoids and phenolic glycosides are included in this list (Bohm, 1989; Drawert et al., 1977; Stevens et al., 1999a, b).

Hops serve as the main source of prenylated flavonoids (predominantly of chalcone type) in beer (Figure 5.9). Xanthohumol, desmethylxanthohumol and 3′-geranylchalconaringenin comprise over 95% of the prenylated flavonoid fraction of hops. In addition, hops contain small quantities of prenylated flavonones that are probably formed in hops upon drying and storage (Stevens et al., 1997). Total content of prenylated flavonoids in hops is between 2000 and 6000 mg/kg (Stevens et al., 1998, 1999b). During the boiling of hops with wort, prenylated chalcones undergo

TABLE 5.17
Classification of Beer Phenolic According to Volatility

Monomeric Monophenols	Monomeric Polyphenols	Polymeric Polyphenols
	Volatility decreases→	

Phenols, cresols, chlorophenols, guaiacols, aromatic alcohols, amines, phenolic and cinnamic acids, lactones, flavonols and glycosides, catechins, anthocyanogens, anthocyanidins, anthocyanins, polymers (tannins)

Source: Adapted from Dadic, M., 1974a, *MBAA Tech. Q.*, 11:140–145.

(a)

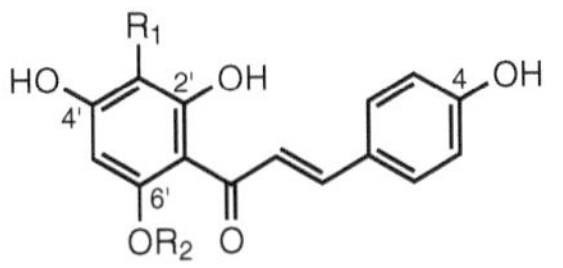

	R_1	R_2
Xanthohumol	isoprenyl	CH_3
Desmethylxanthohumol	isoprenyl	H
3'-Geranylchalconaringenin	geranyl	H

(b)

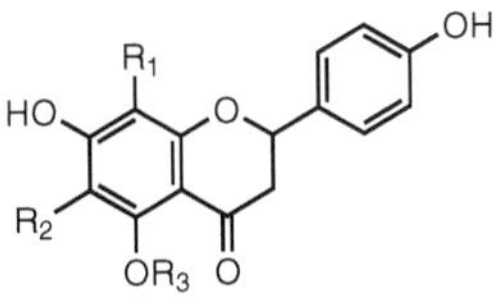

	R_1	R_2	R_3
Isoxanthohumol	isoprenyl	H	CH_3
6-Prenylnaringenin	H	isoprenyl	H
8-Prenylnaringenin	isoprenyl	H	H
6-Geranylnaringenin	H	geranyl	H
8-Geranylnaringenin	geranyl	H	H

Isoprenyl $-CH_2-CH=C(CH_3)_2$ Geranyl

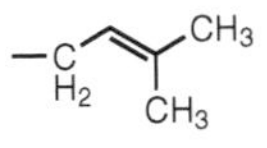

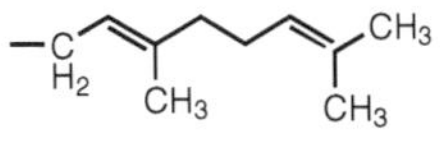

FIGURE 5.9 Chemical structures of some hops flavonoids (A) and their conversion products formed during processing (B).

cyclization into their isomeric flavanones. The presence of free 2′- or 6′-hydroxyl groups on A-ring of prenylated chalcone is a prerequisite for ring closure. Therefore, upon cyclization, xanthohumol yields only one flavanone, namely, isoxanthohumol; cyclization of desmethoxyxanthohumol leads to the formation of 6-prenylnaringenin and 8-prenylnaringenin. Also, 3′-geranylchalconaringenin converts during processing into 6-geranylnaringenin and 8-geranylnaringenin (Figure 5.9) (Stevens et al., 1999a, b).

According to Stevens et al. (1999a), the content of isoxanthohumol in beer accounts for only 22 to 30% of hops' xanthohumol; only 10% of the hops'

TABLE 5.18
Content of Phenolic Acids in Selected Beers[a]

Phenolic Acid	British Export A	American Malt Liquor	German Rauchbier	Irish Ale
Benzoic Acid Derivatives				
Gallic acid	1.8	1.4	3.5	0.3
Protocatechuic acid	1.0	0.5	1.2	0.2
p-Hydroxybenzoic acid	1.8	1.1	0.4	0.3
Vanillic acid	5.4	2.5	12.7	1.5
Syringic acid	1.2	0.8	2.2	0.7
Cinnamic Acid Derivatives				
Caffeic acid	trace	0.1	1.0	0.1
p-Coumaric acid	1.1	1.9	4.6	1.0
Ferulic acid	2.7	4.2	10.8	1.1
Sinapic acid	1.4	1.1	3.0	0.2
Total	16.4	16.6	39.4	5.4

Note: Results are mean values of three samples measured by HPLC on ethyl acetate extracts.

[a] Milligrams per liter.

Source: Adapted from McMurrough, I. et al., 1984, *J. Inst. Brew.*, 90:181–187.

desmethylxanthohumol is completely converted into prenylnaringenin during the brewing process. Therefore, these authors suggest that these prenylated chalcones may also be involved in chemical reactions other than isomerization. Milligan et al. (1999) demonstrated that 8-prenylnaringenin exhibits a greater estrogenic activity than known phytoestrogens such as coumestrol and genistein. Beer may contain up to 20 μg of 8-prenylnaringenin per liter (Tekel et al., 1999). Recently, Promberger et al. (2001) found that average estrogenic activity of beer equals that exerted by 43 ng of 17β-estradiol per liter of beer. Moreover, these authors suggested a negligible human health hazard from drinking beer containing compounds showing activity on the estrogen receptor α. *In vitro* studies also indicate that prenylchalcones and prenylflavones show greater protection for human low-density lipoprotein (LDL) from oxidation than α-tocopherol and genistein (Miranda et al., 2000).

The content of phenolic acids in selected British, American, German and Irish beers is shown in Table 5.18. The lowest content of phenolic acids is found in Irish beer, while German beer contains approximately five times more phenolics. The content of benzoic acid derivatives in the finished beer is higher than that of cinnamic acid derivatives (McMurrough et al., 1984). Italian lager beers contain only 1.187 mg of free phenolic acids per liter (Montanari et al., 1999), while samples of Spanish, Danish and German commercial brands contain 1.759 mg of free phenolic acids per liter (Bartolome et al., 2000). The major free phenolic acids in Italian beers are *m*-, *p*- and *o*-coumaric, and ferulic acids (Montanari et al., 1999), while vanillic, ferulic and *p*-coumaric acids are the predominant free phenolic acids in Spanish, German and Danish brands (Bartolome et al., 2000). Sinapic, vanillic, chlorogenic, homovanillic,

TABLE 5.19
Phenolic Acid Contents of Samples Taken at Stages during Production of Commercial Lager[a]

Phenolic Acid	Unboiled Wort	Fermented Wort	Finished Beer
	Benzoic Acid Derivatives		
Gallic acid	0.1	0.2	0.2
Protocatechuic acid	0.5	0.4	0.3
4-Hydroxybenzoic acid	0.6	1.1	1.1
Vanillic acid	1.4	1.6	0.9
Syringic acid	0.6	0.4	0.3
	Cinnamic Acid Derivatives		
Caffeic acid	0.1	0.4	0.3
p-Coumaric acid	0.6	0.4	0.3
Ferulic acid	1.3	1.7	1.8
Sinapic acid	0.4	0.5	0.4
Total	5.6	8.0	6.7

Note: Results are mean values of three samples in ethyl acetate extracts measured by HPLC.

[a] Milligrams per liter.

Source: Adapted from McMurrough, I. et al., 1984, *J. Inst. Brew.*, 90:181–187.

p-hydroxybenzoic, 2,6- and 3,5-dihydroxybenzoic, syringic, gallic, protocatechuic and caffeic acids are present in small quantities (Bartolome et al., 2000; Montanari et al., 1999). Other monomeric phenolics identified in beer include tyrosol, catechin, 2,3-dihydroxy-1-guaiacoylpropan-1-one and vanillin (Bartolome et al., 2000). Monomeric phenolics account for 10 to 20% of the total content of beer phenolics (Sogawa, 1972). On the other hand, different batches of the same beers may differ in their content and composition of polyphenols (Table 5.19). This difference may be significant for the storage stability of beers (McMurrough et al., 1984).

The contents of catechin, epicatechin and gallocatechin do not generally exceed 15 mg/L (Bellmer, 1977; Drawert et al., 1977; McMurrough et al., 1984); the total content of catechins found in beer is approximately 40 mg/L (Bellmer, 1977). Beer contains less than 1 mg/L hydroxycoumarins (umbelliferon, scopoletin, and daphnetin) as well as anthocyanins and anthocyanidins (cyanin, cyanidin, pelargonin, pelargonidin, and delphidin) (Drawert et al., 1977). The normal concentration of dimeric procyanidins in beer does not exceed 4 mg/L (Bate-Smith, 1973; Haslam, 1974). The contents of leucocyanidin (5,7,3′,4′-tetrahydroxyflavan-3,4-diols) and quercetin are less than 10 mg/L of beer. Moreover, flavones and flavonols such as vitexin, isoquercitin, quercitin, rutin, kaempferol, myricetin and myricitrin have also been found in beer at less than 1 mg/L (Drawert et al., 1977).

Phenolic compounds contribute to bitterness and astringency of beer, play a role in its harshness and intensify its color. Polymerization of phenolics during the mashing and postfermentation periods heightens these latter effects. The flavor thresholds of polyphenols in beer have been found to be 10 to 50 mg/L. The major flavors imparted to beer by pure phenolics, listed in the order of incidence, are sour–acidic, bitter, harsh, bitter–sweet, astringent and aromatic. On the other hand, the oxidation of phenolics during aging brings about the following major notes to beer: harsh, bitter, astringent, sour–acidic, dry and bitter–sweet. Harsh and bitter notes become more pronounced upon oxidation of phenolics due to simultaneous lowering of flavor thresholds (Dadic and Belleau, 1973). Cinnamic acid, after oxidation, gives an old aged taste to beer and *p*-coumaric acid produces a sharp bitter note. Oxidized gallic acid produces a lingering, bitter and astringent note along with depression of the normal flavor characteristics (Charalambous, 1974) while decarboxylation of hydroxycinnamic acids resulting from the formation of 4-vinylguaiacol and 4-vinylphenol gives off-flavor to beer (Madigan et al., 1994).

Tyrosine-derived phenolic amines such as hordenine, N-methyltyramine, and tyrosol possess a bitter taste that is particularly pronounced in lightly hopped beer. These phenolic amines have a taste threshold of 20 mg/L and, according to Charalambous (1974), also have a distinctive influence on the taste and aroma of beer. Presence of some tannin is also necessary to impart desirable qualities to beer. Sometimes an off-flavor described as "smoky" or "apothecary-like" is developed in beer; such beverages contain a higher than usual level of steam-volatile phenols. It is assumed that most of these compounds pass into beer from malt.

The malt obtained from directly fired kilns contains higher levels of steam-volatile phenols than that produced from indirectly heated kilns (Kieninger et al., 1977). A taste panel can detect 25 mg/L of tannins added to beer (Owades et al., 1958). Beer phenolics are also suspected as a common source of some flavor-active carbonyls such as salicylaldehyde and vanillin (Palamand et al., 1971; Dadic et al., 1974). Average content of tannins in malt is 1.1 g/kg compared with 50 to 67 g/kg in hops. However, malt contributes twice as many tannins to beer as hops (Owades et al., 1958).

Effect of Processing

The content of phenolic acids in finished beer depends largely on the extent of their extraction during mashing. Mashing especially increases the content of hydroxycinnamic acid derivatives as a result of their release from previously unextractable combinations. McMurrough et al. (1984) found that mashing malt increases the total content of its phenolic acids from 65 mg/kg before mashing to 191 mg/kg after.

The extraction of phenolics from malt into the wort depends markedly on malt to water ratios. Wort and beer produced from thinner mashes (malt to water ratio of 1:4 and 1:5, respectively) contain higher levels of phenolics than those obtained from more concentrated mashes (1:3). An increase of mashing time from 105 min to 180 and 360 min reduces the level of phenolics in beer, but also adversely affects the taste of the beer (Von Narzib et al., 1979). On the other hand, infusion and decoction-mashing systems produce worts of very different polyphenol compositions with higher levels of oxidizable phenols. The quantities of extracted oxidized and oxidizable polyphenols, however, affect the degree of polyphenol loss during wort boiling and fermentation (Kirby et al., 1977).

The contribution of hops polyphenolics to boiled wort depends on the time of addition to hops; late addition in decoction wort boiling allows inclusion of phenolics but not subsequent removal of oxidizable polyphenols. The late addition of hops to boiling wort may even result in a need for extensive treatment to produce haze-stable beer (Kirby et al., 1977). In total, the water-soluble polyphenols derived from hops account for 20 to 30% of malt polyphenols. Hops polyphenols have a greater affinity to beer proteins; therefore, they tend to precipitate more completely during wort boiling than those extracted from malt (Bellmer, 1981).

The content of polyphenols may decrease during the fermentation process as a consequence of protein precipitation. Sugars present in wort are capable of hydrogen bond formation and compete with polyphenols for available sites. As the fermentation process progresses, the sugar concentration in wort decreases, resulting in increased formation of hydrogen bonding between polyphenols and beer proteins (Kirby et al., 1977).

The content of phenolic acids in the finished beer differs only slightly from that of unboiled wort. Unboiled wort of commercial lager is somewhat deficient in caffeic acid. Boiling wort with hops increases the content of hydroxybenzoic and *p*-coumaric acids. Subsequent unit operations, such as fermentation, cold storage, treatment with adsorbents and filtration, bring about a slight loss of phenolic acids, with the exception of vanillic acid content, which decreases by more than 40% (Table 5.20) (Gardner, 1967; McMurrough et al., 1984).

Nonbrewing strains of *Saccharomyces* yeast often produce undesirable off-flavors referred to as "phenolic," "herbal phenolic," or "clove-like" flavor notes in beer. Decarboxylation of phenol carboxylic acids (*p*-coumaric, ferulic and sinapic acids) may be responsible for production of such flavor notes. Beers with phenolic taste contain significant amounts of 4-vinylguaiacol that may be produced by decarboxylation of ferulic acid present in wort during fermentation (Thurston and Tubb, 1981). Worts fermented with nonbrewing wild-type yeast strains contain 10 times higher

TABLE 5.20
Polyphenol Content in Lager Beer[a]

Beer Sample	Catechin	Epicatechin	Total Monomeric	Total Dimeric
1A	6.3	1.6	7.9	4.3
1B	3.5	1.7	5.2	2.4
2A	3.4	0.9	4.3	3.3
2B	4.9	1.9	6.8	5.0

[a] Milligrams per liter.

Source: Adapted from Kirby, W. and Wheeler, R.E., 1980, *J. Inst. Brew.*, 86:15–17.

TABLE 5.21
Levels of 4-Vinyl Guaiacol in Beers Fermented with Brewing Yeasts and Wild Type Strains[a]

Yeast Strain	Content of 4-guaiacol
Wild yeast	1154
Saccharomyces diastaticus	1914
Saccharomyces pastorianus	1035
Saccharomyces willianus	1333
Saccharomyces uvarum, J-3015	132
Saccharomyces uvarum, J-2270	150
Saccharomyces uvarum, J-2036	115
Saccharomyces uvarum, J-28C4	132

[a] Micrograms per liter.

Source: Adapted from Villarreal, R. et al., 1986, *ASBC J.*, 44:114–117.

levels of 4-vinylguaiacol compared to those found in normal beer (Table 5.21) (Villarreal et al., 1986).

A number of procedures have been developed for production of alcohol-free and low-alcohol beers, including removal of alcohol by (Bartolome et al., 2000):

- Distillation
- Vacuum distillation
- Dialysis
- Reverse osmosis
- Restricted fermentation
- Use of special yeasts
- Use of spent grain
- Carbon dioxide extraction

Alcohol-free beers contain lower levels of phenolics in comparison to the phenolic concentrations found in standard beer. These changes in phenolic contents, according to Bartolome et al. (2000), may be brought about mainly by the differences in fermentation times, the yeast strains employed, and the dealcoholization process employed.

Formation of Haze

Polyphenols in beer may contribute to the formation of haze (Bamforth, 1999; Siebert, 1999; Siebert and Lynn, 1998; Siebert et al., 1996), which may contain up to 55% polyphenols (Table 5.22). According to Siebert (1999), polyphenolics should have at least two binding sites in order to participate in the formation of beer haze, each of which should possess at least two hydroxyl groups on an aromatic ring. The largest amount of haze is formed when the number of polyphenol and protein binding sites are nearly equal (Siebert, 1999). Eleven phenolic acids and two flavonols

TABLE 5.22
Gross Chemical Composition of Beer Haze (%)

Protein	Polyphenol	Carbohydrate	Ash	Reference
40	45	2–4	1–3	Bengough and Harris (1995)
58–77	17–55	2–12.4	2–14	Gramshaw, J.W. (1970)
40–76	17–55	3–13		Clark, A.G. (1960)
15–45	1–3	50–80		Martin et al. (1962)

Source: Adapted from Dadic, M., 1976, *MBAA Tech. Q.*, 13:182–189.

(quercetin and kaempferol), as well as (+)-catechin and (–)-epicatechin, have been identified in ether extracts of acid- and alkali-treated haze (Dadic and Belleau, 1980). The most abundant types of beer haze are chill haze and permanent haze. Chill haze is defined as the haze formed at 0°C and is redissolved upon warming of beer; the haze that remains in beer at or above 20°C is called permanent haze. The chill haze is probably formed through weak intermolecular bonding, while the permanent haze ensues after the formation of a certain number of covalent bonds between the units of the haze particles (Dadic, 1976).

The haze particles are thought to be built from units with a molecular weight of 10,000 to 100,000 Da (Claesson and Sandegren, 1969). Most of the polysaccharides and polypeptides present in beer have the postulated size, while polyphenols reach it through a polymerization process. About 90% of polyphenolics found in aged beer have a molecular weight of 500 to 10,000 Da; only trace quantities of phenolics have a molecular weight over 10,000 Da (Sogawa, 1972). Due to their chemical character and reactivity, polyphenols may undergo oxidative and acid-catalyzed polymerization in beer upon storage (Dadic et al., 1970; Gramshaw, 1968, 1970). Of the phenolics present in beer, catechins in the free form are most sensitive to oxidative changes during the initial stages of aging (Bellmer, 1977). The catalyzed polymerization of phenolics occurs in the later stages of beer storage when most of the air in the headspace has been depleted. Complexation of polymerized polyphenolics with polypeptides, first through hydrogen bonds, produces sufficient hydrophobicity. The insolubility of this complex and consequent formation of haze are largely determined by the nature and distribution of polyphenols in the complex (Figure 5.10). Monomeric phenolics are also found in beer haze; however, their role in haze formation is still disputed because they have usually been identified after alkaline hydrolysis of haze or in beer polyamide adsorbates (Dadic et al., 1971; Gardner and McGuinness, 1977).

The rate of haze formation depends on the degree of polyphenolic polymerization and increases from monomeric to dimeric to trimeric and finally to polymeric catechin units. The structure of polyphenols also plays an important role in haze formation because procyanidin-B3 (catechin–catechin) increases haze formation more quickly than procyanidin-B4 (catechin–epicatechin) (Gardner and McGuinness, 1977). The threshold values for tanning of beer proteins by procyanidins have been found to be 150 mg/L for the trimeric procyanidin and about 850 mg/L for procyanidin B3. On the other hand, the threshold value for tanning by tannic acid is

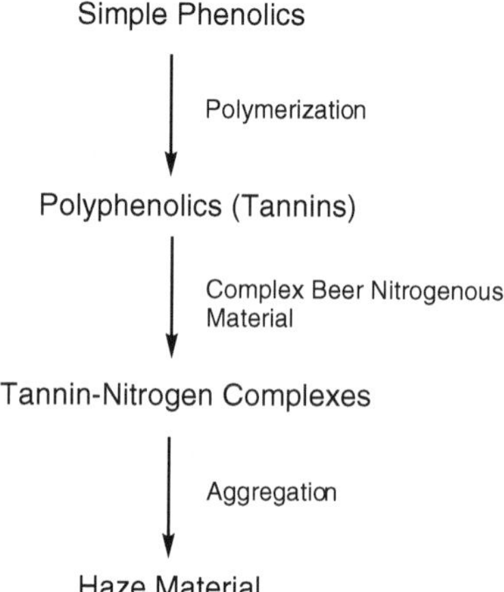

FIGURE 5.10 Participation of phenolics and tannins in haze formation in beverages.

25 mg/L; half of the protein material of this polyphenolic is precipitated at 140 mg/L and total precipitation has been observed at 650 mg/L (Delcour et al., 1983). The normal concentration of dimeric procyanidins in beer does not exceed 4 mg/L (Bate-Smith, 1973; Haslam, 1974); therefore, their contribution to haze formation in beer may be negligible.

Proanthocyanidins from hops and malt have a tremendous influence on haze development. Equal concentrations of hops and malt proanthocyanidins produce similar haze value, but their impact on beer stability depends on their relative concentrations in the beverage (Delcour et al., 1985). Approximately 80% of proanthocyanidins present in regular beer is derived from malt and only 20% from hops (Bellmer, 1981). Thus, malt proanthocyanidins are the main contributors to beer haze.

The permanent and chill haze developed during aging of beer can be effectively reduced by treatment of wort with tannase or laccase before boiling. The action of the enzymes results in some discoloration of wort, but this is totally lost during fermentation and aging of beer (Rossi et al., 1988).

WINE

Phenolic compounds are important components of wine. They contribute to sensory characteristics such as color, flavor, astringency and hardness of wine directly or by interaction with proteins, polysaccharides, or other phenolic compounds (Lee and Jaworski, 1987; Garcia-Viguera et al., 1997). The sensory characteristics of wine are influenced by the grape variety, the vinification process and degradation and polymerization of phenolics during the aging of wine (Auw et al., 1996; Zoecklein et al., 1997). These reactions are influenced by storage temperature (Gomez-Plaza et al., 2000; Somers and Evans, 1986; Somers and Pocock, 1990) and duration (Dallas and Laureno, 1994a, b).

Composition of Polyphenolics

The composition of phenolics in wine depends on the types of fruits used (usually grapes) for vinification and their extraction, procedures employed for wine making and chemical reactions that occur during the aging of wine (Macheix et al., 1990). The wood barrel's contact with must and wine also accounts for the presence of some phenols in wine. Thus, some simple phenols, certain flavonoids and hydrolyzable tannins can be leached out from wood to wine (Maga, 1984). The composition of phenolics is also modified by yeasts as a result of conversion of nonphenolic substances and solubilization and extraction of phenolics by the ethanol produced during fermentation (Singleton and Noble, 1976). Based on a multivariate analysis of midinfrared spectra of wine phenolic extracts, Edelmann et al. (2001) recently developed a method for discrimination of Austrian red wines obtained from different *Vitis vinifera* cultivars.

Phenolics of wine and grapes usually include derivatives of hydroxybenzoic and hydroxycinnamic acids, the trihydroxystilbenes such as *cis*- and *trans*-resveratrol as well as *cis*- and *trans*-piceid, flavonoids such as flavan-3-ols, flavan-3,4-diols, anthocyanins and anthocyanidins, flavonols, flavones and condensed tannins (Burns et al., 2002; Escribano-Bailón et al., 1992; Jaendet et al., 1993; Lamuela-Raventós and Waterhouse, 1993; Lea et al., 1979; Mazza et al., 1999; Siemann and Creasy, 1992; Singleton and Esau, 1969; Soleas et al., 1997; Somers, 1966). The main phenolics identified in *Vitis vinera* berries are given in Table 5.23. White wines tend to be low in their content of phenolics compared to red wines. The proportion of different classes of phenolic substances in wines is influenced by grape variety and harvesting season (Table 5.24) as well as the age of wine (Table 5.25). Malvidin-3-monoglu-

TABLE 5.23
Main Phenolics Identified in *Vitis Vinera* Berries

Phenolic acids
p-Hydroxybenzoic; *o*-hydroxybenzoic; salicylic; gallic; cinnamic; *p*-coumaroylartaric (= coutaric); caffeoyltartaric (= caftaric); feruloylartaric (= fertaric); *p*-coumaroyl glucose; feruloylglucose; glucose ester of coutaric acid

Anthocyanins
Cyanidin 3-glucoside; cy 3-acetylglucoside; cy 3-*p*-coumaryl-glucoside; peonidin 3-glucoside; pn 3-acetylglucoside; pn 3-*p*-coumarylglucoside; pn 3-caffeylglucoside (?); delphinidin 3-glucoside; dp 3-acetylglucoside; dp 3-*p*-coumarylglycoside; petunidin 3-glucoside; pt 3-*p*-coumarylglucoside; malvidin 3-glucoside; mv 3-acetylglucoside; mv 3-*p*-coumaryglucoside; mv 3-caffeylglucoside

Flavonols
Kaempferol 3-glucoside; k 3-glucuronide; k 3-glucosylarabinoside (?); k 3-galactoside; quercetin 3-glucoside; q 3-glucoronide; q 3-rutinoside; q 3-glucosylgalactoside (?); q 3-glucosylxyloside (?); iso-rhamnetic 3-glucoside

Flavan-3-ols and tannins
(+)Catechin; (–)epicatechin; (+)gallocatechin; (–)epigallocatechin; epicatechin-3-O-gallate; procyanidins B1, B2, B3, B4, C1, C2, polymeric forms of condensed tannins

Flavanonols
Dihydroquercetin 3-rhamnoside (= astilbin); dihydrokaempferol 3-rhamnoside (= engeletin)

Source: Adapted from Macheix, J.-J. et al., 1990, *Fruit Phenolics*, CRC Press, Boca Raton, FL.

TABLE 5.24
Effect of Season and Grape Variety on Total Phenolics, Flavonols, and Anthocyanins in Skin, Must, and Wine from Okanagan Valley (B.C., Canada)

Grape Variety	Units	Total Phenolics[a]		Flavonol[b]		Anthocyanins[c]	
		1996	1997	1996	1997	1996	1997
Cabernet Franc							
Skin	mg/kg	548–793	462–1009	82–134	93–200	530–826	289–1048
Must	mg/L	163	326–965	45	96–181	8	42–644
Wine	mg/L	863–982	1095–1193	98–128	121–157	232–420	337–469
Merlot							
Skin	mg/kg	777–1128	633–1028	111–150	118–182	708–1137	485–1094
Must	mg/L	130	196–851	31	47–147	32	81–644
Wine	mg/L	912–974	907–921	128–130	110–113	279–412	338–455
Pinot Noir							
Skin	mg/kg	806–915	955–1074	109–151	167–172	665–803	715–794
Must	mg/L	209	152–650	47	45–109	78	25–314
Wine	mg/L	723–748	913–965	88–94	76–91	223–340	166–219

[a] Expressed as gallic acid equivalents.
[b] Expressed as quercetin equivalents.
[c] Expressed as malvidin-3-glucoside equivalents.

Source: Adapted from Mazza, G. et al., 1999, *J. Agric. Food Chem.*, 47:4009–4017.

TABLE 5.25
Gross Phenol Composition for Typical Wines from *Vitis Vinifera* Grapes[a]

Phenol Class	Source	White Wine Young	White Wine Aged	Red Wine Young	Red Wine Aged
Total nonflavonoids		170	160–260	235	240–500
Cinnamates deriv.	G,D	154	130	165	150
Low volat. benzene deriv.	D,M,G,E	10	15	50	60
Tyrosol	M	10	10	15	15
Volatile phenols	M,D,E,E,	1	5	5	15
Hydrozylable tannins etc.		0	0–100	0	0–260
Macromolecular complexes					
Protein-tannin	G,D,E,	10	5	5	10
Total flavonoids		30	25	1060	705
Catechins	G	25	15	200	150
Flavanols	G,D	tr	tr	50	10
Anthocyanins	G	0	0	200	20
Soluble tannins, deriv.	G,D	5	10	550	450
Other flavonoids, deriv.	G,D,E,M,	?	?	60?	75?
Total phenols		210	190–290	1300	955–1215

Note: D = degradation products; E = environment; G = grapes; M = microbes.

[a] Milligrams of gallic acid equivalents per liter.

Source: Adapted from Singleton, V.L., 1980, *Grape Phenolics: Background and Prospects*, Centenaire, Paris.

coside is the major anthocyanin of Cabernet Franc, Merlot and Pinot Noir wines from British Columbia (Canada). These wines also contain 3-monoglucosides of delphinidin, cyanidin, petunidin and peonidin. In addition, 3-monoglucoside–acetates and 3-monoglucoside–coumarates of anthocyanidins have been found in Cabernet Franc and Merlot wines (Gao et al., 1997; Mazza, 1995; Mazza et al., 1999).

Condensed tannins contribute to astrigency, browning and turbidity in wines and also participate in the aging processes of wine (Haslam and Lilley, 1988). According to Somers (1966), the molecular weight of condensed tannins isolated from wine ranges from 2000 to 4000 Da. Later, Souquet et al. (1996) separated condensed tannins extracted from grape skin (*Vitis vinifera* var. Merlot) into six fractions using normal phase HPLC. The mean degree of polymerization of tannins present in different fractions ranged from 3 (fraction I) to 80 (fraction VI). Catechin comprised 67% of terminal tannin units while 60% of extension units in tannin polymer consisted of epicatechin. Sarni-Manchado et al. (1999) reported that the mean degree of polymerization of tannins precipitated by gelatin from wine is six. The wine used in this study was made from *Vitis venifera* var. Merlot (50%) and var. Carignan (50%). Epicatechin is the major component of precipitable condensed tannins as well as the predominant phenolic of extension units in tannins.

Young wines contain mainly low to medium molecular weight phenolics while aged wines are relatively higher in polymerized phenolics. Of the low molecular weight phenolics found in wines, catechin, epicatechin, procyanidin B1, B2, B3, B4, C1 and T2, quercetin, gallic acid (Carando et al., 1999; Singleton, 1982; Van Dam and Daniel, 1980) and hydroxycinnamates, i.e., *trans*-caftaric, *trans*- and *cis*-coutaric acids (Betes-Saura et al., 1996; Karagiannis et al., 2000), are predominant. Smaller quantities of 2-*S*-glutathionylcaftaric, protocatechuic, *cis*- and *trans*-caffeic, syringic, fertaric, *trans*-coumaric and *trans*-ferulic acids have also been reported (Betes-Saura et al., 1996).

Figure 5.11 shows the structures of some procyanidin dimers and one of the procyanidin trimers commonly found in wine (Lunte et al., 1988). Typical Italian red and white wines contain 203 to 805 and 11 to 49 mg/L of flavanols, respectively. In addition, appreciable amounts of quercetin and rutin have been found in red wines, but only traces are present in white wine (Simonetti et al., 1997). Furthermore, the concentration of (+)catechin and (–)epicatechin in French wines ranges from 32.8 to 209.8 mg/L and from 22.1 to 130.7 mg/L, respectively. The level of procyanidins in these wines is between 7.8 and 39.1 mg/L for B1, 18.3 to 93 mg/L for B2, 21.4 to 215.6 mg/L for B3, 20.2 to 107.2 mg/L for B4, 8.6 to 36.9 mg/L for C1 and 26.7 to 79.3 mg/L for T2 (Carando et al., 1999; Landrault et al., 2001). McDonald et al. (1998) surveyed the flavonol contents of 65 red wines from different geographical origins. The total content in red wines was between 4.6 and 41.6 mg/L, while free flavonols comprised 3 to 67% of total flavonols. The high total flavonol levels were detected in wines produced from thick-skinned grape varieties such as Cabernet Sauvignon.

The total content of phenols, expressed in milligrams of gallic acid equivalents, may range from 50 mg/L in some white wines to 6500 mg/L in certain types of red wine. A typical white wine contains about 250 mg phenols per liter, although some wines may contain even 2000 mg/L. On the other hand, the total content of phenolics in typical red wine ranges from 700 to 4000 mg/L (Landrault et al., 2001; Salunkhe et al., 1990; Sato et al., 1996; Simonetti et al., 1997). The total content of phenolics in wine depends on a number of factors, for example:

- Variety of grapes (red or white)
- Nature of crushing
- Possible inclusion or elimination of crushed grape skins, pulp, and seeds (especially from red grape varieties) prior to fermentation
- Heating of grape skins
- Vinification mode (temperature and maceration time)
- Aging

A longer fermentation period of skins, pulps and seeds allows more phenolics to be extracted into the wine because the ethanol produced acts as an extraction medium for phenolics.

Table 5.26 gives the content of low molecular weight phenolics extractable in ethyl acetate from some young red Bordeaux wines produced from different vine cultivars. Ethyl acetate enables catechins, procyanidins, flavonols, aromatic alcohols and phenolic acids to be extracted from wine. Data presented in Table 5.26 indicate a significant variation in the content of low molecular weight phenolics, depending

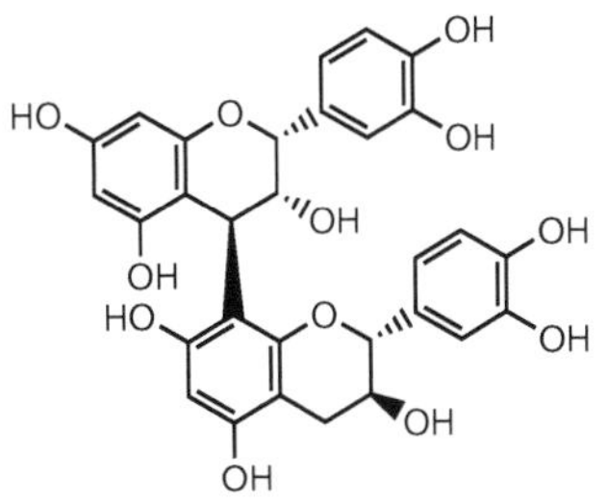

B1

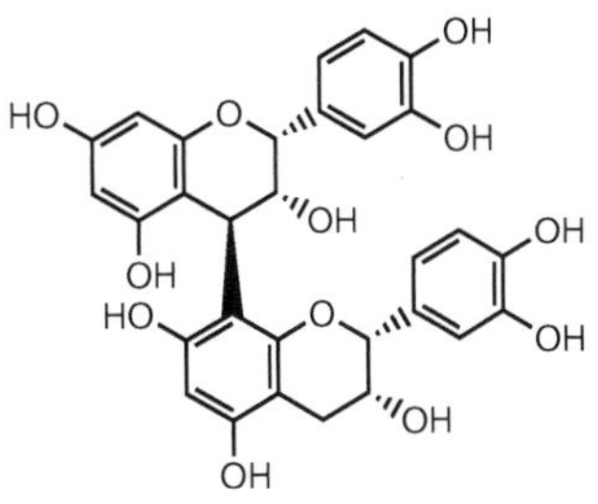

B2

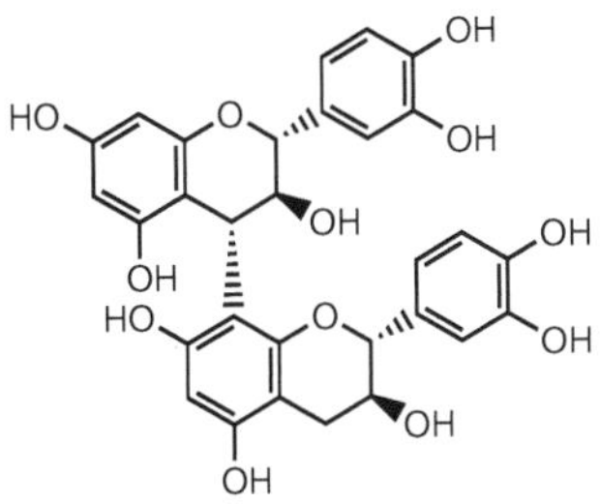

B3

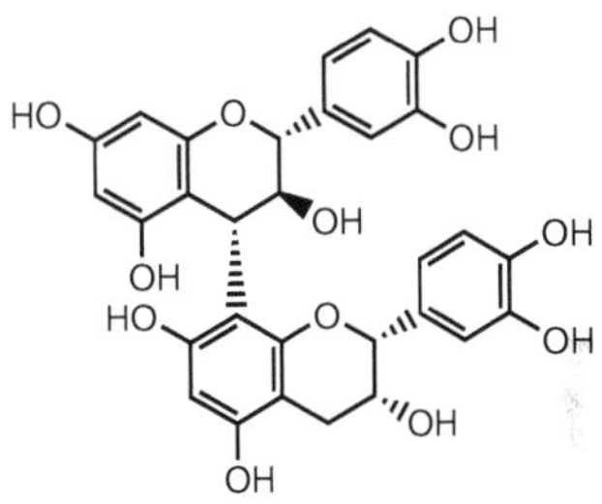

B4

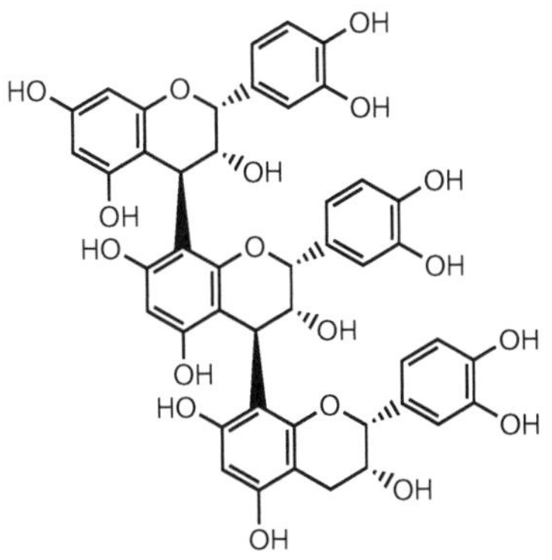

C1

FIGURE 5.11 Some procyanidins found in wine. (Adapted from Lunte, S.M. et al., 1988, *Analyst*, 113:99–102.)

on the vine cultivar. Therefore, Salagoity-Auguste and Bertrand (1984) proposed to use the analysis of low molecular weight phenolics as a fingerprinting method for identification of the cultivar from which red wine is produced. The total content of low molecular weight phenolics ranges from 100 to 200 mg/L, which is low when compared to the total content of phenolics present. Thus, most of the phenolic

TABLE 5.26
Average Concentrations of Phenolic Compounds Extracted from Red Wines of Three Bordeaux Cultivars[a]

Compound	Cultivar		
	Cabernet Sauvignon	Merlot	Malbec
Tyrosol	3.7	3.1	1.7
Tryptophol	1.1	0.9	1.2
Catechin	7.6	11.1	16.3
Epicatechin	11.6	11.3	6.3
Procyanidin A_2	1.5	11.2	1.8
Procyanidin B_2	9.2	6.1	2.4
Procyanidin B_3	38.5	2.9	1.6
Procyanidin B_4	15.7	19.1	5.4
Myricetin	2.6	6.1	3.2
Quercetin	10.4	14.8	7.8
Gallic acid	5.8	3.9	2.7
Protocatechuic acid	0.4	0.2	0.4
p-Hydroxybenzoic acid	2.2	0.9	0.4
Vanillic acid	0.3	0.7	0.6
Caffeic acid	0.9	0.5	0.3
Syringic acid	3.8	1.3	3.0
p-Coumaric acid	1.3	0.4	0.8
Ferulic acid	0.1	0.1	0.1

[a] Milligrams per liter.

Source: Adapted from Salagoity-Auguste, M.-H. and Bertrand, A., 1984, *J. Sci. Food Agric.*, 35:1241–1247.

compounds present in red wine, including anthocyanins and polymerized pigments, are not extracted into ethyl acetate.

Six phenolic acids have been identified in Seyval white wines, including gallic (1.8/L), caffeic (1.1 mg/L), and *p*-coumaric acids (0.6 mg/L) (Jindra and Gallander, 1987). In Vidal white wines, eight acidic phenolics have been identified: gallic acid (1.45 mg/L), *p*-hydroxybenzoic acid (0.17 mg/L), *m*-hydroxybenzoic acid (0.84 mg/L), caffeic acid (12.97 mg/L), salicylic acid (0.66 mg/L), and *p*-coumaric acid (1.94 mg/L) (Woodring et al., 1990). More recently, Soleas et al. (1997) determined the content of 15 polyphenols in a wide range of wine produced in Ontario, Canada. These authors reported that ferulic acid is the predominant phenolic acid in Riesling wines, caffeic acid predominates in Vidal and Chardonnay, and *p*-coumaric acid in Seyval Blanc, while gallic acid is predominant in all red wines.

Baderschneider and Winterhalter (2001) have identified a number of novel phenolic acid derivatives, phenylpropanoids, flavonoids and lignans in Riesling wine. These include 15 new benzoic and cinnamic acid derivatives, two phenylpropanoids, namely, 3-hydroxy-1-(4-hydroxy-3-methoxyphenyl)-propan-1-one and 2,3-dihydroxy-1-(4-hydroxy-3-methoxyphenyl)-propan-1-one, six flavonoids, i.e., dihydrokaempferol, dihydroquercetin and four dihydroflavonol glycosides, three neolignans and six lignans, i.e., derivatives of lariciresinol, isolariciresinol and secoisolariciresinol.

Numerous volatiles present in small quantities in wines are responsible for their specific flavors (Schreier, 1979; Singleton and Noble, 1976). Twelve volatile phenols have been identified in wines stored without contact with oak barrel (Etievant, 1981). Wine aged in wooden barrels contains vanillin, as well as coniferylaldehyde, syringaldehyde, sinapaldehyde, vanillic acid, and ethyl vanillate, most probably extracted from wood by ethanol (Singleton and Noble, 1976). However, some of these compounds may have also been produced from the degradation of hydroxycinnamic acid esters present in juice extracted from grapes. Decarboxylation of *p*-coumaric and ferulic acids by yeasts and bacteria produces 4-vinylphenol and 4-vinylguaiacol, which are volatile and contribute to off-flavors in wine (Schreier, 1979).

Polyphenolics and Vinification

The vinification process brings about qualitative and quantitative changes in wine phenolics. The extraction of phenolics is minimal from grape solids in making white wine, but plays an important role in making red wine. The quantities of pigments and other phenolics may vary greatly with grape variety, region, and cultural practices. The level of phenolics retained during the vinification process determines the type of wine and also has a potential influence on wine quality (Burns et al., 2001; Somers and Pocock, 1986).

Crushing is the first operation in winemaking, enhancing the level of polyphenolics in wine due to the contact of juice with the solid parts of grapes. Crushing increases the surface of the solid parts in contact with the juice. In red wines, it facilitates maceration during the fermentation process and thus increases the extraction of phenolic substances from grape solids. However, excessive crushing with tearing of the solid parts and skins increases the extraction of astringent phenolics into the juice and thus may affect wine quality. Red grapes are usually destemmed prior to crushing, which prevents stems running out into the fermentation vessel and reducing the content of phenolics in wine (Macheix et al., 1991). The crushing of white grape is done without destemming to reduce damage to the stems, which release fewer polyphenols into the juice. Also, tearing solid parts of the fruits and crushing the seeds must be avoided because high levels of phenolics may have a profound and subtle effect on the quality of white wines. Polyphenols are the major contributors to bitterness and astringency in white wines and act as substrates in oxidation processes (Hartnell, 1987; Robichaud and Noble, 1990).

Longer contact periods of skins, pulp, and seeds during fermentation allow more phenolics to be extracted into wine because the ethanol produced enhances the extraction of phenols. Wines with higher levels of phenolics are more astringent and therefore require longer maturation than those with lower phenolic content (Somers and Pocock, 1986). During the initial period of fermentation, phenolic compounds steadily diffuse from skin into the must (Bourzeix, 1971). Thus, 4 to 5 days' maceration is sufficient to produce a colored, low-tannin wine for early drinking, such as Bordeaux wines. However, in the case of wines subjected to aging processes, the skins should be allowed to remain for a longer period in the mixture during the fermentation process to increase the level of tannins (Ribereau-Gayon and Glories, 1987). In addition, Gil-Munoz et al. (1999) demonstrated that cooling the grapes to 10°C prior to crushing decreases the rate of polyphenol extraction from the skin, but only during the first 3 to 4 days of alcoholic fermentation.

The fermentation process brings about qualitative and quantitative changes in phenolic composition of wines. At the beginning, a rapid decrease in the content of hydroxycinnamic acid derivatives occurs, but the rate of degradation of *p*-coumaroyltartaric acid (coutaric acid) is greater than that of caffeoyltartaric acid (caftaric acid) (Macheix et al., 1990). During fermentation, hydroxycinnamic acid derivatives are hydrolyzed to free hydroxycinnamic acid, which may then be oxidized (Cilliers and Singleton, 1989), converted into volatile phenols by yeast decarboxylase (Chatonnet et al., 1993; Dugelay et al., 1993), or adsorbed by yeast (Schreier, 1979; Somers et al., 1987). About 27.6% of hydroxycinnamates are lost in the vinification process (Betes-Saura et al., 1996). Caffeic acid has been found to be the most abundant acidic phenol in Vidal Blanc and white wines from Penedes that are made from grapes high in *trans*-caffeoyl tartrate (Betes-Saura et al., 1996; Woodring et al., 1990). As the fermentation process progresses, due to tannin-anthocyanin polymerization, the level of anthocyanins decreases (Macheix et al., 1990).

Polyphenolics and Aging of Wine

After fermentation, wine usually is matured first in barrels or tanks in the presence of air. This maturation process from end of vinification to bottling typically lasts 12 to 24 months. Following this, the wine is further aged in bottles and thus is protected from air to a considerable extent. During a year of storage in barrels, wine can absorb up to 40 mg of oxygen/L and evidence suggests that oxygen may even penetrate to wine stored in corked bottles (Ribereau-Gayon, 1974; Riberau-Gayon et al., 1983). The presence of oxygen induces the chemical transformation of phenolics and affects the flavor and color of wine. The extent of these changes depends on a number of factors including pH, temperature, presence of oxygen, ethanol content, nature and composition of phenolics, and bisulfite concentration. de Freitas et al. (1998) investigated the oxidative stability of grape seed procyanidins in a model wine solution and found the order of oxidative decomposition for procyanidin dimmers to be B3 ≈ B4 > B7 ≈ B6 > B1 ≈ B2 > B8. These authors also demonstrated that the oxidative decomposition of trimers is affected by their conformation because trimer C1 is more oxidizable than catechins, while trimers (+)catechin-(4a→8)-(+)catechin-(4a→8)-(−)epicatechin and (+)catechin-(4a→8)-(−)epicatechin-(4b→6)-(+)catechin are less oxidizable than catechins.

During barrel aging, composition of phenolics of wine undergoes changes due to an increase in the content of phenols extracted from the wood. Gallic acid, initially low in the control wine (1.8 mg/L), increases to 4.0 mg/L after 12 weeks of aging in American-oak barrels. On the other hand, the content of protocatechuic, vanillic, caffeic, syringic, and *p*-coumaric acids remains relatively constant. Overall, gallic acid represents only 7% of the total increase in nonflavonoid phenolics. However, the remaining 93% increase in phenolic level of aged wine is unaccounted for (Jindra and Gallander, 1987). The increase in gallic acid may be attributed in some cases to the hydrolysis of catechin and epicatechin gallates naturally occurring in wines (Singleton and Essau, 1969). The amount of phenolics extracted into wine depends on aging time, oak type, size and the possibility of the barrel having been used previously (Rous and Alderson, 1983; Singleton, 1974). Some phenols extracted from wood include lignins, hydrolyzable tannins, gallic acid, ellagic acid, aromatic carboxylic acids and aldehydes. Of these, ellagitannins and elagic acid account for 10% of the total phenolic content of Riesling wines treated with oak chips (Quinn and Singleton, 1985).

During the aging of red wine, the content of anthocyanins decreases progressively and irreversibly as a result of formation of polymeric pigments. The color of these pigments is less sensitive to changes of pH (between 2.2 and 5.5) and is quite resistant to decolorization by sulfur dioxide (at a concentration of up to 200 mg/L); however, some of these pigments may be more prone to degradation in aqueous solutions than anthocyanins (Escribano-Bailon et al., 2001). The polymeric forms of pigments account for about 50% of the color density of wine during the first year of aging; this value averages 85% for 10-year-old wines (Somers, 1971). The formation of these pigments has been evaluated in model wine systems. Based on the results of these studies, various mechanisms have been proposed for the formation of polymeric pigments.

According to Mistry et al. (1991) and Figueiredo et al. (1996), the polymeric pigments are the products of intramolecular copigmentation of anthocyanins. Several other researchers (Liao et al., 1992; Santos-Buelga et al., 1996; Somers, 1971) have suggested that the formation of these pigments is the result of direct condensation between anthocyanins and flavanols. Reactions leading to the formation of pigments from anthocyanins are depicted in Figure 5.12. Initially a labile flavone (II) is formed from electrophilic substitution of a carbonium ion (I) into a wine component, RH. The R substituent is most likely a dimeric proanthocyanidin. The flavone is readily oxidized by air or a redox system to condensed anthocyanin species (III); this deprotonates to a more stable quinonoid structure (IV) designated to the chromophores of the polymeric pigment that is then formed. The enhanced stability of the latter pigment is believed to be due to aryl substitution of the anthocyanin at 4-position (Haslam, 1989). Another plausible mechanism suggests that the reactions between anthocyanidins and flavanols are mediated by acetaldehyde (Bakker et al., 1993; Dallas et al., 1996; Escribano-Bailon et al., 2001; Francia-Aricha et al., 1997). This mediation leads to the formation of ethyl bridges between C-8 positions of their phloroglucinol rings (Francia-Aricha et al., 1997). This reaction is favored at more acidic conditions, which promote the formation of the acetaldehyde cation required for mediation of condensation between flavanols and anthocyanins (Garcia-Viguera et al., 1994).

Acetaldehyde is a precursor of ethanol during primary oxidation, but can also be formed during aging as a result of oxidation of ethanol. The latter reaction involves oxidation of *o*-diphenol groups leading to intermediate formation of hydrogen peroxide and subsequent coupled oxidation of ethanol to acetaldehyde (Wildenradt and Singleton, 1974). Francia-Aricha et al. (1997) reported the formation of three novel pigments from malvidin-3-*O*-glucoside (Mv-3-glu) and procyanidin B2 in the

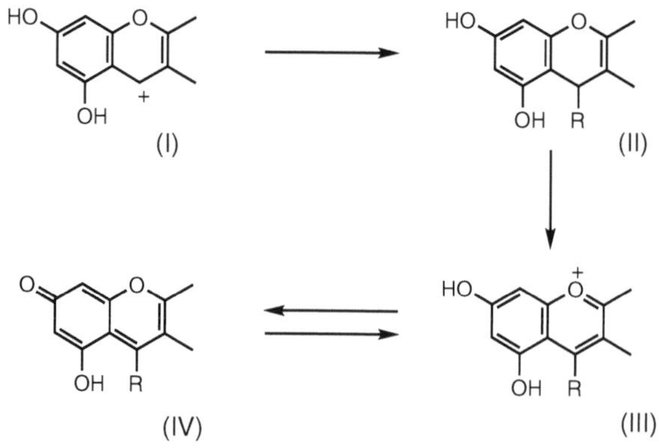

FIGURE 5.12 Reactions involved in the formation of polymeric pigments in red wine. (Adapted from Somers, T.C., 1971, *Phytochemistry*, 10:2175–2186.)

presence of acetaldehyde. Two of these pigments, reddish-blue in color, are identified as enantiomers of Mv-3-glu and procyanidin B2 linked at their C-8 by ethyl bridges. The third pigment is orange-red in color and possibly made of two flavylium mesomeric forms corresponding to malvidin and pelargonidin. This latter pigment does not contain ethyl bridge, but the presence of acetaldehyde is needed for its formation. The mechanism involved in the formation of this pigment remains unknown. Condensed anthocyanin species can be further substituted with additional oligomeric proanthocyanidins. Finally, the polymerization reaction leads to precipitation of polymeric pigments and loss of astringency (Ribereau-Gayon et al., 1983; Timberlake and Bridle, 1976).

Another possible mechanism of the evolution of phenolics in aging red wine involves electrophilic condensation reactions between oligomeric proanthocyanidins and acetaldehyde (Figure 5.13) (Haslam, 1989) as well as between flavanols and acetaldehyde (Es-Safi et al., 1999). Catechin ethyl-bridged dimers and trimers have been identified in red wine samples (Saucier et al., 1997) and it is thought that these reactions may lead to the formation of insoluble polymers, bringing about a decrease in astringency (Tanaka et al., 1994) and changes in color. In addition, the formation of adducts between anthocyanins and pyruvic acid has been reported. Such compounds have been isolated from Port wines (Fulcrand et al., 1998; Mateus and de Freitas, 2001; Mateus et al., 2001; Romero and Bakker, 1999). Using a winelike model system, Vidal et al. (2002) recently investigated possible reactions leading to the loss of astringency in wine. These authors demonstrated that cleavage of the proanthocyanidin interflavanic bond takes place during wine aging. In the absence of (–)epicatechin, the carbocations released upon this cleavage form unknown species (phlobaphenes). However, in the presence of (–)epicatechin, a decrease in mDP (mean degree of polymerization) of proanthocyanidins from 36 to 25 occurred after 53 days of reaction and accumulation of oligomeric proanthocyanidins in the winelike solution was noted.

Storage of wine in contact with yeasts (aging *sur Lie*) alters the flavor and aroma of certain white and red wines and also leads to the production of sparkling wines.

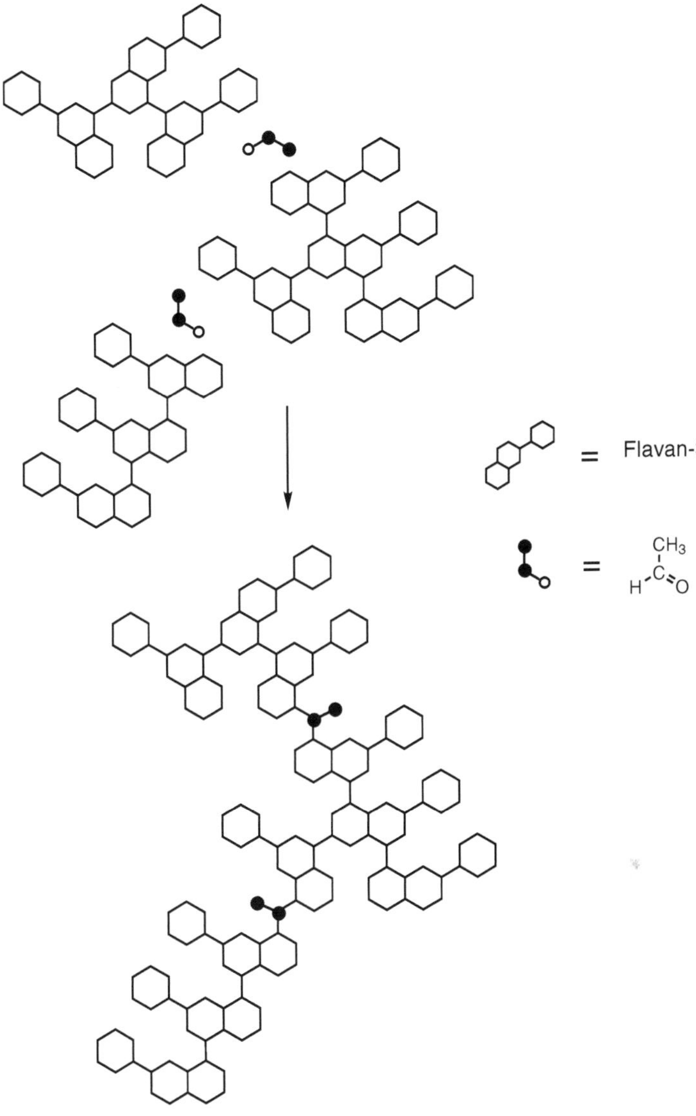

FIGURE 5.13 Proanthocyanidin polymerization in wine via reaction with acetaldehyde. (Adapted from Haslam, E., 1989, *Plant Polyphenols*, Cambridge University Press, Cambridge, U.K.)

Grape-derived glycosides are considered to be the primary precursors of wine aroma (Baltenweck-Guyot et al., 2000; Zoecklein et al., 1997). Aging *sur Lie* decreases the total content of glycosides in Riesling wines by 73.4% (Zoecklein et al., 1997). Vasserot et al. (1997) demonstrated that the yeast lees also have the capacity to adsorb anthocyanins.

Polyphenolics and Browning Reactions in Wines

Phenolic compounds may contribute to oxidative browning of white musts and wines. *Trans*-caffeoyl tartaric acid and other hydroxycinnamic acid derivatives are major phenolic compounds in grape juice (Singleton et al., 1984, 1985) and good substrates for grape polyphenol oxidase (Gunata et al., 1987). The oxidation of phenolics brings about a modification of initial wine phenolics and may result in the formation of condensed brown substances that may negatively affect the color and flavor of white wines. The rate of browning depends on a number of factors; of these, availability of oxygen, the nature and concentration of phenolics, and activity of polyphenol oxidases during the initial phases of wine making are the most important parameters (Macheix et al., 1991).

Susceptibility of grape phenolics to oxidation can be described by the so-called browning potential. This potential is calculated by taking into account the browning index of each phenolic, i.e., phenolic acids, catechin, epicatechin, procyanidins, and catechin–catechin gallate and their concentration in grape juice (Lee and Jaworski, 1987, 1988). However, the browning potential of grapes is not a function of their polyphenol oxidase because the activity of this enzyme is not a limiting factor for browning reactions. The level of oxidizable substrates may be the limiting factor for browning reactions because polyphenol oxidase is present in grapes in excess (Sapis et al., 1983).

The oxidative polymerization leading to the formation of brown pigments depends on the nature and relative concentration of the phenolic compounds present (Cheynier et al., 1988). However, no correlation exists between the level of hydroxycinnamic acid esters and browning potential of grape varieties (Romeyer et al., 1985). Therefore, it has been suggested that caftaric acid/glutathione ratio, which determines the quantities of untrapped quinones, may serve as an important factor to help explain the varietal differences in browning potential (Salgues et al., 1986). On the other hand, a significant correlation has been observed between catechin level and wine browning (Simpson, 1982). However, catechin shows poor affinity towards polyphenol oxidase and is found at low concentrations in wine (Gunata et al., 1987; Singleton, 1982, 1987). Therefore, catechin is probably not much affected by the oxidation process, but may participate in coupled oxidation reactions as demonstrated by Cheynier et al. (1988).

In model systems containing grape polyphenol oxidase, caftaric acid is degraded faster than catechin, epicatechin and epicatechin gallate, while procyanidins B1, B2, B3, B4 and 2-S-glutathionyl caftaric acid (referred to as grape reaction product, GRP) are oxidized slowly (Cheynier et al., 1988). The degradation process is considerably faster for mixtures of monomeric and dimeric flavan-3-ols and gallic acid in proportions naturally found in grape seeds (Oszmianski et al., 1985). Presence of caffeoyl tartaric acids increases the degradation of all flavanols, probably due to a coupled oxidation mechanism. This mechanism involves enzymatic oxidation of tartaric acid followed by chemical oxidation of the second compound by caftaric acid quinone with regeneration of caftaric acid as follows (Cheynier et al., 1988):

- Caftaric acid + 1/2 O_2 → caftaric acid quinone + H_2O
- Caftaric acid quinone + O-diphenol → caftaric acid + o-quinone

The rate of coupled oxidation induced by caftaric acid is influenced by the redox potentials of phenolic compounds involved. The order of redox potential for grape phenolics is: epicatechin gallate < procyanidins < GRP < epicatechin, catechin < caffeoyl tartaric acid (Cheynier et al., 1988). Caftaric acid quinone may also react with hydroquinone to yield a condensation product for the reaction of two semiquinone free radicals (Singleton, 1982, 1987), or by a mechanism similar to a 1,4-Michael addition (Gramshaw, 1970; Singleton, 1982, 1987).

Under normal processing conditions, caftaric acid is oxidized very fast by grape polyphenol oxidase to corresponding orthoquinone, which then reacts primarily with the available glutathione to form the so-called GRP (Cheynier et al., 1986; Singleton et al., 1984). It is thought that formation of GRP limits browning of musts by trapping caftaric acid quinones into a colorless product with a lower affinity for the enzyme as a substrate and thus prevents them from proceeding to brown polymerized products (Singleton et al., 1985). However, capacity of GRP to control browning of grape juice and must depends on the amount of glutathione available, the contribution to coupled oxidation and its oxidation by laccase (Macheix et al., 1991).

The 2-S-glutathionyl caftaric acid may, however, be oxidized by *Botrytis cinerea* laccase if grape juice is prepared from infected fruits. Laccase is an enzyme that attacks a wider range of substrates than polyphenol oxidase (Dubernet and Ribereau-Gayon, 1973). Thus, the oxidation process may contribute to the browning of some white grape juices and wines made from infected grapes. The sequence of reactions involved in the oxidation of 2-S-glutathionyl caftaric acid is depicted in Figure 5.14. First, the laccase catalyzes the oxidation of 2-S-glutathionyl caftaric acid to the corresponding quinone. Following this, in the presence of surplus glutathione, a second thioether linkage is formed. Thus, the vicinal dihydroxyphenol structure is regenerated and 2,5-di-S-glutathionyl caftaric acid is produced. Otherwise, the orthoquinone formed may rapidly proceed to brown polymers. However, glutathionyl derivatives are also oxidizable and therefore can further contribute to the browning reaction (Salgues et al., 1986).

Flavanols may play a more prominent role in the formation of brown pigments than caftaric acid (Cheynier et al., 1989; Fernandez-Zurbano et al., 1995). Results of studies carried out using model systems demonstrate that the formation of catechin oligomers is more rapid than that of caftaric acid oligomers (Cheynier et al., 1989). Fernandez-Zurbano et al. (1998) measured changes in the concentration of hydroxycinnamic acids, hydroxycinnamic acid esters and flavanols brought about by the oxidation process; only the flavanol contents correlated with the degree of wine browning measured as the absorbance change at 420 nm.

OTHER ALCOHOLIC BEVERAGES

Processes involved in production of distilled spirits are specific to each alcoholic beverage (Lembeck, 1983). The composition of phenolic components in these

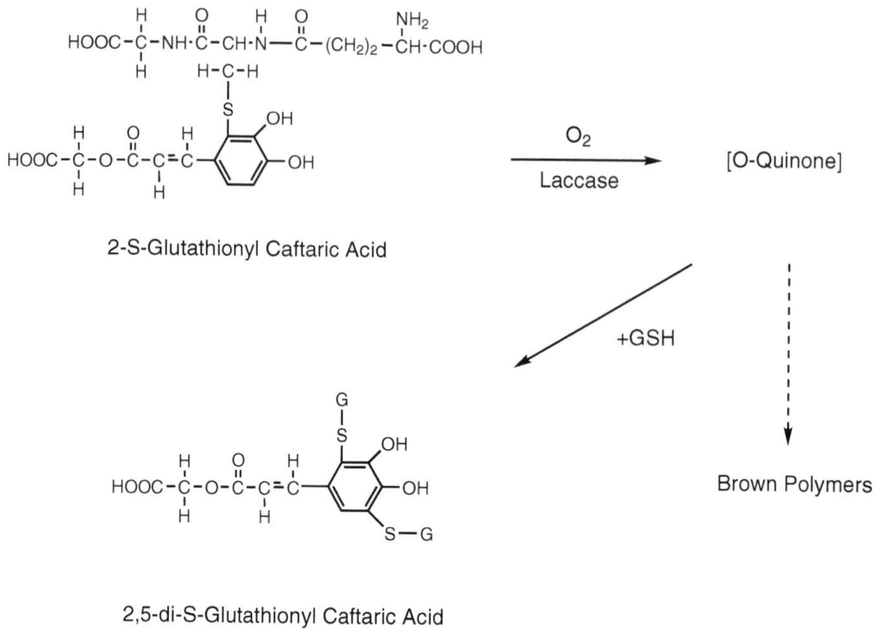

HOOC-CH₂-NH-CO-CH(CH₂-S-)-NH-CO-(CH₂)₂-CH(NH₂)-COOH

2-S-Glutathionyl Caftaric Acid

O_2 / Laccase → [O-Quinone]

+GSH

Brown Polymers

2,5-di-S-Glutathionyl Caftaric Acid

FIGURE 5.14 The proposed reaction sequence for the oxidation of 2-S-glutathionyl caftaric acid (top) by laccase and formation of 2,5-di-S-glutathionyl caftaric acid (bottom); GSH, glutathione. (Adapted from Salgues, M. et al., 1986, *J. Food Sci.*, 51:1191–1194.)

beverages is affected by the starting material and the aging process, if used. Volatile components of alcoholic beverages such as cognac, rum, and whisky contain minute quantities of phenolic compounds; these are also identified in fusel oils obtained by fractional distillation of the beverage. Otsuka et al. (1965) reported the presence of phenolic acids in cognac and whisky stored for a number of years in oak barrels. On the other hand, 4-methylguaiacol and eugenol were first identified in Jamaican rum (Moarse and ten Noever de Brauw, 1966). Later, Timmer et al. (1971) found phenol, *p*-ethylphenol, *o*-cresol, *p*-cresol, guaiacol, *p*-ethylguaiacol and dihydroeugenol. Furthermore, guaiacol, phenol, *o*-cresol, *p*-cresol, 4-methylguaiacol, *o*-ethylphenol and 4-ethylguaiacol have been found in whisky (Kahn, 1969). The presence of 6-methylguaiacol, *o*-isopropylphenol, *p*-ethylphenol, 2,6-xylenol, eugenol, *o*-hydroxyacetophenone and 2-hydroxy-5-methylacetophenone in Japanese and Scotch whisky varieties was reported by Nishimura and Masuda (1971). The content of these compounds ranges from 0.001 to 0.035 mg/L for *p*-ethylguaiacol in Scotch whisky.

Polyphenols represent up to one third of cognac extracts; the total content of polyphenols in aged cognacs ranges from 92 (after 1 year of aging) to 833 mg/L (after 30 years of aging) (Viriot et al., 1993). Distilled spirits aged in oak barrels contain low molecular weight phenolic compounds such as vanillin, syringaldehyde, sinapaldehyde, coniferaldehyde, and scopoletin, as well as vanillic, syringic, gallic, trimetoxyphenylacetic, 3,4-dihydroxybenzoic and ellagic acids (Delgado and Gomez-Cordoves, 1987; Goldberg et al., 1999; Guymon and Crowell, 1968;

Panossian et al., 2001; Puech, 1988; Salagoity-Auguste et al., 1987; Viriot et al., 1993). The presence of these phenolics in beverages is attributed to the lignin and ellagitannin fractions of wood (Guymon and Crowell, 1968; Viriot et al., 1993). The concentration of gallic acid in aged cognacs is of the same magnitude as that of ellagic acid (Viriot et al., 1993). However, Goldberg et al. (1999) found that the content of gallic acid in seven samples of American whiskies (single-malt Scotch, blended Scotch, Canadian rye, American bourbon) and brandies (cognac, armagnac, regular brandy) is 3 to 10 times lower than that of ellagic acid. Panossian et al. (2001) reported that the content of sinapaldehyde in Armenian brandies is between 0.511 and 1.618 mg/L; conferaldehyde is between 0.497 and 2.544 mg/L, syringaldehyde from 2.32 to 34.2 mg/L and vanillin from 1.13 to 12.83 mg/L.

Distilled spirits also contain soluble oligomers of lignins and ellagitannins. Viriot et al. (1993) evaluated the changes in ellagitannins and ellagic acid content during the aging of cognacs in oak barrels. These authors observed that the ellagitannins present in wood are quickly solubilized and degraded, but ellagic acid is released into the beverage at a slower rate. The presence of ellagic acid in cognacs may arise from solubilization of free ellagic acid as well as from the hydrolysis of soluble and insoluble ellagitannins present in woods. Not only ellagitannins but also lignins are chemically altered to soluble forms during aging of distilled spirits in oak barrels. Viriot et al. (1993) have suggested that the solubilization of lignins is brought about by the cleavage of alkyl aryl ether linkages. The content of soluble lignins in distilled spirits increases with duration of aging; one-year-old cognacs contain only 12 mg of lignins per liter while those stored for 30 years in oak barrels have a total lignin content of 219 mg/L (Viriot et al., 1993).

According to Steinke and Paulson (1964), guaiacol and its alkyl derivatives (noted earlier) as well as *p*-ethylphenol and *p*-vinylphenol are produced from coumaric and ferulic acids in grains at the mashing process by heating or by the action of fungi. However, presence of guaiacol, phenol and *p*-cresol may be the result of further degradation of those phenolics. On the other hand, *o*-cresol and 2,6-xylenol may be derived from lignin-like materials (Nishimura and Masuda, 1971).

Tequila is an alcoholic beverage obtained by distillation of fermented juice of cultivated variety of *Agave tequilana* (Valenzuela-Zapata, 1985). The process involves harvesting the stems of the agave plant, cooking them in an oven to hydrolyze inulins to fructose and glucose, and milling and pressing the cooked stems. The juice obtained is first fermented with yeast and then doubly distilled; after adjusting the alcohol content, it is stored in oak barrels (Benn and Peppard, 1996). Guaiacol, cresol, 4-ethylguaiacol, eugenol, vanillin, syringaldehyde and coniferaldehyde have been identified as flavor components of tequila. Of these, vanillin is considered one of the most powerful odorants present; other powerful odorants of tequila are isovalerylaldehyde, 2-phenylethanol and β-damascenone (Benn and Peppard, 1996).

Cider brandy is an alcoholic beverage obtained by distillation of cider followed by a maturation process in American or French oak casks for different time periods, depending on traditional practices (Cabranes et al., 1990). Aging of distilled cider brandies in oak casks increases the total content of phenolics from 12.7 to 25.9 mg/L (Mangas et al., 1997). This increase may be brought about by the extraction and acid ethanolysis of lignin followed by oxidation (Puech, 1988). Syringaldehyde, vanillin and gallic, vanillic and syringic acids are the predominant phenolics present in freshly distilled and aged cider brandies. Small quantities of *p*-hydroxybenzaldehyde and caffeic and protocatechuic acids have also been detected (Mangas et al., 1997).

CIDER AND APPLE JUICE

The traditional method of juice extraction from apples involves crushing and milling the fruit to pulp. The pulp may then be built up in alternate layers with cloths and wooden racks to form a "cheese" from which the juice is squeezed by vertical pack press or extracted by a continuous extraction system such as hot water diffusion. In order to improve pressing characteristics and yield, the pulp may be mixed with pectinolytic enzymes.

Phenolic compounds of apples have aroused considerable interest because of their influence on the sensory characteristics such as color and clarity of juice. Flavons and chlorogenic acid may contribute to the browning of juice because these compounds are good substrates for apple polyphenol oxidase (Stelzig et al., 1972). On the other hand, quercetin glycosides are considered very poor substrates for this enzyme. The content of quercetin glycosides and flavons declines during the storage of juice, but chlorogenic acid is quite stable (Tanner and Rentschler, 1956). Thus, changes in the phenolic composition may play an important role in the overall quality and acceptability of apple juice.

Composition of Polyphenolics

The composition of phenolic compounds in apple juice and cider depends on their initial composition in the apple cultivar, cultural conditions, and the extraction of phenolics during pressing. The composition of phenolics in fruits may vary from one cultivar to another and also depend on the degree of fruit maturity. The contents and relative proportions of phenolics in juices and ciders may also be changed as a result of enzymatic browning reactions and formation of haze. Phenolics have an important role in providing a bitter and astringent flavor to apple juice or cider. Phenolic acids (primarily chlorogenic and *p*-coumaric acids), dihydrochalcones (phloretin xyloglucoside and phloridizin), catechins (epicatechin and catechin) and procyanidins are major groups of phenolic compounds found in apple juice or cider (Lea, 1978; Wilson, 1981). Figure 5.15 depicts the chemical structure of major classes of cider phenolics. Of the phenolics found in apple juice and cider, the polymeric procyanidins are particularly responsible for their bitter/astringent characteristics. Phloridzin is another bitter phenolic found in apple juice and cider; however, due to its low concentration, phloridzin may make only a small contribution to the overall bitterness of these products (Lea and Arnold, 1978; Lea and Timberlake, 1974).

The content of monomeric phenols in ciders, in the order of decreasing concentration, is 5-chlorogenic acid, *p*-coumarylquinic acid, catechin, phloridzin (phloretin glucoside), and phloretin xyloglucoside (Table 5.27) (Whiting and Coggins, 1975). Versari et al. (1997) identified phloridzin and phloretin xyloglucoside as quality indicators of fruit-based beverages. These two compounds are characteristic of apples

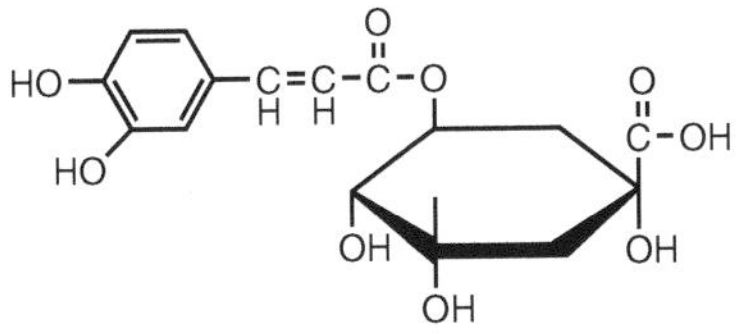

Chlorogenic Acid

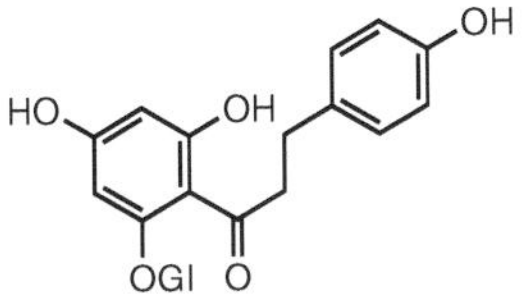

Phloridzin

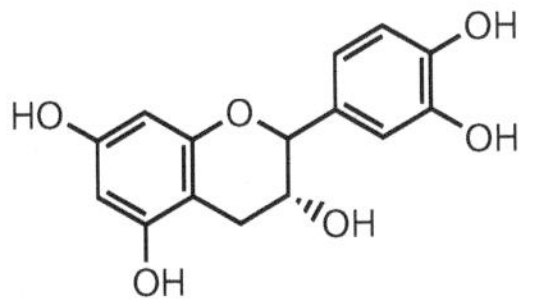

(-)Epicatechin

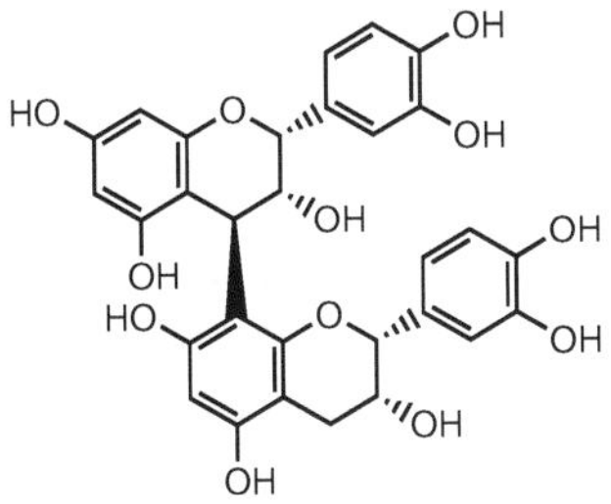

Procyanidin B2

FIGURE 5.15 Phenolic compounds found in cider. (Adapted from Lea, A.G.H., 1978, *J. Sci. Food Agric.*, 29:471–477.)

and have not been detected in any other fruits. The addition of 50 to 60 mL/L of commercial apple juice to pear juice has been detected by the presence of one of these two dihrochalcones (Versari et al., 1997). According to Prabha and Patwardhan (1985), the major monomeric phenols of apple pulp are *cis*-chlorogenic acid followed by epicatechin and *trans*-chlorogenic acid (Table 5.28). However, the contribution of chlorogenic acid, analyzed by HPLC, is 6.2 to 10.7% of the total polyphenols expressed as chlorogenic acid equivalents (Cilliers et al., 1990). The content of monomeric phenolics is also affected by cultural conditions. Cider produced from fruits harvested from trees under reduced supply of nutrients contains 17% higher levels of phenolic compounds (Lea and Beech, 1978).

Chlorogenic acid, a principal component of monomeric phenol fraction in apples, is generally considered a major browning substrate (Hulme and Rhodes, 1971;

TABLE 5.27
Concentration of Phenolics in Ciders from a Range of Cultivars[a]

Cultivar	Chlorogenic Acid	*p*-Coumarylquinic	Catechins	Phloridzin	Phloretin Xyloglucoside
Frederick	0.42	0.07	ND	0.07	0.1
Dabinett	0.4–0.57	0.03	0.3–0.5	0.15–0.16	0.04–0.12
Michelin	0.51	0.06	0.25	0.07	0.04
Vilberie	0.73	0.43	0.4	0.03	0.03
Yarlington	0.68–1.5	6.2–7.9	0.6	0.03	0.02
Kingston	0.25	0.03	0.1	0.05	0.06

[a] Grams per liter.

Source: Adapted from Whiting, G.C. and Coggins, R.A., 1975, *J. Sci. Food Agric.*, 26:1833–1838.

TABLE 5.28
Concentration of Polyphenols in Different Varieties of Apple Pulp[a]

Polyphenol	Aumburi	Royal Delicious	Kesari	Golden Delicious
Epicatechin	250	190	130	70
Quercetin glycoside	38	20	traces	traces
D-Catechin	200	95	80	58
Leucoanthocyanidins	730	340	460	195
Gallocatechin	60	20	traces	10
cis-Chlorogenic acid	370	260	160	140
trans-Chlorogenic acid	180	80	188	50
p-Coumaryl quinic acid	55	12	—	8

[a] Milligrams per kilogram.

Source: Adapted from Prabha, T.N. and Patwardhan, M.V., 1985, *J. Food Sci. Technol.*, 22:404–407.

Stelzig et al., 1972). Thus, changes in chlorogenic acid may contribute to the quality of apple juice. The content of chlorogenic acid in juice obtained from eight apple cultivars grown in Michigan, Washington, Argentina, Mexico and New Zealand ranges from 1.41 mg/L in juice obtained from Granny Smith apples to 214 mg/L in juice from McIntosh apples (Lee and Wrolstad, 1988). Van Buren et al. (1973) reported that the content of chlorogenic acid in juice obtained from different apple varieties ranges from 120 mg/L for juice from Golden Delicious variety to 310 mg/L for Carrel variety juice. Juice obtained from crab apples contains 390 to 1350 mg/L, while fresh fruits have 460 to 2050 mg of chlorogenic acid per kilogram (Kuusi and Pajunen, 1971). However, the pulp obtained from some Indian apple varieties contains from 190 mg of total chlorogenic acid per kilogram for Golden Delicious to

TABLE 5.29
Distribution of Cider Procyanidins[a]

Phenolic Type	Apple Cultivar	
	Tremlett's Bitter	Vilberie
Polymeric procyanidins	1.06	1.23
Pentameric procyanidins	0.51	0.47
Tetrameric procyanidins	0.47	0.42
Trimeric procyanidins C1	0.46	0.39
Dimeric procyanidins B2	0.70	0.61
Epicatechin	0.51	0.40
Total oligomers	2.65	2.29
Total polymers	1.06	1.23
Total procyanidins	3.71	3.52

[a] Grams per liter.

Source: Adapted from Lea, A.G.H. and Arnold, G.M., 1978, *J. Sci. Food Agric.*, 29:478–483.

550 mg/kg for Aumburi variety (Prabha and Patwardhan, 1985) (Table 5.28). Other researchers have reported values of 250 mg chlorogenic acid per kilogram (Macheix, 1968) to 890 mg/kg of fresh mature fruits (Walker, 1963) for different apple varieties.

Cider prepared from true bittersweet apples typically contains 2 to 3 g/L of procyanidins (Lea, 1978; Lea and Timberlake, 1974). Characterization of cider procyanidins shows a range of procyanidin oligomers with varying degrees of polymerization. The main series of procyanidins in cider is based on epicatechin (e.g., procyanidin B2, B5, C1). However, small amounts of a mixed series of catechin–epicatechin (e.g., procyanidin B1) are also present (Lea, 1978; Wilson, 1981). The content of oligomeric and polymeric procyanidins in cider is shown in Table 5.29. The degree of procyanidin polymerization influences the organoleptic characteristics of cider. Thus, cider with a lower ratio of oligomeric to polymeric procyanidins is more astringent while that with a higher ratio is more bitter (Lea and Arnold, 1978).

Effect of Processing

The major loss of phenolics during processing of apples to cider is brought about by oxidation during the milling process and before pressing by incomplete extraction of apple tissues and during clarification of juice. Longer juice contact with air and pomace before pressing results in larger amounts of polyphenolics — in particular, polymeric procyanidins — responsible for sensory characteristics of apple juice and cider. This is probably due to the oxidation of procyanidins to their corresponding quinones, which are then irreversibly bound to the pomace constituents. On the other hand, enzymatically induced oxidative loss of phenolics is much less pronounced after extraction of the juice because the content of soluble polyphenol oxidase is very low in the juice (Lea and Timberlake, 1978).

TABLE 5.30
Content of Polyphenolic in Ciders Obtained from Diffuser and Pack Press Techniques[a]

Phenolics	Diffuser	Pack Press	Pack Press Corrected to Diffuser Dilution
Phenolic acids	850	1170	860
Phloretin xyloglucoside	90	140	100
Epicatechin	200	260	190
Phloridzin	200	150	110
Procyanidin dimers	340	500	360
Procyanidin trimers	100	220	160
Procyanidin tetramers	210	260	190
Procyanidin polymers	850	580	420

[a] Milligrams per liter.

Source: Adapted from Lea, A.G.H. and Timberlake, C.F., 1978, *J. Sci. Food Agric.*, 29:484–492.

Extraction of apple juice by diffusion techniques almost doubles the content of phloridizin and polymeric procyanidins compared to the levels found in the juice obtained by pressing (Table 5.30). Simple warm incubation under nonoxidative conditions also increases the content of polymeric procyanidins and phloridzin. On the other hand, use of pectolytic enzymes to improve the yield of juice extraction does not have any detrimental effect on polyphenols. Clarification of juice with gelatin leads to a significant loss of tetrameric and polymeric procyanidins (Lea and Timberlake, 1978).

Table 5.31 shows the effect of different treatments on the content of polyphenolics in apple juice obtained from Golden Delicious variety. Holding the pulp for up to 3 h under oxidative conditions produces juice with low contents of chlorogenic acid and flavonoids (mainly monomeric and polymeric flavans), but with higher levels of flavonol glycosides. Holding pulp under nonoxidative conditions, however, improves the extraction of all phenolic compounds from pulp into the juice (Van Buren et al., 1976).

OTHER FRUIT JUICES

Orange Juice

Oranges are a good source of flavanones (mainly hesperidin and narirutin) (Albach and Redman, 1969), polymethoxyflavones (e.g., nobiletin, tangeretin and sinensetin) (Park et al., 1983) and hydroxycinnamic acid derivatives (ferulic, *p*-coumaric, sinapic and caffeic acid derivatives) (Clifford, 1999; Risch and Herrmann, 1988). Cinnamoyl-β-D-glucose has been detected only in blood orange juices (Mouly et al., 1997). The total content of phenolics in several juices obtained from five different *Citrus sinensis* (L.)

TABLE 5.31
Phenolic Compounds in Apple Juice[a]

Method of Juice Preparation	Flavonol Glycosides	Other Flavonoids	Chlorogenic Acid	Other Nonflavonoids
Conventional	4	200	140	170
Rapidly pasteurized	2	710	230	80
Oxidized pulp, 1 h	7	90	<10	150
Oxidized pulp, 3 h	60	10	<10	210
Held and nonoxidized pulp	80	1010	320	200

Notes: Flavonol glycosides are expressed in rutin equivalents, other flavonoids are expressed in catechin equivalents, and other nonflavonoids are expressed in chlorogenic acid equivalents.

[a] Milligrams per liter.

Source: Adapted from Van Buren, J. et al., 1976, *J. Agric. Food Chem.*, 24:448–451.

TABLE 5.32
Composition of Phenolics in Orange Juice[a]

Orange Variety	Total Anthocyanins	Total Flavonones	Total Hydroxy-cinnamic Acid	Total Phenols
Moro	97.5–278.4	266.1–444.5	60.1–140.2	673.9–1147.2
Tarocco	1.2–99.4	150.1–179.4	38.0–91.2	387.3–1090.5
Saquinello	6.4–52.6	185.7–300.2	46.4–92.1	382.5–602.9

[a] Milligrams per liter.

Sources: Adapted from Rapisarda et al., 1998, *J. Agric. Food Chem.*, 46:464–470 and 2000, *J. Agric. Food Chem.*, 48:2249–2252.

varieties ranges from 361 (Washington navel variety) to 1147 mg of ferulic acid equivalents per liter (Moro variety) (Table 5.32) (Rapisarda et al., 1998, 2000). According to Gil-Izquierdo et al. (2001), hand-squeezed navel orange juice contains 839 mg of phenolics per liter. Citrus flavonoids possess health-promoting activities; these compounds show activity against myeloid leukemia (Kawaii et al., 1999) and possess anti-inflammatory, antianalgesic (Galati et al., 1994), anticarcinogenic (Bracke et al., 1994; Stavric, 1994), antihypertensive, diuretic (Galati et al., 1996) and hypolipidemic activities (Monforte et al., 1995).

Blood oranges have traditionally been consumed as fresh fruit because processing them into juice is unsatisfactory due to unstable pigments. This problem has been circumvented to a great extent by acidification or addition of antioxidants (Maccarone et al., 1985). Another way of utilizing blood orange juice may be to mix it with other juices for color enhancement (Lee et al., 1990). California and

TABLE 5.33
Effect of Processing on Phenolic Content in Soluble and Cloud Fractions of Orange Juices[a]

Type of Processing	Flavonones		C-glucosylflavone		Total Phenolics
	Soluble	Cloud	Soluble	Cloud	
70°C; 30 s	91.7–116.0	247.0–644.0	338.0–53.3	2.9–8.1	416.5–867.4
92–95°C; 30 s	80.9–127.6	80.9–127.6	32.1–51.1	—	511.0–595.6

[a] Milligrams per liter.

Source: Adapted from Gil-Izquierdo, A. et al., 2001, *J. Agric. Food Chem.*, 49:1035–1041.

Italian blood oranges contain red anthocyanin pigments. The content of anthocyanin pigments in the juice depends on the season and varies from 96 to 166 mg/L of juice obtained from California oranges (Lee et al., 1990) and from 1.2 to 278 mg/L of juice obtained from Italian oranges (Rapisarda et al., 1998, 2000). Comprising over 90% of total anthocyanin pigments, cyanidin is the predominant substance, followed by delphinidin with about 6%, peonidin at 1.2 to 1.7%, and traces of petunidin and pelargonidin. On the other hand, Florida blood oranges contain only 3.7 mg/L of total anthocyanidins, but their profile is similar to those found in California blood oranges (Lee et al., 1990).

Flavanones are the predominant flavonoids found in citrus juices (Table 5.32); of these, hesperidin is the major flavanone glycoside found in orange juices. Other flavanone glycosides identified in citrus fruits include narirutin and didymin (Mouly et al., 1997). Hesperidin is tasteless and therefore does not contribute to the taste of orange juice (Horowitz and Gentili, 1977); its content in orange juice is 350 to 7000 mg/L (Mears and Shenton, 1973). However, much lower amounts of hesperidin have been found in 14 common sweet oranges (Rouseff et al., 1987). Depending on season, California blood oranges contain from 457 to 657 mg of hesperidin per liter of juice (Lee et al., 1990). Narirutin is the other major flavanone glycoside found in orange juices; Rouseff et al. (1987) reported that Florida citrus cultivars contain 18 to 65 mg/L and California blood oranges have 108 to 180 mg/L of narirutin (Lee et al., 1990).

Small quantities of fully methoxylated flavones have also been found in orange juices. According to data reported by Veldhuis et al. (1970), the total content of fully methoxylated flavones in orange juice is less than 7 mg/L. Nobiletin, sinensetin, heptamethoxyflavone, tetra-o-methylscutellarein and tangeretin have been identified in blood orange juices, with nobiletin and heptamethoxyflavone the predominant substances. Commercial orange juices contain only 0.5 mg/L of tangeretin juice, a major fully methoxylated flavone found in tangerine juice (Veldhuis et al., 1970).

Orange juices have a tendency to form cloud upon storage. It has been demonstrated that flavanones are incorporated into the cloud because of their limited solubility in acidic solutions (Table 5.33). The extent of flavanone precipitation is affected by the citrus fruit cultivar, pressure used during juice extraction, holding time before pasteurization, and pasteurization conditions (Baker and Cameron, 1999; Cameron et al., 1997). Recently, Gil-Izquierdo et al. (2001) demonstrated that thermal treatment of citrus juice has a limited effect on precipitation of flavanones.

PEAR JUICE

Pear juice concentrate is used in the production of wine, vinegar and beverages, and as a syrup base for fruit canning. During storage, however, it may become dark brown, which reduces it versatility and commercial value. Discoloration of pear juice is predominantly due to Maillard reactions. Cornwell and Wrolstad (1981) demonstrated that pear juice concentrates with phenolics removed brown at the same rate as those where phenolics are not removed. Thus, oxidation of phenolics does not appear to be a major cause of browning in pear juice. Nonetheless, Anderson (1970) reported the occurrence of oxidation and polymerization of phenolics during storage of fruit juices. Probably the pH of pear juice (pH = 4.2) does not favor oxidation of phenolics that occur at above pH 6 (Gregory and Bendall, 1966). On the other hand, the amber-to-brown color developed during the processing is ascribed to the action of polyphenol oxidase (Sioud and Luh, 1966). Considerable loss of cinnamic acids and total disappearance of procyanidins occur in pear juices processed without addition of SO_2. Arbutin and flavonol glycosides are less affected by SO_2 and processing, but storing juice concentrates for 9 months at 25°C results in extensive degradation of cinnamics and flavonol glycosides and total loss of procyanidins (Spanos and Wrolstad, 1990).

Blending pear juice with anthocyanin-pigmented juices may create a greater market for this type of product, but may also bring about the development of haze and color instability. Anthocyanins degrade during storage and may subsequently condense and polymerize to produce polymeric pigments and sediments during storage. At 25°C, the turbidity and polymeric pigments of juices increase, while anthocyanin levels decrease, during 48 weeks of storing pear juices blended with 5, 10 and 20% of pigmented juices such as Concord grape and black raspberry juice. The changes are more pronounced in blends containing a larger amount of pigmented juice (Spayd et al., 1984).

STRAWBERRY JUICE

Using strawberry juice concentrate as a natural flavorant in food and beverages is still hindered by its susceptibility to color and flavor degradation after a short period of storage at room temperature or longer periods at refrigeration temperatures. Astringent taste, musty or moldy and pungent aromas, and brown colors develop rapidly after 6 days of storage at 20°C (Lundahl et al., 1989). The content of anthocyanins decreases while the level of brown-red pigments increases as the browning process progresses (Wrolstad et al., 1980). The rate of color degradation increases with reduced water activity (Erlandson and Wrolstad, 1972), increased storage temperature (Hassanein, 1982), and lowered concentration of antioxidants in the system (Poei-Langston and Wrolstad, 1981). Phenolics present in the juice may also contribute to nonenzymatic and enzymatic browning reactions. Oxidation of ascorbic acid through a free radical chain reaction produces hydrogen peroxide that, in turn, may react with anthocyanins to form undesirable breakdown products (Sondheimer and Kertesz, 1953; Starr and Francis, 1968).

The flavonols that always accompany anthocyanins in fruits have a protective effect on anthocyanins in fruits by retarding the oxidation of ascorbic acid. Quercetin and quercitin in concentrations of 30 to 60 and 90 mg/L, respectively, retard the oxidation of ascorbic acid by an average of 20% over 72 h. On the other hand, anthocyanins in the absence of flavonols accelerate oxidation of ascorbic acid by an average of 34% over a pure solution of ascorbic acid without anthocyanins (Shrikhande and Francis, 1974). Color degradation can also be accelerated in strawberry juice by polyphenol oxidase. This reaction involves orthoquinones, ascorbic acid and anthocaynins (Wrolstad et al., 1980). However, complexation of anthocyanins with proteins, modified by oxidase or peroxidase, cannot be excluded (Matheis and Whitaker, 1984).

The increase in astringency of strawberry juice can be detected after only 1 day of storage at 20°C (Lundahl et al., 1989). This may be due to polymerization of anthocyanins and other phenolics because this taste attribute has been associated with polymerization of phenolics (Bate-Smith, 1973; Hagerman and Butler, 1981; Lea and Arnold, 1978; Lea and Timberlake, 1974). The formation of protein–phenolic complexes that can elicit an astringent taste cannot be ruled out (Lea and Timberlake, 1974; Lundahl et al., 1989; Matheis and Whitaker, 1984).

Kiwifruit Juice

The total content of phenolics in kiwifruit is much lower than that found in other fruits. One month after polination the fruit contains from 1800 to 2200 mg of total phenolics per kilogram; this level declines by 50% during fruit growth (Fuke and Matsuoka, 1984). Kiwifruit juice concentrate contains only 380 to 440 mg of total phenolics per liter (Wong and Stanton, 1993), but in clarified juice the total content of phenolics is between <1 and 7 mg/L. Webby (1990) identified glucosides, rhamnosides and rutinosides of kaempferol and quercetin. Later, Dawes and Keene (1999) found chlorogenic and protocatechuic acids, derivatives of coumaric and caffeic acids, a derivative of 3,4-dihydroxybenzoic acid, catechin, epicatechin, and procyanidins B2, B3, B4 and oligomers in kiwifruit juices.

REFERENCES

Albach, R.F. and Redman, G.H. 1969. Composition and inheritance of flavanones in citrus fruit. *Phytochemistry*, 8:127–43.

Anderson, H. 1970. Process for improving the stability of fruit juices during storage. South African Patent #69–7002.

Apgar, J.L. and Tarka, S.M. 1998. Methylxanthine composition and consumption patterns of cocoa and chocolate products, in *Caffeine*, Spiller, G.A., Ed., CRC Press, Boca Raton, FL, 162–192.

Arts, I.,C.W., van de Putte, B., and Hollman, P.C.H. 2000. Catechin contents in foods commonly consumed in Netherlands. 2. Tea, wine, fruit juices, and chocolate milk. *J. Agric. Food Chem.*, 48:1752–1757.

Auw, J.M., Blanco, V., O'Keefe, S.F., and Sims, C.A. 1996. Effect of processing on the phenolics and color of Cabernet Sauvignon, Chambourcin and Noble wines and juices. *Am. J. Enol. Vitic.*, 47:279–286.

Baderschneider, B. and Winterhalter, P. 2001. Isolation and characterization of novel benzoates, cinnamates, flavonoids and lignans from Riesling wine and screening for antioxidant activity. *J. Agric. Food Chem.*, 49:2788–2798.

Bailey, R., Nursten, H., and Macdowell, I. 1992. Isolation and analysis of polymeric thearubigin fraction from tea. *J. Sci. Food Agric.*, 59:365–375.

Baker, R.A. and Cameron, R.G. 1999. Clouds of citrus fruits and juice drink. *Food Technol.*, 53(1):64–69.

Bakker, J., Picinelli, A., and Bridle, P. 1993. Model wine solutions: colour and composition changes during aging. *Vitis*, 32:111–118.

Balentine, D.A., Wiseman, S.A., and Bouwens, L.C.M. 1997. The chemistry of tea flavonoids. *CRC Crit. Rev. Food Sci. Nutr.*, 37:693–704.

Balentine, D. 1992. Manufacturing and chemistry of tea, in *Phenolic Compounds in Food and Their Effects on Health,* Ho, C.-T., Lee, C.Y., and Huang, M.T., Eds., American Chemical Society, Washington, D.C., 102–117.

Baltenweck-Guyot, R., Trendel, J.-M., and Albrecht, P. 2000. Glycosides and phenylpropanoid glycerol in *Vitis vinifera* cv. Gewurztraminer wine. *J. Agric. Food Chem.*, 48:6178–6182.

Bamforth, C.W. 1999. Beer haze. *J. Am. Soc. Brew. Chem.*, 57:81–90.

Bartolome, B., Pena-Neira, A., and Gomez-Cordoves, C. 2000. Phenolics and related substances in alcohol-free beer. *Eur. Food Res. Technol.*, 210:419–423.

Bate-Smith, E.D. 1973. Haemanalysis of tannins — the concept of relative astringency. *Phytochemistry*, 12:907–912.

Bellmer, H.-G. 1977. Polyphenole und Alterung des Bieres. *Brauwelt*, 117:660–669.

Bellmer, H.-G. 1981. Hopfenpolyphenole im Brauprozeb. *Brauwelt*, 121:240–245.

Benn, S.M. and Peppard, T.L. 1996. Characterization of tequila flavor by instrumental and sensory analysis. *J. Agric. Food Chem.*, 44:557–566.

Betes-Saura, C., Andres-Lacueva, C., and Lamuela-Raventos, R.M. 1996. Phenolics in white free run juices and wines from Penedes by high-performance liquid chromatography: changes during vinification. *J. Agric. Food Chem.*, 44:3040–3046.

Biehl, B., Voight, J., Henrichs, H., Senjuk, V., and Bytof, G. 1993. pH-dependent enzymatic formation of oligopeptides and amino acids, the aroma precursors in raw cocoa beans, in *Proceedings of XIth International Cocoa Research Conference,* Lafforest, J., Ed., Cocoa Producers' Alliance, Yamassoukro, Ivory Coast, 717–722.

Biehl, B. and Adomako, D. 1983. Fermentation of cocoa (control, acidification, proteolysis). *Lebesmittelchem. Gerichtl. Chem.*, 37:57–63.

Bohm, B.A. 1989. Chalcones and aurones, in *Methods in Plant Biochemistry*, vol.1, *Plant Phenolics,* Harborne, J.B., Ed., Academic Press, London, 237–281.

Bonvehi, S.J. and Coll, V.F. 1997a. Parameters affecting the quality of processed cocoa powder: acidity fraction. *Z. Lebensm. Unters. Forsch.* A., 204:287–292.

Bonvehi, S.J. and Coll, V.F. 1997b. Evaluation of bitterness and astringency of polyphenolic compounds in cocoa powder. *Food Chem.*, 60:365–370.

Bonvehi, S.J. and Coll, V.F. 1998. Evaluation of smoky taste of cocoa powder. *J. Agric. Food Chem.*, 46:620–624.

Bourzeix, M. 1971. La diffusion des composes phenoliques au cours de la fermentation intracellularie. *C.R. J. Maceration Carbonique*, February 10–11, 47.

Bracco, U., Grailhe, N., Rostagno, W., and Egli, R.H. 1969. Analytical evaluation of cocoa curing in the Ivory Coast. *J. Sci. Food Agric.*, 20:713–717.

Bracke, M.E., Bruyneel, E.A., Vermeulen, S.E., Vennekens, K., Van Marck, V., and Mareel, M.M. 1994. Citrus flavonoid effect on tumor invasion and metastasis. *Food Technol.*, 11:121–124.

Bressani, R. and Graham, J.E. 1980. Utilization of coffee pulp as animal feed, in *Neuvieme Colloque Scientifique International sur le Cafe*. ASIC, Paris, 303–323.

Brown, A.G., Eyton, W.B., Holmes, A., and Ollis, W.D. 1969. Identification of thearubigins as polymeric proanthocyanidins. *Nature*, 221:742–745.

Burns, J., Yokota, T., Ashihara, H., Lean, M.E.J., and Crozier, A. 2002. Plant foods and herbal sources of resveratrol. *J. Agric. Food Chem.*, 50:3337–3340.

Burns, J., Gardner, P.T., Matthews, D., Duthie, G.G., Lean, M.E.J., and Crozier, A. 2001. Extraction of phenolics and changes in antioxidant activity of red wine during vinification. *J. Agric. Food Chem.*, 49:5797–5808.

Cabranes, C., Moreno, J., and Mangas, J.J. 1990. Dynamics of yeast populations during cider fermentation in the Asturian region of Spain. *Appl. Environ. Microbiol.*, 56:3881–3884.

Cameron, R.G., Baker, R.A., and Grohman, K. 1997. Citrus tissue extracts affect juice cloud stability. *J. Food Sci.*, 62:242–245.

Carando, S., Teissedre, P.-L., Pascual-Martinez, L., and Cabanis, J.-C. 1999. Levels of flavan-3-ols in French wines. *J. Agric. Food Chem.*, 47:4161–4166.

Charalambous, G. 1974. Progress report of flavor influence and determination of beer polyphenols. *MBAA Tech. Q.*, 11:146–147.

Charley, H. 1982. *Food Science*, 2nd ed., John Wiley & Sons, New York.

Chatonnet, P., Dubourdieu, D., Boidron, J.N., and Lavigne, V. Synthesis of volatile phenols by *Saccharomyces cerevisiae* in wines. *J. Sci. Food Agric.*, 62:191–202.

Chatt, E.M. 1953. *Cocoa: Cultivation, Processing, Analysis,* Interscience Publishing Co., New York.

Chen, Z.-Y., Zhu, Q.Y., Wong, Y.F., Zhang, Z., and Chung, H.Y. 1998. Stabilizing effect of ascorbic acid on green tea catechins. *J. Agric. Food Chem.*, 46:2512–2516.

Chen, Z.-Y., Zhu, Q.Y., Tsang, D., and Huang, Y. 2001. Degradation of green tea catechins in tea drinks. *J. Agric. Food Chem.*, 49:477–482.

Cheynier, V., Trousdale, E., Singleton, V.L., Salgues, M., and Wylde, R. 1986. Characterization of 2-S-glutathionyl caftaric acid and its hydrolysis in relation to grape wines. *J. Agric. Food Chem.*, 34:217–221.

Cheynier, V., Osse, C., and Rigaud, J. 1988. Oxidation of grape juice phenolic compounds in model solutions. *J. Food Sci.*, 53:1729–1732.

Cheynier, V., Basire, N., and Rigaud, J. 1989. Mechanism of *trans*-caffeoyltartaric and catechin oxidation in solutions containing grape polyphenoloxidase. *J. Agric. Food Chem.*, 37:1069–1071.

Cilliers, J.J.L. and Singleton, V.L. 1989. Nonenzymic autooxidative phenolic browning reactions in a caffeic acid model system. *J. Agric. Food Chem.*, 37:890–896.

Cilliers, J.J.L., Singleton, V.L., and Lamuela-Raventos, R.M. 1990. Total polyphenols in apple and ciders; correlation with chlorogenic acid. *J. Food Sci.*, 55:1458–1459.

Claesson, S. and Sandegren, E. 1969. Physical aspects of haze problems, in *Proc. Cong. Eur. Brew. Conv.*, 339–347.

Clark, A.G. 1960. Nonbiological haze in bottled beers: a survey of the literature. *J. Inst. Brewing*, 66:318–330.

Clifford, M.N. and Wight, J. 1976. The measurement of feruloylquinic acids and caffeoylquinic acids in coffee beans. Development of the technique and its preliminary application to green coffee beans. *J. Sci. Food Agric.*, 27:73–84.
Clifford, M.N. and Ohiokpehai, O. 1983. Coffee astringency. *Anal. Proc.* (London), 20:83–86.
Clifford, M.N. 1985. Chemical and physical aspects of green coffee and coffee products, in *Coffee Botany, Biochemistry and Production of Beans and Beverage,* Clifford, M.N. and Wilson, K.C., Eds., AVI Publishing Co., Westport, CT, 305–374.
Clifford, M.N. and Kazi, T. 1987. The influence of coffee bean maturity on the content of chlorogenic acids, caffeine and trigonelline. *Food Chem.*, 26:59–69.
Clifford, M.N. and Ramirez-Martinez, J.R. 1991. Phenols and caffeine in wet-precessed coffee beans and coffee pulp. *Food Chem.*, 40:35–42.
Clifford, M.N. 1999. Chlorogenic acid and other cinnamates — nature, occurrence and dietary burden. *J. Sci. Food Agric.*, 79:362–372.
Cloughley, J.B. 1980. The effect of fermentation temperature on the quality parameters and price evaluation of Central African black teas. *J. Sci. Food Agric.*, 31:911–919.
Cornwell, C.J. and Wrolstad, R.E. 1981. Causes of browning in pear juice concentrate during storage. *J. Food Sci.*, 46:515–518.
Cros, E., Chanliau, S., and Jeanjean, N. 1999. Postharvest processing: a key step in cocoa, in *Confectionary Science II. Proceedings of International Symposium,* Ziegler, G.R., Ed., Pennsylvania State University, State College, PA, 80–95.
Czerny, M., Mayer, F., and Grosch, W. 1999. Sensory study on the character impact odorants in roasted Arabica coffee. *J. Agric. Food Chem.*, 47:695–699.
Czerny, M. and Grosch, W. 2000. Potent odorants in raw arabica coffee. Their changes during roasting. *J. Agric. Food Chem.*, 48:868–872.
Dadic, M. and Belleau, G. 1980. Beer hazes. I. Isolation and preliminary analysis of phenolic and carbohydrate components. *ASBC. J.*, 38:154–158.
Dadic, M. 1976. Current concepts on polyphenols and beer stability. *MBAA Tech. Q.*, 13:182–189.
Dadic, M. and Belleau, G. 1973. Polyphenols and beef flavor. *Proc. Am. Soc. Brew. Chem.*, 107–114.
Dadic, M., Van Gheluwe, J.E.A., and Valyi, Z. 1974. Formation of volatile carbonyls through the action of phenolics. *MBAA Tech. Q.*, 11:164–165.
Dadic, M. 1974a. Role of phenolic compounds in aging of beer. *Brewer's Dig.*, 49:58–70; 101.
Dadic, M. 1974b. Phenolic and beer staling. *MBAA Tech. Q.*, 11:140–145.
Dadic, M., Van Gheluwe, J.E.A., and Valyi, Z. 1971. Chemical constituents of nylon-66 beer adsorbate. II. Thin-layer chromatography of acidic and alkanoline hydrolysis products. *J. Inst. Brewing*, 77:48–56.
Dadic, M., Van Gheluwe, J.E.A., and Valyi, Z. 1970. Chemical constituents of nylon-66 beer adsorbate. I. Preliminary investigations and anthocyanogen contents. *J. Inst. Brewing*, 76:267–280.
Dallas, C. and Laureno, O. 1994a. Effects of SO_2 on the extraction of individual anthcyanins and colored matters of three Portuguese grape varieties during wine making, *Vitis*, 33:41–47.
Dallas, C. and Laureno, O. 1994b. Effects of pH, sulphur dioxide, alcohol content, temperature and storage time on colour composition of young Portuguese red table wine. *J. Sci. Food Agric.*, 65:477–485.
Dallas, C., Ricardo-da-Silva, J.M., and Laureano, O. 1996. Products formed in model wine solutions involving anthocyanins, procyanidin B2, and acetaldehyde. *J. Agric. Food Chem.*, 44:2402–2407.

Dawes, H.M. and Keene, J.B. 1999. Phenolic composition of kiwifruit juice. *J. Agric. Food Chem.*, 47:2398–2403.

de Freitas, V.A.P., Glories, Y., and Laguerre, M. 1998. Incidence of molecular structure in oxidation of grape seed procyanidins. *J. Agric. Food Chem.*, 46:376–382.

Delcour, J.A., Schoetters, M.M., Meysman, E.W., Dondeyne, P., Schrevens, E.L., and Wijnhoven, J. 1985. Flavour and haze stability differences due to hope tannins in all-malt Pilsner beers brewed with proanthocyanidin-free malt. *J. Inst. Brew.*, 91:882.

Delcour, J.A., Verdeyen, J.P., and Dondeyne, P. 1983. The tanning of beer proteins by polyphenols. *Cerevisia*, 8:21–27.

Delgado, T. and Gomez-Cordoves, C. 1987. Content of phenolic acids and aldehydes in Spanish commercial brandies. *Rev. Fr. Oenol.*, 107:39–43.

Drawert, F., Leupold, G., and Lessing, V. 1977. Gaschromatographische Analyse von phenolischen Verbindugen im Bier. *Brauwissenschaft*, 30:13–18.

Dubernet, M. and Ribereau-Gayon, P. 1973. Les polyphenol oxydases du raisin sain et du raisin parasite par *Botrytis cinerea. C.R. Acad. Sci. Paris D.*, 275.

Dugelay, I., Gunata, Z., Sapis, J.C., Baumes, R., and Bayonove, C. 1993. Role of cinnamoyl esterase from enzyme preparations on the formation of volatile phenols during wine-making. *J. Agric. Food Chem.*, 41:2092–2096.

Edelmann, A., Diewok, J., Schuster, K.C., and Lendl, B. 2001. Rapid method for the discrimination of red wine cultivars based on mid-infrared spectroscopy of phenolic wine extracts. *J. Agric. Food Chem.*, 49:1139–1145.

Erlandson, J.A. and Wrolstad, R.E. 1972. Degradation of anthocyanins at limited water concentration. *J. Food Sci.*, 37:592–595.

Escribano-Bailón, T., Alvarez-Garcia, M., Rivas-Gonzalo, J.C., Heredia, F.J., and Santos-Buelga, C. 2001. Color and stability of pigments derived from the acetaldehyde-mediated condensation between malvidin 3-*O*-glucoside and (+) catechin. *J. Agric. Food Chem.*, 49:1213–1217.

Escribano-Bailón, T., Gutiérrez-Fernandez, Y., Rivas-Gonzalo, J.C., and Santos-Buelga, C. 1992. Characterization of procyanidins of *Vitis vinifera* variety Tinta del Pais seeds. *J. Agric. Food Chem.*, 40:1794–1799.

Es-Safi, N.-E., Fulcrand, H., Cheynier, V., and Moutounet, M. 1999. Competition between (+) catechin and (–) epicatechin in acetaldehyde induced polymerization of flavanols. *J. Agric. Food Chem.*, 47:2088–2095.

Etievant, P.X. 1981. Volatile phenol determination in wine. *J. Agric. Food Chem.*, 29:65–67.

Fantozzi, P., Montanari, L., Mancini, F., Gasbarrini, A., Simoncini, M., Nardini, M., Ghiselli, A., and Scaccini, C. 1998. *In vitro* antioxidant capacity from wort to beer. *Lebensm.-Wiss. u-Technol.*, 31:221–227.

Fernandez-Zurbano, P., Ferreira, V., Pena, C., Escudero, A., Serrano, F., and Cacho, J. 1995. Prediction of oxidative browning in white wines as a function of their chemical composition. *J. Agric. Food Chem.*, 43:2813–2817.

Fernandez-Zurbano, P., Ferreira, V., Escudero, A., and Cacho, J. 1998. Role of hydroxy-cinnamic acids and flavanols in the oxidation and browning of white wines. *J. Agric. Food Chem.*, 46:4937–4944.

Figueiredo, P., Elhabiri, M., Toki, K., Saito, N., Dangles, O., and Broullaird, R. 1996. New aspects of anthocyanins complexation. Intramolecular copigmentation as a means of colour loss? *Phytochemistry* 41:301–308.

Finger, A., Engelhardt, U.H., and Wray, V. 1991. Flavonol glycosides in tea-kaempferol and quercetin rhamnodiglucosides. *J. Sci. Food Agric.*, 55:313–321.

Flament, I. 1989. Coffee, cocoa, and tea. *Food Rev. Int.*, 5(3):317–414.

Flament, I. 1995. Coffee, cacao, and tea, in *Volatile Compounds in Foods and Beverages,* Maarse, H., Ed., Marcel Dekker, New York, 617–669.

Forsyth, W.G.C. 1955. Cacao polyphenolic substances. 3. Separation and estimation on paper chromatograms. *Biochem. J.*, 60:1108.

Forsyth, W.G.C. 1963. The mechanism of cacao curing. *Adv. Enz.*, 25:457–492.

Forsyth, W.G.C. 1952. Cacao polyphenolic substances. 2. Changes during fermentation. *Biochem. J.,* 51:516–520.

Forsyth, W.G.C. and Quesnel, V.C. 1957. Cacao polyphenolic substances. 4. The anthocyanin pigments. *Biochem. J.*, 65:177–179.

Francia-Aricha, E.M., Guerra, M.T., Rivas-Gonzalo, J.C., and Santos-Buelga, C. 1997. New anthocyanin pigments formed after condensation with flavanols. *J. Agric. Food Chem.*, 45:2262–2266.

Fuke, Y. and Matsuoka, H. 1984. Changes in content of pectic substances, ascorbic acid and polyphenols, and activity of pectinesterase in kiwifruit during growth and ripening after harvest. *J. Jpn. Soc. Food Sci. Technol.*, 31:31–37.

Fulcrand, H., Benabdeljalil, C., Rigaud, J., Cheynier, V., and Moutounet, M. 1998. A new class of pigments generated by reaction between pyruvic acid and grape anthocyanins. *Phytochemistry*, 47:1401–1407.

Galati, E.M., Monforte, M.T., Kirjavainen, S., Forestieri, A.M., and Trovato, A. 1994. Biological effects of hesperedin, a citrus flavonoid (note I): anti-inflammatory and analgesic activity. *Farmaco*, 49:709–712.

Galati, E.M., Trovato, A., Kiajavainen, S., Foriesteri, A.M., Rossitto, A., and Monforte, M.T. 1996. Biological effects of hesperidin, a citrus flavonoid (note III): anthypertensive and diuretic activity in rat. *Farmaco*, 51:219–221.

Gao, L., Girarrd, B., Mazza, G., and Reynolds, A.G. 1997. Changes in anthocyanins and color characteristics of Pinot Noir wines during different vinification processes. *J. Agric. Food Chem.*, 45:2003–2008.

Garcia-Viguera, C., Bridle, P., and Bakker, J. 1994. The effect of pH on the formation of coloured compounds in model solutions containing anthocyanins, catechins and acetaldehyde. *Vitis*, 33:37–40.

Garcia-Viguera, C., Bakker, S., Bellworthy, S.J., Reader, H.P., Watkins, S.J., and Bridle, P. 1997. The effect of some processing variables on noncoloured phenolic compounds in port wines. *Z. Lebensm. Unters. Forsch A.*, 205:321–324.

Gardner, J.W. 1967. Phenolic constituents of beer and brewing materials. I. Phenolic and nitrogenous components removed from beer by polyamide resins. *J. Inst. Brew.*, 73:258–270.

Gardner, R.J. and McGuiness, J.D. 1977. Complex phenols in brewing — a survey. *Tech. Quant. Master Brew Assoc. Am.*, 14:250–251.

Gil-Izquierdo, A., Gil, M.I., Ferreres, F., and Tomas-Barberan, F.A. 2001. *In vitro* availability of flavonoids and other phenolics in orange juice. *J. Agric. Food Chem.*, 49:1035–1041.

Gil-Munoz, R., Gomez-Plaza, E., Martinez, A., and Lopez-Roca, J.M. 1999. Evolution of phenolic compounds during wine fermentation and postfermentation: influence of grape temperature. *J. Food Compos. Anal.*, 12:259–272.

Goldberg, D.M., Hoffman, B., Yang, J., and Soleas, G.J. 1999. Phenolic constituents, furans and total antioxidant status of distilled spirits. *J. Agric. Food Chem.*, 47:3978–3985.

Gomez-Plaza, E., Gil-Munoz, R., Lopez-Roca, J.M., and Martinez, A. 2000. Color and phenolic compounds of young red wine. Influence of wine-making techniques, storage temperature, and length of storage time. *J. Agric. Food Chem.*, 48:736–741.

Graham, H. 1983. Tea, in *Kirk-OthmerEncyclopedia of Chemical Technology*, vol. 22, Grayson, M. and Eckroth, D., Eds., John Wiley & Sons, New York, 628–44.

Gramshaw, J.W. 1968. Phenolic constituents of beer and beer materials. III. Simple anthocyanogens from beer. *J. Inst. Brew.*, 74:20–38.

Gramshaw, J.W. 1970. Beer polyphenols and chemical basis of haze formation. III. The polymerization of polyphenols and their reactions in beer. *MBAA Tech. Q.*, 7:167–182.

Gregory, R.P.F. and Bendall, D.S. 1966. The purification and some properties of polyphenol oxidase from tea (*Camellia sinensis* L.). *Biochem. J.*, 100:569–581.

Griffiths, L.A. 1957. Detection of the substrate of enzymic browning in cacao by a post-chromatographic enzymatic techniques. *Nature* (London), 180:1373–1374.

Guerrero, G., Suarez, M., and Moreno, G. 2001. Chlorogenic acids as a potential criterion in coffee genotype selections. *J. Agric. Food Chem.*, 49:2454–2458.

Gunata, Y.Z., Sapis, J.C., and Moutounet, M. 1987. Substrates and aromatic carboxylic acid inhibitors of grape polyphenoloxidases. *Phytochemistry*, 26:1573–1575.

Guo, Q., Zha, B., Li, M., Shen, S., and Xin, W. 1996. Studies on protective mechanisms of four components of green tea polyphenols against lipid peroxidation in synapstomes. *Biochim. Biophys. Acta*, 1304:210–222.

Guymon, J.F. and Crowell, E.A. 1968. Separation of vanillin, syringaldehyde, and other aromatic compounds in the extracts of French and American oak woods by brandy and aqueous alcohol solutions. *Qual. Plant. Mater. Veg.*, 16:320–333.

Hagerman, A.E. and Butler, L.G. 1981. The specifity of proanthocyanin-protein interaction. *J. Biol. Chem.*, 256:4494–4497.

Hamboyan, L. and Pink, D. 1990. Ultraviolet spectroscopic studies and the prediction of the feathering of cream in filter coffees. *J. Dairy Sci.*, 57:227–232.

Hamboyan, L., Pink, D., Klapstein, D., MacDonald, L., and Aboud, H. 1989. Ultraviolet spectroscopic studies on the feathering of cream in instant coffee. *J. Dairy Sci.*, 56:1197–1202.

Hammerstone, J.F., Lazarus, S.A., Mitchell, A.E., Rucker, R., and Schmitz, H.H. 1999. Identification of procyanidins in cocoa (*Theobroma caco)* and chocolate using high-performance liquid chromatography/mass spectrophotometry. *J. Agric. Food Chem.*, 47:490–96.

Harbowy, M. and Balentine, D.A. 1997. Tea chemistry. *CRC Crit. Rev. Plant. Sci.*, 16:415–480.

Hartnell, C. 1987. Polyphenols and grape processing: the key to better white wine. *Aust. Grapegrower Winemaker*, April, 13–15.

Hashimoto, F., Nonaka, G., and Nishioka, I. 1988. Tannins and related compounds. LXXIX. Isolation and structure elucidation of B,B′-linked bisflavanoids, theasinensis D-G and oolong theanin from oolong tea. *Chem. Pharm. Bull.*, 36:1676–1684.

Haslam, E. 1974. Polyphenol–protein interactions. *Biochem. J.*, 139:285–288.

Haslam, E. and Lilley, T.H. 1988. Natural astringency in foodstuffs. A molecular interpretation. *CRC Crit. Rev. Food Sci. Nutr.*, 27:1–40.

Haslam, E. 1989. *Plant Polyphenols*, Cambridge University Press, Cambridge, U.K.

Hassanein, S.M. 1982. Color of strawberry juice concentrate as influenced by heating and storage temperature. M.S. thesis, Oregon State University, Corvallis, OR. Cited in Lundahl, D.S., McDaniel, M.R., and Wrolstad, R.E. 1989. Flavor, aroma, and compositional changes in strawberry juice concentrate stored at 20°C. *J. Food Sci.*, 54:1255–1258.

Hilton, P.J. and Ellis, R.T. 1972. Estimation of market value of Central African tea by theaflavin analysis. *J. Sci. Food Agric.*, 23:227–232.

Hor, Y.C., Chin, H.F., and Karim, M.Z. 1984. The effect of seed moisture and storage temperature of storability of cocoa (*Theobroma cacao*) seeds. *Seed Sci. Technol.*, 12:415–420.

Hulme, A.C. and Rhodes, M.J.C. 1971. Pome fruits, in *Biochemistry of Fruits and Their Products,* vol. 2, Hulme, A.C., Ed., Academic Press, New York, 397.

Jaendet, P., Bessis, R., Maume, B.F., and Sbaghi, M. 1993. Analysis of resveratrol in Burgundy wines. *J. Wine Res.*, 4:79–85.

Jalal, M.A.F. and Collin, H.A. 1977. Polyphenols of mature plant, seedling and tissue cultures of *Theobroma cacao. Phytochemistry*, 17:1377–1380.

Jinap, S. 1994. Organic acids in cocoa beans — a review. *ASEAN Food J.,* 9:3–12.

Jindra, J.A. and Gallander, J.F. 1987. Effect of American and French oak barrels on the phenolic composition and sensory quality of Seyval Blanc wines. *Am. J. Enol. Vitic.*, 38:133–138.

Kahn, J.H. 1969. Compounds identified in whisky, wine and beer: a tabulation. *J. Am. Oil Chem. Soc.*, 52:1166–1178.

Karagiannis, S., Economou, A., and Lanaridis, P. 2000. Phenolic and volatile composition of wines made from *Vitis vinifera* cv. Muscat Lefko grapes from island of Samos. *J. Agric Food Chem.*, 48:5369–5375.

Kawaii, S., Tomono, Y., Katase, E., Ogawa, K., and Yano, M. 1999. HL-60 differentiating activity and flavonoid content of the readily extractable fraction prepared from citrus juices. *J. Agric. Food Chem.*, 47:128–135.

Kharlamova, O.A. 1964. The determination of catechols in cocoa beans. *Klebopekar. Konditer. Prom.*, 3:16–318 (Fr), cited in Horowitz, R.M. and Gentili, B. 1977. Flavonoid constituents of citrus, in *Citrus Science and Technology*, Nagy, S., Shaw, P.E., and Veldhuis, M.K., Eds., AVI Publishing Co., Westport, CT, 397–426.

Kieninger, H., Boeck, D., and Schwankl, M. 1977. Über das Auftreten von Flüchtigen Phenolen beim Darren von Malzen, in *Proc. 16th Eur. Brew. Conv.,* Amsterdam, The Netherlands, 129–137.

Kim, H. and Keeney, P.G. 1984. (–)-Epicatechin content in fermented and unfermented cocoa beans. *J. Food Sci.*, 49:1090–1092.

Kirby, W. and Wheeler, R.E. 1980. Extraction of beer polyphenols and their assay by HPLC. *J. Inst. Brew.*, 86:15–17.

Kirby, W., Willias, P.M., Wheeler, R.E., and Jones, M. 1977. The importance of polyphenols of barleys and malts in beer sediment and haze formation, in *Proceeding of the XVIth Eur. Brewery Congr.*, Amsterdam, The Netherlands, 415–427.

Knapp, A.W. 1937. *Cocoa Fermentation.* John Ball, Sons and Curnow Ltd, London, 124.

Kuusi, T. and Pajunen, E. 1971. Less grown apple varieties in juice production and the influence of polyphenols and added ascorbic acid on the juice quality. *J. Sci. Agric. Soc.* (Finland), 43:20.

Lakenbrink, C., Engelhardt, U.H., and Wray, V. 1999. Identification of two novel proanthocyanidins in green tea. *J. Agric. Food Chem.*, 47:4621–4624.

Lakenbrink, C., Lapczynski, S., Mailwald, B., and Engelhardt, U.H. 2000. Flavonoids and other polyphenols in consumer brews of tea and other caffeinated beverages. *J. Agric. Food Agric. Chem.*, 48:2848–2852.

Lamuela-Raventós, R.M. and Waterhouse, A.L. 1993. Occurrence of resveratrol in selected Californian wines by new HPLC method. *J. Agric. Food Chem.*, 41:521–523.

Landrault, N., Poucheret, P., Ravel, P., Gasc, F., Cros, G., and Teissedre, P.-L. 2001. Antioxidant capacities and phenolic levels of French wines from different varieties and vintages. *J. Agric. Food Chem.*, 49:3341–3348.

Lea, A.G.H., Bridle, P., Timberlake, C.F., and Singleton, V. 1979. The procyanidins of white grapes and wines. *Am. J. Enol. Vitic.*, 30:289–300.

Lea, A.G.H. and Timberlake, C.F. 1978. The phenolic of ciders: effect of processing conditions. *J. Sci. Food Agric.*, 29:484–492.

Lea, A.G.H. 1978. The phenolics of ciders: oligomeric and polymeric procyanidins. *J. Sci. Food Agric.*, 29:471–477.

Lea, A.G.H. and Timberlake, C.F. 1974. The phenolic of ciders. I. Procyanidins. *J. Sci. Food Agric.*, 25:1537–1545.

Lea, A.G.H. and Arnold, G.M. 1978. The phenolics of ciders: bitterness and astringency. *J. Sci. Food Agric.*, 29:478–483.

Lea, A.G.H. and Beech, F.W. 1978. The phenolics of cider: effects of cultural conditions. *J. Sci. Food Agric.*, 29:493–496.

Lee, C.Y. and Jaworski, A.W. 1987. Phenolic compounds in white grape grown in New York. *Am. J. Enol. Vitic.*, 38:277–281.

Lee, H.S., Carter, R.D., Barros, S.M., Dezman, D.J., and Castle, W.S. 1990. Chemical characterization by liquid chromatography of Moro blood orange juices. *J. Food Compos. Anal.*, 3:9–19.

Lee, H.S. and Wrolstad, R.E. 1988. Apple juice composition: sugar, nonvolatile acid, and phenolic profiles. *J. Assoc. Off. Anal. Chem.*, 71:789–794.

Lee, C.Y. and Jaworski, A.W. 1988. Phenolics and browning potential of white grapes grown in New York. Am. *J. Enol. Vitic.*, 39:337–340.

Lembeck, H. 1983. *Grossman's Guide to Wines, Beers, and Spirits*. Scribner, New York.

Liao, H., Cai, Y., and Haslam, E. 1992. Polyphenols interactions. Anthocyanidins: copigmentation and color changes in red wine. *J. Sci. Food Agric.*, 59:299–305.

Liebert, M., Licht, U., Bohm, V., and Bitsch, R. 1999. Antioxidant properties and total phenolics content of green tea and black tea under different brewing conditions. *Z. Lebensm. Unters. Forsch. A.*, 208:217–220

Luna, F., Crouzillat, D., Cirou, L., and Buchelli, P. 2002. Chemical composition and flavor of Ecuadorian cocoa liquor. *J. Agric. Food Chem.*, 50:3527–3532.

Lundahl, D.S., McDaniel, M.R., and Wrolstad, R.E. 1989. Flavor, aroma, and compositional changes in strawberry juice concentrate stored at 20°C. *J. Food Sci.*, 54:1255–1258.

Lunte, S.M., Blankenship, K.D., and Scott, A.R. 1988. Detection and identification of procyanidins and flavanols in wine by dual-electrode liquid chromatography. *Analyst*, 113:99–102.

Maccarone, E., Maccarone, A., Perrini, G., and Rapisarda, P. 1985. Stabilization of anthocyanins of blood orange juice. *J. Food Sci.*, 50:901–904.

Macheix, J.-J., Sapis, J.-C., and Fleuriet, A. 1991. Phenolic compounds and polyphenoloxidase in relation to browning in grapes and wines. *CRC Crit. Rev. Food Sci. Nutr.*, 30:441–486.

Macheix, J.-J., Fleuriet, A., and Billot, J. 1990. *Fruit Phenolics*, CRC Press, Boca Raton, FL.

Macheix, J.-J. 1968. Phenolic compounds of apples: preliminary research for a special study of chlorogenic acid. *Fruits*, 23:13–20.

Madigan, D., McMurrough, I., and Smyth, M.R. 1994. Rapid determination of 4-vinylguaiacol and ferulic acid in beers and worts by high-performance liquid chromatography. *J. Am. Soc. Brew. Chem.*, 52:152–155.

Maga, J.A. 1984. Flavor contribution of wood in alcoholic beverages, in *Progress in Flavor Research. Proc. 4th Weurman Flavour Res. Symp.*, Dourdan, France, 409–416.

Mangas, J.J., Rodriguez, R., Moreno, J., Suarez, B., and Blanco, D. 1997. Furanic and phenolic composition of cider brandy. A chemometric study. *J. Agric. Food Chem.*, 45:4076–4079.

Mateus, N. and de Freitas, V. 2001. Evolution and stability of anthocyanin-derived pigments during Port wine aging. *J. Agric. Food Chem.*, 49:5217–5222.

Mateus, N., Silva, A.M.S., Vercauteren, J., and de Freitas, V.A.P. 2001. Occurrence of anthocyanidin-derived pigments in red wines. *J. Agric. Food Chem.*, 49:4836–4840.

Matheis, G. and Whitaker, J.R. 1984. Modification of proteins by polyphenol oxidase and peroxidase and their products. *J. Food Biochem.*, 8:137–162.

Mazza, G., Fukumoto, L., Delaquis, P., Girard, B., and Ewert, B. 1999. Anthocyanins, phenolics, and color of Cabernet Franc, Merlot, and Pinot Noir wines from British Columbia. *J. Agric. Food Chem.*, 47:4009–4017.

McDonald, M.S., Hughes, M., Burns, J., Lean, M.E.J., Matthews, D., and Crozier, A. 1998. Survey of the free and conjugated myrecitin and quercetin content in red wines of different geographical origins. *J. Agric. Food Chem.*, 46:368–375.

Mcdowell, I., Taylor, S., and Gay, C. 1995. The phenolic pigment composition of black tea liquors. Part I. Predicting quality. *J. Sci. Food Agric.*, 69:467–74.

McMurrough, I., Roche, G.P., and Cleary, K.G. 1984. Phenolic acids in beers and worts. *J. Inst. Brew.*, 90:181–187.

Mears, R.G. and Shenton, A.J. 1973. Alteration and characterization of orange and grapefruit juices. *J. Food Sci.*, 50:901–904.

Merrit, M.C. and Proctor, B.E. 1959a. Effect of temperature during the roasting cycle on selected components of different types of whole bean coffee. *Food Res.*, 24:672–680.

Merrit, M.C. and Proctor, B.E. 1959b. Extraction rates for selected components in coffee brew. *Food Res.*, 24:735–743.

Millin, D.J., Swaine, D., and Dix, P.L. 1969. Separation and classification of brown pigments of aqueous infusions of black tea. *J. Sci. Food Agric.*, 20:296–302.

Milligan, S.R., Kalita, J., Heyerick, A., Rong, H., DeCooman, L., and De Keukeleire, D. 1999. Identification of potent phytoestrogen in hops (*Humulus lupulus* L.) and beer. *J. Clin. Endocrinol. Metab.*, 84:2249–2252.

Miranda, C.L., Stevens, J.F., Ivanow, V., McCall, M., Frei, B., Deinzer, M.L., and Buhler, D.B. 2000. Antioxidant and proxidant actions of prenylated and nonprenylated chalcones and flavanones *in vitro*. *J. Agric. Food Chem.*, 48:3876–3884.

Mistry, T.V., Cai, Y., Lilley, T.H., and Haslam E. 1991. Polyphenol interactions. Part 5. Anthocyanin copigmentation. *J. Chem. Soc. Perkin Trans.*, 2:1287–1296.

Moarse, H. and ten Noever de Brauw, M.C. 1966. The analysis of volatile components of Jamaican rum. *J. Food Sci.*, 31:951–955.

Monforte, M.T., Trovato, A., Kirjavainen, S., Foriesteri, A.M., and Galati, E.M. 1995. Biological effects of hesperidin, a citrus flavonoid (note II): hypolipidemic activity on hypercholesterolemia in rat. *Farmaco*, 50:595–599.

Montanari, L., Perretti, G., Natella, F., and Guidi, A. 1999. Organic and phenolic acids in beer. *Lebensm.-Wiss..*, 32:535–539.

Mouly, P.P., Gaydou, E.M., Faure, R., and Estienne, J.M. 1997. Blood orange juice authentication using cinnamic acid derivatives. Variety differentiation associated with flavonone glycoside content. *J. Agric. Food Chem.*, 45:373–377.

Nijssen, L.M., Visscher, C.A., Maarse, H., Willemsens, L.C., and Boelens, M.H. 1996. *Volatile Compounds in Food. Qualitative and Quantitative Data.* 7th ed., TNO Nutrition and Food Research Institute, Zeist, The Netherlands, 72.1–72.23.

Nishimura, K. and Masuda, M. 1971. Minor constituents of whisky fuel oils. I. Basic, phenolic and lactonic compounds. *J. Food Sci.*, 36:819–822.

Nonaka, G., Kawahara, O., and Nishioka, I. 1983. Tannins and related compounds. XXXVI. A new class of dimeric flavan-3-ol gallates, theasinensins A and B, and proanthocyanidin gallates from green tea leaf. *Chem. Pharm. Bull.*, 31:3906–3911.

Oszmianski, J., Ramos, T., and Macheix, J.J. 1985. Changes in grape seed phenols as affected by enzymic and chemical oxidation *in vitro*. *J. Food Sci.*, 50:1505–1506.

Otsuka, K., Imai, S., and Morinega, K. 1965. Studies on the mechanism of aging of distilled liquors. Part II. Distribution of phenolic compounds in aged distilled liquors. *Agric. Biol. Chem.*, 29:27–31.

Owades, J.L., Rubin, G., and Brenner, M.W. 1958. Determination of tannin in beer and brewing materials by ultraviolet spectrophotometry. *Proc. Am. Soc. Brew. Chem.*, 66–73.

Owuor, P.O., Odhiambo, H.O., Robinson, J.M., and Taylor, S.J. 1990. Variations in the leaf standard, chemical composition and quality of black tea (*Camelia sinensis*) due to plucking intervals. *J. Sci. Food Agric.*, 52:63–69.

Ozawa, T., Kataoka, M., Morikawa, K., and Negishi, C. 1996. Elucidation of the partial structure of polymeric thearubigins from black tea by chemical degradation. *Biosci. Biotechnol. Biochem.*, 60:2023–2027.

Palamand, S.R., Markl, K.S., and Hardwick, W.A. 1971. Trace flavor compounds in beer. *Proc. Am. Soc. Brew. Chem.*, 211–218.

Panossian, A., Mamikonyan, G., Torosyan, M., Gabrielyan, E., and Mkhitaryan, S. Analysis of aromatic aldehydes in brandy and wine by high performance capillary electrophoresis. *Anal. Chem.*, 73:4379–4383.

Park, G.L., Avery, S.M., Byers, J.L., and Nelson, D.B. 1983. Identification of bioflavonoids from citrus. *Food Technol.*, 27:98–105.

Pettipher, G.L. 1986. An improved method for extraction and quantification of anthocyanins in cocoa beans and its use as an index of the degree of fermentation. *J. Sci. Food Agric.*, 37:289–296.

Porter, L.J., Ma, Z., and Chan, B.G. 1991. Cacao procyanidins: major flavanoids and identification of minor metabolites. *Phytochemistry*, 30:1657–1663.

Poei-Langston, M.S. and Wrolstad, R.E. 1981. Color degradation in an ascorbic acid-anthocyanin-flavonol model system. *J. Food Sci.*, 46:1218–1225.

Prabha, T.N. and Patwardhan, M.V. 1985. A comparison of polyphenolic patterns in some Indian varieties of apples and their endogenous oxidation. *J. Food Sci. Technol.*, 22:404–407.

Price, K.R., Rhodes, M.J.C., and Barnes, K.A. 1998. Flavonol glycoside content and composition of tea infusions made from commercially available teas and tea products. *J. Agric. Food Chem.*, 46:2517–2522.

Promberger, A., Dornstauder, E., Fruhwirth, C., Schmid, E.R., and Jungbauer, A. 2001. Determination of estrogenic activity in beer by biological and chemical means. *J. Agric. Food Chem.*, 49:633–640.

Puech, J.L. 1988. Phenolic compounds in oak wood extracts used in the aging of brandies. *J. Sci. Food Agric.*, 42:165–172.

Purdon, M.P. and McCamey, D.A. 1987. Use of a 5-caffeoylquinic acid/caffeine ratio to monitor the coffee roasting process. *J. Food Sci.*, 52:1680–1683.

Quinn, M.K. and Singleton, V.L. 1985. Isolation and identification of ellagitannins from white oak wood and estimation of their role in wines. *Am. J. Enol. Vitic.*, 36:148–155.

Rapisarda, P., Carollo, G., Fallico, B., Tomaselli, F., and Maccarone, E. 1998. Hydroxycinnamic acids as markers of Italian blood orange juices. *J. Agric. Food Chem.*, 46:464–470.

Rapisarda, P., Fanella, F., and Maccarone, E. 2000. Reliability of analytical methods for determining anthocyanins in blood orange juices. *J. Agric. Food Chem.*, 48:2249–2252.

Ribereau-Gayon, P., Pontallier, P., and Glories, Y. 1983. Some interpretations of colour changes in young wines during their conservation. *J. Sci. Food Agric.*, 34:505–516.

Ribereau-Gayon, P. 1974. The chemistry of red wine colour, in *Chemistry of Wine-Making,* Dinsmoor-Webb, A., Ed., ACS Symposium Series 137, American Chemical Society, Washington, D.C., 50–87.

Ribereau-Gayon, P. and Glories, Y. 1987. Phenolics in grape and wines, in *Proc. 6th Aust. Wine Ind. Techn. Conf.,* Australian Industrial Publishers, Adelaide, 247.

Risch, B. and Herrmann, K. 1988. Hydroxycinnamic acid derivatives in citrus fruits. *Z. Lebensm. Unters.-Forsch.*, 187:530–534.

Roberts, E.A.H. 1962. Economic importance of flavonoid substances: tea fermentation, in *The Chemistry of Flavonoid Compounds,* Geisman, T.A., Ed., Pergamon Press Ltd., Oxford, U.K., 468.

Roberts, E.A.H. 1952. Chemistry of tea fermentation. *J. Sci. Food Agric.*, 23:193–198.

Robertson, A. 1992. The chemistry and biochemistry of black tea production: the nonvolatiles, in *Tea: Cultivation to Consumption,* Wilson, K.C. and Clifford, M.N., Eds., Chapman & Hall, London, 553–601.

Robichaud, J. and Noble, A. 1990. Astringency and bitterness of selected phenolic in wine. *J. Sci. Food Agric.*, 53:343–353.

Rohan, T.A. 1963. The precursors of chocolate aroma. *J. Sci. Food Agric.*, 14:799–805.

Rohan, T.A. and Connell, M. 1964. The precursors of chocolate aroma: a study of the flavonoids and phenolic acids. *J. Food Sci.*, 29:460–463.

Rohan, T.A. 1964. The precursors of chocolate aroma: a comparative study of fermented and unfermented cocoa beans. *J. Food Sci.*, 29:456–459.

Rohan, T.A. and Stewart, T. 1966a. The precursors of chocolate aroma: changes in free amino acids during the roasting of cocoa beans. *J. Food Sci.*, 31:202–205.

Rohan, T.A. and Stewart, T. 1966b. The precursors of chocolate aroma: changes in the sugars during roasting of cocoa beans. *J. Food Sci.*, 31:202–205.

Rohan, T.A. 1967. The precursors of chocolate aroma: application of gas chromatography in following formation during fermentation of cocoa beans. *J. Food Sci.*, 32:402–404.

Rohan, T.A. and Stewart, T. 1967a. The precursors of chocolate aroma: production of free amino acids during fermentation of cocoa beans. *J. Food Sci.*, 32:395–398.

Rohan, T.A. and Stewart, T. 1967b. The precursors of chocolate aroma: production of reducing sugars during fermentation of cocoa beans. *J. Food Sci.*, 32:399–402.

Rohan, T.A. 1969. The flavor of chocolate, its precursors and a study of their reaction. *Guardian*, 69:443–447; 500–501; 542–544; 587–590.

Romero, C. and Bakker, J. 1999. Interactions between grape anthocyanins and pyruvic acid with effect of pH and acid concentration on anthocyanin composition and color in model solutions. *J. Agric. Food Chem.*, 47:3130–3139.

Romeyer, F.M., Sapis, J.C., and Macheix, J.J. 1985. Hydroxycinnamic esters and browning potential in mature berries of some grape varieties. *J. Sci. Food Agric.*, 36:728–732.

Rossi, M., Giovanelli, G., Cantarelli, C., and Brenna, O. 1988. Effects of laccase and other enzymes on barley wort phenolics as possible treatment to prevent haze in beer. *Bull. de Liason du Groupe Polyphenols*, 14:85–88.

Rous, C. and Alderson, B. 1983. Phenolic extraction curves for white wine aged in French and American oak barrels. *Am. J. Enol. Vitic.*, 34:211–215.

Rouseff, R.L., Martin, S.F., and Youtsey, C.O. 1987. Quantitative survey of narirutin, naringenin, hesperidin, and neohesperidin in citrus. *J. Agric. Food Chem.*, 35:1030–1035.

Salagoity-Auguste, M.-H. and Bertrand, A. 1984. Wine phenolics analysis of low molecular weight components by high performance liquid chromatography. *J. Sci. Food Agric.*, 35:1241–1247.

Salagoity-Auguste, M.-H., Tricard, C., and Sudraud, P. 1987. Estimation of aromatic aldehydes and coumarins by high performance liquid chromatography. Application to wine and spirits aged in oak barrels. *J. Chromatogr.*, 392:379–387.

Salgues, M., Cheynier, V., Gunata, Z., and Wylde, R. 1986. Oxidation of grape juice 2-S-glutathionyl caffeoyl tartaric acid by *Botris cinerea* laccase and characterization of new substance: 2,5-di-S-glutathionyl caffeoyl tartaric acid. *J. Food Sci.*, 51:1191–1194.

Salunkhe, D.K., Chavan, J.K., and Kadam, S.S. 1990. *Dietary Tannins: Consequences and Remedies*. CRC Press, Boca Raton, FL.

Sanbongi, C., Osakabe, N., Natsume, M., Takizawa, T., Gomi, S., and Osawa, T. 1998. Antioxidative polyphenols isolated from *Theobroma cacao*. *J. Agric. Food Chem.*, 46:454–457.

Sanderson, G.W., Ramadive, A.S., Eisenberg, L.S., Farrel, F.J., Simons, R., Manley, C.H., and Coggon, P. 1976. Contribution of phenolic compounds to the taste of tea, in *Phenolic, Sulphur, and Nitrogen Compounds in Food Flavors,* Charalambous, G. and Katz, I., Eds., ACS Symposium Series 26, American Chemical Society, Washington, D.C., 14–46.

Sanderson, G.W. and Graham, H.N. 1973. On the formation of black tea aroma. *J. Agric. Food Chem.*, 21:576–585.

Sanderson, G.W., Berkowitz, J.E., Co, H., and Graham, H.N. 1972. Biochemistry of tea fermentation: products of the oxidation of tea flavanols in model tea fermentation system. *J. Food Sci.*, 37:399–404.

Sanderson, G.W. 1972. The chemistry and manufacturing of tea, in *Structural and Functional Aspects of Phytochemistry, Recent Advances in Phytochemistry,* Runeckles, V.C. and Tso, T.C., Eds., Academic Press, New York, 247–316.

Sanderson, G.W., Co, H., and Gonzalez, J.G. 1971. Biochemistry of tea fermentation: the role of carotenes in black tea formation. *J. Food Sci.*, 36:231–236.

Sanderson, G.W. 1964. The chemical composition of fresh tea flush as affected by clone and climate. *Tea Q.*, 35:101–110.

Sano, M., Suzuki, M., Miyase, T., Yoshino, K., and Maeda-Yamamoto, M. 1999. Novel antiallergic catechin derivatives isolated from oolong tea. *J. Agric. Food Chem.*, 47:1906–1910.

Santos-Buelga, C., Bravo-Haro, S., and Rivas-Gonzalo, J.C. 1996. Role of flavan-3-ol structure on direct condensation with anthocyanidins, in *Polyphenols Communications*, Vercauteren, J., Cheze, C., Dumon, M.C., and Weber, J.F., Eds., Groupe Polyphenols, Bordeaux, vol. 1, 9–10.

Sapis, J.C., Macheix, J.J., and Cordonnier, R.E. 1983. The browning capacity of grapes. 1. Browning potential of polyphenol oxidase during development and maturation of the fruit. *J. Agric. Food Chem.*, 34:342–345.

Sarni-Manchado, P., Deleris, A., Avallone, S., Cheynier, V., and Moutounet, M. 1999. Analysis and characterization of wine condensed tannins precipitated by proteins used as fining agent in enology. *Am. J. Enol. Vitic.*, 50:81–86.

Sato, M., Ramarathnam, N., Suzuki, Y., Ohkubo, T., Takeuchi, M., and Ochi, H. 1996. Varietal differences in phenolic content and superoxide radical scavenging potential of wines from different sources. *J. Agric. Food Chem.*, 44:37–41.

Saucier, C., Little, D., and Glories, Y. 1997. First evidence of acetaldehyde-flavanol condensation products in red wine. *Am. J. Enol. Vitic.*, 48:370–373.

Schreier, P. 1979. Flavor composition of wines: a review. *CRC Crit. Rev. Food Sci. Nutr.*, 12:59–111.

Segall, S. and Proctor, B.E. 1959. The influence of high temperature holding upon the components of coffee brew. *Food Tech.*, 13:266–269.

Shrikhande, A.J. and Francis, F.J. 1974. Effect of flavonols on ascorbic acid and anthocyanin stability in model systems. *J. Food Sci.*, 39:904–906.

Siebert, K.J. 1999. Effects of protein–polyphenol interactions on beverage haze, stabilization, and analysis. *J. Agric. Food Chem.*, 47:353–362.

Siebert, K.J. and Lynn, P.Y. 1998. Comparison of polyphenol interactions with polyvinylpolypyrrolidone and haze-active protein. *J. Am. Soc. Brew. Chem.*, 56:24–31.

Siebert, K.J., Troukhanova, N.V., and Lynn, P.Y. 1996. Nature of polyphenol–protein interactions. *J. Agric. Food Chem.*, 44:80–85.

Siemann, E.H. and Creasy, L.L. 1992. Concentration of phytoalexin resveratrol in wine. *Am. J. Enol. Vitic.*, 43:49–52.

Simonetti, P., Pietta, P., and Testolin, G. 1997. Polyphenol content and total antioxidant potential of selected Italian wines. *J. Agric. Food Chem.*, 45:1152–1155.

Simpson, R.F. 1982. Factors affecting oxidative browning of white wine. *Vitis*, 21:233–239.

Singleton, V.L. and Essau, P. 1969. *Phenolic Substances in Grapes and Wine and Their Significance,* Academic Press, New York.

Singleton, V.L. and Kratzer, F.H. 1973. Plant phenolics, in *Toxicants Occurring Naturally in Foods*, National Academy of Sciences, Washington, D.C., 305–342.

Singleton, V.L. 1974. Some aspects of the wooden containers as a factor in wine maturation, in *Chemistry of Winemaking*, Webb, A.D., Ed., *Adv. Chem. Ser.*, 137:254–277.

Singleton, V.L. and Noble, A.C. 1976. Wine flavor and phenolic substances, in *Phenolic, Sulfur, and Nitrogen Compounds in Food Flavors,* Charalambous, G. and Katz, I., Eds., American Chemical Society, Washington, D.C., 47–70.

Singleton, V.L., Timberlake, C.F., and Lea, A.G.H. 1978. The phenolic cinnamate of white grapes and wines. *J. Sci. Food Agric.*, 29:403–410.

Singleton, V.L. 1980. *Grape Phenolics: Background and Prospects.* Centenaire, Paris.

Singleton, V.L. 1981. Naturally occurring food toxicants: phenolic substances of plant origin common in foods. *Adv. Food Res.*, 27:149–242.

Singleton, V.L. 1982. Grape and wine phenolics: background and prospects, in *Proceedings of University of California, Davis, Grape Wine Centennial Symposium*, Webb, A.D., Ed., University of California Press, Berkeley, 215–227.

Singleton, V.L., Zaya, J., Trousdale, E., and Salgues, M. 1984. Caftaric acid in grapes and conversion to a reaction product during processing. *Vitis*, 23:113–120.

Singleton, V.L., Salgues, M., Zaya, J., and Trousdale, E. 1985. Caftaric acid disappearance and conversion to products of enzymic oxidation in grape must and wine. *Am. J. Enol. Vitic.*, 36:50–56.

Singleton, V.L. 1987. Oxygen with phenols and related reactions in musts, wines and model systems: observations and practical implications. *Am. J. Enol. Vitic.*, 38:67–77.

Sioud, F.B. and Luh, B.S. 1966. Polyphenolic compounds of pear puree. *Food Technol.*, 20:534–538.

Smith, A.W. 1985. Introduction, in *Coffee, Vol. 1. Chemistry,* Clark, R.J. and Macrae, R., Eds., Elsevier Applied Science, London, 1–41.

Sogawa, H. 1972. Polyphenols in brewing. II. Fractionation of polyphenols by ion-exchange chromatography. *Rep. Res. Lab. Kirin Brew. Co.*, 15:17–24.

Soleas, G.J., Dam, J., Carey, M., and Goldberg, D.M. 1997. Towards fingerprinting of wines: cultivar-related patterns of polyphenolic constituents in Ontario wines. *J. Agric. Food Chem.*, 45:3871–3880.

Somers, T.C. and Evans, M.E. 1986. Evolution of red wines. Ambient influences on colour composition during early maturation. *Vitis*, 25:31–39.

Somers, T.C. and Pocock, K.F. 1986. Phenolic harvest criteria for red vinification. *Aust. Grapegrower Winemaker*, 268:24; 26–7; 29–30.

Somers, T.C. and Pocock, F. 1990. Evolution of red wines. III. Promotion of the maturation phase. *Vitis*, 29:109–121.

Somers, T.C., Verette, E., and Pocock, F. 1987. Hydroxycinnamate esters of *Vitis vinefera*: changes during white wine vinification and effects of exogenous enzymic hydrolysis. *J. Sci. Food Agric.*, 40:67–78.

Somers, T.C. 1971. The polymeric nature of wine pigments. *Phytochemistry*, 10:2175–2186.

Somers, T.C. 1966. Wine tannins — isolation of condensed flavonoid pigments by gel filtration. *Nature*, 209:368–370.

Sondheimer, E. and Kertesz, Z.I. 1953. Participation of ascorbic acid in the destruction of anthocyanins in strawberry juice and model systems. *Food Res.*, 18:475–479.

Souquet, J.-M., Cheynier, V., Broussaud, F., and Moutounet, M. 1996. Polymeric proanthocyanidins from grape skins. *Phytochemistry*, 43:509–512.

Spanos, G.A. and Wrolstad, R.E. 1990. Influence of variety, maturity, processing, and storage on phenolic composition of pear juice. *J. Agric. Food Chem.*, 38:817–824.

Spayd, S.E., Nagel, C.W., Hayrynen, L.D., and Drake, S.R. 1984. Color stability of apple and pear juices blended with fruit juices containing anthocyanins. *J. Food Sci.*, 49:411–414.

Stagg, G.V. and Millin, D.J., 1975. The nutritional and therapeutic value of tea — a review. *J. Sci. Food Agric.*, 26:1439–1459.

Starr, M.S. and Francis, F.J. 1968. Oxygen and ascorbic acid effect on the relative stability of four anthocyanin pigments in cranberry juice. *Food Technol.*, 22:1293–1295.

Stavric, B. 1994. Antimutagens and anticarcinogens in foods. *Food Chem. Toxicol.*, 32:79–90.

Steinke, R.D. and Paulson, M.C. 1964. The production of steam-volatile phenols during the cooking and alcoholic fermentation of grains. *J. Agric. Food Chem.*, 12:381–387.

Stelzig, D., Akhtar, S., and Ribeiro, S. 1972. Catechol oxidase of Red Delicious apple peel. *Phytochemistry*, 11:535–539.

Stevens, J.F., Ivancic, M., Hsu, V.L., and Deinzer, M.L. 1997. Prenylflavonoids from *Humulus lupulus. Phytochemistry*, 44:1575–1585.

Stevens, J.F., Taylor, A.W., and Deinzer, M.L. 1998. Chemistry and biology of hop flavonoids. *J. Am. Soc. Brew. Chem.*, 56:136–145.

Stevens, J.F., Taylor, A.W., Clawson, J.E., and Deinzer, M.L. 1999a. Fate of xanthohumol and related prenylflavonoids from hops to beer. *J. Agric. Food Chem.*, 47:2421–2428.

Stevens, J.F., Taylor, A.W., and Deinzer, M.L. 1999b. Quantitative analysis of xanthohumol and related prenylflavonoids in hops and beer by liquid chromatography-tandem mass spectrometry. *J. Chromatogr.*, 832:97–107.

Subramanian, N., Ventkatesh, P., Ganguli, S., and Sinkar, V.P. 1999. Role of polyphenol oxidase and peroxidase in generation of black tea theaflavins. *J. Agric. Food Chem.*, 47:2571–2578.

Tanaka, T., Takahashi, R., Kouno, I., and Nonaka, K. 1994. Chemical evidence for the deastrigency (insolubilization of tannins) of persimmon fruit. *J. Chem. Soc. Perkin. Trans.*, 1:3013–3022.

Tanaka, T., Inoue, K., Betsumiya, Y., Mine, C., and Kouno, I. 2001. Two types of oxidative dimerization of the black tea polyphenol theaflavin. *J. Agric. Food Chem.*, 49:5785–5789.

Tanner, H. and Rentschler, H. 1956. Über Polyphenole der Kernobst-und-Traubensäfte. *Fruchtsaft-Ind.*, 1:231.

Tekel, J., De Keukeleire, D., Rong, H., Daeseleire, E., and Van Peteghem, C. 1999. Determination of the hop-derived phytoestrogen, 8-prenylnaringenin, in beer by gas chromatography/mass spectrometry. *J. Agric. Food Chem.*, 47:5059–5063.

Thompson, R.S., Jacques, D., Haslam, E., and Tanner, R.J.N. 1972. Plant proathocyanidins. Part I. Introduction, the isolation, structure and distribution in nature of plant procyanidins. *J. Chem. Soc. Perkin. Trans.*, I. 11:1387–1399.

Thurston, P.A. and Tubb, R.S. 1981. Screening yeast strains for their ability to produce phenolic off-flavors: a simple method for determining phenols in wort and beer. *J. Inst. Brew.*, 87:177–179.

Timberlake, C.F. and Bridle, P. 1977. Anthocyanins: colour augmentation with catechin and acetaldehyde. *J. Sci. Food Agric.*, 28:539–544.

Timmer, R., ter Heide, R., Wobben, J.J., and de Valois, P.J. 1971. Phenolic compounds in rum. *J. Food Sci.*, 36:462–464.

Trugo, L.C. and Macrae, R. 1984a. A study of the effect of roasting on the chlorogenic acid composition of coffee using HPLC. *Food Chem.*, 15:219–227.

Trugo, L.C. and Macrae, R. 1984b. Chlorogenic acid composition of instant coffees. *Analyst*, 109:263–266.

Ukers, W.H. 1994. Tea in the beginning, in *All about Tea,* Hyperion Press, Westport, CT, 1–22.

Valenzuela-Zapata, A.G. 1985. The tequila industry in Jalisco, Mexico. *Desert Plants*, 7:65–70.

Van Buren, J., De Vos, L., and Pilnik, W. 1976. Polyphenols in Golden Delicious apple juice in relation to the method of preparation. *J. Agric. Food Chem.*, 24:448–451.

Van Buren, J., De Vos, L., and Pilnik, W. 1973. Measurement of chlorogenic acid and flavonol glycosides in apple juice by a chromatographic-fluorometric method. *J. Food Sci.*, 38:656–658.

Van Dam, T.G. and Daniel, R.M. 1980. A method for the separation of coloured components of red wines. *J. Sci. Food Agric.*, 31:267–272.

Vasserot, Y., Caillet, S., and Maujean, A. 1997. Study of anthocyanin adsorption by yeasts Lees. Effect of some physicochemical parameters. *Am. J. Enol. Vitic.*, 47:433–437.

Veldhuis, M.K., Swift, L.J., and Scott, W.C. 1970. Fully methoxylated flavones in Florida orange juices. *J. Agric. Food Chem.*, 18:590–592.

Versari, A., Biesenbruch, D., Barbanti, D., and Farnell, P.J. 1997. Alduration of fruit juices: dihydrochalcones as quality markers for apple juice identification. *Lebensm.-Wiss. u-. Technol.*, 30:585–589.

Vidal, S., Cartalade, D., Souquet, J.-M., Fulcrand, H., and Cheynier, V. 2002. Changes in proanthocyanidin chain length in winlike model solutions. *J. Agric. Food Chem.*, 50:2261–2266.

Villarreal, R., Sierra, J.A., and Cardenas, L. 1986. A rapid and sensitive method for determination of 4-vinyl guaiacol in beer by electron-capture gas-liquid chromatography. *ASBC J.,* 44:114–117.

Villeneuve, F., Cros, E., and Macheix, J.J. 1989. Recherche d'un indice de fermentation du cacao. III Evolution des flavan-3-ols de la féve. *Café Cacao Thé*, 33:165–170.

Villeneuve, F., Cros, E., and Macheix, J.J. 1985. Effects of fermentation on the activity of peroxidases and polyphenol oxidases from cocoa seeds. *Café Cacao Thé*, 29:113–120.

Vinson, J.A., Proch, J., and Zubik, L. 1999. Phenol antioxidant quantity and quality in foods: cocoa, dark chocolate and milk chocolate. *J. Agric. Food Chem.*, 47:4821–4824.

Viriot, C., Scalbert, A., Lapierre, C., and Moutounet, M. 1993. Ellagitannins and lignins in aging of spirits in oak barrels. *J. Agric. Food Chem.*, 41:1872–1879.

Von Narzib, L., Reincheneder, E., and Weihenstephan, R.D. 1979. Über der Einflub der Abläuter-bedingugen auf den Polyphenolgehalt der Würzen und Biere. *Brauwelt*, 119:1009–1014.

Walker, J.R.L. 1963. A note on polyphenol content in ripening apples. *N.Z. J. Sci.*, 6:492–494.

Wang, L.-F., Kim, D.-M., and Lee, C.Y. 2000. Effects of heat processing and storage on flavanols and sensory qualities of green tea beverage. *J. Agric. Food Chem.*, 48:4227–4332.

Webby, R.F. 1990. Flavonoid complement of cultivars of *Actinidia deliciosa* var. *deliciosa*, kiwifruit. *N.Z. J. Crop Hortic. Sci.*, 18:1–4.

Whiting, G.C. and Coggins, R.A. 1975. Estimation of the monomeric phenolics of ciders. *J. Sci. Food Agric.*, 26:1833–1838.

Wickremasinghe, R.L. 1978. Tea, *Adv. Food Res.*, 24:229–286.

Wildenradt, H.L. and Singleton, V.L. 1974. Production of aldehydes as result of oxidation of polyphenolic compounds and its relation to wine aging. *Am. J. Enol. Vitic.*, 25:119–126.

Wilson, E.L. 1981. High-pressure liquid chromatography of apple juice phenolic compounds. *J. Sci. Food Agric.*, 32:257–264.

Wilson, K.C. and Clifford, M.N. 1992. *Tea: Cultivation and Consumption,* Chapman & Hall, London.

Wong, K.M., Dimick, P.S., and Hammerstedt, R.H. 1990. Extraction and high performance liquid chromatographic enrichment of polyphenol oxidase from *Theobroma cacao* seeds. *J. Food Sci.*, 55:1108–1111.

Wong, M. and Stanton, D.W. 1993. Effect of removal of amino acids and phenolic compounds on nonenzymic browning in stored kiwifruit concentrates. *Lebensm. Wiss. u-. Technol.*, 26:138–144.

Woodring, P.J., Edwards, P.A., and Chisholm, M.G. 1990. HPLC determination of non-flavonoid phenols in Vidal Blanc wine by using electrochemical detection. *J. Agric. Food Chem.*, 38:729–732.

Wrolstad, R.E., Lee, D.D., and Poei, M.M. 1980. Effect of microwave blanching on the color and composition of strawberry concentrate. *J. Food Sci.*, 45:1573–1579.

Yamanishi, T., Hara, Y., Luo, S., and Weckremasinghe, R. 1995. Special issue on tea. *Food Rev. Int.*, 11:371–546.

Yoshioka, H., Sugiura, K., Kawahara, R., Fujita, T., Makino, M., Kamiya, M., and Tsuyumu, S. 1991. Formation of radicals and chemiluminescence during the autoxidation of tea catechins. *Agric. Biol. Chem.*, 55:2717–2723.

Zhu, Q.Y., Zhang, A., Tsang, D., Huang, Y., and Chen, Z.-Y. 1997. Stability of green tea catechins. *J. Agric. Food Chem.*, 45:4624–4628.

Ziegleder, G. and Biehl, B. 1988. Analysis of cocoa flavor components and flavor precursors, in *Analysis of Nonalcoholic Beverages,* Linskens, H.F. and Jackson, J.F., Eds., Springer, Berlin, 321–393.

Zoecklein, B.W., Jasinski, Y., and McMahon, H. 1997. Effect of fermentation, aging, and aging *sur Lie* on total and phenol-free Riesling (*Vitis vinifera* L.) glycosides. *J. Food Composition Anal.*, 11:240–248.

6 Phenolics in Herbal and Nutraceutical Products

The fact that phytochemicals occurring in food and natural health products play a significant role in disease prevention and health promotion was first recognized as a result of epidemiological studies using both animal and human subjects. This led to an ever-growing interest in herbs and botanical and nutraceutical products. Bioactives in herbal and nutraceutical products constitute a myriad of chemical compounds, among which phenolic substances often play a primary or a synergistic function. Herbal products are used to enhance immunity, for antiinflammatory, anticancer, and antidepressant effects and for treatment of heart disease, urinary tract infections and prostate conditions, among others. The herbal products of interest include, but are not limited to, echinacea, ginseng, ginkgo biloba, St. John's wort, valerian, kava kava, saw palmetto, black cohosh, devil's claw, goldenseal, hawthorn, ginger, licorice, and milk thistle (Blumenthal, 2000; Lawson and Bauer, 1998). These products serve as a source of the phenolic compounds summarized in this chapter.

ECHINACEA

Echinacea products serve as popular herbal immunostimulants, particularly in North America and Europe (Grunwald and Büttel, 1996). Roots and aerial parts, as well as their tinctures and extracts, are used, mainly in encapsulated forms. Echinacea genus comprises nine species and two varieties (McGregor, 1968), of which only *E. purpurea* (L.), *E. angustifolia* (DC.) and *E. pallida* (Nutt.) are used medicinally. These herbs were used initially by the American Indians for treatment of respiratory ailments, wounds, snake bites, common cold, and headache. Their current use is as antiseptic, antiviral, peripheral vasodialation, healing for skin conditions, carbuncles, boils, wounds, ulcers, burns, bedsores, bites, stings, and poisonous insects, to build the immune system when taken orally and as supportive therapy for cold and influenza (Tyler, 1998).

The active components of Echinacea include polysaccharides, glycoproteins, caffeic acid derivatives, flavonoids, polyacetylenes, and alkylamides. Caffeic acid derivatives are a major group of compounds present (Becker et al., 1982; Facino et al., 1993). Echinacoside is the major polar constituent of the roots of *E. angustifolia* present at 0.3 to 1.7% (Schenk and Franke, 1996); cichoric acid (2,3-*O*-dicaffeoyltartaric acid) occurs in especially high concentrations in the flowerheads and roots (up to 3.1 and 2.1%, respectively). The content of cichoric acid depends on the season and stage of development of the plant (Alhorn, 1992; Bauer and Von Hagen-Plettenberg, 1999). In addition, 1,3-*O*-dicaffeoylquinic acid (cynarine) and 2-*O*-caffeoyltartaric acid (caftaric acid) as well as chlorogenic acid and ferulates

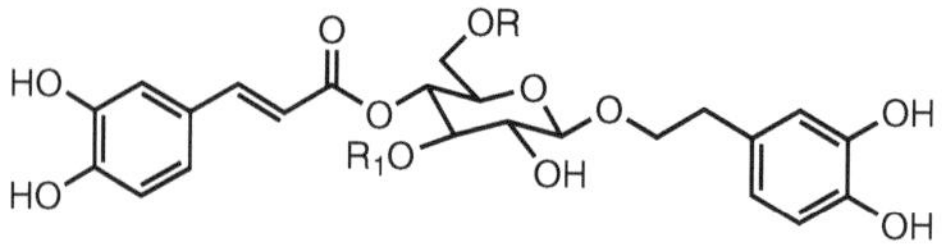

Echinacoside
R = glucose (1, 6-)R_1 = rhamnose (1, 3-)

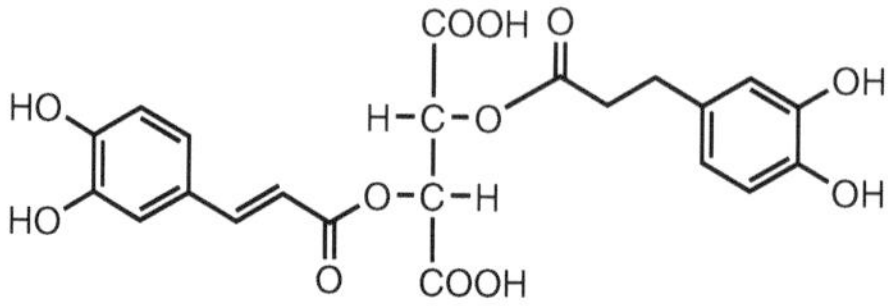

Cichoric Acid
(2,3-dicaffeoyltartaric acid)

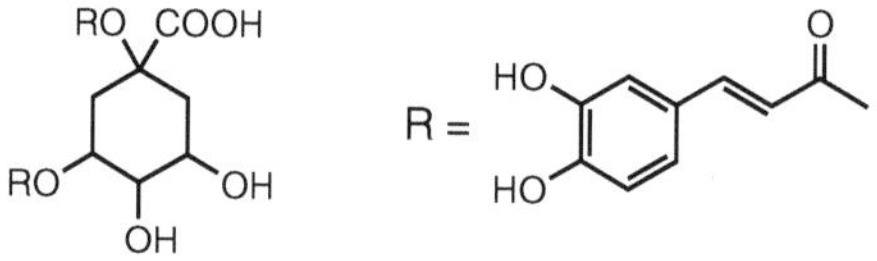

Cynarine
(1,3-dicaffeoylquinic acid)

FIGURE 6.1 Chemical structures of caffeic acid derivatives in Echinacea species.

and tartaric acid may be present (Bisset, 1994; D'Amelio, 1999; Newall, 1996). The chemical structures of some caffeic acid derivatives found in Echinacea products are given in Figure 6.1. Cichoric acid possesses phagocytosis stimulatory activity *in vitro* and *in vivo* (Bauer et al., 1989) and inhibits hyaluronidase (Facino et al., 1993) and free radical degradation of type III collagen (Facino et al., 1995). Cichoric acid also inhibits human immunodeficiency virus type 1 (HIV-I) integrase (Robinson, 1998). Other constituents found in Echinacea include 15 lypophilic alkamides, most of which contain one or two acetylenic bonds. In addition, a number of ketoalkynes have been identified in Echinacea.

GINSENG

Ginseng represents perennial herbs derived from the genus *Panax* grown in China, India, Japan, Korea, Vietnam and North America. The Siberian ginseng is a different plant, referred to as *Eleutherococcus senticosus* ginseng. The beneficial health effects of ginseng relate to its transient regulatory effects on carbohydrates, lipids, and nucleic acid metabolism (Nagasawa et al., 1977; Oura and Hiai, 1973; Sotaniemi et al., 1995). It also has a prophylactic effect that reduces susceptibility to illness and may act as an aphrodisiac. The main active components of ginseng are triterpenoid ginsenosides, and steroidal saponins (panaxosides), which are present at a 2 to 3% level in ginseng roots (Bisset, 1994; D'Amelio, 1999; Newall, 1996). These constituents are linked to different sugar molecules and, together with flavonoids, serve as the main bioactive components of ginseng (Zhang et al., 1979a, b). Presence of caffeic acid, kaempferol, and vanillic acid has also been reported (Kitts, 2000). Ginseng products protect against plasmid DNA scissions and scavenge hydroxyl radicals (Kitts et al., 1999).

GINKGO BILOBA

Ginkgo (*Ginkgo biloba* L.) leaf extracts have served as one of the oldest natural therapeutics; ginkgo's fruits were used as medicinal agents 5000 years ago (Michel, 1986). Its seeds have been used extensively as a remedy for asthma, cough, bladder inflammation and alcohol abuse, among others. The leaves have also been used to alleviate cardiovascular disorders and asthma as well as for the treatment of disorders associated with aging and peripheral circulation insufficiencies (Foster and Chongdi, 1992). The antiinflammatory and antiallergenic effects of ginkgo and its vasodilatory activity have been reported in the literature (D'Amelio, 1999; Juretzek, 1997). The physiological benefits of ginkgo are due to the antioxidive, vasoregulatory, and neuroprotective activities of its constituents. Active components of ginkgo with documented biological activity include ginkgolides, bilobalide, flavonoids, and ginkgolic acids (Tyler, 1998).

Phenolics in ginkgo are aglycones of flavone glycosides. The flavonols present include quercetin, kaempferol, rutin, and isorhamnetin whose glycosidic constituents are mono-, di- or triglycosides of glucose and rhamnose. Thus, dimeric flavonoids, (bilobetin, gingketin, isogingketin, and sciadopitysin), flavonols, flavonol glycosides, and coumaric esters of flavonol glycosides, as well as proanthocyanidins, have been isolated from ginkgo biloba leaves (Van Beek et al., 1998; D'Amelio, 1999; Newall, 1996). Quercetin and kaempferol coumaryl glycosides are essential for the efficacy of extracts of ginkgo. Rutin and quercetin as well as flavonoid mixtures from ginkgo biloba have been found to protect cerebellar cells from oxidative damage and apoptosis by scavenging hydroxyl radicals (Chen et al., 1999). Ginkgo leaves also contain ginkolic acids, which are 6-alkyl salicylic acids and 3-alkyl phenols, known as ginkgols, as well as 5-alkyl resorcins represented by bilobol and hydrobilobol, known as cardols. These latter compounds may cause gastrointestinal disturbance following consumption of ginkgo biloba fruits (Becker and Skipworth, 1975) and may also possess antimicrobial and antitumoral activity (Jaggy and Koch, 1997). Chemical structures of ginkgolic acids, cardanols and cardols, are given in Figure 6.2.

ST. JOHN'S WORT

St. John's wort (*Hypericum perforatum* L.) is a shrubby aromatic perennial herb. It is native to Asia, North Africa, Azores and other parts of Europe and has also been naturalized in Australia and North America. St. John's wort has a long history as a valuable treatment for depression and sleep-related problems (Linde et al., 1996;

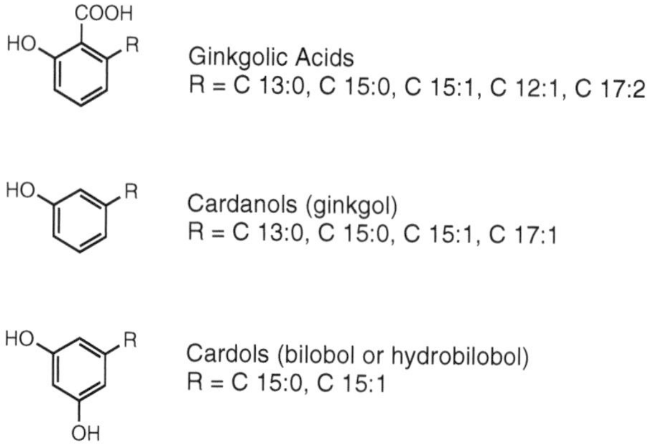

FIGURE 6.2 Chemical structures of ginkgolic acid, cardanols and cardols of ginkgo biloba.

Hippius, 1998; Giese, 1999). It has been found very helpful in easing emotional and stress symptoms and has also been used for the treatment of burns, wounds, insomnia, shocks, concussions, hysteria, gastritis, hemorrhoids, and kidney disorders (Hobbs, 1989, Chavez and Chavez, 1997; Upton, 1997). The main healing qualities of St. John's wort originate from tiny glands on the plant. In addition, extracts of flowering tops of *Hypincum* are known to possess antiinflammatory, wound- and burn-healing effects, and antibacterial, antifungal, antibiotic, and antiviral activities, among others (Hobbs, 1989; Chavez and Chavez, 1997).

St. John's wort contains a complex mixture of bioactive anthraquinones derivatives, mainly 0.05 to 0.3% of hypericin, isohypericin, and protolypericin (Bisset, 1994; Newall, 1996); flavonols (kaempferol, myricetin, and quercetin) (Newall, 1996); flavones (luteolin and rutin) (Newall, 1996); glycosides (hyperoside (hyperin), isoquercitrin, quercitrin and rutin) (Bisset, 1994; Newall, 1996); biflavonoids (biagigenin and amentoflavone) (Bisset, 1994; Newall, 1996) as well as catechins associated with condensed tannins (Newall, 1996). In addition, caffeic, chlorogenic, *p*-coumaric, ferulic, isoferulic, *p*-hydroxybenzoic, genistic, and vanillic acids (Newall, 1996; Hobbs, 1989; Upton, 1997) as well as perenylated phloroglucinol derivatives (Bisset, 1994; Newall, 1996) have been found in St. John's wort. Hyperforin (2 to 4% in the dried herb; *H. perforafum*) has been shown to inhibit the reuptake of serotonin, dopamine and norepinephrine and thus has been identified as the leading candidate for a single antidepressant compound in *Hypericum* (Muller et al., 1998). Flavonoids constitute 11.7 and 7.4% of the leaves and stalks of St. John's wort (Upton, 1997). Proanthocyanidins or condensed tannins are found in large amounts in the areal parts of the plant and are oligomers and polymers of catechin and epicatechin, all of which possess antioxidant, antimicrobial, and cardioprotective effects (Cook and Samman, 1996; Hollman et al., 1996; Baureithel et al., 1997; Kim et al., 1998). The chemical structures of some phenolics of St. John's wort are given in Figure 6.3.

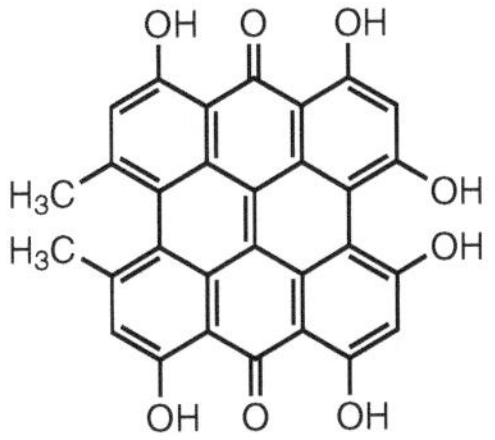

Hypericin

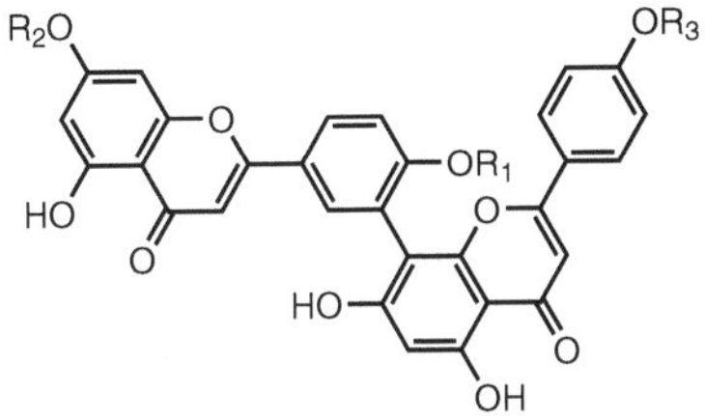

Compound	R_1	R_2	R_3
Amentoflavone	H	H	H
Bilobetin	CH_3	H	H
Ginkgetin	CH_3	CH_3	H
Isoginkgetin	CH_3	H	CH_3
Sciadopitysin	CH_3	CH_3	CH_3

FIGURE 6.3 Chemical structures of hypericin and bioflavonoids of St. John's wort.

VALERIAN

There are approximately 300 species of valeriana, also called valerian, three of which are commonly used for medicinal purposes. These are *Valeriana officinalis* L. or valerian, a Eurasian member of the Valerianceae, *V. wallichii* of India and Pakistan and *V. edulis* of Mexico. The *V. officinalis* is the species of interest in most applications. Roots and rhizomes of valerian have been used for over 2000 years as a sedative and are now used for treating tension, nervous excitability, hysterical state, insomnia, hypothondriasis, migraines, cramps, rheumatic pain, and irritability with falling asleep (Blumenthal et al., 1998; D'Amelio, 1999; Cott, 1995; Houghton, 1988; Newall, 1996).

Active constituents of valerian include pyridine type alkaloids, namely, actinidine, chatinine, shyanthine, valerianine, and valerine (D'Amelio, 1999; Newall, 1996), some of which are phenolic in nature (see Figure 6.4). In addition, they contain a number of iridoids, namely, valtrates and isovaltrates, as well as terpenes and sesquiterpenes. Caffeic and chorogenic acids as well as tannins and other polyphenolics have been identified in roots and rhizomes of valerian (Newall, 1996).

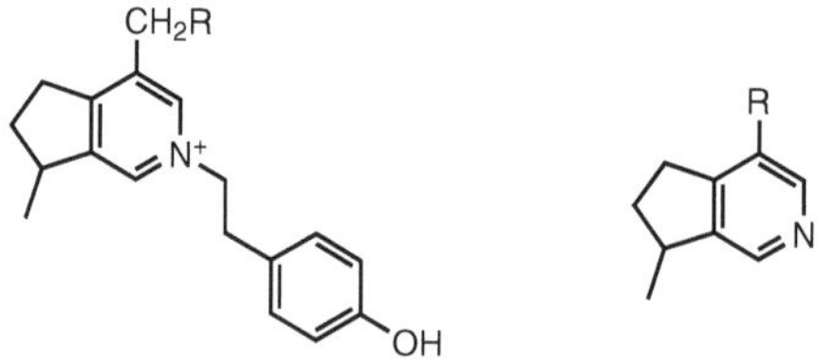

$R = H$ or OH

$R = CH_3$; Actinidine
$R = CH_2OCH_3$; Valeranine

FIGURE 6.4 Chemical structures of pyindine alkaloids of valeriana.

KAVA KAVA

Kava kava (*Piper methysticum*) is an herb originating from Oceania Polynesia, where it has been used as a ceremonial beverage. It is gaining popularity in the United States and Europe as a relatively safe remedy for anxiety and compares favorably with benzodiazopines (Blumenthal et al., 1998). Dried rhizomes are used; these also have a relaxing effect on the central muscular system and are effective in the treatment of urinary tract infections, asthma, anxiety disorders, menopause symptoms, and mental function, as well as acting as anticonvulsant and antispasmodic agents (Backhausß and Krieglstein, 1992; Blumenthal, 2000; Woelk et al., 1993). Kavalactones present at 3.5% or more are the main active components and include methysticin, dihydromethysticin, kavain, 7,8-dihydrokavain, 5,6-dehydrokavain, 5,6-dehydromethysticin and yangonin (Blumenthal, 2000). Chalcones, namely, flavokavains A, B and C, are also present (Blumenthal, 2000)

SAW PALMETTO

Saw palmetto (*Serenoa repens* or *Serenoa serrulata*) is a small shrubby palm grown in the U.S. The dark brown or black single-seeded fruit or berry of saw palmetto looks like an olive in shape and size and turns yellow, orange or purplish-black in color when ripe. Saw palmetto berries are accepted botanical medicine (USP, 1999) and are used for treating chronic and subacute cystitis, laryngitis, bronchitis and, most importantly, benign prostatic hyperplasia (BPH) and enlarged prostate. Saw palmetto stimulates the genitourinary tract and mucous membranes (Brinker, 1994; Bennett and Hicklin, 1998; Wilt et al., 1998), promotes the growth of new flesh and has estrogenic activity (D'Amelio, 1999). The constituents of saw palmetto include steroids, flavonoids and tannins as well as carotenoids and other pigments (Blumenthal, 2000).

BLACK COHOSH

The roots and rhizomes of black cohosh, *Cimicifuga racemosa*, are used for the treatment of depression and hot flashes and is known to possess estrogenic activity (Blumenthal, 2000; D'Amelio, 1999). It has been suggested for treating peripheral arterial diseases because it possesses vasoactive effects. Black cohosh also exhibits antimalarial activity and is effective in relieving the symptoms of menopause. The active components of black cohosh include cimicifugic acids A to E, fukinolic acid and fukiic acid. Cimicifugic acids A to E and fukinolic acid are esters between cinnamic acids and the hydroxyl group of benzyltartaric acids (Noguchi et al., 1998). Tannins are also present in saw palmetto berries (D'Amelio, 1999; Newall, 1996).

DEVIL'S CLAW

Devil's claw (*Harpagophytum procumbens* DC.) is the root (especially its secondary tubers) marketed in Canada and in Europe as a home remedy for the relief of arthritic disease and as an antiinflammatory agent (Bisset, 1994; Blumenthal, 2000; D'Amelio, 1999). When applied externally, it is also effective in the treatment of sores, ulcers, boils and skin lesions (Blumenthal, 2000). The constituents of the plant are primarily sucrose and its galactosides, namely, stachyose, present at 46%, and raffinose. The active constituents present at 0.1 to 2% are primarily iridoiglycosides: harpagoside, harpagide and procumbide (Bisset, 1994; Blumenthal, 2000; D'Amelio, 1999; Newall, 1996) (Figure 6.5). In addition, chlorogenic acid and quinones (D'Amelio, 1999) as well as flavonoids kaempferol and luteolin (Blumenthal, 2000; D'Amelio, 1999; Newall, 1996) are present. Preparations also include verbascosides, acetoside, isoacetoside and abisoide, and glycosylated phytosterols along with oleanic and ursolic acids (Bisset, 1994; Blumenthal, 2000; D'Amelio, 1999; Newall, 1996).

HO O O OCH_3

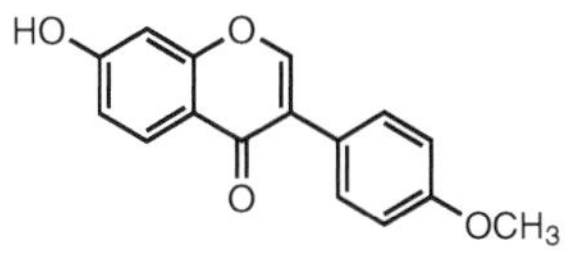

FIGURE 6.5 Chemical structure of bioactive flavonoid of devil's claw.

GOLDENSEAL

Goldenseal (*Hydrastis canadensis* L.) is another medicinal herbaceous perennial that grows in shady woods in Western Ontario and the Eastern U.S. The rhizomes and roots of goldenseal have been used since ancient times in China and India as an antidiarrheal medication. Its liquid extracts may also be used for treatment of colds, flu, sinus infections, and wounds and as an antiinflammatory preparation (Foster, 1991). Duke (1985) has reported its use as a diuretic, echarotic or stimulant and as a cure for boils, hemorrhoids, ulcers, asthma, jaundice, liver ailments and other diseases. These preparations have a bright yellow color due to the presence of the yellow alkaloid berberine and have therefore been used as a dye for clothing (Thornton, 1998). Goldenseal has also been referred to as Indian paint, Indian turmeric or jaundice root (Snow, 1998).

The active components of goldenseal are a number of isoquinoline alkaloids, including berberine, hydrastin, hydrostidine, isohydrastidine, berberostine, canadine, canadaline, corpylpalmine, and isocorpylpalmine (Messana et al., 1980; Leone et al., 1996). At recommended doses berberine and berberine-containing plants are considered nontoxic, but at 2- to 3-g doses goldenseal can lower heartbeat; at higher doses, it can paralyze the central nervous system (Flynn and Roest, 1995).

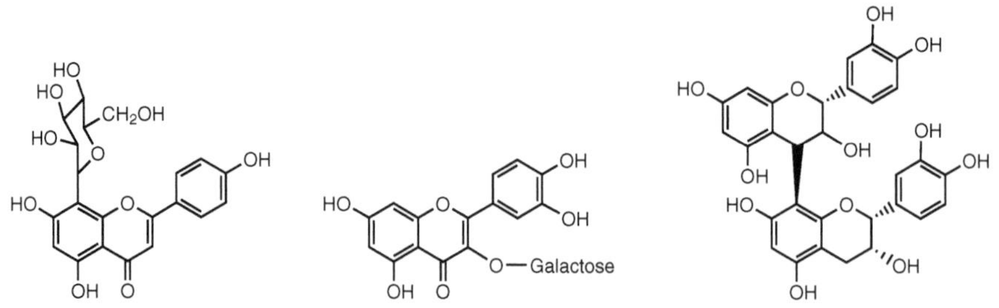

FIGURE 6.6 Chemical structures of some bioactive polyphenolics of hawthorn.

HAWTHORN

The flowers, leaves and berries of hawthorn, crataegus, are used for the treatment of a variety of cardiac diseases (Bisset, 1994; D'Amelio 1999; Newall, 1996), coronary and myocardial circulation, tachycardia, arteriosclerosis and as an antispasmodic and a sedative agent (Bisset, 1994; Blumenthal, 2000; Blumenthal et al., 1998; D'Amelio 1999). The constituents of hawthorn include 1 to 3% oligomeric procyanidins, 1 to 2% flavonoids, hyperosides, vilexins, vilexin glucosides, quercetin and its glucoside. In addition, phenolic acids and catechin are present (Bisset, 1994; Blumenthal, 2000; D'Amelio 1999; Newall, 1996). Figure 6.6 shows structures of some bioactives of hawthorn.

GINGER

An important folk medicine as well as a spice used in Western and Middle Eastern cooking, ginger (*Zingiber officinals*) has been cultivated and used for over 3000 years in India and China (Schulick, 1993). Different parts of the plant, including flower, bud, leaves, fruit and, particularly, rhizome, have been used for treatment of a number of ailments and discomforts. These include stomachache, cough, anorexia, eczema, diarrhea, rheumatism, jaundice, anemia, constipation, asthma, gastritis and can also function as an anticonvulsant and antiamatic agent. The fresh rhizomes of *Zingiber officinale* are dried and used as digestive stimulants and antiinflammatory agents in China and tropical Asia (Kikuzaki et al., 1994).

The bioactive components of ginger include gingerols, which are the main pungent component of the rhizome, and shogaol, a monohydrated gingerol, as well as other gingerol-related compounds formed from phenylalanine via ferulic acid, and diarylheptanoids and terpenoids (Kawakishi et al., 1994; Kikuzaki et al., 1992; 1994; Kikuzaki and Nakatani, 1996; Wu et al., 1998). The antioxidant activity of phenolic constituents of ginger has been documented (Kikuzaki et al., 1994). The compounds involved also served as good antimicrobial agents (Galal, 1996; Yamada et al., 1992). Figure 6.7 and Figure 6.8 summarize the chemical structures of gingerol-related compounds. Figure 6.9 shows the chemical structures of other ginger phenolics.

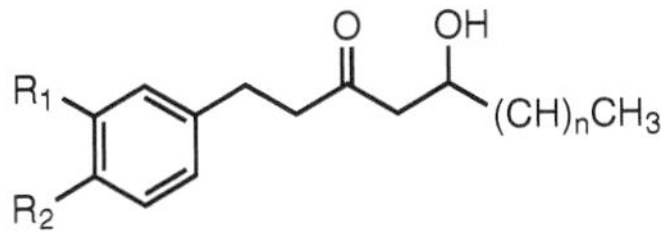

$R_1 = OCH_3$, $R_2 = OH$, n=4, 6, 8; [n+2]-gingerol
$R_1 = R_2 = OCH_3$, n = 4; [6]-methylgingerol
$R_1 = H$, $R_2 = OH$, n = 4; [6]-demethoxygingerol

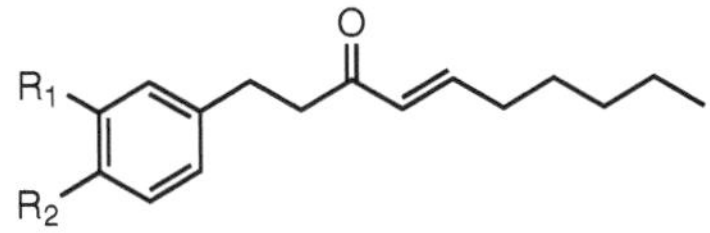

$R_1 = OCH_3$, $R_2 = OH$; [6]-shogaol
$R_1 = R_2 = OCH_3$; [6]-methylshogaol
$R_1 = H$, $R_2 = OH$; [6]-demethoxyshogaol,

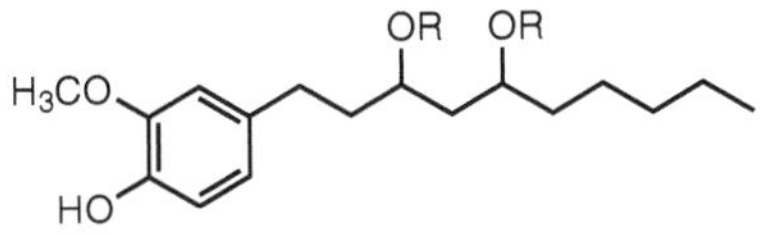

R = H; [6]-gingerdiol
R=Ac; [6]-gingerdiacetate

FIGURE 6.7 Chemical structures of some gingerol-related compounds of ginger.

LICORICE

Licorice is the name used for roots and stolous of some Glycyrrhiza species. It is grown in Spain, Italy, Greece, Russia, Turkey, Iran, Iraq, Syria and China, and also cultivated in South Africa, Australia, New Zealand and the U.S. One of the oldest Chinese medicines, licorice root was also discovered in a Pharaoh's tomb in Egypt. It is used as a flavoring and sweetening agent in chewing gums, candies and beverages. Licorice is also used in certain pharmaceutical products, including antiulcer drugs (Farina et al., 1998), and is reported to have antimutagenic, anti-carcinogenic, antiviral and antiinflammatory properties (Kelloff et al., 1994; Wang et al., 1991). In addition, antihepatitis, anti-HIV, and antiatherogenic effects have been attributed to licorice (Vodovozova et al., 1996). Several triterpenoid saponins, namely, glycyrrhizin (3.63 to 13.06% in dried roots) and glycyrrhetinic acid (α- and β-) have been found in licorice. These are known to inhibit tumorigenesis in mouse lung and liver (Nishino, 1992). Polyphenols found in licorice (Figure 6.10) are also known to serve as important factors responsible for the antiinflammatory, antimutagenic, anticarcinogenic and antiatherogenic effects of licorice (Fuhrman et al., 1997).

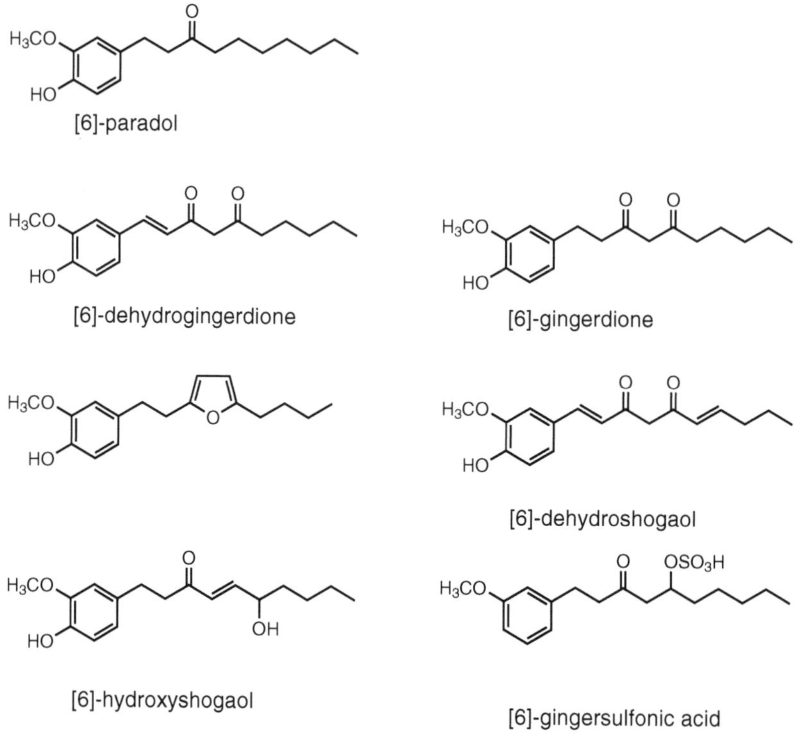

[6]-paradol

[6]-dehydrogingerdione

[6]-gingerdione

[6]-dehydroshogaol

[6]-hydroxyshogaol

[6]-gingersulfonic acid

FIGURE 6.8 Chemical structures of other gingerol-related compounds of ginger.

MILK THISTLE

Milk thistle has been used since antiquity for the treatment of liver disease. Wagner et al. (1971) isolated and identified its active principles as silmaryn, a mixture of flavonolignans, with hepatoprotective effect in laboratory animals exposed to a variety of toxic compounds. It has also been shown to be effective in treating liver cirrhosis in alcoholic subjects and hepatitis (Sonnenbichler et al., 1998). Silibinin makes up approximately 60% of the main active components (Lawson and Bauer, 1998) of milk thistle. Silymarin constituents of milk thistle are shown in Figure 6.11.

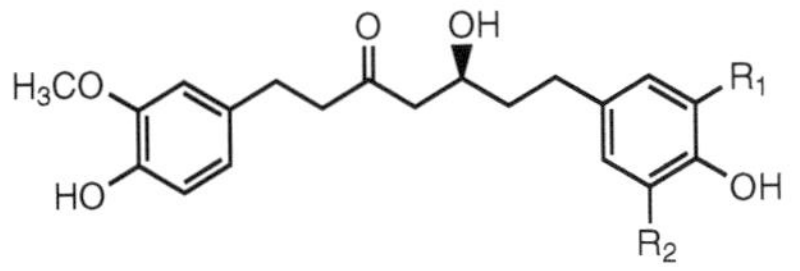

$R_1 = R_2 = H$
$R_1 = R_2 = OCH_3$
$R_1 = OCH_3$, $R_2 = H$; Hexahydrocurcumin

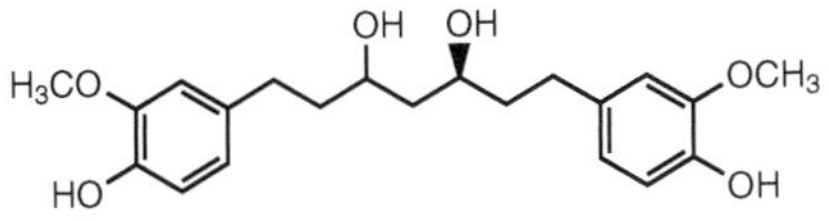

3R or 3S

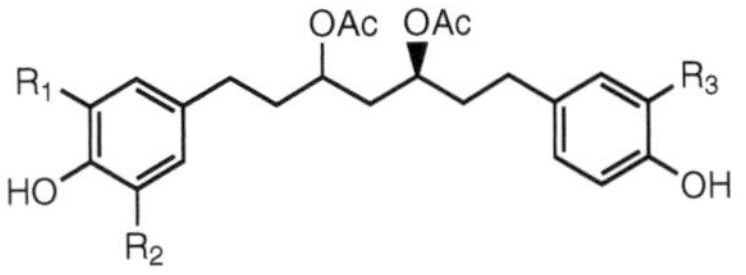

$R_1 = R_3 = OCH_3$, $R_2 = H$; (3R, 5S)
$R_1 = R_2 = R_3 = OCH_3$; (rel RS)
$R_1 = R_2 = R_3 = OCH_3$; (3S, 5S)
$R_1 = OCH_3$, $R_2 = H$, $R_3 = OH$; (rel RS)
$R_1 = R_3 = OH$, $R_2 = H$; (3S 5S)

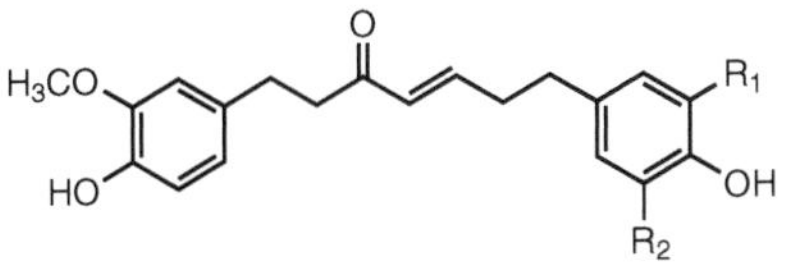

$R_1 = OCH_3$, $R_2 = H$, GingerenoneA
$R_1 = R_2 = OCH_3$
$R_1 = R_2 = H$

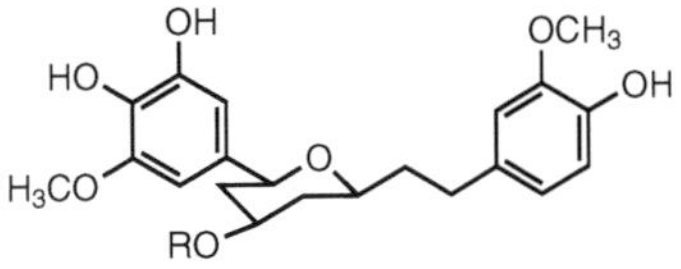

R = H
R = Ac

FIGURE 6.9 Chemical structures of diarylheptanoids of ginger.

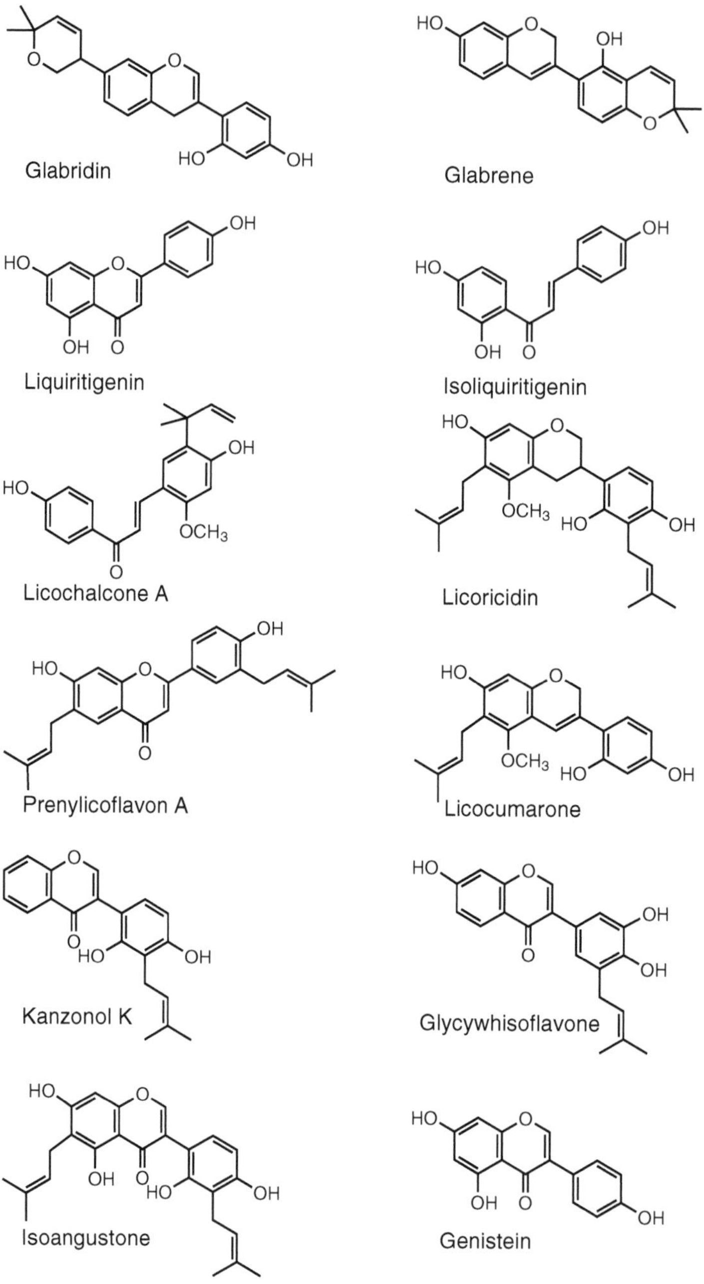

FIGURE 6.10 Chemical structures of polyphenolic compounds of licorice.

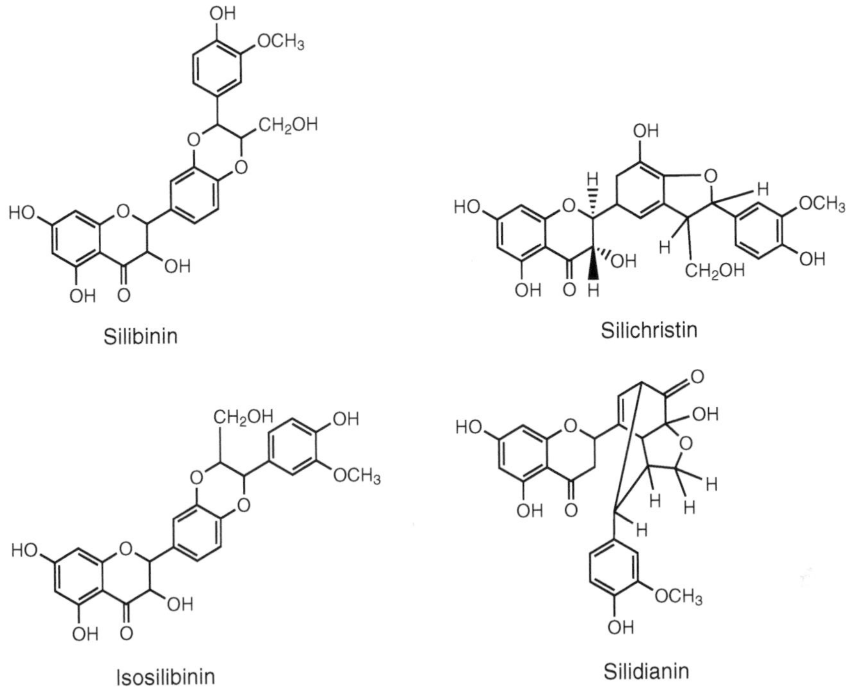

Silibinin

Silichristin

Isosilibinin

Silidianin

FIGURE 6.11 Chemical structures of silymarin constituents of milk thistle.

REFERENCES

Alhorn, R. 1992. Phytochemische und vegetations periodische Undersuchungen von *Edsinacea purpura* (L.) MOENCH unter Berücksichtigungder Kaffeesäurederiuate. Ph.D. Thesis. Universität Marburg/Lahn, Germany.

Backhausß, C. and Krieglstein, J. 1992. Extracts of kava (*Piper methysticum*) and its methysticin constituents protect brain tissue against ischemic damage in rodents. *Eur. J. Pharmacol.*, 215:265–279.

Bauer, R. and Von Hagen-Plettenberg, F. 1999. Der Einfluß von Erntezeitpunkt und Blütenanteil auf die Qualität von *Echinacea purpurea*-Frischpflanzen-Preßsäften, in *Fachtagung Arzneiund Gewürzpflanzen*, Marguard, R. and Schubert, E., Eds., Fachverlag Köhler, Giessen, Germany, 93–100.

Bauer, R., Remiger, P., Jurcic, K., and Wagner, H. 1989. Beeinflussung der Phagozytose-Aktivität durch *Echinacea*-Extrakte. *Z. Phytother*, 10:43–48.

Baureithel, K.H., Buter, K.B., Engesser, A., Burkard, W., and Schaffner, W. 1997. Inhibition of benzodiazepine binding *in vitro* by amentoflavone, a constituent of various species of Hypericum. *Pharm. Act. Helv.*, 72(3):153–157.

Becker, H., Hsieh, W.-C., Wylde, R., Laffite, C., and Andary, C. 1982. Struktur von Echinacosid. *Z. Naturforsch*, 37:351–353.

Becker, L.E. and Skipworth, G.B. 1975. Ginkgo-tree dermatitis, stomatis, and proctitis. *JAMA*, 231:1162–1163.

Bennett, B.C. and Hicklin, J.R. 1998. Uses of saw palmetto (*Serenoa repens*, Arecaceae) in Florida. *Econ. Bot.*, 52:381–393.

Bisset, N.G. 1994. *Herbal Drugs and Phytoharmaceuticals: A Handbook for Practice on a Scientific Basis*, CRC Press, London.

Blumenthal, M. 2000. *Herbal Medicine: Expanded Commission E Monographs,* Integrative Medicine Communications, Newton, MA.

Blumenthal, M., Busse, W., and Goldberg, A. 1998. *The Complete German Commission E Monographs — Therapeutic Guide to Herbal Medicine,* American Botanical Council, Austin, TX.

Brinker, F. 1994. An overview of conventional, experimental and botanical treatments of non-malignant prostate conditions. *Br. J. Phytother.*, 3:154–176.

Chavez, M.L. and Chavez, P.I. 1997. Saint John's Wort. *Hosp. Pharm.*, 32:1621–1632.

Chen, C., Wei, T., Gao, Z., Zhao, B., Hou, J., Xu, H., Xin, W., and Packer, L. 1999. Different effects of the constituents of EGb761 on apoptosis in rat cerebellar granule cells induced by hydroxyl radicals. *Biochem. Mol. Biol. Int.*, 47:397–405.

Cook, N.C. and Samman, S. 1996. Flavonoids — chemistry, metabolism, cardioprotective effects, and dietary sources. *Nutr. Biochem.*, 7:66–76.

Cott, J. 1995. Natural product formulations available in Europe for psychotropic indications. *Psychopharm. Bull.*, 31:745–751.

D'Amelio, F.S. 1999. *Botanicals: A Phytocosmetic Desk Reference,* CRC Press, Boca Raton, FL.

Duke, J.A. 1985. *Handbook of Medicinal Herbs*, CRC Press Inc., Boca Raton, FL, 238–239.

Facino, A.M., Carini, M., Aldini, G., Saibene, L., Piette, P., and Mauri, P. 1995. Echinoside and caffeoyl conjugates protect collagen from free radical-induced degradation: A potential use of Echinasia extracts in prevention of skin photodamage. *Planta Med.*, 61:510:514.

Facino, R.M, Carini, M., Aldini, C., Marinello, C., Arlandini, E., Franzoi, L., Colombo, M., Pietta, P., and Mauri, P. 1993. Direct characterization of caffeoyl esters with anti-hyaluronidase activity in crude extracts from *Echinacea angustifolia* roots by fast atom bombardment tandem mass spectrometry. *Farmaco*, 48:1447–1461.

Farina, C., Pinza, M., and Pifferi, G. 1998. Synthesis and antiulcer activity of new derivatives of glycyrrhetic, oleanolic and ursolic acids. *Farmaco*, 53:22–32.

Flynn, R. and Roest, M. 1995. *Your Guide to Standardized Herbal Products,* One World Press, Prescott, AZ, 38–39.

Foster, S. 1991. Goldenseal: *Hydrastis canadensis*, in American Botanical Council, Botanical Series No. 309, Austin, TX, 1–4.

Foster, S. and Chongdi, Y. 1992. *Herbal Emissaries*. Healing Arts Press, Rochester, VT, 208.

Fuhrman, B., Buch, S., Vaya, J., Belinky, P.A., Coleman, R., Hayek, T., and Aviran, M. 1997. Licorice extract and its major polyphenol glabridin protect low-density lipoprotein against lipid peroxidation: *in vivo* and *ex vivo* studies in humans and in atherosclerotic apolipoprotein E-deficient mice. *Am. J. Chem. Nutr.*, 66:267–275.

Galal, A.M. 1996. Antimicrobial activity of 6-paradol and related compounds. *Int. J. Pharmacogn.*, 34(1):64–69.

Giese, J. 1999. Taste for nutraceutical products. *Food Technol.*, 53(10):43–47.

Grünwald, J. and Büttel, K. 1996. Der europäische Markt für Phytotherapeutika — Zahlen, Trends, Analysen. *Pharm. Ind.*, 58:209–214.

Hippius, H. 1998. St. John's wort (*Hypericum perforatum*) a herbal antidepressant. *Curr. Med. Res. Opin.*, 14:171–184.

Hobbs, C. 1989. St. John's wort (*Hypericum perforatum* L.). A review. *HerbalGram*, 18/19:24–33. http://www.healthy,net/library/articles/hobbs/hypericm.htm.

Hollman, P.C., Hertog, M.G.L., and Katan, M.B. 1996. Analysis and health effects of flavonoids. *Food Chem.*, 57:43–46.

Houghton, P.J. 1988. The biological activity of valerian and related plants. *J. Ethnopharmacol.*, 22:121–142.

Jaggy, H. and Koch, E. 1997. Chemistry and biology of alkylphenols from ginkgo biloba L. *Pharmazie*, 52:735–738.

Juretzek, W. 1997. *Recent Advances in Ginkgo Biloba Extract (EGb 761). Ginkgo Biloba – A Global Treasure*, Springer-Verlag, Tokyo.

Kawakishi, S., Morimitsu, Y., and Osawa, T. 1994. Chemistry of ginger components and inhibitory factors of the arachidonic acid cascade, in *Food Phytochemicals for Cancer Prevention*. II. *Teas, Speices and Herbs*. Ho, C-T., Osawa, T., Huang, M-T., and Rosen, R.T., Eds., ACS Symposium Series 547, American Chemical Society, Washington, D.C., 244–250.

Kelloff, G.J., Crowell, J.A., Boone, C.W., Steele, V.E., Lubet, R.A., Greenwald, P., Alberts, D.S., Covey, J.M., Doody, L.A., and Knapp, G.G. 1994. Clinical development plan: 18 beta-glycyrrhetinic acid. *J. Cell Biochem. Suppl.*, 20:166–175.

Kikuzaki, H., Kawakishi, S., and Nakatani, N. 1994. Structure of antioxidative compounds in ginger, in *Food Phytochemicals for Cancer Prevention*. II. *Teas, Spices and Herbs*, Ho, C-T., Osawa, T., Huang, M-T., and Rosen, R.T., Eds., ACS Symposium Series 547, American Chemical Society, Washington, D.C., 237–243.

Kikuzaki, H. and Nakatani, N. 1996. Cyclic diarylheptanoids from *Rhizomes* of *Zingiber officinale. Phytochemistry*, 43:273–277.

Kikuzaki, H., Tsai, S.M., and Nakatani, N. 1992. Gingerdiol related compounds from the *Rhizomes* of *Zingiber* officinale. *Phytochemistry*, 31:1783–1786.

Kim, H.K., Son, K.H., Chang, H.W., Kang, S.S., and Kim, H.P. 1998. Amentoflavone, a plant biflavone: a new potential antiinflammatory agent. *Arch. Pharm. Res.*, 21: 406–410.

Kitts, D.D. 2000. Chemistry and pharmacology of ginseng and ginseng products, in *Herbs, Botanicals and Teas*, Mazza, G. and Domah, B.D., Eds., Technomic Publishing Co., Inc. Lancaster, PA, 23–44.

Kitts, D.D., Hu, C., and Wijewickreme, A.N. 1999. Antioxidant activity of a North American ginseng extract, in *Food for Health in the Pacific Rim, 3rd Int. Conf. Food Sci. Technol.*, Whitaker, J.R., Haard, N.F., Shoemaker, C.F., and Singh, P.P., Eds., Food and Nutrition Press, Trumbull, CT, 232–242.

Lawson, L.D. and Bauer, R. 1998. *Phytomedicines of Europe: Chemistry and Biological Activity*. ACS Symposium Series 691. American Chemical Society, Washington, D.C.

Leone, M.G., Cometa, M.F., Palmery, M., and Saso, L. 1996. HPLC determination of major alkaloids extracted from *Hydrastis canadensis* L. *Phytother. Res.*, 10:545–546.

Linde, K., Ramirez, G., Mulrow, C.D., Pauls, A., Weidenhammer, W., and Melchart, D. 1996. St. John's wort for depression — an overview and meta-analysis of randomized clinical trials. *Bri. Med. J.*, 313:253–258.

McGregor, R.L. 1968. The taxonomy of the genus *Echinacea* (Compositae). *Univ. Kansas Sci. Bull.*, 48:113–142.

Messana, I., La Bua, R., and Galeffi, C. 1980. The alkaloids of *Hydrastis canadensis* L. (*Ranunculaceae*). Two new alkaloids: hydrastidine and isohydrastinidine. *Gazz. Chim. Ital.*, 110:539–543.

Michel, P.F. 1986. *Ginkgo Biloba: L'Arbre Qui à Vaincu Le Temps*, Felin, Paris.

Muller, W.E., Singer, A., Wonnemann, M., Hafner, U., Rolli, M., and Schafer, C. 1998. Hyperforin represents the neurotransmitter reuptake inhibiting constituent of hypericum extract. *Pharmacopsychiatry*, 31 (Supp. 1):16–21.

Nagasawa, T., Oura, H., Hiai, S., and Nishinaga, K. 1977. Effect of ginseng extract on ribonucleic acid and protein synthesis in rat kidney. *Chem. Pharm. Bull.* (Tokyo), 25:1665–1670.

Newall, C.A. 1996. *Herbal Medicines: A Guide for Health Care Professionals*, Pharmaceutical Press, London.

Nishino, H. 1992. Antitumor-promoting activity of glycyrrhetinic acid and its related compounds, in *Cancer Chemoprevention,* Wattenberg, L., Lipkin, M., Boone, C.W., and Kelloff, G.J., Eds., CRC Press, Boca Raton, FL, 457–467.

Noguchi, M., Nagai, M., Koeda, M., Nakayama, S., Sakurai, N., Takahisa, M., and Kusano, G. 1998. Vasoactive effects of cumicifugic acids (and 1) and fukinolic acid *incimicifuga rhizome. Biol. Pharm. Bull.*, 21:1163–1168.

Oura, H. and Hiai, S. 1973. Cited in Sakakibara, K., Shibata, Y., Higashi, T., Sanada, S., and Shoji, J. 1975. Effect of ginseng saponins on cholesterol metabolism. I. The level and synthesis of oerum and liver cholesterol in rats treated with ginsenosides. *Chem. Pharm. Bull.* (Tokyo), 23:1017–1024.

Robinson, W.E. 1998. L-chicoric acid, an inhibitor of human immunodeficiency virus type 1 (HIV-1) integrase, improves on the *in vitro* anti-HIV-1 effect of Zidovudine plus a protease inhibitor (AG1350). *Antiviral Res.*, 39:101–111.

Schenk, R. and Franke, R. 1996. Content of echinacoside in *Echinacea* roots of different origin. *Beitr. Züchtungsforsch.*, 2:64–67.

Schulik, P. 1993. *Ginger Common Spice and Wondr Drug*. Herbal Free Press Ltd., Brattleboro, UT, 7.

Snow, J.M. 1998. *Hydrastis canadensis* L. (*Ranunculaceae*). *Prot. Bot. Med.*, 2:25–28.

Sonnenbichler, J., Sonnenbichler, I., and Scalera, F. 1998. Influence of flavonolignan silibinin of milk thistle on hepatocytes and kidney cells, in *Phytomedicines of Europe: Chemistry and Biological Activity*, Lawson, L.D. and Bauer, K., Eds., ACS Symposium Series 691. American Chemical Society, Washington, D.C., 263–277.

Sotaniemi, E.A., Haapakoski, E., and Rautio, A. 1995. Ginseng therapy in noninsulin dependent diabetic patients. *Diabetes Care*, 18:1373–1375.

Thornton, L. 1998. The ethics of wildcrafting. *The Herb Quarterly*, Fall Issue: 41–42.

Tyler, V.E. 1998. Importance of European phytomedicinals in the American market: an overview, in *Phytomedicines of Europe: Chemistry and Biological Activity,* Lawson, L.D. and Bauer, R. Eds., ACS Symposium Series 691. American Chemical Society, Washington, D.C., 2–12.

Upton, R. 1997. St. John's wort monograph. American herbal pharmacoepea and therapeutic compendium. *HerbalGram*, 40(5):1–32.

USP 23 – NF 18. 1999. The United States Pharmacopeia XXIII and the National Formulary XVIII. United States Pharmacopeial Convention, Inc., Rockville, MD, 4454–4455.

Van Beek, T.A., Bombardelli, E., Morazzoni, P., and Peterlongo, F. 1998. Ginkgo biloba L. *Fitoterapia*, 69:195–243.

Vodovozova, E.L., Pavlova, I.B., Pobshkina, M.A., Rzhaninova, A.A., Garaev, M.M., and Molotkovskii, I.G. 1996. New phospholipids inhibitors of human immunodeficiency virus reproduction. Synthesis and antiviral activity. *Bioorg. Khim.*, 22:451–457.

Wagner, H., Seligmann, O., Hörhammer, L., Seitz, M., and Sonnenbichler, J. 1971. Zur Struktur von Silychristin Einem Zweiten silymarin-isomeren aus *Silybum marianum. Tetrahedron Lett.*, 22:1895–1899.

Wang, Z.Y., Agarwal, R., Zhou, Z.C., Bickers, D.R., and Mukhtar, H. 1991. Inhibition of mutagenicity in *Salmonella typhimurium* and skin tumor initiating and tumor promoting activities in SENCAR mice by glycyrrhetinic acid: comparison of 18 alpha- and 18 beta-stereoisomers. *Carcinogenesis*, 12:187–192.

Wilt, T.J., Ishani, A., Stark, G., MacDonald, R., Lau, J., and Mulrow, C. 1998. Saw palmetto extracts for treatment of benign prostatic hyperplasia. A systematic review. *JAMA*, 280:1604–1609.

Woelk, H., Kapoula, S., and Lehrl, S. 1993. Treatment of patients suffering from anxiety, double blind study: kava special extract vs. benzodiazopines. Z. *Allegmeinmed.*, 69:271–277.

Wu, T.S., Wu, Y.C., Wu, P.L., Chen, C.Y., Leu, Y.L., and Chan, Y.Y. 1998. Structure and synthesis of [n]-dehydroshogaols from *Zingiber officinale. Phytochemistry*, 48:889–891.

Yamada, Y., Kikuzaki, H., and Nakatani, N. 1992. Identification of antimicrobial gingerols from ginger (*Zingiber officinale* Roscoe). *J. Antibact. Antifung. Agents*, 20:309–311.

Zhang, G.D., Zhou, Z.H., Wang, M.Z., and Gao, F.Y. 1979a. Analysis of ginseng II. *Acta Pharmaceut. Sin.*, 15:175–181.

Zhang, G.D., Zhou, Z.H., Wang, M.Z., and Gao, F.Y. 1979b. Analysis of ginseng I. *Acta Pharmcaeut. Sin.*, 14:309–314.

7 Nutritional and Pharmacological Effects of Food Phenolics

Antinutritional effects of polyphenolics have been demonstrated using laboratory animals (Glick and Joslyn, 1970; Myer and Gorbet, 1985; Rostagno et al., 1973) and humans (Hussein and Abbas, 1985; Stavric and Matula, 1992). These effects can be manifested as a decrease in growth rate or feed conversion (Mehansho et al., 1985; Myer and Gorbet, 1985; Rostagno et al., 1973), inhibition of enzymes (Ahmed et al., 1991; Guyot et al., 1996; Oh and Hoff, 1986), and lower egg production (Sell et al., 1983). Diets containing high levels of tannins have been reported to interfere with pancreatic digestion (Driedger and Hatfield, 1972) as well as to cause methionine deficiency by requiring active methyl groups derived from methionine in the detoxification process (Singleton and Kratzer, 1969). A number of reviews have been published on dietary intake and bioavailability of polyphenols (Bravo, 1998; Clifford, 2000; Heider and Fuchs, 1997; Heim et al., 2002; Hollman, 1997; Hollman et al., 1997; Hollman and Katan, 1997, 1999; Peterson and Dwyer, 1998; Remesy et al., 1996; Ren et al., 2001; Santos-Buelga and Scalbert, 2000; Scalbert and Williamson, 2000; Scalbert et al., 2002; Stavric and Matula, 1992; Teissedre and Landrault, 2000; Waterhouse and Walzem, 1998). Therefore, this chapter discusses only topics related to dietary intake and metabolism of polyphenols.

DIETARY INTAKE

Dietary intake of phenolics is greatly affected by the eating habits and preferences of individuals. The average daily intake of dietary polyphenols is about 1 g per person; the main sources are beverages, fruits and, to a lesser extent, vegetables and legumes (Scalbert and Williamson, 2000). Furthermore, factors such as structural diversity of phenolics, lack of standardized and reliable methods for quantification of various classes of phenolics, variations in phenolic content within each food, uneven distribution of phenolics in plant food, and food processing make accurate assessment of dietary intake of food phenolics cumbersome and difficult (Scalbert and Williamson, 2000). Thus, more precise data on the dietary intake of food phenolics could be obtained by carrying out a comprehensive measurement of phenolic contents in a wide range of foods.

Hertog et al. (1993a) reported that the total daily intake of flavonoids by men participating in the Zutphen Elderly Study ranged from 12 to 41.9 mg/day; however, according to Linseisen et al. (1997), the daily Western diet provides approximately

50 mg of different flavonoids. Black tea is the richest source of flavonoids, contributing 61% of total dietary intake, while onions and apples comprise only 13 and 10% of total dietary flavonoid intake, respectively. On the other hand, 2 and 21 mg of flavones and flavonols are consumed daily by the Dutch population (Hertog et al., 1993b).

According to Deschner et al. (1991), the daily intake of quercetin with food is 50 to 500 mg. Kühnau (1976) estimated the average daily intake of anthocyanins in the U.S. population to be 185 to 215 mg, while Hollman and Katan (1999) and Scalbert and Williamson (2000) have found the average daily intake of anthocyanins by humans to range from 25 to 1000 mg. Subsequently, using a representative sample of the Dutch population (Dutch National Food Consumption Survey), Arts et al. (1999) estimated the contribution of tea and chocolate to the daily intake of catechins. These authors reported that chocolate provides 20% of the daily intake of catechins and that about 55% of daily catechin intake is contributed by tea. The highest amount of tea consumption per person (3.16 kg/year) is in Northern Ireland, followed by the U.K. (2.53 kg/year) and Kuwait (2.52 kg/year) (Weisburger, 1999). The daily intake of isoflavones by Japanese ranges from 30 to 40 mg (Kimira et al., 1998; Wakai et al., 1999) compared to a maximum of 5 mg/day in the U.S. (Ho et al., 2000).

ABSORPTION AND METABOLISM

Information on the bioavailability and absorption of food phenolics is diverse, fragmentary and controversial. Figure 7.1 shows possible routes for metabolism of polyphenols ingested by humans; enzymes, phenol sulfotransferases, catechol-*O*-transferases, β-glucosidases, lactase-phloridzin oxidases, and UDP glucuronosyl transferases are involved in polyphenol metabolism (Scalbert and Williamson, 2000). Absorption and bioavailability of polyphenols in the body depend on their metabolism in the small intestine (Gee et al., 2000; Piskula and Terao, 1998; Ren et al., 2001; Sfakianos et al., 1997; Spencer et al., 1999; Walle et al., 1999). Polyphenol metabolism is influenced by factors such as molecular size, lipophilicity, solubility, and pK_a, as well as gastric and intestinal transit time, membrane permeability, lumen pH, and first-pass metabolism (Higuchi et al., 1981; Lin et al., 1999). Only polyphenols not absorbed in the stomach and small intestine and polyphenol metabolites excreted back to the small intestine are degraded by gut microflora (Scalbert and Williamson, 2000; Setchell, 2000).

Glycosylated polyphenols may be absorbed as such or after hydrolysis by intestinal enzymes (Scalbert and Williamson, 2000). Hawksworth et al. (1971) demonstrated β-glucosidase activity by *Lactobacilli, Bacteroides* and *Bifidobacteria* of human gut. Using an *in-situ* perfusion model, Crespy et al. (1999) showed that quercetin is absorbed, metabolized and partially reexcreted by the small intestine. Donovan et al. (2001) have reported that after absorption into intestinal wall, catechins are extensively metabolized. Only glucuronide conjugates of catechin and 3-*O*-methylcatechin (no intact catechin) have been detected in mesenteric plasma. Additional metabolism of polyphenols involving their conjugation by glucuronidation, *O*-methylation, sulfation, or a combination may occur in the liver (Donovan

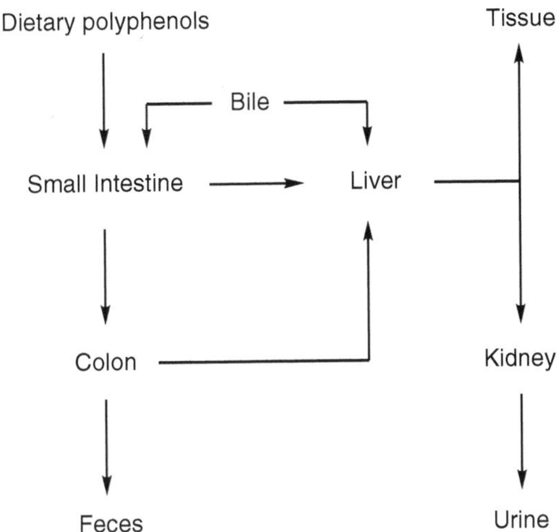

FIGURE 7.1 Possible routes for dietary polyphenols in humans. (Adapted from Scalbert, A. and Williamson, G., 2000, *J. Nutr.*, 130:2073S-2085S.)

et al., 2001; Ren et al., 2001; Scalbert and Williamson, 2000). The conjugated metabolites are then transferred to tissues or excreted in urine or bile (Scalbert and Williamson, 2000; Zhang et al., 1999b).

Phenolic Acids

Olthof et al. (2001) investigated the absorption of chlorogenic and caffeic acids in ileostomy human subjects, in which the degradation of phenolics by gut microflora is minimal. The absorption of chlorogenic and caffeic acids in the small intestine was 33% for chlorogenic acid and 95% for caffeic acid. Only traces of chlorogenic acid and about 11% of ingested caffeic acid were excreted in the urine. According to these authors, chlorogenic acid is absorbed in the intact form and subsequently metabolized in the liver. On the other hand, the absorption of caffeic acid may have occurred by passive nonionic diffusion in the stomach as well as via an active Na^+-dependent transport mechanism in the small intestine. Subsequently, Simonetti et al. (2001) examined the bioavailability of caffeic acid in plasma following the intake of red wine. The concentration of caffeic acid in plasma was dose dependent and the maximum level of caffeic acid was reached approximately 60 min after red wine intake. Recently, Nardini et al. (2002) evaluated the absorption of phenolic acids in humans after the consumption of 200 mL of freshly brewed coffee. Only caffeic acid was detected in plasma, with an absorption peak at 60 min following the intake of coffee.

Gallic acid is one the main phenolics of black tea. Shahrzad and Bitsch (1998) reported that 4-*O*-methylgallic acid is a major metabolite of gallic acid in humans. Later, Hodgson et al. (2000) identified the metabolites of gallic acid in human urine 24 h after 4-week ingestion of five cups of black tea per day. They identified three metabolites in human urine: 4-*O*-methylgallic acid, 3-*O*-methylgallic acid and

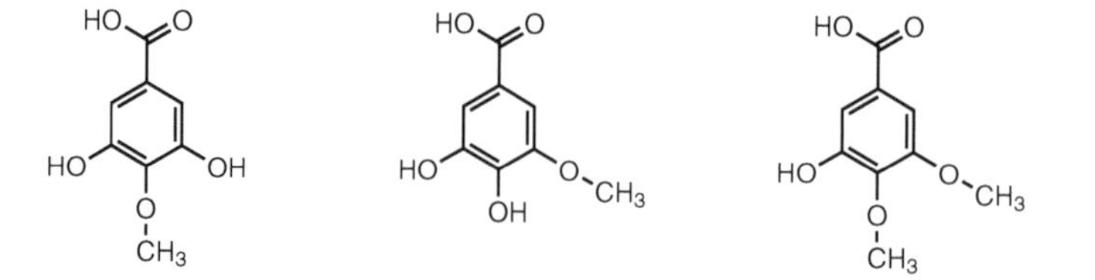

4-O-Methylgallic Acid 3-O-Methylgallic Acid 3,4-O-Dimethylgallic Acid

FIGURE 7.2 Chemical structures of some metabolites of gallic acid.

3,4-*O*-dimethylgallic acid (Figure 7.2). Total excretion of these metabolites was 1.5 mg/day, accounting for only 7.5% of ingested free gallic acid and 1.5% of ingested total gallic acid. According to these authors, the phenolics examined may be used as potential markers of black tea phenolic intake.

Anthocyanins

Little is known about the metabolic fate of anthocyanins; however, published data suggest that they are poorly absorbed. Tsuda et al. (1996) reported that protocatechuic acid may be considered an oxidative degradation product of cyanidin-3-glucoside. Later, Tsuda (1999) also suggested that protocatechuic acid may contribute to the *in vivo* antioxidative activity of cyanidin-3-glucoside. Morazzoni et al. (1991) investigated the pharmokinetics of anthocyanins in rats fed with a mixture of 15 anthocyanins at a level of 400 mg/kg of body weight. Only 2.5 mg/L of anthocyanins were transported into plasma 15 min after ingestion and about 5% of the administered anthocyanins were found intact in urine within 24 h. Later, Lapidot et al. (1998) detected only 1.5 to 5.1% of ingested red wine anthocyanins in the urine of human subjects within 12 h of wine consumption. These authors suggested that anthocyanins absorbed through the guts undergo molecular modification brought about by their antioxidative action.

Yamasaki et al. (1996) demonstrated bleaching of anthocyanins by superoxide ($O_2\bullet^-$) in *in vitro* studies. Furthermore, Cao and Prior (1999) reported poor absorption of cyanidin glucosides from elderberry extract into human plasma and Miyazawa et al. (1999) demonstrated that intact cyanidin glucosides are absorbed across the intestinal wall into plasma and liver of rats as well as into human plasma. The amounts of cyanidins incorporated into plasma were similar to those estimated by Marazzoni et al. (1991). Subsequently, Youdim et al. (2000) demonstrated an incorporation of elderberry anthocyanins into the membrane and cytosol of vascular endothelial cells. Later, Bub et al. (2001) investigated the absorption of malvidin-3-glucoside from red wine, dealcoholized red wine and grape juice; only small amounts of malvidin-3-glucoside were detected in the plasma and urine. Matsumoto et al. (2001) also reported poor absorption of blackcurrant anthocyanins — only 0.11% of anthocyanin intake was excreted 0 to 8 h after ingestion in the intact form. Recently, Milbury et al. (2002) examined the bioavailability of elderberry anthocyanins in humans. These researchers detected anthocyanins in their unchanged glycosylated forms in plasma and urine samples and demonstrated that elimination

TABLE 7.1
Concentrations of Catechin Metabolites[a] in Mesenteric Plasma, Abdominal Aortic Plasma and Bile of Rats after Perfusion with 30 μmol Catechin/L

Catechin Metabolite	Mesenteric Plasma	Aortic Plasma	Bile
Catechin glucuronide	0.7 ± 0.2	0.2 ± 0.1	0.2 ± 0.0
Catechin glucuronide + sulfate	—	—	0.8 ± 0.2
3′-*O*-methyl catechin glucuronide	0.6 ± 0.2	0.3 ± 0.2	9.0 ± 2.2
3′-*O*-methyl catechin glucuronide + sulfate	—	—	16.8 ± 0.4
Total metabolites	1.2 ± 0.1	0.5 ± 0.1	26.8 ± 2.6

Note: Values are means ± standard deviations; n = 4.

[a] Micromoles per liter.

Source: Adapted from Donovan et al., 2001, *J. Nutr.*, 130:1753–1757.

of anthocyanins in plasma follows first-order kinetics. Thus, more research is needed to investigate the metabolic fate of anthocyanins in the human body.

Flavanols and Proanthocyanidins

Kuhnle et al. (2000a) demonstrated that epicatechin and catechin are extensively metabolized and conjugated during transfer from the gut lumen to the serosal surface. In the serosal fluid, glucuronidated, *O*-methylated, and *O*-methylglucuronidated flavanol derivatives account for 45, 30 and 20% of the total flavanols consumed, respectively. Furthermore, Donovan et al. (1999) reported that catechin metabolites are predominantly found in the plasma of human subjects after consumption of red wine. Later, Rein et al. (2000a) detected a 12-fold increase in the plasma epicatechin, from 22 to 257 nmol/L, after consumption of 80 g of procyanidin-rich chocolate containing 557 mg of procyanidin per meal. Unmethylated glucuronide, glucuronide sulfate, unmethylated sulfate of epicatechin and free epicatechin contributed to the increase in plasma epicatechin. Subsequently, using an *in situ* perfusion rat model, Donovan et al. (2001) showed that approximately 35% of catechin is absorbed into the intestinal wall during the first 30-min period. Only glucuronidated and *O*-methylated metabolites of catechin were detected in the mesenteric plasma and abdominal aortic plasma. On the other hand, glucuronidates or sulfates of *O*-methylated catechin metabolites were identified in the bile (Table 7.1).

Recently, Warden et al. (2001) found that only 1.68% of catechins ingested from black tea are detected in the plasma, urine and feces in the free form. Rapid metabolism of catechins by gut bacteria, quick absorption and metabolism of catechins in the gastrointestinal tract or their quick distribution in the tissues may be responsible for a low content of catechins detected in plasma, urine and feces. Li et al. (2000) identified two metabolites of catechins, (–)-5-(3′,4′,5′-trihydroxyphenyl)-γ-valerolactone and (–)-5-(3′,4′-dihydroxyphenyl)-γ-valerolactone, in the

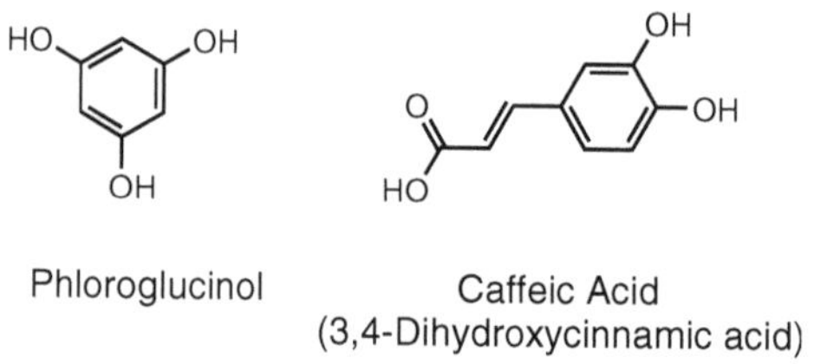

HO OH OH Phloroglucinol

O HO OH OH Caffeic Acid (3,4-Dihydroxycinnamic acid)

FIGURE 7.3 Chemical structures of some metabolites of eriocitrin.

plasma and urine of healthy human subjects consuming green tea. These metabolites may be the products of catechin metabolism by anaerobic gut bacteria. On the other hand, Clifford et al. (2000) recently detected an increase in hippuric acid in the urine of subjects consuming black tea. The rise in hippuric acid may be due to the metabolism of tea polyphenols by gut microflora into 3-phenylpropionic acid followed by an endogenous β-oxidation and conjugation to glycine.

Moridani et al. (2001) reported that enzymatic oxidation of catechin by tyrosinase in the presence of glutathione results in the formation of catechin–glutathione conjugates, such as mono-, di-, and triglutathione conjugates of catechin and mono- and diglutathione conjugates of catechin dimer. When tyrosinase is replaced with a peroxidase/hydrogen peroxide system only different monoglutathione conjugates of catechin are detected. Catechin–glutathione conjugates are also formed in the presence of rat liver microsomes and NADPH.

Spencer et al. (2000) examined the stability of procyanidins isolated from *Theobroma cacao* in simulated gastric juice of pH 2 at 37°C for up to 3.5 h. Dimers were least susceptible to decomposition while higher molecular weight procyanidins (trimers to hexamers) were cleaved to monomers and dimers. Up to 80% of procyanidins were hydrolyzed during the first 90 min of incubation. According to these authors, the decomposition of procyanidins in the gastric milieu may enhance their potential for absorption in the small intestine.

Flavanones and Flavanonols

Miyake et al. (1997) investigated the metabolism of eriocitrin (eriodictyol-7-rutinoside), a flavonoid glycoside in lemon fruits, by 18 different intestinal bacteria cultured from human feces. Eriocitrin was metabolized by bacteria belonging to *Bacteroides, Bifidobacterium, Propionibacterium, Enterbacter, Enterococcus, Lactobacillus* and *Streptococcus* genera. The primary eriocitrin metabolite was identified as eriodictyol, which was then further metabolized into 3,4-dihydroxycinnamic acid and phloroglucinol by *Clostridium butyricum* (Figure 7.3). No eriocitrin was detected after 15 h of culturing, while the presence of eriodictyol and 3,4-dihydroxycinnamic acid was detected after 6 and 9 h of incubation, respectively.

Flavones and Flavonols

Olthof et al. (2000) demonstrated that quercetin-3-glucoside and quercetin-4′-glucoside are rapidly absorbed by healthy human subjects. The peak concentrations of

TABLE 7.2
Kinetics of Quercetin Absorption and Elimination in Human Plasma after One-Time Ingestion of Quercetin-3-Glucoside or Quercetin-4-Glucoside

Property	Quercetin-3-glucoside	Quercetin-4-glucoside
Peak concentration (μg/L)	1526 ± 315	1345 ± 212
Time to reach peak concentration (min)	37 ± 12	27 ± 5
Elimination half-life (h)	18.5 ± 0.8	17.7 ± 0.9

Note: Values are means ± standard deviations; n = 9.

Source: Adapted from Olthof, M.R. et al., 2000, *J. Nutr.*, 130:1200–1203.

quercetin in the blood were reached 37 ± 12 and 27 ± 5 min after ingestion of quercetin-3-glucoside and quercetin-4′-glucoside, respectively. The half-life of elimination of quercetin from the blood was 18.5 ± 0.8 h and 17.7 ± 0.9 h after ingestion of quercetin-3-glucoside and quercetin-4′-glucoside, respectively (Table 7.2). According to Walle et al. (2000), quercetin glucosides are completely hydrolyzed in ileostomy patients before absorption. Only 19.5 to 35.2% of total ingested quercetin glucosides was detected in the ileostomy fluid. Using intestinal Caco-2 cell monolayers as a model of human intestinal absorption, Walgren et al. (1998, 2000a, b) demonstrated that quercetin is absorbed more rapidly than its glycosides. These data are in contrast to those reported by Gee et al. (2000) and Hollman et al. (1995, 1996), who demonstrated that quercetin monoglucosides are transported more rapidly than quercetin aglycone in the small intestine.

It has been proposed that sugar moiety may be involved in transporting flavonoid glucoside by the sodium-dependent glucose transporter-1 (SGLT1) (Lostao et al., 1994; Panayotova-Heiermann et al., 1996). Gee et al. (2000) observed rapid deglycosylation and then glucuronidation of flavonoid glycosides during passage across the rat small intestine epithelium. However, because the location of hydrolysis is still unknown, several possible locations have been suggested. Recently, de Vries et al. (2001) investigated the absorption of quercetin in humans after consumption of similar quantities of quercetin from red wine, yellow onions or tea. The absorption of quercetin was estimated based on its concentration in the plasma and urine. Results suggest that the bioavailability of quercetin from red wine is lower than that from onions, but slightly higher than that from tea.

Booth et al. (1956) identified hydroxyphenylacetic and phenylpropionic acids in the urine of rats fed with rutin (quercetin-3-O-rhamnoglucoside). Later, Bokkenheuser et al. (1987) and Winter et al. (1989) reported that rutin is metabolized by the intestinal microflora to quercetin, phloroglucinol and 3,4-dihydroxyphenylacetic acid. Schneider et al. (1999) estimated that 10^7 to 10^9 fecal bacteria per gram of dry mass are able to degrade rutin and suggested that the degradation of quercetin by *Eubacterum ramulus* requires the presence of sugar liberated from rutin. Justesen et al. (2000) investigated the metabolism of rutin and quercetin by human fecal microflora using an *in vitro* model. Rutin was almost completely degraded within

48 h of culturing and only 5% of rutin originally present in fermenting medium was detected. On the other hand, quercetin was completely metabolized into six metabolites after 8 h of incubation. Five of these metabolites were identified: 3,4-dihydoxytoluol, 3,4-dihydroxybenzaldehyde, 3- or 4-hydroxyphenylacetic acid, 3- or 4-hydroxypropionic acid and 3,4-dihydroxyphenylacetic acid.

Isoflavones

Joannou et al. (1995) postulated that isoflavone β-glycosides are hydrolyzed by β-glucosidase present in the jejunum and then further metabolized in the distal intestine. Meanwhile, Lundh (1995) reported that isoflavones are predominantly (95%) conjugated in the liver with glucuronic acid. Isoflavone aglycones and their bacterial metabolites undergo mainly glucuronidation during uptake by the enterocyte (Axelson and Setchell, 1981; Setchell, 2000; Sfakianos et al., 1997; Zhang et al., 1999a). The extent of this metabolism is highly variable among individuals and affected by dietary factors as well (Setchell and Cassidy, 1999). Xu et al. (1995) reported that isoflavone bioavailability varies from 13 to 35% depending on the ability of individual gut microflora to degrade isoflavones. According to Andlauer et al. (2000), the bioavailability of isoflavones is significantly affected by the food matrix. Subsequently, Izumi et al. (2000) observed that isoflavone aglycones in fermented soy food products are absorbed more rapidly than their corresponding glycosides.

Piskula et al. (1999) examined the absorption of orally administered isoflavone aglycones and glucosides in surgically treated rats with absorption site restricted to the stomach. The absorption was measured by monitoring the level of isoflavone metabolites in blood plasma for 30 min. Results indicated that isoflavones are rapidly absorbed in the rat stomach. A few minutes' delay in the detection of metabolites of isoflavone glucosides in plasma compared to those of aglycones was observed. This, according to the authors, indicates that isoflavone aglycones, but not their glucosides, are aborbed from the rat stomach.

Using the simulator of the human intestinal microbial ecosystem, de Boever et al. (2000) demonstrated the cleavage of β-glycosidic bond of conjugated isoflavones under the conditions prevailing in the large intestine. Hur et al. (2000) isolated two strains of human microflora, *Escherichia coli* HGH21 and an unindentified Gram-positive strain HGH6, metabolizing daidzin and genistin to their respective aglycones. Administration of antibiotics significantly blocks the metabolism of isoflavonones. Also, limited isoflavone metabolism has been observed in infants of up to 4 months old, when gut microflora is underdeveloped. These data suggest that gut microflora plays an important role in isoflavone metabolism (Axelson and Setchell, 1981; Setchell et al., 1997, 1998).

Setchell et al. (2001) have suggested that, based on pK_a values, isoflavone aglycones are absorbed from jejunum by the nonionic passive diffusion mechanism. A maximum concentration of isoflavones in blood plasma is observed between 4 and 7 h and 8 to 11 h after ingesting isoflavone aglycones and their β-glycosides, respectively (Setchell et al., 2001; Watanabe et al., 1998). Presence of isoflavones

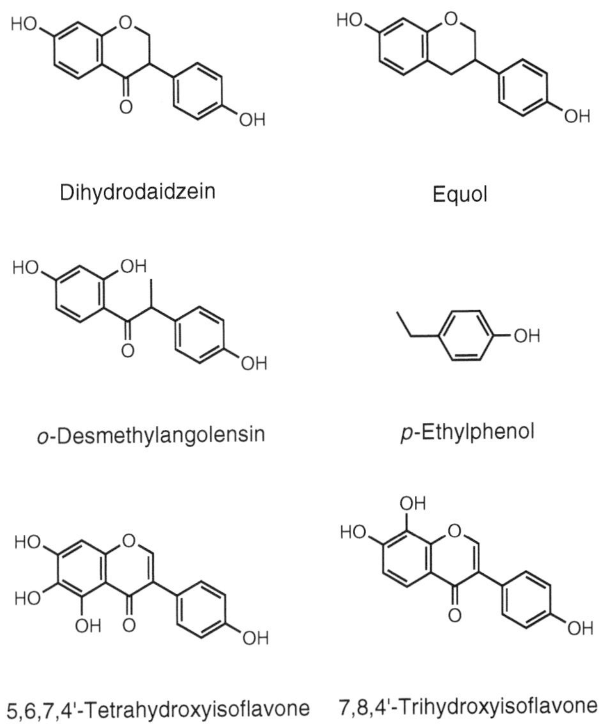

Dihydrodaidzein Equol

o-Desmethylangolensin *p*-Ethylphenol

5,6,7,4'-Tetrahydroxyisoflavone 7,8,4'-Trihydroxyisoflavone

FIGURE 7.4 Chemical structures of some metabolites of isoflavones.

in plasma has been detected up to 24 h after consumption (Setchell, 2000; Zhang et al., 1999b).

Figure 7.4 shows that aglycones, genistein and daidzein may be further metabolized to equol, 7-hydroxy-3-(4′-hydroxyphenyl)chroman, dihydrodaidzein, *O*-desmethylangolensin and *p*-ethylphenol (Chang and Nair, 1995; Setchell and Cassidy, 1999). Daidzein is first transformed by gut microflora to dihydrodaidzein, which is then metabolized to equol and *O*-desmethylangolensin. On the other hand, genistein is first transformed by gut microflora to dihydrogenistein and then converted to 6′-hydroxy-*O*-desmethylangolensin, which is degraded to 4-hydroxyphenyl-2-propionic acid. Decarboxylation of this acid leads to the formation of *p*-ethylphenol (Coldham et al., 1999; Joannou et al., 1995; Kurzer and Xu, 1997). The amounts of aglycones and their metabolites recovered in urine and feces are summarized in Table 7.3. Of the isoflavone metabolites, equol possesses an estrogenic activity of an order higher than that displayed by daidzein (Axelson et al., 1982a; Shutt and Cox, 1972). Higher conversion of daidzein to equol has been observed with diets higher in carbohydrates (Lampe et al., 1998; Shutt and Cox, 1972).

Kulling et al. (2001) studied the oxidative metabolites of isoflavones excreted in the urine of human subjects ingesting soya products for 2 days. The following compounds were identified as oxidative metabolites of daidzein:

TABLE 7.3
Recovery of Daidzein, Genistein, *O*-Desmethylangolensin and Equol in Urine and Feces of Men for 3 Days after Ingestion of 60 g of Kinako (Baked Soybean Powder)[a]

	Urine		Feces	
	Range	Mean ± SD	Range	Mean ± SD
Daidzein	27.28–64.56	36.97 ± 13.90	1.07–12.59	4.57 ± 4.04
Genistein	9.68–47.91	19.71 ± 14.03	0.20–7.14	2.85 ± 2.82
O-desmethylangolensin	0.59–11.70	4.80 ± 4.10	0.08–6.12	1.97 ± 1.97
Equol	0.07–33.51	7.60 ± 13.37	0.08–8.12	1.67 ± 3.07

Note: SD = standard; n = 7.

[a] Micromoles.

Source: Adapted from Watanabe, S. et al., 1998, *J. Nutr.*, 128:1710–1715.

- 6,7,4′-Trihydroxyisoflavone
- 7,3′,4′-Trihydroxyisoflavone
- 7,8,4′-Trihydroxyisoflavone
- 7,8,3′,4′-Tetrahydroxyisoflavone
- 6,7,8,4′-Tetrahydroxyisoflavone
- 6,7,3′,4′-Tetrahydroxyisoflavone

Of these, only 6,7,8,4′-tetrahydroxyisoflavone is detected in trace amounts. On the other hand, the main oxidative metabolites of genistein are 5,6,7,4′-tetrahydroxy-isoflavone, 5,7,8,4′-tetrahydroxyisoflavone, and 5,7,3′4′-tetrahydroxyisoflavone; 5,7,8,3′4′-pentahydroxyisoflavone and 5,6,7,3′,4′-pentahydroxyisoflavone are identified as minor metabolites.

Fang et al. (2002) identified 17 metabolites of isoflavones in the urine of female rats fed on a diet containing soy protein isolates. The detected isoflavone metabolites indicated that any of the hydroxyl groups of the isoflavone molecule may be involved in the formation of glucuronides. In addition, glucuronides of equol and *O*-desmetylangolensin were also identified in rat urine.

Other Phenolics

Kuhnle et al. (2000b) used an *in vitro* model system to investigate the absorption and metabolism of resveratrol. These authors noted that only small amounts unmetabolized resveratrol are absorbed across the enterocytes of jejenum and ileum and identified the glucuronide conjugate of resveratrol as a major metabolite on the serosal side. Based on mass spectrometry data, these authors suggest that the 4′-hydroxyl site of resveratrol serves as the preferential site for glucuronidation.

Malathi and Crane (1969) detected lactase–phloridzin hydrolase activity in the brush border of hamster small intestine. This enzyme hydrolyzes phloridzin (phloretin 2′-*O*-glucose) to phloretin and glucose. Later, Monge et al. (1984) demonstrated that phloretin is metabolized by fecal microflora to phloretic acid and phloroglucinol. Recently, Crespy et al. (2002) reported that phloretin is absorbed more rapidly than phloridzin. However, only free phloretin and its glucuronidated or sulfated conjugates have been detected in plasma up to 24 h after ingesting meal containing phloretin or phloridzin and only 10.4% of the ingested phloretin and phloridzin recovered in the urine after 24 h.

Nakazawa and Ohsawa (1998) investigated the metabolism of rosmarinic acid administered orally to rats. They found metabolites attributed to this acid only in the urine and plasma of rats, detecting four in the plasma: *trans*-caffeic acid 4-*O*-sulfate, *trans*-*m*-coumaric acid 3-*O*-sulfate, *trans*-ferulic acid 4-*O*-sulfate, *m*-hydroxypropionic acid and *trans*-*m*-coumaric acid. In addition, unchanged rosmarinic acid and *trans*-caffeic acid were identified in the urine. The total amount of rosmarinic acid and its metabolites excreted in the urine 48 h after the oral administration accounted for 31.8% of the administered dose.

For many years, plant lignans matairesinol and secoisolariciresinol were considered the only precursors of mammalian lignans enterodiol and enterolactone (Axelson et al., 1982b; Borriello et al., 1985). Recently, Heinonen et al. (2001) demonstrated that lariciresinol and pinoresinol are also effectively converted by gut microflora to mammalian lignans. It has been suggested that the microbial metabolism of pinoresinol involves the formation of secoisolariciresinol via lariciresinol. Other known plant lignans are not effectively converted to mammalian lignans.

Only limited data have been published on the metabolism of hydroxytyrosol, the principal phenolic component in olive oil. Manna et al. (2000) reported that, in Caco-2-cells, hydroxytyrosol is converted to homovanillic alcohol. Subsequently, while investigating the metabolism of hydroxytyrosol after oral consumption of olive oil by human subjects, Visioli et al. (2000) identified only unmetabolized hydroxytyrosol and its glucuronide conjugate in the urine. Reexamination of these urine samples led to the identification of homovanillic acid (4-hydroxy-3-methoxyphenylacetic acid) and homovanillic alcohol (Figure 7.5) (Caruso et al., 2001). Following this, Tuck et al. (2001) observed extensive absorption of hydroxytyrosol administered to rats in the form of olive oil solution. Five unidentified hydroxytyrosol metabolites were detected in the urine, three of which were later identified: a monosulfate conjugate, monoglucuronide conjugate and 3-hydroxy-4-methoxyphenylacetic acid (Figure 7.5) (Tuck et al., 2002).

Because a number of coumarin metabolites have been identified, namely, 3-, 4-, 5-, 6-, 7-, and 8-hydroxycoumarins, *o*-hydroxyphenylacetyladehyde (*o*-HPA) and *o*-hydroxyphenylacetic acid (*o*-HPAA) (Fentem and Fry, 1993; Pelkonen et al., 1997), different pathways may be involved in the metabolism of coumarins in various tissues (Pelkonen et al., 1997). Of these, 8-hydroxycoumarin, *o*-HPAA, and *o*-HPA are the major metabolites of coumarin formed by rat esophageal microsomes and cytochrome P450 enzymes (von Weymarn and Murphy, 2001). According to Born et al. (1997, 2000), *o*-HPA is formed from the unstable coumarin 3,4-epoxide, while *o*-HPAA is a product of *o*-HPA oxidation.

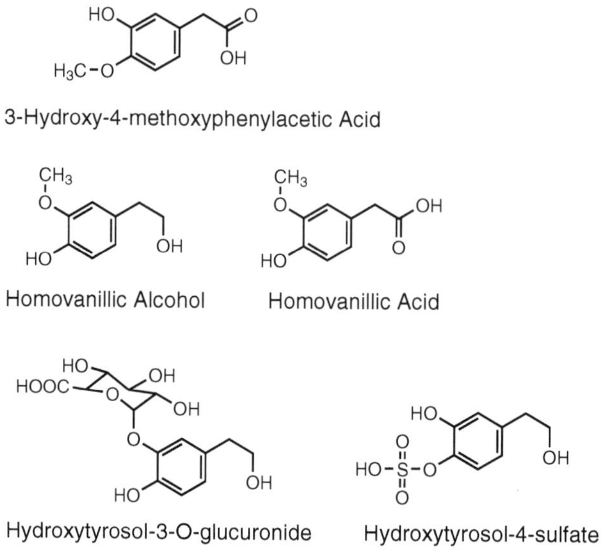

3-Hydroxy-4-methoxyphenylacetic Acid

Homovanillic Alcohol

Homovanillic Acid

Hydroxytyrosol-3-O-glucuronide

Hydroxytyrosol-4-sulfate

FIGURE 7.5 Chemical structures of some metabolites of hydroxytyrosol.

NUTRITIONAL IMPLICATIONS

Tannin–protein complexes may account for the antinutritional effects of tannin-containing feeds in nonruminants (Clandinin and Robblee, 1981; Martin-Tanguy et al., 1977) and ruminants (Kumar and Singh, 1984). Detrimental effects, including a decrease in food palatability, voluntary intake and growth of experimental animals and a decrease in fiber and protein digestibility (Barry and Duncan, 1984; Barry and Manley, 1984; Barry et al., 1986; Jung and Fahey, 1983; Mitaru et al., 1984a), have been associated with ingestion of polyphenols (Glick, 1981; Reddy et al., 1985). Recently, Carbonaro et al. (2001) demonstrated partial binding of catechins and tannic acid by endogenous proteins in the intestinal lumen. These authors postulated that this binding is responsible for antinutritional effects displayed by tannic acid because it may interfere with the function of gut.

Mitjavila et al. (1977) reported increased fecal calcium loss by rats fed 3% tannic acid in their diet as a result of increased endogenous gut secretion. Marquardt and Ward (1979) observed that tannins account for about 50% of the growth depression of chicks fed a faba bean diet. The tannin content was highly correlated with decreased weight gain, decreased feed intake, and decreased protein and increased fat retention. Utilization of barley proteins by rats has also been negatively correlated with the level of tannins (0.55% to 1.23%) in samples (Eggum and Christensen, 1975). The reduction in utilization of proteins may be due to the binding of tannins to digestive enzymes or to dietary proteins. Condensed tannins may also bind to methionine, thus making it unavailable and, in turn, lowering the utilization of dietary proteins (Armstrong et al., 1974; Ford and Hewitt, 1974).

Tannins in rapeseed meal are reported to cause tainting of eggs. It has been postulated that they block the metabolism of trimethylamine (TMA) by inhibiting

TMA oxidase (Butler et al., 1982; Fenwick et al., 1981, 1984). This enzyme converts TMA to odorless, water-soluble TMA oxide. Addition of tannins extracted from rapeseed meal to soybean-containing diets for chicks results in a reduction of their metabolizable energy; however, it has no apparent effect on the absorption of proteins by chicks (Yapar and Clandinin, 1972). In another study, Mitaru et al. (1982) reported that condensed tannins isolated from rapeseed hulls do not inhibit the activity of α-amylase *in vitro*. However, other *in vitro* studies have indicated that tannins may show stimulatory, rather than inhibitory, effects on protein digestion (Mole and Waterman, 1985; Oh and Hoff, 1986), perhaps due to partial denaturation of protein substrates.

On the other hand, it has been shown that water extracts of field bean testa inhibit the *in vitro* activity of α–amylase, lipase, and trypsin (Griffith, 1979). Reporting that sorghum tannins inhibit the activity of α-amylase, Davis and Hoseney (1979) found a near-linear inhibition of enzymes that is quite sensitive to tannin concentration in the range between 0.0115 and 0.014 mg of tannins/mL (Figure 7.6). Later, Griffith and Moseley (1980) found that trypsin and α–amylase activities in the intestine of rats fed on a diet containing high levels of tannins (field bean testa) are significantly reduced. However, trypsin activity is recovered when guts were extracted with a polyvinylpyrrolidone–saline solution. These results indicate that tannins of field bean testa are responsible for the inhibition of trypsin activity.

Moreover, under *in vivo* conditions, digestive enzymes are less susceptible to inhibition by tannins due to competition of other proteins, such as salivary proteins for binding (Mehansho et al., 1987; Robbins et al., 1987). Phospholipids and bile acids in the intestine may also cause breakdown of some tannin–protein complexes (Blytt et al., 1988).

INTERACTIONS WITH PROTEINS

The complexation of polyphenols and their enzymatic and nonenzymatic oxidation products with proteins in seeds, meals or flours may reduce the nutritive value of the proteins involved. Oxidized phenolics may react with amino acids and proteins and inhibit the activity of enzymes such as trypsin and lipase (Milic et al., 1968). The ability of polyphenols to form insoluble complexes with proteins interferes with utilization of dietary proteins (Butler et al., 1986). In addition, a decrease in the antioxidant activity of phenolic–protein systems has also been associated with phenolic–protein interactions (Hainonen et al., 1998).

Formation of Phenolic Acid–Protein Complexes

Phenolic acids can form complexes with food proteins, thus lowering nutritional value of proteins. Loomis and Battaile (1966) suggested that phenols may reversibly complex with proteins by a hydrogen-bonding mechanism or irreversibly by oxidation to quinones that combine with reactive groups of protein molecules. Wade et al. (1969) found that the binding of serum albumin correlated well with the pK_a of simple phenols. Thus, the hydrogen bond from phenol to protein is stronger for more acidic phenols. Products of enzymatic and nonenzymatic oxidation of phenolics in

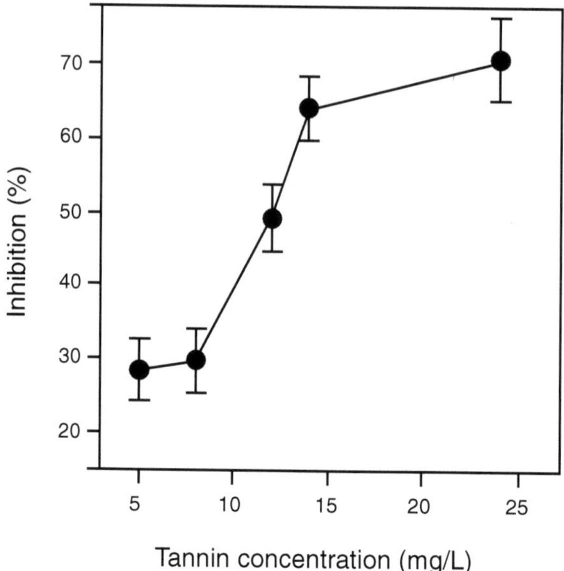

FIGURE 7.6 Percent α-amylase inhibition vs. the concentration of condensed tannins (mg/mL). Standard deviation at 0.012 mg tannins/mL = 4.3. (Adapted from Davis, A.B. and Hoseney, R.C., 1979, *Cereal Chem.*, 56:310–314.)

seeds, meals or flours may readily react with the ε-NH_2 group of lysine and CH_3S group of methionine of enzymes and other proteins to form complexes, thus rendering them nutritionally unavailable to monogastric animals (Rutkowski et al., 1977).

Bartolome et al. (2000) used model systems to study interactions between a mixture of low molecular weight phenolics (*p*-hydroxybenzoic, protocatechuic, *p*-coumaric and caffeic acids, and catechin) and bovine serum albumin (BSA). This mixture was separated using Sephadex G-50. Change in the elution profile was used as an indicator of affinity of phenolics for BSA. Protocatechuic and caffeic acids exhibited the strongest affinity for BSA while *p*-hydroxybenzoic acid did not interact with it at all. Crude extracts of lentil phenolics displayed interactions with BSA similar to those observed in the model systems.

The possibility of phenolic–protein complex formation has also been indirectly evidenced from the amount of soluble matter extracted into 80% ethanol. Kozlowska and Zadernowski (1988) reported that the quantity of extracted matter increases with increasing pH of the 80% ethanol used for its extraction (Figure 7.7). The formation of these complexes has also been investigated in model systems consisting of sinapic acid and BSA by using a fluorescence spectrophotometric technique (Figure 7.8). These studies indicated that complex formation is favored under neutral and basic conditions (Smyk and Drabent, 1989).

Sunflower protein solutions under neutral and alkaline conditions develop dark-green and brown colors, presumably due to the formation of complexes between oxidized polyphenols and proteins. Sabir et al. (1973) reported bonding of chlorogenic acid to low molecular weight sunflower proteins, which constitute about 15%

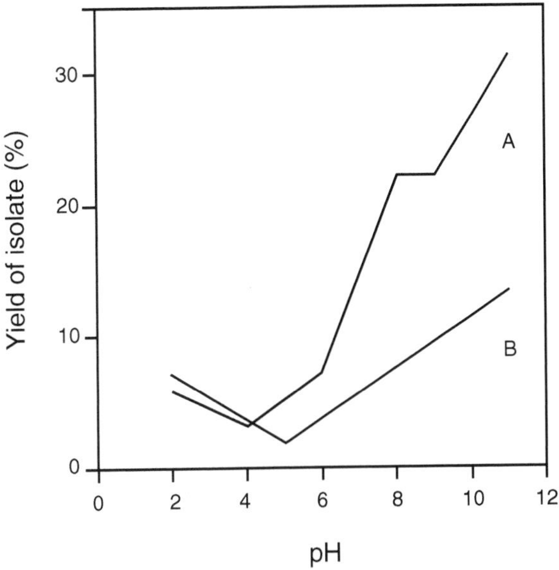

FIGURE 7.7 Effect of pH on the yield of isolate (A) and on the amount of substances extracted from rapeseed with 80% ethanol (B). (Adapted from Kozlowska, H. et al., 1990, in *Rapeseed and Canola: Production, Chemistry, Nutrition and Processing Technology*, Shahidi, F., Ed., Van Nostrand Reinhold, New York, 193–210.)

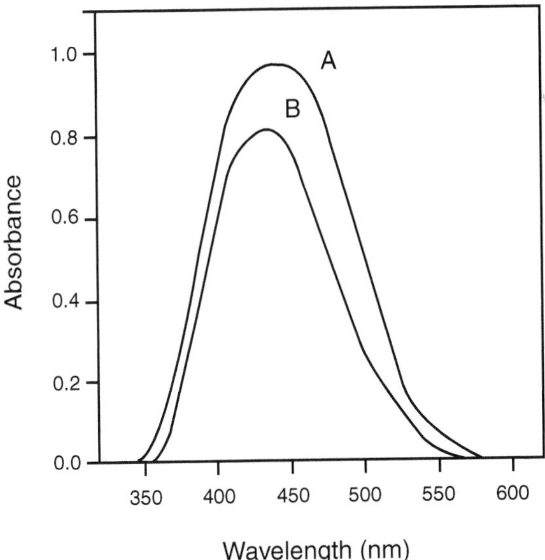

FIGURE 7.8 The fluorescence spectra of sinopic acid (A) and sinopic acid-bovine serum albumin. (Adapted from Kozlowska, H. et al., 1990, in *Rapeseed and Canola: Production, Chemistry, Nutrition and Processing Technology*, Shahidi, F., Ed., Van Nostrand Reinhold, New York, 193–210.)

of the salt-extractable proteins in sunflower flour. Later, Sabir et al. (1974) reported that 30% of chlorogenic acid is nondialyzable and remains bound to the flour constituents. These authors also found that aproximately 50% of the polyphenolic compounds is extracted with the fraction of sunflower proteins soluble in neutral salt solutions. Fractionation of the neutral salt soluble protein on Sephadex G-200 revealed that soluble chlorogenic acid is associated with low molecular weight proteins (MW < 5000 Da). In addition, these researchers found that 68% of the soluble chlorogenic acid is hydrogen bonded and 32% covalent bonded to the salt-soluble sunflower proteins.

It has been reported that polyphenols found in beans and peas inhibit the activity of amylases *in vitro* and may also be responsible for diminishing the overall digestibility of carbohydrates in the intestinal tract of rats (Griffith, 1979, 1981). This may be due to phenol–enzyme interaction as well as starch–phenol association.

Dick and Bearne (1988) studied the inhibition of purified β-glucosidase from "Spartan" apples by flavonoids, flavonoid glycosides, phenolic acids and polyphenols. They found that quercetin, fustin, rutin, *p*-coumaric acid, and catechol cause classical noncompetitive inhibition, while keampferol and chlorogenic acid do not. The *trans*-3-(4-hydroxyphenyl)-propenoyl group common to 4′-hydroxyflavones and *p*-hydroxyphenolic acids is an important structural feature for effective inhibition of β-glucosidase activity that has been implicated in the softening process of fruits (Bartley, 1974).

Formation of Tannin–Protein Complex

Tannins may form soluble or insoluble complexes with proteins (Calderon et al., 1968; Hagerman, 1989, 1992; Mole and Waterman, 1982; Siebert, 1999). The specificity of tannin–protein interactions depends on the size, conformation, and charge of the protein molecules (Hagerman and Butler, 1981; Hagerman, 1989, 1992). Relative affinity of proteins for tannins, estimated by using a competitive binding assay, may vary by four orders of magnitude or more (Table 7.4). Thus, proteins with a high affinity for tannins may be selectively bound out of a large excess of proteins with a lesser affinity for tannins (Butler, 1989a, b). Proteins with compact globular structure, such as ribonuclease, lysozyme or cytochrome C, exhibit low affinity for tannins, whereas conformationally open proteins such as gelatin and polyproline readily form complexes with tannins.

Precipitation of tannin–protein complex is due to the formation of a sufficient hydrophobic surface on the complex (McManus et al., 1981). Haslam (1989) suggested nonspecific coating of the protein surface with hydrophobic polyphenols. Later, Kawamoto et al. (1996) and Luck et al. (1994) postulated that the formation of polyphenol–protein precipitate is a two-stage process: (1) soluble polyphenol–protein complexes are formed and (2) subsequent interaction between these complexes leads to the formation of insoluble high molecular weight aggregates. Binding of polyphenols by proteins does not affect their abilities to self-associate, i.e., to form cross links with the protein and other polyphenol molecules. Therefore, aggregation of soluble tannin–protein complexes can be accomplished by noncovalent cross linking mediated by polyphenols (Baxter et al., 1997; Charlton et al., 2002).

TABLE 7.4
Relative Affinity of Proteins for Sorghum Tannins

Protein	Relative Affinity
Gelatin	14.0
Proline-rich salivary	6.8
Pepsin	1.1
Bovine serum albumin	1.0
Bovine hemoglobin	0.068
Ovalbumin	0.016
α-Lactoglobulin	0.0087
Lysozyme	0.0048
Soybean trypsin inhibitor	<0.001

Sources: Adapted from Butler, L.G., 1989a, in *Food Products*, AOCS, Champaign, IL and Hagerman, A.E. and Butler, L.G., 1981, *J. Biol. Chem.*, 256:4494–4497.,

According to Kawamoto and Nakatsubo (1997) pH, temperature, and ionic strength have a greater effect on the formation of insoluble than soluble tannin–protein complexes. These authors also suggested that the absence of precipitation does not indicate lack of binding between tannins and proteins. Subsequently, Silber et al. (1998) postulated that the precipitation of tannin–protein occurs when a critical number of tannin molecules are associated with a protein molecule. At low concentrations of proteins, the precipitation is due to the formation of a hydrophobic monolayer of polyphenols on the protein surface. However, at higher concentrations, the hydrophobic surface results from complexing of polyphenols on the protein surface and cross linking of different protein molecules with polyphenols. Thus, the stoichiometry of protein–phenol complex depends on the protein concentration in the solution.

In the presence of an excess amount of tannins, precipitates with a fixed ratio of tannin to protein are formed (20 mol of polyphenol to 1 mol of protein for epicatechin (4→8) catechin (EC_{16}-C) and 40 mol of polyphenol to 1 mol of protein for pentagalloylglucose) (Hagerman et al., 1998). The formation of insoluble protein–phenol complexes may be reversed by the addition of an excess amount of protein (Haslam, 1989). Binding a tannin to a protein molecule may also bring about changes in the conformational structure of both molecules, thus lowering their solubility (Asquith and Butler, 1986).

The lowest solubility of tannin–protein complex occurs at a pH near the isoelectric point of the protein (Hagerman and Butler, 1981; Van Buren and Robinson, 1969). Figure 7.9 shows the effect of pH on the amount of canola tannins precipitating with selected proteins. Bovine serum albumin (BSA), fetuin, gelatin and pepsin were precipitated at pH values between 3.0 and 5.0, but the maximum precipitation of lysozyme occurred at pH > 8.0. The precipitation of proteins by tannins depends on the availability of unionized phenolic groups for hydrogen bonding (Naczk et al., 1996).

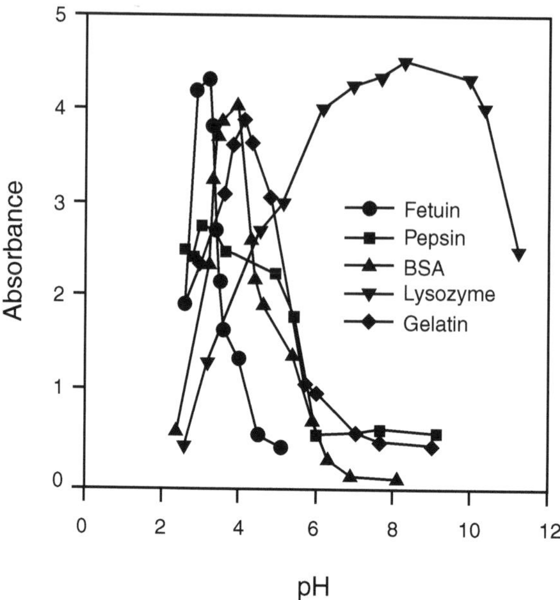

FIGURE 7.9 pH dependence of complex formation between crude condensed tannins of canola hulls and several proteins. (Adapted from Naczk, M. et al., 1996, *J. Agric. Food Chem.*, 44:2144–2148.)

The tannin–protein interaction also depends on the initial concentration of tannins and proteins. All proteins are precipitated in the presence of excess tannins (Hagerman and Robbins, 1987); however, an excess of protein leads to the formation of soluble protein–tannin complexes (Calderon et al., 1968; Hagerman and Robbins, 1987; McManus et al., 1981). The relationship between the amount of tannin–protein complex precipitated and the quantity of protein added to the reaction mixture is described by a bell-shaped curve (Hagerman and Robinson, 1987; Naczk et al., 1996; Silber et al., 1998). Silber et al. (1998) developed a quantitative model describing the dependence of the yield of protein–tannic acid precipitate as a function of protein and tannic acid concentrations. This model was derived based on the assumption that the random binding of tannic acid to proteins follows a Poisson-type distribution and that the protein-binding sites are identical and noninteractive. According to these authors, the proposed model described reasonably well the experimental data obtained for BSA and gelatin.

According to Hagerman (1980) and Hagerman and Butler (1981), proteins with a high affinity for tannins usually have a high molecular weight and open and loose conformational structure, and are rich in proline and other hydrophobic amino acids. Siebert (1999) also indicated that haze-forming activity of polypetides in beer is affected by the content of tannins present. Asquith et al. (1987) reported that the glycosylated salivary proteins have enhanced affinity and selectivity towards binding of tannins. Removing the carbohydrate portion from a salivary proline-rich protein (GP66sm) results in reduction of the affinity of deglycosylated protein for sorghum tannins by a four-fold factor and for quebracho tannins by an eight-fold factor

(Asquith et al., 1985; Mehensho et al., 1987). The enhanced affinity of glycosylated proteins for tannins may be due to the ability of the oligosaccharide portion of the proteins to maintain the protein structure in a relatively open conformation. Thus, this kind of protein structure would more readily form hydrogen bonds with tannins.

Subsequently, Lu and Bennick (1998) demonstrated that the fraction of basic proline-rich salivary proteins displays a greater affinity for condensed and hydrolyzable tannins than the acidic and glycolysated proline-rich salivary protein fractions. Later, Naurato et al. (1999) reported that human salivary histatins are more effective tannin precipitants than human salivary proline-rich proteins. Histatins are a proline-free group of peptides that are high in histidine and exhibit antibacterial and antifungal activities (Azen et al., 1978; Yan and Bennick, 1995). Perez-Madonaldo (1994) observed that soluble complexes between salivary proteins from sheep and goats and tannic acid and condensed tannins are formed only at 38°C. The formation of insoluble complexes was noticed after the reaction mixture was kept for 12 h at 5°C. Similar observations have been reported for salivary proteins of cattle (Jones and Mangan, 1977) and cattle and sheep (Austin et al., 1989).

Binding of proteins to tannins is also affected by the presence of surfactants. Martin and Martin (1984) observed that higher concentrations of tannins are required for precipitation of rubilose-1,5-bisphosphate carboxylase/oxygenase (RuBPC) from *Menduca sexta* gut fluid adjusted to pH 6.5. These authors suggested that interference of surfactants present in the insect gut fluids might be responsible for this observation. Later, Martin et al. (1985) demonstrated that, between pH 6.15 and 7.55, lysolecithin (0.06%), a surfactant detected in the gut fluids of *Pierris brassicae* (Turunen and Kastari, 1979), prevents the precipitation of RuBPC by tannic acid. However, at pH 8.3 and 9.3, lysolecithin decreases the amount of RuBPC precipitated from 49 to 16% and from 54 to 21%, respectively.

Binding of proteins to tannins may be due to the formation of multiple hydrogen bonds between the hydroxyl groups of tannins and the carbonyl functionalities of peptide bonds of proteins (Gustavson, 1954; Haslam, 1974; Loomis and Battaile, 1966; Spencer et al., 1988). The tannin–protein complex may also be stabilized by other types of molecular interactions such as ionic bonds between the phenolate anion and the cationic site of protein molecules (Loomis, 1974) or covalent links formed as a result of condensation of oxidized phenolic groups of tannins with nucleophilic groups (SH, OH, NH_2) of proteins (Haslam, 1974; Loomis, 1974; Pierpoint, 1969). Figure 7.10 shows the hydrophobic interactions between the aromatic rings of tannins and hydrophobic regions of proteins that may also be involved (Goldstein and Swain, 1965; Hagerman and Butler, 1980; Hagerman et al., 1998; Loomis, 1974; Naczk et al., 2001; Oh et al., 1980; Spencer et al., 1988).

The 1,2-di- (or 1,2,3-tri-) hydroxyphenyl residue is considered the prime binding site of tannins. It is believed, however, that tannin–protein complexation is usually the result of formation of hydrogen bonds and hydrophobic interactions (Hagerman and Butler, 1981; Oh et al., 1980), particularly under acidic conditions (McManus et al., 1985). Hagerman and Butler (1978) did not observe any precipitation of proteins at pH values above pK_a of the phenolic groups. Based on these results, it has been suggested that the ionic bonds between protein and tannin moieties are less important.

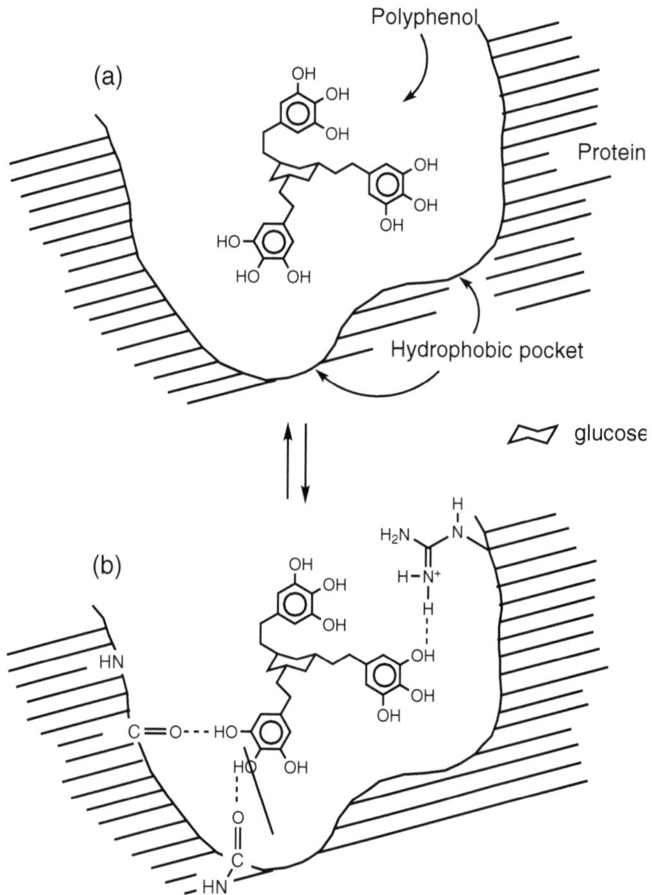

FIGURE 7.10 Polyphenol–protein complexation: (A) docking of the polyphenol; (B) hydrogen bonding to the protein surface. (Adapted from Spencer, C.M. et al., 1988, *Phytochemistry*, 27:2397–2409.)

Hoff and Singleton (1977) demonstrated that BSA immobilized on Sepharose binds tannins selectively at pH 4. Monomeric polyphenolic substances such as gallic acid, chlorogenic acid and catechol do not complex with proteins under these conditions. They also reported that protein–tannin complexes can be dissociated by organic solvents such as methanol and dimethylformamide. Later, Serafini et al. (1997) postulated that ethanol is responsible for the increase in bioavalibility of polyphenols in red wine by inhibiting the interactions between proteins and polyphenols.

The formation of tannin–protein complex is affected not only by the composition and structure of proteins but also by the size, length and flexibility of tannin molecules and the number and stereospecificity of binding sites on the tannin molecule. Haslam et al. (1992) suggested the existence of an inverse correlation between the solubility of polyphenols in water and their ability to precipitate proteins. It has been demonstrated that tannins should possess at least three flavonol subunits to be effective protein-precipitating agents. Dimers precipitate proteins, but are much less effective; simple flavonols do not precipitate proteins at all (Artz et al., 1987; Bate-Smith, 1973; Roux, 1972). Similarly, Porter and Woodruffe (1984) found that the ability of condensed tannins to form insoluble complexes with proteins (hemoglobin) depends more on the molecular weight of tannins than on the configuration and number of hydroxyl groups on the B-ring of tannins. Later, Sarni-Manchado et al. (1999) reported that high molecular weight procyanidins (average degree of polymerization of 7.4) are selectively precipitated by salivary proteins from solutions that also contain dimers and trimers.

The configuration and number of hydroxyl groups on the B-ring may also affect the ability of tannins to precipitate proteins. It has been demonstrated that flavonols with three orthohydroxyl groups, like prodelphinidins, bind proteins more tightly than those with two orthohydroxyl groups on their B-ring, such as procyanidins (Asano et al., 1984; Bate-Smith, 1975). Later, Bacon and Rhodes (1998) found a relationship between the degree of galloylation of flavan-3-ol monomers from tea and their affinity to human parotid saliva proteins. The compounds were ranked in the order of their ability to bind to salivary proteins:

(–)-epigallocatechin gallate > (–)-gallocatechin gallate > (–)-epicatechin gallate > (–)-epigallocatechin = (–)-epicatechin = (+)-catechin

Subsequently, de Freitas and Mateus (2001) showed that tannin–protein interactions are affected by the stereochemistry of interflavonoid linkage. Procyanidins with a 4→8 interflavonoid linkage have a greater affinity for proteins than those with a 4→6 linkage. In addition, esterification of (–)-epicatechin or (–)-epicatechin moiety of procyanidin dimer with gallic acid at C3 carbon enhances the affinity of these polyphenols for proteins.

The protein-precipitating capacity of hydrolyzable tannins is influenced by the degree of galloylation, hexahydroxydiphenol esterification, and polymerization (Bacon and Rhodes, 2000; Hagerman et al., 1992; Kawamoto et al., 1996), as well as the flexibility of tannin molecules (Haslam et al., 1992). Beart et al. (1985) observed that the affinity of galloyl esters of D-glucose for bovine serum albumin is enhanced with the addition of every galloyl ester group. Similarly, Ozawa et al. (1987) reported that the binding of β-glucosidase is affected by the degree of glucose galloylation. Subsequently, based on analysis of the galloylglucose–protein co-precipitates, Kawamoto et al. (1995, 1996) found a positive linear relationship between the degree of galloylation and the affinity of tannins for BSA. The compounds were ranked in order of their ability to bind to BSA:

penta- > tetra- > 2,3,6-tri- > 2,3,4-tri- >> di- >> monogalloylglucose

Furthermore, Baxter et al. (1997) reported that more than one galloyl ester group of hydrolyzable tannins may be involved in association of polyphenol with proline-rich peptide, while Haslam (1998) postulated that the ability of polyphenols to form multiple bonds with salivary proteins is affected by their molecular weights. Thus, tannins are able to form numerous bonds with adjacent proline residues of protein molecules but simple phenolics may only bind to a single proline residue.

Based on evaluation of protein-precipitating activities exhibited by pentagalloylglucose and procyanidin (epicatechin$_{16}$ (4→8) catechin; EC$_{16}$-C), Hagerman et al. (1998) recently reported that tannin–protein interactions are strongly influenced by the structure of tannins. Hydrogen bonds are primarily involved in interactions between EC$_{16}$-C and protein, while hydrophobic interactions are the predominant mode of interaction between pentagalloylglucose and proteins. Hydrophobic interactions also play an important role in complexation of proteins with condensed tannins isolated from grape (Oh et al., 1980) and canola hulls (Naczk et al., 2001), and in the formation of haze in beer (Siebert et al., 1996).

INTERACTIONS WITH CARBOHYDRATES

Polyphenols may complex with carbohydrates; the affinity of polyphenols for polysaccharides depends strongly on molecular size, conformational mobility and shape, as well as solubility of polyphenols in water. Thus, an increase in the molecular size and conformational flexibility of tannins enhances their affinity for carbohydrates (Cai et al., 1989).

The affinity of phenolics for carbohydrates is well represented by their association with Sephadex chromatographic gels. The complexation may be due to inclusion of phenolics within the pores of Sephadex, or interaction between oxygen atoms from ether groups that cross link gels and phenolic hydroxyl groups or between a phenyl ring acting as an electron donor and the hydroxyl groups of gels (Brook and Housley, 1969; deLigny, 1979; Haglund, 1978). According to Cai et al. (1989), even subtle changes in the structure of polyphenols may drastically affect the retention of polyphenols by Sephadex gel.

Polyphenolics show a strong affinity toward cyclodextrins. Ya et al. (1989) suggested that a "key and lock" type matching is probably responsible for the association of phenolics within the cavity of a cyclodextrin molecule. This association depends on the size of the cavity, as well as the configuration of the phenolic compound involved. Later, Bianco et al. (1997) applied NMR methodology to investigate the interactions of 3,4-dihydroxyphenylacetic acid (DHPA) with β-cyclodextrin. The thermodynamic data indicated that these interactions are controlled by enthalpy with an opposing entropy contribution.

Davis and Harbers (1974) found that starch prepared by wet milling of bird-resistant sorghum is less susceptible to attack by enzymes than other starches. They suggested that absorption and retention of condensed tannins on starch might be responsible for this phenomenon. Later, Davis and Hoseney (1979) reported that, depending on the source of tannins as well as the original type of starch, 40 to 60% of tannins are bound to starch (Figure 7.11). These researchers also found two α-amylase-inhibiting fractions; one is absorbed on starch and the other one is not.

Deshpande and Salunkhe (1982) studied the interaction of tannic acid and catechin with different legume and potato starches in model systems. Starches associated up to 652 μg of tannic acid and up to 586 μg of catechin per 100-mg

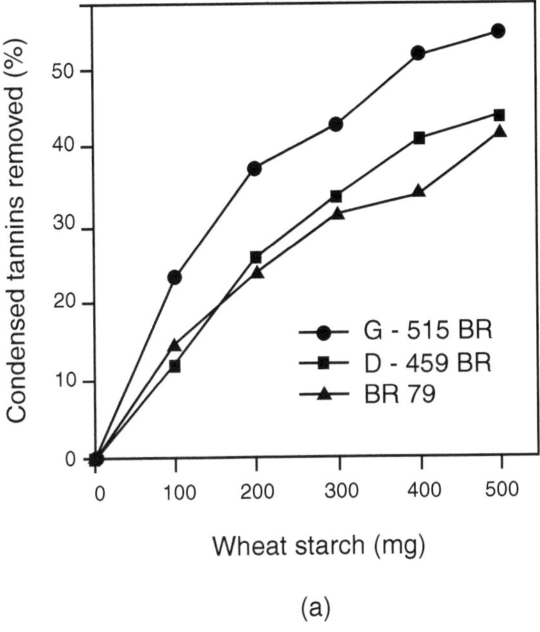

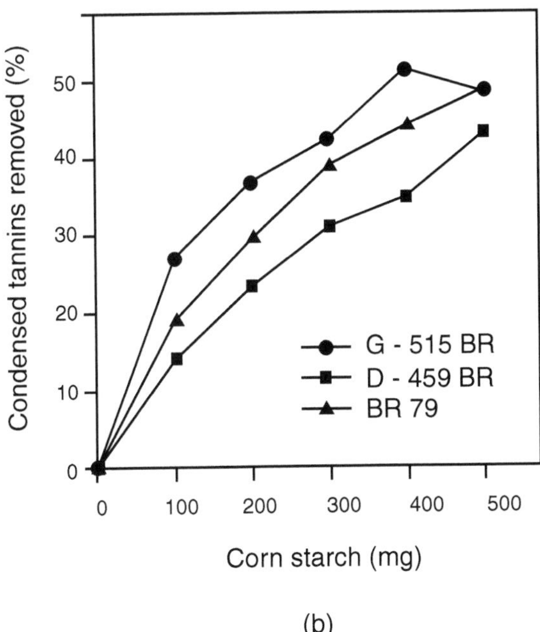

FIGURE 7.11 Percent condensed tannins removed by certain amounts of wheat (A) and corn (B) starch. (Adapted from Davis, A.B. and Hoseney, R.C., 1979, *Cereal Chem.*, 56:310–314.)

TABLE 7.5
Binding of Tannic Acid to Different Starch/Starch Fractions[a]

	Starch/Starch Fraction				
Conditions	Split Yellow Pea	Small Red Bean	Potato	Amylose	Amylopectin
21°C for 4 h	522	261	358	652	587
95°C for 30 min	394	278	267	186	214

[a] Milligrams of catechin equivalents per 100 g of starch.

Source: Adapted from Deshpande, S.S. and Salunkhe, D.K., 1982, *J. Food Sci.*, 47:2080–2081, 2083.

sample. A 30-min heating at 95°C substantially decreased the ability of starch molecules to associate with tannic acid and catechin (Table 7.5).

It has been reported that some phenolic acids may have flatulence-inhibiting properties. Rackis et al. (1970) found that phenolic acids such as syringic and ferulic acids inhibit flatulence activity of soybean meals in *in vitro* and *in vivo* studies.

INTERACTIONS WITH MINERALS AND VITAMINS

Tannins containing *o*-dihyhroxyphenyl groups may form a stable complex with a wide range of essential minerals, thus lowering their bioavailibility (Conrad, 1970; Faithful, 1984; Kennedy and Powell, 1985; Powell and Rate, 1987; Slabbert, 1991). Scalbert (1991) and Mila and Scalbert (1994) suggested that the formation of tannin–metal complexes may play a similar role for plant defense against microorganisms as the complexation of tannins with proteins does. Furthermore, many industrial applications of tannins involve their complexation with metal ions. These include using tannins as modifiers of rheological properties of minerals and clays (Chang and Anderson, 1968), chelators for metal recovery from waste waters (Randall et al., 1974), components of anticorrosive primers (Seavell, 1978), and writing inks (Grimshaw, 1976).

McDonald et al. (1996) used model systems for studying the precipitation of copper (II) and zinc (II) by low molecular weight phenols, penta-*O*-galloyl-β-D-glucose, and commercial tannin at pH 5. They found that the precipitation of polyphenol–metal complexes is affected by initial concentrations of the metal ion and phenol in the reaction mixture and the control of acidification of the reaction mixture brought about by this complexation.

Low availability of iron in the diet is considered an important factor contributing to the high prevalence of iron deficiency anemia in developing countries (Tatala et al., 1998). Phenolic compounds in tea (Disler et al., 1975a; Hallberg and Rossander, 1982; Hurrell et al., 1998, 1999; South et al., 1997), vegetables (Gillooly et al., 1983), wines (Bezwoda et al., 1985; Cook et al., 1995), sorghum (Gillooly et al., 1984) and coffee (Brune et al., 1989; Derman et al., 1977; Hallberg and Rossander, 1982; Hurrell et al., 1998; Morck et al., 1983) have been identified as possible inhibitors of iron absorption through the gut barrier. This inhibition may be due to formation of insoluble iron (III)–phenol complexes, thus making the iron unavailable for absorption in the gastrointestinal tract (Gust and Suwalski, 1994; Porter, 1992; Slabbert, 1992). Phenolic compounds bearing galloyl and catechol groups are mainly responsible for inhibition of iron absorption (Brune et al., 1989, 1991; Gust and Suwalski, 1994; Slabbert, 1992; Zijp et al., 2000).

Gallooly et al. (1983) found an inverse relationship between the content of polyphenols in foods and the degree of inhibition of iron absorption. Later, Brune et al. (1989) reported that the inhibition of iron absorption by tannic acid is a function of its dose. Gallic acid decreases the iron absorption to the same extent that tannic acid does, but no inhibition is observed in the presence of catechin. On the other hand, addition of ascorbic acid increases the absorption of nonheme iron in the presence of polyphenols due to the reduction of Fe(III) to Fe(II) (Siengeberg et al., 1991). Germination of finger millet also improves the accessible iron due to a reduction in phytate and tannin contents (Mbithi-Mwikya et al., 2000; Sripriya et al., 1997; Udayasekhara and Deostahale, 1988). In addition, Matuschek et al. (2001) have demonstrated that reduction of phytate in high-tannin cereals (sorghum and finger millet), followed by oxidation of polyphenols with polyphenol oxidase (mushroom tyrosinase), increases the amount of iron available for absorption approximately 1.5 and 3 times in finger millet and sorghum, respectively.

A number of researchers have reported that adding tea to test meals for a period of 1 to 2 weeks lowers the retention of iron in the blood of human subjects and that chronic tea drinking is related to the incidence of anemia. It has been suggested that flavonoids present in tea may be responsible for decreasing iron absorption from test meal by 63 to 91%, thus lowering its incorporation into blood cells (Brune et al., 1989; Derman et al., 1977; Disler et al., 1975a, b; Morck et al., 1983; Reddy et al., 1991; Rossander et al., 1979). The results of these studies are quite diversified: the consumption of a cup of tea with bread as a test meal (Disler et al., 1975a) and a cup of tea dextrimaltose as a test meal (Reddy and Cook, 1991) bring about inhibition of iron absorption by 64 and 91%, respectively. Later, Zijp et al. (2000) reported that strength of the tea consumed does not have a major impact on iron absorption and that the consumption of black tea infusion with iron-containing foods decreases iron absorption by 60 to 70%. However, drinking black tea between meals inhibits iron absorption only by about 20%.

Caffeic acid and tea flavonoids have been reported to possess an antithiamine effect. Formation of thiamine oxidation products is probably implicated because addition of ascorbic acid prevents this antithiamine effect (Rungruangsak et al., 1977; Somogyi, 1978). Tannic acid has been found to precipitate vitamin B_{12}, which makes it unavailable and results in anemia (Carrera et al., 1973).

REMEDIES FOR ANTINUTRITIONAL EFFECTS OF FOOD PHENOLICS

Several methods for lowering the potential antinutritional effect of phenolics in the diet have been evaluated. These include supplementing the diet with phenol-binding

TABLE 7.6
Effects of Dehulling on Removal of Tannins and Proteins from Sorghum Seeds

Variety	Dehulled Fraction (% of Grain)	Tannins (%)	
		In Dehulled Grain	Removed
BR 64 (high tannin)	0	4.54	—
	12.3	3.99	33.0
	24.3	1.58	73.6
	37.0	0.17	97.6
BR 626 (low tannin)	0	0.54	—
	11.3	0.35	42.5
	23.4	0.30	57.5
	36.0	0.19	77.5

Source: Adapted from Chibber, B.A.K. et al., 1978, *J. Agric. Food Chem.*, 26:679–683.

materials, dehulling, cooking, soaking in water and chemical solutions, and promoting metabolic detoxification (Butler, 1989b; Salunkhe et al., 1989).

The antinutritional effects of tannins for poultry can be overcome by supplementing diet with methionine and choline (Chang and Fuller, 1964). Methionine may participitate as a methyl group donor in the detoxification methylation of tannin breakdown products (Salunkhe et al., 1989). The growth-depressing effect of tannins can also be alleviated by increasing the level of proteins in the diet. The supplementary proteins, such as gelatin, serve as tannin-binding agents and not as a source of amino acids (Rogler et al., 1985; Schaffert et al., 1974). Weight gain and feed-to-weight ratio can be significantly improved by adding tannin-binding polymers such as polyvinylpyrrolidone to the diet (Armstrong et al., 1973; Marquardt et al., 1977).

Because tannins are mostly located in the pericarp and testa of seed coats, tannin content in cereals and legumes can be significantly lowered by dehulling grains. According to Chibber et al. (1978), the efficiency of the dehulling process in the removal of tannins depends on the amount of dry matter removed from seeds. The tannin content of high-tannin sorghum can also be reduced with cooking (Mitaru et al., 1984b), steaming (Ekpenyong, 1985), soaking and germination (Elmaki et al., 1999; Iwuoha and Aina, 1997; Mukuru et al., 1992; Obizoba and Atii, 1991).

Approximately 37% of dry matter of BR64 sorghum must be removed in order to lower its tannin content by 98% (Table 7.6). A poor recovery of sorghum grains dehulled with a mechanized milling process has been noted (Reichert et al., 1988); however, wet and dry milling techniques are commonly used in Asia for commercial dehulling of grain legumes. The dehulling removes from 24.4 to 98.2% of tannins initially present in soybean and black gram beans, respectively (Salunkhe et al., 1989).

Heat processing seeds or their meals may lower their level of assayable tannins, perhaps because of the formation of insoluble complexes between tannins and other seed components (Butler, 1989a). The amount of assayable tannins and loss of bound

TABLE 7.7
Effects of Soaking on Content of Tannins in Dry *Phaseolus* Beans[a]

Treatment	Pinto Beans	Cranberry Beans	Viva Pink Beans
Raw beans	264.7	76.3	122.1
Beans soaked in			
Distilled water			
6 h	208.6	67.0	100.0
12 h	96.1	57.0	80.0
18 h	69.9	51.0	73.0
Sodium bicarbonate (2%) 12 h	24.1	25.0	27.0
Mixed salt solution[b]	18.0	20.0	20.0

[a] Milligrams of catechin equivalents per 100 g of starch.
[b] Mixed aqueous salt solution contained 2.5% (w/v) sodium chloride, 1% (w/v) sodium tripolyphosphate, and 1.5% (w/v) sodium carbonate.

Source: Adapted from Deshpande, S.S. and Cheryan, M., 1983, *Nutr. Rep. Int.*, 27:371–377.

tannins in cooked black beans account for 60.4% compared to 20.5% in raw black beans (Bressani et al., 1982). Furthermore, Price et al. (1980) have reported that cooking high-tannin sorghum grains does not improve their nutritional value, but instead depresses the growth rate of rats fed on such diets.

Several chemical treatments have been proposed to improve the nutritional value of cereals and legumes. These include ammoniation, treatment with alkaline solutions, acids and formaldehyde (Butler, 1989b). Ammoniation of high-tannin sorghum grains at 80 psi for 1 h significantly decreases the content of assayable tannins and improves the growth of chickens fed on such grains; however, similar treatment of low-tannin sorghum has a detrimental effect on its nutritional value. The mechanism responsible for this detrimental effect in low-tannin sorghum has not been elucidated (Price et al., 1978, 1979). On the other hand, soaking cowpea seeds in an alkaline or acidic solution decreases the content of assayble tannins by approximately 84%. Soaking in a 0.5-M solution of NaOH, KOH or acetic acid is most effective in removing polyphenols from cowpeas (Laurena et al., 1986). Similarly, as shown in Table 7.7, up to 93% of tannins are removed from dry *Phaseolus* beans after 12 h of soaking in a 2% sodium bicarbonate or a mixed-salt solution (Deshpande and Cheryan, 1983). Makkar and Becker (1996) have also reported that tannins are effectively inactivated in the presence oxidizing agents under alkaline conditions.

TOXICITY OF ABSORBED POLYPHENOLS

Morton (1978) attributed esophageal cancer to the use of tannin-rich beverages such as tea. A significant correlation exists between the incidence of cancer in some parts of the world and excessive drinking of tea (Morton, 1989) and apple-based drinks. Apple ciders are made of fruits rich in tannins, which give the beverage its bitterness and astringency (Haslam, 1989). According to Haslam (1989), frequently drinking beverages with high tannin content may damage the esophageal epithelium, thus increasing its susceptibility to known carcinogens. Singleton (1981) attributed the effect of tannins to irritation and cellular damage to DNA mutagenic-type cancer. Salunkhe et al. (1989) have also associated the possibility of high risk of cancer incidence to the chronic ingestion of high quantities of tannins in the diet. However, they suggested that regular consumption of tannins may induce the development of a defensive mechanism by animals and human beings to lower the risk of cancer.

When Singleton and Kratzer (1969) studied the acute toxicity of tannins, they found that LD_{50} values for rats, mice and rabbits for a single dose of orally administered tannins range from 2.25 to 6.00 g/kg of body weight. They suggested that phenols may become toxic if natural barriers or detoxification mechanisms are overloaded by the amount of ingested phenols. The toxicity level of phenols depends also on the manner of their administration and is affected by the presence of substances containing diether or isopropenoid structures. According to Singleton (1981), small- and large-size phenolics may cause nutritional implications as a result of consuming the metabolized energy in their detoxification process or by lowering the contribution of methyls or glucuronic acid to more useful metabolisms. Phenolic compounds can produce a long-term toxic effect at the magnitude of common phenol intakes ranging from 1 to 5% in the diet. These values are much higher for insoluble phenolics such as lignin or ellagic acid. General agreement is that phenolics from common foodstuffs at the regular intake have very low toxicities. This is attributed to their low absorption and rapid metabolism, as well as the presence of an efficient defense mechanism in mammals (Lake, 1999; Singleton, 1981; Williams et al., 1999).

Wren et al. (2002) assessed the safety of grape extract containing less than 5.5% catechin monomers. The grape seed extract was administered to rats in their feed at levels of 0, 0.5, 1.0, and 2.0% for 90 days. A significant increase in body weight in rats consuming grape seed extract compared to that of the control rats was observed. No significant toxicological effects in rats were noticed during this study. Furthermore, potential adverse effects of excessive intake of flavonoids should not be overlooked as high intakes of food phenolics may potentially lead to mutagenesis, prooxidant and inhibition of key enzymes (Galati et al., 2002; Skibola and Smith, 2000). More research is still required to establish toxicological effects associated with high intake of food phenolics.

PHARMACOLOGICAL EFFECTS

ANTIMICROBIAL PROPERTIES

Many preservatives are added to foods as antioxidants or antimicrobial agents. Phenolic compounds are known to possess both of these properties (Atroshi et al., 2003; Haslam, 1989; Nychas, 1995; Nychas et al., 1990; Scalbert, 1991; Schmitz et al., 1993; Takenaka et al., 1997); however, the lipophilic nature of phenols may reduce their antimicrobial properties (Baranowski and Nagel, 1984).

Antimicrobial action of phenolic compounds was first related to the inactivation of cellular enzymes, which depended on the rate of penetration of the substance into the cell (Fogg and Lodge, 1945). Later, Judis (1963) suggested that inhibition of

TABLE 7.8
Effect of BHA on Aflatoxin Production of *Aspergillus parasiticus* Spores[a]

Treatment	Aflatoxins[b]			
	B_1	B_2	G_1	Total
Control	3.56	0.54	2.02	6.12
BHA (0.1 mg/kg)	2.34	0.62	1.76	4.72
BHA (10 mg/kg)	10.5	0.84	4.34	15.68
BHA (1000 mg/kg)	ND	ND	ND	ND

Note: ND = no toxin detected.

[a] Molds were grown in 50 mL of glucose-salts medium in 125-mL flask for 7 days at room temperature.
[b] Milligrams per liter of broth.

Source: Adapted from Chang, H.C. and Branen, A.L., 1975, *J. Food Sci.*, 40:349–351.

microbial growth is due to weakening or destruction of the permeability of cell membranes. Subsequently, Juven et al. (1972) demonstrated that phenolics cause membrane permeability changes. They showed that oleuropein, a phenolic glycoside found in olives, causes leakage of the radioactively labeled glutamate, potassium, and phosphorus from *Lactobacillus plantarum*. According to Bernheim (1972), changes in the permeability of cell membrane may be due to the interaction of phenols with phospholipid components of membranes. However, using intracellular UV-absorbing materials and C^{14} labeled compounds, Davidson and Branen (1980a, b) proposed that disruption of cytoplasmic membranes is at least partially responsible for inhibitory action of butylated hydroxyanisole (BHA) on *Pseudomonas fluorescens* and *Pseudomonas fragi*. Moreover, they speculated that BHA may also affect the activity of cell-bound enzyme systems.

Chang and Branen (1975) found that BHA at 1000 mg/kg totally prevents the growth and production of aflatoxin by *Aspergillus parasiticus* (Table 7.8). They also demonstrated that BHA has a profound antimicrobial effect against *Staphylococcus aureus, Escherichia coli* and *Salmonella typhimurium*. Later, Davidson and Branen (1980a, b) examined the antimicrobial activity of BHA against psychorotopic bacteria, namely, *P. fluorescens* and *P. fragi*. They found that 100 mg/kg of BHA delays the growth of *P. fluorescens* at 22°C and totally inhibits it at 7°C. On the other hand, *P. fragi* is more resistant to BHA: delayed growth has been observed at ≥300 mg/kg. According to Shelef and Liang (1982), BHA acts as an effective bacteriostatic agent against *Bacillus cereus, Bacillus subtilis* and *Bacillus megaterium* at ≤200 mg/kg in laboratory media and at ≤10,000 mg/kg in food systems. They also suggested that including BHA to control food microorganisms is premature because BHA is fat soluble; thus its inhibitory effects in lipid-containing foods may be influenced by the partition coefficient between the lipid and the bacteria.

Klindworth et al. (1979) observed reduced antibacterial activity for BHA against *Clostridium perfringens* in the presence of lipids. On the other hand, Davidson et al. (1981) reported that potassium sorbate and BHA have little or no delaying effect in initiating growth of *S. aureus* or *S. typhimurium*. However, potassium sorbate (500 to 1000 mg/kg) and BHA (50 to 100 mg/kg) show a synergistic effect in supressing the growth of both microorganisms, thus allowing a reduced level of potassium sorbate to be used when BHA is present.

Pierson and Reddy (1982) examined the activity of some phenolic antioxidants against the growth and toxin production of *Clostridium* types A and B. They found that inoculation with 8000 spores/g of sample of meat-containing esters of gallic acid, *p*-hydroxybenzoic acid, and related phenolics delays the time of first toxic swell for 3 days beyond the control. The inhibitory activity of phenolics increases as the spore level per gram of meat decreases.

Kubo et al. (2002) have examined a series of synthetic alkyl gallates as potential antifungal agents against *Saccharomyces cerevisiae*, which is responsible for spoilage of sauerkraut, syrups, honey, fruit juices, meats, wine and beer (Fleet, 1992). Of these synthetic alkyl gallates, nonyl gallate, followed by octyl gallate, is the most effective antifungal agent tested. The nonyl and octyl gallates display fungicidal activities at a minimum concentration of 12.5 and 25 mg/L, respectively. The antifungal activity of these gallates is not affected by pH and the growing stage of *S. cerevisiae*; the antifugicidal activity of nonyl gallate is 64 times greater than that of sorbate (Kubo et al., 2002).

Widely distributed as secondary metabolities in plants, hydroxycinnamates exhibit considerable antimicrobial effect under appropriate conditions. Gupta et al. (1976) found that naturally occurring ethyl *p*-methoxycinnamate inhibits growth of selected molds (*Trichophyton rubrum, Aspergillus niger, S. cerevisiae, Epidermophyton floccosum, Aspergillus fumigatus, Penicillium purpurogenum, Trignoposis variabilis, Microsporum gypseum, Sclerotium rolifsii, Geotricular candidae, Fusarium oxysporum, Helminthosporium oryzale, Candida cruseum* and *Trichophyton mentagrophytes*) at concentrations of 10 to 50 mg/kg. Later, Baranowski et al. (1980) reported that *p*-coumaric acid at a 100 mg/kg-level increases the lag phase of *S. cerevisiae*; at concentrations >250 mg/kg the inhibition after 72 h of growth is proportional to its concentration. They also found that ferulic acid at 50 mg/kg brings about an increase in the lag phase and that as little as 250 mg/kg results in complete inhibition.

Baranowski and Nagel (1982) examined the effectiveness of alkyl esters of six hydroxycinnamic acids and cinnamic acid (so-called alkacins), as well as free acids, in inhibiting the growth of *P. fluorescens* (a bacterium commonly implicated in food spoilage). They found that free caffeic, *p*-coumaric, ferulic, sinapic, cinnamic, *p*-methoxycinnamic and 3,4-dimethoxycinnamic acids are largely ineffective at 400 mg/kg. On the other hand, alkyl esters of *p*-coumaric and caffeic acids have a more inhibitory effect than parabens (alkyl esters of *p*-hydroxybenzoic acid) at 125 mg/kg (Table 7.9). Later, Baranowski and Nagel (1984) demonstrated the inhibitory effect of alkacins in a wide range of bacteria in model systems and showed it to be comparable to the effects of methyl and propyl parabens. They also found that, although these compounds have fair to very good antimicrobial activity in model systems, their relative effect in foods depends on the system under investigation.

Marvan and Nagel (1986) demonstrated that ethanolic extracts of cranberry exert a significant antimicrobial effect on *Saccharomyces bayanus* and *P. fluorescens*. They

TABLE 7.9
Relative Effectiveness of Alkacins and Parabens against *Pseudomonas Fluorescens* in Trypticase Soy Broth[a]

Compound	Concentration[b] (mg/kg)			
	25	50	125	250
Methyl caffeoate	1.00	1.06	1.71	2.84
Ethyl caffeoate	1.03	1.07	1.66	2.24
Propyl caffeoate	1.00	1.24	1.84	2.84
Methyl cinnamate	0.96	1.03	1.28	2.46
Methyl *p*-coumarate	1.03	1.15	1.72	2.11
Methyl ferulate	1.02	1.00	1.02	1.20
Methyl sinapate	1.02	1.01	1.02	1.06
Methyl *p*-methoxycinnamate	1.02	1.02	1.02	1.02
Methyl 3,4-dimethoxycinnamate	1.00	0.99	1.03	1.02
Methyl paraben	1.00	1.07	1.31	2.18
Propyl paraben	1.05	1.09	1.38	2.40

[a] pH 7.0 at 20°C.
[b] Milligram per kilogram.

Source: Adapted from Baranowski, J.D. and Nagel, C.W., 1982, *J. Food Sci.*, 47:1587–1589.

showed that proanthocyanidins account for 21.3% of total cranberry inhibition, while the inhibitory effects of flavonols and benzoic acid are 18.5 and 15.6%, respectively, when tested at pH 4.0. A synergistic effect between benzoic acid and proanthocyanidins or flavonols was also observed, but the activities of proanthocyanidins and flavonols are additive. At higher pH values, the dominant inhibitors are proanthocyanidins. The proanthocyanidin fraction completely inhibits the growth of microorganisms at pH 5.0 and 6.0 for 1 month. This may be brought about by stronger interaction between proanthocyanidins and microorganisms' cell membrane proteins or oxidation and polymerization of proanthocyanidins. Waage et al. (1984) showed that procyanidin of higher molecular weight (MW) has a broader activity than that of those with lower MW.

Cranberry proanthocyanidins have been linked to the maintenance of urinary tract health (Gibson et al., 1991; Howell, 2002; Papas et al., 1996). These compounds may be responsible for inhibition of cellular adherence of uropathogenic strains of P-type (mannose-resistant) *E. coli* to mucosal cells in the urinary tract (Howell, 2002; Howell et al., 1998; Ofek et al., 1991; Zafriri et al., 1989) because the urinary tract infection is initiated by bacteria adhering to the mucosal cells (Beachey, 1981). Foo et al. (2000a, b) demonstrated that cranberry proanthocyanidins with A-type linkage exhibit greater antiadhesion activities than those with B-type linkage. Howell (2002) suggested that cranberry proanthocyanidins competitively inhibit the adhesion of *E. coli* to mucosal cells through receptor–ligand interactions.

Oleuropein, a phenolic compound found in olive, and its hydrolysis products, namely, aglycone and elenolic acids, possess antimicrobial properties (Fleming et al.,

1973; Juven and Henis, 1970; Juven et al., 1972). Pradhan et al. (1999) evaluated the antimicrobial properties of two major phenolic compounds present in green pepper berries (*Piper nigrum* L.): 3,4-dihydroxyphenylethanol glucoside and 3,4-dihydroxy-6-(N-ethylamino) benzamide. These phenolics are not detected in black pepper because of their oxidation to black pigment by polyphenoloxidase during sun drying of green pepper berries (Bondopadhyay et al., 1990; Variyar et al., 1988). Of these phenolic compounds, 3,4-dihydroxyphenyl ethanol glucoside displays greater bacteriostatic activity against *S. typhimurium, S. aureus, Bacillus aureus* and *E. coli* than 3,4-dihydroxy-6-(N-ethylamino) benzamide (Pradhan et al., 1999).

Norton (1999) examined the inhibitory effects of anthocyanidins (cyanidin, pelargonidin, malvidin, peonidin, apigenidin, and luteolinidin) and their mono- and diglucosides against the growth and biosynthesis of aflatoxin B_1 by *Aspergillus flavus.* All anthocyanidins tested inhibited the synthesis of aflatoxin B_1; those with a 3-hydroxyl group (peonidin, delphinidin, pelargonidin, and cyanidin) displayed approximately three times greater inhibitory effect than the corresponding 3-deoxy anthocyanidins. Kaempferol and naringenin also display strong inhibitory effects on the growth of *A. flavus* and formation of aflatoxin (Mallozzi et al., 1996), while catechin and luteolin are weak inhibitors (Norton, 1999). On the other hand, the steryl ferulate and *p*-coumarate ester fraction isolated from corn does not inhibit the growth of *A. flavus*, but at a concentration range of 0.33 to 1.0 g/L, displays a stimulatory effect on the biosynthesis of aflatoxin B_1 (Norton and Dowd, 1996).

Kubo et al. (1999) characterized antibacterial activity of several anarcadic acids against *Helicobacter pylori*, an infectious bacterium responsible for gastric ulcer and other gastrointestinal diseases. Adhesion of bacteria to epithelial cells is an essential step in the deveolpment of these diseases (Dorell et al., 1998; Dunn et al., 1997; Lingwood et al., 1992). Maximum antibacterial activity is displayed by anarcadic acid with saturated $C_{12:0}$ and unsaturated $C_{15:3}$ and $C_{15:2}$ alkyl side chains, while anarcadic acid with saturated $C_{15:0}$ alkyl side chain is inactive against *H. pylori* (Kubo et al., 1999). The antiulcer activity of anarcadic acids may be from their action as detergents (Kubo et al., 1995) and urease inhibitors (Kubo et al., 1999). Urease neutralizes the acid environment of the stomach by hydrolyzing urea to ammonia and carbon dioxide (Blaser, 1993) and thus is thought to help in colonization and survival of *H. pylori*. According to Kubo et al. (1999), regular consumption of fresh cashew apple and its processed products may effectively control *H. pylori.* Burger et al. (2002) have demonstrated the inhibitory effect of high molecular weight constituents of cranberry (proanthocyanidins) on a specific adhesion of *H. pylori* human gastric cell lines *in vitro*. Cranberry phenolics are thought to act on sialyllactose-specific adhesion of *H. pylori*; in turn, this prevents adhesion of pathogen to the stomach (Burger et al., 2002).

Tea polyphenols substantially inhibit the growth of cariogenic bacteria, namely, *Streptococcus mutans* and *Streptococcus sobrinus*, and the synthesis of extracellular glucans from sucrose by the action of their glucosyltransferases (Hamilton-Miler, 2001; Hattori et al., 1990; Matsumoto et al., 1999; Ooshima et al., 1998; Sakanaka and Kim, 1998; Sakanaka et al., 1992). The glucans are important cariogenic factors mediating the aggregation of cariogenic bacteria on tooth surfaces — an accumulation that leads to formation of dental plaque and caries (Freedman and Tanzer, 1974;

TABLE 7.10
Comparison of 50% Inhibitory Doses (ID_{50}) of Selected Phenolics on Glucosyltransferase Activity[a]

Compound	Glucosyltransferase from	
	S. mutants	*S. sobrinus*
Crude apple polyphenol extract	120	25
Caffeic acid	>1000	>1000
Chlorogenic acid	>1000	>1000
Epicatechin	>1000	>1000
Procyanidin B2	>1000	>1000
Apple condensed tannins	5	1.5
Phlorizidin	>1000	>1000
Epigallocatechin gallate	>1000	>1000
Quercetin	ND	120

Note: ND = no data.

[a] Milligrams per liter.

Source: Adapted from Yanagida, A. et al., 2000, *J. Agric. Food Chem.*, 48:5666–5671.

Hamada and Torii, 1978). Bactericidal activity for anarcadic acid against *S. mutans* has also been reported (Muroi and Kubo, 1993). Moreover, apple polyphenols (Matsudaira et al., 1998) and cranberry juice constituents (Weiss et al., 1999) effectively inhibit the formation of dental plaque in human subjects. Yanagida et al. (2000) have demonstrated that apple polyphenols are effective in inhibiting the glucosyltransferase activity (Table 7.10), but have no significant effect on *in vitro* growth of cariogenic bacteria. Recently, Daglia et al. (2002) observed the inhibition of adsorption of *S. mutants* to hydroxyapatite beads coated with saliva by a roasted coffee solution. The antiadhesive property of the roasted coffee is thought to be due to absorption of coffee components (trigonelline, nicotinic acid, chlorogenic acids) on tooth sufaces, thus preventing adhesion of bacteria to these surfaces.

Antiviral Properties

A number of flavonoids present in foods of plant origin possess antiviral activity. Quercetin, taxifolin and flavonoids from red wine are able to inactivate a wide range of viruses and prevent their infectivity (Bakay et al., 1968; Konowalchuk and Speirs, 1976a). Tannins from strawberry are able to inactivate polio, enteric, and herpes viruses (Konowalchuk and Speirs, 1976a, b). Quercetin, a flavonol aglycone found in a number of fruits such as apple, apricot, fig, plum, strawberry, and tomato, shows antiviral activities against herpes simplex virus type 1, parainfluenza virus type 3 and polio virus type 1 in *in vivo* and *in vitro* studies (Middleton, 1986; Musci, 1986; Veckenstedt et al., 1986).

There have also been reports that quercetin synergistically enhances the beneficial effect of interferon (Musci, 1986). Tea polyphenols are able to inhibit influenza

A and B infections by preventing the virus from adsorbing to cells (Nakamaya et al., 1993), while hesperetin inhibits the infectivity of herpes simplex type viruses, polio viruses, and parainfluenza viruses (Kaul et al., 1985). Furthermore, curcumin and its analogs exhibit inhibitory activities against human immunodeficiency virus type 1 (HIV-1) integrase by preventing binding of the enzyme to viral DNA (Mazumder et al., 1995, 1997), and tannins from pericarp of pomegranate display antiviral activity against the genital herpes virus (Zhang et al., 1995). King et al. (1999) examined 34 analogues of dicaffeoylquinic and dicaffeoyltartaric acids as potential inhibitors of HIV-1 integrase and replication. Biscatechol group was found to be essential for the inhibition of integrase while at least one free carboxyl group was required for anti-HIV activity. Of the evaluated phenolic acid analogues, D-cichoric, *meso*-cichoric, *bis*(3,4-dihydroxybenzoyl)-L-tartaric acid, *bis*(3,4-dihydroxydihydrocinnamoyl)-L-tartaric acid, digalloyl-L-tartaric acid, dicaffeoylglyceric acid and *bis*(3,4-dihydroxyphenylacetyl)-L-tartaric acid were the most potent inhibitors of HIV-1 integrase.

Other Pharmacological Potentials

A number of books and reviews have been published to describe the pharmacological potentials of food phenolics (Adams and Lewis, 1977; Bagchi et al., 2000; Benavente-Garcia et al., 1997; Bidlack et al., 2000; Block et al., 1992; Blot et al., 1996; Böhm et al., 1998; Brandi, 1997; Butrum, 1996; Cody et al., 1986; Cook and Samman, 1996; De Bruyne et al., 1999; Dubnick and Omaye, 2001; Dugas et al., 2000; Facino et al., 1998; Gilman et al., 1995; González de Mejía et al., 1999; Habtemariam, 1997; Haslam et al., 1989; Haslam, 1996; Hertog et al., 1997a, b, 1993a, b; Hertog, 1996; Hollman and Katan, 1998; Huang et al., 1994a, b; Ielpo et al., 2000; Ito and Hirose, 1989; Jang et al., 1997; Katiyar and Mukhtar, 1996; Keli et al., 1996; Knekt et al., 1997; Konneh and Caen, 1996; Kuntz et al., 1999; Kuo, 1997; Langseth, 1995; Leake, 1997; Mabry and Ulubelen, 1980; Messina et al., 1994; Meyers et al., 1996; Middleton and Kandaswami, 1993; Morton et al., 2000; Murakami et al., 1996; Nehlig and Debry et al., 1994; Noguchi, 2002; Pathak et al., 1991; Reed, 2002; Rice-Evans and Packer, 1997; Stavric and Matula, 1988, 1992; Steinmetz and Potter, 1996; Stoner and Mukhtar, 1995; Suschetet et al., 1996; Tapiero et al., 2002; Temple, 2000; Tijburg et al., 1997; Ursini et al., 1999; Verstraeten et al., 2003; Waladkhani and Clemens, 1998; Willet, 2002; Wollgast and Anklam, 2000; Yang and Wang, 1993; Yochum et al., 1999). Therefore, this section provides a discussion of only some aspects of pharmacological potentials of food phenolics.

Saito et al. (1998) used rats to examine the antiulcer property of grape seed extracts. Rich in flavanol, these extracts strongly suppress the stomach mucosal injury caused by acidified ethanol. Active oxygen species have been implicated in the development of gastric mucosal damage caused by acidified ethanol (Itoh and Guth, 1985; Matsumoto et al., 1993; Pihan et al., 1987). It has been suggested that the antiulcer property of grape procyanidins may be due to their radical scavenging activity and their ability to bind to proteins. However, the exact antiulcer mechanism of procyanindins against stomach mucosa injury is still not well understood (Saito et al., 1998). Procyanidin-rich extracts from grape seeds also display anticataract activity in hereditary cataractous (ICR/f) rats fed a standard diet containing 0.213% (w/w) of grape seed extract (Yamakoshi et al., 2002), as well as in streptozotosin-induced diabetic rats fed a diet containing 0.5% (w/w) (Nguyen et al., 1999).

According to Spector (1995) and Taylor et al. (1995), oxidative stress is responsible for cataractogenesis; therefore, the retardation of cataract progression may be due to the antioxidant activity of grape seed procyanidins and their metabolites (Yamakoshi et al., 2002). Other reported biological activities of grape seed extracts include antiatherosclerotic (Yamakoshi et al., 1999), antidiabetic (Nguyen et al., 1999), and anticarcinogenic (Alam et al., 2002; Arii et al., 1998; Bagchi et al., 2000; Krohn et al., 1999; Ye et al., 1999; Zhao et al., 1999) properties, production of vasorelaxation by inducing nitric oxide release from endothelium (Fitzpatrick et al., 2000), control of LDL-cholesterol level in high cholesterol-fed rats (Tebib et al., 1994a, b), and prevention of plasma postprandial oxidative stress in humans (Natella et al., 2002).

Cocoa procyanidins are also effective scavengers of reactive oxygen and nitrogen species (Arteel and Sies, 1999; Kondo et al., 1996; Steinberg et al., 2002; Waterhouse et al., 1996; Wollgast and Anklam, 2000) and have the ability to modulate immune function and platelet aggregation (Mao et al., 1999; Rein et al., 2000b; Sanbongi et al., 1997; Wollgast and Anklam, 2000). Yamagishi et al. (2000) demonstrated that cacao liquor polyphenols exhibit a strong antimutagenic activity on the action of heterocyclic amines (HCA) *in vitro* and *ex vivo*. The supression of antimutagenic activity of HCA may involve not only inhibition of the metabolic activation of HCA in the liver but also adsorption of HCA by some phenolic component of cacao liquor such as procyanidins. Pomegranate polyphenolics exhibit strong antiatherogenic (Aviram et al., 2000; Kaplan et al., 2001) and antitumoral (Kashiwada et al., 1992) activities. Moreover, pomegranate juice has been recommended in the treatment of acquired immune deficiency syndrome (AIDS) due to pomegranate bioflavonoids' inhibitory effect on lipoxygenase (Lee and Watson, 1998).

Phenolic compounds extracted from bilberry (*Vaccinium myrtillus)* juice after fermentation constitute the active principle of a drug used for vascular protection. Anthocyanins present in the extract act on capillary permeability and fragility (Azar et al., 1987; Wagner, 1985). Pool-Zobel et al. (1999) have demonstrated that the anthocyanins from *Aronia melanocarpa,* elderberry, and Macqui and Tintorera fruits exhibit stronger antioxidant activities *in vitro* than those displayed by Trolox and vitamin C in a ferric-reducing ability assay. Anthocyanins are not able to prevent endogenous oxidative DNA damage in human colon cells; thus, according to Pool-Zobel et al. (1999), the cancer preventive potential of anthocyanins within specific tissue is questionable. On the other hand, anthocyanins exhibit inhibitory effects on the formation of nitric oxide in lipopolysaccharide (LPS)- and interferon-γ (INF-γ)-activated mouse macrophage cell line RAW 264.7 (American type culture collection).

Anthocyanins at a concentration ≥250 mg/L have a stimulatory effect on secretion of tumor necrosis factor-α (TNF-α) in microphages (Wang and Mazza, 2002a, b), which display cytostatic and cytotoxic activities on malignant cells (Camussi et al., 1991; Wadsworth and Koop, 1999). Frank et al. (2002) have reported that cyanidin-3-glucoside and anthocyanin mixtures from elderberry and blackcurrant, when consumed at normal dietary levels, have little effect on the cholesterol and liver fatty acid profiles in healthy rats. However, cyanidin-3-glucoside elevates the level of tocopherols in the liver and lungs.

Tea polyphenols exhibit a very broad spectrum of biological activities (Erland et al., 2001; Gao et al., 1994; Katiyar and Mukhtar, 1996; Stensvold et al., 1992; Trevisanato and Kim, 2000; Yang and Chung, 2000; Yang and Wang, 1993; Yang et al., 1996). Green tea polyphenols inhibit intestinal uptake of glucose through the sodium-dependent glucose transporter (SGLT1) of the rabbit intestinal epithelial cells. Presence of galloyl group enhances the inhibitory effect of tea polyphenols. It has been suggested that tea polyphenols act as antagonist-like molecules (Kobayashi et al., 2000) and thus may contribute to the reduction of blood glucose level. Aqueous extracts of green, oolong, pouchong, and black tea inhibit over 90% of mutagenicity of the 2-amino-3-methylimidazo(4,5-f)quinoline, 3-amino-1,4-dimethyl-5H-pyrodyl(4,3-b)indole, 2-amino-6-methylpyrido(1,2-a:3′,2′-d)imidazole, benzo[a]pyrene and aflatoxin B1 toward *S. typhimurium* TA 98 and TA100 (Yen and Chen, 1994).

Imai et al. (1997) and Nakachi et al. (1998) have reported that increased consumption of green tea is responsible for lowering the risk for breast cancer metastasis and recurrence among Saitama, Japan, women. The (–)-epigallocatechin-3-gallate (EGCG), a polyphenolic component of green tea, has been found to reduce the incidence of spontaneously and chemically induced tumors in experimental animals, as observed for tumors of liver, stomach, skin, lungs and esophagus (Huang et al., 1992). EGCG causes cell cycle dysregulation and apoptosis of cancer cells mediated through nuclear factor-κB (NF-κB). The inhibition of NF-κB constitutive expression and activation in cancer cells occurs at lower concentrations of EGCG compared to those of normal cells (Ahmad et al., 2000a, b). Pan et al. (2000) have reported that, isolated from oolong tea, theasinensin A and mixtures of theaflavin-3-gallate and theaflavin-3′-gallate exhibit strong growth inhibitory effects against human hystolytic lymphoma U937, but are less effective against human acute T cell leukemia Jurkat. Induced by tea phenolics, the apoptosis of these cells involves the caspase-3 mediated mechanism.

Yang and Chung (2000) demonstrated that effective levels of tea polyphenols required for imparting signal transduction pathways, inhibiting cell proliferation and inducing apoptosis of cancer cells are higher than those detected in blood and tissues. In addition, Maeda-Yamamoto (1999) reported that tea polyphenols, namely, epicatechin gallate, epigallocatechin gallate, and theaflavin, strongly prevent the invasion of highly metastatic human fibrosarcoma HT1080 cells into the monolayer of human umbilical vein endothelial cells/gelatin membrane *in vitro*. Of these phenolics, only epicatechin gallate inhibits invasion in the absence of cytotoxicity. Thus, more studies are needed to determine the mechanisms involved in prevention of cancer by tea phenolics.

Low incidence of coronary heart disease among populations consuming red wine has been linked to the presence of resveratrol in wine (Das et al., 1999; Goldberg, 1995; Goldberg et al., 1995; Gronbaek, 1999; Hegsted and Ausman, 1988; Renaud and de Lorgeril, 1992; Sharpe et al., 1995; Waterhouse et al., 1999). Animal studies indicated that resveratrol and piceid lower the accumulation of cholesterol and synthesis of triacylglycerols in rat liver (Arichi et al., 1982). Waffo-Teguo et al.

(2001) have reported that two products of oxidative degradation of stilbene, namely, resveratrol (*E*)-dehydrodimer 11-*O*-β-D-glucopyranoside and resveratrol (*E*)-dehydrodimer, exhibit a nonspecific inhibitory activity against cyclooxygenase-1 and -2. These enzymes play an important role in the development of tumorigenesis and inflammation (Subbaramaiah et al., 1998).

Pterostilbenes display moderate inhibitory activity against cyclooxygenase-1 and weak inhibition of cycloxygenase-2 (Rimando et al., 2002). Furthermore, *trans*- and *cis*-resveratrols display inhibitory effects against kinases (Jayatilake et al., 1993), platelet aggregation (Chung et al., 1992), and LDL oxidation (Belguendouz et al., 1997; Fauconneau et al., 1997; Stivala et al., 2001), while piceids are effective against platelet aggregation (Chung et al., 1992; Orsini et al., 1997) and LDL oxidation (Frankel et al., 1995; Fremont et al., 1999).

Resveratrol also prevents cell profileration in tumor cell lines *in vitro* (Jang et al., 1997; Hsieh and Wu, 1999) and decreased tumor growth in a rat tumor model (Carbo et al., 1999). In addition, resveratrol and its oxidation product are a potent inhibitor of dioxygenase activity of lipoxygenase, but not its hydroperoxidase activity (Papatheofanis and Land, 1985; Pinto et al., 1999). However, by its hydroperoxidase activity, lipoxygenase is involved in the oxidation of resveratrol in the presence of hydroperoxy derivatives of polyunsaturated fatty acids or hydrogen peroxide (Pinto et al., 1999). Lipoxygenase has been implicated in inflammatory responses in allergy, asthma, and arthritis (Kulkarni et al., 1990), and in the formation of atherosclerotic lessions (Prigge et al., 1997) and carcinogenic processes (Furstenberger et al., 1991; Kamitani et al., 1998).

Frankel et al. (1993) suggested that the beneficial effect of red wine may be explained by the inhibition of oxidation of low-density lipoproteins (LDL) by wine phenolics. Later, Satué-Gracia et al. (1999) reported that phenolics present in Spanish sparkling wines (cava) also inhibit oxidation of LDL. This activity correlates positively with the total phenolic content, quercetin 3-glucuronide, *trans*-caffeic acid, protocatechuic acid, and coumaric acid. Subsequently, Gómez-Cordovés et al. (2001) demonstrated that all fractions of wine phenolics (anthocyanins, other flavonoids, phenolic acids) show potential as therapeutic agents in the treatment of melanoma.

Citrus flavonoids display a broad spectrum of biological activities (Rouseff and Nagy, 1994). Of these, auraptene (7-geranyloxycoumarin) found in citrus fruit peel exhibits chemopreventive activity in mouse skin (Murakami et al., 1997), rat colon (Tanaka et al., 1997, 1998a) and rat tongue (Tanaka et al., 1998b) carcinogenesis models; 8-geranyloxypsolaren, bergamottin, and 5-geranyloxy-7-methoxycoumarin are effective inhibitors of tumor promoter 12-*O*-tetradecanoyl-13-phorbol acetate-induced Espstein-Barr virus activation in Raji cells (Miyake et al., 1999). Flavanones display protective effects against cancer (Koyuncu et al., 1999), as well as rendering antiallergic and antiinflammatory properties (Gabor, 1986; Noguchi et al., 1999) and a cholesterol-lowering activity in the rat (Bok et al., 1999; Galati et al., 1994). A number of polymethoxylated flavones suppress the production of tumor necrosis factor-α in culture of human monocytes (Manthey et al., 1999). These flavones are also effective inducers of HL-60 (human acute promyelocytic leukemia cells) differentiation (Kawaii et al., 1999a, b; Sugiyama et al., 1993) and therefore may have great therapeutic potential (Koeffler, 1983). Diosmin is an active component of some drugs used for the treatment of circulatory system illnesses (Galley and Thiollet, 1993; Tsouderos, 1991), inflammatory disorders (Jean and Bodinier, 1994) and severe hemorroidal disease (Godeberg, 1994). Furthermore, diosmin and its aglycone (diosmetin) prevent the activation of carcinogenesis by heterocyclic amines (Ciolino et al., 1998).

Isoflavones exert a broad spectrum of biological activities. Besides estrogenic activities, isoflavones protect against several chronic diseases. Results of epidemiological studies indicate that consumption of soybean isoflavones lowers the incidence of breast, prostate, urinary tract and colon cancers; they also provide protection against coronary heart diseases and osteoporosis (Brandi, 1997; Clarkson et al., 1995; Cline and Hughes, 1998; Miyazawa et al., 2001; Moyad, 1999; Peterson and Dwyer, 1991; Sheu et al., 2001; Su et al., 2000). Peterson and Dwyer (1991) examined the effect of isoflavones on the growth of breast carcinoma cell lines, MDA-468 (estrogen receptor negative), MCF-7 and MCF-7-D-40 (estrogen receptor positive). Genistein is the most potent inhibitor of cancer cell growth, while daidzein and biochanin A display weaker inhibitory activity. However, isoflavone glucosides, genistin and daidzin have little effect on the growth of breast cancer cells (Peterson and Dwyer, 1991). In addition, isoflavones exhibit marked inhibitory activity against oxidation of lipoprotein in serum (Hodgson et al., 1996; Kerry and Abbey, 1998). Sheu et al. (2001) have reported that genistein, daidzein and glycetein suppress the formation of nitric oxide in RAW 264.7 macrophages by inhibiting activity and expression of the inducible nitric oxide synthase in lipopolysaccharide-activated macrophages.

The pharmacological effects of anarcadic acids, cardols and cardanols, the major phenolics of cashew nuts, include cytotoxic activity against BT-20 breast carcinoma cells (Kubo et al., 1993) and antiacne (Kubo et al., 1994a) and antibacterial (Gellerman et al., 1969; Himejima and Kubo, 1991; Kubo et al., 2002) activity. Anarcadic acids also exhibit inhibitory effects against a number of enzymes, including aldose reductase (Toyomizu et al., 1993), glycerol-3-phosphate dehydrogenase (Irie et al., 1996), lipoxygenase (Grazzini et al., 1991; Shobha et al., 1994), prostaglandin synthase (Bhattacharya et al., 1987; Grazzini et al., 1991; Kubo et al., 1987) and tyrosinase (Kubo et al., 1994b).

Curcuminoids, namely, curcumin, demethoxycurcumin and bisdemethoxycurcumin, are the major phenolics found in turmeric (*Curcuma longa* L.) (Figure 7.12) (Govindarajan, 1980; Jayaprakasha et al., 2002). Of these, curcumin is the predominant phenolic constituent of turmeric (Ashan et al., 1999; Govidarayan, 1980). The total content of curcuminoids in four samples of commercially available varieties of turmeric ranged between 23.4 and 91.8 g/kg (Jayaprakasha et al., 2002). Curcuminoids display antioxidant (Ashan et al., 1999; Kim et al., 2001; Osawa et al., 1994; Reddy and Lokesh, 1992; Schaich et al., 1994; Wright, 2002), anticancer, antimutagenic, anti-inflammatory (Brouet and Ohshima, 1995; Chang and Fong, 1994; Lin et al., 1994; Srimal, 1997), and antiviral (Mazumder et al., 1995, 1997) activities. Simon et al. (1998) evaluated curcumin, demethylcurcumin and bisdemethylcurcumin as potential inhibitors of proliferation of MCF-7 human breast cancer cells. Of these, demethylcurcumin displayed the best inhibitory effect, followed by curcumin and bisdemethylcurcumin. Lin et al. (1994) demonstrated that curcumin

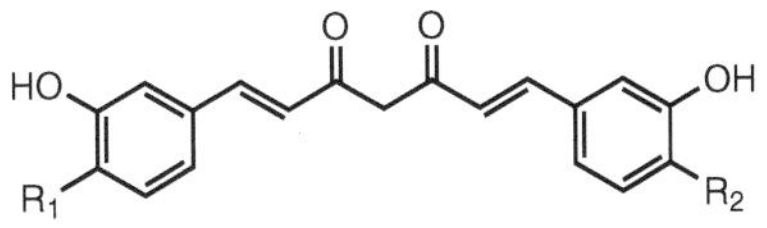

Compound	R_1	R_2
Curcumin	OCH_3	OCH_3
Demethoxycurcumin	H	OCH_3
Bisdemethoxycurcumin	H	H

FIGURE 7.12 Chemical structures of some curcuminoids.

is an effective inhibitor of tumor promotion induced by 12-*O*-tetradecanoylphorbol-13-acetate (TPA). Curcumin suppresses the TPA-induced synthesis of protein kinase C in mouse fibroblast cells and 8-hydroxydeoxyguanosine in cellular DNA. In addition, curcumin displays immunomodulating acivities by enhancing the production of interleukin-4 in T helper-2 cells (Chang and Fong, 1994).

The ellagic, protocatechuic and chlorogenic acids present in fruits and vegetables have been found to serve as potential chemopreventers against several carcinogens (Nakamura et al., 2001; Tanaka et al., 1992). Repeated oral administration of protocatechuic acid at a 100 μg/kg-level inhibits the growth of colon and oral cancers in rats (Tanaka et al.1993; Ueda et al., 1996), while topical administration of protocatechuic acid at doses less than 1/1000 of toxic level effectively inhibits growth of tumor in mouse skin (Nakamura et al., 2000). However, topical pretreatment of mouse skin with a high dose (>150 mg) of protocatechuic acid enhances tumor promotion and contact hypersensitivity in mouse skin (Nakamura et al., 2000). Meanwhile a toxic oral dose of protocatechuic acid (500 mg/kg) causes depletion of glutathione, the major cellular antioxidant, in mouse liver and kidney (Nakamura et al., 2001).

The colonic neoplasia induced by azoxymethanol can be inhibited by quercetin and rutin found in vegetables and fruits (Deschner, 1992; Deschner et al., 1991). Quercetin is also known for its vasoactive properties (Alarcon de la Lastra et al., 1994; Formica and Regelson, 1995) and gastroprotective effect (Kahraman et al., 2003; Martin et al., 1998). In addition, quercetin displays an inhibitory effect on the mutagenic activity of heterocyclic amines (HCA). Several mutagenic HCAs have been isolated from cooked foods (Ohgaki et al., 1991; Sugimura, 1988; Wakabayashi et al., 1993). The suppression of HCA activity is due to inhibition of metabolic activation of HCA in the liver (Alldrick et al., 1986, 1989). On the other hand, quercetin may also act as a tumor promoter and carcinogenic agent (Formica and Regelson, 1995; Stavric, 1994). According to Nagabhushan (1990), catechin from sorghum and faba beans acts as an antimutagen of benzo[α]pyrene and dimethyl[α]anthracene in *Salmonella.*

Pearson et al. (1997) demonstrated that carnosic acid, carnosol, and rosmarinic acid present in rosemary extract and thymol, carvacrol, and zingerone (Figure 7.13) found in thyme, oregano, and ginger display inhibitory effects on endothelial cell-mediated oxidation of LDL. Their inhibitory activities decrease as follows:

$$\text{carnasol} > \text{carnosic acid} \approx \text{rosmarinic acid} >>> \text{thymol} > \text{carvacrol} > \text{zingerone}$$

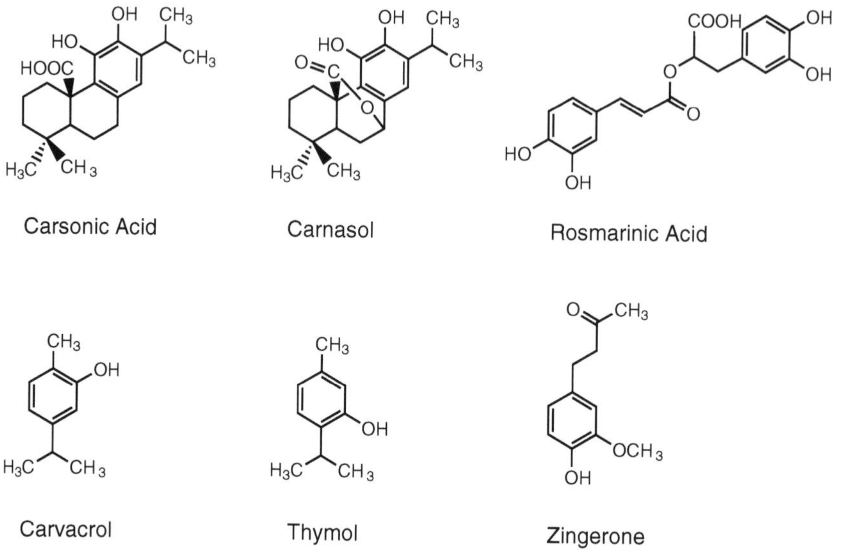

Carsonic Acid Carnasol Rosmarinic Acid

Carvacrol Thymol Zingerone

FIGURE 7.13 Chemical structure of some phenolics found in rosemary, thyme, oregano, and ginger.

The oxidation of LDL is considered an important step in atherogenesis, i.e., the deposition of plaques containing cholesterol and lipids in arterial walls (Esterbauer et al., 1992; Grundy, 1996; Kane, 1996; Reed, 2002). Sawa et al. (1999) reported that rutin, chlorogenic acid, vanillin, vanillic acid, neohesperidin, gallic acid, rhamnetin and kaempferol exhibit strong alkylperoxyl radical scavenging activity *in vitro.* The alkylperoxyl radicals have been found to enhance carcinogenesis in rats treated with carcinogens (Sawa et al., 1998). Oxidation of xanthine and hypoxanthine to uric acid plays a crucial role in the pathogenesis of gout (Hatano et al., 1991). Cos et al. (1998) examined a series of flavonoids as potential inhibitors of xanthine oxidase.

Flavones exhibit a somewhat higher inhibitory activity than flavonols. Flavonoids possessing hydroxyl groups at C-5 and C-7 positions and a double bond between C-2 and C-3 positions display strong inhibitory activity on xanthine oxidase. The presence of hydroxyl groups at C-3 and C-3′ positions is essential for flavonoids to show a strong superoxide-scavenging activity.

REFERENCES

Adams, J.H. and Lewis, J.R. 1977. Eupatorin a constituent of *Merilla caloxylon. Planta Med.*, 32:86–87.

Ahmad, N., Gupta, S., and Mukhtar, H. 2000a. Cell cycle dysregulation by green tea polyphenol epigallocatechin-3-gallate. *Biochem. Biophys. Res. Commun.*, 275:328–334.

Ahmad, N., Gupta, S., and Mukhtar, H. 2000b. Green tea polyphenol epigallocatechin-3-gallate differentially modulates nuclear factor κb in cancer cells vs. normal cells. *Arch. Biochem. Biophys.*, 376:338–346.

Ahmed, A.E., Smithard, R., and Ellis, M. 1991. Activities of the enzymes of the pancreas, and the lumen and mucosa of the small intestine in growing broiler cockerelsfed on tannin-containing diet. *Brit. J. Nutr.*, 65:189–197.

Alam, A., Khan, N., Sharma, S., Saleem, M., and Sultana, S. 2002. Chemopreventive effect of *vitis vinifera* extract on 12-*o*-tetradecanoyl-13-phorbol acetate-induced cutaneous oxidative stress and tumor promotion in murine skin. *Pharmacol. Res.*, 46:557–564.

Alarcon de La Lastra, C., Martin, M.J., and Motilva, V. 1994. Antiulcer and gastroprotective effects of quercetin: a gross and histologic study. *Pharmacology*, 48:56–62.

Alldrick, A.J., Lake, B.G., and Rowland, I.R. 1989. Modification of *in vivo* heterpocyclic amine genotoxicity by dietary flavonoids. *Mutagenesis*, 4:365–370.

Alldrick, A.J., Flynn, J., and Rowland, I.R. 1986. Effects of plant-derived flavonoids and polyphenolic acids on the activity of mutagens from cooked food. *Mutat. Res.*, 163:225–232.

Andlauer, W., Kilb, J., and Furst, P. 2000. Isoflavones from tofu are absorbed and metabolized in the isolated rat small intestine. *J. Nutr.*, 130:3021–3027.

Arichi, H., Kimura, Y., Okuda, H., Baba, M., Kazawa, M., and Arichi, S. 1982. Effects of stilbene components of roots of *Polygonum Cuspidatum Zieb et Zucc.* on lipid metabolism. *Chem. Pharm. Bull.*, 30:1766–1770.

Arii, M., Miki, R., Hosoyama, H., Ariga, T., Yamaji, N., and Kataoka, S. 1998. Chemopreventive effect of grape seed extract on intestinal carcinogenesis in APC^{Min} mouse. *Proc. 89th Annu. Meeting Am. Assoc. Cancer Res.*, 39:20.

Armstrong, W.D., Featherston, W.R., and Rogler, J.C. 1973. Influence of methionine and other dietary additions on the performance of chicks fed bird resistance sorghum grain diets. *Poultry Sci.*, 52:1592–1599.

Armstrong, W.D., Featherston, W.R., and Rogler, J.C. 1974. Effect of bird resistant sorghum grain and various commercial tannins on chick performance. *Poultry Sci.*, 53:2137–2142.

Arteel, G.E. and Sies, H. 1999. Protection against peroxynitrite by cocoa polyphenol oligomers. *FEBS Lett.*, 462:167–170.

Arts, I.C., Hollman, P.C.H., and Kromhout, D. 1999. Chocolate as a source of tea flavonoids. *Lancet*, 354:488.

Artz, W.E., Bishop, P.D., Dunker, A.K., Schanus, E.G., and Swanson, B.G. 1987. Interaction of synthetic proanthocyanidin dimer and trimer with bovine serum albumin and purified bean globulin fraction G-1. *J. Agric. Food Chem.*, 35:417–421.

Asano, K., Ohtsu, K., Shinigawa, K., and Hashimoto, N. 1984. Affinity of proanthocyanidins and their oxidation products for haze-forming proteins of beer and the formation of chill haze. *Agric. Biol. Chem.*, 48:1139–1146.

Ashan, H., Parveen, N., Khan, N.U., and Hadi, S.M. 1999. Pro-oxidant and antioxidant and cleavage activities on DNA of curcumin and its derivatives demethoxycurcumin and bismethoxycurcumin. *Chem. Biol. Interact.*, 121:161–175.

Asquith, T.N. and Butler, L.G. 1986. Interactions of condensed tannins with selected proteins. *Phytochemistry*, 25:1591–1593.

Asquith, T.N., Mehansho, M., Rogler, J.C., Butler, L.G., and Carlson, D.M. 1985. Induction of proline-rich protein biosynthesis in salivary glands by tannins. *Fed. Proc.*, 44:1097.

Asquith, T.N., Uhlig, J., Mehansho, M., Putman, L., Carlson, D.M., and Butler, L. 1987. Binding of condensed tannins to salivary proline-rich glycoproteins: the role of carbohydrate. *J. Agric. Food Chem.*, 35:331–334.

Atroshi, F., Rizzo, A., Watermarck, T., and Ali-Vehmas, T. 2003. Antioxidant nutrients and mycotoxins. *Toxicology*, 180:151–167.

Austin, P.J., Suchar, L.A., Robbins, C.H., and Hagerman, A.E. 1989. Tannin-binding proteins in saliva of deer and their absence in saliva of sheep and cattle. *J. Chem. Ecol.*, 15:1335–1347.

Aviram, M., Dornfield, L., Roseblat, M., Volkova, N., Kaplan, M., Coleman, R., Hayek, T., Presser, D., and Fuhrman, B. 2000. Pomegranate juice consumption reduces oxidative stress, atherogenic modification to LDL, and platelet aggregation. *J. Clin. Nutr.*, 71:1062–1076.

Azar, M., Verette, E., and Brun, S. 1987. Identification of some phenolic compounds in bilberry juice *Vaccinium myrtillus*. *J. Food Sci.*, 52:1255–1257.

Azen, E.A., Leutenegger, W., and Peters, E.H. 1978. Evolutionary and dietary aspects of salivary basic (Pb) and post Pb (PPb) proteins in anthropoid primates. *Nature*, 273:775–778.

Axelson, M. and Setchell.K.D. 1981. The excretion of lignans in rats: evidence for an intestinal bacterial source for this new group of compounds. *FEBS Lett.*, 123:49–53.

Axelson, M., Kirk, D.N., Farrant, R.D., Cooley, G., Lawson, A.M., and Setchell, K.D. 1982a.The identification of the weak oestrogen equol (7-hydroxy-3-(4′-hydroxyphenyl)chroman) in human urine. *Biochem. J.*, 201:353–357.

Axelson, M., Sjovall, J., Gustafsson, B.E., and Setchell, K. 1982b. Origin of lignans in mammals and identification of precursors from plants. *Nature*, 298:659–660.

Bacon, J.R. and Rhodes, M.J.C. 2000. Binding affinity of hydrolyzable tannins to parotid saliva and to proline-rich proteins derived from it. *J. Agric. Food Chem.*, 48:838–843.

Bacon, J.R. and Rhodes, M.J.C. 1998. Development of a competition assay for the evaluation of the binding of human parotid salivary proteins to dietary complex phenols and tannins using a peroxidase-labeled tannin. *J. Agric. Food Chem.*, 46:5083–5088.

Bagchi, D., Bagchi, M., Stohs, S.J., Das, D.K., Ray, S.D., Kruszynski, C.A., Joshi, S.S., and Pruess, H.G. 2000. Free radicals and grape seed proanthocyanidin extract: importance in human health and disease prevention. *Toxicology*, 148:187–197.

Bakay, M., Musci, I., Beladi, I., and Gabor, M. 1968. Effect of flavonoids and related substances. II. Antiviral effect of quercetin, dihydroquercetin, and dihydrofisetin. *Acta Microbiol. Acad. Hung.*, 15:223–227.

Baranowski, J.D., Davidson, P.M., Nagel, C.W., and Branen, A.L. 1980. Inhibition of *Saccharomyces cerevisiae* by naturally occurring hydroxycinnamates. *J. Food Sci.*, 45:592–594.

Baranowski, J.D. and Nagel, C.W. 1982. Inhibition of *Pseudomonas fluorescens* by hydroxycinnamic acids and their alkyl esters. *J. Food Sci.*, 47:1587–1589.

Baranowski, J.D. and Nagel, C.W. 1984. Antimicrobial and antioxidant activities of alkyl hydroxycinnamates (alkacins) in model systems. *Can. Inst. Food Sci. Technol. J.*, 17:79–85.

Barry, T.N. and Duncan, S.J. 1984. The role of condensed tannins in nutritional value of *Lotus penduculatus* for sheep. *Br. J. Nutr.*, 51:485–491.

Barry, T.N. and Manley, T.R. 1984. The role of condensed tannins in nutritional value of *Lotus penduculatus* for sheep. *Br. J. Nutr.*, 51:493–504.

Barry, T.N., Manley, T.R., and Duncan, S.J. 1986. The role of condensed tannins in the nutritional value of *Lotus penduculatus* for sheep. *Br. J. Nutr.*, 55:123–137.

Bartley, I.M. 1974. α-Glucosidase activity in ripening apples. *Phytochemistry*, 13:2107–2111.

Bartolome, B., Estrella, I., and Hernandez, M.T. 2000. Interaction of low molecular weight phenolics with proteins (BSA). *J. Food Sci.*, 65:617–621.

Bate-Smith, E.C. 1973. Haemanolysis of tannins: the concept of relative astringency. *Phytochemistry*, 12:907–912.

Bate-Smith, E.C. 1975. Phytochemistry of proanthocyanidins. *Phytochemistry*, 14:1107–1113.

Baxter, N.J., Lilley, T.H., Haslam, E., and Williamson, M.P. 1997. Multiple interactions between polyphenols and a salivary proline-rich protein repeat result in complexation and precipitation. *Biochemistry*, 36:5566–5577.

Beachey, E.H. 1981. Bacterial adherence: adhesion-receptor interactions mediating the attachment of bacteria to mucosal surface. *J. Infect. Dis.*, 143:325–345.

Beart, J.E., Lilley, T.H., and Haslam, E. 1985. Secondary metabolism and chemical defense: some observations. *Phytochemistry*, 24:33–38.

Belguendouz, L., Fremont, L., and Linard, A. 1997. Resveratrol inhibits metal ion-dependent and independent peroxidation of porcine low-density lipoproteins. *Biochem. Pharmacol.*, 53:1103–1104.

Benavente-Garcia, O., Castillo, J., Marin, F.R., Ortuño, A., and Del Rio, J. 1997. Uses and properties of *Citrus* flavonoids. *J. Agric. Food Chem.*, 45:4505–4515.

Bernays, E.A. 1981. Plant tannins and insect herbivores: an appraisal. *Ecol. Entomol.*, 6:353–360.

Bernheim, F. 1972. The effect of chloroform, phenols, alcohols and cyanogen iodide on the swelling of *Pseudomonas aeruginosa* in various salts. *Microbios*, 5:143–149.

Bezwoda, J.R., Torrance, J.D., Bothwell, T.H., Macphail, A.P., Graham, B., and Mills, W. 1985. Iron absorption from red and white wines. *Scand. J. Haematol.*, 34:121–127.

Bhattacharya, S.K., Mukhopadhyay, M., Mohan Rao, P.J., Bagchi, A., and Ray A.B. 1987. Pharmacological investigation on sodium salt and acetyl derivatives of anarcadic acid. *Phytother. Res.*, 1:127–134.

Bianco, A., Chicchio, U., Rescifina, A., Romeo, G., and Uccella, N. 1997. Biomimetic supramolecular biophenol-carbohydrate and biophenol-protein models by NMR experiments. *J. Agric. Food Chem.*, 45:4281–4285.

Bidlack, W.R., Omaye, S.T., Meskin, M.S., and Topham, D.K.W. 2000. *Phytochemicals as Bioactive Agents*. Technomic Publishing Co., Inc., Lancaster, PA.

Blaser, M.J. 1993. *H. pylori*: microbiology of a 'slow' bacterial infection. *Trends Microbiol.*, 1:255–260.

Block, G., Patterson, B., and Subar, A. 1992. Fruit, vegetables, and cancer prevention: a review of the epidemiological evidence. *Nutr. Cancer*, 18:1–29.

Blot, W.J., Chow, W.-H., and McLaughlin, J.K. 1996. Tea and cancer: a review of the epidemiological evidence. *Eur. J. Cancer. Prev.*, 5:425–438.

Blytt, H.J., Guscar, T.K., and Butler, L.G. 1988. Antinutritional effects and ecological significance of dietary condensed tannins may not be due to binding and inhibiting digestive enzymes. *J. Chem. Ecol.*, 15:1455–1465.

Böhm, H., Boeing, H., Hempel, J., Raab, B., and Kroke, A. 1998. Flavonols, flavones, and anthocyanins as native antioxidants of food and their possible role in the prevention of chronic diseases. *Z. Ernährungswiss*, 37:147–163.

Bok, S.H., Lee, S.H., Park, Y.B., Bae, K.H., Son, K.H., Jeong, T.S., and Choi, M.S. 1999. Plasma and hepatic activities of 3-hydroxy-3-methyl-glutaryl coenzyme A reductase and acyl coenzyme A: cholesterol acyltransferase are lower in rats fed citrus peel extract or a mixture of citrus bioflavonoids. *J. Nutr.*, 128:1182–1185.

Bokkenheuser, V.D., Schackleton, C.H.L., and Winter, J. 1987. Hydrolysis of dietary flavonoid glycosides by strains of intestinal *Bacteroides* from humans. *Biochemistry*, 248:953–956.

Booth, A.N., Murray, C.W., Jones, F.F., and DeEds, F. 1956. The metabolic fate of rutin and quercetin in the animal body. *J. Biol. Chem.*, 233:251–260.

Born, S.L., Rodriguez, P.A., Eddy, C.L., and Lehmann-McKeeman, L.D. 1997. Synthesis and reactivity of coumarin 3,4-epoxide. *Drug Metab. Dispos.*, 25:1318–1323.

Born, S.L., Hu, J.K., and Lehmann-McKeenan, L.D. 2000. *O*-Hydroxyphenylacetylaldehyde as a hepatoxic metabolite of coumarin. *Drug Metab. Dispos.*, 28:218–223.

Borriello, S.P., Setchell, K.D.R., Axelson, M., and Lawson, A.M. 1985. Production and metabolism of lignans by the human faecal flora. *J. Appl. Bacteriol.*, 58:37–43.

Brandi, M.L. 1997. Natural and synthetic isoflavones in the prevention and treatment of chronic diseases. *Calcif. Tissue Int.*, 61:S5-S8.

Bravo, L. 1998. Polyphenols: chemistry, dietary sources, metabolism and nutritional significance. *Nutr. Rev.*, 56:317–333.

Bressani, R., Elias, L.G., and Braham, J.E. 1982. Reduction of digestibility of legume proteins by tannins. *J. Plant Foods*, 4:43–55.

Brook, A.J.W. and Housley, S. 1969. Interactions of phenols with Sephadex gels. *J. Chromatogr.*, 41:200–204.

Brouet, I. and Ohshima, H. 1995. Curcumin, an antitumor promoter and antiinflammatory agent, inhibits induction of nitric oxide synthase in activated macrophages. *Biochem. Biophys. Res. Commun.*, 206:533–540.

Brune, M., Rossander, L., and Hallberg, L. 1989. Iron absorption and phenolic compounds. Importance of different phenolic structure. *Eur. J. Clin. Nutr.*, 43:547–558.

Bruno, M., Hallberg, L., and Skanberg, A.B. 1991. Determination of iron-binding phenolic groups in foods. *J. Food Sci.*, 56:128–131, 167.

Bub, A., Watzl, B., Heeb, D., Rechkemmer, G., and Briviba, K. 2001. Malvidin-3-glucoside bioavalibility in humans after ingestion of red wine, dealcoholized red wine and red grape juice. *Eur. J. Nutr.*, 40:113–120.

Burger, O., Weiss, E., Sharon, N., Tabak, M., Neeman, I., and Ofek, I. 2002. Inhibition of *Helicobacter pylori* adhesion to human gastric mucus by high-molecular-weight constituent of cranberry juice. *CRC Crit. Rev. Food Sci. Nutr.*, 42(Suppl.):279–284.

Butler, E.J., Pearson, A.W., and Fenwick, G.R. 1982. Problems which limit the use of rapeseed meal as a protein source in poultry diets. *J. Sci. Food Agric.*, 33:866–875.

Butler, L.G. 1989a. New perspectives on the antinutritional effects of tannins, in *Food Products*, Kinsella, J.E. and Soucie, W.G., Eds., AOCS, Champaign, IL, 402–409.

Butler, L.G. 1989b. Effects of condensed tannin on animal nutrition, in *Chemistry and Significance of Condensed Tannins,* Hemingway, R.W. and Karchesy, J.J., Eds., Plenum Press, New York, 391–403.

Butler, L.G., Rogler, J.C., Mahansho, H., and Carlson, D.M. 1986. Dietary effects of tannins, in *Plant Flavonoids in Biology and Medicine: Biochemical, Pharmacological and Structure-Activity Relationship*, Cody, V., Harborne, J.B., and Middleton, E., Eds., Alan R. Liss, New York, 141–157.

Butrum, R.R., Ed. 1996. *Dietary Phytochemicals in Cancer Prevention and Treatment*, Plenum Press, New York.

Cai, Y., Gaffney, S.H., Lilley, T.H., and Haslam, E. 1989. Carbohydrate-polyphenol complexation, in *Chemistry and Significance of Condensed Tannins*, Hemingway, R.W. and Karchesy, J.J., Eds., Plenum Press, New York, 307–322.

Calderon, P., Van Buren, J., and Robinson, W.B. 1968. Factors influencing the formation of precipitates and hazes by gelatin and condensed and hydrozable tannins. *J. Agric. Food Chem.*, 16:479–482.

Camussi, G., Albano, E., Tetta, C., and Bussolino, F. 1991. The molecular action of tumor necrosis factor-α. *Eur. J. Biochem.*, 202:3–14.

Cao, G. and Prior, R.L. 1999. Anthocyanins are detailed in human plasma after oral administration of an elderberry extract. *Clin. Chem.*, 45:574–576.

Carbo, N., Costelli, P., Baccino, F.M., Lopez-Soriano, F.J., and Argilles, J.M. 1999. Resveratrol, a natural product present in wine, decreases tumour growth in a rat tumour model. *Biochem. Biophys. Res. Commun.*, 254:739–743.

Carbonaro, M., Grant, G., and Pusztai, A. 2001. Evaluation of polyphenol bioavailability in rat small intestine. *Eur. J. Nutr.*, 40:84–90.

Carrera, G., Mitjavila, S., and Derache, R. 1973. Effect de l'acide tannique sur l'absorption de la vit. B_{12} chex le rat. *C.R. Hebd. Seances Acad. Sci. Ser. D*, 276:239–242. Cited in Singleton, V. 1981. Naturally occurring food toxicants: phenolic substances of plant origin common in foods. *Adv. Food Res.*, 27:149–243.

Caruso, D., Visioli, F., Patelli, R., Galli, C., and Galli, G. 2001. Urinary excretion of olive oil phenols and their metabolites in humans. *Metab. Clin. Exp.*, 50:1426–1428.

Chang, Y.-C. and Nair, M.G. 1995. Metabolism of daidzein and genistein by intestinal bacteria. *J. Nat. Prod.*, 58:1892–1896.

Chang, M.M.-Y. and Fong, D. 1994. Antiinflammatory and cancer-preventive immunomodulation through diet: effects of curcumin on T-lymphocytes, in *Food Phytochemicals for Cancer Prevention II. Teas, Spices, and Herbs,* Ho, C.-T., Osawa, T., Huang, M.-T., and Rosen, R.T., Eds., ACS Symposium Series 547, American Chemical Society, Washington, D.C., 222–230.

Chang, H.C. and Branen, A.L. 1975. Antimicrobial effects of butylated hydroxyanisole (BHA). *J. Food Sci.*, 40:349–351.

Chang, C.W. and Anderson, J.U. 1968. Flocculation of clays and soils by organic compounds. *Soil Sci. Soc. Am. Proc.*, 32:23–27.

Chang, S.I. and Fuller, H.L. 1964. Effect of tannin content of grain sorghum on their feeding value of growing chicks. *Poultry Sci.*, 43:30–36.

Charlton, A.J., Baxter, N.J., Khan, M.L., Moir, A.J.G., Haslam, E., Davies, A.P., and Williamson, M.P. 2002. Polyphenol/peptide binding and precipitation. *J. Agric. Food Chem.*, 50:1593–1601.

Chibber, B.A.K., Mertz, E.T., and Axtell, J.D. 1978. The effect of dehulling on tannin content, protein distribution and quality of high and low tannin sorghum. *J. Agric. Food Chem.*, 26:679–683.

Chung, M.I., Teng, C.M., Cheng, K.L., Ko, F.N., and Lin, C.N. 1992. An antiplatelet principle of *Veratrum formosanum. Planta Med.*, 58:274–276.

Ciolino, H.P., Wang, T.T.Y., and Yeh, C.C. 1998. Diosmin and diosmetin are agonists of the aryl hydrocarbon receptor that differentially affect cytochrome P450 1A1 activity. *Cancer Res.*, 58:2754–2760.

Clandinin, D.R. and Robblee, A.R. 1981. Rapeseed meal in animal nutrition. II. Nonruminant animals. *J. Am. Oil Chem. Soc.*, 58:682–686.

Clarkson, T.B., Anthony, M.S., and Hughes, C.L. 1995. Estrogenic soybean isoflavones and chronic disease. Risks and benefits. *Trend End. Met.*, 6:11–16.

Clifford, M.M., Copeland, E.L., Bloxsidge, J.P., and Mitchell, L.A. 2000. Hippuric acid as a major excretion product associated with black tea consumption. *Xenobiotica*, 30:317–326.

Clifford, M.N. 2000. Anthocyanins — nature, occurrence and dietary burden. *J. Sci. Food Agric.*, 80:1063–1072.

Cline, J.M. and Hughes, C.L., Jr. 1998. Phytochemicals for the prevention of breast and endometrial cancer. *Cancer Treat. Res.*, 94:107–134.

Cody, V., Middleton, E., and Harborne, J. 1986. *Plant Flavonoids in Biology and Medicine. Biochemical, Pharmacological, Structure–Activity Relationships*. Alan R. Liss, New York.

Coldham, N.G., Howells, L.C., Santi, A., Montesissa, C., Langlais, C., King, L.J., MacPherson, D.D., and Sauer, M.J. 1999. Biotransformation of genistein in the rat: elucidation of metabolite structure by product ion mass fragmentology. *J. Steroid Biochem. Mol. Biol.*, 70:169–184.

Conrad, M.E. 1970. Factors affecting iron absorption, in *Iron Deficiency, Pathogenesis: Clinical Aspects: Therapy,* Hallberg, L., Harworth, G., and Vannotti, A., Eds., Academic Press, New York, 87–120.

Cook, N.C. and Samman, S. 1996. Flavonoids-chemistry, metabolism, cardioprotective effects, and dietary sources. *J. Nutr. Biochem.*, 7:66–76.

Cook, J.D., Reddy, M.B., and Hurrell, R.F. 1995. The effects of red and white wines on non heme-iron absorption in humans. *Am. J. Clin. Nutr.*, 61:800–804.

Cos, P., Ying, L., Calomme, M., Hu, J.P., Cimanga, K., Van Poel, B., Pieters, L., Vlientinck, A.J., and Berghe, D.V. 1998. Structure–activity relationship and classification of flavonoids as inhibitors of xanthine oxidase and superoxide scavengers. *J. Nat. Prod.*, 61:71–76.

Crespy, V., Morand, C., Manach, C., Besson, C., Demigne, C., and Remesy, C. 1999. Part of quercetin absorbed in the small intestine is conjugated and further secreted in the intestinal lumen. *Am. J. Physiol.*, 277:120–126.

Crespy, V., Aprikian, O., Morand, C., Besson, C., Manach, C., Demigne, C., and Remesy, C. 2002. Bioavailability of phloretin and phlorizidin in rats. *J. Nutr.*, 132:3227–3230.

Daglia, M., Tarsi, R., Papetti, A., Grisoli, P., Dacarro, C., Pruzzo, C., and Gazzani, G. 2002. Antiadhesive effect of green and roasted coffee on *Streptococcus mutans* adhesive properties on saliva-coated hydroxyapatite beads. *J. Agric. Food Chem.*, 50:1225–1229.

Das, D.K., Sato, M., Ray, P.S., Maulik, G., Engelman, R.M., Bertelli, A.A.E., and Bertelli, A. 1999. Cardioprotection of red wine: role of polyphenolic antioxidants. *Drugs Exp. Clin. Res.*, 25:115–120.

Davidson, P.M. and Branen, A.L. 1980a. Inhibition of two psychrotopic *Pseudomonas* species by butylated hydroxyanisole. *J. Food Sci.*, 45:1603–1606.

Davidson, P.M. and Branen, A.L. 1980b. Antimicrobial mechanisms of butylated hydroxyanisole against two *Pseudomonas* species. *J. Food Sci.*, 45:1607–1613.

Davidson, P.M., Brekke, C.J., and Branen, A.L. 1981. Antimicrobial activity of butylated hydroxyanisole, tertiary butylhydroquinone, and potassium sorbate in combination. *J. Food Sci.*, 46:314–316.

Davis, A.B. and Harbers, L.H. 1974. Hydrolysis of sorghum grain starch by rumen microorganisms and purified alpha-amylase as observed by scanning electron microscopy. *J. Animal Sci.*, 38:900–907.

Davis, A.B. and Hoseney, R.C. 1979. Grain sorghum condensed tannins. I. Isolation, estimation, and selective absorption by starch. *Cereal Chem.*, 56:310–314.

de Boever, P., Deplancke, B., and Verstraete, W. 2000. Fermentation by gut microbiota cultured in a similator of the human intestinal microbial ecosystem is improved by supplementing a soy germ powder. *J. Nutr.*, 130:2599–2606.

De Bruyne, T., Pieters, L., Witvrouw, M., De Clercq, E., Berghe, D.V., and Vlietinck, A.J. 1999. Biological evaluation of proanthocyanidin dimers and related polyphenols. *J. Nat. Prod.*, 62:954–958.

de Freitas, V. and Mateus, N. 2001. Structural features of procyanidin interaction with salivary proteins. *J. Agric. Food Chem.*, 49:940–945.

deLigny, C.L. 1979. Absorption of monosubstituted phenols on Sephadex gel G-15. *J. Chromatogr.*, 172:397–398.

Derman, D., Sayers, M., Lynch, S.R., Charlton, R., Bothwell, T.H., and Mayet, F. 1977. Iron absorption from a cereal-based meal containing cane sugar fortified with ascorbic acid. *Br. J. Nutr.*, 38:261–269.

Deschner, E.E. 1992. Dietary quercetin (QU) and rutin (RU) as inhibitors of experimental colonic neoplasia, in *Phenolic Compounds in Food and Their Effects on Health II. Antioxidant and Cancer Prevention*, Huang, M.-T., Ho, C.-T., and Lee, C.Y., Eds., ACS Symposium Series 507, American Chemical Society, Washington, D.C., 265–268.

Deschner, E.E., Ruperto, J., Wong, G., and Newmark, H.L. 1991. Quercetin and rutin as inhibitors of azoxymethanol-induced colonic neoplasia. *Carcinogenesis*, 12:1193–1196.

Deshpande, S.S. and Cheryan, M. 1983. Changes in phytic acid, tannins and trypsin inhibitory activity on soaking of dry beans (*Phaseolus vulgaris* L.). *Nutr. Rep. Int.*, 27:371–377.

Deshpande, S.S. and Salunkhe, D.K. 1982. Interactions of tannic acid and catechin with legume starches. *J. Food Sci.*, 47:2080–2081, 2083.

de Vries, J.H.M., Hollman, P.C.H., van Amersfoot, I., Olthof, M.R., and Katan, M.B. 2001. Red wine is a poor source of bioavailable flavonols in men. *J. Nutr.*, 131:745–748.

Dick, A.J. and Bearne, S.L. 1988. Inhibition of β-glucosidase of apple by flavonoids and other polyphenols. *J. Food Biochem.*, 12:97–108.

Disler, P.B., Lynch, S.R., Charlton, R.W., Torrance, J.D., Bothwell, T.H., Walker, R.B., and Mayet, F. 1975a. The effect of tea on iron absorption. *Gut*, 16:193–200.

Disler, P.B., Lynch, S.R., Torrance, J.D., Sayers, M.H., Bothwell, T.H., and Charlton, R.W. 1975b. The mechanism of inhibition of iron absorption by tea. *S. Afr. J. Med. Sci.*, 40:109–116.

Donovan, J.L., Bell, J.R., Kasim-Karakas, S., German, J.B., Walzem, R.L., Hansen, R.J., and Waterhouse, A.L. 1999. Catechin is present as metabolites in human plasma after consumption of red wine. *J. Nutr.*, 129; 1662–1668.

Donovan, J.L., Crespy, V., Manach, C., Morand, C., Besson, C., Scalbert, A., and Remesy, C. 2001. Catechin is metabolized by both the small intestine and liver of rats. *J. Nutr.*, 130:1753–1757.

Dorell, N., Crabtree, J.E., and Wren, B.W. 1998. Host-bacterial interactions and the pathogenesis of *Helicobacter pylori* infection. *Trends. Microbiol.*, 6:379–381.

Driedger, A. and Hatfield, E.E. 1972. Influence of tannins on the nutritive value of soybean meals for ruminants. *J. Animal Sci.*, 34:465–468.

Dubnick, M.A. and Omaye, S.T. 2001. Evidence for grape, wine and tea polyphenols as modulators of atherosclerosis and heart disease in humans. *J. Nutr. Func. Med. Foods*, 3:67–93.

Dugas, A.J., Castañeda-Acosta, J., Bonin, G.C., Price, K.L., Fischer, N.H., and Winston, G.W. 2000. Evaluation of the total peroxyl radical-scavenging capacity of flavonoids: structure-activity relationships. *J. Nat. Prod.*, 63:327–331.

Dunn, B.E., Cohen, H., and Blaser, M.J. 1997. *Helicobacter pylori. Clin. Microbiol. Rev.*, 10:720–741.

Eggum, B.O. and Christensen, K.D. 1975. Influence of tannin on protein utilization in feedstuffs with special reference to barley, in *Breed, Seed Protein Improv. Using Nucl. Tech. Proc. Res. Co-ord. Meet.*, 2nd, 135–143.

Ekpenyong, T.E. 1985. Effect of cooking on polyphenolic content of some Nigerian legumes and cereals. *Nutr. Rep. Int.*, 21:561–565.

Elmaki, H.B., Babiker, E.E., and El Tinay, A.H. 1999. Changes in chemical composition, grain malting, starch and tannin contents and *in vitro* protein digestibility during germination of sorghum cultivars. *Food Chem.*, 64:331–336.

Erlund, I., Alfthan, G., Maenpaa, J., and Aro, A. 2001. Tea and coronary heart disease: the flavonoid quercetin is more bioavailable from rutin in women than in men. *Arch. Intern. Med.*, 161:1919–1920.

Esterbauer, H., Janusz, G., Puhl, H., and Jurgens, G. 1992. The role of lipid peroxidation and antioxidants in oxidative modification of LDL. *Free Radical Biol. Med.*, 13:341–390.

Facino, R.M., Carini, M., Aldini, G., Berti, F., Rossoni, G., Bombadelli, E., and Morazzoni, P. 1998. Diet enriched with procyanidins enhances antioxidant activity and reduces myocardial post-ischaemic damage in rats. *Life Sci.*, 64:627–642.

Faithful, N.T. 1984. The *in vitro* digestibility of feedstuffs — a century of ferment. *J. Sci. Food Agric.*, 35:819–826.

Fang, N., Yu, S., and Badger, T.M. 2002. Characterization of isoflavones and their conjugates in female rat urine using LC/MS/MS. *J. Agric. Food Chem.*, 50:2700–2707.

Fauconneau, B., Waffo-Teguo, P., Huguet, F., Barrier, L., Decendit, A., and Merillon, J.-M. 1997. Comparative study of radical scavenger and antioxidant properties of phenolic compounds from *Vitis vinifera* cell cultures using *in vitro* tests. *Life Sci.*, 61:2103–2110.

Fentem, J.H. and Fry, J.R. 1993. Species differences in the metabolism and hepatoxicity of coumarin. *Comp. Biochem. Physiol.*, 104C:1–8.

Fenwick, G.R., Pearson, A.W., Greenwood, N.M., and Butler, E.G., 1981. Rapeseed meal tannins and egg taint. *Animal Feed Sci. Technol.*, 6:421–431.

Fenwick, G.R., Curl, C.L., Butler, E.J., Greenwood, N.M., and Pearson, A.W. 1984. Rapeseed meal and egg taint: effects of low glucosinolate *Brassica napus* meal, dehulled meal and hulls, and of neomycin. *J. Sci. Food Agric.*, 35:749–756.

Fitzpatrick, D.F., Fleming, R.C., Bing, B., Maggi, D.A., and O'Malley, R.M. 1999. Isolation and characterization of endothelium-dependent vasorelaxing compounds from grape seeds. *J. Agric. Food Chem.*, 48:6384–6390.

Fleet, G. 1992. Spoilage yeasts. *CRC Crit. Rev. Biotechnol.*, 12:1–44.

Fleming, H.P., Walter, W.H., Jr., and Etchells, J.L. 1973. Antimicrobial properties of oleuropein and products of its hydrolysis from green olives. *Appl. Microbiol.*, 26:777–782.

Fogg, A.H. and Lodge, R.M. 1945. The mode of antimicrobial action of phenols in relation to drug fastness. *Trans. Faraday Soc.*, 41:359–365.

Foo, L.Y., Lu, Y., Howell, A.B., and Vorsa, N. 2000a. The structure of cranberry proanthocyanidins which inhibit adherence of uropathogenic P-fimbriated *Escherichia coli in vitro*. *Phytochemistry*, 54:173–181.

Foo, L.Y., Lu, Y., Howell, A.B., and Vorsa, N. 2000b. A-type proanthocyanidin trimers from cranberry that inhibit adherence of uropathogenic P-fimbriated *Escherichia coli*. *J. Nat. Prod. Chem.*, 63:1225–1228.

Ford, J.E. and Hewitt, D. 1974. Protein quality in cereals and pulses. 2. Influence of poly ethylene glycol on nutritional availability of methionine in sorghum (*Sorghum vulgare pers*), field beans (*Vicia faba* L.) and barley. *Br. J. Nutr.*, 42:317–323.

Formica, J.V. and Regelson, W. 1995. Review of biology of quercetin and related bioflavonoids. *Food Chem. Toxicol.*, 33:1061–1080.

Frank, J., Kamal-Eldin, A., Lundh, T., Määttä, K., Törrönen, R., and Vessby, B. 2002. Effects of dietary anthocyanins on tocopherols and lipids in rats. *J. Agric. Food Chem.*, 50:7226–7230.

Frankel, E.N., Waterhouse, A.L., and Teissedre, P.L. 1995. Principal phenolic phytochemicals in selected California wines and their antioxidant activity in inhibiting oxidation of human low-density lipoproteins. *J. Agric. Food Chem.*, 43:890–894.

Frankel, E.N., Waterhouse, A.L., and Teissedre, P. 1993. Inhibition of oxidation of human low-density protein by phenolic substances in red wine. *Lancet*, 341:1103–1104.

Freedman, M.L. and Tanzer, J.M. 1974. Dissociation of plaque formation from glucan-induced agglutination in mutants of *Streptococcus mutans*. *Infect. Immun.*, 10:189–196.

Fremont, L., Belguendouz, L., and Delpal, S. 1999. Antioxidant activity of resveratrol and alcohol-free wine polyphenols related to LDL oxidation and polyunsaturated fatty acids. *Life Sci.*, 64:2511–2521.

Furstenberger, G., Hagedorn, H., Jacobi, T., Besemfelder, E., Lehman, W.D., and Marks, F. 1991. Characterization of an 8-lipoxygenase activity induced by the phorbol ester tumor promoter 12-*O*-tetradecanoylphorbol-13-acetate in mouse skin *in vivo*. *J. Biol. Chem.*, 266:15738–15745.

Gabor, M. 1986. Antiinflammatory and antiallergic properties of flavonoids, in *Plant Flavonoids in Biology and Medicine: Biochemical, Pharmacological, and Structure-Activity Relantionships*, vol. 213, Cody, V., Middleton, E., Harborne, J.V., and Beretz, A., Eds., Alan R. Liss, New York, 471–480.

Galati, E.M., Monforte, M.T., Kirjainen, S., Forestieri, A.M., Trovato, A., and Tripodo, M.M. 1994. Biological effect of hesperidin, a citrus flavonoid. (Note I): antiinflammatory and analgesic activity. *Farmaco*, 49:709–712.

Galati, G., Sabzevari, O., Wilson, J.K., and O'Brien, P.J. 2002. Prooxidant activity and cellular effects of the phenoxyl radical of dietary flavonoids and other polyophenolics. *Toxicology*, 177:91–104.

Galley, P. and Thiollet, M.A. 1993. A double-blind placebo-controlled trial of a new venoactive flavonoid fraction in the treatment of symptomatic capillary fragility. *Int. Angiol.*, 12:69–72.

Gao, Y.T., McLaughlin, J.K., Blot, W.J., Ji, B.T., Dai, A., and Fraumeni, J.F., Jr. 1994. Reduced risk of esophageal cancer associated with green tea consumption. *J. Nat. Cancer Inst.*, 86:855–858.

Gee, J.M., DuPont, S.M., Day, A.J., Plumb, G.W., Williamson, G., and Johnson, I.T. 2000. Intestinal transport of quercetin glycosides in rats involves both the deglycosylation and interaction with hexose transport pathway. *J. Nutr.*, 130:2765–2771.

Gellerman, J.L., Wash, N.J., Werner, N.K., and Schlenk, H. 1969. Antimicrobial effects of anarcadic acids. *Can. J. Microbiol.*, 15:1219–1223.

Gibson, L., Pike, L., and Kilbourn, J.P. 1991. Clinical study: effectiveness of cranberry juice in preventing urinary tract infections in long-term care facility patients. *J. Naturapathic Med.*, 2:45–47.

Gillooly, M., Bothwell, T.H., Charlton, R.W., Torrance, J.D., Bezwoda, W.R., MacPhail, A.P., Derman, D.P., Novelli, L., Morrall, P., and Mayet, F. 1984. Factors affecting the absorption of iron from cereals. *Br. J. Nutr.*, 51:37–46.

Gillooly, M., Bothwell, T.H., Torrance, J.D., MacPhail, A.P., Derman, D.P., Bezwoda, W.R., Mills, W., Charlton, R.W., and Mayet, F. 1983. The effects of organic acids, phytates and polyphenols on the absorption of iron from vegetables. *Br. J. Nutr.*, 49:331–342.

Gilman, M.W., Cupples, L.A., Gagnon, D., Posner, B.M., Ellison, R.C., Castelli, W.P., and Wolf, P.A. 1995. Protective effect fruits and vegetables on development of stroke in men. *J. Am. Med. Assoc.*, 273:1113–1117.

Glick, Z. and Joslyn, M.A. 1970. Food intake depression and other metabolic effects of tannic acid in the rat. *J. Nutr.*, 100:509–515.

Glick, Z. 1981. Modes of action of gallic acid in suppressing food intake by rats. *J. Nutr.*, 111:1910–1916.

Godeberg, P. 1994. Daflon 500 mg in the treatment of hemorroidal disease: a demonstration of efficacy in comparison with placebo. *Angiology*, 45:574–578.

Goldberg, D.M. 1995. Does wine work? *Clin. Chem.*, 41:14–16.

Goldberg, D.M., Hahn, S.E., and Parkes, J.G., 1995. Beyond alcohol: beverage consumption and cardiovascular mortality. *Clin. Chem. Acta*, 237:155–187.

Goldstein, J. and Swain, T. 1965. The inhibition of enzymes by tannins. *Phytochemistry*, 4:185–192.

Gómez-Cordovés, C., Bartolomé, B., Viera, W., and Virador, V.M. 2001. Effects of wine phenolics and sorghum tannins on tyrosinase activity and growth of melanoma cells. *J. Agric. Food Chem.*, 49:1620–1624.

González de Mejía, E., Castaño-Tostado, E., and Loarca-Piña, G. 1999. Antimutagenic wffwcts of natural phenolic compounds in beans. *Mutation Res.*, 441:1–9.

Govindarajan, V.S. 1980. Turmeric chemistry, technology, and quality. *CRC Crit. Rev. Food Sci. Nutr.*, 12:199–301.

Grazzini, R., Hesk, D., Heiminger, E., Hildenbrandt, G., Reddy, C.C., Cox-Foster, D., Medford, J., Craig, R., and Mumma R.O. 1991. Inhibition of lipoxygenase and prostaglandin endoperoxide synthase by anarcadic acids. *Biochem. Biophys. Res. Commun.*, 176:775–780.

Griffith, D.W. 1981. The polyphenolic content and enzyme inhibitory activity of testas from faba bean (*Vicia faba*) and pea (*Pisum spp.*) varieties. *J. Sci. Food Agric.*, 32:797–804.

Griffith, D.W. and Moseley, G. 1980. The effect of diet containing field beans of high and low polyphenolic content on the activity of digestive enzymes in the intestines of rats. *J. Sci. Food Agric.*, 31:255–259.

Griffith, D.W. 1979. The inhibition of digestive enzymes by extracts of field beans (*Vicia fabia*). *J. Sci. Food Agric.*, 30:458–462.

Griffiths, L.A. 1982. Mammalian metabolism of flavonoids, in *The Flavonoids: Advances in Research,* Harborne, J.B and Marby, T.S., Eds., Chapman & Hall, London, 681–718.

Grimshaw, J. 1976. Phenolic aralkylamines, monoketones and monocarboxylic acids, in *Rodd's Chemistry*, 2nd ed., Coffey, S., Ed., Elsevier, Amsterdam, 141–202.

Gronbaek, M. 1999. Type of alcohol and mortality from cardiovascular disease. *Food Chem. Toxicol.*, 37:921–924.

Grundy, S.M. 1996. Lipids, nutrition, and coronary heart disease, in *Atherosclerosis and Coronary Artery Disease,* Fuster, V., Ross, R., and Topol, E.J., Eds., Lippincott-Raven Publishers, Philadelphia, 45–68.

Gupta, S.K., Banerjee, A.B., and Achari, B. 1976. Isolation of ethyl *p*-methoxycinnamate, the major antifungal principle of *Curcumba zedoaria. Lloydia*, 39:218–222.

Gust, J. and Suwalski, J. 1994. Use of Mossbauer spectroscopy to study reaction products of polyphenols and iron compounds. *Corrosion*, 50:355–365.

Gustavson, K.H. 1954. Interaction of vegetable tannins with polyamides as proof of the dominant function of the peptide bond of collagen for its binding of tannins. *J. Poly. Sci.*, 12:317–324.

Guyot, S., Pellerin, P., Brillouet, J.-M., and Cheynier, V. 1996. Inhibition of β-glucosidase (*Amygdalae dulces*) by (+) catechin oxidation products and procyanidin dimers. *Biosci. Biotechnol. Biochem.*, 60:1131–1135.

Habtemariam, S. 1997. Flavonoids as inhibitors or enhancers of cytotoxicity of tumor necrosis factor-α in L-929 tumor cells. *J. Nat. Prod.*, 60:775–778.

Hagerman, A.E. and Butler, L.G. 1978. Protein precipitation method for the quantitative determination of tannins. *J. Agric. Food Chem.*, 26:809–812.

Hagerman, A.E. 1980. Ph.D. Thesis, Purdue University, in Asquith, T.N. and Butler, L.G. 1986. Interactions of condensed tannins with selected proteins. *Phytochemistry*, 25:1591–1593.

Hagerman, A.E. and Butler, L.G. 1980. Condensed tannin purification and characterization of tannin associated proteins. *J. Agric. Food Chem.*, 28:947–952.

Hagerman, A.E. and Butler, L.G. 1981. The specifity of proanthocyanidin-protein interaction. *J. Biol. Chem.*, 256:4494–4497.

Hagerman, A.E. and Robbins, C.T. 1987. Implications of soluble tannin-protein complexes for tannin analysis and plant defense mechanisms. *J. Chem. Ecol.*, 13:1243–1259.

Hagerman, A.E. 1989. Chemistry of tannin-protein complexation, in *Chemistry and Significance of Condensed Tannins*, Hemingway, R.W. and Karhesy J.J., Eds., Plenum Press, New York, 307–322.

Hagerman, A.E. 1992. Tannin–protein interactions, in *Phenolic Compounds in Food and Their Effects on Health I. Analysis, Occurrence, and Chemistry*; Ho, C.-T., Lee, C.Y., and Huang, M.-T., Eds., ACS Symposium Series 506, American Chemical Society, Washington, D.C., 236–248.

Hagerman, A.E., Robbins, C.T., Weerasurija, Y., Wilson, T.C., and McArthur, C. 1992. Tannin chemistry in relation to digestion. *J. Range Manag.*, 45:57–62.

Hagerman, A.E., Rice, M.E., and Ritchard, N.T. 1998. Mechanisms of protein precipitation for two tannins, pentagalloyl glucose and epicatechin$_{16}$ (4→8) catechin (procyanidin). *J. Agric. Food Chem.*, 46:2590–2595.

Haglund, A.C. 1978. Absorption of monosubstituted phenols on Spehadex G-15. *J. Chromatogr.*, 156:317–322.

Hainonen, M., Rein, D., Satue-Gracia, M.T., Huang, S.W., German, J.B., and Frankel, E.N. 1998. Effect of protein on the antioxidant activity of phenolic compounds in a lecithin liposome oxidation system. *J. Agric. Food Chem.*, 46:917–922.

Hallberg, L. and Rossander, L. 1982. Effect of different drinks on the absorption of nonheme iron from composite meals. *Human Nutr. Appl. Nutr.*, 36A:116–123.

Hamada, S. and Torii, M. 1978. Effect of sucrose in culture media on the location of glucosyltransferase from *Streptococcus mutants* and cell adherence to glass surfaces. *Infect. Immun.*, 20:592–599.

Hamilton-Miler, J.M. 2001. Anticariogenic properties of tea (*Camellia sinensis*). *J. Med. Microbiol.*, 50:299–302.

Haslam, E. 1974. Polyphenol–protein interactions. *Biochem. J.*, 139:285–288.

Haslam, E. 1989. *Plant Polyphenols. Vegetable Tannins Revisited,* Cambridge University Press, Cambridge, U.K.

Haslam, E., Lilley, T.H., Cai, Y., Martin, R., and Magnolato, D. 1989. Traditional herbal medicines — the role of polyphenols. *Planta Med.*, 55:1–8.

Haslam, E., Lilley, T.H., Warminski, E., Liao, H., Cai, Y., Martin, R., Gaffney, S.H., Goulding, P.N., and Luck, G. 1992. Polyphenol complexation: a study of molecular recognition, in *Phenolic Compounds in Foods and Their Effects on Health I. Analysis, Occurrence and Chemistry*, Ho, C.-T., Lee, C.Y., and Huang, M.-T., Eds., ACS Symposium Series 506, American Chemical Society, Washington, D.C., 5–80.

Haslam, E. 1996. Natural polyphenols (vegetable tannins) as drugs: possible modes of action. *J. Nat. Prod.*, 59:205–215.

Haslam, E. 1998. *Practical Polyphenolics*, Cambridge University Press, Cambridge, U.K.

Hatano, T., Yasuhara, T., Yoshihara, R., Ikegami, Y., Matsuda, M., Yazaki, K., Agata, K., Nishibe, S., Noro, T., Yoshizaki, M., and Takuo, O. 1991. Inhibitory effects of galloylated flavonoids on xanthine oxide. *Planta Med.*, 57:83–84.

Hattori, M., Kusumoto, I.T., Namba, T., Ishigami, T., and Hara, Y. 1990. Effect of tea polyphenols on glucan synthesis by glucotransferase from *Streptococcus mutants. Chem. Pharm. Bull.*, 38:717–720.

Hawksworth, G., Drasar, B.S., and Hill, M. 1971. Intestinal bacteria and the hydrolysis of glycosidic bonds. *J. Med. Microbiol.*, 4:451–459.

Hegsted, D.M. and Ausman, L.M. 1988. Diet, alcohol, and coronary heart disease in man. *J. Nutr.*, 118:1184–1189.

Heider, J. and Fuchs, G. 1997. Microbial anaerobic aromatic metabolism. *Anaerobe*, 3:1–22.

Heim, K.E., Tagliaferro, A.R., and Bobilya, D.J. 2002. Flavonoid antioxidants: chemistry, metabolism and structure–activity relationship. *J. Nutr. Biochem.*, 13:572–584.

Heinonen, S., Nurmi, T., Liukkonen, K., Poutanen, K., Wahala, K., Deyama, T., Nishibe, S., and Adlercreutz, H. 2001. *In vitro* metabolism of plant lignans: new precursors of mammalian lignans enterolactone and eneterodiol. *J. Agric. Food Chem.*, 49:3178–3186.

Hertog, M.L.G., Sweetnam, P.M., Fehily, A.M., Elwood, P.C., and Kromhout, D. 1997a. Antioxidant flavonols and ischaemic heart disease in Welsh population of men. The Cearphilly study. *Am. J. Clin. Nutr.*, 65:1489–1494.

Hertog, M.G.L., van Poppel, G., and Verhoeven, D. 1997b. Potentially anticarcinogenic secondary metabolites from fruits and vegetables, in *Phytochemistry of Fruit and Vegetables, Proceedings of the Phytochemical Society of Europe 41*, Tomás-Barberán, F.A. and Robins, R.J., Eds., Clarendon Press, Oxford, U.K., 313–330.

Hertog, M.G.L., 1996. Flavonols in wine and tea and prevention of coronary heart disease, in *Polyphenols 96,* Vercauteren, J., Cheze, C., and Triaud, J., Eds., INRA Editions 87, INRA, Paris, 117–132.

Hertog, M., Feskens, E., Hollman, P., Katan, M., and Kromhout, D. 1993a. Dietary antioxidant flavonoids and risk of coronary heart disease: the Zutphen elderly study. *Lancet*, 342:1007–1011.

Hertog, M., Hollman, P., Katan, M., and Kromhout, D. 1993b. Estimation of daily intake of potentially anticarcinogenic flavonoids and their determinants in adults in the Netherlands. *Nutr. Cancer*, 20:21–29.

Higuchi, W.I., Ho, N.F., Park, J.Y., and Komiya, I. 1981. Rate-limiting steps and factors in drug absorption, in *Drug Absorption,* Prescott, L.F. and Nimno, W.S., Eds., ADIS Press, New York, 35–60.

Himejima, M. and Kubo, I. 1991. Antibacterial agents from the cashew *Anacardium occidentale* (Anacardiacae) nut shell oil. *J. Agric Food Chem.*, 39:418–421.

Hodgson, J.M., Morton, L.W., Puddey, I.B., Beilin, L.J., and Croft, K.D. 2000. Gallic acid metabolites are markers of black tea intake in humans. *J. Agric. Food Chem.*, 48:2276–2280.

Hodgson, J.M., Croft, K.D., Puddey, I.B., Mori, T.A., and Beilin, L.J. 1996. Soybean isoflavonoids and their metabolic products inhibit *in vitro* lipoprotein oxidation in serum. *J. Nutr. Biochem.*, 7:664–669.

Hoff, J.E. and Singleton, K.I. 1977. A method for determination of tannins in foods by means of immobilized protein. *J. Food Sci.*, 42:1566–1569.

Hollman, P.C. 1997. Bioavailability of flavonoids. *Eur. J. Clin. Nutr.*, 51:S66-S69.

Hollman, P.C., de Vries, J.H., van Leeuwen, S.D., Mengelers, M.J., and Katan, M.B. 1995. Absorption of dietary quercetin glycosides and quercetin in healthy ileostomy volunteers. *Am. J. Clin. Nutr.*, 62:1276–1282.

Hollman, P.C., Gaag, M., Mengelers, M.J., van Trip, J.M., de Vries, J.H., and Katan, M.B. 1996. Absorption and disposition kinetics of the dietary antioxidant quercetin in man. *Free Radical Biol. Med.*, 21:703–707.

Hollman, P.C. and Katan, M.B. 1999. Dietary flavonoids: intake, health effects and bioavailabilty. *Food Chem. Toxicol.*, 37:937–942.

Hollman, P.C. and Katan, M.B. 1998. Bioavailability and health effects of dietary flavanols in man. *Arch. Toxicol. Supl.*, 20:237–248.

Hollman, P.C. and Katan, M.B. 1997. Absorption, metabolism, and bioavailability of flavonoids, in *Flavonoids in Health and Disease,* Rice-Evans, C. and Packer, L., Eds., Marcel Dekker, New York, 483–522.

Hollman, P.C., Tijburg, L.B.M., and Yang, C.S. 1997. Bioavailability of flavonoids from tea. *CRC Crit. Rev. Food Sci. Nutr.*, 37:719–738.

Howell, A.B. 2002. Cranberry proanthocyanidins and the maintenance of urinary tract health. *CRC Crit. Rev. Food Sci. Nutr.*, 42 (Suppl.):273–278.

Howell, A.B., Vorsa, N., Der Marderosian, A., and Foo, L.Y. 1998. Inhibition of the adherence of P-fimbriated *Escherichia coli* to uroepithelial-cell surfaces by proanthocyanidin extracts from cranberry. *N. Engl. J. Med.*, 339:1085–1086.

Hsieh, T.C. and Wu, J.M. 1999. Differential effects on growth, cell cycle arrest, and induction of apoptosis by resveratrol in human prostate cancer cell lines. *Exp. Cell Res.*, 249:109–115.

Huang, M.-T., Osawa, T. Ho, C.-T., and Rosen, R.T., Eds. 1994a. *Food Phytochemicals for Cancer Prevention I. Fruits and Vegetables,* ACS Symposium Series 546, American Chemical Society, Washington, D.C.

Huang, M.-T., Osawa, T., Ho, C.-T., and Rosen, R.T., Eds. 1994b. *Food Phytochemicals for Cancer Prevention II. Teas, Spices, and Herbs,* ACS Symposium Series 547, American Chemical Society, Washington, D.C.

Huang, M.-T., Ho, C.-T., and Lee, C.Y., Eds. 1992. *Phenolic Compounds in Food and Their Effects on Health. II.* ACS Symposium Series 507, American Chemical Society, Washington, D.C.

Hur, H.-G., Lay Jr., J.O., Beger, R.D., Freeman, J.P., and Rafii, F. 2000. Isolation of human intestinal bacteria metabolizing the natural isoflavone glycosides daidzin and genistin. *Arch. Microbiol.*, 174:422–428.

Hurrell, R.F., Reddy, M.B., and Cook, J.D. 1999. Inhibition of nonheme iron absorption in man by polyphenolic-containing beverages. *Br. J. Nutr.*, 81:289–295.

Hurrell, R.F., Reddy, M.B., and Cook, J.D. 1998. Inhibition of nonheme iron absorption in man by herb teas, coffee and black tea. *Br. J. Nutr.*, 61:800–804.

Hussein, L. and Abbas, H. 1985. Nitrogen balance studies among boys fed combinations of faba beans and wheat differing in polyphenolic contents. *Nutr. Rep. Int.*, 31:67–81.

Ielpo, M.T.L., Basile, A., Miranda, R., Moscatiello, V., Nappo, C., Sorbo, S., Laghi, E., Ricciardi, M.M., Ricciardi, L., and Vuotto, M.L. 2000. Immunopharmacological properties of flavonoids. *Fitoterapia*, 71:S101-S109.

Imai, K., Suga, K., and Nakachi, K. 1997. Cancer-preventive effects of drinking green tea among a Japanese population. *Prev. Med.*, 26:769–775.

Irie, J., Murata, M., and Homma, S. 1996, Glycerol-3-phosphate dehydrogenase inhibitors, anarcadic acids, from *Gingko biloba. Biosci. Biotechnol. Biochem.*, 60:240–243.

Ito, N. and Hirose, M. 1989. Antioxidants-carcinogenic and chemopreventive properties. *Adv. Cancer Res.*, 53:247–302.

Itoh, M. and Guth, P.H. 1985. Role of oxygen-derived free radicals in hemorrhagic shock-induced gastric lessions in rats. *Gastroenterology*, 88:1162–1167.

Iwuoha, C.I. and Aina, J.O. 1997. Effects of steeping condition and germination time on the alpha-amylase activity, phenolics content and malting loss of Nigerian local red and hibryd short Kaura sorghum malts. *Food Chem.*, 58:289–295.

Izumi, T., Piskula, M.K., Osawa, S., Obata, A., Saito, M., and Kataoka, S. 2000. Soy isoflavone aglycones are absorbed faster and in higher amounts than their glycosides. *J. Nutr.*, 130:1695–1699.

Jang, M, Cai, L., Udeani, G.O., Slowing, K.V., Thomas, C.F., Beecher, C.W.W., Fong, H.H.S., Farnsworth, N.R., Kinghorn, A.D., Mehta, R.G., Moon, R., and Pezzuto, J.M. 1997. Cancer chemopreventive activity of resveratrol, a natural product derived from grape. *Science*, 275:218–220.

Jayaprakasha, G.K., Rao, L.J.M., and Sakariah, K.K. 2002. Improved HPLC method for determination of curcumin, desmethylcurcumin, and bismethylcurcumin. *J. Agric. Food Chem.*, 50:3668–3672.

Jayatilake, G.S., Jayasuriya, H., Lee, E.S., Koonchanok, N.M., Geahlen, R.L., Ashendel, C.L., McLaughlin, J.L., and Chang, C.L. 1993. Kinase inhibitors from *Polygonum cuspidatum. J. Nat. Prod.*, 56:1805–1810.

Jean, T. and Bodinier, M.C. 1994. Mediators involved in inflammation: effect of Daflon 500 mg on their release. *Angiology*, 45:554–559.

Joannou, G.E., Kelly, G.E., Reeder, A.Y., Waring, M.A., and Nelson, C. 1995. A urinary profile study of dietary phytoestrogens. The identification and mode of metabolism of new isoflavonoids. *J. Steroid Biochem. Mol. Biol.*, 54:167–184.

Jones, W.T. and Mangan, J.L. 1977. Complexes of the condensed tannins of sainfoin (*Onobrychis viciifolia* Scop) with fraction 1 leaf protein and with submaxilar mucoprotein, and their reversal by polyethylene glycol and pH. *J. Sci. Food Agric.*, 28:126–136.

Judis, J. 1963. Studies on the mechanism of action of phenolic disinfectants. 2. Patterns of release of radioactivity from *Escherichia coli* labeled by growth on various compounds. *J. Pharm. Sci.*, 52:126–131.

Jung, H.G. and Fahey, G.C. 1983. Effects of phenolic monomers on rat performance and metabolism. *J. Nutr.*, 113:546–556.

Justesen, U., Arrigoni, E., Amado, R., and Larsen, B.R. 2000. Degradation of flavonoid glycosides *in vitro* fermentation with human faecal flora. *Lebensm.-Wiss. u.-Technol.*, 33:424–430.

Juven, B. and Henis, Y. 1970. Studies on the antimicrobial activity of olive phenolic compounds. *J. Appl. Bacteriol.*, 33:721–732.

Juven, B., Henis, Y., and Jacoby, B. 1972. Studies on the mechanism of the antimicrobial action of oleuropein. *J. Appl. Bacteriol.*, 35:559–567.

Kahraman, A., Erkasap, N., Köken, T., Serteser, M., Aktepe, F., and Erkasap, S. 2003. The antioxidative and antihistaminic properties of quercetin in ethanol-induced gastric lesions. *Toxicology*, 183:133–142.

Kamitani, H., Geller, M., and Thomas, E. 1998. Expression of 15-lipoxygenase by human colorectal carcinoma Caco-2 cells during apoptosis and cell differentiation. *J. Biol. Chem.*, 273:21569–21577.

Kane, J.P. 1996. Structure and function of the plasma lipoproteins and their receptors, in *Atherosclerosis and Coronary Artery Disease,* Fuster, V., Ross, R., and Topol, E.J., Eds., Lippincott-Raven Publishers, Philadelphia, 89–103.

Kaplan, M., Hayek, T., Raz, A., Coleman, R., Dornfield, L., Vaya, J., and Aviram, M. 2001. Pomegranate juice supplementation to atheroclerotic mic reduce micropaphage lipid peroxidation, cellular cholesterol accumulation and development of atherosclerosis. *J. Nutr.*, 131:2082–2089.

Kashiwada, Y., Nonaka, G.I., Nishioka, I., Chang, J.J., and Lee, K.H. 1992. Antitumor agents, 129. Tannins and related compounds as selective cytotoxic agents. *J. Nat. Prod.*, 1033–1043.

Katiyar, S.K. and Mukhtar, H. 1996. Tea in chemoprevention of cancer: epidemiological and experimental studies (review). *Int. J. Oncology*, 8:221–238.

Kaul, T.N., Middleton, E., and Ogra, P.L. 1985. Antiviral effects of flavonoids on human viruses. *J. Med. Virol.*, 15:71–79.

Kawaii, S., Tomono, Y., Katase, E., Ogawa, K., and Yano, M. 1999a. HL-60 differentiating activity and flavonoid content of readily extractable fraction prepared from *Citrus* juices. *J. Agric. Food Chem.*, 47:128–135.

Kawaii, S., Tomono, Y., Katase, E., Ogawa, K., and Yano, M. 1999b. Effect of citrus flavonoids on HL-60 cell differentiation. *Anticancer Res.*, 19:1261–1269.

Kawamoto, H., Nakatsubo, F., and Murakami, K. 1995. Quantitative determination of tannin and protein precipitates by high-performance liquid chromatography. *Phytochemistry*, 40:1503–1505.

Kawamoto, H., Nakayama, M., and Murakami, K. 1996. Stoichiometric studies of tannin protein co-precipitation. *Phytochemistry*, 41:1427–1431.

Kawamoto, H. and Nakatsubo, F. 1997. Effects of environmental factors on two stage tannin protein coprecipitation. *Phytochemistry*, 46:479–483.

Keli, S.O., Hertog, M.G.L., Feskens, E.J.M., and Kromhout, D. 1996. Dietary flavonoids, antioxidant vitamins and incidence of stroke. *Arch. Intern. Med.*, 156:637–642.

Kennedy, J.A. and Powell, K.J. 1985. Polyphenol interactions with aluminium (III) and iron (III): their possible involvement in the podzolization process. *Aust. J. Chem.*, 38:879 888.

Kerry, N. and Abbey, M. 1998. The isoflavone genistein inhibits copper and peroxyl radical mediated low-density lipoprotein oxidation *in vitro*. *Atherosclerosis*, 140:341–347.

Kim, D.S.H.L., Park, S.Y., and Kim, J.Y. 2001. Curcuminoids from *Curcuma longa* L. (Zingiberaceae) that protect PC12 rat pheochromocytoma and normal human umbilical vein endothelial cells from βA (1–42) insult. *Neurosci. Lett.*, 303:57–61.

Kimira, M., Arai, Y., Shimoi, K., and Watanabe, S. 1998. Japanese intake of flavonoids and isoflavonoids from foods. *J. Epidemiol.*, 8:168–175.

King, P.J., Ma, G., Miao, W., Jia, Q., McDougall, B., Reinecke, M.G., Cornell, C., Kuan, J., Kim, T.R., and Robinson Jr., E. 1999. Structure-activity relationships: analogues of the dicaffeoylquinic and dicaffeoyltartaric acids as potent inhibitors of human immunodeficiency virus type 1 integrase and replication. *J. Med. Chem.*, 42:497–509.

Klindworth, K.J., Davidson, P.M., Brekke, C.J., and Branen, A.L. 1979. Inhibition of *Clostridium perfringens* by butylated hydroxyanisole. *J. Food Sci.*, 44:564–567.

Knekt, P., Jarvinen, R., Seppanen, R., Hellovaara, M., Teppo, L., Pukkala, E., and Aromas, A. 1997. Dietary flavonoids and the risk of lung cancer and other malignant neoplasms. *Am. J. Epidemiol.*, 146:223–230.

Kobayashi, Y., Suzuki, M., Satsu, H., Arai, S., Hara, Y., Suzuki, K., Miyamoto, Y., and Shimizu, M. 2000. Green tea polyphenols inhibit the sodium-dependent glucose transporter of intestinal epithelial cells by a competitive mechanism. *J. Agric. Food Chem.*, 48:5618–5623.

Koeffler, H.P. 1983. Induction of differentiation of human acute myelogenous leukemia cell: therapeutic implications. *Blood*, 62:709–721.

Kondo, K., Hirano, R., Matsumoto, A., Igarashi, O., and Itakura, H. 1996. Inhibition of LDL oxidation by cocoa. *Lancet*, 348:1514.

Konneh, M. and Caen, J. 1996. Red wine derived compounds and their putative antiatherogenic properties, in *Polyphenols 96,* Vercauteren, J., Cheze, C., and Triaud, J., Eds., INRA Editions 87, INRA, Paris, 105–106.

Konowalchuk, J. and Speirs, J. 1976a. Antiviral activity of fruit extracts. *J. Food Sci.*, 41:1013–1017.

Konowalchuk, J. and Speirs, J. 1976b. Virus inactivation by grapes and wines. *Appl. Environ. Microbiol.*, 32:757–763.

Koyuncu, H., Bekarda, B., Baykut, F., Soybir, G., Alatli, C., Gül, H., and Altun, M. 1999. Preventive effect of hesperidin against inflammation in CD-1 mouse skin caused by tumor promoter. *Anticancer Res.*, 19:3237–3242.

Kozlowska, H. and Zadernowski, B. 1988. Phenolic compounds of rapeseed as factors limiting the utilization of protein in nutrition. Presented at the 3rd Chemical Congress of North America, Toronto, Canada, June 5 –10.

Kozlowska, H., Naczk, M., Shahidi, F., and Zadernowski, R. 1990. Phenolic acids and tannins of rapeseed and canola, in *Rapeseed and Canola: Production, Chemistry, Nutrition and Processing Technology,* Shahidi, F., Ed., Van Nostrand Reinhold, New York, 193–210.

Krohn, R.L., Ye, X., Liu, W., Joshi, S.S., Bagchi, M., Preuss, H.G., Stohs, S.J., and Bagchi, D. 1999. Differential effect of a novel grape seed extract on cultured human normal and malignant cells, in *Natural Antioxidants and Anticarcinogens in Nutrition, Health, and Disease*, Kumpulainen, J.T. and Salonen, J.T., Eds., Royal Society of Chemistry, Cambridge, U.K., 443–450.

Kubo, I., Xiao, P., Nichei, K.-I., Fujita, K.-I., Yamagiwa, Y., and Kamikawa, T. 2002. Molecular design of antifungal agents. *J. Agric. Food Chem.*, 50:3992–3998.

Kubo, I., Muroi, H., and Kubo, A. 1995. Structural functions of antimicrobial long chain alcohols and phenolics. *Bioorg. Med. Chem.*, 3:873–880.

Kubo, I., Muroi, H., and Kubo, A. 1994a. Naturally occurring antiacne agents. *J. Nat. Prod.*, 57:59–62.

Kubo, I., Kinst-Hori, I., and Yokokawa, Y. 1994b. Tyrosinase inhibitors from *Anarcadium occidentale* fruits. *J. Nat. Prod.*, 57:545–551.

Kubo, I., Ochi, M., Vieira, P.C., and Komatsu, S. 1993. Antitumor agents from the cashew (*Anarcardium occidentale)* apple juice. *J. Agric. Food Chem.*, 41:1012–1015.

Kubo, J., Lee, J.R., and Kubo, I. 1999. Antihelicobacter pyroli agents from the cashew apple. *J. Agric. Food Chem.*, 47:533–537.

Kubo, S., Kim, M., Naya, K., Komatsu, S., Yamagiwa, Y., Ohashi, K., Sakamoto, Y., Harakawa, S., and Kamikawa, T. 1987. Prostoglandin synthetase inhibitors from the African medicinal plant *Ozoroa mucronata. Chem. Lett.*, 1101–1104.

Kühnau, J. 1976. The flavonoids: a class of semiessential food components; their role in human nutrition. *Wld. Rev. Nutr. Diet*, 24:117–191.

Kuhnle, G., Spencer, J.P.E., Schroeter, H., Shenoy, B., Debnam, E.S., Kaila, S., Srai, S., Rice-Evans, C., and Hahn, U. 2000a. *Biochem. Biophys. Res. Commun.*, 277:507–512.

Kuhnle, G., Spencer, J.P.E., Chowrimootoo, G., Schroeter, H., Debnam, E.S., Kaila, S., Rice-Evans, C., and Hahn, U. 2000b. Resveratrol is absorbed in the small intestine as resveratrol glucuronide. *Biochem. Biophys. Res. Commun.*, 272:212–217.

Kulkarni, A.P., Mitra, A., Chaudhuri, J., Byczkowski, J., and Richards, I. 1990. Hydrogen peroxide: a potent activator of dioxygenase activity of soybean lipoxygenase. *Biochem. Biophys. Res. Commun.*, 166:417–423.

Kulling, S.E., Honig, D.M., and Metzler, M. 2001. Oxidative metabolism of the soy isoflavones daidzein and genestein in humans *in vitro* and *in vivo. J. Agric. Food Chem.*, 49:3024–3033.

Kumar, R. and Singh, M. 1984. Tannins: their adverse role in ruminant nutrition. Review. *J. Agric. Food Chem.*, 32:447–453.

Kuntz, S., Wenzel, U., and Daniel, H. 1999. Comparative analysis of the effects of flavonoids on profileration, cytotoxicity, and apoptosis in human colon cancer cell lines. *Eur. J. Nutr.*, 38:133–142.

Kuo, S.M. 1997. Dietary flavonoid and cancer prevention: evidence and potential mechanism. *Crit. Rev. Oncog.*, 8:47–69.

Kurzer, M.S. and Xu, X. 1997. Dietary phytoestrogens. *Annu. Rev. Nutr.*, 17:353–381.

Lake, B.G. 1999. Coumarin metabolism. Toxicity carcinogenicity relevance for human risk assessment. *Food Chem. Toxicol.*, 37:423–453.

Lampe, J.W., Karr, S.C., Hutchins, A.M., and Slavin, J.L. 1998. Urinary equol excretion with a soy challenge: influence of habitual diet. *Proc. Soc. Exp. Biol. Med.*, 217:335–339.

Langseth, L. 1995. *Oxidants, Antioxidants, and Disease Prevention,* ILSI Press, Brussels, Belgium.

Lapidot, T., Harel, S., Granit, R., and Kanner, J. 1998. Bioavailability of red wine anthoyanins as detected in human urine. *J. Agric. Food Chem.*, 46:4297–4302.

Laurena, A.C., Garcia, V.V., and Mendoza, E.T. 1986. Effect of soaking in aqueous acidic and alkali solutions on removal of polyphenols and *in vitro* digestibility of lowpea. *Qual. Plant. Plant Foods Hum. Nutri.*, 36:107–118.

Leake, D. 1997. The possible role of antioxidants in fruit and vegetables in protecting against coronary heart disease, in *Phytochemistry of Fruit and Vegetables, Proceedings of the Phytochemical Society of Europe 41*, Tomás-Barberán, F.A. and Robins, R.J., Eds., Clarendon Press, Oxford, U.K., 287–312.

Lee, J. and Watson, R.R. 1998. Pomegranate: role in health promotion and AIDS? in *Nutrition Food and AIDS,* Watson, R.R., Ed., CRC Press, Boca Raton, FL, 179–192.

Li, C., Lee, M.-J., Sheng, S., Meng, X., Prabhu, S., Winnik, B., Huang, B., Chung, J.Y., Yan, S., Ho, C.-T., and Yang, C.S. 2000. Structural identification of two metabolites of catechins and their kinetics in human urine and blood after tea ingestion. *Chem. Res. Toxicol.*, 13:177–184.

Lin, J.H., Chiba, M., and Baillie, T.A. 1999. Is the role of the small intestine in the first-pass metabolism over-emphasized? *Pharmacol. Rev.*, 51:135–157.

Lin, J.K., Huang, T.S., Shih, C.A., and Liu, J.Y. 1994. Molecular mechanism of action of curcumin. Inhibition of 12-*O*-tetradecanoyl-13-phorbol acetate-induced responses associated with tumor promotion, in *Food Phytochemicals for Cancer Prevention II. Teas, Spices, and Herbs,* Ho, C.-T., Osawa, T., Huang, M.-T., and Rosen, R.T., Eds., ACS Symposium Series 547, American Chemical Society, Washington, D.C., 196–203.

Lingwood, C.A., Huesca, M., and Kuksis, A. 1992. The glycerolipid receptor for *Helicobacter pylori* (and exoenzyme S) is phosphatidylethanolamine. *Infect. Immun.*, 60:2470–2474.

Linseisen, J., Radtke, J., and Wolfram, G. 1997. Flavonoid intake of adults in a Bavarian subgroup of the national food consumption survey. *Z. Ernährungswiss*, 36:403–412.

Loomis, W.D. and Battaile, J. 1966. Plant phenolic compounds and isolation of plant enzymes. *Phytochemistry*, 5:423–438.

Loomis, W.D. 1974. Overcoming problems of phenolics and quinones in the isolation of plant enzymes and organelles. *Methods Enzymol.*, 31:528–544.

Lostao, M.P., Hirayama, B.A., Loo, D.D., and Wright, E.M. 1994. Phenylglucosides and the Na^+/glucose cotransporter (SLGT1): analysis of interactions. *J. Membr. Biol.*, 142:161–170.

Lu, Y. and Bennick, A. 1998. Interaction of tannins with salivary proline-rich proteins. *Arch. Oral. Biol.*, 43:717–728.

Luck, G., Liao, H., Murray, N.J., Grimmer, H.R., Warminski, E.E., Williamson, M.P., Lilley, T.E., and Haslam, E. 1994. Polyphenols, astrigency and proline-rich proteins. *Phytochemistry*, 37:357–371.

Lundh, T. 1995. Metabolism of estrogenic isoflavones in domestic animals. *Proc. Soc. Exp. Biol. Med.*, 208:33–39.

Mabry, T.J. and Ulubelen, A. 1980. Chemistry and utilization of phenylpropanoids including flavonoids, coumarins and lignans. *J. Agric. Food Chem.*, 28:188–196.

Maeda-Yamamoto, M., Kawahara, H., Tahara, N., Tsuji, K., Hara, Y., and Isemura, M. 1999. Effects of tea polyphenols on the invasion and matrix metalloproteinases activities of human fibrosarcoma HT1080 cells. *J. Agric. Food Chem.*, 47:2350–2354.

Makkar, H.P.S. and Becker, K. 1996. Effect of pH, temperature, and time on inactivation of tannins and possible implications in detannification studies. *J. Agric. Food Chem.*, 44:1291–1295.

Malathi, P. and Crane, R.K. 1969. Phloridzin hydrolase: a beta-glucosidase of hamster intestinal brush border membrane. *Biochim. Biophys. Acta*, 173:245–256.

Mallozzi, M.A.B., Correa, B., Haraguchi, M., and Neto, F.B. 1996. Effect of flavonoids on *Aspergillus flavus* growth and aflatoxin production. *Rev. Microbiol.*, 27:161–165.

Manna, C., Galletti, P., Maisto, G., Ciciolla, V., D'Angelo, S., and Zappia, V. 2000. Transport mechanism and metabolism of olive oil hydroxytyrosol in Caco-2 cells. *FEBS Lett.*, 470:341–344.

Manthey, J.A., Grohmann, K., Monatanari, A., Ash, K., and Manthey, C.L. 1999. Polymethoxylated flavones derived from citrus suppress tumor necrosis factor-α expression by human monocytes. *J. Nat. Prod.*, 62:441–444.

Mao, T.K., Powell, J.J., van de Water, J., Keen, C.L., Schmitz, H.H., and Gershwin, M.E. 1999. The influence of cocoa procyanidins on transcription of interleukin-2 in peripheral blood molecular cells. *Int. J. Immunother.*, 15:23–29.

Marquardt, R.R. and Ward, A.T. 1979. Chick performance as affected by autoclave treatment of tannin-containing and tannin-free cultivars of faba beans. *Can. J. Animal Sci.*, 59:781–788.

Marquardt, R.R., Ward, T., Campbell, L.D., and Cansfield, P.E. 1977. Purification, identification and characterization of growth inhibitors in faba beans (*Vicia faba* L. var. *Minor*). *J. Nutr.*, 107:1313–1324.

Martin, M.M. and Martin, J.S. 1984. Surfactants: their role in preventing the precipitation of proteins by tannins in insect guts. *Oecologia*, 61:342–345.

Martin, M.M., Rockholm, D.C., and Martin, J.C. 1985. Effects of surfactants, pH, and certain cations on precipitation of proteins by tannins. *J. Chem. Ecol.*, 11:485–493.

Martin, M.C., La Casa, C., Alarcon de la Lastra, C., Cabeza, J., Villegas, I., and Motilva, V. 1998. Antioxidant mechanisms involved in gastroprotective effects of quercetin. *Z. Naturforsch.* (C), 53:82–88.

Martin-Tanguy, J., Guillaume, J., and Kossa, A. 1977. Condensed tannins in horse bean seeds: chemical structure and apparent effects on poultry. *J. Sci. Food Agric.*, 28:757–765.

Marvan, A.G. and Nagel, C.W. 1986. Microbial inhibitors of cranberries. *J. Food Sci.*, 51:1009–1013.

Matsudaira, F., Kitamura, T., Yamada, H., Fujimoto, I., Arai, M., Karube, H., and Yanagida, A. 1998. Inhibitory effect of polyphenol extracted from immature apples on dental plague formation. *J. Dent. Health*, 48:230–235 (in Japanese), cited in Yanagida, A., Kanda, T., Tanabe, M., Matsudaira, F., and Cordeiro, J.G.O. 2000. Inhibitory effects of apple polyphenols and related compounds on cariogenic factors of mutants *Streptococci. J. Agric. Food Chem.*, 48:5666–5671.

Matsumoto, H., Inaba, H., Kishi, M., Tominaga, S., Hirayama, M., and Tsuda, T. 2001. Orally administered delphinidin 3-rutinoside and cyanidin 3-rutinoside are directly absorbed in rats and humans and appear in the blood as the intact forms. *J. Agric. Food Chem.*, 49:1546–1551.

Matsumoto, M., Minami, T., Sasaki, H., Sobue, S., Hamada, S., and Ooshima, T. 1999. Inhibitory effects of olong tea extract on caries inducing properties of mutants streptococci. *Caries Res.*, 33:441–445.

Matsumoto, T., Moriguchi, R., and Yamada, H. 1993. Role of polymorphonuclear leucocytes and oxygen-derived free radicals in the formation of gastric lessions induced by HCl/ethanol, and possible mechanism of protection by antiulcer polysaccharides. *J. Pharm. Pharmacol.*, 45:535–539.

Matuschek, E., Towo, E., and Svanberg, U. 2001. Oxidation of polyphenols in phytate-reduced high-tannin cereals: effect on different phenolic groups and on *in vitro* accessible iron. *J. Agric. Food Chem.*, 49:5630–5638.

Mazumder, A., Neamati, N., Sunder, S., Schultz, J., Pertz, H., Eich, E., and Pommier, Y. 1997. Curcumin analogs with altered potencies against HIV-1 integrase as probes for biochemical mechanisms of drug action. *J. Med. Chem.*, 40:3057–3063.

Mazumder, A., Raghavan, K., Weinstein, J.N., Kohn, K.W., and Pommier, Y. 1995. Inhibition of human immunodeficiency virus type 1 integrase by curcumin. *Biochem. Pharmacol.*, 49:1165–1170.

Mbithi-Mwikya, S., van Camp, J., Yiru, Y., and Huyghebaert, A. 2000. Nutrient and anti-nutrient changes in finger millet (*Elusine coracana)* during sprouting. *Lebensm.-Wiss.u.-Technol.*, 33:9–14.

McDonald, M., Mila, I., and Scalbert, A. 1996. Precipitation of metal ions by plant polyphenols: optimal conditions and origin of precipitation. *J. Agric. Food Chem.*, 44:599–606.

McManus, J.P., Davis, K.G., Lilley, T.H., and Haslam, E. 1981. The association of proteins with polyphenols. *J. Chem. Soc. Chem. Comm.*, 309–311.

McManus, J.P., Davis, K.G., Beart, J.E., Gaffney, S.H., Lilley, T.H., and Haslam, E. 1985. Polyphenol interactions. Part 1. Introduction: some observations on the reversible complexation of polyphenols with proteins and polysaccharides. *J. Chem. Soc. Perkin Trans.*, 2:1429–1438.

Mehansho, H., Clements, S., Sheares, B.T., Smith, S., and Carlson, D.M. 1985. Induction of proline-rich glycoprotein synthesis in mouse salivary glands by isoproterenol and by tannins. *J. Biol. Chem.*, 260:4418–4423.

Mehansho, H., Butler, L.G., and Carlson, D.M. 1987. Dietary tannins and salivary proline-rich proteins: interactions, induction, and defense mechanisms, *Ann. Rev. Nutr.*, 7:423–440.

Messina, M.J., Parsky, V., Setchell, K.D.R., and Barnes, S. 1994. Soy intake and cancer risk: a review of *in vitro* and *in vivo* data. *Nutr. Cancer*, 21:333–340.

Meyers, D.G., Maloley, P.A., and Weeks, D. 1996. Safety of antioxidant vitamins. *Arch. Intern. Med.*, 156:925–935.

Middleton, E.J. and Kandaswami, C., 1993. Plant flavonoid modulation of immune and inflammatory cell functions. Nutrition and immunology, in *Human Nutrition. A Comprehensive Treatise,* Klurfeld, D.M., Alfin-Slater, R.B., and Kritchevsky, D., Eds., Plenum Press, New York, 239–266.

Middleton, E.J. 1986. Some effects of flavonoids on mammalian cell systems, in *Flavonoids and Bioflavonoids,* Farkas, L., Gabor, M., and Kallay, F., Eds., Elsevier, Amsterdam, 381–388.

Mila, I. and Scalbert, A. 1994. Tannin antimicrobial properties through iron deprivation: a new hypothesis. *Acta Hortic.*, 381:749–755.

Milbury, P.E., Cao, G., Prior, R.L., and Blumberg, J. 2002. Bioavailability of elderberry anthocyanins. *Mechan. Ageing Develop.*, 123:997–1006.

Milic, B., Stojanovic, S., Vucurevic, N., and Turcic, M. 1968. Chlorogenic and quinic acids in sunflower meal. *J. Sci. Food Agric.*, 19:108–113.

Mitaru, B.N., Reichert, R.D., and Blair, B. 1984a. The binding of dietary protein by sorghum tannins in digestive tract of pigs. *J. Nutr.*, 114:1787–1796.

Mitaru, B.N., Reichert, R.D., and Blair, B. 1984b. Kinetics of tannin deactivation during anaerobic storage and boiling treatments of high tannin sorghums. *J. Food Sci.*, 49:1566–8, 1583.

Mitaru, B.N., Blair, R., Bell, J.M., and Reichert, R.D. 1982. Tannin and fiber contents of rapeseed and canola hulls. Notes. *Can. J. Animal Sci.*, 62:661–663.

Mitjavila, S., Lacombe, C., Carrera, G., and Dererache, R. 1977. Tannic acid and oxidized tannic acid on the functional state of rat intestinal epithelium. *J. Nutr.*, 107:2113–2121.

Miyake, Y., Yamamoto, K., and Osawa, T. 1997. Metabolism of antioxidant in lemon fruit (*Citrus limon* Burm.f.) by human intestinal bacteria. *J. Agric. Food Chem.*, 45:3738–3742.

Miyake, Y., Murakami, A., Sugiyama, Y., Isobe, M., Koshimizu, K., and Ohigashi, H. 1999. Identification of coumarins from lemon fruit (*Citrus limon*) as inhibitors of *in vitro* tumor promotion and superoxide and nitric oxide generation. *J. Agric. Food Chem.*, 47:3151–3157.

Miyazawa, T., Nakagawa, K., Kudo, M., Muraishi, K., and Someya, K. 1999. Direct intestinal absorption of red fruit anthocyanins, cyanidin-3-glucoside and cyanidin-3,5-diglucoside, into rats and humans. *J. Agric. Food Chem.*, 47:1083–1091.

Miyazawa, M., Sakano, K., Nakamura, S.-I., and Kosaka, H. 2001. Antimutagenic activity of isoflavone from *Pueraria lobata. J. Agric. Food Chem.*, 49:336–341.

Mole, S. and Waterman, P.G. 1985. Stimulary effects of tannic and cholic acid on tryptic hydrolysis of proteins: ecological implications. *J. Chem. Ecol.*, 11:1323–1332.

Monge, P., Solheim, E., and Scheline, R.R. 1984. Dihydrochalcone metabolism in the rat: phloretin. *Xenobiotica*, 14:917–924.

Morazzoni, P., Livio, S., Scillingo, A., and Maladrino, S. 1991. *Vaccinum myrtillus* anthocyanosides pharmokinetics in rats. *Arzneim. Forsch.*, 41:128–131.

Morck, T.A., Lynch, S.R., and Cook, J.D. 1983. Inhibition of food iron absorption by coffee. *Am. J. Clin. Nutr.*, 37:416–420.

Moridani, M.Y., Scobie, H., Salehi, P., and O'Brien, P.J. 2001. Catechin metabolism: glutathione conjugate formation catalyzed by tyrosinase, peroxidase, and cytochrome P450. *Chem. Res. Toxicol.*, 14:841–848.

Morton, J.E. 1978. Economic botany in epidemiology. *Econ. Bot.*, 32:111–116.

Morton, J.F. 1989. Tannin as carcinogen in bush-tea: tea, mate, and khat, in *Chemistry and Significance of Condensed Tannins,* Hemingway, R.W. and Karchesy, J.J., Eds., Phenum Press, New York, 403–416.

Morton, L.W., Abu-Amsha Cacetta, R., Puddey, I.B., and Croft, K.D. 2000. Chemistry and biological effects of dietary phenolic compounds: relevance to cardiovascular disease. *Clin. Exp. Pharmacol. Physiol.*, 27:152–159.

Moyad, M.A. 1999. Soy, disease prevention and prostate cancer. *Semin. Urol. Oncol.*, 17:97–102.

Mukuru, S.Z., Butler, L.G., Rogler, J.C., Kirleis, A.W., Ejeta, G., Axtell, J.D., and Mertz, E.T. 1992. Traditional processing of high-tannin sorghum grain in Uganda and its effect on tannin, protein digestibility, and rat growth. *J. Agric. Food Chem.*, 40:1172–1175.

Murakami, A., Kuki, W., Takahashi, Y., Yonei, H., Nakamura, Y., Ohto, Y., Ohigashi, H., and Koshimizu, K. 1997. Auraptene, a citrus coumarin, inhibits 12-*O*-tetradecanoylphorbol-13-acetate-induced tumor promotion in ICR mouse skin, possibly through suppression of superoxide generation in leukocytes. *Jpn. J. Cancer Res.*, 88:443–452.

Murakami, A., Ohigashi, H., and Koshimizu, K. 1996. Antitumor promotion with food phytochemicals: a strategy for cancer chemoprevention. *Biosci. Biotech. Biochem.*, 60:1–8.

Muroi, H. and Kubo, I. 1993. Bactericidal activity of anarcadic acid against *Streptococcus mutans* and their potentiation. *J. Agric. Food Chem.*, 41:1780–1783.

Musci, I. 1986. Combined antiviral effect of quercetin and interferon on the multiplication of herpes simplex virus in cell cultures, in *Flavonoids and Bioflavonoids,* Farkas, L., Gabor, M., and Kallay, F., Eds., Elsevier, Amsterdam, 333–338.

Myer, R.O. and Gorbet, D.W. 1985. Waxy and normal grain sorghums with varying tannin contents in diets for young pigs. *Animal Feed Sci. Technol.*, 12:179–186.

Naczk, M., Oickle, D., Pink, D., and Shahidi, F. 1996. Protein precipitating capacity of crude canola tannins: effect of pH, tannin and protein concentration. *J. Agric. Food Chem.*, 44:2144–2148.

Naczk, M., Amarowicz, R., Zadernowski, R., and Shahidi, F. 2001. Protein-precipitating capacity of crude condensed tannins of canola and rapeseed hulls. *J. Am. Oil Chem. Soc.*, 78:1173–1178.

Nagabhushan, M. 1990. Catechin as antimutagen and anticarcinogen. *Toxicologist*, 10:166–172.

Nakachi, K., Suemasu, K., Suga, K., Takeo, T., Imai, K., and Higashi, Y. 1998. Influence of drinking green tea on breast cancer malignancy among Japanese patients. *Jpn. J. Cancer Res.*, 89:254–261.

Nakamura, Y., Koji, T., and Ohigashi, H. 2001. Toxic dose of a simple phenolic antioxidant, protocatechuic acid, attenuates the glutathione level in ICR mouse liver and kidney. *J. Agric. Food Chem.*, 49:5674–5678.

Nakamura, Y., Torikai, K., Ohto, Y., Murakami, A., Tanaka, T., and Ohigashi, H. 2000. A simple phenolic acid antioxidant protocatechuic acid enhances tumor promotion and oxidative stress in female ICR mouse skin: dose- and timing-dependent enhancement and involment of bioactivation by tyrosinase. *Carcinogenesis*, 21:1899–1907.

Nakayama, M., Suzuki, K., Toda, M., Okubo, S., Hara, Y., and Shimamura, T. 1993. Inhibition of the infectivity of influenza virus by tea polyphenols. *Antiviral Res.*, 21:289–299.

Nakazawa, T. and Ohsawa, K. 1998. Metabolism of rosmarinic acid in rats. *J. Nat. Prod.*, 61:993–996.

Nardini, M., Cirillo, E., Natella, F., and Scaccini, C. 2002. Absorption of phenolic acids in humans after coffee consumption. *J. Agric. Food Chem.*, 50:5735–5741.

Natella, F., Belelli, F., Gentili, V., Ursini, F., and Scaccini, C. 2002. Grape seed proanthocyanidins prevent plasma postprandial oxidative stress in humans. *J. Agric. Food Chem.*, 50:7720–7725.

Naurato, N., Wong, P., Lu, Y., Wroblewski, K., and Bennick, A. 1999. Interaction of tannin with human salivary histatins. *J. Agric. Food Chem.*, 47:2229–2234.

Nehlig, A. and Debry, G. 1994. Potential genotoxic, mutagenic and antimutagenic effects of coffee: a review. *Mut. Res./Rev. Genet. Toxicol.*, 317:145–162.

Nguyen, V.C., Lako, J.V., Oizumi, A., Ariga, T., and Kataoka, S. 1999. Anticataract activity of proanthocyanidin-rich grape seed extract in streptozotosin-induced diabetic rats. *Proc. Jpn. Soc. Biotechnol. Agrochem. 1999 Annu. Meeting*, 73:133.

Noguchi, Y., Fukuda, K., Matsushima, A., Haishi, D., Hiroto, M., Kodera, Y., Nishimura, H., and Inada, Y. 1999. Inhibition of Df-protease associated with allergic diseases by polyphenol. *J. Agric. Food Chem.*, 47:2969–2972.

Noguchi, N. 2002. Novel insights into the molecular mechanisms of the antiatherosclerotic properties of antioxidants: the alternatives to radical scavenging. *Free Radical Biol. Med.*, 33:1480–1489.

Norton, R.A. 1999. Inhibition of aflatoxin B_1 biosynthesis in *Aspergillus flavus* by anthocyanidins and related flavonoids. *J. Agric. Food Chem.*, 47:1230–1235.

Norton, R.A. and Dowd, P.F. 1996. Effect of steryl cinnmaic acid derivatives from corn bran on *Aspergillus flavus,* corn earworm larvae, and dried fruit beetle larvae and adults. *J. Agric. Food Chem.*, 44:2412–2416.

Nychas, G.J.E. 1995. Natural antimicrobials from plants, in *New Methods of Food Preservation,* Gould, G.W., Ed., Blackie Academic and Professional, London, 58–89.

Nychas, G.J.E., Tassou, S.C., and Board, R.G. 1990. Phenolic extract from olives: inhibition of *Staphylococcus aureus. Lett. Appl. Microbiol.*, 10:27–220.

Obizoba, I.C. and Atii, J.V. 1991. Effect of soaking, sprouting, fermentation and cooking on nutrient composition and some antinutritional factors of sorghum (Guinesia) seeds. *Plant Foods Hum. Nutr.*, 41:203–212.

Ofek, I., Goldhar, J., Zafriri, D., Lis, H., Adar, R., and Sharon, N. 1991. Anti-*Escherichia* adhesion activity of cranberry and blueberry juices. *N. Engl. J. Med.*, 324:1599.

Oh, H.I., Hoff, J.E., Armstrong, G.S., and Haff, L.A. 1980. Hydrophobic interaction in tannin-protein complexes. *J. Agric. Food Chem.*, 28:394–398.

Oh, H.I. and Hoff, J.E. 1986. Effect of condensed grape tannins on the *in vitro* activity of digestive proteases and activation of their zymogens. *J. Food Sci.*, 51:577–580.

Ohgaki, H., Takayama, S., and Sugimura, T. 1991. Carcinogenicities of heterocyclic amines in cooked food. *Mutat. Res.*, 259:399–410.

Olthof, M.R., Hollman, P.C.H., and Katan, M.B. 2001. Chlorogenic acid and caffeic acid are absorbed in humans. *J. Nutr.*, 131:66–71.

Olthof, M.R., Hollman, P.C.H., Vree, T.B., and Katan, M.B. 2000. Bioavailabilities of quercetin 3-glucoside and quercetin-4′-glucoside do not differ in humans. *J. Nutr.*, 130:1200–1203.

Ooshima, T., Minami, T., Matsumoto, M., Fujiwara, T., Sobue, S., and Hamada, S. 1998. Comparison of the cariostatic effects between regiments to administer oolong tea polyphenols in SPF rat. *Caries Res.*, 32:75–80.

Orsini, F., Pelizzoni, F., Verotta, L., and Aburjai, T. 1997. Isolation, synthesis, and antiplatelet aggregation activity of resveratrol 3-*O*-D-glucopyranoside and related compounds. *J. Nat. Prod.*, 60:1082–1087.

Osawa, T., Sugiyama, Y., Inayoshi, M., and Kawakischi, S. 1994. Chemistry and antioxidative mechanisms of β-diketones, in *Food Phytochemicals for Cancer Prevention II. Teas, Spices, and Herbs,* Ho, C.-T., Osawa, T., Huang, M.-T., and Rosen, R.T., Eds., ACS Symposium Series 547, American Chemical Society, Washington, D.C., 196–203.

Ozawa, T., Lilley, T.H., and Haslam, E. 1987. Polyphenol interactions: astringency and the loss of astringency in ripening fruit. *Phytochemistry*, 26:2937–2942.

Pan, M.-H., Liang, Y.-C., Lin-Shiau, S.-Y., Zhu, N.-Q., Ho, C.-T., and Lin, J.-K. 2000. Induction of apoptosis by the oolong tea polyphenol theasinen A through cytochrome C release and activation of caspase-9 and caspase-3 in human U937 cells. *J. Agric. Food Chem.*, 48:6337–6346.

Panayotova-Heiermann, M., Loo, D.D., Kong, C.T., Lever, J.E., and Wright, E.M. 1996. Sugar binding to Na-/glucose cotransporters is determined by carboxyl-terminal half of protein. *J. Biol. Chem.*, 271:1029–1034.

Papatheofanis, F.J. and Land, W.E.M. 1985. Lipoxygenase mechanism, in *Biochemistry of Arachidonic Acid Metabolism,* Lands, W.E.M., Ed., Martinus Nijhoff Publishing, Boston, 9–39.

Pathak, D., Pathak, K., and Singla, A.K. 1991. Flavonoids as medicinal agents-recent advances. *Fitoterapia*, 62:371–389.

Pearson, D.A., Frankel, E.N., Aeschbach, R., and Bruce, J.B. 1997. Inhibition of endothelial cell-mediated oxidation of low-density lipoprotein by rosemary and plant phenolics. *J. Agric. Food Chem.*, 45:578–582.

Pelkonen, O., Raunio, H., Rautio, A., Pasanen, M., and Lang, M.A. 1997. The metabolism of coumarin, in *Coumarins: Biology, Applications and Mode of Action*, O'Kennedy, R. and Thornes, R.D., Eds., John Wiley & Sons Ltd., New York, 67–92.

Perez-Madonaldo, R.A., Norton, B.W., and Kerven, G.L. 1995. Factors affecting *in vitro* formation of tannin–protein complexes. *J. Sci. Food Agric.*, 69:291–298.

Peterson, J. and Dwyer, J. 1998. Flavonoids: dietary occurrence and biochemical activity. *Nutr. Res.*, 18:1995–2018.

Peterson, G. and Dwyer, S. 1991. Genistein inhibition of the growth of human breast cancer cells: independence from estrogen receptors and multi-drug resistance gene. *Biochem. Biophys. Res. Commun.*, 179:661–667.

Pierpoint, W.S. 1969. *o*-Quinones formed in plant extracts. Their reactions with bovine serum albumin. *Biochem. J.*, 112:619–629.

Pierson, M.D. and Reddy, N.R. 1982. Inhibition of *Clostridium botulinum* by antioxidants and related phenolic compounds in comminuted pork. *J. Food Sci.*, 47:1926–1929; 1935.

Pihan, G., Regillo, C., and Szabo, S. 1987. Free radicals and lipid peroxidation in ethanol- and aspirin-induced gastric mucosal injury. *Dig. Dis. Sci.*, 32:1395–1401.

Pinto, M.C., Garcia-Barrado, J.A., and Macias, P. 1999. Resveratrol is a potent inhibitor of the dioxygenase activity of lipoxygenase. *J. Agric. Food Chem.*, 47:4842–4846.

Piskula, M.K. and Terao, J. 1998. Accumulation of (–)-epicatechin metabolites in rat plasma after oral administration and distribution of conjugation enzymes in rat tissues. *J. Nutr.*, 128:1172–1178.

Piskula, M.K., Yamakoshi, J., and Iwai, Y. 1999. Daidzein and genistein but not their glucosides are absorbed from the rat stomach. *FEBS Lett.*, 447:287–291.

Pool-Zobel, B.L., Bub, A., Schröder, N., and Rechkemmer, G. 1999. Anthocyanins are potent antioxidants in model systems but do not reduce endogenous oxidative DNA damage in human colon system. *Eur. J. Nutr.*, 38:227–234.

Porter L.J. 1992. Structure and chemical properties of the condensed tannins, in *Plant Polyphenols-Synthesis, Properties and Significance*, Hemingway, R.W. and Laks, P.E., Eds., Plenum Press, New York, 245–258.

Porter, L.J. and Woodruffe, J. 1984. Haemanalysis: the relative astringency of proanthocyanidin polymers, *Phytochemistry*, 23:1255–1256.

Powell, H.K.J. and Rate, A.W. 1987. Aluminium–tannin equilibria: a potentiometric study. *Aust. J. Chem.*, 40:2015–2022.

Pradhan, K.J., Variyar, P.S., and Bandekar, J.R. 1999. Antimicrobial activity of novel phenolic compounds from green pepper (*Piper nigrum* L.). *Lebensm.-Wiss.u.-Technol.*, 31:121–123.

Price, M.L., Butler, L.G., Rogler, J.C., and Featherston, W.R. 1978. Detoxification of high tannin sorghum grain. *Nutr. Rep. Int.*, 17:229–236.

Price, M.L., Butler, L.G., Rogler, J.C., and Featherston, W.R. 1979. Overcoming the nutritionally harmful effects of tannin in sorghum grain by treatment with inexpensive chemicals. *J. Agric. Food Chem.*, 27:441–445.

Price, M.L., Hagerman, A.E., and Butler, L.G. 1980. Tannin in sorghum grain: effect of cooking on antinutritional properties in rats. *Nutr. Rep. Int.*, 21:761–767.

Prigge, S.T., Boyington, J.C., Faig, M., Doctor, K.S., Gaffney, B., and Amzel, L.M. 1997. Structure and mechanism of lipoxygenases. *Biochimie*, 79:629–636.

Rackis, J.J., Honig, D.H., Sessa, D.J., and Steggerda, F.R. 1970. Flavor and flatulence factors in soybean protein products. *J. Agric. Food Chem.*, 18:977–982.

Randall, J.M., Bermann, R.L., Garrett, V., and Waiss, A.C.J. 1974. Use of bark to remove heavy metal ions from waste solutions. *For. Prod. J.*, 24:80–84.

Reddy, A.C.P. and Lokesh, B.R. 1992. Studies on spice principles as antioxidants in the inhibition of lipid oxidation of rat liver microsomes. *Mol. Cell. Biochem.*, 111:117–124.

Reddy, N.R., Pierson, M.D., Sathe, S.K., and Salunkhe, D.K. 1985. Dry bean tannins: a review of nutritional implications. *J. Am. Oil Chem. Soc.*, 62:541–549.

Reddy, M.B. and Cook, J.D. 1991. Assessment of dietary determinants of nonheme iron absorption in humans and rats. *Am. J. Clin. Nutr.*, 54:723–728.

Reed, J. 2002. Cranberry flavonoids, atherosclerosis and cardiovascular health. *CRC Crit. Rev. Food Sci. Nutr.*, 42 (Suppl.):301–316.

Reichert, R.D., Mwasaru, M.A., and Mukuru, S.Z. 1988. Characterization of colored-grain sorghum lines and identification of high-tannin lines with good dehulling characteristics. *Cereal Chem.*, 65:165–170.

Rein, D., Lotito, S., Holt, R.R., Keen, C.L., Schmitz, H.H., and Fraga, C.G. 2000a. Epicatechin in human plasma: *in vivo* determination and effect of chocolate consumption on plasma oxidation status. *J. Nutr.*, 130:2109S-2114S.

Rein, D., Paglieroni, T.G., Wun, T., Pearson, D.A., Achmitz, H.H., Gosselin, R., and Keen, C.L. 2000b. Cocoa inhibits platelet activation and function. *Am. J. Clin. Nutr.*, 72:30–35.

Remesy, C., Manach, C., Demigne, C., Texier, O., and Regerat, F. 1996. Interest of polyphenols in preventive nutrition, in *Polyphenols 96;* Vercauteren, J., Cheze, C., and Triaud, J., Eds., INRA Editions 87, INRA, Paris, 251–266.

Ren, M.Q., Kuhn, G., Wegner, J., and Chen J. 2001. Isoflavones, substances with multibiological and clinical properties. *Eur. J. Nutr.*, 40:135–146.

Renaud, S. and De Lorgeril, M. 1992. Wine, alcohol, platelets, and French paradox for coronary heart disease. *Lancet*, 339:1523.

Rice-Evans, C. and Packer, L., Eds. 1997. *Flavonoids in Health and Disease,* Marcel Dekker, New York.

Rimando, A.M., Cuendet, M., Desmarchelier, C., Mehta, R.G., Pezzuto, J.M., and Duke, S.O. 2002. Cancer chemopreventive and antioxidant activities of pterostilbene, a naturally occurring analogue of resveratrol. *J. Agric. Food Chem.*, 50:3453–3457.

Robbins, C.T., Mole, S., Hagerman, A.E., and Hanley, T.A. 1987. Role of tannins in defending plants against ruminants: reduction of dry matter digestibility. *Ecology*, 68:1606–1615.

Rogler, J.C., Ganduglia, H.R.R., and Elkin, R.G. 1985 Effects of nitrogen source and level on the performance of chicks and rats fed low and high tannin sorghum. *Nutr. Res.*, 5:1143–1152.

Rossander, L., Hallberg, L., and Bjorn-Rasmussen, E. 1979. Absorption of iron from breakfast meals. *Am. J. Clin. Nutr.*, 32:2484–2489.

Rostagno, H.S., Featherston, W.R., and Rogler, J.C. 1973. Nutritional value of sorghum grains with varying tannin contents for chicks. I. Growth studies. *Poultry Sci.*, 52:765–772.

Rouseff, R.L. and Nagy, S. 1994. Health and nutritional benefits of citrus fruit component. *Food Technol.*, (11):125–132.

Roux, D.G. 1972. Recent advances in chemistry and utilization of the natural condensed tannins. *Phytochemistry*, 11:1219–1230.

Rungruangsak, K., Tosukhowong, P., Panijpan, B., and Vimokesant, S.L. 1977. Chemical interactions between thamine and tannic acid. I. Kinetics, oxygen dependence and inhibition by ascorbic acid. *Am. J. Clin. Nutr.*, 30:1680–1685.

Rutkowski, A., Barylko-Pikielna, N., Kozlowska, H., Borowski, J., and Zawadzka, L. 1977. Evaluation of soybean protein isolates and concentrates as meat additives to provide a basis for increasing utilization of soybean. USDA-ARS Grant No. 13, Final Report, Olsztyn, Poland.

Sabir, M.A., Sosulski, F.W., and Finlayson, A.J. 1974. Chlorogenic acid-protein interactions in sunflower. *J. Agric. Food Chem.*, 22:575–578.

Sabir, M.A., Sosulski, F.W., and MacKenzie, S.L. 1973. Gel chromatography of sunflower proteins. *J. Agric. Food Chem.*, 21:988–993.

Saito, M., Hosoyama, H., Ariga, T., Kataoka, S., and Yamaji, N. 1998. Antiulcer activity of grape seed extract and procyanidins. *J. Agric. Food Chem.*, 46:1460–1464.

Sakanaka, S., Shimura, N., Aizawa, M., Kim, M., and Yamamoto, T. 1992. Preventive effect of green tea polyphenols against dental caries in conventional rats. *Biosci. Biotechnol. Biochem.*, 56:592–594.

Sakanaka, S. and Kim, M. 1998. Green tea polyphenols to prevent dental caries. *Polyphenols Actualities*, 18(2):18–22.

Sanbongi, C., Suzuki, N., and Sakane, T. 1997. Polyphenols in chocolate, which have antioxidant activity, modulate immune functions in humans *in vitro*. *Cell Immunol.*, 177:129–136.

Santos-Buelga, C. and Scalbert, A. 2000. Review. Proanthocyanidins and tannin-like compounds — nature, occurrence, dietary intake and effects on nutrition and health. *J. Sci. Food Agric.*, 80:1094–1117.

Salunkhe, D.K., Chavan, J.K., and Kadam, S.S. 1989. *Dietary Tannins: Consequences and Remedies,* CRC Press, Boca Raton, FL.

Sarni-Manchado, P., Cheynier, V., and Moutounet, M. 1999. Interactions of grape seed tannins with salivary proteins. *J. Agric. Food Chem.*, 47:42–47.

Satué-Gracia, M.T., Andrés-Lacueva, C., Lamuela-Raventós, R.M., and Frankel, E.N. 1999. Spanish sparkling wines (cavas) as inhibitors of *in vitro* human low-density lipoprotein oxidation. *J. Agric. Food Chem.*, 47:2198–2202.

Sawa, T., Nakao, M., Akaike, T., Ono, K., and Maeda, H. 1999. Alkylperoxyl radical scavenging activity of various flavonoids and other phenolic compounds: implications for the antitumor-promoter effect of vegetables. *J. Agric. Food Chem.*, 47:397–402.

Sawa, T., Akaike, T., Kida, K., Fukushima, Y., Takagi, K., and Maeda, H. 1998. Lipid peroxyl radicals from oxidized oils and heme iron: implication of fat diet in colon carcinogenesis. *Cancer Epidemiol. Biomarkers Prev.*, 7:1007–1012.

Scalbert, A. 1991. Antimicrobial properties of tannins. *Phytochemistry*, 30:3875–3883.

Scalbert, A. and Williamson, G. 2000. Dietary intake and bioavailability of polyphenols. *J. Nutr.*, 130:2073S-2085S.

Scalbert, A., Morand, C., Manach, C., and Rémésy, C. 2002. Absorption and metabolism of polyphenols in the gut and impact on health. *Biomed. Pharmacother.*, 56:276–282.

Schaffert, R.E., Oswalt, D.L., and Axtell, J.D. 1974. Effect of supplemental protein on nutritive value of high and low tannin sorghum bicolor (L.) Moench grain for the growing rat. *J. Animal Sci.*, 39:500–505.

Schaich, K.M., Fischer, C., and King, R. 1994. Formation of reactivity of free radicals in curcuminoids: an electron paramagnetic resonance study, in *Food Phytochemicals for Cancer Prevention II. Teas, Spices, and Herbs,* Ho, C.-T., Osawa, T., Huang, M.-T., and Rosen, R.T., Eds., ACS Symposium Series 547, American Chemical Society, Washington, D.C., 196–203.

Schmitz, S., Weidenboerner, M., and Kunz, B. 1993. Herbs and spices as selective inhibitors of mould growth. *Chem. Microbiol. Technol. Lebesm.*, 15(5/6):175–177.

Schneider, H., Schwiertz, A., Collins, M.D., and Blaut, M. 1999. Anaerobic transformation of quercetin-3-glucoside by bacteria from the human intestinal tract. *Arch. Microbiol.*, 171:81–91.

Seavell, A.J. 1978. Anticorrosive properties of mimosa (wattle) tannin. *J. Oil Col. Chem. Assoc.*, 61:439–462.

Seeram, N.P., Bourquin, L.D., and Nair, M.G. 2001. Degradation products of cyanidin glycosides from tart cherries and their bioactivities. *J. Agric. Food Chem.*, 49:4924–4929.

Sell, D.R., Rogler, J.C., and Featherston, W.R. 1983. The effects of sorghum tannin and protein level on the performance of laying hens maintained in two temperature environments. *Poultry Sci.*, 62:2420–2428.

Serafini, M., Maiani, G., and Ferro-Luzzi, A. 1997. Effect of ethanol on red wine tannin-protein (BSA) interactions. *J. Agric. Food Chem.*, 45:3148–3151.

Setchell, K.D.R., Nechemias-Zimmer, L., Cai, J., and Heubi, J.E. 1997. Exposure of infants to phytoestrogens from soy infant formulas. *Lancet*, 350:23–27.

Setchell, K.D.R., Nechemias-Zimmer, L., Cai, J., and Heubi, J.E. 1998. The isoflavone content of infant formulas and the metabolic fate of these phytoestrogens in early life. *Am. J. Clin. Nutr.*, 68:1453S-1461S.

Setchell.K.D.R. and Cassidy, A. 1999. Dietary isoflavones: biological effects and relevance to human health. *J. Nutr.*, 129:758S-767S.

Setchell, K.D.R. 2000. Absorption and metabolism of soy isoflavones — from food to dietary supplements and adults to infants. *J. Nutr.,* 130:654S-655S.

Setchell, K.D.R., Brown, N.M., Desai, P., Zimmer-Nechemias, L., Wolfe, B.E., Brashear, W.T., Kirschner, A.S., Cassidy, A., and Heubi, J.E. 2001. Bioavailability of pure isoflavones in healthy humans and analysis of commercial soy isoflavone supplements. *J. Nutr.*, 131:1362S-1375S.

Sfakianos, J., Coward, L., Kirk, M., and Barnes, S. 1997. Intestinal uptake and biliary excretion of the isoflavone genistein in rats. *J. Nutr.*, 128:1172–1178.

Shahrzad, S. and Bitsch, I. 1998. Determination of gallic acid and its metabolites in human plasma and urine by high-performance liquid chromatography. *J. Chromatogr. B.* 705:87–95.

Sharpe, P.C., McGrath, L.T., McClean, E., Young, I.S., and Archbold, G.P.R. 1995. Effect of red wine consumption on lipoprotein (a) and other risk factors for aetherosclerosis. *Q. J. Med.*, 88:101–108.

Shelef, L.A. and Liang, P. 1982. Antibacterial effects of butyated hydroxyanisole (BHA) against *Bacillus* species. *J. Food Sci.*, 47:796–799.

Sheu, F., Lai, H.-H., and Yen, G.-C. 2001. Suppression effect of soy isoflavones on nitric oxide production in RAW 264.7 macrophages. *J. Agric. Food Chem.*, 49:1767–1772.

Shobha, S.V., Ramadoss, C.S., and Ravindranath, B. 1994. Inhibition of soybean lipoxygenase-I by anarcadic acids, cardols, and cardanols. *J. Nat. Prod.*, 57:1755–1757.

Shutt, D.A. and Cox, R.I. 1972. Steroid and phytoestrogen binding to sheep uterine receptors *in vitro*. *J. Endocrinol.*, 52:299–310.

Siebert, K.J., Troukhanova, N.V., and Lynn, P.Y. 1996. Nature of polyphenol-protein interactions. *J. Agric. Food Chem.*, 44:80–85.

Siebert, K.J. 1999. Effects of protein–polyphenol interactions on beverage haze. Stabilization and analysis. *J. Agric. Food Chem.*, 47:353–362.

Siegenberg, D., Baynes, R.D., Bothwell, T.H., MacFarlane, B.J., Lamparelli, R.D., Car, N.G., MacPhail, P, Schmidt, U., Tal, A., and Mayet, F. 1991. Ascorbic acid prevents the dose-dependent inhibitory effects of polyphenols and phytates on non-heme absorption. *Am. J. Clin. Nutr.*, 53:537–541.

Silber, M.L., Davitt, B.B., Khairutdinov, R.F., and Hurst, J.K. 1998. A mathematical model describing tannin–protein association. *Anal. Biochem.*, 263:46–50.

Simon, A., Allais, D.P., Duroux, J.L., Basly, J.P., Durand-Fontanier, S., and Delage, C. 1998. Inhibitory effect of curcuminoids on MCF-7 cell proliferation and structure-activity. *Cancer Lett.*, 129:111–116.

Simonetti, P., Gardana, C., and Pietta, P. 2001. Plasma level of caffeic acid and antioxidant status after red wine intake. *J. Agric. Food Chem.*, 49:5964–5968.

Singleton, V.L. and Kratzer, F.H. 1969. Toxicity and related physiological activity of phenolic substances of plant origin. *J. Agric. Food Chem.*, 17:497–512.

Singleton, V.L. 1981. Naturally occurring food toxicants: phenolic substances of plant origin common in foods. *Adv. Food Res.*, 27:149–242.

Skibola, C.F. and Smith, M.T. 2000. Potential health impacts of excessive flavonoid intake. *Tree Rad. Biol. Med.*, 29:375–383.

Slabbert, N. 1992. Complexation of condensed tannins with metal ions, in *Plant Polyphenols, Synthesis, Properties, Significance,* Hemingway, R.W. and Laks, P.E., Eds., Plenum Press, New York, 421–436.

Smyk, B. and Drabent, R. 1989. Spectroscopic investigation of the equilibria of the ionic form of sinapic acid. *Analyst*, 114:723–726.

Somogyi, C. 1978. Natural toxic substances in food. *World Rev. Nutr. Diet.*, 29:42–49.

South, P.K., House, W.A., and Miller, D.D. 1997. Tea consumption does not affect iron absorption in rats unless tea and iron are consumed together. *Nutr. Res.*, 17:1303–1310.

Spector, A. 1995. Oxidative stress-induced cataract: mechanism of action. *FASEB J.*, 9:1172–1182.

Spencer, C.M., Cai, Y., Martin, R., Gaffney, S.H., Goulding, P.N., Magnoloto, D., Lilley, T.H., and Haslam, E. 1988. Polyphenol complexation. Some thoughts and observations. *Phytochemistry*, 27:2397–2409.

Spencer, J.P., Chowrimootoo, G., Choudhury, R., Debnam, E.S., Srai, S.K., and Rice-Evans, C. 1999. The small intestine can both absorb and glucuronidate luminal flavonoids. *FEBS Lett.*, 458:224–230.

Spencer, J.P.E., Chaudry, F., Pannala, A.S., Srai, S.K., Debnam, E., and Rice-Evans, C. 2000. Decomposition of cocoa procyandins in gastric milieu. *Biochem. Biophys. Res. Commun.*, 272:236–241.

Srimal, R.C. 1997. Turmeric: a brief review of medicinal properties. *Fitoterapia*, 68:483–493.

Sripriya, G., Anthony, U., and Chandra, T.S. 1997. Changes in carbohydrate, free amino acids, organic acids, phytate and HCl extractability of minerals during germination and fermentation of finger millet (*Eleusine coracana*). *Food Chem.*, 58:345–350.

Stavric, B. 1994. Quercetin in our diet: from poteny mutagen to probable anticarcinogen. *Clin. Biochem.*, 27:245–248.

Stavric, B. and Matula, T.I. 1992. Flavonoids in foods: their significance for nutrition and health, in *Lipid-Soluble Antioxidants: Biochemistry and Clinical Applications*, Ong, A.S.H. and Packer, L., Eds., Birkhäuser Verlag, Basel, Switzerland, 274–294.

Stavric, B. and Matula, T.I. 1988. Biological significance of flavonoids in foods: a review of some recent issues and problems. *Bull. Liaison Groupe Polyphenols*, 14:95–104.

Steinberg, F.M., Holt, R.R., Schmitz, H.H., and Keen, C.L. 2002. Cocoa procyanidin chain length does not determine ability to protect LDL from oxidation when monomer units are controlled. *J. Nutr. Biochem.*, 13:645–652.

Steinmetz, K.A. and Potter, J.D. 1996. Vegetable, fruit and cancer prevention. A review. *J. Am. Diet. Assoc.*, 96:1027–1039.

Stensvold, I., Tverdal, A., Solvoll, K., and Per Foss, O. 1992. Tea consumption. Relationship to cholesterol, blood pressure, and coronary and total mortality. *Prev. Med.*, 21:546–553.

Stivala, L.A., Savio, M., Fedarico, C., Perucca, P., Bianchi, L., Magas, G., Forti, L., Pagnoni, U.M., Albini, A., Prosperi, E., and Vannini, V. 2001. Specific structural determinants are responsible for the antioxidant activity and the cell cycle effects of resveratrol. *J. Biol. Chem.*, 276:22586–22594.

Stoner, G.D. and Mukhtar, H. 1995. Polyphenols as cancer chemopreventive agents. *J. Cell. Biochem.*, 22(Suppl.):168–180.

Su, S.J., Yeh, T.M., Lei, H.Y., and Chow, N.H. 2000. The potential of soybean foods as a chemoprevention approach for human urinary tract cancer. *Clin. Cancer Res.*, 6:230–236.

Subbaramaiah, K., Chung, W.J., Michaluart, P., Telang, N., Tanabe, T., Inoue, H., Jang, M., Pezzuto, J.M., and Dannenberg, A.J. 1998. Resveratrol inhibits cyclooxygenase-2 transcription and activity in phorbol ester-treated human mammary epithelial cells. *J. Biol.Chem.*, 273:21975–21882.

Sugimura, T. 1988. New environmental carcinogens in daily life. *Trends Pharmacol. Sci.*, 9:205–209.

Sugiyama, S., Umehara, K., Kuroyanagi, M., Ueno, A., and Taki, T. 1993. Studies on differentiation inducers of myeloid leukemic cells from *Citrus* species. *Chem. Pharm. Bull.*, 41:714–719.

Suschetet, M., Siess, M.-H., Le Bon, A.-M., and Canivenc-Lavier, M.-C. 1996. Anticarcinogenic properties of some flavonoids, in *Polyphenols 96,* Vercauteren, J., Cheze, C., and Triaud, J., Eds., INRA Editions 87, INRA, Paris, 165–204.

Takenaka, M., Watanabe, T., Sugahara, K., Harada, Y., Yoshida, S., and Sugawara, F. 1997. New antimicrobial substances against *Streptococcus scabies* from rosemary (*Rosmarinus officinalis* L.) *Biosci. Biotechnol. Biochem.*, 61:1440–1444.

Tanaka, T., Kawabata, K., Kakumoto, M., Hara, A., Murakami, A., Kuki, W., Takahashi, Y., Maeda, M., Ota, T., Odashima, S., Yamane, T., Koshimizu, K., and Ohigashi, H. 1998a. Citrus auraptene exerts dose-dependent chemopreventive activity in rat large bowel tumorigenesis: the inhibition correlates with suppression of cell profileration and lipid peroxidation and with induction of phase II drug metabolising enzymes. *Cancer Res.*, 58:2550–2556.

Tanaka, T., Kawabata, K., Kakumoto, M., Matsunaga, K., Mori, H., Murakami, A., Kuki, W., Takahashi, Y., Yonei, H., Satoh, K., Hara, A., Maeda, M., Ota, T., Odashima, S., Koshimizu, K., and Ohigashi, H. 1998b. Chemoprevention of 4-nitriquinoline 1-oxide-induced oral carcinogenesis by citrus auraptene in rats. *Carcinogenesis*, 19:425–431.

Tanaka, T., Kawabata, K., Kakumoto, M., Makita, H., Hara, A., Mori, H., Satoh, K., Murakami, A., Kuki, W., Takahashi, Y., Yonei, H., Koshimizu, K., and Ohigashi, H. 1997. Citrus auraptene inhibits chemically induced colonic aberrant crypt foci in male F344 rats. *Carcinogenesis*, 18:2155–2161.

Tanaka, T. Kojima, T., Suzui, M., and Mori, H. 1993. Chemoprevention of colon carcinogenesis by natural product of a simple phenolic compound protocatechuic acid: suppressing effects on tumor development and biomarkers expression of colon tumorigenesis. *Cancer Res.*, 53:3908–3913.

Tanaka, T., Yoshimi, N., Sugie, S., and Mori, H. 1992. Protective effects against liver, colon, and tongue carcinogenesis by plant polyphenols, in *Phenolic Compounds in Food and Their Effects on Health II. Antioxidant and Cancer Prevention,* Huang, M.-T., Ho, C.-T., and Lee, C.Y., Eds., ACS Symposium Series 507, American Chemical Society, Washington, D.C., 326–337.

Tatala, S., Svanberg, U., and Mduma, B. 1998. Low dietary iron availability is a major cause of anemia. A nutrition survey in the Lindi district of Tanzania. *Am. J. Clin. Nutr.*, 68:171–178.

Tapiero, H., Tew, K.D., Nguyen Ba, G., and Mathé, G. 2002. Polyphenols: do they play a role in the prevention of human pathologies? *Biomed. Pharmacother.*, 56:200–207.

Taylor, A., Jacques, P.F., and Epstein, E.M. 1995. Relationship among aging, antioxidant status and cataract. *Am. J. Clin. Nutr.*, 62:1439S-1447S.

Tebib, K., Bitri, L., Besanncon, P., and Rouanet, J.-M. 1994a. Polymeric grape seed tannins prevent plasma cholesterol changes in high-cholesterol-fed rats. *Food Chem.*, 49:403–406.

Tebib, K., Rouanet, J.M., and Besancon, P. 1994b. Effect of grape seed tannins on the activity of some rat intestinal enzymes activities. *Enzyme Protein*, 48:51–60.

Teissedre, P.-L. and Landrault, N. 2000. Wine phenolics: contribution to dietary intake and bioavalability. *Food Res. Intern.*, 33:461–467.

Temple, N.J. 2000. Antioxidants and disease: more questions than answers. *Nutr. Res.*, 20:449–459.

Tijburg, L.B.M., Mattern, T., Folts, J.D., Weisgerber, U.M., and Katan, M.B. 1997. Tea flavonoids and cardiovascular diseases: a review. *CRC Crit. Rev. Food Sci. Nutr.*, 37:771–785.

Toyomizu, M., Sugiyama, S., Jin, R.L., and Nakatsu, T. 1993. α-Glucosidase and aldose reductase inhibitors — constituents of cashew, *Anarcadium occidentale,* nut shell liquids. *Phytother. Res.*, 7:252–254.

Trevisanato, S.I. and Kim, Y.-I. 2000. Tea and health. *Nutr. Rev.,* 58:1–10.

Tsouderos, Y. 1991. Venous tone: are the phlebotonic properties of a therapeutic benefit? A comprehensive view of our experience with Daflon 500 mg. *Z. Kardiol.*, 80:S95-S101.

Tsuda, T., Horio, F., and Osawa, T. 1999. Absorption and metabolism of cyanidin 3-*O*-β-D-glucoside in rats. *FEBS Lett.*, 449, 179–182.

Tsuda, T., Ohshima, K., Kawakishi, S., and Osawa, T. 1996. Oxidation products of cyanidin 3-*O*-β-D-glucoside with a free radical initiator. *Lipids*, 31:1259–1263.

Tuck, K.L., Freeman, M.P., Hayball, P.J., Stretch, G.L., and Stupans, I. 2001. The *in vivo* fate of hydroxytyrosol and tyrosol, antioxidant phenolic constituents of olive oil, following intravenous and oral dosing of labeled compounds to rats. *J. Nutr.*, 131:1993–1996.

Tuck, K.L., Hayball, P.J., and Stupans, I. 2002. Structural characterization of the metabolites of hydroxytyrosol, the principal phenolic component in olive oil, in rats. *J. Agric. Food Chem.*, 50:2404–2409.

Turunen, S. and Kastari, T. 1979. Digestion and absorption of lecithin in larvae of cabbage butterfly, *Pierris brassicae*. *Comp. Biochem. Physiol.*, 62A:933–937.

Udayasekhara, R.P. and Deostahale, Y.G. 1988. *In vitro* availability of iron and zinc in white and coloured ragi (*Eleusine coracana*): role of tannin and phytate. *Plant Foods Hum. Nutr.*, 38:35–41.

Ueda, J., Saito, N., Shimazu, Y., and Ozawa, T. 1996. A comparison of scavenging abilities of antioxidants against hydroxyl radicals. *Arch. Biochem. Biophys.*, 333:377–384.

Ursini, F., Tubaro, F., Rong, J., and Sevanian, A. 1999. Optimization of nutrition: polyphenols and vascular protection. *Nutr. Rev.*, 57:241–249.

Van Buren, J.P. and Robinson, W.B. 1969. Formation of complexes between protein and tannic acid. *J. Agric. Food Chem.*, 17:772–777.

Variyar, P.S., Pendharkar, M.B., Banergee, A., and Bandopadhyay, C. 1988. Blackening of green pepper berries. *Phytochemistry*, 27:715–717.

Veckenstedt, A., Buttner, J., Pusztai, R., Heinecke, H., Hartl, A., and Beladi, I. 1986. Antiviral, immunologic and toxicologic studies with quercetin in mice, in *Flavonoids and Bioflavonoids,* Farkas, L., Bagor, M., and Kallay, F., Eds., Elsevier, Amsterdam, 339.

Verstraeten, S.V., Keen, C.L., Schmitz, H.H., Fraga, C.G., and Oteiza, P.I. 2003. Flavan-3-ols and procyanidins protect liposomes against lipid oxodation and disruption of the bilayer structure. *Free Radical Biol. Med.*, 34:84–92.

Visioli, F., Galli, C., Bornet, F., Mattei, A., Patelli, R., Galli, G., and Caruso, D. 2000. Olive oil phenolics are dose-dependently absorbed in humans. *FEBS Lett.*, 468:159–160.

von Weymarn, L.B. and Murphy, S.E. 2001. Coumarin metabolism by rat esophageal microsomes in cytochrome P450 2A3. *Chem. Res. Toxicol.*, 14:1386–1392.

Wade, S., Tomioka, S., and Moriguchi, I. 1969. Protein binding, 6. Binding phenols to bovine serum albumin. *Chem. Pharm. Bull.*, 17:320–323.

Waage, S.K., Hedin, P.A., and Grimley, E. 1984. A biologically active procyanidin from *Machaerium floribundum*. *Phytochemistry*, 23:2785–2787.

Wadsworth, T. and Koop, D.R. 1999. Effects of the wine polyphenolics quercetin and resveratrol on pro-inflammatory cytokine expression in RAW 264.7 macrophages. *Biochem. Pharmacol.*, 57:941–949.

Waffo-Teguo, P., Lee, D., Cuendet, M., Merillon, J.-M., Pezzuto, J.M., and Kinghorn, A.D. Two new stilbene dimer glucosides from grape (*Vitis vinifera*) cell cultures. *J. Nat. Prod.*, 64:136–138.

Wagner, H. 1985. New plant phenolics of pharmaceutical interest, in *Ann. Proc. Phytochem. Soc. Eur.*, vol. 25, Van Sumere, C.F. and Lea, P., Eds., Clarendon Press, Oxford, U.K., 409.

Wakabayashi, K., Ushiyama, H., Takahashi, M., Nukaya, H., Kim, S.-B., Hirose, M., Ochiai, M., Sugimura, T., and Nagao, M. 1993. Exposure of heterocyclic amines. *Environ. Health Perspect.*, 99:129–134.

Wakai, K., Egami, I., Kato, K., Kawamura, T., Tamakoshi, A., Lin, Y., Nakayama, T., Wada, M., and Ohno, Y. 1999. Dietary intake and sources of isoflavones among Japanese. *Nutr. Cancer*, 33:139–145.

Walgren, R.A., Kamaky, K.J., Lindenmayer, G.E., and Walle, T. 2000a. Efflux of dietary of dietary flavonoid quercetin 4′-β-glucoside across human intestinal Caco-2 cell monolayers by apical multidrug resistance-associated protein-2. *J. Pharmacol. Exp. Ther.*, 294:830–836.

Walgren, R.A., Lin, J.-T., Kinne, R.K.-H., and Walle, T. 2000b. Cellular uptake of dietary flavonoid quercetin 4′-β-glucosidase by sodium dependent glucose transporter SGLT1. *J. Pharmacol. Exp. Ther.*, 294:837–843.

Walgren, R.A., Walle, U.K., and Walle, T. 1998. Transport of quercetin and its glucosides across human intestinal epithelial Caco-2 cells. *Biochem. Pharmacol.*, 55:1721–1727.

Waladkhani, A.R. and Clemens, M.R. 1998. Effect of dietary phytochemicals on cancer development (review). *Int. J. Molec. Med.*, 1:747–753.

Walle, U.K., Galijatovic, A., and Walle, T. 1999. Transport of flavonoid chrysin and its conjugated metabolites by the human intestinal cell line Caco-2. *Biochem. Pharmacol.*, 58:431–438.

Walle, T., Otake, Y., Walle, K., and Wilson, F.A. 2000. Quercetin glucosides are completely hydrolysed in ileostomy patients before absorption. *J. Nutr.,* 130:2658–2661.

Wang, J. and Mazza, G. 2002a. Inhibitory effects of anthocyanins and other phenolic compounds on nitric oxide production in LPS/INF-γ-activated RAW 264.7 macrophages. *J. Agric. Food Chem.*, 50:850–857.

Wang, J. and Mazza, G. 2002b. Effects of anthocyanins and other phenolic compounds on the production of tumor necrosis factor α in LPS/INF-γ-activated RAW 264.7 macrophages. *J. Agric. Food Chem.*, 50:4183–4189.

Warden, B.A., Smith, L.S., Beecher, G.R., Balentine, D.A., and Clevidence, B.A. 2001. Catechins are bioavailable in men and women drinking black tea throughout the day. *J. Nutr.*, 131:1731–1737.

Watanabe, S., Yamaguchi, M., Soube, T., Takahashi, T., Miura, T., Arai, Y., Mazur, W., Wahala, K., and Adlercreutz, H. 1998. Pharmacokinetics of soybean isoflavones in plasma, urine and feces of men after ingestion of 60 g baked soybean powder (kinako). *J. Nutr.*, 128:1710–1715.

Waterhouse, A.L., German, J.B., Walzem, R.L., Hansen, R.J., and Kasin-Karakas, S.E. 1999. Is it time for a wine trial? *Am. J. Clin. Nutr.*, 68:220–221.

Waterhouse, A.L. and Walzem, R.L. 1998. Nutrition of grape phenolics, in *Flavonoids in Health and Disease,* Rice-Evans, C. and Parker, L., Eds., Marcel Dekker, New York, 349–387.

Waterhouse, A.L., Sirley, J.R., and Donovan, J.L. 1996. Antioxidants in chocolate. *Lancet*, 348:834.

Weisburger, J.H. 1999. Second International Scientific Symposium on Tea and Health: an introduction. *Proc. Soc. Exp. Biol. Med.*, 220:193–194.

Weiss, E., Lev-Dor, R., Kashamn, Y., Goldhar, J., Sharon, N., and Ofek, I. 1999. Inhibiting interspecies aggregation of plague bacteria with cranberry juice constituents. *J. Am. Dent. Assoc.*, 129:1719–1723.

Willet, W.C. 2002. Balancing life-style and genomics research for disease prevention. *Science*, 296:695–698.

Williams, G.M., Iatropoulos, M.J., and Whysner, J. 1999. Safety assessment of butylated hydroxyanisole and butylated hydroxytoluene as antioxidant food additives. *Food Chem. Toxicol.*, 37:1027–1038.

Winter, J., Lilian, H.M., Dowell, V.R. Jr., and Bokkenheuser, V.D. 1989. C-ring cleavage of flavonoids by human intestinal bacteria. *Appl. Environ. Microbiol.*, 55:1203–1208.

Wollgast, J. and Anklam, E. 2000. Polyphenols in chocolate: is there a contribution to human health? *Food Res. Intern.*, 33:449–459.

Wren, A.F., Cleary, M., Frantz, C., Melton, S., and Norris, L. 2002. 90-day oral toxicity study of grape seed extract (IH636) in rats. *J. Agric. Food Chem.*, 50:2180–2192.

Wright, J.S. 2002. Predicting the antioxidant activity of curcumin and curcuminoids. *J. Mol. Struct.*, (Theochem) 591:207–217.

Yamagishi, M., Natsume, M., Nagaki, A., Adachi, T., Osakabe, N., Takizawa, T., Kumon, H., and Osawa, T. 2000. Antimutagenic activity of cacao: inhibitory effect of cacao liquor polyphenols on the mutagenic action of heterocyclic amines. *J. Agric. Food Chem.*, 48:5074–5078.

Yamakoshi, J., Saito, M., Kataoka, S., and Tokutake, S. 2002. Procyanidin-rich extract from grape seeds prevents cataract formation in hereditary cataractous (ICR/f) rats. *J. Agric. Food Chem.*, 50:4983–4988.

Yamakoshi, J., Kataoka, S., Koga, T., and Ariga, T. 1999. Proantocyanidin-rich extract of grape seeds attenuates the development of aortic atherosclerosis in cholesterol-fed rabbits. *Atherosclerosis*, 142:139–149.

Yamasaki, H., Uefuji, H., and Sakihama, Y. 1996. Bleaching of red antocyanin induced by superoxide radical. *Arch. Biochem. Biophys.*, 332:183–186.

Yan, Q. and Bennick, A. 1995. Identification of histatins and tannin-binding protein in human saliva. *Biochem. J.*, 311:341–347.

Yanagida, A., Kanda, T., Tanabe, M., Matsudaira, F., and Cordeiro, J.G.O. 2000. Inhibitory effects of apple polyphenols and related compounds on cariogenic factors of mutants *Streptococci. J. Agric. Food Chem.*, 48:5666–5671.

Yang, S.Y. and Chung, J.E. 2000. Tea and tea polyphenols in cancer prevention. *J. Nutr.*, 130:472S-478S.

Yang, C.S., Laihshun, C., Lee, M.-J., and Landau, J.M. 1996. Effects of tea on carcinogenesis in animal models and humans, in *Dietary Phytochemicals in Cancer Prevention and Treatment,* Back, N., Kritchevsky, I.R., Kritchevsky, D., Lajtha, A., and Paoletti, R., Eds., Plenum Press, New York, 51–61.

Yang, C.S. and Wang, Z.-Y. 1993. Tea and cancer: a review. *J. Natl. Cancer Inst.*, 58:1038–1049.

Yapar, Z. and Clandinin, D.R. 1972. Effect of tannins in rapeseed meal on its nutritional value for chicks. *Poultry Sci.*, 51:222–228.

Ye, X., Krohn, R.L., Liu, W., Joshi, S.S., Kuszynski, C.A., McGinn, T.R., Bagchi, M., Preuss, H.G., Stohs, S.J., and Bagchi, D. 1999. The cytotoxic effects of novel IH636 grape seed proanthocyanidin extract on cultured human cancer cells. *Mol. Cell Biochem.*, 196:99–108.

Yen, G.-C. and Chen, H.-Y. 1994. Comparison of antimutagenic effect of various tea extracts (green, oolong, pouchong, and black tea). *J. Food Protect.*, 57:54–58.

Yochum, L., Kushi, L.H., Meyer, K., and Folsom, A.R. 1999. Dietary flavonoid intake and risk of cardiovascular disease in postmenopausal women. *Am. J. Epidemiol.*, 149:943–949.

Youdim, K.A., Martin, A., and Joseph, J.A. 2000. Incorporation of the elderberry anthocyanidins by endothelial cells increases protection against oxidative stress. *Free Radic. Biol. Med.*, 29:51–60.

Xu, X., Harris, K.S., Wang, H.-J., Murphy, P.A., and Hendrich, S. 1995. Bioavailability of soybean isoflavones depends upon gut microflora in women. *J. Nutr.*, 125; 2307–2315.

Zafriri, D., Ofek, I., Pocino, A.R., and Sharon, N. 1989. Inhibitory activity of cranberry juice on adherence of type 1 and type P fimbriated *Escherichia coli* to eucaryotic cells. *Antimicrob. Agents Chemother.*, 33:92–98.

Zhang, J., Zhan, B., Yao, X., and Song, J. 1995. Antiviral activity of tannin from the pericarp of *Punica granatum* L. against genital herpes virus *in vitro*. *Zhongguo Zhongyao Zazhi*, 20:556–558.

Zhang, Y., Song, T.T., Cunnick, J.E., Murphy, P.A., and Hendrich, S. 1999a. Daidzein and genistein glucuronides *in vitro* are weakly estrogenic and activate human natural killer cells at nutritionally relevant concentrations. *J. Nutr.*, 129:399–405.

Zhang, Y., Wang, G.-J., Song, T.T., Murphy, P.A., and Hendrich, S. 1999b. Urinary disposition of soybean isoflavones daidzein, genistein and glycitein differs among humans with moderate fecal isoflavones degradation activity. *J. Nutr.*, 129:957–962.

Zhao, J., Wang, J., Chen, Y., and Agarwal, R. 1999. Antitumor-promoting activity of polyphenolic fraction isolated from grape seeds in the mouse skin two-stage initiation-promotion protocol and identification of B5–3′-gallate as the most effective antioxidant constituent. *Carcinogenesis*, 20:1737–1745.

Zijp, I.M., Korver, O., and Tijburg, B.M. 2000. Effect of tea and other dietary factors on iron absorption. *CRC Crit. Rev. Food Sci. Nutr.*, 40:371–398.

8 Antioxidant Properties of Food Phenolics

INTRODUCTION

Antioxidants markedly delay or prevent oxidation of the substrate (Halliwell, 1999; Shahidi, 2000), when they are present in foods or in the body at low concentrations compared to that of an oxidizable substrate. Food manufacturers have used food-grade antioxidants, mainly of a phenolic nature, to prevent quality deterioration of products and to maintain their nutritional value. Antioxidants have also been of interest to health professionals because they help the body to protect itself against damage caused by reactive oxygen species (ROS) as well as reactive nitrogen species (RNS) and reactive chlorine species (RCS) associated with degenerative diseases (Figure 8.1).

Antioxidants act at different levels in the oxidative sequence involving lipid molecules. They may decrease oxygen concentration, intercept singlet oxygen, prevent first-chain initiation by scavenging initial radicals such as hydroxyl radicals, bind metal ion catalysts, decompose primary products of oxidation to nonradical species and break chains to prevent continued hydrogen abstraction from substrates (Shahidi, 2000, 2002).

Natural antioxidants from dietary sources include phenolic and polyphenolic compounds, among others. The mechanism by which these antioxidants exert their effects may vary depending on the compositional characteristics of the food, including its minor components. Furthermore, the beneficial health effects of consuming plant foods have been ascribed, in part, to the presence of phenolics, which are associated with counteracting the risk of cardiovascular diseases, cancer and cataract as well as a number of other degenerative diseases. This is achieved by preventing lipid oxidation, protein cross linking and DNA mutation and, at later stages, tissue damage (Figure 8.2).

Although phenolic compounds and some of their derivatives are very efficient in preventing autoxidation, only a few phenolic compounds are currently allowed as food antioxidants. The major considerations for acceptability of such antioxidants are their activity and potential toxicity and/or carcinogenicity. The approved phenolic antioxidants have been extensively studied, but the toxicology of their degradation products still is not clear.

The process of autoxidation of polyunsaturated lipids in food involves a free radical chain reaction that is generally initiated by exposure of lipids to light, heat, ionizing radiation, metal ions or metalloprotein catalysts. Enzyme lipoxygenase can also initiate oxidation. The classical route of autoxidation includes initiation (production of lipid free radicals), propagation and termination (production of nonradical products) reactions (Reaction 8.1 to Reaction 8.4). Figure 8.3 represents a general

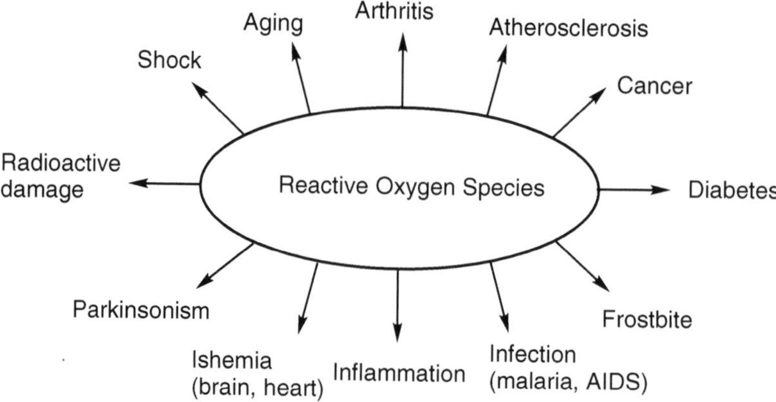

FIGURE 8.1 Diseases and damages caused by reactive oxygen species.

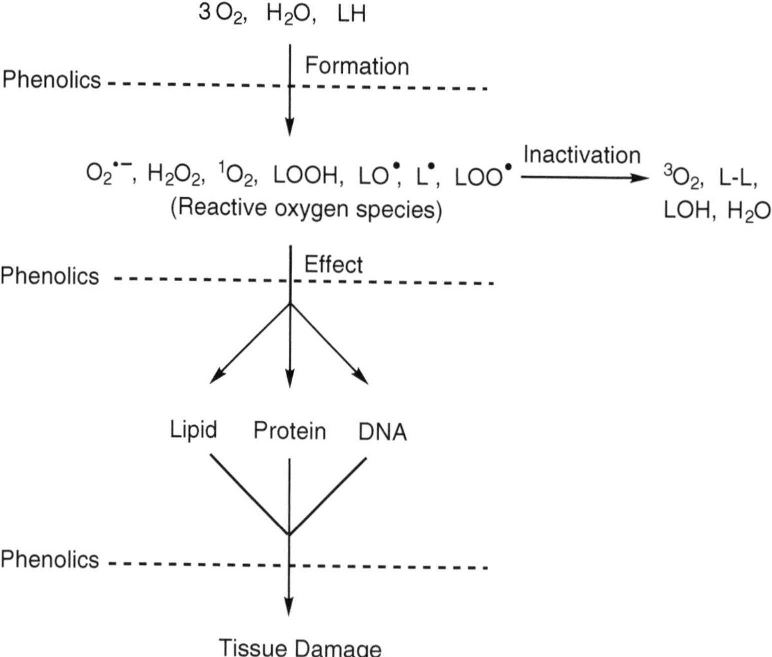

FIGURE 8.2 Consequences of reactive oxygen species in diseases and preventive role of phenolics.

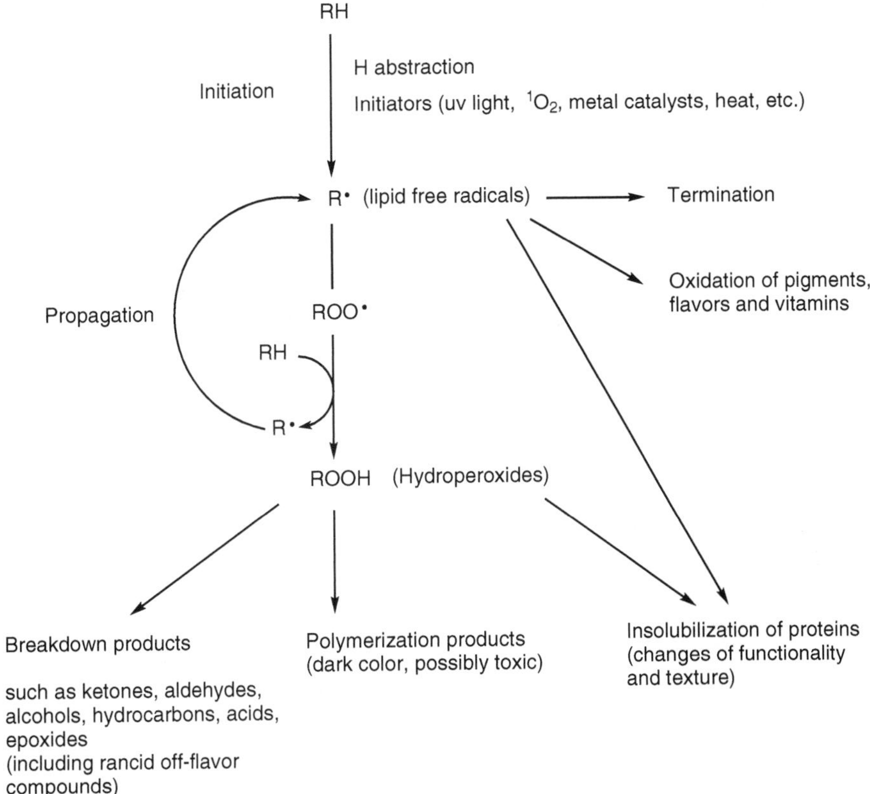

FIGURE 8.3 General scheme for autoxidation of lipids containing polyunsaturated fatty acids (RH) and their consequences.

scheme for autoxidation of polyunsaturated lipids and their consequence in quality deterioration of food:

Initiation $$RH \rightarrow R^{\bullet} + H^{\bullet} \tag{8.1}$$

Propagation $$R^{\bullet} + O_2 \rightarrow ROO^{\bullet} \tag{8.2}$$

$$ROO^{\bullet} + RH \rightarrow R^{\bullet} + ROOH \tag{8.3}$$

Termination $$\left.\begin{array}{l} R^{\bullet} + R^{\bullet} \\ R^{\bullet} + ROO^{\bullet} \\ ROO^{\bullet} + ROO^{\bullet} \end{array}\right\} \rightarrow \text{nonradical products} \tag{8.4}$$

Some commercially produced plant phenolic compounds have recently been considered as antioxidants. Flavonoids, the most potent antioxidative compounds of plant phenolics, need further investigation in order to explore the feasibility of their use in foods and to determine their toxicological implications.

PREVENTION OF AUTOXIDATION AND USE OF ANTIOXIDANTS

Vacuum (or *sous vide*) packaging or packaging under an inert gas (i.e., modified atmosphere packaging) to exclude oxygen, as well as refrigeration or freezing, can reduce the rate of autoxidation (Dziedzic and Hudson, 1983a). However, these means are not always practical because very little oxygen is needed to initiate and maintain the oxidation process. It is neither economical nor practical to remove traces of oxygen from foods; therefore, it is quite common to combine such methods with the use of antioxidants. The main justification for using an antioxidant is to extend the shelf life of foodstuffs and to reduce waste and nutritional losses by inhibiting and delaying oxidation. However, antioxidants cannot improve the quality of an already oxidized food product (Sherwin, 1978; Coppen, 1983).

Antioxidants are regarded as compounds capable of delaying, retarding or preventing autoxidation processes. According to the USDA Code of Federal Regulations, "antioxidants are substances used to preserve food by retarding deterioration, rancidity or discoloration due to oxidation" (21, CFR 170.3 (0) (3); Dziezak, 1986). Synergists are substances that enhance the activity of antioxidants without possessing antioxidant activity of their own (Nawar, 1996).

It has been suggested that an ideal food-grade antioxidant should be safe, not impart color, odor or flavor, be effective at low concentrations, be easy to incorporate, survive after processing, and be stable in the finished product (carry through) as well as available at a low cost (Coppen, 1983).

ESTIMATION OF ANTIOXIDANT ACTIVITY

The activity of an antioxidant can be estimated by quantitatively determining primary or secondary products of autoxidation of lipids or by monitoring other variables. Generally, the delay in hydroperoxide formation or production of secondary products of autoxidation by chemical or sensory methods can be used. These procedures can be applied to intact foods, their extracts or to model systems. Studies on foods can be performed under normal storage conditions or under accelerated oxidation such as active oxygen method (AOM), Schaal oven test, oxygen uptake/absorption, and oxygen bomb calorimetry, or by using a fully automated oxidative stability instrument (OSI), a Rancimat apparatus, or an oxidograph, among others (Kahl and Hildebrandt, 1986; Kochhar and Rossell, 1990). The extension of the induction period by addition of an antioxidant has been related to antioxidant efficacy, which is sometimes expressed as antioxidant index or protection factor. It is also possible to use a luminescence apparatus, also known as PHOTOCHEM (Analytik Jena, Delaware, OH), which measures antioxidant activity of hydrophilic and lipophilic compounds (Amarowicz et al., 2003). ORAC (oxygen radical absorbance capacity) and TEAC (Trolox equivalent antioxidant capacity) tests have also been used in the recent literature; artificial radicals such as DPPH (2,2-diphenyl-1-picrylhydrazyl) radical have been employed. All of these offer means of evaluating antioxidant activity of food phenolics and other constituents.

The ORAC test measures the scavenging capacity at 37°C of peroxyl radical induced by 2,2′-azobis- (2-amidinopropane) dihydrochloride (AAPH) using fluorescence at a wavelength of 565 nm with excitation at 540 nm (Cao et al., 1993). The results are calculated and expressed as millimoles of Trolox equivalents per gram. The TEAC assay determines the capacity of an antioxidant, antioxidant mixtures or an extract to scavenge ABTS (2,2′-azino-bis-(3-ethylbenzthiazoline-6-sulfonic acid) radical and thus disappearance of the blue/green color of this radical upon the action of antioxidants (Arts et al., 2003). Rice-Evans et al. (1995, 1996) have reported TEAC values for selected flavonoids. In addition, DPPH assay measures the scavenging of this radical with antioxidants using spectrophotomeric or EPR (electron paramagnetic resonance) techniques (Wettasinghe and Shahidi, 2000; Yu et al., 2002).

Formation of hydroperoxides is measured by an iodometric titration of released iodine by hydroperoxides and is generally expressed as the peroxide value (PV). The 2-thiobarbituric acid (TBA) test or para-anisidine test, measurement of total carbonyls, or a selected carbonyl compound and assessment of off-flavors and off-odors due to the formation of volatile decomposition products of hydroperoxides by objective and subjective means are used extensively (Shahidi, 1998) to measure hydroperoxide decomposition products and determine aldehydic compounds (e.g., spectrophotometric determination of malonaldehyde and related compounds) collectively referred to as malonaldehyde equivalents. Sensory evaluation provides information on the overall acceptability of foods in addition to the results provided by chemical determinations (Kahl and Hildebrandt, 1986). Application of Fourier transform infrared (FT-IR) spectroscopy (van de Voort et al., 1994) and nuclear magnetic resonance (NMR) spectroscopy (Wanasundara and Shahidi, 1993) for the evaluation of the oxidative state of food lipids have recently been reported (van de Voort et al., 1994). These techniques reflect the overall changes in primary and secondary oxidation products.

Model systems for testing the antioxidant activity of food components and additives have been used extensively. Peroxidation catalyzed by metmyoglobin or a Fe^{2+}-EDTA system, as well as AOM tests, may be performed on linoleic acid instead of food in evaluation of the activity of antioxidants. However, model systems have a disadvantage in that intact foods also contain natural compounds that may possess antioxidant or synergistic properties.

TYPES OF ANTIOXIDANTS

According to their mode of action, antioxidants may be classified as free radical terminators, chelators of metal ions, or oxygen scavengers that react with oxygen in closed systems. Thus, primary antioxidants react with high-energy lipid radicals to convert them to thermodynamically more stable products. Secondary antioxidants, also known as preventive antioxidants, function by retarding the rate of chain initiation by breaking down hydroperoxides. Examples of these secondary antioxidants include dilauryl thiodipropionate and thiodipropionic acid (Dziezak, 1986; Gordon, 1990); phenolic antioxidants are included in the category of free radical terminators. Because the main emphasis is on phenolic antioxidants, the mechanism of the action of free radical terminators will be discussed in detail.

MECHANISM OF ACTION OF PHENOLIC ANTIOXIDANTS

The first detailed kinetic study of antioxidant activity was conducted by Boland and ten-Have (1947) who postulated Reaction 8.5 and Reaction 8.6 for free radical terminators. Phenolic antioxidants (AH) interfere with lipid oxidation by rapid donation of a hydrogen atom to lipid radicals (Reaction 8.5 and Reaction 8.6). The latter reactions compete with chain propagation Reaction 8.3 and Reaction 8.9:

$$ROO^{\bullet} + AH \rightarrow ROOH + A^{\bullet} \quad (8.5)$$

$$RO^{\bullet} + AH \rightarrow ROH + A^{\bullet} \quad (8.6)$$

$$ROO^{\bullet} + A^{\bullet} \rightarrow ROOA \quad (8.7)$$

$$RO^{\bullet} + A^{\bullet} \rightarrow ROA \quad (8.8)$$

$$RO^{\bullet} + RH \rightarrow ROOH + R^{\bullet} \quad (8.9)$$

These reactions are exothermic in nature. The activation energy increases with increasing A–H and R–H bond dissociation energy. Therefore, the efficiency of the antioxidants (AH) increases with decreasing A–H bond strength. The resulting phenoxy radical must not initiate a new free radical reaction or be subject to rapid oxidation by a chain reaction. In this regard, phenolic antioxidants are excellent hydrogen or electron donors; in addition, their radical intermediates are relatively stable due to resonance delocalization and lack of suitable sites for attack by molecular oxygen (Sherwin, 1978; Nawar, 1986; Belitz and Grosch, 1987).

In the body, free radicals may be involved in a number of diseases and tissue injuries such as those of the lung, heart, cardiovascular system, kidneys, liver, eye, skin, muscle and brain, as well as the process of ageing. Oxidants and radicals are known to mediate such disorders, but are generally neutralized by antioxidant enzymes in healthy individuals. However, with age and in individuals with certain ailments, the endogenous antioxidants may require exogenous assistance from dietary antioxidants in order to maintain the integrity of cell membranes.

The phenoxy radical formed by reaction of a phenol with a lipid radical is stabilized by delocalization of unpaired electrons around the aromatic ring as indicated by the valence bond isomers in Reaction 8.10.

OH / OH → (ROO•, ROOH) → O• / OH → (ROO•, ROOH; O• / OH, OH / OH) → O / O (8.10)

However, phenol is inactive as an antioxidant. Substitution of the hydrogen atoms in the ortho and para positions with alkyl groups increases the electron density of the OH moiety by an inductive effect and thus enhances its reactivity toward lipid radicals. Substitution at the para position with an ethyl or *n*-butyl group rather than a methyl group improves the activity of the phenolic antioxidant; however, presence of chain or branched alkyl groups in this position decreases the antioxidant activity (Gordon, 1990).

The stability of the phenoxy radical is increased by bulky groups at the ortho positions as in 2,6-di-tertiary-butyl, 4-methoxyphenol or butylated hydroxyanisole (BHA) (Miller and Quackenbush, 1957). Because these substituents increase the steric hindrance in the region of the radicals, they further reduce the rate of possible propagation reactions that may occur, as in Reaction 8.11 to Reaction 8.13, involving antioxidant free radicals (Gordon, 1990):

$$A^{\bullet} + O_2 \rightarrow AOO^{\bullet} \quad (8.11)$$

$$AOO^{\bullet} + RH \rightarrow AOOH + R^{\bullet} \quad (8.12)$$

$$A^{\bullet} + RH \rightarrow AH + R^{\bullet} \quad (8.13)$$

The introduction of a second hydroxy group at the ortho or para position of the hydroxy group of a phenol increases its antioxidant activity. The effectiveness of a 1,2-dihydroxybenzene derivative is increased by the stabilization of the phenoxy radical through intramolecular hydrogen bond. Thus, catechol and hydroquinone are much more effective in their peroxynitrite scavenging activity than phenol (Heignen et al., 2001). Similarly, flavonols containing a catechol moiety (3′- and 4′-OH) in ring B (rutin and monohydroxyethyl rutinoside) or an AC-ring with three OH groups (3-, 5-, and 7-OH) are potent scavengers. The 3-OH group is found to be the active center, its activity influenced by electron-donating groups at the 5- and 7-positions (galangin, kaempferol, and trihydroxyethyl quercetin). In another study, Heim et al. (2002) found that multiple hydroxyl groups conferred substantial antioxidant, chelating, and, in some cases, pro-oxidant activity to the molecule. Methoxy groups introduce unfavorable steric effects, but presence of a double band and a carbonyl functionality in the C ring increases the activity by affording a more stable flavonoid radical through conjugation and electron delocalization.

Finally, the antioxidant activity of hydroxyflavones is influenced by pH. The antioxidant potential increases, as determined by the TEAC assay, upon deprotonation of the hydroxyl group. This indicates that the mechanism of action of flavonoids is variable and, although abstraction of the hydrogen atom is involved for underprotonated species, electron (not hydrogen) atom donation is involved in the deprotonated species (Lemanska et al., 2001). Furthermore, the hydroxyl radical scavenging activity of phenolics involves multiple mechanisms, including hydroxyl bond strength, electron donating ability, enthalpy of single electron transfer and spin distribution of the phenoxy radical after hydrogen abstraction (Cheng et al., 2002). More recently, Arts et al. (2003) have reported a critical evolution of the use of antioxidant capacity in defining optimal antioxidant structures.

The antioxidant activity of dihydroxybenzene derivatives is partly due to the fact that the semiquinoid radical produced initially can be further oxidized to a quinone by reaction with another lipid radical. It can also form into a quinone or hydroquinone molecule, as in Reaction 8.14:

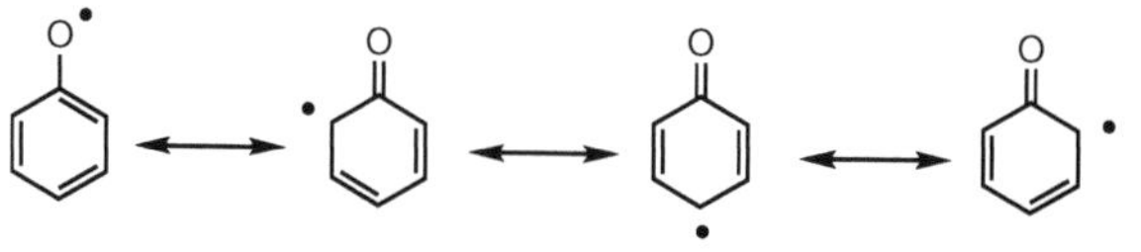

(8.14)

The activity of 2-methoxyphenol is, in general, much lower than that of catechol, which possesses two free hydroxy groups (Rosenwald and Chenicek, 1951), because 2-methoxyphenols are unable to stabilize the phenoxy radical by hydrogen bonding as in Reaction 8.14 (Gordon, 1990). The effect of antioxidant concentration on autoxidation rates depends on many factors, including the structure of the antioxidant, oxidation conditions, and nature of the sample oxidized. Often phenolic antioxidants lose their activity at high concentrations and behave as pro-oxidants (Cillard et al., 1980) by involvement in initiation reactions such as those in Reaction 8.11) to Reaction 8.13 (Gordon, 1990). Antioxidant activity by donation of a hydrogen atom is unlikely to be limited to phenols. Endo et al. (1985) have suggested that the antioxidant effect of chlorophyll in the dark occurs by the same mechanism as phenolic antioxidants.

Phenolic antioxidants are more effective in extending the induction period when added to an oil that has not deteriorated to any great extent. However, they are ineffective in retarding decomposition of already deteriorated lipids (Mabrouk and Dugan, 1961). Thus, antioxidants should be added to foodstuffs as early as possible to achieve maximum protection against oxidation (Coppen, 1983).

SYNTHETIC PHENOLIC ANTIOXIDANTS

The application of antioxidants in foods is governed by federal regulations. FDA regulations require that the ingredient labels of products declare antioxidants and their carriers followed by an explanation of their intended purpose (Dziezak, 1986). Table 8.1 and Table 8.2 summarize the permitted food phenolic antioxidants, some of their properties and amounts of their allowable usage. Synthetic food antioxidants currently permitted for use in foods are butylated hydroxytoluene (BHT), butylated

TABLE 8.1
Synthetic Food Antioxidants and Their Properties

Characteristic	BHA	BHT	Gallates: Propyl	Gallates: Dodecyl	TBHQ
Melting point °C	50–52	69–70	146–148	95–98	126–128
Carry-through properties	Very good	Fair–good	Poor	Fair–good	Good
Synergism	BHT and gallates	BHA	BHA	BHA	—
Solubility (w/w%)					
Water	0	0	<1	<1	<1
Propylene glycol	50	0	6.5	4	30
Lard	30–40	50	1	—	05–10
Corn oil	30	40	0	0	10
Glycerol	1	0	25	—	<1
Methyl linoleate	Very soluble	Very soluble	1	1	>10

Sources: Adapted from Coppen, P.P., 1983, in *Rancidity in Foods*, Allen, J.C. and Hamilton, R.J., Eds., Applied Science Publishing Company, London, 67–87 and Dziezak, J.D., 1986, *Food Technol.*, 9, 94–102.

hydroxyanisole (BHA), propyl gallate (PG), dodecyl gallate (DG) and tertiary-butylhydroquinone (TBHQ).

BHA and BHT are monohydric phenolic antioxidants (Figure 8.4) originally used to protect petroleum from oxidative degumming before their introduction and acceptance in the food industry (Porter, 1980). Chemically, BHA is a mixture of 3-tertiary-butyl-4-hydroxyanisole (90%) and 2-tertiary-butyl-4-hydroxyanisole (10%). BHA is commercially available as white waxy flakes and BHT is available as a white crystalline compound; both are extremely soluble in fats and insoluble in water (Table 8.2). Furthermore, both assert a good carry-through effect; however, BHA is slightly better than BHT in this respect (Dziezak, 1986).

BHT is more effective in suppressing oxidation of animal fats than vegetable oils. Among its multiple applications, BHA is particularly useful in protecting the flavor and color of essential oils and is considered the most effective of all food-approved antioxidants for this application (Stuckey, 1972). BHA is particularly effective in controlling the oxidation of short-chain fatty acids such as those found in the coconut and palm kernel oils typically used in cereal and confectionary products (Dziezak, 1986).

As a monophenol, BHT can produce radical intermediates with moderate resonance delocalization. The tertiary butyl groups of BHT do not generally allow involvement of the radical formed from it after hydrogen abstraction in other reactions. Thus, a lipid peroxy radical may join the molecule of BHT in the para position to the phenoxy group (Dziezak, 1986).

The volatile nature of BHA and BHT makes them important additives in packaging materials because they can migrate into foods. For this purpose, these antioxidants are added directly to the wax used in making inner liners or applied to the packaging board as an emulsion (Porter, 1980; Dziezak, 1986). A synergistic effect

TABLE 8.2
Maximum Usage Levels Permitted by U.S. FDA in Specific Application of Antioxidants (from Code of Federal Regulations)

Food	Maximum Usage Levels[a]			
	BHA[b,c]	BHT[c,d]	PG[e]	TBHQ[b]
Dehydrated potato shreds	50	50	—	—
Active dry yeast	1000[g]	—	—	—
Beverages prepared from dry mixes	2[g]	—	—	—
Dry breakfast cereals	50[g]	50	—	—
Dry diced glazed fruits	32[g]	—	—	—
Dry mixes for beverages and desserts	50	—	—	—
Emulsion stabilizers for shortenings	200	200	—	—
Potato flakes	50	50	—	—
Potato granules	10	10	—	—
Sweet potato flakes	50	50	—	—
Poultry products[h]	100[g]	100[j]	100	100
Dry sausage[i]	30[g]	30[j]	30	30
Fresh sausage[i]	100[g]	100[j]	100	100
Dried meat[i]	100[g]	100[j]	100	100

[a] Milligrams per kilogram.
[b] 21 CFR 172. 110.
[c] Given levels are for total BHT and BHA.
[d] 21 CFR 172. 115.
[e] 21 CFR 184. 1660.
[f] 21CFR 172. 185.
[g] BHA only.
[h] 9 CFR 381. 147(f) (3).
[i] 9 CFR 318.7 © (4).
[j] BHT only.

Source: Adapted from Dziezak, J.D., 1986, *Food Technol.*, 9, 94–102.

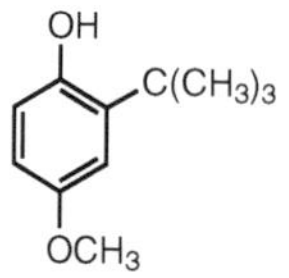

BHA

3-tertiary-butyl-4-hydroxyanisole
2-tertiary-butyl-4-methoxyphenol

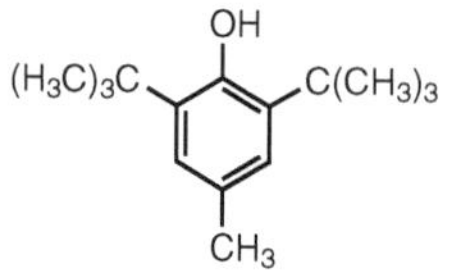

BHT

3,5-di-tertiary-butyl-4-hydroxytoluene
2,6-di-tertiary-butyl-4-methylphenol

FIGURE 8.4 Chemical structures of butylated hydroxyanisole (BHA) and butylated hydroxytoluene (BHT).

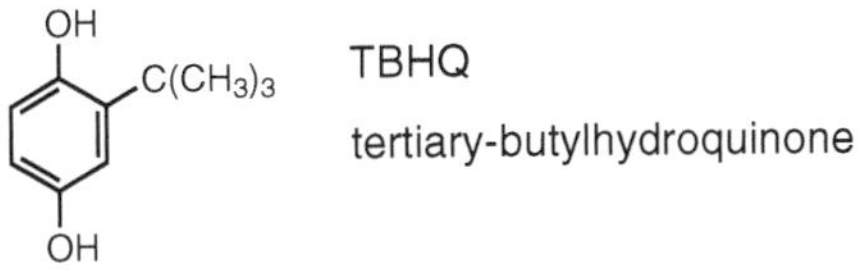

OH
$C(CH_3)_3$ TBHQ
tertiary-butylhydroquinone
OH

FIGURE 8.5 Chemical structure of tertiary-butylhydroquinone (TBHQ).

has been shown to exist when BHT and BHA are used in combination. The oxidative reactions of nut and nut products are very responsive to the combination of these two antioxidants (Dziezak, 1986).

Tertiary-butylhydroquinone, shown in Figure 8.5, is regarded as the best antioxidant for protecting frying oils against oxidation (Khan and Shahidi, 2001) and provides good carry-through protection to fried products similar to those of BHA and BHT. TBHQ may be considered as an alternative to hydrogenation for increasing oxidative stability (Dziezak, 1986). TBHQ is adequately soluble in fats and does not complex with iron or copper; therefore, it does not discolor the treated products. TBHQ is available as a beige-colored powder to be used alone or in combination with BHA or BHT at a maximum amount of 0.02% or 200 mg/kg, based on the fat content of foods, including essential oils. TBHQ is not permitted in combination with propyl gallate (Dziezak, 1986). Coppen (1983) has reported that TBHQ shows good performance in stabilizing crude oils.

Chelating agents such as citric acid and monoacylglycerol citrate can further enhance lipid-stabilizing properties of TBHQ. This combination is primarily used in vegetable oils and shortenings but not extensively for animal fats. Confectioneries, including nuts and candies, also benefit from the use of TBHQ or its mixtures (Buck, 1984). As a diphenolic antioxidant, TBHQ reacts with peroxy radicals to form a semiquinone resonance hybrid. The semiquinone radical intermediates may undergo different reactions to form more stable products. They can also react with one another to produce dimmers or react with one another to produce a quinone and a hydroquinone molecule or add to a lipid peroxy radical to produce a semiquinone.

Propyl gallate is commercially prepared by esterification of gallic acid with propyl alcohol followed by distillation to remove the excess alcohol (Figure 8.6). PG is available as a white crystalline powder and is sparingly soluble in water; it functions particularly well in stabilizing animal fats and vegetable oils. With a melting point of 148°C, PG loses its effectiveness during heat processing and is therefore not suitable in frying applications that involve temperatures exceeding 190°C. PG chelates iron ions and forms an unappealing blue–black complex. Hence, PG is always used with chelators such as citric acid to eliminate the pro-oxidative iron and copper catalysts. Good synergism is obtained with BHA and BHT; however, their coapplication with TBHQ is not permitted (Buck, 1984).

PG may be used to inhibit the oxidation of vegetable oils, animal fats and meat products, including fresh and frozen sausages and snacks. Its usage has been permitted in chewing gum base at <0.1% and with BHA and/or BHT at a total concentration of <0.1%. Moreover, the amphiphilic nature of PG makes it a very effective antioxidant for dry vegetable oils. Gallates have lower volatility and thus

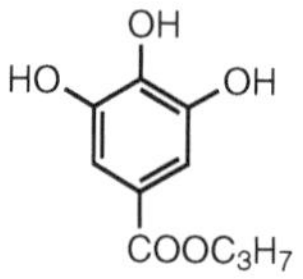

PG

3,4,5-trihydroxypropylbenzoate

FIGURE 8.6 Chemical structure of propyl gallate (PG).

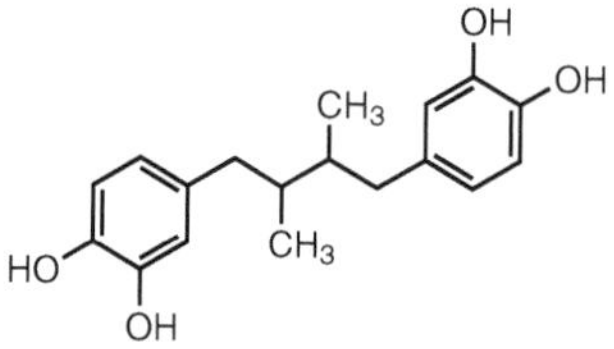

NDGA

4,4'-(2,3-dimethyl-1,4-butanediyl)bis[1,2-benzenediol]

FIGURE 8.7 Chemical structure of nordihydroguaiaretic acid (NDGA).

have less phenolic odor than monohydric phenols such as BHA and BHT (Dziezak, 1986).

Nordihydroguaiaretic acid (NDGA) is a grayish-white crystalline compound that was widely used as an antioxidant in animal fats in the 1950s and 1960s (Figure 8.7). It possesses phenolic properties similar to gallates, including their advantages and disadvantages. Besides the isolation of natural material (resinous exudate of creosote bush), NDGA has also been chemically synthesized. Due to unfavorable toxicological findings, NDGA is no longer of practical importance in the food industry (Gordon, 1990; Schuler, 1990).

Degradation Products of Phenolic Antioxidants

Phenolic antioxidants generally undergo degradation during the course of oxidation of fats and oils and produce a range of products, especially dimers of the antioxidants. Most oxidation products of antioxidants retain some antioxidant activity. Among the breakdown products of BHT shown in Figure 8.8, products (2) through (4) possess antioxidant properties while products (1) and (5) are not effective as antioxidants.The degradation products of BHA shown in Figure 8.9 are less effective than BHA itself, in the order of BHA > (8) > (7), as reported by Kikugawa et al. (1990).

The mechanism and breakdown products of TBHQ due to irradiation have been reviewed by Kikugawa et al. (1990). All the oxidation products of TBHQ retain some antioxidant effect, but compounds (10) and (12) exhibit antioxidant activities greater than that of TBHQ (Figure 8.10). As shown in Figure 8.11, irradiation of propyl gallate in ethanol produces ellagic acid with excellent antioxidant activity (Kikugawa et al., 1990).

Degradation of mixtures of BHA, BHT and PG produced heterodimers between different antioxidant components (Figure 8.12). Although product (15) was new, the heterodimer (16) and heterodimer (17) exhibited activities comparable to that of PG (Kikugawa et al., 1990).

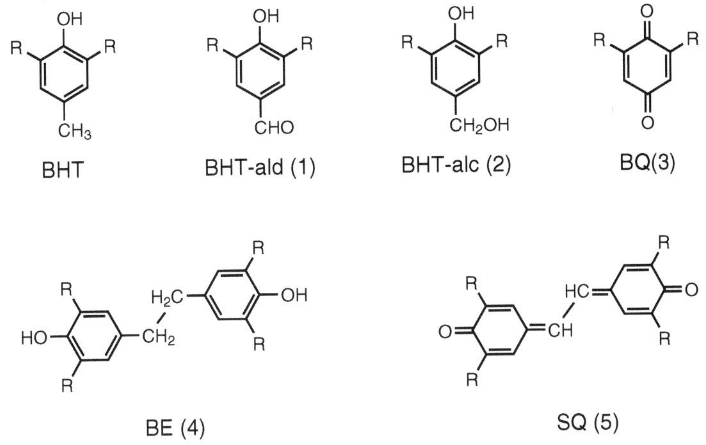

R = tertiary-butyl

(1) 3,5-ditertiary-butyl-4-hydroxybenzaldehyde
(2) 3,5-ditertiary-butyl-4-hydroxybenzylalcohol
(3) 2,6-ditertiary-butyl-benzoquinone
(4) 3,5,3',5'-tetra-tertiary-butyl-4,4'-dihydroxy-1,2-diphenylethane
(5) 3,5,3',5'-tetra-tertiary-butyl-stilbenequinone

FIGURE 8.8 Degradation products of BHT. (Adapted from Kikugawa, K. et al., 1990, in *Food Antioxidants*, Hudson, B.J.F., Ed., Elsevier, London, 65–98.)

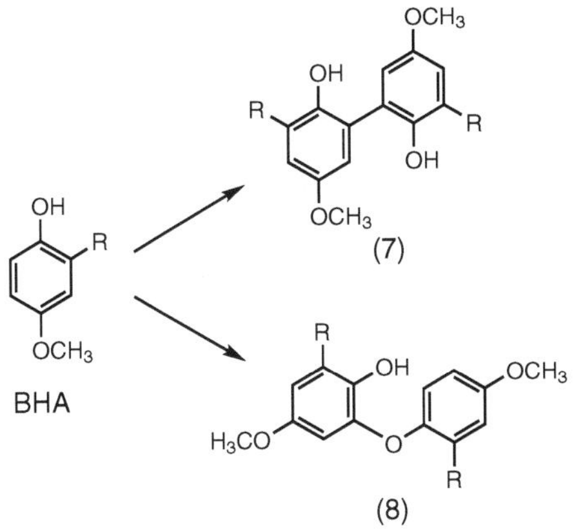

R = tertiary butyl

(7) 2,2'-dihydroxy-5,5'dimethoxy-3,3'-di-tertiary-butyl biphenyl
(8) 2',3-ditertiary-butyl-2-hydroxy-4',5-dimethoxy biphenyl ether

FIGURE 8.9 Degradation products of BHA. (Adapted from Kikugawa, K. et al., 1990, in *Food Antioxidants*, Hudson, B.J.F., Ed., Elsevier, London, 65–98.)

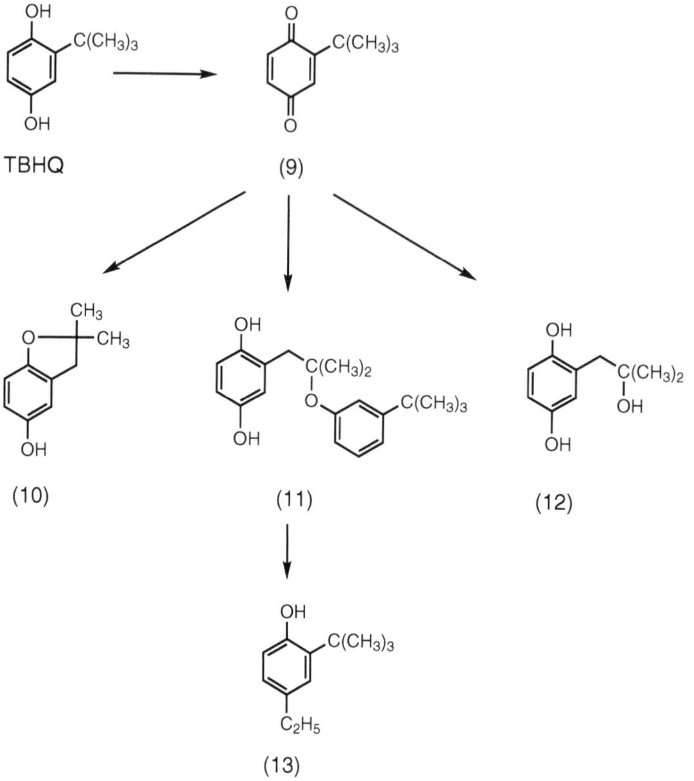

(9) 2-tertiary-butyl-*p*-benzoquinone
(10) 2,2'-dimethyl-5-hydroxy-2,3-dihydrobenzo[6]furan
(11) 2-[2-(3'-tertiary-butyl-4'-hydroxy-phenoxy-2-methyl-1-propyl]hyd roquinone
(12) 2-(2-hydroxy-2-methyl-1-propyl)hydroquinone
(13) 2-tertiary-butyl-4-ethoxyphenol

FIGURE 8.10 Degradation products of TBHQ. (Adapted from Kikugawa, K. et al., 1990, in *Food Antioxidants*, Hudson, B.J.F., Ed., Elsevier, London, 65–98.)

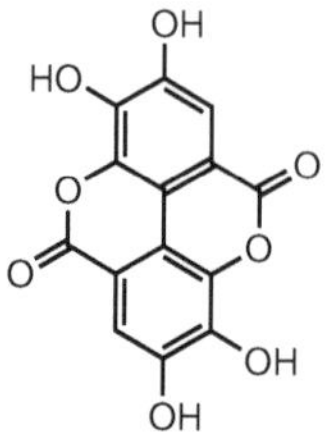

(14) ellagic acid

FIGURE 8.11 Degradation of PG and production of ellagic acid. (Adapted from Kikugawa, K. et al., 1990, in *Food Antioxidants*, Hudson, B.J.F., Ed., Elsevier, London, 65–98.)

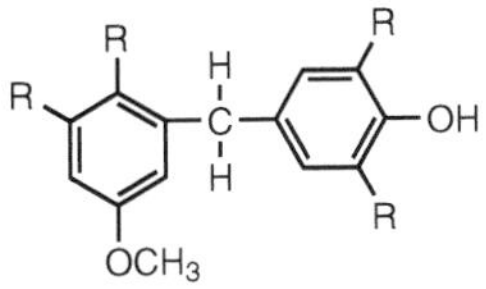

(15)

3,3',5'-tri-tertiary-butyl-5-methoxy-2,4'-dihydroxy diphenylmethane

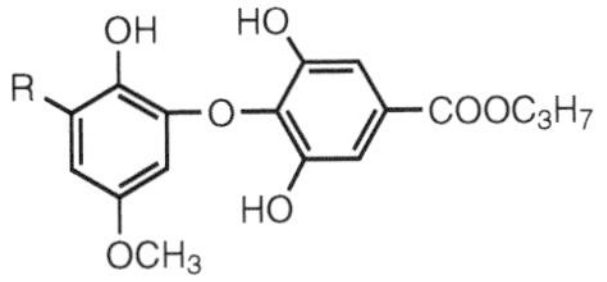

(16)

propyl-3,5-dihydroxy-4-(2'-hydroxy-5-methoxy-3'-tertiary-butyl phenoxy) benzoate

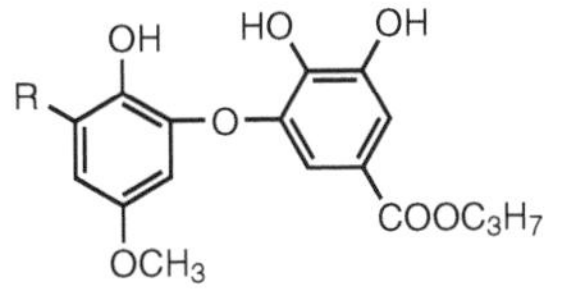

(17)

propyl-3.4-dihydroxy-5-(2'-hydroxy-5'-methoxy-3'-tertiary-butyl phenoxy) benzoate

FIGURE 8.12 Degradation products of a mixture of BHA and PG. (Adapted from Kikugawa, K. et al., 1990, in *Food Antioxidants*, Hudson, B.J.F., Ed., Elsevier, London, 65–98.)

NATURAL ANTIOXIDANTS

Antioxidants in foods may originate from compounds that occur naturally in the foodstuff or from substances formed during its processing (Shahidi, 1997; Shahidi and Wanasundara, 1992). Natural antioxidants are primarily plant-phenolic and polyphenolic compounds that may occur in all parts of the plant. Plant phenolics are multifunctional and can act as reducing agents (free radical terminators), metal chelators, and singlet oxygen quenchers. Examples of common plant phenolic antioxidants include flavonoid compounds, cinnamic acid derivatives, coumarins, tocopherols and polyfunctional organic acids (Löliger, 1983; Pratt and Hudson, 1990). Several studies have been carried out in order to identify natural phenolics that possess antioxidant activity. Some natural antioxidants have already been extracted from plant sources and are produced commercially (Schuler, 1990).

Tocopherols and Their Antioxidant Activity

Monophenolic antioxidants that help to stabilize most vegetable oils, tocopherols occur widely in nature. Tocopherols are composed of eight different compounds belonging to two families, namely, tocols and tocotrienols, referred to as α, β, γ, or δ, depending on the number and position of methyl groups attached to the chromane rings. Tocopherols also possess vitamin E activity. In tocols, the side chain is

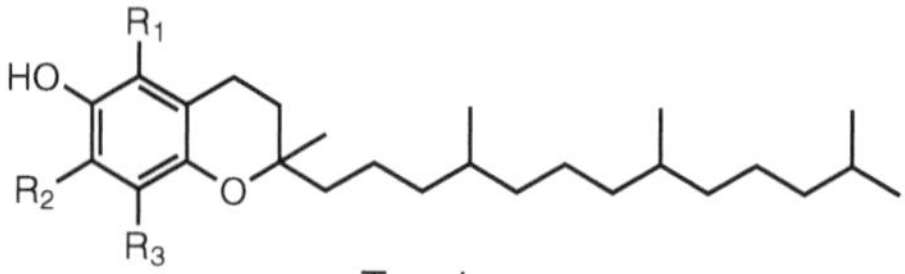

Tocols

Compound	R_1	R_2	R_3
5, 7, 8-Trimethyl tocol (α-tocopherol)	CH_3	CH_3	CH_3
7, 8-Dimethyl tocol (β-tocopherol)	CH_3	H	CH_3
5, 8-Dimethyl tocol (γ-tocopherol)	H	CH_3	CH_3
8-Methyl tocol (δ-tocopherol)	H	H	CH_3

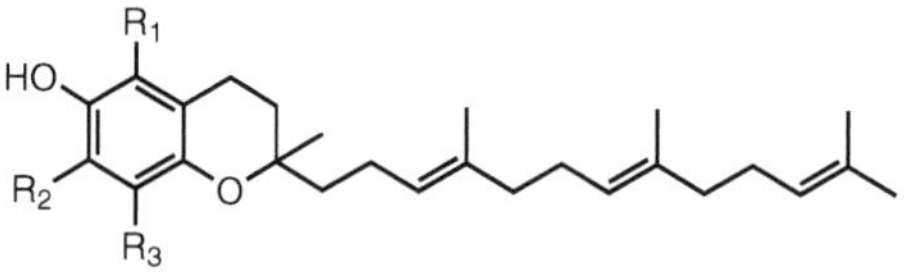

Tocotrienols

Compound	R_1	R_2	R_3
5, 7, 8-Trimethyl tocotrienol (α-tocotrienol)	CH_3	CH_3	CH_3
7, 8-Dimethyl tocotrienol (β-tocotrienol)	CH_3	H	CH_3
5, 8-Dimethyl tocotrienol (γ-tocotrienol)	H	CH_3	CH_3
8-Methyl tocotrienol (δ-tocotrienol)	H	H	CH_3

FIGURE 8.13 Chemical structures of vitamin E and related compounds.

saturated, while in tocotrienols it is unsaturated (Figure 8.13). With regard to vitamin E activity, α-tocopherol is the most potent member of this family. The antioxidant activity decreases from δ to α (Dziezak, 1986).

Vegetable foods contain considerable amounts of different tocopherols and tocotrienols in their lipid fraction (Table 8.3). Cereals and cereal products, oilseeds, nuts and vegetables are rich sources of tocopherols (Table 8.4); however, in the animal kingdom, tocopherols are only found in trace quantities. During manufacturing of oils, 30 to 40% of tocols and tocotrienols are lost (Dziezak, 1986; Schuler, 1990).

Tocopherols are important biological antioxidants. Alpha-tocopherol, or vitamin E, prevents oxidation of body lipids including polyunsaturated fatty acids and lipid components of cells and organelle membranes. Tocopherols are produced commercially and used as food antioxidants. The antioxidant activity of tocopherol is based mainly on the tocopherol–tocopheryl quinone redox system (Figure 8.14).

Tocopherols (AH_2) are radical scavengers and quench lipid radicals ($R^\bullet$), thus regenerating RH molecules as well as producing a tocopheryl semiquinone radical. Two tocopheryl semiquinone radicals (AH; Figure 8.15) may form a molecule of

TABLE 8.3
Approximate Content of Tocopherol and Tocotrienol Found in Vegetable Oils[a]

	Tocopherols				Tocotrienols			
Oil	**α**	**β**	**γ**	**δ**	**α**	**β**	**γ**	**δ**
Coconut	5–10	—	5	5	5	Trace	1–20	—
Cottonseed	40–560	—	270–410	0	—	—	—	—
Maize, grain	60–260	0	400–900	1–50	—	0	0–240	0
Maize, germ	300–430	1–20	450–790	5–60	—	—	—	—
Olive	1–240	0	0	0	—	—	—	—
Palm	180–260	Trace	320	70	120–150	20–40	260–300	70
Peanut	80–330	—	130–590	10–20	—	—	—	—
Rapeseed/canola	180–280	—	380	10–20	—	—	—	—
Safflower	340–450	—	70–190	230–240	—	—	—	—
Soybean	30–120	0–20	250–930	50–450	0	0	0	—
Sunflower	350–700	—	10–50	1–10	—	—	—	—
Walnut	560	20–40	590	450	—	—	—	—
Wheat germ	560–1200	660–810	260	270	20–90	80–190	—	—

[a] Milligrams per kilogram.

Source: Adapted from Schuler, P., 1990, in *Food Antioxidants*, Hudson, B.J.F., Ed., Elsevier, London, 99–170.

TABLE 8.4
Tocopherol Content of Cereal Grains

	Tocopherols (mg/100 g)			
Product	**α**	**β**	**γ**	**α-Tocotrienol**
Whole yellow corn	1.5	—	5.1	0.5
Yellow corn meal	0.4	—	0.9	—
Whole wheat	0.9	2.1	—	0.1
Wheat flour	0.1	1.2	—	—
Whole oats	1.5	—	0.05	0.3
Oat meal	1.3	—	0.2	0.5
Whole rice	0.4	—	0.4	—
Milled rice	0.1	—	0.3	—

Source: Adapted from Herting, D.C. and Durry, E.J.E., 1969, *J. Agric. Food Chem.*, 17, 785–790.

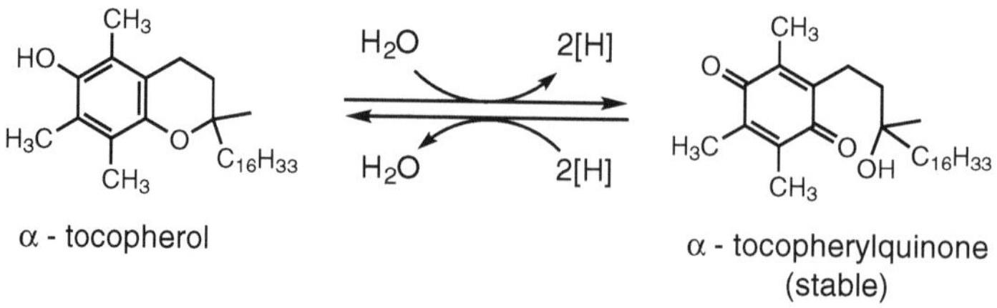

α - tocopherol

α - tocopherylquinone (stable)

FIGURE 8.14 Alpha-tocopherol/alpha-tocopheryl quinine redox system. (Adapted from Schuler, P., 1990, in *Food Antioxidants*, Hudson, B.J.F., Ed., Elsevier, London, 99–170.)

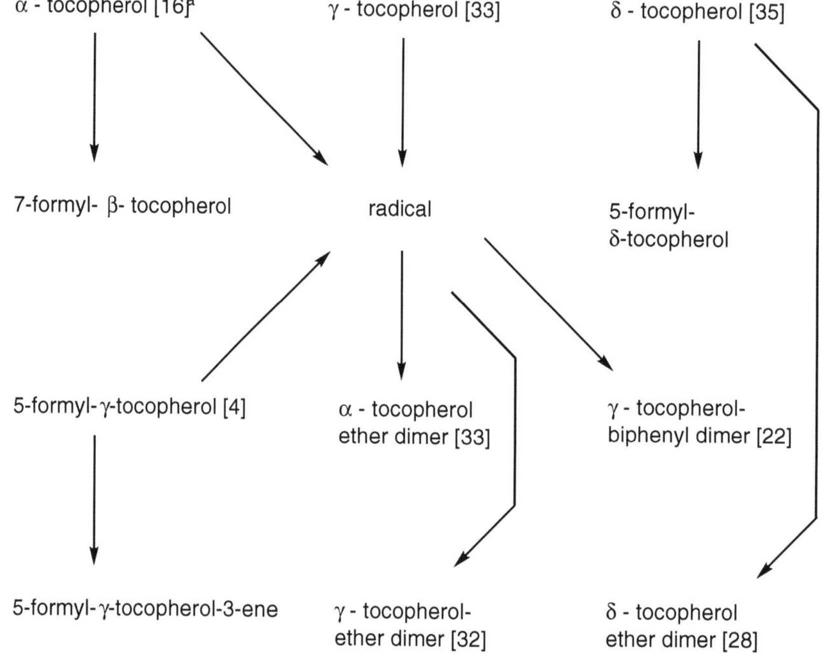

FIGURE 8.15 Antioxidant activity of selected tocopherols and their decomposition products. Length of induction period is given in days in brackets. Control without antioxidants had an induction period of 2 days. (Adapted from Ishikawa, Y. and Yuki, E., 1975, *Agric. Biol. Chem.*, 39, 851–857.)

tocopheryl quinone (A) and a regenerated molecule of tocopherol, as can be seen in Reaction 8.20 and Reaction 8.21:

$$R^{\bullet} + AH_2 \rightarrow RH + AH^{\bullet} \tag{8.20}$$

$$2AH^{\bullet} \rightarrow A^{\bullet} + AH_2 \tag{8.21}$$

The mechanism of oxidation of α-tocopherol with linoleate hydroperoxides has been studied in detail (Tappel, 1972). After releasing one H atom, the α-tocopheryl radical formed releases another H atom to produce methyl tocopheryl quinone, which is unstable and gives rise to α-tocopheryl quinone as its main product. The reaction between two semiquinoid radicals may also lead to the formation of α-tocopherol dimer, which possesses antioxidant properties (Schuler, 1990). Ishikawa and Yuki (1975) described the antioxidant effect of the products of tocopherols as oxidized α-, γ-, and δ-tocopherol with trimethylamine oxide. Some of the oxidation products formed were isolated and tested for their antioxidant activity; the results are summarized in Figure 8.15.

Tocopherols are commercially extracted from deodorizer sludge obtained in the deodorization of vegetable oils. Various tocols and tocotrienols of such extracts contain sterols, esters of sterols, free fatty acids and triacylglycerols. The separation of tocopherols from other compounds is possible by esterification with lower alcohol, washing and vacuum distillation or by saponification or fractional liquid–liquid extraction. Further purification may be achieved using molecular distillation, extraction, crystallization or a combination of these processes (Schuler, 1990). The total tocopherol content of the extracts is usually between 30 and 80%, but higher in γ- and δ-tocopherols. To obtain stable ∀-tocopheryl acetate, the mixture is methylated and subsequently acetylated. The α-tocopheryl acetate is the commercially available form of vitamin E, which is not an antioxidant because its active OH group is blocked. However, under acidic aqueous conditions, tocopherol is released by hydrolysis and the released tocopherol may then act as an antioxidant (Schuler, 1990).

The commercial synthesis of α-tocopherol involves condensation of 2,3,5-trimethyl hydroquinone with phytol, isophytol or phytol halogenides. The crude tocopherol product is purified by vacuum distillation. Because use of isophytol is preferred, the distillation produces a racemic mixture of all possible tocopherol isomers. Products intended for antioxidant applications are generally marketed in oil forms. Pure, all-racemic α-tocopherol, mixed tocopherols having various contents of α-, γ- and/or δ-tocopherols; synergistic mixtures composed of tocopherols, ascorbyl palmitate or other antioxidants, and synergists such as lecithin, citric acid and carriers are also available (Schuler, 1990).

Phenolic Acids, Flavonoids and Their Antioxidant Activity

Flavonoids usually occur in living cells as glycosides and may break down to their respective aglycone and sugar by enzymes or acid-heat treatments. Pratt and Watts (1964) and Pratt (1965, 1972) have considered flavonoids as primary antioxidants. Many of the flavonoid and related phenolic acids have shown marked antioxidant characteristics (see Table 8.5 through Table 8.7) (Mehta and Seshadri, 1959). Structures of these flavonoid and related compounds are given in Chapter 1.

Flavonoids and cinnamic acids are known as primary antioxidants and act as free radical acceptors and chain breakers. Flavonols are known to chelate metal ions at the 3-hydroxy-4-keto group and/or 5-hydroxy-4-keto group (when the A-ring is hydroxylated at the fifth position). An *o*-quinol group at the B-ring can also demonstrate metal chelating activity (Pratt and Hudson, 1990).

TABLE 8.5
Antioxidant Activity of Flavones

Compound	Time to Reach Peroxide Value of 50 (h)[a]	Induction Period by Rancimat (h)[b]
Control		
Stripped corn oil	105	—
Lard	—	1.4
Aglycones		
Quercetin (3,5,7,3′-4′-pentahydroxy)	475	7.1
Fisetin (3,7,3′,4′-tetrahydroxy)	450	8.5
Lueteolin (5,7,3′,4′-tetrahydroxy)	—	4.3
Myricetin (3,5,7,3′,4′,5′-hexahydroxy)	552	—
Robinetin (3,7,3′,4′,5′-pentahydroxy)	750	—
Rhamnetein (3,5,3′,4′-tetrahydroxy-7-methoxy)	375	—
Glycosides		
Quercetrin (3-rhamnoside)	475	1.9
Rutin (3-rhamnoglucoside)	195	—

[a] 5×10^{-4} M in stripped corn oil.
[b] 2.3×10^{-4} M in lard.

Source: Adapted from Pratt, D.E. and Hudson, B.J.F., 1990, in *Food Antioxidants*, Hudson, B.J.F., Ed., Elsevier Applied Science, London, 171–192.

It has been established that the position and degree of hydroxylation are of primary importance in determining antioxidant activity of flavonoids. The *o*-dihydroxylation of the B-ring contributes to the antioxidant activity. The *p*-quinol structure of the B-ring has been shown to impart an even greater activity than *o*-quinol; however, para and meta hydroxylation of the B-ring do not occur naturally (Pratt and Hudson, 1990). All flavonoids with 3′,4′-dihydroxy configuration possess antioxidant activity (Dziedzic and Hudson, 1983b).

Robinetin and myricetin have an additional hydroxy group at their 5′ position, thus leading to enhanced antioxidant activities over those of their corresponding flavones that do not possess the 5′ hydroxy group, namely, fisetin and quercetin (Figure 1.9 and Table 8.5). Two flavanones (naringenin and hesperitin) have only one hydroxy group on the B-ring and possess little antioxidant activity (Figure 1.9 and Table 8.5). Hydroxylation of the B-ring is the major consideration for antioxidant activity (Pratt and Hudson, 1990). Other important features include a carbonyl group at position 4 and a free hydroxy group at position 3 and/or 5 (Dziedzic and Hudson, 1983b).

Uri (1961) has investigated the importance of other sites of hydroxylation. It has been shown that the *o*-dihydroxy grouping on one ring and *p*-dihydroxy grouping on the other (e.g., 3,5,8,3′,4′- and 3,7,8,2′,5′-pentahydroxy flavones) produce very potent antioxidants, while 5,7 hydroxylation of the A-ring apparently has little

TABLE 8.6
Antioxidant Activity of Flavanones

Compound	Time to Reach Peroxide Value of 50 (h)[a]	Induction Period by Rancimat (h)[b]
Control		
Stripped corn oil	105	—
Lard	—	1.4
Aglycones		
Taxifolin (dihydroquercetin)	470	8.2
(3,5,7,3′, 4′-pentahydroxy)		6.7
Fustin (3,7,3′,4′-tetrahydroxy)	—	6.7
Eriodictyol (3,7,3′,4′-tetrahydroxy)	—	—
Naringenin (5,7,4′-tetrahydroxy)	198	—
Hesperitin (5,7,3′-trihydroxy-4 methoxy)	125	—
Glycosides		
Hesperidin (7-rhamnoglucoside)	125	—
Neohesperidin (7-glucoside)	135	—

[a] 5×10^{-4} M in stripped corn oil.
[b] 2.3×10^{-4} M in lard.

Source: Adapted from Pratt D.E. and Hudson, B.J.F., 1990, in *Food Antioxidants*, Hudson, B.J.F., Ed., Elsevier Applied Science, London, 171–192. With permission.

influence on the antioxidant activity of the compounds (Pratt and Hudson, 1990). Thus, quercetin and fisetin have almost the same activity, while myricetin possesses an activity similar to that of robetin (Figure 1.9 and Table 8.5). The 3-glycosylation of flavonoids with monosaccharides/disaccharides reduces their activity compared with that of the corresponding aglycones (e.g., rutin is less active than quercetin; Figure 1.9).

The ability of flavonoids to form complexes with cupric ion (Figure 8.16) has also been demonstrated by UV spectral studies. Such complexations may contribute to the antioxidative action of flavonoids (Hudson and Lewis, 1983). Chelation of metal ions renders them catalytically inactive.

Chalcones, the natural precursors of flavones and flavanones, are readily cyclized under acidic conditions (see Chapter 1) and have been shown to possess potent antioxidant activity. The 3,4-dihydroxychalcones are particularly effective and chalcones are more effective than their corresponding flavanones. Effectiveness of 3,4-dihydroxychalcones, namely, butein and okanin (Figure 1.9 and Table 8.7), depends on the formation of resonance-stabilized free radicals (Dziedzic and Hudson, 1983b) (Figure 8.17).

In the isoflavone, it is clear that both hydroxyl groups in 4′ and 5 positions are needed for significant antioxidant activity as genistein (Figure 8.18). Even 6,7,4′-trihydroxyisoflavone is marginally active when compared to analogous flavone

TABLE 8.7
Antioxidant Activity of Some Flavonoid-Related Compounds

Compound	Time to Reach Peroxide Value of 50 (h)[a]	Induction Period by Rancimat (h)[b]	
Control			
Stripped corn oil	110	A	B
Lard	—	1.3	0.35
Isoflavones			
Daidzein (7,4′-dihydroxy)	—	1.4	—
Genistein (5,7,4′-trihydroxy)	—	2.6	—
Chalcones			
Butein (2′,4′,3,4-tetrahydroxy)	—	94.0	—
Okanin (2′,3′,4′,3,4-pentahydroxy)	—	97.0	—
Phenolic acids			
Protocatechuic acid (3,4-dihydroxybenzoic acid)	—	—	4.8
Gallic acid (3,4,5-trihydroxybenzoic acid)	—	—	28.6
Coumaric acid (*p*-hydroxycinnamic acid)	—	120	0.8
Ferulic acid (4-hydroxy-3-methoxy-cinnamic acid)	—	145	2.0
Caffeic acid (3,4-dihydroxycinnamic acid)	—	495	23.3
Dihydrocaffeic acid (3,4-dihydroxyphenyl-propionic acid)	—	—	31.4
Chlorogenic acid (caffeoyl quinic ester)	—	505	—
Quinic acid	—	105	—
Phenolic ester			
Propyl gallate	—	435	21.8
Miscellaneous			
D-Catechin	—	410	—
Hesperidin methychalcone	—	135	—
Asculetin (6,7-dihydroxycoumarin)	—	—	15.5

Note: A and B denote different batches of oil.

[a] 5×10^{-4} M in stripped corn oil.
[b] 2.3×10^{-4} M in lard.

Source: Adapted from Pratt D.E. and Hudson, B.J.F., 1990, in *Food Antioxidants*, Hudson, B.J.F., Ed., Elsevier Applied Science, London, 171–192.

apigenin, which is inactive as an antioxidant (Figure 1.9). Genistein is particularly active. The resonance-stabilized quinoid structures show that for isoflavone the carbonyl group at position four remains intact and can interact with the 5-hydroxy group, if present; however, in flavone, the carbonyl group at position four loses its functionality. This may explain the superior antioxidant activity of genistein compared with that of apigenin (Dziedzic and Hudson, 1983a).

Naturally occurring isoflavones are found mainly in the Leguminoseae family and are much less widespread than flavones. Pronounced synergism was evidenced

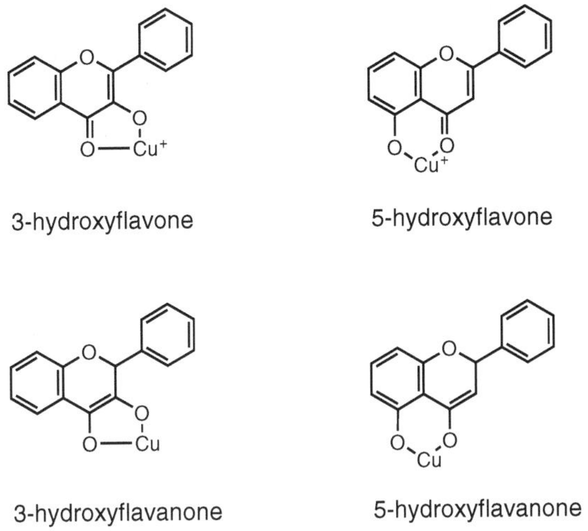

3-hydroxyflavone 5-hydroxyflavone

3-hydroxyflavanone 5-hydroxyflavanone

FIGURE 8.16 Possible copper complexes of flavones and flavonones. (Adapted from Hudson, B.J.F. and Lewis, I.F., 1983, *Food Chem.*, 10, 47–55.)

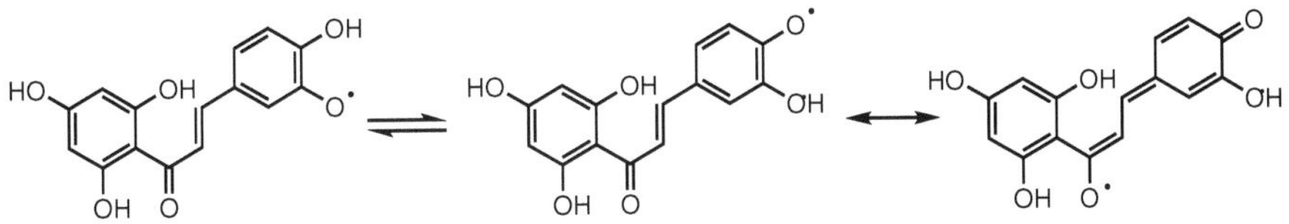

FIGURE 8.17 Resonance-stabilized free radicals of 3′,4′-dihydroxychalcones. (Adapted from Dziedzic, S.Z. and Hudson, B.J.F., 1983, *Food Chem.*, 12, 205–212.)

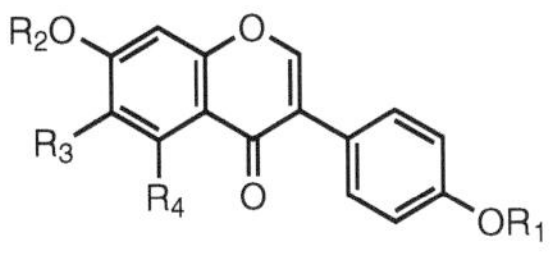

Compound	R_1	R_2	R_3	R_4
Daidzein	H	H	H	H
Formononetin	CH_3	H	H	H
Genestein	H	H	H	OH
Prunetin	H	CH_3	H	OH
4',6',7'-trihydroxyisoflavone	H	H	OH	H
glycitein (7-O-glucoside)	H	Gluc	OCH_3	H

FIGURE 8.18 Chemical structures of isoflavone and isoflavone glycosides of soybean.

when isoflavones were supplemented with phospholipids such as phosphatidyl ethanolamine (Dziedzic and Hudson, 1983a).

The antioxidant activity of phenolic acids and their esters depends on the number of hydroxy groups in the molecule; this will be strengthened by steric hindrance (Dziedzic and Hudson, 1983b). Hydroxylated cinnamic acids have been found to be more effective than their benzoic acid counterparts. It has been mentioned that at least two and, even better, three neighboring phenolic hydroxy groups (catechol or pyrogallol structure) and a carbonyl group in the form of an aromatic ester or lactone or a chalcone, flavonone or flavone are the essential molecular features required to achieve a high level of antioxidant activity (Dziedzic and Hudson, 1983b).

NATURAL SOURCES OF PLANT ANTIOXIDANTS

Although plant-based antioxidants are found in different plant parts, including fruits, seeds and leaves among others, this chapter provides examples of selected oilseeds, teas, several herbs and spices, and wood smoke only .

SOYBEAN

Soybean (*Glycine max* L.) flour has shown antioxidant property in various food products (Table 8.8). In soybean oil, the active antioxidant is tocopherol (mainly α-tocopherol) and, to a lesser extent, δ-tocopherol (Schuler, 1990). Soybean flour and other soybean derivatives are sources of a large variety of antioxidant compounds (Table 8.9) belonging to the family of isoflavone glycosides and their derivatives, phospholipids, tocopherols, amino acids and peptides (Gyorgy et al., 1964; Naim et al., 1973, 1974; Herrmann, 1976; Hayes et al., 1977; Murakani et al., 1984). Isolated isoflavone glycosides from soy flour are genistein, daidzein and glycitein; 7,4′-dihydroxy-6-methylisoflavone (Figure 8.18; Rackis, 1972; Naim et al., 1973).

It has been reported that 99% of the isoflavones are present as glycosides, of which 64% are genistein, 23% daidzein and 13% glycitein 7-*O*-β-glycoside (Naim et al., 1973). Murakani et al. (1984) have shown that, in tempeh (a fermented soybean product), the isoflavones are liberated from glycosides by acids formed during the fermentation. The main antioxidants in tempeh were determined to be daidzein and genistein; glycitein had only a slight antioxidant activity. The reported isoflavone and isoflavone glycosides in soybean meal, soy protein products and soy isoflavones are summarized in Table 8.8 through Table 8.10, respectively. The antioxidant activity of aglycones is superior compared to the parent isoflavone glycoside (Hayes et al., 1977). As discussed earlier, the antioxidant activity is not as pronounced as that of their corresponding flavones possessing dihydroxy substitutes in the A- or the B-ring.

Phenolic acids, namely, syringic, vanillic, caffeic, ferulic, *p*-coumaric and *p*-hydroxybenzoic acids, possessing antioxidative activity have also been found in soybean and soybean products (Arai et al., 1966; Hammerschmidt and Pratt, 1978; Pratt and Birac, 1979). Among polyphenolic antioxidants extracted by methanol or water from dried and fresh soybean, chlorogenic, isochlorogenic, caffeic and ferulic acids were identified in addition to isoflavones (Pratt and Birac, 1979). The antioxidant activity of soy isoflavones in a β-carotene/linoleate model system in the increasing order was glycetin, diadzein, genistein, quercetin and 6,7,4′-trihydroxy isoflavones, respectively. Meanwhile, for the phenolic acids, the increasing order of antioxidant activity was *p*-coumaric, ferulic, chlorogenic and caffeic acids, respectively.

TABLE 8.8
Effective Antioxidant Levels of Soybean in Various Products

Product	Effective level (% of flour)	Comment
Lard	5–10	—
Premier jus	2–6 (unspecified type)	Original condition of fat strongly influenced soy effectiveness
Ghee (buffalo butter)	0.5–1.0 (full-fat)	About equally effective
Frozen pastry	5–20 (full-fat)	All concentrations were about equally effective
Raw pastry mixes and baked pastry	10 (low-fat)	50°C storage
Ration biscuits	4–20 (defatted)	Progressively lower peroxide values with increasing conc.
Dehydrated pork-corn meal scrapple	2.8 (full-fat)	—
Frozen, raw ground pork and frozen, precooked ground pork	2.5–7.5 (full-fat)	No significant difference in peroxide values with conc.
Degermed, uncooked corn meal–soy flour blend	15, 20 (toasted, defatted) or (commercial-process full-fat) or (extrusion-cooked full-fat)	No flavor difference between 15 and 20% products
Degermed, uncooked corn meal-soy flour blend plus ferrous sulfate	15–25 (toasted, defatted)	No rancidity after storage
Instant, fully cooked corn meal–soy flour-milk blend (plus 5% soybean oil)	27.5 (toasted, defatted)	Low peroxide values after storage

Source: Adapted from Pratt D.E. and Hudson, B.J.F., 1990, in *Food Antioxidants*, Hudson, B.J.F., Ed., Elsevier Applied Science, London, 171–192.

Peanut and Cottonseed

The methanolic extracts of protein ingredients of peanut (*Arachis hypogea*, variety: Spanish) and glandless cottonseed (*Gossypium hirsutum*, variety: McNair) have demonstrated antioxidant activity in lipid peroxidation model systems catalyzed by metmyoglobin and Fe^{2+}- EDTA, and also in fresh beef homogenate (Rhee et al., 1979). The same protein ingredients in beef patties (replaced beef up to 10%) were able to retard rancidity development when cooked and refrigerated (Ziprin et al., 1981). Cottonseed protein ingredients have been shown to have a higher activity in retarding development of rancidity than peanut meals.

Pratt and Miller (1984) have identified dihydroquercetin taxifolin as an antioxidant in hot methanolic extracts of peanut (variety: Spanish; Figure 1.9). Quercetin and rutin are identified as the major flavonoids in the methanolic extracts of delinted

TABLE 8.9
Probable Principal Antioxidants Compounds of Soybean Products and Derivatives

Derivative	Antioxidant Component
Aqueous extract	Isoflavone glycosides, their aglycones, phenolic acids
Organic solvent extract	Flavonoids, tocopherols, phospholipids
Soy protein concentrate/isolate	Isoflavone glycosides, their aglycones
Protein hydrolysate	Amino acids and peptides
Textured vegetable proteins	Phospholipids

Sources: Adapted from Pratt, D.E., 1972, *J. Food Sci.*, 37, 322–323; Hammerschmidt, P.A. and Pratt, D.E., 1978, *J. Food Sci.*, 43, 556–559; Hayes, R.E. et al., 1977, *J. Food Sci.*, 42, 1527–1532; Pratt, D.E. and Birac, P., 1979, *J. Food Sci.*, 44, 1720–1722; Pratt, D.E. et al., 1981; Murakani, H. et al., 1984, *Agric. Biol. Chem.*, 48, 2971–2975.

TABLE 8.10
Isoflavone and Isoflavone Glycosides Content of Soybean Meal[a]

Component	Full-Fat Flakes (fat-free basis)	Defatted Flakes
Daidzin	118.5	114.5
Glycitin-7-β-glucoside	0.9	0.8
Genistin	204.1	188.5
Daidzein	2.0	2.5
Glycitein	1.0	1.2
Genistin	4.4	4.4
Total	330.9	311.4

[a] Milligrams per 100 grams.

Source: Adapted from Hudson, B.J.F. and Lewis, J.I., 1983, *J. Food Chem.*, 10, 47–55.

cottonseed (Whiltern et al., 1984). Both possess potent antioxidant activity; however, rutin is comparatively inferior (Yousseff and Rahman, 1985).

The antioxidant activity of phenolics of rapeseed and mustard has been reported in model systems of β-carotene/linoleate and muscle foods (Takagi and Iida, 1980; Ramanathan and Das, 1992; Shahidi et al., 1992). Canola and mustard flour, or their extracts, possess strong antioxidative activity. The effectiveness in retarding meat flavor deterioration of an enzyme-deactivated mustard flour, referred to as deheated mustard flour (DMF), at 1 to 2% level of addition is superior to that of BHT at 200 mg/kg. Extracts of DMF, however, are less effective. Furthermore, the activity of these extracts is proportional to their total content of phenolics (Shahidi et al., 1992).

Borage, Evening Primrose, and Blackcurrant

The low molecular weight phenolics of borage (*Borago officinalis* L.) meal have been recently identified (Wettasinghe et al., 2001). Syringic acid, sinapic acid and rosmarinic acid are important phenolics found in defatted borage meal. Following chromatographic separation on Sephadex LH-20, the crude extracts of borage as well as their fractions possess antioxidant activity by quenching free radicals or by chelation (Wettasinghe and Shahidi, 1999a, 2002).

In evening primrose (*Oenothera biennis* L.), Wettasinghe et al. (2002) identified gallic acid as well as (+)-catechin and (–)-epicatechin in the extracts. In addition, a substantial amount of high molecular weight compounds, including oenothin B, are tentatively identified in the extracts of evening primrose. The latter compound is an oligomeric hydrolyzable tannin first reported by Okuda et al. (1993) as present in green parts of the evening primrose plant. Other polymeric phenolics have also been found in evening primrose extracts (Lu and Foo, 1995). The extracts of evening primrose, or their fractions, are antioxidative in nature, as exhibited by their effective quenching of free radicals as well as chelation of metal ions (Wettasinghe and Shahidi, 1999b, 2002).

Blackcurrant (*Ribes nigrum*) seeds provide a good source of an array of polyphenols with potentially excellent antioxidant activity. The polyphenolics present are dominated by rutinosides and glucosides of delphinidin and cyanidin, as well as 1-cinnamoyl- and 1-*p*-coumaroyl-β-D-glycosides, as shown in Figure 8.19 (Lu and Foo, 2003).

Rice

The methanolic extracts of rice hulls (*Oryza sativa Linn.*) from long life (Katakutara) and short life (Kusabue) varieties exhibit superior antioxidant activity compared with α-tocopherol (Ramarathnam et al., 1989). The HPLC separation of methanol–water (50:50, v/v) extracts of both varieties showed that the short-life variety includes compounds similar to α-tocopherol. The long-life variety had a stronger antioxidant activity and contained compounds other than α-tocopherol. The active compound has been characterized as isovitexin, a C-glycosyl flavonoid (Ramarathnam et al., 1989; Figure 8.20).

It has been observed and established that storability and longevity of rice grains is related to the antioxidative activity of the seed coat or husk (Osawa et al., 1985a; Ramarathnam et al., 1986). Presence of isovitexin has confirmed the antioxidative activity of long-life rice (Ramarathnam et al., 1989).

Sesame Seed

The oil from sesame seed (*Sesamum indicum* L.) has a superior oxidative stability compared to other vegetable oils; this special property has been attributed to the presence of sesamol in the seeds (Lyon, 1972). Sesamol is usually present in traces, but may also be released from sesamolin by hydrogenation, bleaching earth, or other conditions of processing. Sesame seed contains 0.4 to 1.1% sesamin, 0.3 to 0.65%

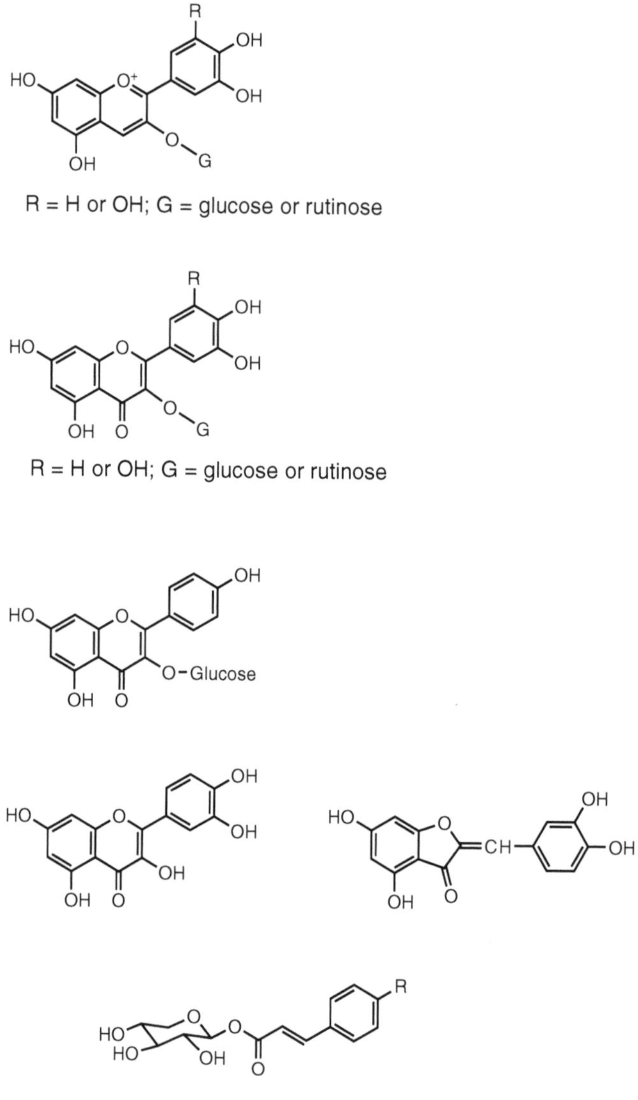

R = H or OH; G = glucose or rutinose

R = H or OH; G = glucose or rutinose

FIGURE 8.19 Chemical structures of polyphenols of blackcurrant seeds.

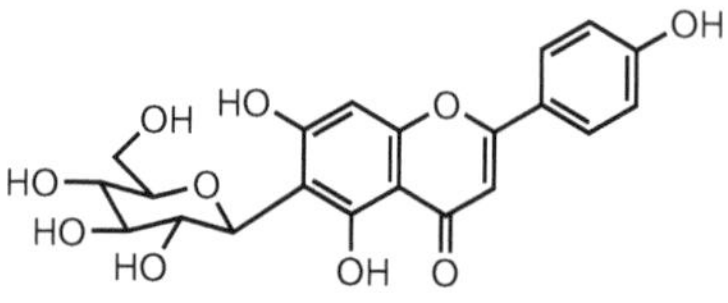

Isovitexin

FIGURE 8.20 Chemical structure of isovitexin.

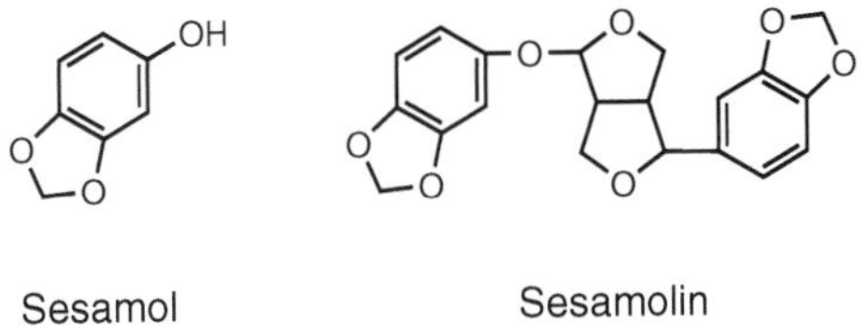

FIGURE 8.21 Chemical structures of sesamol and sesamolin.

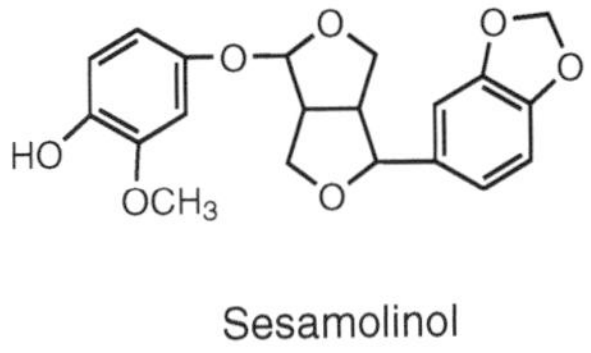

FIGURE 8.22 Chemical structures of sesamolinol.

sesamolin and traces of sesamol (Figure 8.21). Sesamol is a free 3,4-methylenediphenoxy phenol; sesamolin is an acetal type derivative of sesamol and sesamin. The 3,4-methylenediphenoxy phenol is attached to the 2,7-dioxabicyclo-(3,3,0)-octane nucleus directly.

Sesamol is as effective as BHT and BHA and more effective than PG (Lyon, 1972). Fukuda et al. (1985) have reported that acetone extract of sesame seed produces a strong antioxidant activity. The active compounds were identified as bisepoxylignan or sesamolinol (Osawa et al., 1985b), as given in Figure 8.22. Sesamole is readily oxidized to sesamol dimers and then to sesamol dimer semiquinone by treatment with H_2O_2/horseradish peroxidase (Kikugawa et al., 1990).

Canary Seeds

Canary seeds (*Phalaris canariensis*) are a popular bird feed. Takagi and Iida (1980) have shown that extracts (i.e., successive extraction of seeds with hexane, ether and methanol) of canary seeds exhibit excellent antioxidant activity in lard and sardine oil. The active components are identified by gas chromatography to be esters of caffeic acid as well as cycloartenol, gramisterol, sitosterol and campesterol, and lesser quantities of 24-methylenecycloartanol, obtusifoliol, brassicasterol and Δ,7-stigmasterol (Takagi and Iida, 1980). These sterols and triterpene alcohol esters of caffeic acid are lipid soluble and have higher melting points. Lipid solubility and lack of bitter taste may enhance the potential use of this source of natural antioxidant.

Tea

Presence of catechin and its derivatives in tea has been well documented. Matsuzaki and Hara (1985) have reported the antioxidative efficiency of isolated catechins from green tea leaves. The extracts included (–)-epicatechin (EC), (–)-epigallocatechin

TABLE 8.11
Antioxidant Activity of Spices in Oils and Fats[a]

Spice/Antioxidant 30 mg/kg	Soybean	Linseed	Olive	Sesame
Fennel	2.3	2.2	8.0	5.1
Ginger	1.9	2.6	7.8	3.8
Capsicum	2.0	2.6	7.3	5.4
Clove	5.1	6.0	23.8	8.5
Garlic	2.3	2.2	6.6	7.0
Turmeric	2.0	2.1	8.5	3.3
BHA (20 mg/100 g)	1.1	1.1	2.5	1.1
Control	9.5	3.7	16.0	8.7

Note: Antioxidant Index = $\frac{\text{Time for treated sample to reach PV of 100}}{\text{Time for control to reach PV of 10}}$

[a] Active oxygen method (AOM) h.

Sources: Adapted from Gordon, M.H., 1990, in *Food Antioxidants*, Hudson, B.J.F., Ed., Elsevier, London, 1–18 and Gyorgy, P. et al., 1964,. *Nature* (London), 203, 870–872.

(EGC), (–)-epicatechin gallate (ECG), and (–)-epigallocatechin gallate (EGCG) (Figure 1.7). The activity of catechins in model systems is in the following order:

$$\text{EC} < \text{ECG} < \text{EGC} < \text{EGCG}$$

At similar molar concentrations, the activity of these compounds is superior to those of BHA and di-α-tocopherol in lard (Matsuzaki and Hara, 1985; Namiki, 1990).

Herbs and Spices

Chipault et al. (1952, 1956) investigated antioxidant activity of spices in various fats. In general, alcoholic and ether extracts of spices are less active than native spices (Table 8.11 and Table 8.12). Allspice, clove, sage, oregano, rosemary, and thyme possess antioxidant activity in all types of fats examined. Clove appears to be the most active antioxidant in vegetable oils; however, extracts of rosemary and sage are the most effective (Chipault et al., 1952, 1956). Spice extracts have attracted a lot of interest in recent years because they can be easily added to fats and oils in bulk. Many extracts possess a strong odor and bitter taste and thus are of limited use in many food products. Rosemary and sage provide the most potent antioxidant spice extracts with little odor of the original spice.

Chang et al. (1977) were able to prepare an odorless and flavorless natural antioxidant from rosemary (*Rosmarinus officinalis* L.) and sage (*Salvia officinalis* L.). These antioxidants can be successfully extracted into different organic solvents such as benzene, chloroform, diethyl ether and methanol. The diethyl ether extract of rosemary was purified and evaluated for its antioxidant activity (peroxide value) at 0.02% (w/w) in potato chips, sunflower oil and corn oil. It showed a very low

TABLE 8.12
Antioxidative Activity of Ground Spices, Distilled Water Soluble Fraction and Ethanol Soluble Fraction from Spices in Lard[a]

Spice	Ground Spice (0.25%)	Distilled Water Soluble Fraction[b]	Ethanol Soluble Fraction[b]
Control	5.7	5.7, 21.9	5.7, 21.9
Allspice	11.8	9.7, 32.8	8.7, 15.5
Black pepper	7.7	5.8, 28.9	6.4, 24.7
Capsicum	8.0	6.8,31.7	7.0, 21.8
Clove	19.6	13.9, 33.4	18.9, 27.5
Ginger	14.7	6.7, 30.8	11.7, 24.7
Mace	24.0	6.3, 27.1	21.7, 34.6
Nutmeg	21.7	6.1, 27.3	18.5, 21.7
Rosemary	67.0	8.2, 29.7	58.5, 58.1
Sage	54.8	6.9, 26.6	42.1, 42.7
Turmeric	16.8	7.0, 27.7	12.4, 24.5

[a] Active oxygen method (AOM) h.
[b] First entry = 0.25% ground spice equivalent; second entry = soluble fraction + 100 mg/kg of α-tocopherol.

Source: Adapted from Watanabe Y. and Ayano, Y., 1974, *J. Jpn. Soc. Food Nutr.*, 27, 181–188.

peroxide value and provided excellent flavor stability to the products tested (Chang et al., 1977).

The extracts of rosemary leaves contain a phenolic diterpene, namely, carnosol (Houlihan et al., 1984). Furthermore, rosmanol, another phenolic diterpene that has a closely related structure to carnosol, and rosmaridiphenol have also been identified in rosemary leaves (Figure 8.23). Rosmaridiphenol (at 0.02%) has shown antioxidant activity similar to BHT at the same level in prime steam lard. Carnosic acid and rosmaric acid are the most active antioxidant constituents of rosemary and rosmaric acid possesses an activity comparable to that of caffeic acid (Schuler, 1990). In animal fats, carnosic acid has been described as the most active antioxidative constituent of rosemary (Schuler, 1990). Commercial antioxidant extracts (molecular or vacuum distilled) from rosemary are available as a fine powder. Depending on their content of active antioxidants, they are recommended for use at concentrations ranging between 200 and 1000 mg/kg of the processed product. However, pure carnosic acid is not available as a food antioxidant (Schuler, 1990). It has also been observed that ascorbic acid (500 mg/kg of ascorbic acid + 200 mg/kg of rosemary extract) enhances the antioxidant activity of rosemary extract in lard (Chang et al., 1977).

The extracts of dried leaves of oregano (*Origanum vulgare* L.) in dichloromethane and methanol have been isolated as reported for rosemary. The main antioxidative compound of oregano is identified as a phenolic glycoside (Nakatani

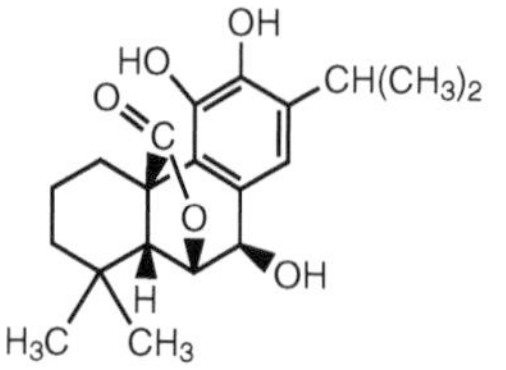

Rosmanol

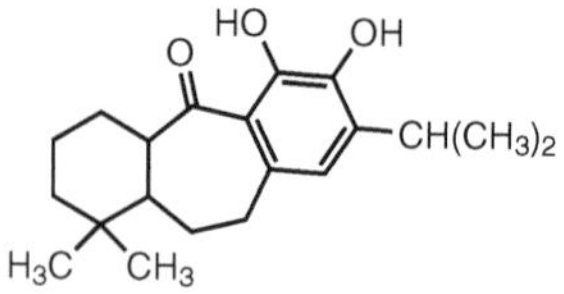

Rosmaridiphenol

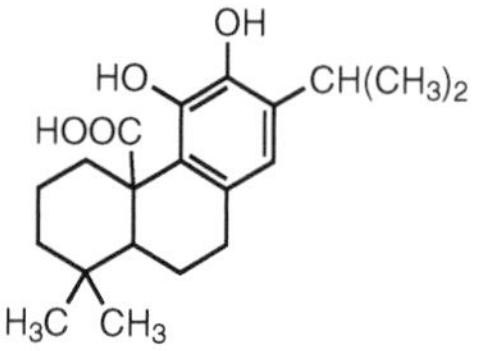

Carnosic acid

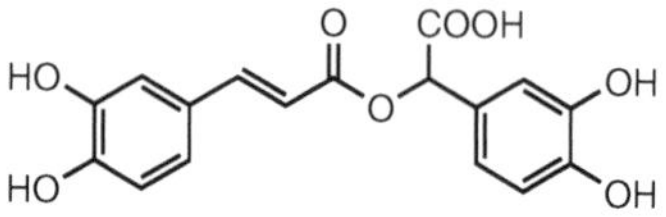

Rosmaric acid

FIGURE 8.23 Chemical structures of selected autioxidant compounds in rosemary.

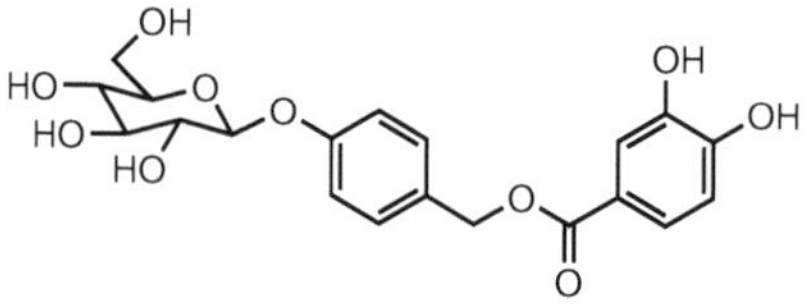

Antioxidative glycoside from Oregano

FIGURE 8.24 Chemical structure of an antioxidative compound in oregano.

and Kikuzaki, 1987). Further investigations have suggested that the compound is 2-caffeoyloxy-3 [2-(4-hydroxybenzoyl)-4,5-dihydroxy]phenyl propionic acid (Kikuzaki and Nakatani, 1989; Figure 8.24).

The strong antioxidant activity of Papua mace (*Myristica argentea*) is related to the presence of 2-allylphenols and a number of lignans (Nakatani and Ikeda, 1984). Capsicin, a pungent antioxidant component of *Capsicum frutescens*, and a new strong antioxidant (Figure 8.25) with no pungency have also been isolated from this spice (Nakatani et al., 1988). Meanwhile, a ferulic acid amide of tyramine (Figure 8.25) and piperine-related compounds with an open methylenedioxy ring found in black pepper (*Riper nigum*) have stronger antioxidant activity than tocopherol. These compounds are fat soluble, odorless and tasteless (Nakatani et al., 1986). In addition, tetrahydrocurcumin, a colorless, heat-resistant, antioxidative compound, has been found in turmeric (*Curama longa* L.) (Figure 8.25; Osawa et al., 1989).

OTHER SOURCES OF ANTIOXIDANTS

Olives and their leaves contain polyphenolic compounds that may be extracted into methanol and ethyl acetate, successively. Fractions of this polyphenolic extract were

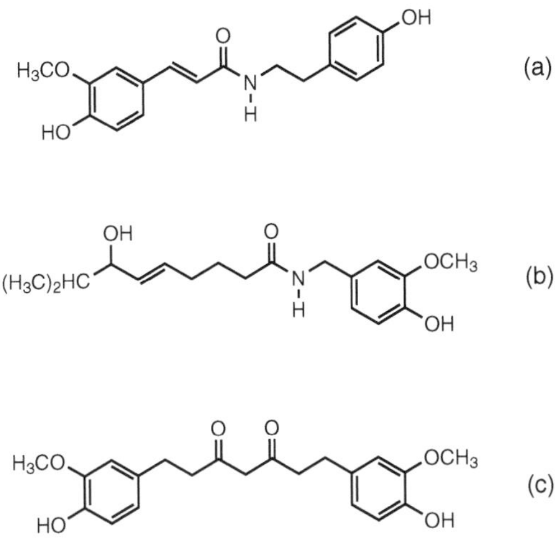

FIGURE 8.25 Chemical structure of antioxidative compounds in (a) black pepper, (b) capsicum and (c) turmeric.

eluted on TLC and the compounds with the highest *o*-diphenol concentration separated and evaluated for their inhibition of oxidative deterioration of vegetable oils. Oxidative deterioration of soybean and olive oils was inhibited when the extracts were used at a concentration of 100 mg/kg. Nergiz and Unal (1991) have reported that virgin olive oil contains high amounts of *p*-coumaric acid (0.73 to 10.37 mg/kg) compared to the syringic, vanillic, and ferulic acids that are also present. The active antioxidant compounds in olive have not been studied in detail.

Varieties of onion (*Allium cepa* L.) with colored skin have exceptionally high flavonol content, mainly 2.5 to 6.5% quercetin (as aglycone) (Herrmann, 1976). Akaranta and Odozi (1986) have shown that acetone extracts of red onion skin (0.3%) reduce the peroxide value of oils as effectively as the commercial antioxidant 1,2-dihydro-2,2,4-trimethyl-quinoline at a 0.1% level of addition. Meanwhile, various phenolic compounds of waste fluid from the starch and ethanol manufacturing process of sweet potato (*Iopomea batatas*) have been determined (Hayes and Kato, 1984). The antioxidative activity of a 70% methanol extract of sweet potato was evaluated in a linoleate aqueous system and its strong antioxidant activity was attributed to the presence of chlorogenic and isochlorogenic acids. Caffeic and 4-*O*-caffeoyl quinic acids were also present. The synergistic effect of these phenolic compounds with amino acids has also been contemplated (Hayes and Kato, 1984).

Oat (*Avena sativa* L.) products have also been suggested as potential food antioxidants. Their antioxidant activity is supposedly based on their high content of dihydrocaffeic acid and phospholipids (Schuler, 1990).

Flavoglaucin, a phenolic compound isolated from the mycelial mat of *Eurotium chevalieri*, has been found to be an excellent antioxidant in vegetable oils at a 0.05% (w/w) level of addition (Figure 8.26; Ishikawa et al., 1984). It stabilized lard when used in combination with α-tocopherol at a 0.04% concentration. However, this fungal antioxidative compound has not shown mutagenic activity to *Salmonella*

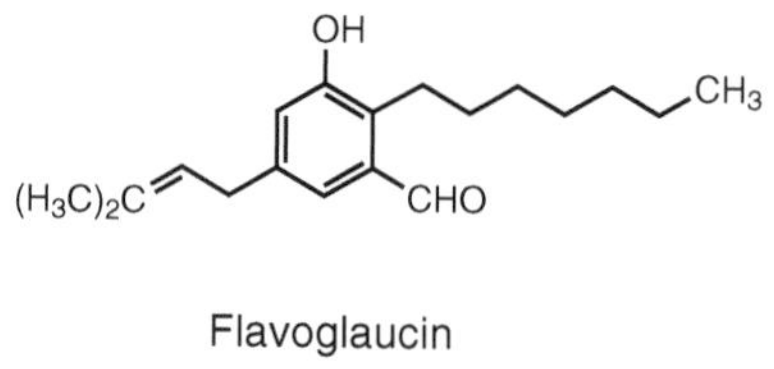

Flavoglaucin

FIGURE 8.26 Chemical structure of flavoglaucin.

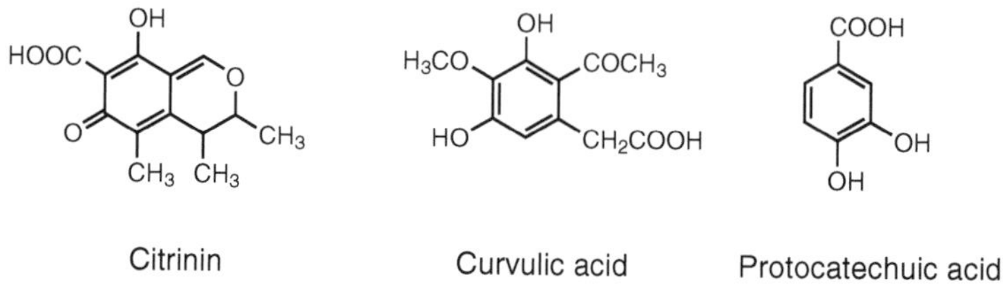

Citrinin Curvulic acid Protocatechuic acid

FIGURE 8.27 Chemical structures of citrinin, curvulic acid and protocatechnic acid.

typhimurium, TA 100 and TA 98, and seems to have limited potential use in foods (Ishikawa et al., 1984). Aoyama et al. (1982) screened 750 strains of filamentous fungi from soil for their antioxidant activity against linoleate. Fourteen strains were positive for the test and the two superior strains were studied for their microbial products. The curvulic acid produced by these two fungi strains possess significant antioxidant activity (Figure 8.27). Furthermore, they were able to find protocatechuic acid and citrinin as other antioxidative components present (Figure 8.28). The antioxidative activity of these compounds was compared with BHA and α-tocopherol. It was concluded that citrinin was as effective as α-tocopherol, while curvulic acid and protocatechuic acid showed an activity between α-tocopherol and BHA. Citrinin is a known mycotoxin and has limited potential in foods; protocatechuic acid is a known antioxidant and curvulic acid has shown low toxicity at 150 mg/kg in rats (Aoyama et al., 1982).

WOOD SMOKE

Wood smoke is a complex system consisting of dispersed and particulate phases. Absorption of vapor (dispersed phase) by the foodstuff results in the characteristic color, flavor and preservative properties of smoked foods. It has been shown that certain levels and types of specific wood smoke have antioxidative properties (Maga, 1988). Wood smoke can retard oxidative rancidity in smoked foods. The neutral fraction of wood smoke, which contains most of the phenols, has been shown to possess the highest antioxidative properties while the basic fraction actually promotes lipid oxidation (Maga, 1988). During pyrolysis of wood, its lignins produce the most important compounds in smoke, namely, phenols and phenol ethers, typified by guaiacol (2-methoxyphenol) and syringol (2,6-dimethoxyphenol) and their homologous derivatives. Soft woods form guaiacols predominantly, and hard woods

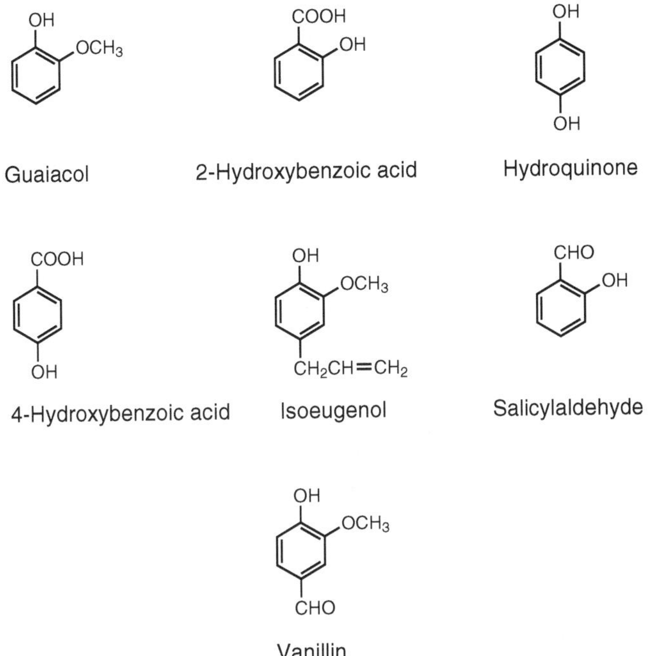

FIGURE 8.28 Selected phenolic compounds of wood smoke.

produce a mixture of guaiacols and syringols (Maga, 1988). It is interesting to note that lignin has antioxidant properties; inclusion of 1 to 10% lignin in the rat diet results in increased deposition of retinal as compared to cellulose-fed controls (Catignani and Carter, 1982). Studies on lignin have shown the presence of phenolic compounds such as ferulic, vanillic, syringic, and *p*-hydroxybenzoic acids (Catignani and Carter, 1982) — compounds that may contribute to the antioxidant properties of lignin.

It has been demonstrated that phenols are the primary antioxidant-related components associated with smoke (Miller and Quackenbush, 1957). Phenols with a high boiling point are the key compounds responsible for antioxidant properties, while low-boiling phenols show weak antioxidant activity (Barylko-Pikielna, 1977; Maga, 1988). The 2,6-dimethoxyphenols exhibit strong antioxidant effect (Kjällstrand and Petersson, 2001). The chemical structures of some antioxidative phenolics of wood smoke are given in Figure 8.28.

REFERENCES

Akaranta, O. and Odozi, T.O. 1986. Antioxidant properties of red onion skin (*Allium cepa*) tannin extracts. *Agric. Wastes*, 18:299–303.

Amarowicz, R., Raab, B. and Shahidi, F. 2003. Antioxidant activity of phenolic fractions of rapeseed. *J. Food Lipids*, 10:51–62.

Aoyama, T., Nakakita, Y., Nakagawa, M., and Sakai, H. 1982. Screening for antioxidants of microbiol origin. *Agric. Biol. Chem.*, 46:2369–2371.

Arai, S., Suzuki, H., Fujimaki, M., and Sukrai, Y. 1966. Studies on flavor components in soybean. II. Phenolic acids in defatted soybean flour. *Agric. Biol. Chem.*, 30:364–369.

Arts, M.J.I.J., Dallinga, J.S., Voss, H-P., Haenen, G.R.M.M., and Bast, A. 2003. A critical appraisal of the use of the antioxidant capacity (TEAC) assay in defining optimal antioxidant structures. *Food Chem.*, 80:409–414.

Barylko-Pikielna, N. 1976. Contribution of smoke components to sensory, bacteriostatic and antioxidative effects in smoke foods, in *Advances in Smoking of Foods*, Rutkowski, A., Ed., Pergamon Press, Oxford, U.K., 1667–1671.

Belitz, H.D. and Grosch, W. 1987. *Food Chemistry*. Springer-Verlag, New York.

Boland, J.L. and ten-Have, P. 1947. Kinetics in the chemistry of rubber and related materials; the inhibitory effect of hydroquinone on the thermal oxidation of ethyl linoleate. *Trans. Faraday Soc.*, 43:201–204.

Buck, D.F. 1984. Food antioxidants — applications and uses in snack foods. *Cereal Foods World*, 29:301–303.

Cao, G., Alessio, H.M., and Cutler, R. 1993. Oxygen radical absorbance capacity assay for antioxidants. *Free Rad. Biol. Med.*, 14:303–311.

Catignani, G.L. and Carter, M.E. 1982. Antioxidant properties of lignin. *J. Food Sci.*, 47:1745–1748.

Chang, S.S., Ostric-Matijasevic, B., Hsieh, O.A.L., and Huang, C.L. 1977. Natural antioxidants from rosemary and sage. *J. Food Sci.*, 42:1102–1106.

Cheng, I., Ren, J., Li, Y., Chang, W., and Chen, Z. 2002. Study on the multiple mechanisms underlying the reaction between hydroxyl radical and phenolic compounds by qualitative structure and activity relationship. *Bioorg. Med. Chem.*, 10:4067–4073.

Chipault, J.R., Mizuno, G.R., Hawkins, J.M., and Lundberg, W.P. 1952. The antioxidant properties of natural spices. *Food Res.*, 17:46–49.

Chipault, J.R., Mizuno, G.R., and Lundberg, W.P. 1956. The antioxidant properties of spices in foods. *Food Technol.*, 10:209–212.

Cillard, J., Cillard, P., and Cormier, M. 1980. Effect of experimental factors on the pro-oxidant behaviour of tocopherol. *J. Am. Oil Chem. Soc.*, 57:255–261.

Coppen, P.P. 1983. Use of antioxidants, in *Rancidity in Foods*, Allen, J.C. and Hamilton, R.J., Eds., Applied Science Publishing Company, London, 67–87.

Dziezak, J.D. 1986. Preservatives: antioxidants. *Food Technol.*, 9:94–102.

Dziedzic, S.Z. and Hudson, B.J.F. 1983a. Hydroxyisoflavones as antioxidants for edible oils. *Food Chem.*, 11:161–166.

Dziedzic, S.Z. and Hudson, B.J.F. 1983b. Polyhydroxy chalcones and flavonones as antioxidants for edible oils. *Food Chem.*, 12:205–212.

Endo, Y., Usuki, R., and Kareda, T. 1985. Antioxidant effects on chlorophyll and pheophytin on the autoxidation of oils in the dark II. *J. Am. Oil Chem. Soc.*, 62:1375–1378.

Fukuda, Y., Osawa, T., Namiki, M., and Ozaki, T. 1985. Studies on antioxidative substances in sesame seeds. *Agric. Biol. Chem.*, 49:301–306.

Gordon, M.H. 1990. The mechanism of antioxidant action *in vitro*, in *Food Antioxidants*, Hudson, B.J.F., Ed., Elsevier, London, 1–18.

Gyorgy, P., Murata, K., and Ikehata, H. 1964. Antioxidants isolated from fermented soybean (tempeh). *Nature* (London), 203:870–872.

Halliwell, B. 1999. Food-derived antioxidants. Evaluating their importance in food and *in vivo*. *Food Sci. Agric. Chem.*, 18:1–29.

Hammerschmidt, P.A. and Pratt, D.E. 1978. Phenolic antioxidants of dried soybeans. *J. Food Sci.*, 43:556–559.

Hayes, F. and Kato, H. 1984. Antioxidative compounds of sweet potatoes. *J. Nutr. Sci. Vitaminol.*, 30:37–46.

Hayes, R.E., Bookwalter, G.N., and Bagley, E.B. 1977. Antioxidant activities of soybean flour and derivatives — a review. *J. Food Sci.*, 42:1527–1532.

Heignen, C.G.M., Haenon, G.R.M.M., Vekemans, J.A.J.M., and Bast, A. 2001. Peroxynitrite scavenging of flavonoids: structure activity relationship. *Environ. Toxicol. Pharmacol.*, 10:199–206.

Heim, K.E., Tagliaferro, A.R., and Bobilya, D.J. 2002. Flavonoid antioxidants: chemistry, metabolism and structure-activity relationship. *J. Nutr. Biochem.*, 13:572–584.

Herrmann, K. 1976. Flavanols and flavones in food plants — a review. *J. Food Technol.*, 11:433–448.

Herting, D.C. and Durry, E.J.E. 1969. ∀-Tocopherol content of cereal grains and processed cereals. *J. Agric. Food Chem.*, 17:785–790.

Houlihan, C.M., Ho, C.T., and Chang, S.S. 1984. Elucidation of the chemical structure of novel antioxidant, rosmaridiphenol isolated from rosemary. *J. Am. Oil Chem. Soc.*, 61:1036–1039.

Hudson, B.J.F. and Lewis, J.I. 1983. Polyhydroxy flavonoid antioxidants for edible oils: structural criteria for activity. *Food Chem.*, 10:47–55.

Ishikawa, Y., Marimoto, K., and Hamasaki, T. 1984. Flavoglaucin, a metabolite of *Eurotium chavalieri*, its oxidation and synergism with tocopherol. *J. Am. Oil Chem. Soc.*, 61:1864–1868.

Ishikawa, Y. and Yuki, E. 1975. Reaction products from various tocopherols with trimethyl oxide and their antioxidative activities. *Agric. Biol. Chem.*, 39:851–857.

Kahl, R. and Hildebrandt, A.G. 1986. Methodology for studying antioxidant activity and mechanisms of action of antioxidants. *Food Chem. Toxicol.*, 24:1007–1014.

Khan, M.R. and Shahidi, R. 2001. Effects of natural and synthetic antioxidants on the oxidative stability of borage and evening primrose triacylglycerols. *Food Chem.*, 75:431–437.

Kikugawa, K., Kunugi, A., and Kurechi, T. 1990. Chemistry and implication of degradation of phenolic antioxidants, in *Food Antioxidants*, Hudson, B.J.F., Ed., Elsevier, London, 65–98.

Kikuzaki, H. and Nakatani, N. 1989. Structure of a new antioxidative phenolic acid from oregano. *Agric. Biol. Chem.*, 53:519–524.

Kjällstrand, J. and Petersson, G. 2001. Phenolic antioxidants in wood smoke. *Sci. Total Environ.*, 277:69–75.

Kochhar, S.P. and Rossell, J.D. 1990. Detection, estimation and evaluation of antioxidants in food systems, in *Food Antioxidants*, Hudson, B.J.F., Ed., Elsevier, London, 19–64.

Lemanska, K., Szymusiak, H., Tyrakowska, B., Zielinski, R., Soffers, A.E.M.F., and Reitjens, I.M.C.M. 2001. The influence of pH on antioxidant properties and the mechanism of antioxidant action of hydroxyflavones. *Free Rad. Biochem.*, 31:572–584.

Löliger, J. 1983. *Natural Antioxidants in Rancidity in Foods*, Allen, J.C. and Hamilton, R.J., Eds., Applied Science, London, 89–107.

Lu, F. and Foo, L.Y. 1995. Phenolic antioxidant components of evening primrose, in *Nutrition, Lipids, Health and Disease*, Ong, A.S.H., Niki, E., and Packer, L., Eds., AOCS Press, Champaign, IL, 86–95.

Lu, F. and Foo, L.Y. 2003. Polyphenolic constituents of blackcurrant seed residues. *Food Chem.*, 80:71–76.

Lyon, C.K. 1972. Sesame, present knowledge of composition and use. *J. Am. Oil Chem. Soc.*, 49:245–249.

Mabrouk, A.F. and Dugan, L.R. 1961. Kinetic investigation into glucose, fructose and sucrose activated oxidation of methyl linoleate emulsion. *J. Am. Oil Chem. Soc.*, 53:572–576.

Maga, J.A. 1988. *Smoke in Food Processing*. CRC Press, Boca Raton, FL, 99–101.

Matsuzaki, T. and Hara, M. 1985. Antioxidative activity of tea leaf catechins. *Nippon Nogeikagatu Kai Shi.*, 59:129–132.

Mehta, A.C. and Seshadri, T.R. 1959. Flavonoids as antioxidants. *J. Sci. Ind. Res.*, 18B, 24.

Miller, G.J. and Quackenbush, F.W. 1957. A comparison of alkylated phenols as antioxidants for lard. *J. Am. Oil Chem. Soc.*, 34:249–252.

Murakani, H. Asakawa, T., Terao, J., and Matsushita, S. 1984. Antioxidative stability of tempeh and liberation of isoflavones by fermentation. *Agric. Biol. Chem.*, 48:2971–2975.

Naim, M., Gestetner, B., Zilkah, S., Birk, Y., and Bondi, A. 1973. A new isoflavone from soybean. *Phytochemistry*, 12:169–170.

Naim, M., Gestetner, B., Zilkah, S., Birk, Y., and Bondi, A. 1974. Soybean isoflavones characterization, determination and detection and antifungal activity. *J. Agric. Food Chem.*, 22:806–810.

Nakatani, N. and Ikeda, K. 1984. Isolation of antioxidative lignan from papua mace. *Nippon Kasei Giakkaishi*, 58:79–83.

Nakatani, N., Inatani, R., Ohta, H., and Nishioka, A. 1986. Chemical constituents of pepper (*Piper* spp.) and application of food preservation for naturally occurring antioxidative compounds. *Eviron. Health. Perspect.*, 67:135–142.

Nakatani, N. and Kikuzaki, H. 1987. A new antioxidative glycoside isolated from oregano. *Agric. Biol. Chem.*, 51:2727–2732.

Nakatani, N., Tachibana, Y., and Kikuzaki, H. 1988. Medical, biochemical and chemical aspects of free radicals, in *Proceedings of the 4th Biennial General Meeting of the Society for Free Radical Research*, Kyoto, 453–456.

Namiki, M. 1990. Antioxidants/antimutagens in food. *CRC Crit. Rev. Food Sci. Nutr.*, 29:273–300.

Nawar, W.W. 1996. Lipids, in *Food Chemistry*, 3rd Ed., Fennema, O.R., Ed., Marcel Dekker Inc., New York, 225–319.

Nergiz, C. and Unal, K. 1991. Determination of phenolic acids in virgin olive oil. *Food Chem.*, 39:237–240.

Okuda, T., Yoshida, T., and Hatano, T. 1993. Classification of oligomeric hydrolyzable tannins and specificity of their occurrence in plants. *Phytochemistry*, 32:507–521.

Osawa, T., Inayoshi, M., Kawakishi, S., Mimura, Y., and Tahara, T. 1989. A new antioxidant, tetrahydrocurcumin. *Nippon Nogeikagku Kaishi*, 63:345–347.

Osawa, T., Ramarathnam, N., Kawakishi, S., Namiki, M., and Tashiro, T. 1985a. Autoxidative defense system in rice hull against damage caused by oxygen radicals. *Agric. Biol. Chem.*, 49:3085–3087.

Osawa, T., Ramarathnam, N., Kawakishi, S., Namiki, M., and Tashiro, T. 1985b. Sesamolinol: a novel antioxidant isolated from sesame seeds. *Agric. Biol. Chem.*, 49:3351–3352.

Porter, W.L. 1980. Recent trends in food applications of antioxidants, in *Autoxidation in Food and Biological Systems*, Simic, M.G. and Karel, M., Eds., Plenum Press, New York, 295–365.

Pratt, D.E. 1965. Lipid antioxidants in plant tissues. *J. Food Sci.*, 30:737–741.

Pratt, D.E. 1972. Water soluble antioxidant activity of soybeans. *J. Food Sci.*, 37:322–323.

Pratt, D.E. and Birac, P. 1979. Source of antioxidant activity of soybean and soy products. *J. Food Sci.*, 44:1720–1722.

Pratt, D.E. and Hudson, B.J.F. 1990. Natural antioxidants not exploited commercially, in *Food Antioxidants*, Hudson, B.J.F., Ed., Elsevier Applied Science, London, 171–192.

Pratt, D.E. and Miller, E.E. 1984. A flavonoid antioxidant in Spanish peanuts. *J. Am. Oil Chem. Soc.*, 61:1065–1067.

Pratt, D.E. and Watts, B.M. 1964. The antioxidant activity of vegetable extracts. I. Flavone aglycones. *J. Food Sci.*, 29:27–30.

Pratt, D.E., Pietor, C.D., Porter, W.L., and Giffee, J.W. 1981. Phenolic antioxidants of soy protein hydrolyzate. *J. Food Sci.*, 47:24–25, 35.

Rackis, J.J. 1972. Biologically active compounds, in *Soybean Chemistry and Technology*, Smith, A.K. and Gircler, S.J., Eds., AVI Publishing Company, West Port, CT, 158–202.

Ramanathan, L. and Das, N.P. 1992. Studies on the control of lipid oxidation in ground fish by some polyphenolic natural products. *J. Agric. Food Chem.*, 40:17–21.

Ramarathnam, N., Osawa, T., Namiki, M., and Tashiron, T. 1986. Studies on the relationship between antioxidantive activity of rice hull and germination ability of rice seeds. *J. Sci. Food Agric.*, 37:719–726.

Ramarathnam, N., Osawa, T., Namiki, M., and Kawakishi, S. 1989. Chemical studies of novel rice antioxidants. 2. Identification of isovitexin; a C-glycosyl flavonoid. *J. Agric. Food Chem.*, 37:316–319.

Rhee, K.S., Ziprin, Y.A., and Rhee, K.C. 1979. Water soluble antioxidant activity of oilseed protein derivatives in model lipid peroxidation system of meat. *J. Food Sci.*, 44:1132–1135.

Rice-Evans, C.A., Miller, N.J., Bolwell, P.G., Bramley, P.M., and Pridham, J.B. 1995. The relative activities of plant-derived polyphenolic flavonoids. *Free Rad. Res.*, 22:375–383.

Rice-Evans, C.A., Miller, N.J., and Paganga, G. 1996. Structure–antioxidant activity relationship of flavonoids and phenolic acids. *Free Rad. Biol. Med.*, 20:933–956.

Rosenwald, R.H. and Chenicek, J.A. 1951. Alkyl hydroxyanisoles as antioxidants. *J. Am. Oil Chem. Soc.*, 28:185–187.

Schuler, P. 1990. Natural antioxidants exploited commercially, in *Food Antioxidants*, Hudson, B.J.F., Ed., Elsevier, London, 99–170.

Shahidi, F. 1997. *Natural Antioxidants: Chemistry, Health Effects and Applications*. AOCS Press, Champaign, IL.

Shahidi, F. 1998. Assessment of lipid oxidation and off-flavor development in meat, meat products and seafoods, in *Flavor of Meat, Meat Products and Seafoods*. 2nd Ed., Shahidi, F., Ed., Blackie Academic & Professional, London, 373–394.

Shahidi, F. 2000. Antioxidants in food and food antioxidants. *Nahrung*, 44:158–163.

Shahidi, F. 2002. Antioxidants in plants and oleaginous seeds, in *Free Radicals in Food: Chemistry, Nutrition and Health Effects*. Morello, M.J., Shahidi, F., and Ho, C-T., Eds., ACS Symposium Series 807. American Chemical Society, Washington, D.C., 162–175.

Shahidi, F. and Wanasundara, P.K.J.P.D. 1992. Phenolic antioxidants. *CRC Crit. Rev. Food Sci. Nutr.*, 32:67–103.

Shahidi, F., Wanasundara, P.K.J.P.D., and Hong, C. 1992. Antioxidant activity of phenolic compounds in meat model systems, in *Phenolic Compounds in Food and Their Effect on Health, I.* Ho, C.-T., Lee, C.Y., and Huang, M.-T., Eds., ACS Symposium Series 506, American Chemical Society, Washington, D.C., 214–222.

Sherwin, E.R. 1978. Oxidation and antioxidants in fat and oil processing. *J. Am. Oil Chem.* Soc., 55:809–814.

Stuckey, B.N. 1972. Antioxidants as food stabilizers, in *CRC Handbook of Food Additives*, Furia, T.E., Ed., The Chemical Rubber Co., Ohio, 209–245.

Takagi, T. and Iida, T. 1980. Antioxidant for fats and oils from canary seeds; sterol and triterpin alcohol esters of caffeic acid. *J. Am. Oil Chem. Soc.*, 57:326–330.

Tappel, A.L. 1972. Vitamin E and free radical peroxidation of lipids. *Ann. New York Acad. Sci.*, 203:12–19.

Uri, N. 1961. Mechanism of antioxidation, in *Autoxidation and Antioxidants*, Lundbert, W.O., Ed., Interscience Publishers, New York, 133–169.

van de Voort, F.R., Ismail, A.A., Sedman, J., and Emo, G. 1994. Monitoring the oxidation of edible oils by Fourier transform infrared spectroscopy. *J. Am. Oil Chem. Soc.*, 71:243–253.

Wanasundara, U.N. and Shahidi, F. 1993. Application of NMR spectroscopy to assess oxidative stability of canola and soybean oils. *J. Food Lipids*, 1:15–26.

Watanabe, Y. and Ayano, Y. 1974. The antioxidative activities of distilled water and ethanol soluble fraction from ground spice. *J. Jpn. Soc. Food Nutr.*, 27:181–188.

Wettasinghe, M. and Shahidi, F. 1999a. Antioxidant and free radical-scavenging properties of ethanolic extract of defatted borage (*Borago offinalis* L.) seeds. *Food Chem.*, 67:399–414.

Wettasinghe, M. and Shahidi, F. 1999b. Evening primrose meal: a source of natural antioxidants and scavengers of hydrogen peroxide and oxygen-derived free radicals. *J. Agric. Food Chem.*, 47:1801–1812.

Wettasinghe, M. and Shahidi, F. 2000. Scavenging of reactive-oxygen species and DPPH free radicals by extracts of borage and evening primrose meals. *Food Chem.*, 70:17–26.

Wettasinghe, M. and Shahidi, F. 2002. Iron (II) chelation activity of extracts of borage and evening primrose meals. *Food Res. Int.*, 35:65–71.

Wettasinghe, M., Shahidi, F., Amarowicz, R., and Abou-Zaid, M. 2001. Phenolic acids in defatted seeds of borage (*Borago officinalis* L.). *Food Chem.*, 75:49–56.

Wettasinghe, M., Amarowicz, R., and Shahidi, F. 2002. Identification and quantitation of low-molecular-weight phenolic antioxidants in seeds of evening primrose (*Oenothera biennis* L.). *J. Agric. Food Chem.,* 50:1267–1271.

Whiltern, C.C., Miller, E.E., and Pratt, D.E. 1984. Cotton seed flavonoid as lipid antioxidants. *J. Am. Oil Chem. Soc.*, 61:1075–1078.

Youssef, A.H. and Rahman, A. 1985. Flavonoids of cotton seed and peanut as antioxidants. *La Rivista Italiana Dele Sostanze Grasse*, LXII, 147–149.

Yu, L., Haley, S., Perret, J., Harris, M., Wilson, J., and Qian, M. 2002. Free radical scavenging properties of wheat extracts. *J. Agric. Food Chem.*, 50:1619–1624.

Ziprin, Y.A., Rhee, K.S., Carpenter, Z.L., Hostetler, R.L., Terrel, R.N., and Rhee, K.C. 1981. Glandless cotton seed, peanut and soyprotein ingredients in ground beef patties, effect on rancidity and other quality factors. *J. Food Sci.*, 46:58–61.

9 Contribution of Phenolic Compounds to Flavor and Color Characteristics of Foods

CONTRIBUTION TO FLAVOR

In general, phenolic compounds contribute to the aroma and taste of numerous food products of plant and animal origin (Crouzet et al., 1997; Teranishi et al., 1989). According to Arai et al. (1966), the objectionable flavor of some oilseeds is due to the presence of phenolics that possess sour, bitter, astringent and/or phenolic-like flavor characteristics. Huang and Zayas (1991) found that the presence of ferulic and coumaric acids together may allow prediction of the intensity of sourness of corn germ protein flour products (Table 9.1). On the other hand, a number of alkylphenols contribute to the sheep/mutton flavor of ovine fat and aged cheese (Ha and Lindsay, 1991a, b). A mixture of pure alkylphenols in concentrations equivalent to those found in ram depot fat results in a sheep-like aroma (Ha and Lindsay, 1990).

The detection thresholds and odor description of some volatile phenols are given in Table 9.2 and Table 9.3, respectively. Taste thresholds for some individual phenolic acids present in foods ranges from 30 mg/kg (protocatechuic acid) to 240 mg/kg (syringic acid). The taste threshold for sinapic acid has not been determined due to its insolubility in water at concentrations required for testing. Combining phenolic acids results in much more sensitive detection thresholds than in the individual acids (Maga and Lorenz, 1973). On the other hand, the recognition threshold of the acid taste of *p*-coumaric as bitter, astringent and unpleasant was perceived at 48 mg/kg. Ferulic acid can be perceived as having a sour taste at 90 mg/kg. The combination of *p*-coumaric and ferulic acids results in a sour–bitter taste at 20 mg/kg (Huang and Zayas, 1991). The lower taste thresholds for combined phenolics are due to a synergistic effect. Studies, therefore, have indicated that free phenolic acids contribute to the taste of foods.

Sinapine is a bitter phenolic compound present in rapeseed meals at quite high concentrations (0.39 to 2.85%) (Naczk et al., 1998); therefore, it also contributes to the reported unpleasant and bitter flavor of glucosinolate-free rapeseed flour (Sosulski et al., 1977). Moreover, it may have adverse effects on the palatability of rapeseed products (Appelqvist, 1972). Sinapine is also linked to a crabby or fishy taint noted in eggs from some brown-egg laying hens (Butler et al., 1982; Fenwick et al., 1984; Hobson-Frohock et al., 1973) and is responsible for production of

TABLE 9.1
Regression Models of Free Phenolic Acids for Sensory Attributes in Tested Corn Germ Protein Products

Equation of Model	R^2
S.I. = 1.18 + 0.083 (Ferulic) – 0.019 (*o*-Coumaric)	0.989
B.I. = 0.326 – 1.638 (Syringic) + 0.353 (*p*-Coumaric)	0.514
A.I. = 0.428 + 0.069 (Vanillic) + 0.112 (*p*-Coumaric)	0.750

Note: S.I., sourness intensity; B.I., bitterness intensity; A.I., astringency intensity.

Source: Adapted from Huang, C.A. and Zayas, J.F., 1991, *J. Food Sci.*, 56:1308–1313; 1315.

TABLE 9.2
Effect of Phenolic Acid Combinations on Flavor Thresholds[a]

Compounds	Individual Thresholds	Combination Thresholds
Salicylic + *p*-hydroxybenzoic	(90); (40)	35
Salicylic + *p*-hydroxybenzoic + gentisic	(90); (40); (90)	40
Vanillic + *p*-hydroxybenzoic	(30); (40)	10
Vanillic + syringic	(30); (240)	90
Ferulic + *p*-coumaric	(90); (40)	25
Ferulic + gentisic	(90); (90)	80
Ferulic + gentisic + caffeic	(90); (90); (90)	60
Ferulic + gentisic + caffeic + syringic	(90); (90); (90); (240)	95

[a] Milligrams per kilogram.

Source: Adapted from Maga, J.A. and Lorenz, K., 1973, *Cereal Sci. Today*, 18:326–329.

trimethylamine (TMA) (Pearson et al., 1980) at low concentrations of about 1 mg/kg (Hobson-Frohock et al., 1973).

Singlenton and Nobel (1976) related the presence of chlorogenic acid, other hydroxycinnamates and, in particular, oligomeric proanthocyanidins to the bitterness and astringency of wine and cider. Similarly, Dadic and Bellau (1973) associated bitterness of beer to the presence of phenolics. They found bitterness threshold values of chlorogenic acid to be 20 mg/L in beer and 10 mg/L in 5% ethanol. On the other hand, Nagel et al. (1987) reported that chlorogenic acid even at 100 mg/L in aqueous 2% potassium acid tartrate solution is not bitter when its acid character is masked. It has also been postulated that hydroxycinnamic acids and their derivatives are responsible for the sour–bitter taste of cranberries (Marwan and Nagel, 1982).

Phenolic compounds are responsible for the bitterness and astringency of potatoes. The cortex tissue of potatoes cooked with peel is more bitter than that of potatoes cooked without the peel (Mondy and Gosselin, 1988). Mondy et al. (1971)

TABLE 9.3
Odor Description of Some Volatile Phenolic Compounds

Compounds	Aroma Description
p-, *m*-, and/or *o*-Thiocresol	Sulfurous, stench, burnt rubber, medicinal, meaty, old fruit, cooked meat, brothy
Phenol	Phenolic, medicinal, smoky
2,6-Diisopropylphenol	Phenolic, medicinal, chemical
o-Cresol	Phenolic, indol-like, medicinal
p-Cresol	Phenolic, wood preservative-like smoky, very animal-like, sheep wool-like
m-Cresol	Phenolic, wood preservative-like, smoky
3,4-Dimethylphenol	Phenolic, medicinal, animal-like
2-Isopropylphenol	Must, medicinal, wood cedar-like, licorice-like
Thymol	Phenolic, medicinal, musty
Carvacrol	Phenolic, medicinal
3-and/or 4-Isopropylphenol	Phenolic, medicinal

Source: Adapted from Ha, J.K. and Lindsay, R.C., 1991b, *J. Food Sci.*, 56:1197–1202.

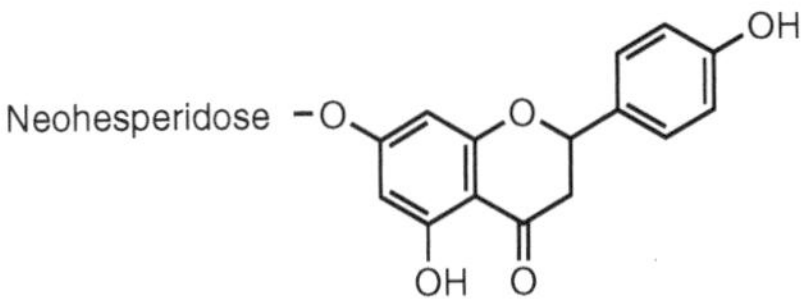

Neohesperidose –O ... OH ... OH O

(1)

Neohesperidose –O ... OH ... OCH_3 ... OH O

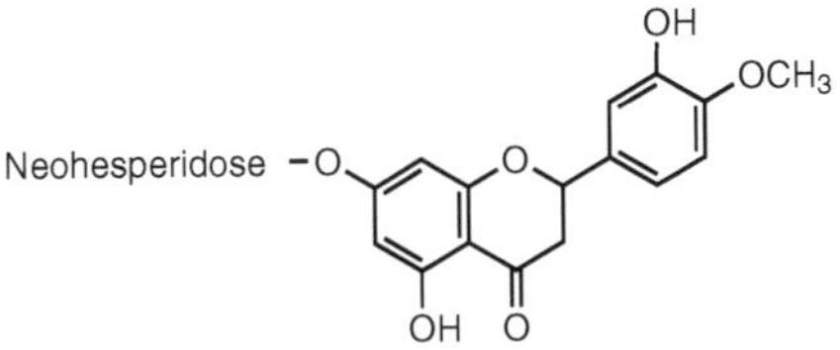

(2)

FIGURE 9.1 Structures of naringin (1) and neohesperidin (2).

found a high positive correlation between phenol content and bitterness (+0.72) and astringency (+0.82) of potatoes. Sinden et al. (1976) reported similar results. Increased bitterness of potatoes with peel may be due to the migration of phenolics from peel into the cortex tissue.

Citrus fruits contain a number of flavanone glycosides that contribute to their bitter taste. Naringin (naringenin 7-neohesperidoside) is the predominant bitter flavanone found in grapefruits; naringin and neohesperidin (hesperetin 7-neohesperidoside) are dominant in sour oranges (Figure 9.1) (Rouseff et al., 1987). The sugar moiety of these glycosides is linked to the 7-hydroxy group of aglycone and is a disaccharide neohesperidose composed of L-rhamnose linked (1→2) to D-glucose. The bitterness of flavanone glycosides depends on the structure of sugar moiety because the point of attachment of rhamnose to glucose affects the taste. Thus, a 1→2 linkage between L-rhamnose and D-glucose is associated with bitterness. The bitter taste of pure naringin can be detected at 10^{-4} to 10^{-5} M in aqueus solutions (Horowitz and Gentilli, 1969); however, narirutin (naringenin 7-rutinoside) with a 1→6 linkage between L-rhamnose and D-glucose is not bitter. Free ferulic acid is a precursor of the objectionable flavor in stored orange juice caused by the presence of *p*-vinylguaiacol produced by decarboxylation of ferulic acid (Naim et al., 1988). This phenol is also present in processed alcoholic beverages, cooked beef, and dried mushroom (Maga, 1978); however, its presence in popcorn is considered desirable (Walradt et al., 1970).

Phyllodulcin, a 3,4-dihydroisocoumarin compound, is reported to be 200 to 300 times sweeter than sucrose (Figure 9.2). This sweet-tasting phenol was first isolated in 1916 by Asahina and Ueno from the *Hydrangea Macrophylla Seringe* and structurally identified by Arkawa and Nakazari in 1959 (Crosby, 1976). In comparison to sucrose, the phyllodulcin gives a delayed onset of sweetness and a lingering licrorice-like aftertaste. The bitter flavonones in citrus, naringin and neohesperidin can be hydrolyzed to chalcones and then converted by simple hydrogenation into the corresponding dihydrochalcones; chalcones and dihydrochalcones are sweet (Horwitz and Gentili, 1969 and 1974). Some B-ring-substituted neohesperidin dihydrochalcones are reported to possess up to 2000 times the sweetness of sucrose (Figure 9.3) (Inglett, 1969). The

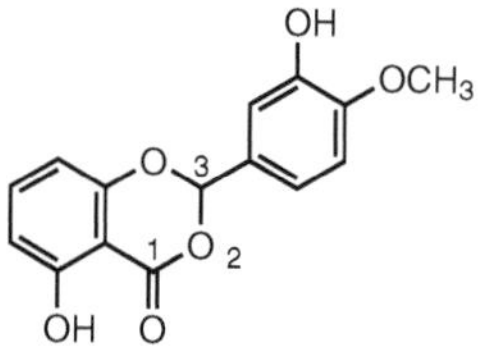

FIGURE 9.2 Structure of phyllodulcin.

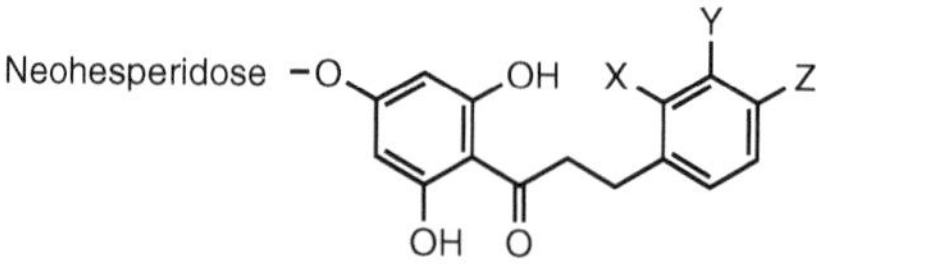

Sweetness (Sucrose=1)	X	Y	Z
0	H	H	H
0	OH	H	H
110	H	OH	H
110	H	H	OH
0	H	OH	OH
0	H	OCH_3	OH
950	H	OH	OCH_3
1100	H	OH	OC_2H_5
2000	H	OH	OC_3H_7

FIGURE 9.3 Sweetness of neohesperidin dihydrochalcone modifies with various substituents in ring B. (Adapted from Inglett, G.E., 1969, *J. Food Sci.*, 34:101–103.)

glycoside portion of dihydrochalcone is unnecessary for sweetness, but its removal lowers the water solubility of the resultant aglycone. On the other hand, replacement of the glycoside portion by carboxymethyl or carboxypropyl groups produces dihydrochalcone derivatives sweeter than saccharin and cyclamate. However, all dihydrochalcone sweeteners are notably slow in onset of sweetness and have a considerable lingering effect that may render them suitable for many food uses (DuBois et al., 1977).

Phenolic substances contribute to the flavor of vanilla pod and vanilla extracts. Vanillin, *p*-hydroxybenzaldehyde and *p*-hydroxybenzyl methyl ester are the most abundant volatiles present. Simple phenolics such as *p*-cresol, eugenol, guaiacol, *p*-vinylguaiacol and *p*-vinylphenol, as well as aromatic acids such as vanillic and salicylic acids, are also identified as minor components (Purseglove et al., 1981).

Ripe banana contains volatile phenolics such as eugenol, methyleugenol, elimicin, and vanillin. Strawberry volatiles contain esters of some phenolic acids such as ethylsalicylic, methylcinnamic and ethylbenzoic acids (Nurstern, 1970); thymol, eugenol and carvacrol are found in plum mirabelle (Walker, 1975a). Thymol is also a major contributor to the flavor of essential oils from tangerine and mandarin orange (Wilson and Shaw, 1981). Moreover, several coumarins such as citral, nooketone and citropten are found in the oil of lime and thymol in the peel oil of mandarin orange. A number of synthetic fruits and chocolate flavorings contain vanillin and ethylvanillin (Table 9.4).

Phenolic substances may be responsible for the flavor of a number of spices and herbs. Anethole, estragole, eugenol, thymol and carvacrol are common constituents of many herbs and spices, thus contributing to their overall sensory characteristics.

TABLE 9.4
Composition of Some Synthetic Flavoring Formulations

Flavoring	Component	Amount (mg)
Raspberry	Vanillin	20
	Ethylvanillin	8
	α-Ionone	1
	Maltol	30
	1-(*p*-hydroxyphenol)-3-butanone	100
	Dimethyl sulfide	1
	2,5-Dimethyl-N-(2-pyrazinyl) pyrrole	1
Chocolate	Dimethyltrisulphide	1
	2,6-Dimethylpyrazine	3324
	Ethylvanillin	143
	Isovaleraldehyde	100

Note: Carriers and solvents are not included.

Source: Adapted from Coultate, T.P., 1984, *Food — the Chemistry of its Components*, The Royal Society of Chemistry, London, using U.S. Patents 3886289 and 3619210.

Eugenol is one of the main flavor compounds found in essential oils obtained by distillation of bark or leaf oil from cinnamon. Presence of thymol is reported in the headspace volatiles of living and picked thyme plant, *Thymus vulgaris*, as well as in the commercial oil produced from this herb. Thymol constitutes 15.7, 9.0, and 39.0% of the headspace volatiles found in the living plant, picked plant and commercial oil of thyme, respectively (Mookherjee et al., 1989; Russell and Olson, 1972). Wyllie et al. (1990) identified thymol and carvacrol in the volatile components of the fruit of *Pistacia Lentiscus*, a fruit used to aromatize olives, vinegar, and salad dressings.

Some vegetables and spices contain derivatives of phenolics that cause a characteristic hot, sharp and stinging sensation in the mouth referred to as pungency. The pungent compounds found in *Capsicum* species such as red and green chilies are known as capsaicinoids. These substances are vanillylamides of monocarboxylic acids with varying degrees of unsaturation and chain length (Figure 9.4) and are ascribed mainly to alkaloids. The pungent characteristic of fresh ginger, as well as ginger oleoresin, is brought about by phenylalkyl ketones that are derivatives of vanillin. This group of compounds is known as gingerols. The predominant gingerol homologues are those with n = 4 (Figure 9.5); Govindarajan (1979) reported that (6)-gingerol is the most pungent active member of these compounds. Processing of ginger, such as drying, can result in the loss of a water molecule from gingerols, producing shogoals that have a greater pungency than the gingerols (Figure 9.6). The essential pungent substance found in cloves is eugenol.

Millin et al. (1969) examined the effect of nonvolatile compounds of black tea on its flavor. They studied the taste of pure compounds as well as the nonvolatile fractions isolated from tea. Based on this study, they concluded that monomeric phenolics (flavanols, flavonols, theogallin, chlorogenic acid, *p*-coumarylquinic acid, and caffeic acid) do not contribute significantly to the taste of tea.

A number of polyphenols are found in the methanolic extracts of cocoa beans that produce chocolate aroma. Rohan and Connell (1964) identified two flavonols (quercetin and quercetrin) and three phenolic acids (*p*-coumaric, caffeic and chlorogenic) in cocoa beans. Stewart (1970) studied the development of chocolate aroma in model systems and found that phenolics enhanced the production of chocolate aroma. Phenolics may also contribute to the quality defects found in some cocoa beans and chocolate liquors, thus imparting a smoke- or ham-like flavor note. This type of flavor defect has been associated with the method used to dry beans (Lehrian et al., 1978).

Phenolic compounds are known to contribute to the flavor of fermented soy sauce. Of these, *p*-ethylguaiacol and *p*-ethylphenol are considered the most important flavor components of soy sauce; a difference of 0.5% in *p*-ethylguaiacol content is easily detectable. Compounds *p*-ethylguaiacol and *p*-ethylphenol are products of metabolism of ferulic and *p*-hydroxycinnamic acids during the fermentation process (Yokotsuka, 1986).

Nine phenolics have been identified in the volatile constituents of yellow passion fruit juice: 2-, 3-, and 4-methylphenols, 4-ethyl and 4-allylphenol, and 2,4-dimethylphenol, 3,4,5-trimethylphenol, 2-methoxyphenol and 4-allyl-2-methoxyphenol. These phenolics are found in quantities below 0.01 mg/kg (Casimir et al., 1981).

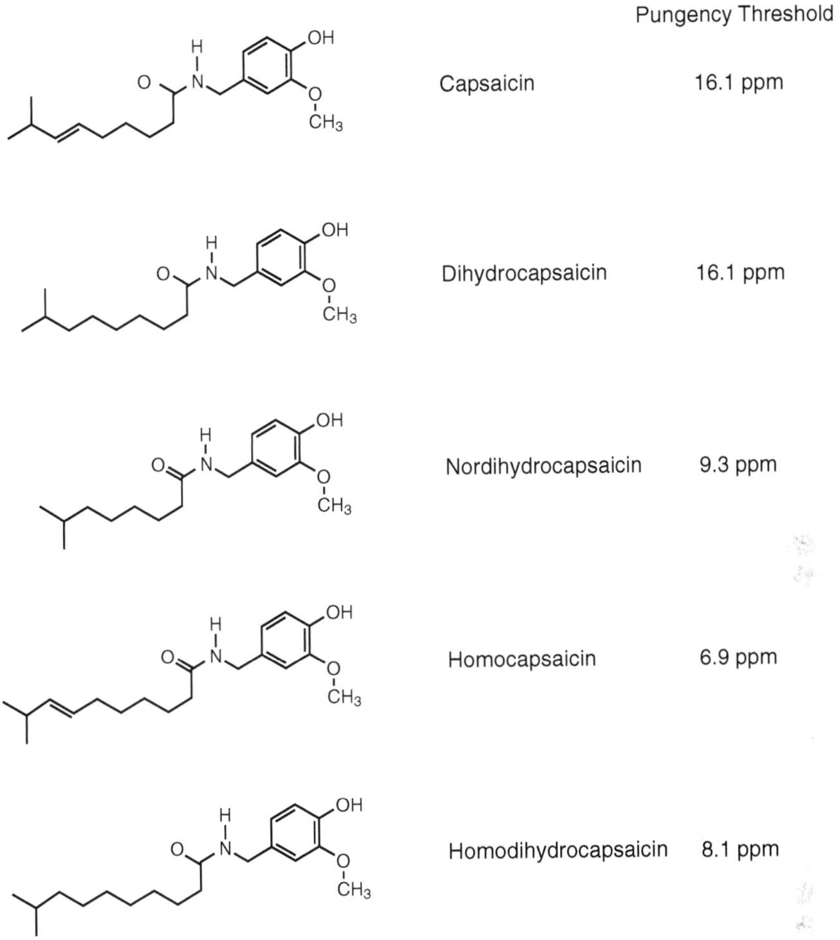

	Pungency Threshold
Capsaicin	16.1 ppm
Dihydrocapsaicin	16.1 ppm
Nordihydrocapsaicin	9.3 ppm
Homocapsaicin	6.9 ppm
Homodihydrocapsaicin	8.1 ppm

FIGURE 9.4 Structure of capsaicinoids and their pungency thresholds. (Adapted from Wong, D.W.S., 1989, in *Mechanism and Theory in Food Chemistry,* AVI Publishing Co., Westport, CT, 147–189.)

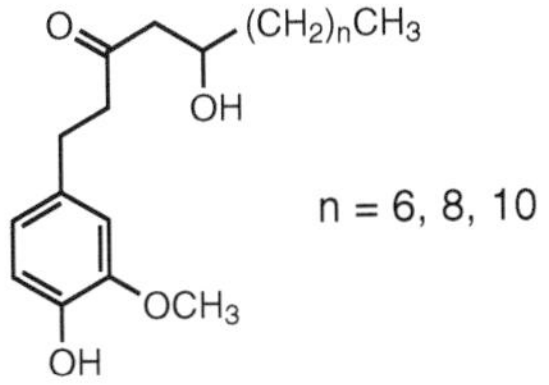

$(CH_2)_nCH_3$

n = 6, 8, 10

FIGURE 9.5 Chemical structures of gingerols.

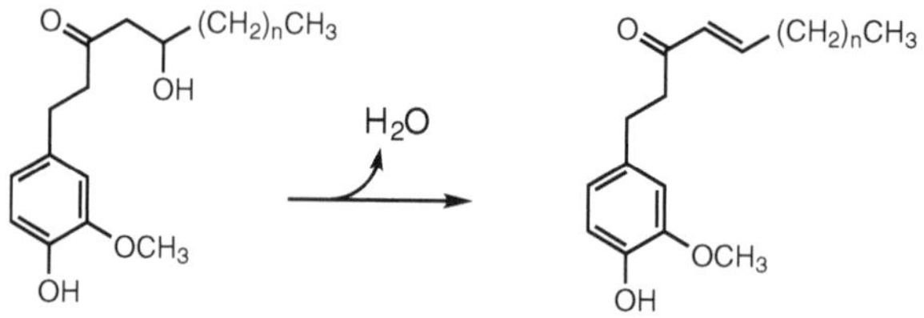

FIGURE 9.6 Reaction of gingerol dehydration. (Adapted from Wong, D.W.S., 1989, in *Mechanism and Theory in Food Chemistry*, AVI Publishing Co., Westport, CT, 147–189.)

TABLE 9.5
Content of Volatile Phenolics in Some Animal Fats[a]

Phenols	Caprine	Ovine		
	Male Goat	Ram	Lamb	Beef
Thiophenol	5	1766	8	—
Phenol	3	7	0.6	109
o-Cresol	1	901	5	14
2-Ethylphenol	0.6	200	26	—
p-Cresol	5	173	33	25
m-Cresol	0.5	5	2	68
2-Isopropylphenol	—	1493	156	—
3,4-Dimethylphenol	3	65	16	—
3- and(or) 4-Ethylphenol	—	—	—	2
Thymol	—	1139	170	—
Carvacrol	0.4	480	36	—
3- and(or) 4-Isopropylphenol	—	1065	70	—
3,5-Diisopropylphenol	—	26	0.7	—
2,6-Diisopropylphenol	—	—	—	0.3

[a] Parts per billion (ppb).

Source: Adapted from Ha, J.K. and Lindsay, R.C., 1991b, *J. Food Sci.*, 56:1197–1202.

Alkylphenols such as *p*-cresol, 2-isopropylphenol, 3,4-dimethylphenol, thymol, carvacrol, 3-isopropylphenol and 4-isopropylphenol contribute distinguishing phenolic-sheep odors to ovine fat. On the other hand, *p*-cresol and 3- and/or 4-ethylphenol combined with 3-methylbutanoic acid are possibly responsible for a pork-like flavor of swine fat. Presence of cresols and, particularly, *m*-cresol may have an influence on the beefy flavor note of fats (Ha and Lindsay, 1991b). The content of volatile phenolic compounds in animal fats is summarized in Table 9.5. Presence of phenolics in animal muscles may be derived from plant phenolics ingested from feeds, as well as from phenolics derived from microbial intestinal fermentation of tyrosine (Brot et al., 1965; Enei et al., 1972). A schematic representation for the formation of *p*-cresol, *p*-ethylphenol, and phenol from tyrosine by microbial metabolism is shown in Figure 9.7.

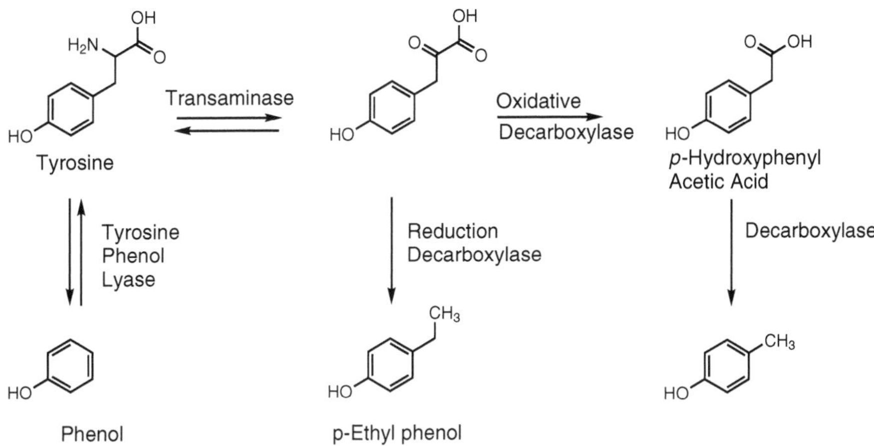

FIGURE 9.7 Schematic for the formation of *p*-cresol, 4-ethylphenol and phenol from tyrosine by microbial metabolism. (Adapted from Ha, J.K. and Lindsay, R.C., 1991a, *J. Food Sci.*, 56:1241–1247.)

Phenolic compounds also contribute to the flavor of dairy products. Presence of flavor notes such as phenolic, medicinal, cow and sheep has been noticed in varieties of aged Italian cheese such as provolone, Romano peccorino, and parmesan. This is attributed to the presence of phenol, *m*-cresol, *p*-cresol, *p*-ethylphenol, *p*-ethylphenol, 3,4-dimethylphenol, *o*-isopropylphenol, thymol and carvacrol (Ha and Lindsay, 1991a; Ney, 1973). Phenol and *o*-cresol have also been found as flavor components of butter (Urbach et al., 1972) and have been included as a part of flavoring mix used to simulate a cheese flavor (Urbach et al., 1972).

Tilgner (1977) estimated that smoke used in meat and fish curing may contain up to 10,000 different compounds; of these, probably about 500 are responsible for a smoky flavor note. A number of studies have indicated that phenolic compounds present in the vapor phase of smoke may contribute to imparting a smoky flavor to foods (Baryłko-Pikielna, 1977; Bratzler et al., 1969; Daun, 1972; Hamm, 1977). Some 85 different phenolics have been identified in smoke condensate, but only 20 phenols in smoked food product (Toth and Potthast, 1984). The flavor characteristics of some simple phenols found in wood smoke are summarized in Table 9.6. Kurko (1963) reported that only phenolics with a boiling point of 76 to 89°C at 5.33 hPa (4 mm Hg) carry a smoke-like flavor. Syringol, guaiacol, 4-methylguaiacol, 4-methylsyringol and eugenol may be the dominant contributors to the pleasant smoke-like flavor (Fiddler et al., 1970). Guaiacols contribute to a smoky taste and syringols impart a smoky odor; however, a much more complex mixture of compounds is responsible for the characteristic aroma of smoked products (Daun, 1972).

Presence of simple bromophenols has been detected in Australian ocean prawns (Whitfield et al., 1988), North American crustaceans and molluscs (Boyle et al., 1992a), Pacific salmon, molluscs (Boyle et al., 1992a), and North Atlantic shrimp (Anthoni et al., 1990). Of these, 2- and 4-bromophenols, 2,4- and 2,6-dibromophenols and 2,4,6-tribromophenol have been identified as key flavor components in

TABLE 9.6
Sensory Characteristics of Phenols from Smoke

Phenol	Optimum Concentration (mg/100 mL H_2O)	Smell	Taste
Guaiacol	3.75	Phenolic, smoky aromatic, hot, sweet	Phenolic, hot spicy, flavor of smoked ham, sweet, dry
Isoeugenol	9.80	Sweet, fruity, vanilla-like, phenolic	Sweet-fruity, mild smoke flavor, dry, hot
o-Cresol	7.50	Phenolic, sweet, fruity, aromatic caramel-like, like smoked ham	Sweet, hot, unpleasant smoky, burning

Source: Adapted from Toth, L. and Potthast, K., 1984, *Adv. Food Res.*, 29:87–158.

seafoods (Boyle et al., 1992a, b; Whitfield et al., 1988, 1997, 1998). Flavor thresholds for 2-bromophenol, 2,6-dibromophenol and 2,4,6-tribromophenol in water are 3×10^{-2}, 5×10^{-4} and 0.6 µg/kg, respectively. The flavor of 2,4,6-tribromophenol at 0.6 µg/kg is described as iodoform-like while that of 2,6-dibromophenol at its threshold level is phenolic/iodine-like (Whitfield et al., 1998). It has been postulated that bromophenols at concentrations below their flavor thresholds contribute the marine- or ocean-like flavor to seafoods and may enhance existing seafood flavors (Boyle et al., 1992a; Whitfield et al., 1997, 1998). Polychaetes and bryozoans have been implicated as the main sources of bromophenols in seafoods (Whitfield et al., 1999).

CONTRIBUTION OF FOOD PHENOLICS TO ASTRINGENCY

Some phenolic substances present in foods are able to bring about a puckering and drying sensation over the whole surface of the tongue and the buccal mucosa (Lea and Arnold, 1978). This sensation is referred to as astringency and is related to the ability of the substance to precipitate salivary proteins (Bate-Smith, 1973; Clifford, 1997). The presence of astringent phenols may affect the acceptability of a number of foods. The most significant examples of astringent foods are cider apples, Japanese persimmons, sloes, perry pears, cranberries, and wines.

Astringent phenols are substances of moderate molecular size whose phenolic groups are oriented into 1,2-dihydroxy or 1,2,3-trihydroxy configurations. At least two such orientations of phenolic groups are required to impart an astringent characteristic to a phenolic compound (Haslam, 1981). These phenolic substances bind to proteins more strongly than phenols with isolated hydroxyl groups (McManus et al., 1981). The phenol–protein complex can only precipitate when the complex becomes sufficiently hydrophobic. Molecular interpretation of chemical reactions responsible for the formation of astringency has been reported (Haslam et al., 1986; Haslam and Lilley, 1988).

Delcour et al. (1984) determined the astringency thresholds for tannic acid, (+)-catechin, procyanidin B-3, and mixtures of trimeric and tetrameric proanthocyanidins dissolved in deionized water. These threshold values range from 4.1 to 46.1 mg/L; greater molecular weight substances have lower threshold values. The recognition taste threshold of *p*-coumaric acid perceived as astringent is 48 mg/kg (Huang and Zayas, 1991). Good correlations exist between the astringent note of corn-germ, protein-flour products and the content of free *p*-coumaric acid (r = 0.850) and *p*-hydroxybenzoic acid (r = 0.815).

Millin et al. (1969) examined the effect of nonvolatile compounds on the flavor of black tea. Monomeric phenolics, such as flavanols, flavonols, theogallin, chlorogenic acid, *p*-coumarylquinic acid, and caffeic acid do not significantly contribute to the taste of tea. However, intermediate molecular weight oxidation products of flavanols formed during the fermentation process, such as theaflavin, are astringent. Later, Sanderson et al. (1976) reported that astringency of tea is related to the quantity of polyphenols and the degree of oxidation of flavanols as well as the degree of substitution of flavanols with galloyl groups. The astringency of theaflavins in tea is attenuated by the formation of complexes with caffeine (Stagg and Millin, 1975).

The astringency of unripened fruits is related to the presence of condensed tannins. According to Haslam (1975), only tannins with molecular weights ranging from 500 to 3000 Da may bring about the astringency sensation. The process of fruit ripening brings about changes in the composition and quantity of tannins, thus lowering astringency and improving palatability of fruits. Tannin polymerization is thought to be responsible for reduced astringency of ripened fruits (Goldstein and Swain, 1963). Increase in the molecular size of tannins is a factor that inhibits the interaction of tannins with salivary proteins (Mehansho et al., 1987). On the other hand, according to Ozawa et al. (1987), loss of astringency during ripening of fruits may also be due, in part, to the interaction of the involved polyphenols and proteins. However, the interaction of tannins with acetaldehyde is attributed to the loss of astringency upon ripening of the persimmon fruit (Matsuo and Itoo, 1982).

Polyphenolic compounds also contribute to the remarkably astringent taste of semidried banana products obtained by dehydration of ripe fruits following their blanching or sulfiting. This is due to the diffusion of tannin-like materials from the latex cells to the surrounding tissues. Light microscopy observations show that blanching and sulfiting disrupt the cell walls and result in extensive leakage of active tannins (Ramirez-Martinez et al., 1977).

PHENOLICS AS NATURAL FOOD PIGMENTS

A large and diversified group of phenolic substances, known as flavonoids, are responsible for the color and flavor of foods. Over 4000 flavonoids have been identified in vascular plants (Harborne, 1988a, b). Among them, the anthocyanins are responsible for the pink, scarlet, red, mauve, blue and violet colors of vegetables, fruits, fruit juices and wines. Other flavonoids may contribute to the ivory and pale yellow colors or act as copigments to anthocyanins (Harborne, 1967, 1988b). Most flavonoids are present in plant cells in the form of glycosides. The sugar part usually consists of mono-, di-, and trisaccharides of glucose, rhamnose, galactose, xylose and arabinose. The sugar moiety can be additionally acylated with phenolic acids such as *p*-coumaric, caffeic, or ferulic acids. The role of flavonoids in foods has been extensively reported and several comprehensive reviews have been published (Brouillard et al., 1997; Harborne et al., 1975, 1988a, b; Jackman et al., 1987; Kuhnau, 1976; Markakis, 1974, 1982; Timberlake, 1980). Foods of plant origin may contain up to several grams of flavonoids per kilogram of fresh weight. It is estimated that daily average consumption of flavonoids may range from 50 mg to about 1 g per person. Nearly half of the total flavonoid intake is anthocyanins, catechins and 4-oxoflavonoids (Kuhnau, 1976).

ANTHOCYANINS

Anthocyanins are water-soluble pigments found in most species in the plant kingdom; they are responsible for the bright red skin of radishes, the red skin of potatoes, and the dark skin of eggplants. Blackberries, red and black raspberries, nectarines, peaches, blueberries, bilberries, cherries, currants, grapes, pomegranates, cranberries, and ripe gooseberries contain anthocyanins. The visual detection thresholds for selected anthocyanins are shown in Table 9.7. The total content of anthocyanins varies among fruits and in different cultivars of the same fruit and is also affected by genetic make-up, light, temperature, and agronomic factors (Delgado-Vargas et al., 2000; Francis, 1989; Harborne and Grayer, 1988; Harborne and Williams, 2001; Ju et al., 1995; Mazza and Miniati, 1993; Wrolstad, 2000). The exposure of apples (Faragher, 1983), cranberries (Hall and Stark, 1972), and nectarines (Basiouny and Buchanan, 1977) to low temperature promotes the formation of anthocyanins. However, blackberries ripened at low temperatures contained fewer anthocyanins compared to those ripened at warm weather conditions (Naumann and Wittemberg, 1980).

The content of anthocyanins is also affected by the intensity and quality of light. For example, light is the predominant factor affecting the content of anthocyanins in olives; olive fruits ripened in the dark contain 10 times fewer anthocyanins than those ripened under normal conditions (Vazquez-Roncero et al., 1970). In apples,

TABLE 9.7
Visual Detection Thresholds (VDT) for Selected Anthocyanins

Anthocyanin	VDT (mg/L)
Cyanidin 3-glucoside	1.3
Cyanidin 3-xylosyl-galactoside	0.9
Cyanidin 3-xylosyl-glucosyl-galactoside	2.4
Cyanidin 3-sinapoyl-xylosyl-glucosyl-galactoside	0.9
Cyanidin 3-feruoyl-xylosyl-glucosyl-galactoside	0.4
Cyanidin 3-sophoroside-5-glucoside	3.6
Cyanidin 3-coumaroyl-sinapoyl-sophoroside-5-glucoside	2.0

Source: Adapted from Stintzing, F.C. et al., 2002, *J. Agric. Food Chem.*, 50:6172–6182.

the production of anthocyanins is stimulated by light in the blue–violet region (Siegelman and Hendricks, 1958) as well as light in the UV region (Chalmers and Faragher, 1977). The presence of sucrose, glucose, fructose, lactose, and maltose also stimulates the formation of anthocyanins (Vestrheim, 1970). Endogenous growth promoters such as auxins, cytokinins and gibberellins affect the synthesis of anthocyanins (Mazza and Miniati, 1993). Morever, treating plants with excess nitrogen reduces the level of anthocyanins in fruits. Diversion of products of photosynthesis toward the formation of amino acids and proteins may be contemplated (Macheix et al., 1989).

Over 250 naturally occurring anthocyanins have been identified (Strack and Wray, 1993); of these, approximately 70 have been reported in fruits. The composition and content of anthocyanins allow differentiation of fruits from one another (Clifford, 2000; Delgado-Vargas et al., 2000; Jackman and Smith, 1996; Mazza and Miniati, 1993; Strack and Wray, 1993). These pigments usually are found in nature in the form of 3-monoglycosides, 3,5-diglycosides and acylglycosides of anthocyanidins; however, glycosylated anthocyanins in positions 7, 3′, 4′ or 5′ may also occur (Brouillard, 1982). Presence of a second glycosyl group in anthocyanins brings about a characteristic bathochromic shift in their absorption maxima. This is used as a means for spectral differentiation of anthocyanins with a glycosyl group at position C-3 from those with glycosyl groups at position C-3 and position C-5 (Harborne, 1958). The sugar moiety can be glucose, arabinose, xylose, galactose, rhamnose, and fructose as well as rutinose (6-*O*-α-L-rhamnosyl-D-glucose), sophorose (2-*O*-β-D-xylosyl-D-glucose), gentobiose (6-*O*-β-D-glucosyl-D-glucose), sambubiose (2-*O*-β-D-xylosyl-D-glucose), xylosylrutinose and glycosylrutinose. Common names of some naturally occurring anthocyanins are listed in Table 9.8. The aglycone remaining after hydrolysis of anthocyanins is known as anthocyanidin.

Most anthocyanins found in nature are the glycosides of perlargonidin, cyanidin, peonidin, delphinidin, petunidin and malvidin. Figure 9.8 shows structures of some important anthocyanidins such as pelargonidin found in strawberry and radish, as well as cyanidin present in apple and red cabbage, and malvidin and delphinidin in

TABLE 9.8
Common Names of Major Anthocyanins Found in Fruits and Vegetables

Common Names	Compound
Kuromanin	Cyanidin-3-glucoside
Keracyanin	Cyanidin-3-rhamnoglucoside
Ideain	Cyanidin-3-galactoside
Cyanin	Cyanidin-3,5-diglucoside
Callistephin	Pelargonidin-3-glucoside
Pelargonin	Pelargonidin-3,5-diglucoside
Oenin	Malvidin-3-glucoside
Malvin	Malvidin-3,5-diglucoside

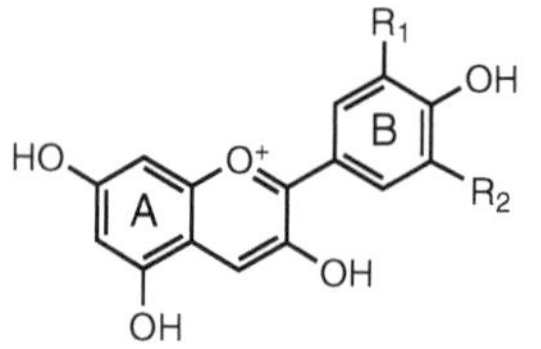

Compound	R_1	R_2
Cyanidin	OH	H
Delphinidin	OH	OH
Malvidin	OCH_3	OCH_3
Pelargonidin	HH	

FIGURE 9.8 Chemical structures of some anthocyanidins found in foods.

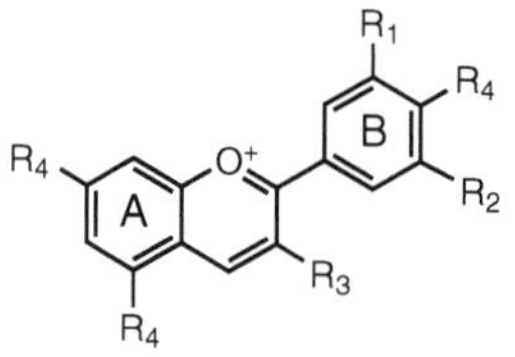

R_1	H, OH or OCH_3
R_2	H, OH or OCH_3
R_3	H or Glycosyl
R_4	OH or Glycosyl

FIGURE 9.9 Chemical structure of flavylium cation.

grapes. The basic structure of these anthocyanidins is the flavylium (2-phenylbenzopyrilium) cation substituted with a number of hydroxyl and methoxyl groups (Figure 9.9). The patterns of hydroxylation and methoxylation affect the color of anthocyanidins (Table 9.9). Presence of larger numbers of hydroxyl groups in the molecule tends to deepen the blue hue, while an increase in the number of methoxyl groups enhances the redness (Braverman, 1963; Delgado-Vargas et al., 2000; Francis, 1989; Iacobucci and Sweeny, 1983; Mazza and Miniati, 1993). The intensity of color, however, depends also on the pH, presence of metal ions, self-association of anthocyanins (Figuerido et al., 1996a, b; Hoshino et al., 1980; Mazza and Miniati, 1993; Wrolsdat, 2000), pigment mixtures and copigments such as other colorless phenolic compounds (Brouillard, 1983; Dangles et al., 1993; Mazza and Miniati, 1993; Ohta et al., 1980), as well as processing and storage conditions (temperature, sugar content, presence of ascorbic acid, and presence of oxygen, among others) (Dao et al., 1998; Debicki-Pospisil et al., 1982; Delgado-Vargas et al., 2000; Francis, 1989; Markakis, 1974). Depending on the conditions, ascorbic acid may have positive or detrimental effects on anthocyanin stability (Sarma et al., 1997).

TABLE 9.9
Naturally Occurring Anthocyanidins

Name	Substitution Pattern[a]							Color
	3	5	6	7	3′	4′	5′	
Apigenidin	H	OH	H	OH	H	OH	H	Orange
Auratinidin	OH	OH	OH	OH	H	OH	H	Orange
Cyanidin	OH	OH	H	OH	OH	OH	H	Orange-red
Delphinidin	OH	OH	H	OH	OH	OH	OH	Bluish-red
Europinidin	OH	OMe	H	OH	OMe	OH	OH	Bluish-red
Hirsutidin	OH	OH	H	OMe	OMe	OH	OMe	Bluish-red
Luteolinidin	H	OH	H	OH	OH	OH	H	Orange
Malvidin	OH	OH	H	OH	OMe	OMe	OMe	Bluish-red
Pelargonidin	OH	OH	H	OH	H	OH	H	Orange
Peonidin	OH	OH	H	OH	OMe	OH	H	Orange-red
Petunidin	OH	OH	H	OH	OMe	OH	OH	Bluish-red
Puchellidin	OH	OMe	H	OH	OH	OH	OH	Bluish-red
Rosinidin	OH	OH	H	OMe	OMe	OH	H	Red
Tricetinidin	H	OH	H	OH	OH	OH	OH	Red

[a] For substitution patterns see Figure 9.9.

Source: Adapted from Mazza, G. and Miniati, E., 1993, *Anthocyanins in Fruits, Vegetables and Grains*, CRC Press, Boca Raton, FL.

Acylation of the sugar moiety of anthocyanins contributes to their color stability. Acylation usually occurs at the C-3 position of the sugar (Harborne, 1964). Presence of two or more acyl groups also increases the color stability of anthocyanins in aqueous solutions (Brouillard, 1982). For example, stabilization of anthocyanins containing two acyl groups is brought about by the formation of sandwich-type stacking due to hydrophobic interaction between the anthocyanidin ring and the acyl groups (intramolecular interactions) (Brouillard, 1982; Mazza and Miniati, 1993; Wrolstad, 2000). The sugar is usually acylated with *p*-coumaric, caffeic, ferulic or sinapic acids (Francis and Harborne, 1966; Harborne, 1964; Somers, 1966). The stability of anthocyanins in plants is also associated with their ability to form complexes with other phenolics, nucleic acids, sugars, amino acids and metal ions such as calcium, magnesium and potassium (Brouillard, 1983). The colored forms of anthocyanidins are stabilized by the presence of hydroxyl groups at the C-5 position and substitution at the C-4 position. This inhibits the addition of water and subsequent formation of colorless species (Mazza and Miniati, 1993). Moreover, the addition of a methyl or phenyl group at the C-4 position (Timberlake and Bridle, 1968) or the conversion to deoxyanthocyanidins (Sweeny and Iacobucci, 1983) improves the stability of anthocyanidins in the presence of sulfur dioxide or ascorbic acid.

Anthocyanins are very sensitive to changes in pH (Dao et al., 1998; Mazza and Brouillard, 1987; Ohta et al., 1980; Sweeny and Iacobucci, 1980). In solution, they exist in four different forms: neutral and ionized quinonoidal base, flavylium cation

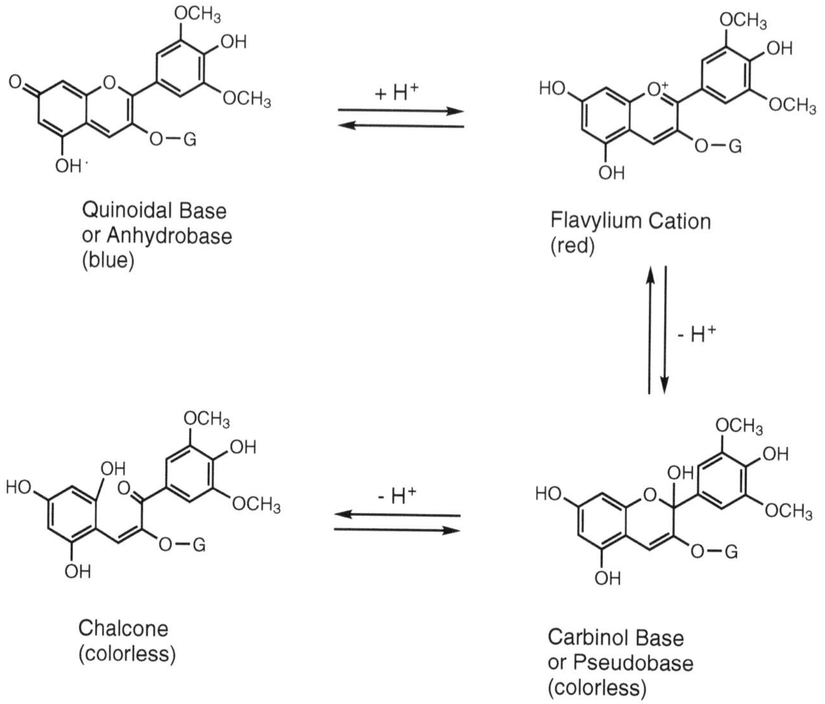

FIGURE 9.10 Chemical structures of anthocyanin chromophores. (Adapted from Timberlake, C.F., 1980, *Food Chem.*, 5:69–90 and Wong, D.W.S., 1989, in *Mechanism and Theory in Food Chemistry,* AVI Publishing Co., Westport, CT, 147–189.)

or oxonium salt, the colorless pseudobase and chalcone. The flavylium cation is weak acid, and a neutral quinoidal base behaves as a weak acid and a weak base (Brouillard, 1983; Mazza and Miniati, 1993). The distribution of these forms at equilibrium depends on pH and structure of anthocyanidin. The effect of pH on the predominant forms of anthocyanin chromophores is given in Figure 9.10. Due to the equilibrium between flavylium cation and colorless carbinol structures, most intense red coloration of anthocyanins occurs in the pH range of 1 to 3 (Hrazdina, 1981).

In slightly acidic solutions (pH 4 to 6.5), anthocyanins undergo color fading. In this range of pH, anthocyanins rapidly transform to red or blue quinonoidal bases as a result of rapid proton loss of any hydroxyl groups at C-4′, C-5 or C-7. Following this, nucleophilic addition of water to anthocyanins at C-2 or C-4 results in the formation of colorless carbinol structures and this equilibrates to the open, colorless chalcone forms. The reaction involves transferring the proton as well as breaking or forming the C–O bond (Brouillard, 1982, 1988; Brouillard and Dangles, 1994; Dao et al., 1998; Mazza and Miniati, 1993; Timberlake and Bridle, 1967a; Wrolsdat, 2000). Anthocyanins also show a minimum coloration at their isoelectric point (Markakis, 1960). Shifting the pH of anthocyanins to higher values may even bring about complete loss of color. Thus, alkaline conditions, in particular at high temperatures, should be avoided in the processing of anthocyanin-containing foods (Brouillard, 1982).

The change in coloration of anthocyanins may be due not only to pH but also to the formation of a weak complex with another anthocyanin molecule (self-association) or with a colorless phenolic compound such as gallotannin, quercetin, rutin or hydroxycinnamic acid derivatives (intermolecular copigmentation). Self-association brings about a greater than linear increase in the color intensity with increasing pigment concentration. The formation of anthocyanin–anthocyanin complex is primarily due to the hydrophobic interaction between aromatic nuclei of the molecules involved (Goto, 1987). The intermolecular copigmentation process brings about a shift from red to blue and increases the absorbance of the visible band, thus enhancing the color (Asen et al., 1972; Brouillard, 1982; Markakis, 1974; Mazza and Miniati, 1993; Osawa, 1982; Wrolsdat, 2000).

Anthocyanins are very unstable compounds. Loss of their color can occur during thawing of frozen vegetables and fruits as well as during processing and storage of plant products. This loss of color may be due to the hydrolysis of unstable aglycones (Shrikhande, 1976). On the hand, high temperature and pH bring about the formation of an unstable quinoidal base, carbinol pseudobase and chalcones (Brouillard, 1982, 1988; Brouillard and Dangles, 1994; Dao et al., 1998; Timberlake, 1981; Wrolsdat, 2000). Anthocyanins may also degrade by intermediate products of nonenzymatic browning reactions and by formation of copolymers with melanins (Debicki-Pospisil et al., 1982); they can form complexes with flavonoids, heavy metals and other compounds present in foods (Asen et al., 1972; Markakis, 1974; Mazza and Miniati, 1993).

Stability of anthocyanins during processing depends on the composition of food (anthocyanins, enzymatic systems, etc.) as well as temperature, pH, level of sugar, light, organic chemicals (ascorbic acid, tartaric acid, etc.) and contaminations with metal ions (de Ancos et al., 2000; Francis, 1989; Gil et al., 1997; Macheix et al., 1989; Markakis, 1982; Wong, 1989; Wrolstad et al., 1990; Wrolstad, 2000; Zabetakis et al., 2000). The mechanism of thermal degradation of anthocyanin is not understood well, although a number of pathways have been proposed. In the presence of oxygen a brown-colored degradation product is formed. Degradation of anthocyanins depends on the temperature and duration of heat treatment. Application of HTST (high temperature-short time) processes improves the retention of anthocyanins. It has been demonstrated that loss of anthocyanins in red fruit juices is neglegible when heat processing is carried out for less than 12 min at 100°C (Markakis, 1974, 1982). However, thermal treatments such as pasteurization, sterilization and concentration bring about degradation of anthocyanins (Mazza and Miniati, 1993).

The rate of anthocyanin degradation is also affected by storage temperature. A linear relationship was found between the logarithm of the rate of color loss of anthocyanins in strawberry preserves and storage temperature (Meschter, 1953). Thus, the half-life of anthocyanin color in strawberry preserves is 1 h, 240 h, 1300 h and 11 months when stored at 100, 38, 20, and 0°C. Similarly, higher storage temperatures also increased the loss of anthocyanin color in orange and blackberry juices (Figure 9.11). On the other hand, storing apples at 2°C and 70 to 80% humidity slows the loss of anthocyanins (Lin et al., 1989). Thus, the retention of anthocyanin color may be enhanced by lowering storage temperature of foods.

The loss of anthocyanins can also be minimized by storing fruits under a modified atmosphere (Gil et al., 1997; Kalitka and Skrypnik, 1983; Lin et al., 1989), high hydrostatic pressure treatment (Zabetakis et al., 2000) or freezing (de Ancos et al., 2000; Polessello and Bonzini, 1977; Wrolstad et al., 1990). Gil et al. (1997) reported that moderate carbon dioxide concentrations have little effect on the external color of strawberries; however, changes in the color of strawberry skin have been noted at carbon dioxide levels greater than 40% (Gil et al., 1997; Kader, 1986; Ke et al., 1991; Li and Kader, 1989). Furthermore, Zabetakis et al. (2000) have demonstrated that high hydrostatic pressure treatment (800 Mpa) of strawberries for 15 min reduces the loss of anthocyanins in fruits stored for 8 days at 4°C. Drying food at high temperatures brings about rapid degradation of anthocyanins (Macheix et al., 1989). Anthocyanins can be bleached by sulfur dioxide; the colorless sulfonic acid derivatives are formed as a result of addition of sulfur dioxide to the 4-position of the anthocyanin molecule, but this reaction is reversible (Timberlake and Bridle, 1967b).

OTHER FLAVONOIDS

Other flavonoids may also contribute to the color of food products. The group of yellow or ivory flavonoid pigments that exists in plants includes flavonols, flavones, chalcones, aurones, flavanones, isoflavanones and biflavonols. Although many of these compounds are colorless, under conditions of food handling and processing they can be converted to colored products.

Flavones are characterized by the presence of a double bond between C-2 and C-3, whereas flavonols are flavones with a hydroxy group in the 3-position (Figure 9.12). Many plants contain appreciable quantities of flavones or flavonols in the form of glycosides. These phenolics may act in plant as copigments to modify the color of anthocyanins. Figure 9.13 shows four flavonol aglycons commonly found in fruits: kaempferol, quercetin, myricetin and isorhamnetin (Macheix et al., 1990). The flavonols kaempferol, quercetin and myricetin also contribute to the color of green tea (Roberts et al., 1956). Flavones and flavonols do not contribute much to the coloration of plants unless they are present in high concentrations such as in onion skins. Outer dry skin of onions contains 2.5 to 6.5% quercetin, mostly in the free aglycone form (Herrmann, 1976). Flavones and flavonols possessing orthodihydroxy groups (quercetin, luteolin) may be responsible for discoloration of foods because they are able to chelate metal ions. Iron chelates are dark in color while aluminum chelates are bright yellow or brown (Bate-Smith, 1954; Swain, 1962). In the form of bright yellow crystals, rutin, a 3-rhamnoglucoside of quercetin, occasionally precipitates on asparagus packed in glass containers (Dame et al., 1959). This is due to its limited solubility in water at room temperature. Rutin may also complex with iron to give a dark discoloration in canned asparagus.

Flavan-3-ols and flavan-3,4-diol are commonly found in plant foods (Figure 9.14). Flavan-3-ols participate in the structure of protoanthocyanidins as their monomers. Although they are colorless, under certain conditions they may

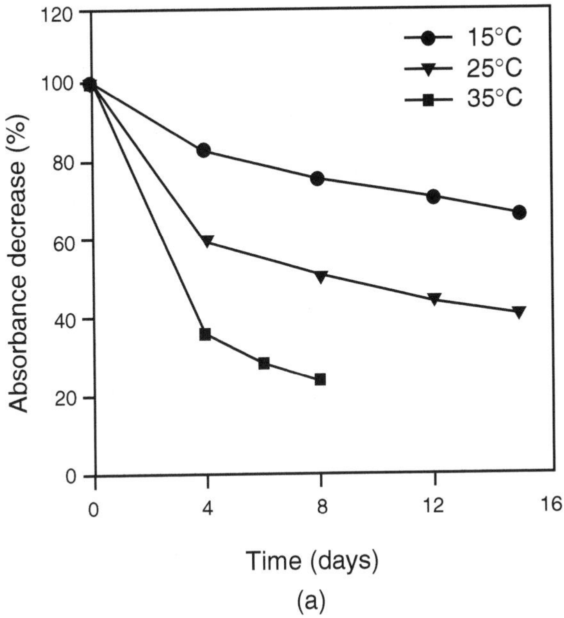

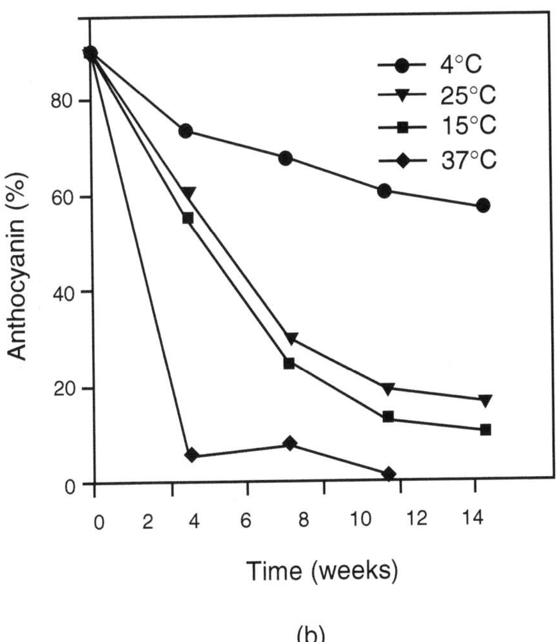

FIGURE 9.11 Effect of storage temperature on the color degradation of orange juice (a) and blackcurrant juice (b). (Adapted from Macheix, J.-J. et al., 1989, *Fruit Phenolics,* CRC Press, Boca Raton, FL.)

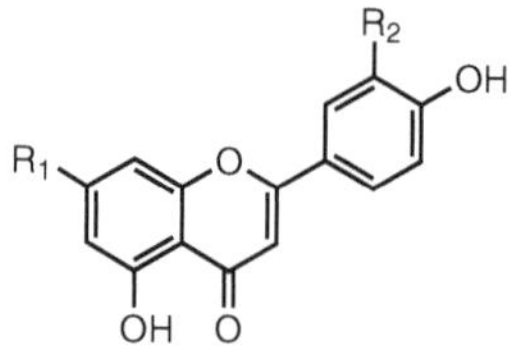

Compound	R_1	R_2
Apigenin	OH	H
Luteolin	OH	OH
Tangeretin	H	OCH_3

FIGURE 9.12 Chemical structures of some flavones.

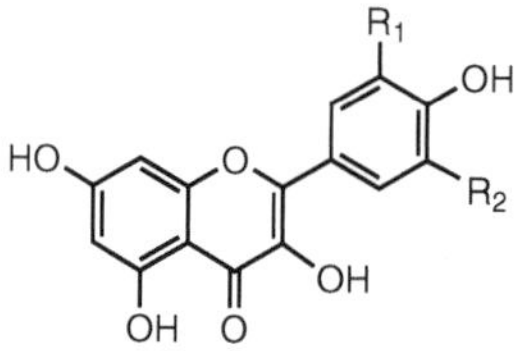

Compound	R_1	R_2
Kaempferol	H	H
Myricetin	OH	OH
Quercetin	OH	H
Isorhamnetin	OCH_3	H

FIGURE 9.13 Chemical structures of some flavonols.

cause discoloration of fruits and vegetables. They may also form dark-colored chelates with metals as well as participate in enzymatic browning reactions. Bate-Smith (1954) suggested that proanthocyanidins might participate in the formation of pinkish discoloration sometimes observed in overcooked stewed pears. Joslyn and Peterson (1956) demonstrated the presence of proanthocyanidins in Bartlett pears that yielded cyanidin. A pinkish color is often observed in pears with a higher proanthocyanidin content, especially at lower pH values (Luh et al., 1960).

ENZYMATIC REACTION OF DISCOLORATION

A major concern to food processors and researchers working with plant foods is the presence of polyphenol oxidase (monophenol monoxygenase or tyrosinase: EC 1.14.18.1). Diphenol oxidase or catechol oxidase (EC 1.10.3.2) and laccase (EC 1.10.3.1) are also involved in enzymatic browning reactions. Producing insoluble polymers that serve as a barrier to spread of infection in living plants, these enzymes play an important role in resistance to infection by viruses, fungi, bacteria or mechanical damage (Eskin, 1990). In food products polyphenol oxidase is mainly

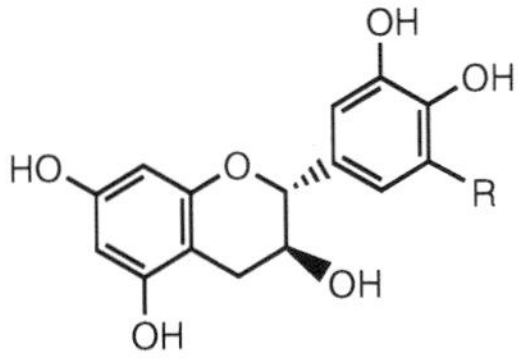

R = H [+] Catechin

R = OH [+] Gallocatechin

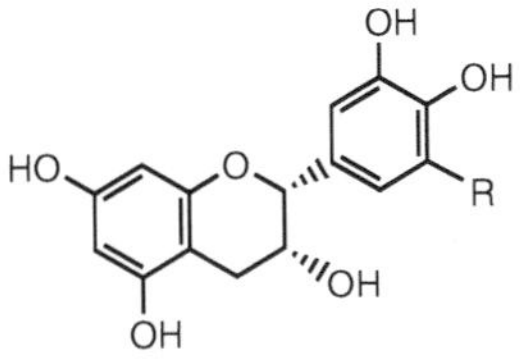

R = H [-] Epicatechin

R = OH [-] Epigallocatechin

FIGURE 9.14 Chemical structures of catechins.

responsible for spoilage because it catalyzes oxidation of phenols to highly active quinones. Subsequently, quinones may react with amino and sulfhydryl groups of proteins and enzymes, as well as with anthocyanins. These secondary reactions may bring about changes in physical, chemical and nutritional characteristics of food proteins and may also affect sensory properties of food products (Mayer and Harel, 1979). Quinones may also contribute to the formation of brown pigments due to participation in polymerization and condensation reactions with proteins (Mathew and Parpia, 1971).

ENZYMATIC BROWNING REACTIONS

Some enzymes present in fruits and vegetables have the ability to catalyze oxidation of phenols to quinones with subsequent rapid nonenzymatic polymerization. The *o*-quinones may also modify proteins as a result of reaction with their amino and sulfhydryl groups (Matheis and Whitaker, 1984). Moreover, *o*-quinones may oxidize compounds of lower oxidation-reduction potential such as anthocyanins to colorless products (Friedman, 1996; Mathew and Parpia, 1971; Peng and Markikis, 1963). The enzyme-catalyzed oxidation of phenols leads to browning of foods such as potatoes, peaches, bananas, tea, and coffee. The enzymatic browning reaction is considered wholesome in the processing of coffee, leaf tea and dates.

The enzymatic browning reaction occurs during the ageing or senescence of fruits and vegetables or as a result of injury to plant product. Polyphenol oxidase is activated as a result of disruption of cell integrity and when the contents of plastid and vacuole are mixed (Amiot et al., 1997; Eskin, 1990). The enzymatic oxidation of flavonols plays an important role in the formation of color of fermented tea. Oxidized flavonols readily condense to theaflavins and subsequently to thearubigins. Theaflavins are yellow-orange in color whereas thearubigins, the polymeric form of oxidized flavonols, are black-brown in color and participate in the formation of bulk color of tea (Sanderson and Graham, 1973).

The enzymatic browning reaction is considered harmful when it causes degradation of food quality due to browning discoloration or development of off-flavor and off-odor (Burnette, 1977; Friedman, 1996). A number of technological treatments, such as crushing, slicing, cutting, extraction, handling, storage conditions, low temperature, and thawing, may cause unwanted browning discoloration in fruits and plants. The "black ring" defect in canned beet root slices is associated with endogenous polyphenol oxidase activity. According to Im et al. (1990), the initiation of this discoloration requires exposure of beet slices to oxygen for a minimum time (20 min) under high-temperature treatment. Thus, thermal processing may bring about quinone participation in secondary reactions to yield dark pigmented products. Enzymes may also be responsible for dark discoloration or melanosis of shrimp and other crustacean species, thus reducing the market value of these products (Savagaon and Sreenivasan, 1978; Simpson et al., 1988).

The enzymes involved in catalyzing the reaction of *o*-diphenol oxidation are copper-containing enzymes called phenolases, polyphenolases, polyphenol oxidases, or phenol oxidases (Mathew and Parpia, 1976; Sanchez-Ferrer et al., 1995). The reaction can be represented as

$$o\text{-diphenol} + \text{oxygen} \rightarrow o\text{-quinone} + \text{water}$$

Depending on the type of substrate, this enzyme can show cresolase (monophenolase) activity when monophenol is converted to *o*-diphenol and catecholase (diphenolase) activity when *o*-diphenol is oxidized to *o*-quinone (Espin et al., 1997; Itoh et al., 2001; Sanchez-Ferrer et al., 1995) (Figure 9.15). It has been demonstrated that monophenolase activity of polyphenol oxidase displays a pH-dependent lag period prior to reaching a steady state rate (Escribano et al., 1997; Espin et al., 1997). In many vegetables and fruits, polyphenol oxidase exists in the inactive or latent form (Fraignier et al., 1995; Laveda et al., 2001; Nunez-Delicado et al., 1996; Sanchez-Ferrer et al., 1993). *In vitro* activation of the latent enzyme can be brought about by treatments with surfactants (Moore and Flurkey, 1990; Nunez-Delicado et al., 1996; Sanchez-Ferrer et al., 1993), fatty acids (Golbeck and Cammarata, 1981) and proteases (Golbeck and Cammarata, 1981; Harel et al., 1973; King and Flurkey, 1987; Laveda et al., 2001), as well as exposure to low pH (Kenten, 1957). It has been postulated that the activation of the latent polyphenol oxidase may by due to its solubilization (Mayer and Harel, 1979), limited conformational changes in enzyme (Swain et al., 1966) or dissociation of an enzyme-inhibitor complex (Kenten, 1958).

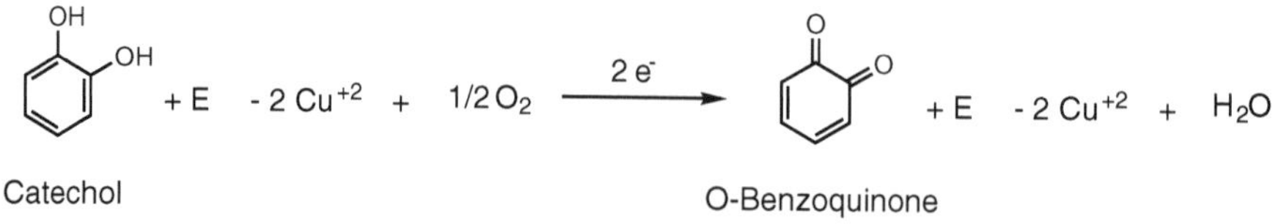

$$\text{Catechol} + E - 2\,Cu^{+2} + 1/2\,O_2 \xrightarrow{2e^-} \text{O-Benzoquinone} + E - 2\,Cu^{+2} + H_2O$$

FIGURE 9.15 Enzymatic oxidation of *O*-diphenol by catecholase.

The oxidation of food phenolics may also be catalyzed by another group of enzymes in foods called peroxidases (Burnette, 1977; Nicolas et al., 1994; Williams et al., 1985). These enzymes may use polyphenols as hydrogen donors and are able to oxidize hydroxycinnamic derivatives and flavans (Nicolas et al., 1994; Robinson, 1991). Richard-Forget and Gaullard (1997) reported that in the presence of polyphenol oxidase, peroxidase enhances the oxidation of polyphenol. This enhancement may be due to the formation of hydrogen peroxide during oxidation of phenolics by polyphenol oxidase. The hydrogen peroxide may then be used by peroxidase as electron acceptor to oxidize the phenolic compounds further. The possible use of quinonoid forms by peroxidase as peroxide substrate has also been postulated.

The rate of enzymatic browning reaction depends on the nature and content of phenolic compounds, activity of phenol oxidases present in foods, presence of oxygen-reducing substances and metal ions, pH and temperature. The enzymatic browning reaction can be controlled by inactivating or inhibiting phenol oxidases, excluding oxygen, modifying or lowering the phenolic content, adding reducing substances, and interacting with the copper prosthetic group, as well as reducing or trapping quinones and even removing products of browning reactions (Friedman, 1996).

Some phenolic compounds do not serve as substrates of enzymatic browning reactions but may act as synergists or inhibitors of polyphenol oxidases (PPO) from certain sources. Shannon and Pratt (1967) reported that *p*-coumaric acid is a noncompetitive inhibitor of apple PPO that catalyzes the oxidation of chlorogenic acid. Protocatechuic acid completely inhibits the browning reaction in broad beans (*Vicia faba* L.) (Robb et al., 1966), but resorcinol and phloroglucinol act as synergists of PPO-mediated browning reactions (Pratt, 1972; Shannon and Pratt, 1967). Similarly, Kahn (1976) reported that *p*-coumaric acid, *p*-cresol and protocatechuic acid arrest browning of 4-methylcatechol by avocado PPO, while phloroglucinol, orcinol and resorcinol markedly increase the browning reaction rate.

Polyphenol oxidase can be inactived by application of heat for a sufficient time period to denature the protein. However, thermal treatment may have detrimental effects on sensorial and nutritional quality of vegetables and fruits (Lund, 1977; Sapers, 1993). Development of bitter off-flavors has been reported in avocadoes heated to inactivate the enzyme (Benet et al., 1973). Dimick et al. (1951) found that a temperature of 90°C is necessary for almost total inactivation of polyphenolase activity in pear puree. Arslan et al. (1998) reported approximately 50% loss in enzymatic activity by polyphenol oxidase isolated from Malaya apricot after heating the enzyme for 47 and 16 min at 60 and 80°C, respectively. Hot water or steam blanching are usually used as pretreatment of plant foods prior to freezing, canning and drying. This process retards enzymatic browning but results in the leaching of soluble materials. The time and temperature used for blanching should be carefully selected in order to avoid undesirable changes in flavor and texture (Table 9.10). Weemaes et al. (1998) investigated the effect of high hydrostatic pressures and thermal treatments on inactivation of avocado polyphenol oxidase. Inactivation of enzyme followed first-order kinetics in the pH range between 5 and 8, but displayed a deviation from first order at pH 4. The minimum pressure required to inactivate PPO ranged from 450 to 850 MPa at pH 4 and 8, respectively.

TABLE 9.10
Blanching Time in Water at 100°C for Vegetables for Freezing

Vegetable	Blanching Time (min)
Asparagus	
Small (5/16 in. diam. or less at butt)	2
Medium (6/16 to 9/16 in. diam. at butt)	3
Large (10/16 in. diam. or larger at butt)	4
Peas	1–1.5
Spinach	1.5
Swiss chard	2
Beets	
Small, whole (1.25 in. diam. or less)	3–5
Beans	
Small (less than 5/16 in. diam. or sieve no. 2 or smaller)	1–1.5
Medium (5/16–6/16 in. diam. or sieve no. 3 and 4)	2–3
Large (6/16 in. diam. and larger or sieve no. 5 or larger)	3–4

Source: Adapted from Potter, N.N., 1986, *Food Science*, 4th ed., Van Nostrand Reinhold, New York.

Another method of retarding enzymatic browning reaction is to lower the pH of plant food tissue because the optimum pH for most polyphenol oxidases ranges from 4.0 to 7.0 (Eskin, 1990). However, polyphenol oxidase isolated from Malaya apricot (*Prunus armeniaca* L.) toward catechol, *L*-dopa and gallic acid displays a maximum activity at pH 8.5 (Arslan et al., 1998). Citric, ascorbic, malic and phosphoric acids are usually employed. Recently, Soliva-Fortuny et al. (2001) reported a 62% depletion of polyphenol oxidase activity in ready-to-eat apples stored under a modified atmosphere (90.5% nitrogen + 7% carbon dioxide + 2.5% oxygen), which extended their shelf-life by several weeks.

Numerous chemical species have been identified as effective inhibitors of browning reaction, including

- Ascorbic acid and its derivatives (Sapers, 1993)
- 2-Mercaptoethanol, sodium metabisulfite, and thiourea (Arslan et al., 1998)
- Cysteine and cysteine derivatives (Molnar-Perl and Friedman, 1990a, b)
- Honey (Chen et al., 2000; Oszmianski and Lee, 1990)
- 4-Hexylresorcinol (Dawley and Flurkey, 1993; Gonzalez-Aguilar et al., 2000; Monsalve-Gonzalez et al., 1995)
- Oxalic acid (Son et al., 2000)
- Kojic acid (Chen et al., 1991a,b)
- Tropolone (Kahn and Andrawis, 1985)
- β-Cyclodextrin (Hicks et al., 1996)
- Protein hydrolyzates (Kahn and Andrawis, 1985)
- Proteases (Labuza et al., 1992)
- Calcium chloride (Tomas-Barberan et al., 1997)
- 2,4-Dichlorophenoxyacetic acid (Tomas-Barberan et al., 1997)
- Acetic acid (Tomas-Barberan et al., 1997)

Inhibitors may prevent browning by directly inhibiting polyphenol oxidase activity, removing phenolic substrate or reducing *o*-quinones back to *o*-diphenols (Hicks et al., 1996).

Irwin et al. (1994) demonstrated that cyclodextrins protect chlorogenic acid from oxidation by forming a 1:1 inclusion complex with the acid. Later, Hicks et al. (1996) showed that in the presence of 1 to 1.5% of soluble β–cyclodextrin, browning of apple juice held at room temperature is efficiently inhibited for 6 h. Addition of 0.25 to 0.5% of sodium hexametaphosphate inhibits the browning of apple juice for 1 day at room temperature or for up to 3 weeks at 4°C. Furthermore, 10-min stirring of apple, pear, green grape and celery juice with an insoluble β-cyclodextrin or elution of juices through the column filled up with insoluble β-cyclodextrin results in a product that is β-cyclodextrin free and also resistant to browning (Irwin et al., 1994).

Walker (1975b, 1977) found that cinnamic and *p*-coumaric acids are potent inhibitors of browning reactions in fruit juices. Less than 0.01% of cinnamic acid is required to prevent the browning reaction. Addition of sodium chloride also effectively inhibits the formation of browning reaction products in apples (Rouet-Mayer and Philippon, 1986), cherries (Benjamin and Montgomery, 1973) and grapes (Interesse et al., 1984). Sapers et al. (1990) investigated the use of vacuum and pressure infiltration as a means of applying ascorbate- and erythorbate- based enzymatic browning inhibitors to apple and potato surfaces. They found that pressure treatment brings about a more uniform uptake of solution and less extensive water logging than vacuum treatment. The apple cuts treated by pressure infiltration gained 3 to 7 days storage life, while potato plugs gained 2 to 4 days, compared to dipping.

Enzymatic browning reactions in foods of plant origin as well as discoloration (melanosis) of shrimp and other crustacean species can be prevented by sulfiting agents. Sulfites are also used to delay browning reactions in salad bar items and in prepeeled potatoes, cut apples and other fruits supplied to the baking industry (Ponting et al., 1971). The mechanism of sulfite action may be due to direct inhibition of enzymes by sulfite, or interaction of sulfites with intermediates of enzymatic browning reactions, or action of sulfite as a reducing agent, thus converting quinones back to the original phenolic compounds. The level of sulfite addition to foods depends on the nature of the phenolics present and the length of time that browning reaction should be inhibited (Taylor et al., 1986). However, increased consumer concern about the safety of sulfites has brought about a search for suitable alternatives. So far the potential substitutes are less effective and more costly (Otwell and Marshall, 1986; Taylor et al., 1986).

OTHER ENZYMATIC DISCOLORATIONS

Anthocyanins in food products are poor substrates for oxidation reactions catalyzed by phenolases. However, they may undergo other types of enzymatic degradation resulting in discoloration. Degradation of anthocyanins may be due to their hydrolysis to anthocyanidins by anthocyanases (Forsyth and Quensel, 1957; Huang, 1955, 1956).

Phenolases may indirectly accelerate degradation of anthocyanins as a result of contribution of *o*-quinones to colorless products. Peng and Markikis (1963) studied

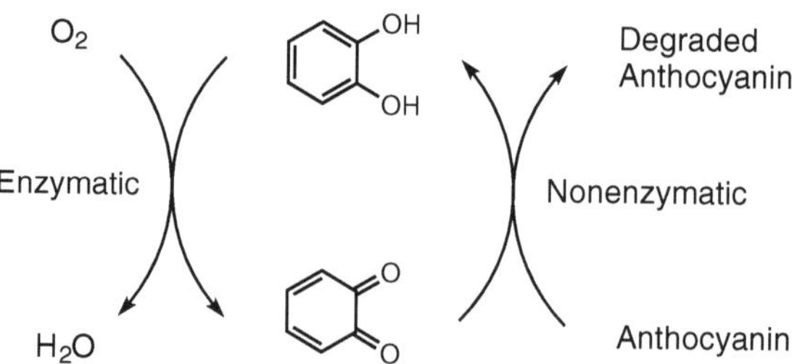

FIGURE 9.16 Possible mechanism of anthocyanin degradation by *O*-diphenol peroxidase. (Adapted from Pifferi, P.G. and Cultrera, R., 1974, *J. Food Sci.*, 39:786–791.)

the degradation of anthocyanins isolated from red tart cherries by mushroom phenolase and found that the presence of catechol accelerates their degradation. Based on the results of their study, these researchers proposed a mechanism of sequential degradation of anthocyanins involving phenolases and catechol (Figure 9.16). This reaction scheme was also confirmed by Sakamura et al. (1965), who found that eggplant anthocyanins are rapidly degraded by phenolases in the presence of catechin and chlorogenic acid. Degradation of anthocyanins may be due to their reaction with *o*-quinones or polymeric products of *o*-quinones condensation. Wesche-Ebeling and Montgomery (1990) studied the degradation of anthocyanins by strawberry polyphenolase in model systems containing D-catechin alone or in combination with pelargonidin or cyanidin. Polyphenol oxidase and D-catechin together caused the loss of 50 to 60% of anthocyanin pigments after 24 h at room temperature with the formation of a precipitate.

Similarly, flavone and flavonol glycosides remain unoxidized or are oxidized relatively slowly in the presence of oxidative enzymes such as *o*-polyphenoloxidase. However, they undergo a more rapid oxidization in the presence of transfer substances such as chlorogenic acid and catechin (Baruah and Swain, 1959).

NONENZYMATIC DISCOLORATION REACTIONS

Alkaline treatment of sunflower seeds brings about the development of a distinct color that progresses from a cream yellow to light green, dark green, and eventually brown. Thus, the conventional process of producing protein isolates from sunflower would result in dark and discolored products; this discoloration is due to the complexation of sunflower proteins with chlorogenic acid (Cater et al., 1972; Smith and Johnsen, 1948). Addition of sunflower meal to baked products also results in the formation of gray-colored crumbs (Clandinin, 1958).

Phenols are involved in discoloration of certain potato cultivars shortly after cooking (Table 9.11). This discoloration is often referred to as "after-cooking darkening." This is due to the formation of phenol-ferrous complexes that, upon exposure to air, turn to bluish-gray phenol-ferric complexes as cooking eliminates reducing conditions in the tuber (Heisler et al., 1964; Hughes et al., 1962; Hughes and Swain, 1962). The discoloration occurs more often in the stem end of cooked potatoes where phenols are more concentrated (Reeve et al., 1969). Mondy and Gosselin (1988)

TABLE 9.11
Effect of Irradiation[a] on Relative Tendency of Certain Potato Cultivars to Darken after Cooking

Cultivar	Type of Flesh Discoloration	
	Soon after Boiling	After 4-h Exposure to Air
Kufri Alankar	Pale gray	Dark gray
Kufri Deva	White	Pale gray
Kufri Kuber	Straw yellow	Straw yellow with pale gray background
Kufri Sheetman	Pale gray	Medium gray
Kufri Sinduri	Straw yellow	Straw yellow with pale medium-gray background

[a] kRad.

Source: Adapted from Thomas, P., 1981, *J. Food Sci.*, 46:1620–1621.

found that cortex tissue of potatoes cooked with the peel discolor significantly more ($p < 0.05$) than cortex tissues cooked without peel.

The appearance of a blue discoloration on the surface of the meat or in the coagulated blood released from the meat can greatly decrease the quality of canned or cooked crab. Discolorations ranging from blue-gray to black have been observed in canned Dungeness crab (Babbit et al., 1973a, b; Elliot and Harvey, 1951). Phenolics (Babbit et al., 1973b), iron (Waters, 1971) and copper (Babbit et al., 1975; Elliot and Harvey, 1951) are involved in the process of blueing canned crab meat. The discoloration is not considered purely enzymatic because the thermal processing of canned crab denatures the enzymes present. However, phenol oxidases present in live crab may participate in the initiation of blueing because precooking freshly caught crab greatly improves the overall quality of canned crabs (Babbit et al., 1975). According to Babbit et al. (1973b), the mechanism of blueing may involve initiation of phenol oxidation by enzymes followed by further oxidation and polymerization of phenols to colored melanins; this may proceed nonenzymatically, particularly in the presence of metals (Cu, Fe) and under alkaline conditions.

Canned green asparagus is often subjected to greenish-black discoloration of the spears and the liquid in the can — a phenomenon of great economic significance. The color change usually occurs from a few minutes to 3 hours after opening the can and is due to the formation of a pigmented rutin-ferric ion complex. The ferric ion is formed via oxidation and dissolution of iron from the can. The black discoloration is pH dependent and does not occur when sufficient amounts of stannic ions are present in the solution (Hernandez and Vosti, 1963; Lueck, 1970).

OTHER UNDESIRABLE EFFECTS

The reaction between polyphenolic compounds and proteins may produce haze and precipitates in stored beverages such as wine, beer and fruit juice. Ferulic acid is also considered a potential off-flavor precursor in stored orange juice. Under mild acidic conditions, ferulic acid may decarboxylate to *p*-vinylguaiacol (PVG), which is considered the most detrimental component of "old-fruit" or "rotten" flavor of the juice. Ferulic acid is generally present in orange juice in the bound form (feruoylglucose, feruloyl putrescine) (Reschke and Herrmann, 1981; Wheaton and Stewart, 1965). However, the juice contains up to 200 μg/L of free ferulic acid and pasteurization may increase its content to approximately 300 μg/L. It is also possible that some free ferulic acid may be generated from the bound forms during juice storage. These small amounts of free ferulic acid are sufficient to bring about off-flavor development in the juice (Naim et al., 1988).

Phenolic acids, notably chlorogenic and caffeic acids, are also implicated in the curding of unstabilized coffee cream in hot coffee known as feathering (Charley, 1982). Hamboyan et al. (1989) reported that the extent of feathering is not uniquely determined by the content of chlorogenic acid. However, when the amount of chlorogenic acid is considered together with a parameter that described the shape of ultraviolet spectrum in the 230- to 340-nm range, a reliable criterion for the occurrence of feathering can be obtained.

REFERENCES

Amiot, M.J., Fleuriet, A., Cheynier, V., and Nicolas, J. 1997. Phenolic compounds and oxidative mechanisms in fruits, in *Phytochemistry of Fruit and Vegetables,* Proceedings of the Phytochemical Society of Europe 41, Tomás-Barberán, F.A. and Robins, R.J., Eds., Clarendon Press, Oxford, U.K., 51–86.

Anthoni, U., Larsen, C., Nielsen, P.H., and Christophersen, C. 1990. Off-flavor from commercial crustaceans from North Atlantic zone. *Biochem. Syst. Ecol.*, 18:377–379.

Appelqvist, L.-A. 1972. Chemical constituents of rapeseed, in *Rapeseed,* Appelquist, L.A. and Olson, R., Eds., Elsevier Publishing Co., Amsterdam, 168–180.

Arai, S.H., Suzuki, H., Fujimaki, M., and Sakurai, Y. 1966. Flavor components in soybean. II. Phenolic acids in defatted soybean flour. *Agric. Biol. Chem.*, 30:364–369.

Arslan, O., Temur, A., and Tozlu, I. 1998. Polyphenol oxidase from Malaya apricot (*Prunus armeniaca* L.). *J. Agric. Food Chem.*, 46:1239–1241.

Asen, S., Stewart, R.N., and Norris, K.H. 1972. Co-pigmentation of anthocyanins in plant tissues and its effect on color. *Phytochemistry*, 11:1139–1144.

Babbit, J.K., Law, D.K., and Crawford, D.L. 1973a. Phenolases and blue discoloration in whole cooked Dungeness crab (*Cancer magister*). *J. Food Sci.*, 38:1089–1090.

Babbit, J.K., Law, D.K., and Crawford, D.L. 1973b. Blueing discoloration in canned crab meat (*Cancer magister*), *J. Food Sci.*, 38:1101–1103.

Babbit, J.K., Law, D.K., and Crawford, D.L. 1975. Effect of precooking on copper content, phenolic content and blueing of canned Dungeness crab meat. *J. Food Sci.*, 40:649–650.

Baruah, P. and Swain, T. 1959. Action of potato phenolase on flavonoid compounds. *J. Sci. Food Agric.*, 10:125–129.

Baryłko-Pikielna, N. 1977. Contribution of smoke compounds to sensory, bacteriostatic and antioxidative effects of smoked foods. *Pure Appl. Chem.*, 49:1667–1671.

Basiouny, F.M. and Buchanan, D.W. 1977. Fruit quality of sungold nectarine as influenced by shade and sprinkling. *Soil Crop Sci. Fla. Proc.*, 36:130.

Bate-Smith, E.C. 1954. Flavonoid compounds in foods. *Adv. Food Res.*, 5:261–300.

Bate-Smith, E.C. 1973. Haemanalysis of tannins: the concept of relative astringency. *Phytochemistry*, 12:907–912.

Benet, G., Dolev, A., and Tatarsk, D. 1973. Compounds contributing to heat-induced bitter off-flavor in avocado. *J. Food Sci.*, 38:546–547.

Benjamin, N.D. and Montgomery, M.W. 1973. Polyphenol oxidase of Royal Ann cherries. Purification and characterization. *J. Food Sci.*, 38:799–806.

Boyle, J.L., Lindsay, R.C., and Stuiber, D.A. 1992a. Bromophenol distribution in salmon and selected seafoods of fresh- and saltwater origin. *J. Food Sci.*, 57:918–922.

Boyle, J.L., Lindsay, R.C., and Stuiber, D.A. 1992b. Contributions of bromophenols to marine-associated flavors of fish and seafood. *J. Aquat. Food Prod. Technol.*, 1:43–63.

Bratzler, L.J., Spooner, M.E., Weatherspoon, J.B., and Maxey, J.A. 1969. Smoke flavor as related to phenol, carbonyl and acid content of bologna. *J. Food Sci.*, 34:146–148.

Braverman, J.B.S. 1963. *Introduction to the Biochemistry of Foods,* Elsevier Publishing Co., New York.

Brot, N., Smit, Z., and Weissbach, H. 1965. Conversion of L-tyrosine to phenol by *Clostridium tetanomorphum. Arch. Biochem. Biophys.*, 112:1–6.

Brouillard, R. 1982. Chemical structure of anthocyanins, in *Anthocyanins as Food Colors,* Markakis, P., Ed., Academic Press, New York, 1–40.

Brouillard, R. 1983. The *in vivo* expression of anthocyanins colour in plants. *Phytochemistry*, 22:1311–1323.

Brouillard, R. 1988. Flavonoids and flower colour, in *The Flavonoids: Advances in Research Since 1980,* Harborne, J.B., Ed., Chapman & Hall, London, 525–538.

Brouillard, R. and Dangles, O. 1994. Flavonoids and flower colour, in *The Flavonoids: Advances in Research Since 1986,* Harborne, J.B., Ed., Chapman & Hall, London, 565–588.

Brouilard, R., Figuerido, P., Elhabire, M., and Dangles, O. 1997. Molecular interactions of phenolic compounds in relation to colour of fruits and vegetables, in *Phytochemistry of Fruit and Vegetables, Proceedings of the Phytochemical Society of Europe 41*, Tomás-Barberán, F.A. and Robins, R.J., Eds., Clarendon Press, Oxford, U.K., 29–50.

Burnette, F.S. 1977. Peroxidase and its relationship to food flavor and quality: a review. *J. Food Sci.*, 42:1–6.

Butler, E.J., Pearson, A.W., and Fenwick, G.R. 1982. Problems which limit the use of rapeseed meal as a protein source in poultry diets. *J. Sci. Food Agric.*, 33:866–875.

Casimir, D.J., Keffort, J.F., and Whitfield, F.B. 1981. Technology and flavor chemistry of passion fruit juices and concentrates. *Adv. Food Res.*, 27:243–296.

Cater, C.M., Gheyasuddin, S., and Mattil K.F. 1972. Effect of chlorogenic, quinic, and caffeic acids on the solubility and color of protein isolates especially from sunflower seed. *Cereal Chem.*, 49:508–514.

Chalmers, D.J. and Faragher, J.D. 1977. Regulation of anthocyanin synthesis in apple skin. I. Comparison of the effects of cycloheximide, ultraviolet light, wounding and maturity. *Aust. J. Plant Physiol.*, 4:111–121.

Charley, H. 1982. *Food Science*, 2nd ed., John Wiley & Sons, New York.

Chen, L., Mehta, A., Berenbaum, M., Zangerl, A.R., and Engeseth, N.J. 2000. Honeys from different floral sources as inhibitors of enzymatic browning in fruit and vegetable homogenates. *J. Agric. Food Chem.*, 48:4997–5000.

Chen, J.S., Wei, C., Rolle, R.S., Otwell, W.S., Balaban, M.O., and Marshall, M.R. 1991a. Inhibitory effect of kojic acid on some plant and crustacean polyphenol oxidases. *J. Agric. Food Chem.*, 39:1396–1401.

Chen, J.S., Wei, C., and Marshall, M.R. 1991b. Inhibition mechanism of kojic acid on polyphenol oxidase. *J. Agric. Food Chem.*, 39:1897–1901.

Clandinin, D.R. 1958. Sunflower seed oil meal, in *Processed Plant Protein Foodstuffs,* Altchul, A.M., Ed., Academic Press, New York, 557.

Clifford, M.N. 1997. Astringency, in *Phytochemistry of Fruit and Vegetables, Proceedings of the Phytochemical Society of Europe 41*, Tomás-Barberán, F.A. and Robins, R.J., Eds., Clarendon Press, Oxford, U.K., 87–108.

Clifford, M.N. 2000. Nature, occurrence and dietary burden. *J. Sci. Food Agric.*, 80:1063–1072.

Coultate, T.P. 1984. *Food — the Chemistry of its Components*, The Royal Society of Chemistry, London.

Crosby, G.A. 1976. New sweeteners. *CRC Crit. Rev. Food Sci. Nutr.*, 7:297–323.

Crouzet, J., Sakho, M., and Chassagne, D. 1997. Fruit aroma precursors with special reference to phenolics, in *Phytochemistry of Fruit and Vegetables, Proceedings of the Phytochemical Society of Europe 41*, Tomás-Barberán, F.A. and Robins, R.J., Eds., Clarendon Press, Oxford, U.K., 109–124.

Dadic, M. and Belleau, G. 1973. Polyphenolics and beer flavor. *Am. Soc. Brew. Chem. Proc.*, 107–114.

Dame, C., Chichester, C.O., and Marsh, G.L. 1959. Studies of processed all-green asparagus. IV. Studies on the influence of tin on the solubility of rutin and the concentration of rutin present in the brines of asparagus processed in glass and tin containers. *Food Res.*, 24:28–36.

Dangles, O., Saito, N., and Brouillard, R. 1993. Kinetic and thermodynamic control of flavylium hydration of the pelargonidin-cinnamic acid complexation. Origin of extraordinary color diversity of *Pharbitis nil. J. Am. Chem. Soc.*, 115:3125–3132.

Dao, L.T., Tekeoka, G.R., Edwards, R.H., and Berrios, J.J. 1998. Improved method for stabilization of anthocyanidins. *J. Agric. Food Chem.*, 46:3564–3569.

Daun, H. 1972. Sensory properties of selected phenolic compounds isolated from curing smoke as influenced by its generation parameters. *Lebensm.-Wiss.Technol.*, 5:102–105.

Dawley, R.M. and Flurkey, W.H. 1993. 4-Hexylresorcinol, a potent inhibitor of mushroom tyrosinase. *J. Food Sci.*, 58:609–610; 670.

Debicki-Pospisil, J., Lovric, T., Trinajstic, N., and Sabljic, A. 1982. Anthocyanin degradation in the presence of furfural and 5-hydroxymethylfurfural. *J. Food Sci.*, 48:411–416.

de Ancos, B., Ibañez, E., Reglero, G., and Cano, M.P. 2000. Frozen storage effects on anthocyanins and volatile compounds of raspberry fruit. *J. Agric. Food Chem.*, 48:873–879.

Delcour, J.A., Vandenberghe, M.M., Corten, P.F., and Dondeyne, P.C. 1984. Flavor thresholds of polyphenolics in water. *Am. J. Enol. Vitic*, 35:134–136.

Delgado-Vargas, F., Jiménez, A.R., and Parades-López, O. 2000. Natural pigments: carotenoids, anthocyanins, and betalains-characteristics, biosynthesis, processing and stability. *CRC Crit. Rev. Food Sci. Nutr.*, 40:173–289.

Dimick, K.P., Ponting, J.D., and Makover, B. 1951. Heat inactivation of polyphenolase in fruit puree. *Food Technol.*, 5:237–241.

DuBois, G.E., Crosby, G.A., and Saffron, P. 1977. Nonnutritive sweetners: taste-structure relationships for new simple dihydrochalcones. *Science*, 195:397–366.

Elliott, H.H. and Harvey, E.W. 1951. Biological methods of blood removal and their effectiveness in reducing discoloration in canned Dungeness crab meat. *Food Technol.*, 5:163–166.

Enei, H., Matsui, H., Yamashita, K., Okumura, S., and Yamada, H. 1972. Microbial synthesis of L-tyrosine and 3,4-di-hydroxyphenyl-L-alanine I. Distribution of tyrosine phenol lyase in microorganisms. *Agric. Biol. Chem.*, 36:1861–1868.

Escribano, J., Cabanes, J., Chazarra, S., and Garcia-Carmona, F. 1997. Characterization of monophenolase activity of table beet polyphenol oxidase. Determination of kinetic parameters on the tyramine/dopamine pair. *J. Agric. Food Chem.*, 45:4209–4214.

Eskin, N.A.M. 1990. *Biochemistry of Foods*, 2nd ed., Academic Press Inc., San Diego, CA.

Espin, J.C., Trujano, M.F., Tudela, J., and Garcia-Canovas, F. 1997. Monophenolase activity of polyphenol oxidase from Haas avocado. *J. Agric. Food Chem.*, 45:1091–1096.

Faragher, J.D. 1983. Temperature regulation of anthocyanin accumulation in apple skin. *J. Exp. Bot.*, 34:1291–1298.

Fenwick, G.R., Curl, C.L., Butler, E.J., Greenwood, N.M., and Pearson, A.W. 1984. Rapeseed meal and egg taint: effects of low glucosinolate *Brassica napus* meal, dehulled meal and hulls, and of neomycin. *J. Sci. Food Agric.*, 35:749–756.

Fiddler, W., Waserman, A.E., and Doerr, R.C. 1970. "Smoke" flavour fraction of a liquid smoke solution. *J. Agric. Food Chem.*, 18:934–936.

Figueiredo, P., Elhabiri, M., Toki, K., Saito, N., and Brouillard, R. 1996a. Anthocyanin intramolecular interactions. A new mathematical approach to account for the remarkable colorant properties of the pigments extracted from *Matthiola incana*. *J. Am. Chem. Soc.*, 118:4788–4793.

Figueiredo, P., Elhabiri, M., Toki, K., Saito, N., and Brouillard, R. 1996b. A new aspect of anthocyanin complexation. Intramolecular copigmentation as a means for color loss? *Phytochemistry*, 41:301–308.

Forsyth, W.G.C. and Quesnel, V.C. 1957. Cacao phenolic substances. 4. The anthocyanins pigments. *Biochem. J.,* 65:177–179.

Fraignier, M.P., Marques, L., Fleuriet, A., and Macheix, J. 1995. Biochemical and immunochemical characterics of polyphenol oxidase from different fruits of *Prunus*. *J. Agric. Food Chem.*, 43:2375–2380.

Francis, F.J. and Harborne, J.B. 1966. Anthocyanins of the garden huckleberry, *Solanum guineese*. *J. Food Sci.*, 31:524–528.

Francis, F.J. 1989. Food colorants: anthocyanins. *CRC Crit. Rev. Food Sci. Nutr.*, 28:273–314.

Friedman, M. 1996. Food browning and its prevention: an overview. *J. Agric. Food Chem.*, 44:631–653.

Gil, M.I., Holcroft, D.M., and Kader, A.A. 1997. Changes in strawberry anthocyanins and other polyphenols in response to carbon dioxide treatments. *J. Agric. Food Chem.*, 45:1662–1667.

Golbeck, J.H. and Cammarata, K. 1981. Spinach polyphenol oxidase. Isolation, activation and properties of the native chloroplast enzyme. *Plant Physiol.*, 67:977–984.

Goldstein, J.L. and Swain, T. 1963. Changes in tannins in ripening fruit. *Phytochemistry*, 2:371–383.

Gonzalez-Aguilar, G.A., Wang, C.Y., and Buta, J.G. 2000. Maintaining quality of fresh-cut mangoes using antibrowning agents and modified atmosphere packaging. *J. Agric. Food Chem.*, 48:4204–4208.

Govindarajan, V.S. 1979. Pungency: the stimuli and their evaluation, in *Food Taste Chemistry*, Boudreau J.C., Ed., American Chemical Society, Washington, D.C., 52–91.

Goto, T. 1987. Structure, stability and color variation of natural anthocyanins. *Prog. Chem. Org. Nat. Prod.*, 52:113–158.

Ha, J.K. and Lindsay, R.C. 1990. Distribution of branched-chain fatty acids in perinephric fats of various red meat species. *Lebensm. Wissen. u- Technol.*, 23:433–440.

Ha, J.K. and Lindsay, R.C. 1991a. Volatile branched-chain fatty acids and phenolic compounds in aged Italian cheese flavors. *J. Food Sci.*, 56:1241–1247.

Ha, J.K. and Lindsay, R.C. 1991b. Volatile alkylphenols and tiophenols in species-related characterizing flavors of red meats. *J. Food Sci.*, 56:1197–1202.

Hall, I.V. and Stark, R. 1972. Anthocyanin production in cranberry leaves and fruit related to cool temperature at a low light intensity. *Hort. Res.*, 12:183–187.

Hamboyan, L., Pink, D., Klapstein, D., MacDonald, L., and Aboud, H. 1989. Ultraviolet spectroscopic studies on the feathering of cream in instant coffee. *J. Dairy Sci.*, 56:741–748.

Hamm, R., 1977. Analysis of smoke and smoke products. *Pure Appl. Chem.*, 49:1655–1666.

Harborne, J.B. 1958. Spectral methods of characterizing anthocyanins. *Biochem. J.*, 70:22–28.

Harborne, J.B. 1964. Plant polyphenols. XI. The structure of acetylated anthocyanins. *Phytochemistry*, 3:151–160.

Harborne, J.B. 1967. *Comparative Biochemistry of Flavonoids,* Academic Press, New York.

Harborne, J.B., Mabry, T.J., and Mabry, H. 1975. *The Flavonoids*, Chapman & Hall, London.

Harborne, J.B., Ed., 1988a. *The Flavonoids: Advances in Research Since 1980*, Chapman & Hall, London.

Harborne, J.B. 1988b. The flavonoids: recent advances, in *Plant Pigments,* Goodwin, T.W., Ed., Academic Press, London, 298–343.

Harborne, J.B. and Grayer, R.J. 1988. The anthocyanins, in *The Flavonoids: Advances in Research Since 1980*, Harborne, J.B., Ed., Chapman & Hall, London, 1–20.

Harborne, J.B. and Williams, C.A. 2001. Anthocyanins and other flavonoids. *Nat. Prod. Rep.*, 18:310–333.

Harel, E., Mayer, A.M., and Lehman, E. 1973. Multiple forms of *Vitis vinifera* catechol oxidase. *Phytochemistry*, 12:2649–2654.

Haslam, E. 1975. The chemistry and biochemistry of plant proanthocyanins, in *Top. Flavonoids Chem. Biochem. Proc. Hung. Bioflavonoid Symp. 4th* (1973), Farkas, L., Gabor, M., and Kelley, F. Eds., Elsevier Publishing Co., Amsterdam, 77–97.

Haslam, E. 1981. Vegetable tannins, in *The Biochemistry of Plants,* Vol. 7, Stumpf, P.K. and Conn., E.E., Eds., Academic Press, London, 527–556.

Haslam, E. and Lilley, T.H. 1988. Natural astringency in foodstuffs. A molecular intrepretation. *CRC Crit. Rev. Food Sci. Nutr.*, 27:1–40.

Haslam, E., Lilley, T.H., and Azawa, T. 1986. Polyphenol complexation. Astringency in fruits and beverages. *Bull. Liaison Groupe Polyphenols*, 13:352–353.

Heisler, E.G., Siciliano, J., Treadway, R.H., and Porter, W.L. 1964. After-cooking darkening of potatoes. Role of organic acids. *J. Food Sci.*, 29:555–564.

Hernandez, H.H. and Vosti, D.C. 1963. Dark discoloration of canned all-green asparagus. I. Chemistry and related factors. *Food Technol.*, 17:95–99.

Herrmann, K. 1976. Flavonols and flavones in food plants: a review. *J. Food. Technol.*, 11:433–448.

Hicks, K.B., Haines, R.M., Tong, C.B.S., Sapers, G.M., El-Atawy, Irwin, P.L., and Seib, P.A. 1996. Inhibition of enzymatic browning in fresh fruit and vegetable juices by soluble and insoluble forms of β-cyclodextrin alone or in combination with phosphates. *J. Agric. Food Chem.*, 44:2591–2594.

Hobson-Frohock, A., Land, D.G., Griffiths, N.M., and Curtis, R.F. 1973. Egg taints: association with trimethylamine. *Nature*, 243:303–305.

Horowitz, R.M. and Gentili, B. 1969. Taste and structure of in phenolic glycosides. *J. Sci. Food Agric.*, 17:696–700.

Horowitz, R.M. and Gentili, B. 1974. Dihydrochalcone sweeteners, in *Sweeteners*, Inglett, G.E., Ed., AVI Publishing Co., Westport, CT, 182–193.

Hoshino, T., Matsumoto, U., Goto, T., and Harada, N. 1980. Evidence for self-association of anthocyanins. IV. PMR spectroscopic evidence for the vertical stacking of anthocyanin molecules. *Tetrahedron Lett.*, 23:433–436.

Hrazdina, G. 1981. Anthocyanins and their role in food products. *Lebensm. Wissen.u- Technol.*, 14:283–286.

Huang, H.T. 1955. Decolorization of anthocyanins by fungal enzymes. *J. Agric. Food Chem.*, 3:141–146.

Huang, H.T. 1956. The kinetics of the decolorization of anthocyanins by "anthocyanase." *J. Am. Chem. Soc.*, 78:2390–2393.

Huang, C.A. and Zayas, J.F. 1991. Phenolic acid contribution to taste characteristics of corn germ protein flour products. *J. Food Sci.*, 56:1308–1313; 1315.

Hughes, J.C., Ayers, J.E., and Swain, T. 1962. After-cooking blackening in potatoes. I. Introduction and analytical methods. *J. Sci. Food Agric.*, 13:224–228.

Hughes, J.C. and Swain, T. 1962. After-cooking blackening in potatoes. II. Core experiments. *J. Sci. Food Agric.*, 13:229–236.

Iacobucci, G.A. and Sweeny, J.G. 1983. The chemistry of anthocyanins, anthocyanidins and related flavylium salts. *Tetrahedron*, 39:3005–3038.

Im, J.-S., Parkin, K.L., and Von Elbe, J.H. 1990. Endogenous polyphenoloxidase activity associated with the "black ring" defect in canned beet (*Beta vulgaris* L.) root slices. *J. Food Sci.*, 55:1042–1045 and 1059.

Inglett, G.E. 1969. Dihydrochalcone sweetners — sensory and stability evaluation. *J. Food Sci.*, 34:101–103.

Interesse, F.S., Alloggio, V., Lamparelli, F., and Davella, G. 1984. Characterization of the oxidative enzymatic system of the phenolic compounds from muscat grapes. *Lebensm. Wiss. u-Technol.*, 17:5–10.

Irwin, P.L., Pfeffer, P.E., Doner, L.W., Sapers, G.M., Brewster, J.D., Nagahashi, G., and Hicks, K.B. 1994. Binding geometry, stoichiometry, and thermodynamics of cyclomaltooligosaccharide (cyclodextrin) inclusion complex formation with chlorogenic acid, the major substrate of apple polyphenol oxidase. *Carbohydr. Res.*, 256:13–27.

Itoh, S., Kumei, H., Taki, M., Nagatomo, S., Kitagawa, T., and Fukuzumi, S. 2001. Oxygenation of phenols to catechols by A (μ-η^2:η^2-peroxo)dicopper(II) complex: mechanistic insight into the phenolase activity of tyrosinase. *J. Am. Chem. Soc.*, 123:6708–6709.

Jackman, R.L., Yada, R.Y., Tung, M.A., and Speers, A. 1987. Anthocyanins as a food colorants — a review. *J. Food Biochem.*, 11:201–247.

Jackman, R.L. and Smith, J.L. 1996. Anthocyanins and betalains, in *Natural Food Colorants,* Hendry, G.F. and Houghton, J.D., Eds., Blackie Academic & Professional, London, 244–309.

Joslyn, M.A. and Peterson, R. 1956. Occurrence of leucoanthocyanin in pears. *Nature*, 178:318.

Ju, Z., Liu, C., and Yuan, Y. 1995. Activities of chalcone synthase and UDPGal:flavonoid-3-*O*-glycosyltransferase in relation to anthocyanin synthesis in apple. *Sci. Hortic.*, 63:175–185.

Kader, A.A. 1986. Biochemical and physiological basis for effects of controlled and modified atmospheres on fruits and vegetables. *Food Technol.*, 40:99–100, 102–104.

Kahn, V. 1976. Effect of some phenolic compounds on the oxidation of 4-methyl catechol catalyzed by avocado polyphenol oxidase. *J. Food Sci.*, 41:1011–1012.

Kahn, V. and Andrawis, A. 1985. Inhibition of mushroom tyrosinase by tropolone. *Phytochemistry*, 24:905–908.

Kalitka, V.V. and Skrypnik, V.V. 1983. Changes in the anthocyanin content of cherries during controlled athmosphere storage. *Konservn. Ovoshchesush. Prom-St.*, 11:46–47.

Ke, D., Goldstein, L., O'Mahony, M., and Kader, A.A. 1991. Effects of short term exposure to low O_2 and high CO_2 atmospheres on quality attributes of strawberries. *J. Food Sci.*, 56:50–54.

Kenten, R.H. 1958. Latent phenolase in extracts of broad bean (*Vicia faba* L.) 2. Activation by anionic wetting agents. *Biochem. J.*, 68:244–251.

Kenten, R.H. 1957. Latent phenolase in extracts of broad bean (*Vicia faba* L.) leaves. 1. Activation by acid and alkali. *Biochem. J.*, 67:300–307.

King, R.S. and Flurkey, W.H. 1987. Effect of limited proteolysis on broad bean polyphenol oxidase. *J. Sci. Food Agric.*, 41:231–240.

Kuhnau, J. 1976. The flavonoids — a class of semiessential food components: their role in human nutrition. *World Rev. Nutr. Diet.*, 24:117–191.

Kurko, V.I. 1963. *The Chemical and Physiochemical Foundations of the Smoking Process*, Wydawnictwo Przemysłu Lekkiego i Spozywcżego, Warsaw, Poland (in Polish).

Labuza, T.P., Lillemo, H.H., and Paoukis, P.S. 1992. Inhibition of polyphenol oxidase by proteolytic enzymes. *Fruit Proc.*, 2:9–13.

Laveda, F., Nunez-Delicado, E., Garcia-Carmona, F., and Sanchez-Ferrer, A. 2001. Proteolytic activation of latent Paraguaya peach PPO. Characterization of monophenolase activity. *J. Agric. Food Chem.*, 49:1003–1008.

Lea, A.G.H. and Arnold, G.M. 1978. The phenolics of ciders: bitterness and astringency. *J. Sci. Food Agric.*, 29:478–483.

Lehrian, D.W., Keeney, P.G., and Lopez, A.S. Method for the measurement of phenols associated with the smoky/hammy flavor defect of cocoa beans and chocolate liquor. *J. Food Sci.*, 43:743–745.

Li, C. and Kader, A.A. 1989. Residual effects of controlled atmospheres on postharvest physiology and quality of strawberries. *J. Am. Soc. Hort. Sci.*, 114:629–634.

Lin, T.Y., Koehler, P.E., and Schewfelt, R.L. 1989. Stability of anthocyanins in the skin of Starkrimson apples stored unpacked, under heat shrinkable wrap and in package modified athmosphere. *J. Food Sci.*, 54:405–407.

Lueck, R.H. 1970. Black discoloration in canned asparagus: interrelations of iron, tin, oxygen, and rutin. *J. Agric. Food Chem.*, 18:607–612.

Luh, B.S., Leonard, S.J., and Patel, D.S. 1960. Pink discoloration in canned Bartlett pears. *Food Technol.*, 14:53–56.

Lund, D.B. 1977. Design of thermal processes for maximizing nutrient retention. *Food Technol.*, 31:532–538.

Macheix, J.-J., Fleuriet, A., and Billot, J. 1989. *Fruit Phenolics,* CRC Press, Boca Raton, FL.

Maga, J.A. and Lorenz, K. 1973. Taste thresholds values for phenolic acids which can influence flavor properties of certain flours, grains, and oilseeds. *Cereal Sci. Today*, 18:326–329.

Maga, J.A. 1978. Simple phenol and phenolic compounds in food flavor. *CRC Crit. Rev. Food Sci. Nutr.*, 10:323–372.

Markakis, P. 1960. Zone electrophoresis of anthocyanins. *Nature*, 187:1092–1093.

Markakis, P. 1974. Anthocyanins and their stability in foods. *CRC Crit. Rev. Food Technol.*, 4:437–456.

Markakis, P. 1982. *Anthocyanins as Food Colors*, Academic Press, New York.

Marwan, A.G. and Nagel, C.W. 1982. Separation and purification of hydroxycinnamic acid derivatives in cranberries. *J. Food Sci.*, 47:585–588.

Matheis, G. and Whitaker, J.R. 1984. Modification of proteins by polyphenoloxidase and peroxidase and their products. *J. Food Biochem.*, 8:137–162.

Mathew, A.G., and Parpia, H.A.B. 1971. Food browning as a polyphenol reaction. *Adv. Food Res.*, 19:75–145.

Matsuo, T. and Itoo, S. 1982. A model experiment for deastringency of persimmon fruit with high carbon dioxide treatment, *in vitro* gelation of kaki-tannin by reacting with acetylaldehyde. *Agric. Biol. Chem.*, 46:683–689.

Mazza, G. and Brouillard, R. 1987. Recent developments in the stabilization of anthocyanins in food products. *Food Chem.*, 25:207–225.

Mazza, G. and Miniati, E. 1993. *Anthocyanins in Fruits, Vegetables and Grains,* CRC Press, Boca Raton, FL.

Mayer, A.M. and Harel, E. 1979. Review: polyphenol oxidase in plants. *Phytochemistry,* 18:193–225.

McManus, J.P., Davis, K.G., Lilley, T.H., and Haslam, E. 1981. The association of proteins with polyphenols. *J. Chem. Soc. Chem. Commun.*, 309–311.

Mehansho, H., Butler, L.G., and Carlson, D.M. 1987. Dietary tannins and salivary proline-rich proteins: interactions, induction and defense mechanisms. *Ann. Rev. Nutr.*, 7:423–440.

Meschter, E.E. 1953. Fruit color loss, effects of carbohydrates and other factors on strawberry products. *J. Agric. Food Chem.*, 1:574–579.

Millin, D.J., Sinclair, D.S., and Swaine, D. 1969. Nonvolatile components of black tea and their contribution to the character of the beverage. *J. Agric. Food Chem.*, 17:717–722.

Molnar-Perl, I. and Friedman, M. 1990a. Inhibition of browning reaction by sulfur amino acids. 2. Fruit juices and protein containing foods. *J. Agric. Food Chem.*, 38:1648–1651.

Molnar-Perl, I. and Friedman, M. 1990b. Inhibition of browning reaction by sulfur amino acids. 3. Apples and potatoes. *J. Agric. Food Chem.*, 38:1652–1656.

Mondy, N.I., Metcalf, C., and Plaisted, R.L. 1971. Potato flavor as related to chemical composition. I. Polyphenols and ascorbic acids. *J. Food Sci.*, 41:459–461.

Mondy, N.I. and Gosselin, B. 1988. Effect of peeling on total phenols, total glycoalkaloids, discoloration and flavor of cooked potatoes. *J. Food Sci.*, 53:756–759.

Monsalve-Gonzalez, A., Barbosa-Canovas, G.V., McEvily, A., and Iyengar, A.J. 1995. Inhibition of enzymatic browning in apple products by 4-hexylresorcinol. *Food Technol.*, 49:110–118.

Mookherjee, B.D., Wilson, R.A., Trenkle, R.W., Zampino, M.J., and Sands, K.P. 1989. New trends in flavor research: herbs and spices, in *Flavor Chemistry: Trends and Developments,* Teranishi, R., Buttery, R.G., and Shahidi, F., Eds., ACS Symposium Series 388, American Chemical Society, Washington, D.C., 176–187.

Moore, B.M. and Flurkey, W.H. 1990. Sodium dodecyl sulfate activation of a plant polyphenol oxidase. *J. Biol. Chem.*, 265:4982–4988.

Naczk, M., Amarowicz, R., Sullivan, A., and Shahidi, F. 1998. Current research developments on polyphenolics of rapeseed/canola: a review. *Food Chem.*, 62:489–502.

Nagel, C.W., Herrick, I.W., and Graber, W. 1987. Is chlorogenic acid bitter? *J. Food Sci.*, 52:213.

Naim, M., Striem, B.J., Kanner, J., and Peleg, H. 1988. Potential of ferulic acid as a precursor to off-flavors in stored orange juice. *J. Food Sci.*, 53:500–503; 512.

Naumann, W.D. and Wittemburg, U. 1980. Anthocyanins, soluble solids and titrable acidity in blackberries as influenced by preharvest temperature. *Acta Hortic.*, 112:183.

Ney, K.H. 1973. Technique of flavor research. *Guardian*, 73:380, 382, 384, 387.

Nicolas, J.J., Richard-Forget, F., Goupy, P., Amiot, M.J., and Aubert, S. 1994. Enzymatic browning reactions in apple and apple products. *CRC Crit. Rev. Food Sci. Nutr.*, 34:109–157.

Nunez- Delicado, E., Bru, R., Sanchez-Ferrer, A., and Garcia-Carmona, F. 1996. Triton X-114-aided purification of latent tyrosinase. *J. Chromatogr.*, 680:105–112.

Nurstern, H.E. 1970. Volatile compounds: the aroma of fruits, in *The Biochemistry of Fruits and Their Products*, Vol. 1, Hulme, A.C., Ed., Academic Press, London, 239.

Ohta, H., Akuta, S., and Osajima, Y. 1980. Stability of anthocyanin pigments and related compounds in acidic solutions. *Nippon Shokuhin Gakkaishi*, 27:81–85.

Osawa, Y. 1982. Copigmentation of anthocyanins, in *Anthocyanins as Food Colors*, Markikis, P., Ed., Academic Press, New York, 41–86.

Oszmianski, J. and Lee, C.Y. 1990. Inhibition of polyphenol oxidase activity and browning by honey. *J. Agric. Food Chem.*, 38:1892–1895.

Otwell, W.S. and Marshall, M.R. 1986. Screening alternatives to sulfiting agents to control shrimp melanosis, in *Proceedings of the 11th Annual Trop. Subtrop. Fish Technol. Conference of the Americas,* Ward and Smith, compilers, TAMU-SG-86–102, Texas A & M University, College Station, TX, 35–44.

Ozawa, T., Lilley, T.H., and Haslam, E. 1987. Polyphenol interactions. Part 3. Astringency and the loss of astringency in ripening fruits. *Phytochemistry*, 26:2937–2942.

Pearson, A.W., Bulter, E.J., and Fenwick, G.R. 1980. Rapeseed meal and egg taint: the role of sinapine. *J. Sci. Food Agric.*, 31:898–904.

Peng, C.Y. and Markakis, P. 1963. Effect of phenolase on anthocyanins. *Nature*, 199:597–598.

Pifferi, P.G. and Cultrera, R. 1974. Enzymatic degradation of anthocyanins: the role of sweet cherry polyphenoloxidase. *J. Food Sci.*, 39:786–791.

Polesello, A. and Bonzini, C. 1977. Observations on pigments of sweet cherries and on pigment stability during frozen storage. *Confructa*, 22: 170–175.

Ponting, J.D., Jackson, R., and Watters, G. 1971. Refrigerated apple slices: effects of pH, sulfites and calcium on texture. *J. Food Sci.*, 36:349–350.

Potter, N.N. 1986. *Food Science*, 4th ed., Van Nostrand Reinhold, New York.

Pratt, D.E. 1972. The role of phloroglucinol in browning of chlorogenic acid catalyzed by apple polyphenol oxidase. *Can. Inst. Food Sci. Technol. J.*, 5:207–209.

Purseglove, J.W., Brown, E.G., Green, C.L., and Robbins, S.R.J., 1981. *Spices* (*Vanilla*), vol. 2, Longman, London.

Ramirez-Martinez, J.R., Levi, A., Padua, H., and Bakal, A. 1977. Astringency in an intermediate moisture banana products. *J. Food Sci.*, 42:1201–1203; 1217.

Reeve, R.M., Hautala, E., and Weaver, M.L. 1969. Anatomy and compositional variation within the potatoes. II. Phenolics, enzymes, and other minor components. *Am. Potato J.*, 46:374–386.

Reschke, A. and Herrmann, K. 1981.Occurrence of 1-*O*-hydroxycinnamoyl-β-D-glucoses in fruits. 15. Phenolics of fruits. *Lebensm. -Unters. Forsch.*, 173:458–463.

Richard-Forget, F.C. and Gaullaird, F.A. 1997. Oxidation of chlorogenic acid, catechins, and 4-methylcatechol in model solutions by combinations of pear (*Pyrus communis* Cv. Williams) polyphenol oxidase and peroxidase in enzymatic browning. *J. Agric. Food Chem.*, 45:2472–2476.

Robb, D.A., Swain, T., and Mapson, L.W. 1966. Substrates and inhibitors of the activated tyrosinase of broad bean (*Vicia faba* L.). *Phytochemistry*, 5:665–675.

Roberts, E.A.H., Cartwright, R.A., and Wood, D.J. 1956. The flavonols of tea. *J. Sci. Food Agric.*, 7:637–646.

Robinson, D.S. 1991. Peroxidases and catalases in foods, in *Oxidative Enzymes in Foods*, Robinson, D.S. and Eskin, N.A.M., Eds., Elsevier Publishing Co., London, p. 1.

Rohan, T.A. and Connell M. 1964. The precursor of chocolate aroma: a study of the flavonoids and phenolic acids. *J. Food Sci.*, 29:460–464.

Rouet-Mayer, M.A. and Philippon, J. 1986. Inhibition of catechol oxidases from apples by sodium chloride. *Phytochemistry*, 25:2717–2719.

Rouseff, R.L., Martin, S.F., and Youtsey, C.O. 1987. Quantitative survey of narirutin, naringin, hesperidin and neohesperidin in *Citrus*. *J. Agric. Food Chem.*, 35:1027–1030.

Russell, G.F. and Oslon, K.V. 1972. The volatile constituents of oil of thyme. *J. Food Sci.*, 37:405–407.

Sakamura S., Watanabe S., and Obata T. 1965. Anthocyanase and anthocyanins occurring in eggplant (*Solanum molegana*). III. Oxidative decolorization of anthocyanins by polyphenol oxidase. *Agric. Biol. Chem.*, 29:181–190.

Sanchez-Ferrer, A., Rodriguez-Lopez, J.N., Garcia-Canovas, F., and Garcia-Carmona, F. 1995. Tyrosinase: a comprehensive review of its mechanism. *Biochim. Biophys. Acta*, 1247:1–11.

Sanchez-Ferrer, A., Laveda, F., and Garcia-Carmona, F. 1993. Substrate-dependent activation of latent potato leaf polyphenol oxidase by anionic surfactants. *J. Agric. Food Chem.*, 41:1219–1223.

Sanderson, G.W. and Graham, H.N. 1973. On the formation of black tea aroma. *J. Agric. Food Chem.*, 21:576–585.

Sanderson, G.W., Ranadive, A.S., Eisenberg, L.S., Farrell, F.J., Simons, R., Manley, C.H., and Coggon, P. 1976. Contribution of polyphenolic compounds to the taste of tea, in *Sulfur and Nitrogen Compounds in Food Flavors*, Charalambous, G. and Katz, I., Eds., ACS Symposium Series 26, American Chemical Society, Washington, D.C., 14–16.

Sapers, G.M., Garzarella, L., and Pilizota, V. 1990. Application of browning inhibitors to cut apple and potato by vacuum and pressure infiltration. *J. Food Sci.*, 55:1049–1053.

Sapers, G.M. 1993. Browning of foods: control by sulfites, antioxidants and other means. *Food Technol.*, 47:75–84.

Sarma, A.D., Sreelakshmi, Y., and Sharma, R. 1997. Antioxidant ability of anthocyanins against ascorbic acid oxidation. *Phytochemistry*, 45:671–674.

Savagaon, K.A. and Sreenivasan, A. 1978. Activation mechanism of prephenolase in lobster and shrimp. *Fish Technol.*, 15:49–55.

Shannon, C.T. and Pratt, D.E. 1967. Apple polyphenol oxidase activity in relation to various phenolic compounds. *J. Food Sci.*, 32:479–482.

Shrikhande, A.J. 1976. Anthocyanins in foods. *CRC Crit. Rev. Food Technol.*, 24:169–218.

Siegelman, H.W. and Hendricks, S.B. 1958. Photocontrol of anthocyanin synthesis in apple skin. *Plant Physiol.*, 33:185–190.

Simpson, B.K., Marshall, M.R., and Otwell, W.S. 1988. Phenoloxidases from pink and white shrimp: kinetic and other properties. *J. Food Biochem.*, 12:205–217.

Sinden, S.L., Deahl, R.L., and Aulenback, B.B. 1976. Effect of glycoalkaloids and phenolics on potato flavor. *J. Food Sci.*, 41:520–523.

Singlenton, V.L. and Nobel, A.C. 1976. Wine flavor and phenolic substances, in *Phenolic Sulphur and Nitrogen Compounds in Food Flavors*, ACS Symposium Series 26, Charalambous, G. and Katz, I., Eds., American Chemical Society, Washington, D.C., 47–70.

Smith, A.K. and Johnsen, V.L. 1948. Sunflower seed protein. *Cereal Chem.*, 25:399–406.

Soliva-Fortuny, R.C., Grigelmo-Miguel, N., Odriozola-Serrano, I., Gorinstein, S., and Martin-Belloso, O. 2001. Browning evaluation of ready-to-eat apples as affected by modified atmosphere. *J. Agric. Food Chem.*, 49:3685–3690.

Somers, T.C. 1966. Grape phenolics: the anthocyanins of *Vitis vinifera*, variety Shiraz. *J. Sci. Food Agric.*, 17:215–219.

Son, S.M., Moon, K.D., and Lee, C.Y. 2000. Kinetic study of oxalic acid inhibition on enzymatic browning. *J. Agric. Food Chem.*, 48:2071–2074.

Sosulski, F.W., Humbert, E.S., Lin, M.J.Y., and Card, J.W. 1977. Rapeseed-supplemented wieners. *Can. Inst. Food Sci. Technol. J.*, 10:9–12.

Stagg, G.V. and Millin, D.J. 1975. The nutritional and therapeutic value of tea — a review. *J. Sci. Food Agric.*, 26:1439–1459.

Stewart, T.F. 1970. The precursor of chocolate aroma. *Br. Food Manuf. Ind. Res. Assoc. Res. Rep.*, 157, cited in Martin, R.A., Jr. 1987. *Chocolate Adv. Food Res.*, 31:211–342.

Stintzing, F.C., Stintzing, A.S., Carle, R., Frei, B., and Wrolsdat, R.E. 2002. Color and antioxidant properties of cyanidin-based anthocyanin pigments. *J. Agric. Food Chem.*, 50:6172–6182.

Strack, D. and Wray, V. 1993. The anthocyanins, in *The Flavonoids: Advances in Research Since 1986*, Harborne, J.B., Ed., Chapman & Hall, London, 1–22.

Swain, T. 1962. Economic importance of flavonoid compounds: foodstuffs, in *The Chemistry of Flavonoid Compounds*, Geissman, T.A., Ed., Pergamon Press, Oxford, U.K., 513–552.

Swain, I., Mapson, L.W., and Robb, D.A. 1966. Activation of *Vicia faba* tyrosinase as affected by denaturating agents, *Phytochemistry*, 5:469–482.

Sweeny, J.G. and Iacobucci, G.A. Effect of substitution on the stability of 3-deoxyanthocyanidins in aqueous solutions. *J. Agric. Food Chem.*, 31:531–533.

Taylor, S.L., Higley, N.A., and Bush, R.K. 1986. Sulfites in foods: uses, analytical methods, residues, fate, exposure assessment metabolism, toxity, and hypersensitivity. *Adv. Food Res.*, 30:1–76.

Teranishi, R., Buttery, R.G., and Shahidi, F., Eds. 1989. *Flavor Chemistry: Trends and Developments*, ACS Symposium Series 388. American Chemical Society, Washington, D.C.

Thomas, P. 1981. Involvement of polyphenols in the after-cooking darkening of gamma irradiated potatoes. *J. Food Sci.*, 46:1620–1621.

Tilgner, D.J. 1977. The phenomena of quality in the smoke curing. *Pure Appl. Chem.*, 49:1629–1638.

Timberlake, C.F. and Bridle, P. 1967a. Flavylium salts, anthocyanidins and anthocyanins. I. Structural transformation in acid solutions. *J. Sci. Food Agric.*, 18:473–478.

Timberlake, C.F. and Bridle, P. 1967b. Flavylium salts, anthocyanidins and anthocyanins. II. Reactions with sulphur dioxide. *J. Sci. Food Agric.*, 18:479–485.

Timberlake, C.F. and Bridle, P. 1968. Flavylium salts resistant to sulfur dioxide. *Chem. Ind.* (London), 1489.

Timberlake, C.F. 1980. Anthocyanins — occurrence, extraction, and chemistry. *Food Chem.*, 5:69–90.

Timberlake, C.F. 1981. Anthocyanins in fruits and vegetables, in *Recent Advances in the Biochemistry of Fruits and Vegetables*, Friend, J. and Rhodes, M.J.C., Eds., Academic Press, New York, 221.

Tomas-Barberan, F., Gil, M.I., Casatner, M., Artes, F., and Saltveit, M.E. 1997. Effect of selected browning inhibitors on phenolic metabolism in stem tissue of harvested lettuce. *J. Agric. Food Chem.*, 45:583–589.

Toth, L. and Potthast, K. 1984. Chemical aspects of the smoking of meat and meat products. *Adv. Food Res.*, 29:87–158.

Urbach, G., Stark, W., and Fross, D.A. 1972. Volatile compounds in butter oil. *J. Dairy Res.*, 39:35.

Vasquez Roncero, A., Maestro Duran, R., and Janer der Valle, M.L. 1970. Los colorantes de la ceituna madura. II. Varanciones durante la maduracion. *Grasas Aceites*, 21:337, cited in Mazza, G. and Miniati, E. 1993. *Anthocyanins in Fruits, Vegetables, and Grains.* CRC Press, Boca Raton, FL.

Vestrheim, S. 1970. Effects of chemical compounds on anthocyanin formation in 'McIntosh' apple skin. *J. Am. Soc. Hortic. Sci.*, 95:712–715.

Walker, J.R.L. 1975a. Enzymatic browning in foods. *Enzyme Technol. Dig.*, 3:89–100.

Walker, J.R.L. 1975b. Studies on the enzymatic browning of apples. Inhibition of apple *O*-diphenol oxidase by phenolic acids. *J. Sci. Food Agric.*, 26:1825–1831.

Walker, J.R.L. 1977. Enzymic browning in foods. Its chemistry and control. *Food Technol. N.Z.*, 12:19; 21–22.

Walradt, J.P., Lindsay, R.C., and Libbey, L.M. 1970. Popcorn flavor: identification of volatile compounds. *J. Agric. Food Chem.*, 18:926–928.

Waters, M.E. 1971. Blueing of processed crab meat. 2. Identification of some factors involved in the blue discoloration of canned meat (*Callinectes sapidus*). USDC Special Scientific Report-Fisheries No. 633, cited in Babbit, J.K., Law, D.K., and Crawford, D.L. 1975. Effect of precooking on copper content, phenolic content and blueing of canned Dungeness crab meat. *J. Food Sci.*, 40:649–650.

Weemaes, C.A., Ludikhuyze, L.R., Van de Broeck, I., and Hendrickx, M.E. 1998. Effect of pH on pressure and thermal inactivation of avocado polyphenol oxidase: a kinetic study. *J. Agric. Food Chem.*, 46:2785–2792.

Wesche-Ebeling, P. and Montgomery, M.W. 1990. Strawberry polyphenoloxidase: its role in anthocyanin degradation. *J. Food Sci.*, 55:731–734; 745.

Wheaton, T.A. and Stewart, I. 1965. Feruoylputrescine: isolation and identification from citrus leaves and fruits. *Nature*, 206:620–621.

Whitfield, F.B., Last, J.H., Shaw, K.J., and Tindale, C.R. 1988. 2,6-Dibromophenol: the cause of an iodoform-like off-flavor in some Australian crustaceans. *J. Sci. Food Agric.*, 46:29–42.

Whitfield, F.B., Helidoniotis, F., Shaw, K.J., and Svoronos, D. 1997. Distribution of bromophenols in Australian wild-harvested and cultivated prawns (shrimp). *J. Agric. Food Chem.*, 45:4398–4405.

Whitfield, F.B., Helidoniotis, F., Shaw, K.J., and Svoronos, D. 1998. Distribution of bromophenols in species of ocean fish from eastern Australia. *J. Agric. Food Chem.*, 46:3750–3757.

Whitfield, F.B., Drew, M., Helidoniotis, F., and Svoronos, D. 1999. Distribution of bromophenols in species of marine polychaetes and bryozoans from eastern Australia and the role of such animals in flavor of edible ocean fish and prawns (shrimp). *J. Agric. Food Chem.*, 47:4756–4762.

Williams, D.C, Lim, M.H., Chen, O.A., Pangborn, R.M., and Whitaker, J.R. 1985. Blanching of vegetables for freezing. Which indicator to choose? *Food Technol.*, 40:130–140.

Wilson, C.W. and Shaw, P.E. 1981. Importance of thymol, methyl N-methylanthranalate and monoterpene hydrocarbons to the aroma and flavor of mandarin cold-pressed oils. *J. Agric. Food Chem.*, 29:494–496.

Wong, D.W.S. 1989. Colorants, in *Mechanism and Theory in Food Chemistry,* AVI Publishing Co., Westport, CT, 147–189.

Wrolsdat, R.E. 2000. Anthocyanins, in *Natural Food Colorants: Science and Technology,* Lauro, G.J. and Francis, F.J., Eds., Marcel Dekker Inc., New York, 237–252.

Wrolsdat, R.E., Skrede, G., Lea, P., and Enersen, G. 1990. Influence of sugar on anthocyanin pigment stability in frozen strawberries. *J. Food Sci.*, 55:1064–1065; 1070.

Wyllie, S.G., Brophy, J.J., Sarafis, V., and Hobbs, M. 1990. Volatile components of the fruit of *Pistachia Lentiscus*. *J. Food Sci.*, 55:1325–1326.

Yokotsuka, T. 1986. Soy sauce biochemistry. *Adv. Food Res.*, 30:196–220.

Zabetakis, I., Leclerc, D., and Kajda, P. 2000. The effect of high hydrostatic pressure on strawberry anthocyanins. *J. Agric. Food Chem.*, 48:2749–2754.

10 Methods of Analysis and Quantification of Phenolic Compounds

INTRODUCTION

Nutritional and sensory effects of food phenolics have become the center of attention in recent years. Although these compounds generally possess beneficial health effects, they may also lower the availability of dietary proteins and minerals and affect the flavor and color of foods. No reliable method for quantification of phenolic compounds that could be applied to all food products exists; moreover, not all phenolics present in foods are of nutritional or sensory concern. Therefore, determining their total content in food cannot be considered a reliable index for phenolics' nutritional and sensory qualities.

Various assay methods for phenolic compounds have been developed and a number of reviews on the analysis of polyphenolics published (Antolovich et al., 2000; Deshpande et al., 1986; Hagerman et al., 1997; Jackman et al., 1987; Makkar, 1989; Markham, 1975; Porter, 1989; Scalbert et al., 1989; Scalbert, 1992; Tempel, 1982). These assays can be classified as those that determine total phenolic content or those quantifying a specific group or class of phenolic compounds. Folin-Denis assay and Prussian blue test are examples of methods used for total phenol determination (Salunkhe et al., 1989). The vanillin test is used as an assay method for catechins and proanthocyanidins (Price et al., 1978) while protein precipitation methods are employed for biologically active phenols (Makkar et al., 1988). A number of chromatographic techniques have also been developed to identify and quantify specific phenolics. However, only extractable polyphenolics under selected conditions are measured by all the methods described (Antolovich et al., 2000). Hillis and Swain (1959) found that, after exhaustive extraction, some additional phenolics could be extracted with change of solvent. The content of phenolic compounds, determined by various assay methods, is summarized in Table 10.1 and Table 10.2.

Analysis of polyphenolic compounds is influenced by their chemical nature, the extraction method employed, sample particle size, and storage time and conditions, as well as assay method, selection of standards and presence of interfering substances such as waxes, fats, terpenes and chlorophylls. Therefore, no completely satisfactory extraction procedure is suitable for extraction of all phenolics or a specific class of phenolic substances in foods. Solubility of polyphenolics is governed by the type of solvent (polarity) used, degree of polymerization of phenolics, interaction of phenolics with other food constituents and formation of insoluble complexes. Solvents frequently used for the extraction of polyphenolic compounds

TABLE 10.1
Phenolic Compound[a] Content in Some Foods and Beverages as Determined by Different Assays

Foods	FAS Assay[b]	F-C Assay[c]	Vanillin Assay[d]
Spinach	0.15	0.43	0.15
Tomato	ND	0.23	0.03
Peanut	ND	4.3	ND
Oregano	20.5	70.5	1.2
Cinnamon	38.7	27.5	27.0
Black tea leaf	105.0	67.7	92.0
Green tea leaf	84.0	62.3	70.0
Coffee powder	5.6	33.2	33.0
Red wine	87.0	1345.0	990.0
White wine	ND	181.0	35.0

Note: ND = not detected.

[a] Milligrams per fresh sample or milligrams per liter.
[b] FAS assay = assay for determination of iron-binding phenolic groups according to Brune, M. et al., 1991.
[c] F-C assay = Folin-Ciocalteu assay according to Hoff, J.F. and Singleton, K.I., 1977.
[d] Vanillin assay determined according to Price, M.L. et al., 1978.

Source: Adapted from Brune, M. et al., 1991, *J. Food Sci.*, 56:128–131; 167.

TABLE 10.2
Phenolic Content of Dry Beans as Measured by Prussian Blue and 0.5% Vanillin Methods[a]

Legume Variety	Prussian Blue	Vanillin
Cranberry	106	94
Viva pink	114	181
Pinto	153	291
Small red	136	239
Black beauty	113	26
Dark red kidney	136	143
Light red kidney	112	191

[a] Milligrams of catechin equivalents per 100 g of beans.

Source: Adapted from Deshpande, S.S. and Cheryan, M., 1987, *J. Food Sci.*, 52:332–334.

include methanol, ethanol, acetone, water, ethyl acetate, propanol, dimethylformamide and their combinations (Antolovich et al., 2000). It is difficult to find a specific and suitable standard for quantification of polyphenolics due to the complexity of food phenolic substances as well as existing differences in the reactivity of phenols toward reagents used for their assay.

EXTRACTION PROCEDURES

Because of the chemical nature of food phenolics, no satisfactory solvent extraction systems are suitable for isolation of all classes of food phenolics or even a specific class of phenolics. These vary from simple to very highly polymerized substances that include varying quantities of phenolic acids, phenylpropanoids, anthocyanins and tannins, and possible interactions of phenolics with carbohydrates, proteins and other food components. Because some high molecular weight phenolics and their complexes are quite insoluble, the extracts always contain a mixture of different classes of phenolic substances soluble in the solvent system of choice. Usually, additional steps are required in order to purify the isolates and to remove unwanted phenolics and nonphenolic substances.

Krygier et al. (1982) extracted free and esterified phenolic acids from oilseeds using a mixture of methanol–acetone–water (7:7:6, v/v/v) at room temperature. First, the free phenolics were extracted with diethyl ether, then the extract was treated with 4M NaOH under nitrogen. The hydrolyzate was acidified and the liberated phenolic acids were extracted with diethyl ether. After exhaustive extraction with a mixture of methanol–acetone–water, the left-over sample was treated with 4M NaOH under nitrogen to liberate insoluble bound phenolic acids. On the other hand, Kozlowska et al. (1983) extracted phenolic acids with hot 80% methanol (See Figure 10.1). Other solvents, such as ethanol and acetone, have also been used for the extraction of phenolics, often with different proportions of water (Antolovich et al., 2000; Naczk et al., 1992a).

Anthocyanins are usually extracted with an acidified organic solvent, most commonly methanol. This solvent system destroys cell membranes and simultaneously dissolves the anthocyanins. The acid also provides favorable conditions for formation of flavylium chloride salt, thus stabilizing the anthocyanins. However, according to Moore et al. (1982a, b), the acid may change the native form of anthocyanins by breaking down their complexes with metals and co-pigments. The acidic extracts of anthocyanins are first concentrated under vacuum and then extracted with petroleum ether, ethyl acetate or diethyl ether in order to remove lipids and unwanted polyphenols (Fuleki and Francis, 1968a, b). The pigment can also be partially purified using ion exchange resins, as described by Fuleki and Francis (1968c).

Concentration of extracts of pigments in acidic solutions before purification, however, may bring about losses of labile acyl and sugar residues (Adams, 1972). In order to avoid this detrimental effect, several researchers have proposed reducing the acid contact (Du and Francis, 1975), using neutral organic solvents or boiling water (Adams, 1972), or using weak organic acids such as formic or acetic acid to acidify the solvents (Antolovich et al., 2000; Moore et al., 1982a, b). Anthocyanins

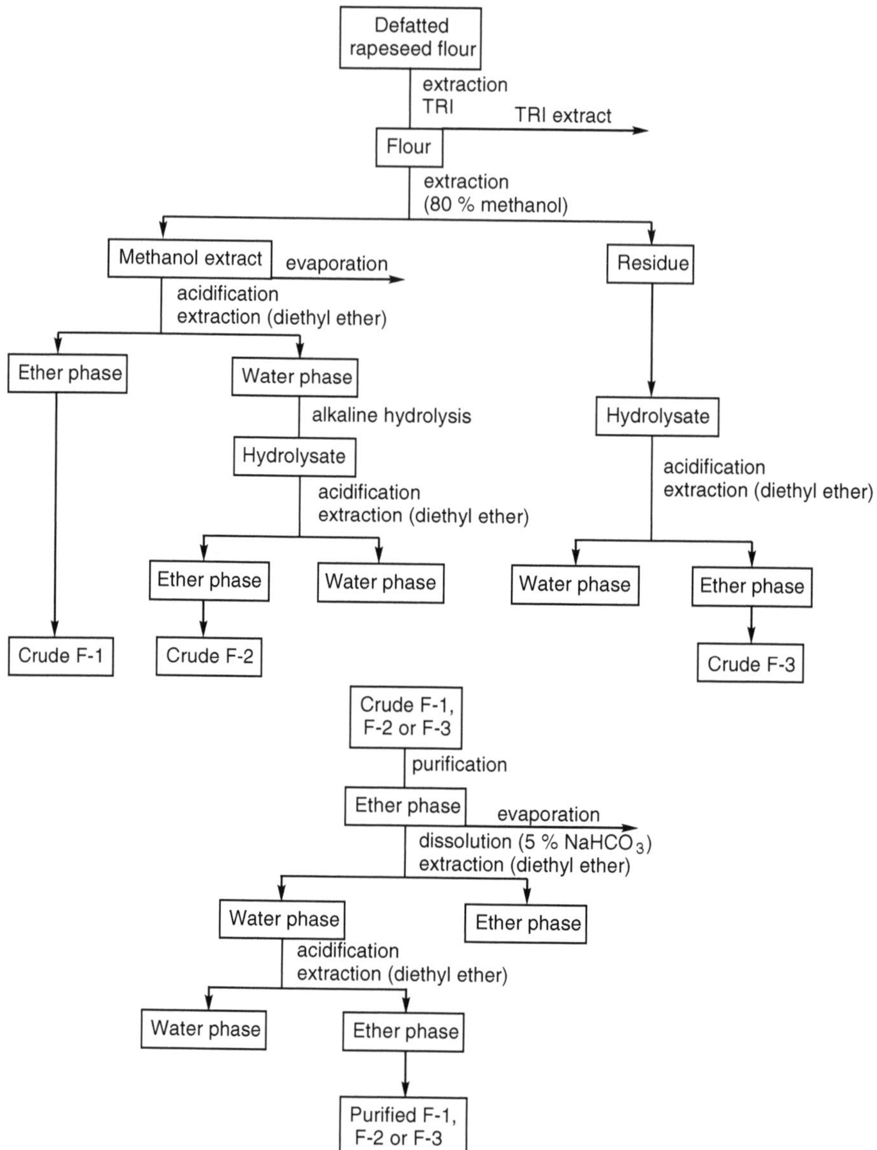

FIGURE 10.1 A simplified flow diagram for fractionation of free, esterified and bound phenolic compounds and their hydrolysis to free phenolic acids. TRI = trichloroethylene; F-1 = free phenolic acids; F-2 = esterified phenolic acids; and F-3 = insoluble phenolic acids. (Adapted from Krygier, K. et al., 1982, *J. Agric. Food Chem.*, 30:330–333.)

may also be recovered using solid-phase extraction (SPE) or solid-phase micro-extraction (SPME) on C_{18} cartridges. Hong and Wrolstad (1990) fractionated anthocyanins by elution through an SPE cartridge with an alkaline borate solution. Only anthocyanins with *o*-dihydroxy groups (cyanidin, delphinidin and petunidin) were

TABLE 10.3
Effect of HCl Addition and Boiling on Recovering Total Phenolics and Tannins from Canola Meal

Solvent System	Total Phenolics	Tannins
70% Methanol	847.0	241.3
1% HCl in methanol	892.1	73.0
1% HCl in 70% methanol	1079.8	225.9
1% HCl in methanol + 4 min of boiling	1053.3	112.5
70% Acetone	805.8	321.3
1% HCl in 70% acetone	1010.7	216.9
1% HCl in 70% acetone + 4 min of boiling	1051.6	338.7

Notes: Total phenolics are expressed as milligrams of sinapic acid equivalents per 100 g of meal; tannin content is expressed as milligrams of catechin equivalents per 100 g of meal.

Source: Adapted from Naczk, M. and Shahidi, F., 1991, presented at 8th International Rapeseed Congress, Saskatoon, Canada, July 9–11, 1991.

preferentially eluted from the SPE cartridge with borate solution due to the formation of hydrophilic borate–anthocyanin complex. On the other hand, Wang and Sporns (1999) isolated anthocyanins from fruit juices and wine using an SPE cartridge and elutions with a methanol–formic acid–water (70:2:28, v/v/v) solvent system.

For the extraction of condensed tannins, several solvent systems have been used. Absolute methanol, ethanol, acidified methanol, acetone, water, and their combinations are among the most commonly used solvent systems. For example, 1% HCl in methanol has been used for the extraction of tannins from sorghum (Price et al., 1978) and dry beans (Deshpande and Cheryan, 1985). Acetone–water (70:30, v/v) solution is the best solvent system for the extraction of tannins from rapeseed and canola (Table 10.3) (Leung et al., 1979; Shahidi and Naczk, 1989, Naczk and Shahidi, 1991), beach pea (*Lathyrus maritimus* L.) (Chavan et al., 2001; Shahidi et al., 2001) and blueberries (Prior et al., 2001). Acetone–water (60:40, v/v) solution is used for extraction of tannins from cider apple (Guyot et al., 2001) and grape skins (Labarbe et al., 1999). Tannins of cloves and allspice have been extracted with boiling water over a 2-h period (AOAC, 1980), a method also used to determine the content of tannins in rapeseed meal (Kozlowska et al., 1990).

The extraction period is another factor that affects the recovery of polyphenolics. Extraction periods varying from 1 min (Price and Butler, 1977) to 24 h (Burns, 1971; Maxson and Rooney, 1972) have been reported. However, longer extraction times increase the possibility of oxidation of phenolics unless reducing agents are added to the solvent system (Khanna et al., 1968). Price et al. (1978) reported that rotating the test tube containing suspensions of samples in a given solvent system is sufficient to extract tannins from sorghum. However, Deshpande (1985) has demonstrated that the optimum extraction time required for dry bean phenolics is 50 to 60 min. On the other hand, Naczk and Shahidi (1991) and Naczk et al. (1992a) found that a two-stage extraction with 70% (v/v) acetone, 1 min each at 10,000 rpm using a

TABLE 10.4
Effect of Number of Extraction Steps on Recovery of Total Phenolics and Tannins

Number of Extraction Steps	70% Acetone		70% Methanol	
	Total Phenolics	Tannins	Total Phenolics	Tannins
1	720.0	268.1	837.4	127.7
2	805.8	321.3	847.0	241.3
4	972.1	328.0	1025.7	243.8
6	1075.0	331.3	1081.0	235.5

Notes: Total phenolics are expressed as milligrams of sinapic acid equivalents per 100 g of meal; tannin content is expressed as milligrams of catechin equivalents per 100 g of meal.

Source: Adapted from Naczk, M. and Shahidi, F., 1991, presented at 8th International Rapeseed Congress, Saskatoon, Canada, July 9–11, 1991.

TABLE 10.5
Effect of the Canola Meal-to-Solvent Ratio (R) on Recovery of Total Phenolics and Tannins

R	70% Acetone		70% Methanol	
	Total Phenolics	Tannins	Total Phenolics	Tannins
1:5	773.5	257.3	844.3	188.3
1: 0	805.8	321.3	874.0	241.3
1:20	948.0	324.6	886.7	275.4

Note: Total phenolics are expressed as milligrams of sinapic acid equivalents per 100 g of meal; tannin content is expressed as milligrams of catechin equivalents per 100 g of meal.

Source: Adapted from Naczk, M. and Shahidi, F., 1991, presented at 8th International Rapeseed Congress, Saskatoon, Canada, July 9–11, 1991.

Polytron homogenizer, is sufficient for extraction of tannins from commercial canola meals. Further extraction (up to six stages) only marginally enhances the yield of extraction of other phenolic compounds (Table 10.4).

The recovery of polyphenols from food products is also influenced by the ratio of sample to solvent (R) (Table 10.5). Naczk and Shahidi (1991) and Naczk et al. (1992a) found that changing R from 1:5 to 1:10 increases the extraction of condensed tannins from commercial canola meals from 257.3 to 321.3 mg/100 g of meal and total phenolics from 773.5 to 805.8/100 g of meal when using 70% acetone.

Deshpande and Cheryan (1985) demonstrated that the yield of tannin recovery from dry beans is strongly influenced by variations in the sample particle size. They found that 0.5% vanillin-assayable tannins (Price et al., 1978) decrease by about 25 to 49% as the minimum size is reduced from 820 to 250 μ (Table 10.6).

TABLE 10.6
Effects of Particle Size on Tannin Analysis of Dry Beans Using 0.5% Vanillin Assay[a]

Minimum Particle Size (μ)	Bean Variety			
	Pinto	Small Red	Viva Pink	Black Beauty
820	337	304	169	154
420	269	262	111	172
350	245	250	105	213
250	223	227	87	258

[a] Milligrams of catechin equivalents per 100 g of beans, on dry basis.

Source: Adapted from Deshpande, S.S. and Cheryan, M., 1985, *J. Food Sci.*, 50:905–910.

SPECTROPHOTOMETRIC ASSAYS

A number of spectrophotometric methods for quantification of phenolic compounds in plant materials have been developed. Based on different principles, these assays are used to determine various structural groups present in phenolic compounds. Spectrophotometric methods may quantify all extractable phenolics as a group (Earp et al., 1981; Price and Butler, 1977; Swain and Hillis, 1959) or may determine a specific phenolic substance such as sinapine (Tzagoloff, 1963), or sinapic acid (Naczk et al., 1992b), or a given class of phenolics such as phenolic acids (Brune et al., 1991; Mole and Waterman, 1987a, b; Naczk and Shahidi, 1989; Price et al., 1978). The modified vanillin test (Price et al., 1978), Folin-Denis assay (Swain and Hillis, 1959), Prussian blue test (Price and Butler, 1977) and Folin-Ciocalteu assay (Deshpande and Cheryan, 1987; Earp et al., 1981; Hoff and Singleton, 1977; Maxson and Rooney, 1972) are most frequently used.

DETERMINATION OF TOTAL PHENOLICS

Folin-Denis Assay

The Folin-Denis assay is the most widely used procedure for quantification of total phenolics in plant materials and beverages. Reduction of phosphomolybdic-phosphotungstic acid (Folin-Denis) reagent to a blue-colored complex in an alkaline solution occurs in the presence of phenolic compounds (Folin and Denis, 1912). Swain and Hillis (1959) modified this method for routine analysis of a large number of samples. The Folin-Denis assay is also used for determination of total phenolics using the AOAC method (1980). This procedure is outlined in Figure 10.3.

Total phenolics are assayed colorimetrically by the method of Swain and Hillis (1959): to 0.5 to 1.0 mL of phenolics solution, 7 mL of distilled water are added and mixed well, followed by the addition of 0.5 mL of Folin-Denis reagent. After 3 min standing, 1 mL of saturated sodium bicarbonate solution is added and the total volume is made up to 10 mL with distilled water. After thorough mixing, the

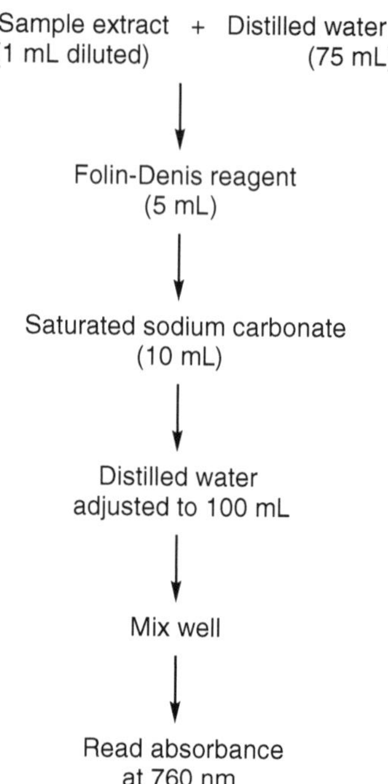

FIGURE 10.2 Procedure for tannin analysis by the Folin-Denis method. (Adapted from Salunkhe, O.K. et al., 1989, *Dietary Tannins: Consequences and Remedies*, CRC Press, Boca Raton, FL.)

absorbance is read at 725 nm after 1 h standing at room temperature. A mixture of water and reagents is used as a blank.

Folin-Ciocalteu Assay

The Folin-Ciocalteu assay is often used to determine the total content of food phenolics (Brune et al., 1991; Deshpande and Cheryan, 1987; Earp et al., 1981; Hoff and Singleton, 1977; Maxson and Rooney, 1972). Folin-Ciocalteu reagent is not specific and detects all phenolic groups found in extracts including those found in the extractable proteins. A disadvantage of this assay is the interference of reducing substances such as ascorbic acid with the determinations.

The total phenolics are assayed colorimetrically as modified by Singleton and Rossi (1965) and Hoff and Singleton (1977): 2.5 mL of 10-fold diluted Folin-Ciocalteu reagent, 2 mL of a 7.5% solution of sodium carbonate, and 0.5 mL of phenolics solution are mixed well. The absorbance is measured at 765 nm after a 15-min heating at 45°C; a mixture of water and reagents is used as a blank. The content of phenolics is expressed as gallic acid or catechin equivalents.

Prussian Blue Assay

The Prussian blue method was suggested by Price and Butler (1977) for determination of phenolics from sorghum grain and was later reevaluated by Deshpande and

Cheryan (1987) for dry beans. Reduction of ferric to ferrous ion by polyphenolic compounds and formation of ferricyanide–ferrous ion complex, also called Prussian blue, is involved. The ability of polyphenolic compounds to reduce ferric ion depends on their hydroxylation pattern and the degree of polymerization. According to Deshpande and Cheryan (1987), tannic acid is about 25% more effective reducing ferric ion than catechin because tannic acid has more phenolic hydroxyl groups than catechin does. This method's sensitivity towards flavonoids is sufficient to measure concentrations of less than 10^{-4} M. The possibility of interference from sulfides, aromatic amines, unsaturated aliphatics, glucose and ascorbic acid that may reduce ferricyanide (Snell and Snell, 1953), as well as organic solvents (Price and Butler, 1977), should be considered.

Price and Butler (1977) and Deshpande and Cheryan (1987) reported that the content of polyphenolics obtained by the Prussian test is always somewhat lower than that obtained by the vanillin assay (Table 10.2). These authors suggested that catechin used as a standard in the vanillin test may lead to overestimation of results. Furthermore, the Prussian blue test may also underestimate the results if polyphenolics are more oxidized than catechin or less reactive than the standard monomer.

The Prussian blue procedure involves a 1-min rapid extraction of phenolic compounds with water (Price and Butler, 1977). Subsequently, an aliquot of the extract is diluted with distilled water in order to eliminate any interfering color. Then an excess of ferric chloride in 0.1 M HCl (added to increase stability) is added to ensure rapid and complete reaction. Due to inherent instability of diluted ferric chloride solution, ferricyanide should be added immediately.

The polyphenolic compounds are assayed colorimetrically by the method of Price and Butler (1977): to 6 mL of aqueous solution of phenolics, 50 mL of distilled water is added followed by mixing. To this, 3 mL 0.1 M of ferric chloride is added followed immediately by timed addition of 3 mL of 0.008 M potassium ferricyanide solution. The absorbance at 720 nm is read after 10 min standing at room temperature. Distilled water is used as a blank.

Determination of Condensed Tannins

Vanillin Assay

The vanillin method is widely used for quantification of proanthocyanins (condensed tannins) in fruits (Goldstein and Swain, 1963a) and grains (Burns, 1971). This assay has been critically reevaluated for detection of tannins in sorghum (Price et al., 1978), dry beans (Deshpande and Cheryan, 1985), forage legumes (Broadhurst and Jones, 1978) and canola meals (Naczk and Shahidi, 1991), as well as catechins and proanthocyanidins (Sun et al., 1998a). The vanillin test is specific for flavan-3-ols, dihydrochalcones and proanthocyanins, which have a single bond at the 2,3-position and possess free metahydroxy groups on the B-ring (Sarkar and Howarth, 1976; Gupta and Haslam, 1980). Catechin, a monomeric flavan-3-ol, is often used as a standard in the vanillin assay. According to Price et al. (1978) and Gupta and Haslam (1980), this may lead to overestimation of tannin contents.

TABLE 10.7
Response of Some Flavonoid and Chromone Compounds to Vanillin Test

Class	Compound	Reaction with Vanillin
Flavonols	Catechin (5,7,3′,4′-tetrahydroxyflavan-e-ol	+ + +
Dihydrochalcones	Phoretin (4,2′,4′,6′-tetrahydroxydihydrochalcone	+ + +
Chalcone	Butein (3,4,2′,4′-tetrahydroxychalcone)	–
Flavanones	Naringenin (5,7,4′-trihydroxyflavavones)	–
Flavones	7-Hydroxyflavone	–
Flavanonol	Dihydroxyquercitin	+
Flavanols	Kaempferol (5,7,4′-trihydroxyflavonol)	–
	Quercitin (5,7,3′,4′-tetrahydroxyflavanol)	
Chromones	Eugenin (2-methyl-5-hydroxy-7-methoxychromone)	–

Notes: +++ = very positive; + = positive; and – = negative.

Source: Adapted from Sarkar, S.K. and Howarth, R.E., 1976, *J. Agric. Food Chem.*, 24:317–320. With permission.

Methanol, a usual solvent used for vanillin test, may affect the kinetics of catechin and tannin reactions with vanillin differently. The vanillin assay in methanol is more sensitive to polymeric tannins than monomeric flavan-3-ols; due to its simplicity, sensitivity and specificity, it is generally recognized as a useful method for the detection and quantification of condensed tannins in plant materials. The method can be used for quantifying condensed tannins in the range of 5 to 500 μg with precision and accuracy of greater than 1 μg when the optimum concentrations of reactants and solvents are selected (Broadhurst and Jones, 1978).

The possibility of interference with dihydrochalcones and anthocyanins (Sarkar and Howarth, 1976) when determining tannins, as well as ascorbic acid or ascorbate and chlorophylls (Broadhurst and Jones, 1978; Sun et al., 1998a), should be considered (Table 10.7). At usual concentrations present in plant extracts, anthocyanins do not react with vanillin (Broadhurst and Jones, 1978; Sun et al., 1998a). However, their presence in phenolic extract may lead to overestimation of tannins because maximum absorption of tannin–vanillin adducts coincides with that of anthocyanins. This interference can be eliminated by the use of an appropriate blank (Broadhurst and Jones, 1978).

Ascorbic acid or ascorbate interferes with the vanillin assay in the presence of sulfuric acid, which may be due to the oxidative nature of sulfuric acid (Broadhurst and Jones, 1978). Sun et al. (1998b) described a simple procedure for the removal of ascorbic acid and ascorbates from wine and grape juice by fractionation with C_{18} Sep-Pak cartridges. Chlorophyll-like anthocyanins also absorb in the region that coincides with the maximum absorption displayed by tannin–vanillin pigment. Therefore, presence of chlorophylls in plant phenolic extracts may lead to overestimation of tannins. This interference may be eliminated by hexane extraction of chlorophylls from the aqueous phenolic solution (Sun et al., 1998a).

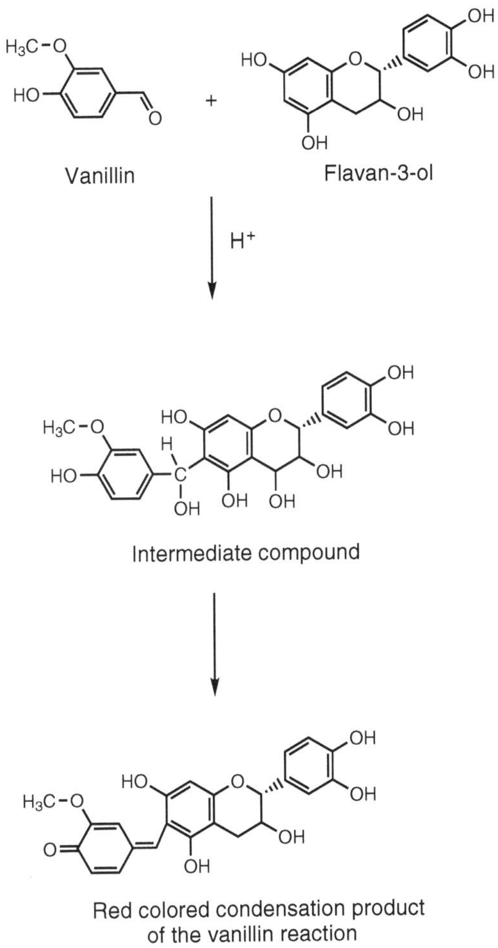

Vanillin + Flavan-3-ol

H^+

Intermediate compound

Red colored condensation product of the vanillin reaction

FIGURE 10.3 Condensation reactions of vanillin with leucocyanidin. (Adapted from Salunkhe, O.K. et al., 1989, *Dietary Tannins: Consequences and Remedies*, CRC Press, Boca Raton, FL.)

The vanillin method is based on condensation of the vanillin reagent with proanthocyanins in acidic solutions. Protonated vanillin, a weak electrophilic radical, reacts with the flavonoid ring at the 6- or 8-position. The intermediate product of this reaction dehydrates readily to give a light pink to deep red cherry-colored product (Figure 10.2). The vanillin reaction is affected by the acidic nature and concentrations of substrate, the reaction time, the temperature, the vanillin concentration and water content (Sun et al., 1998a). Improvement in carbonium ion stability may be responsible for an increase in sensitivity (Scalbert, 1992) of vanillin assay when sulfuric acid is used rather than hydrochloric acid (Scalbert et al., 1989; Sun et al., 1998a). However, high acid concentrations should be avoided because they may promote self-reaction of vanillin and decomposition of proanthocyanidins (Broadhurst and Jones, 1978; Beart et al., 1985; Scalbert, 1992). According to Sun et al. (1998a), the

TABLE 10.8
Molar Extinction Coefficient of Phenolic Complexes with Vanillin

Compound	$\epsilon_{500} \times 10^{-3}$ in Methanol	$\epsilon_{500} \times 10^{-3}$ in Acetic Acid
Monomers:		
Epicatechin	0.43	9.8
Catechin	0.30	8.5
Dimers:		
B-2 (epicatechin)$_2$	2.52	11.2
B-3 (catechin)$_2$	1.82	12.3
Trimers:		
B-9 (epicatechin)$_3$	4.16	12.2

Source: Adapted from Butler, L.G. et al., 1982, *J. Agric. Food Chem.*, 30:1087–1089.

vanillin concentration used in the vanillin assay should be between 10 and 12 g/L. At higher vanillin concentrations self-reaction of vanillin may lead to the formation of colored products (Broadhurst and Jones, 1978). Furthermore, the color stability of vanillin–tannin adducts may increase when light is excluded and the temperature of the reaction is controlled (Broadhurst and Jones, 1978). Thus, to obtain accurate and reproducible results, the vanillin assay must be conducted at a fixed temperature (Gupta and Haslam, 1980; Sun et al., 1998a).

The reaction of vanillin with phenolics containing meta-oriented di- and trihydroxy groups on the benzene rings is approximately stoichiometric (Goldstein and Swain, 1963b). The solvent used in the vanillin assay may also affect the reaction kinetics. In methanol, the time course of the vanillin reaction with catechin is different from that with tannins; however, use of glacial acetic acid in place of methanol affords similar kinetics for the monomers and polymers. This, according to Butler et al. (1982), suggests that vanillin reacts only with the terminal groups in glacial acetic acid and thus the concentration of oligomeric molecules rather than the total content of flavan-3-ols is measured. Table 10.8 depicts the molar extinction coefficients of vanillin monomer and vanillin oligomer adducts.

The aforementioned authors proposed that vanillin assay in glacial acetic acid may be used for estimating the degree of polymerization of condensed tannins. The procedure involves extraction of phenolic compounds with methanol (Burns, 1971), 1% concentrated HCl in methanol (Price et al., 1978), and 70% acetone (Broadhurst and Jones, 1978; Shahidi and Naczk, 1989). The sample should be extracted within 1 day after grinding. Extracts are directly assayed or the supernatants are first evaporated to dryness at 30°C under vacuum and the residues then dissolved in methanol. The vanillin assay is performed at 30°C; vanillin reagent must be prepared freshly each day.

The condensed tannins are assayed colorimetrically by the method of Price et al. (1978): to 1 mL of methanolic solution of condensed tannins, 5 mL of freshly prepared 0.5% vanillin solution in methanol containing 4% concentrated HCl (sample) or 5 mL of 4% concentrated HCl in methanol (blank) are added and mixed well. The absorbance of the sample or the blank is then read at 500 nm, after 20 min standing at 30°C. The condensed tannins are expressed as milligrams of catechin equivalents per 100 g of sample.

DMCA (4-(Dimethylamino)-Cinnamaldhyde) Assay

The formation of a green chromophore between catechin and 4-(dimethylamino)-cinnamaldehyde (DMCA) was first reported by Thies and Fisher (1971). Subsequently, McMurrough and McDowell (1978), Delcour and de Varebeke (1985) and Treutter (1989) demonstrated that DMCA does not react with a wide range of flavonoids including dihydrochalcones, flavanones and flavononols, and phenolic acids. Weak responses are detected for resorcinol, orcinol, naphtoresorcinol, and phloretin. Treutter (1989) also reported that DMCA reacts with indoles and terpenes. Later, Li et al. (1996) reexamined the use of the DMCA assay for determination of proanthocyanidins in plants. According to these authors, on a molar basis, the sensitivity of DMCA toward proanthocyanidins is 4 and 30,000 times greater than that of indole and thymol, respectively.

In comparison to the vanillin reaction carried out in methanol, the DMCA reagent reacts only with the terminal groups of condensed tannins. The presence of methanol, acetone, ethyl acetate, and dimethylformamide does not have any detrimental effect on the rate and intensity of the color development. However, the DMCA reagent is sensitive to monomeric and to polymeric units (McMurrough and McDowell, 1978), which may lead to overestimation of condensed tannin content.

The colorimetric assay of condensed tannins using the DMCA reagent described by McMurrough and McDowell (1978) and Thies and Fisher (1971) is: 1 mL of diluted (up to 1:10, v/v) methanolic solution of condensed tannins is mixed with 5 mL of DMCA reagent. The absorbance of the sample is read at 635 nm, after 15 min standing at room temperature, using an appropriate blank. The DMCA reagent is prepared by dissolving 1 g of DMCA in a precooled solution containing 250 mL of concentrated HCl and 700 mL of methanol. Following this, the solution is made up to 1 L with methanol. The DMCA reagent is stable for at least 1 week when stored in the dark.

Proanthocyanidin Assay

The proanthocyanidin assay is carried out in a solution of butanol-concentrated hydrochloric acid (95:5, v/v). In the presence of this hot HCl, proanthocyanidins (condensed tannins) are converted to anthocyanidins that are products of autoxidation of carbocations formed by cleavage of interflavanoid bonds (Porter et al., 1986). The yield of reaction depends on the content of HCl and water, temperature and length of the reaction period, presence of transition metals, and length of the proanthocyanidin chain (Porter et al., 1986; Scalbert et al., 1989). Govindarajan and Mathew (1965) reported that water content of up to 20% does not affect the yield of anthocyanidins. Later, Porter et al. (1986) demonstrated that the maximum yield of anthocyanidins is obtained at a 6% (v/v) level of water in the solvent.

Both lower and higher contents of water have a detrimental effect on the formation of anthocyanidins. The reproducibility and the yield of conversion of proanthocyanidin to anthocyanidins are enhanced in the presence of transition metals. Porter et al. (1986) investigated the effect of several transition metals on the formation of anthocyanidins. These authors found that Fe^{+2} and Fe^{+3} are the most effective catalysts. The maximum yield of anthocyanidin is achieved at Fe^{+3} concentration of 7×10^{-4} M. Scalbert et al. (1989) reported that the reproducibility of the method is significantly improved if the reaction temperature is strictly controlled.

The colorimetric assay of condensed tannins by the proanthocyanidin procedure descibed by Mole and Waterman (1987a) is: to 1 mL of methanolic solution of condensed tannins, add 10 mL of 1-butanol–HCl reagent. This mixture is heated in sealed ampules for 2 h in a boiling water bath and then allowed to cool in an ice-water bath. The absorbance of the solution is read at 550 nm against a reagent blank. For $A > 0.75$, the reaction mixture is diluted with 1-butanol. The 1-butanol–HCl reagent is prepared by dissolving 0.7 g of ferrous chloride heptahydrate in 25 mL of concentrated HCl containing a small amount of 1-butanol; this solution is then made up to 1 L with 1-butanol.

Determination of Insoluble Condensed Tannins

Insoluble condensed tannins were first identified in herbs by Bate-Smith (1973a). Subsequently, Terrill et al. (1992) fractionated condensed tannins into tannins soluble in organic solvents, tannins soluble in solution of sodium dodecyl sulfate (SDS), and insoluble tannins. The procedure proposed by these authors included the extraction of soluble tannins with a mixture of acetone–water–diethyl ether (4.7:2.0:3.3, v/v/v), followed by the extraction of SDS-soluble tannins with a boiling aqueous solution of SDS containing 2-mercaptoethanol. The insoluble tannins were then determined directly on the remaining residue after extraction of SDS-soluble tannins by using the proanthocyanidin assay. The residue was subjected to one-step treatment with butanol–HCl; this treatment with butanol–HCl was also used by Makkar et al. (1997), Degen et al. (1995) and Makkar and Singh (1991) for determining insoluble condensed tannins in forages. Later, Makkar et al. (1999) reported incomplete recovery of insoluble condensed tannins after one-step treatment of sample with butanol–HCl. These authors used solid state ^{13}C NMR as a probe for the measurement of bound condensed tannins. Recently, Naczk et al. (2000) investigated a multistep treatment of canola hulls with butanol–HCl and reported that, with butanol–HCl. a six-step extraction of hulls free of soluble tannins removes most of the insoluble condensed tannins as anthocyanidins.

Determination of Hydrolyzable Tannins

Various approaches have been used for convenient screening of a large number of plant samples for their hydrolyzable tannins. Of these, the most widely used method is based on the reaction between potassium iodate and hydrolyzable tannins. This reaction was first described by Haslam (1965) and later utilized by Bate-Smith (1977) for the development of an analytical assay for estimation of hydrolyzable tannins in plant materials. The original assay included mixing the sample and potassium iodate, cooling the mixture in an ice-water bath for 40 min (Haslam, 1965) or 90 min (Bate-Smith, 1977) and measuring the absorbance at 550 nm. Later, Willis and Allen (1998) demonstrated that chilling the samples after addition of potassium iodate is unnecessary. These authors recommended carrying out the reaction at 25°C for a time unique for each type of plant material.

Several limitations have been reported for the original assay, including variability in the reaction time required for different tannins for maximum color development, variability in spectral properties of chromophores formed from different hydrolyzable tannins, formation of yellow oxidation products by nontannin phenolics, the possibility of precipitate formation and difficulties in reproducibility of the results (Inoue and Hagerman, 1988; Hartzfeld et al., 2002; Waterman and Mole, 1994).

Recently, Hartzfeld et al. (2002) modified this assay by including a methanolysis step followed by oxidation with potassium iodate. The assay is based on gallic acid, which is a common structural component found in gallotannins and ellagitannins. Methyl gallate, formed upon methanolysis of hydrolyzable tannins in the presence of strong acids, reacts with potassium iodate to produce a red chromophore with a maximum absorbance at 525 nm. The oxidation reaction is carried out at 30°C for 50 min at pH 5.5. The detection limit of the method is 1.5 µg of methyl gallate. Methanol and acetone stabilize the chromophore, while the presence of water accelerates degradation of the resultant pigment. According to Hartzfeld et al. (2002), the modified potassium iodate method provides a good estimate for gallotannins, but underestimates the content of ellagitannins.

Other analytical assays proposed for quantification of hydrolyzable tannins in plant materials include the rhodanine (Inoue and Hagerman, 1988) and the sodium nitrate (Wilson and Hagerman, 1990) methods. The rhodanine assay may be used for estimating gallotannins and is based on determination of gallic acid in a sample subjected to acid hydrolysis under anaerobic conditions. The sodium nitrate assay was developed for quantitative determination of ellagic acid in sample hydrolysates, but requires large quantities of pyridine as a solvent.

DETERMINATION OF SINAPINE

Sinapine, a choline ester of sinapic acid, has been reported in seeds of *Brassica napus, Brassica campestris* and *Crambe abyssinica* (Austin and Wolff, 1968; Clandinin, 1961). Methods for estimating sinapine are based on the formation of a water-insoluble complex between Reinecke salt and quaternary nitrogen base (Austin and Wolff, 1968; Tzagoloff, 1963) and formation of a colored complex between sinapine and titanium chloride (Ismail and Eskin, 1979; Eskin, 1980).

Reinecke Salt Assay

This procedure involves three extractions with hot methanol (25, 15, and 15 mL per extraction, respectively) of 1 g of ground, defatted meal. The extracts are then combined, the solvent evaporated under reduced pressure, and the residue suspended in water and filtered through analytical grade Celite (Austin and Wolff, 1968). A 5-mL clear aliquot of the extract is chilled in an ice bath for a few minutes, then

1 mL of a 50% saturated aqueous solution of Reinecke salt (not more than 2 days old) is added and the mixture is allowed to stand in an ice bath for 1 h. The precipitate is then centrifuged at 1750 × g for 15 min. Following this, the precipitate is dissolved in 5 mL of slightly acidified methanol. This methanol solution is alkalized with 1 mL of 0.2 M NaOH and its absorbance read immediately (within 5 to 7 sec after addition of alkali because the color fades quickly) at 400 nm (Austin and Wolff, 1968; Tzagoloff, 1963).

Titanium Tetrachloride Assay

This procedure involves extracting 5 g of defatted product with acetone–water (60:40, v/v; acidified to pH 3 with trichloroacetic acid) three times for 30 min under reflux conditions. The combined acidified extracts are evaporated under reduced pressure to remove acetone, and washed with diethyl ether to remove the trichloroacetic acid. Following this, an aqueous extract is passed through a neutral aluminum oxide column to remove phenols, and sinapine is eluted with water. Subsequently, the aliquot (0.2 to 2.0 mL) is dried (50°C and 700 mm pressure). The dried extract is then dissolved in 5 mL of concentrated HCl containing 0.25 mL of titanium tetrachloride; the absorbance of the solution is read at 485 nm. The concentrated HCl is the best medium for the formation of titanium tetrachloride complex with sinapine, whereas acetone is best for other phenolic compounds (Ismail and Eskin, 1979).

Determination of Chlorogenic Acid

A number of UV spectrophotometric methods have been used for determination of chlorogenic acid in foodstuff (AOAC, 1980; Corse et al., 1962; Merrit and Proctor, 1959; Moores et al., 1948). A colorimetric method for quantification of chlorogenic acid in green coffee beans was proposed by Clifford and Wright (1976), who found that metaperiodate reagent provides a rapid and simple means of measuring the total chlorogenic acid. On the other hand, successive use of molybdate and metaperiodate reagents enables the estimation of total content of feruloylquinic acids as well as accurate measurement of caffeoylquinic acids and total chlorogenic acids. Molybdate reagent is specific for orthodihydrophenols while metaperiodate reagent reacts with *ortho-* and *para-*dihydroxyphenols and their monomethyl ethers to produce a yellow-orange color (Clifford and Wright, 1973).

The molybdate reagent can be prepared by dissolving the following in 800 mL of distilled water, one by one: sodium molybdate dihydrate (16.5 g), disodium hydrogen phosphate dihydrate (8.04 g) and sodium dihydrogen phosphate dihydrate (7.93 g). Following this, the solution is made up of 1000 mL with distilled water (Swain and Hillis, 1959); the phenolics are assayed by adding 10 mL of molybdate reagent to 0.1 mL aliquot of phenolic extract and subsequent mixing. This solution is then examined spectrophotometrically against reagent blank at 370 nm for caffeoylquinic acid and 353 nm for caffeic acid (Clifford and Wright, 1976).

The determination of phenolics with metaperiodate reagent (0.25% aqueous solution of sodium metaperiodate) involves addiing 1 mL of aliquot of phenolic solution to 10 mL of metaperiodate reagent, mixing thoroughly and allowing to stand for 10 min at room temperature. Following this the absorbance of the solution is read at 406 nm for chlorogenic acids and 423 nm for caffeic and ferulic acids (Clifford and Wright, 1976).

Determination of Anthocyanins

Quantification of anthocyanins takes advantage of their characteristic behavior in acidic media; anthocyanins exist in these media as an equilibrium between the colored oxonium ion and the colorless pseudobase form. The oxonium form comprises about 15% of the equilibrium mixture at pH 3.9 and 100% at pH 1 (Timberlake and Bridle, 1967). Sondheimer and Kertesz (1948) first developed the analytical procedure for quantification of anthocyanins. Swain and Hillis (1959) later modified this procedure by suggesting expression of the concentration of pigments in terms of the change in the absorbance at λ_{max} between pH 3.5 and pH ≤ 1.0. Fuleki and Francis (1968a-c) suggested extending the pH differential to between pH 4.5 and 1.0. They proposed to determine the optical density of the two samples at 515 nm for aliquots buffered to the preceding pH values. Using an average extinction coefficient (Moskowitz and Hrazdina, 1981), the total content of anthocyanins may also be estimated from the absorption of the total extracts at 520 nm.

Anthocyanins may also be assayed colorimetrically by the differential method of Fuleki and Francis (1968a, c). In this method, the anthocyanins are extracted twice with acidified ethanol (95% v/v ethanol:1.5M HCl = 85:15; v/v); 1 mL of extract is then diluted in a buffer pH 1.0 (0.2 M KCl:0.2 M HCl = 25:67; v/v) until an absorbance reading of 0.6 to 0.8 is obtained. Subsequently, the same amount of extract is diluted in a buffer pH 4.5 (1 M sodium acetate:1 M HCl:water = 100:60:90; v/v/v). The diluted samples are allowed to equilibrate in the dark for 2 h and then the absorbance of sample is read at its λ_{max}. The difference in the absorbance is then calculated by subtracting the reading at pH 4.5 from that at pH 1.0 and dividing it by the average extinction coefficients for the four major anthocyanins or by the extinction coefficient of the principal anthocyanin of the sample in order to yield the total anthocyanin content. However, Little (1977) reported that downward and upward pH manipulations offer no advantage for products such as wine, with a natural pH of around 3.4. The use of transreflectometric thin layer measurements at two pH levels, natural (about 3.5) and pH 1.0, was suggested in order to obtain the content of anthocyanins.

Determination of Theaflavins

Theaflavins in black tea can be determined spectrophotmetrically using the Flavognost method described by Hilton (1973). Flavognost reagent is believed to form a green complex with *cis*-2,2-dihydrobenzene rings associated with theaflavins, so the color intensity may be used to calculate the total theaflavin content (Figure 10.4) (Spiro and Price, 1986). The procedure consists of three critical stages: (1) sample preparation, (2) infusion, and (3) partitioning and color development. According to Reeves et al. (1985), control of infusion temperature is of prime importance in

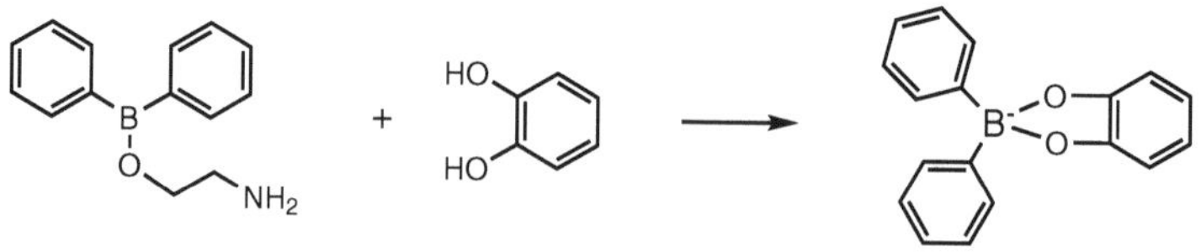

FIGURE 10.4 Reaction of complexation of the Flavognost reagent with theaflavin. (Adapted from Spiro, M. and Price, W.E., 1986, *Analyst*, 3:331–333.)

theaflavin analysis. However, the results of critical investigation into this method by Robertson and Hall (1989) indicate that failure to control conditions at each of these critical stages can give a four-fold variation in the results.

Extraction of theaflavins is influenced by the particle size of ground sample and the type of grinder used to produce it. The temperature of water used and time over which the infusion is carried out are also very significant. The extraction of theaflavins decreases rapidly after 12.5 min of infusion, which may be due to the oxidation of theaflavins to thearubigins. Moreover, it has been found that shaking the infused leaf may affect the extraction of theaflavins. Thus, occasional but regular inversion of flask by hand gives results similar to those obtained using a horizontal shaker. On the other hand, the use of vibratory shakers may result in poor extraction of theaflavins.

The theaflavin analysis according to the Flavognost method is: the black tea must be first ground to particles of less than 0.5-mm diameter. Next, 9 g of ground tea is infused in 175 g of boiling water; the infusion temperature should be kept close to boiling. The infusion is carried out in a thermos flask with continuous mechanical shaking. After infusion, the hot liquor is filtered and quickly cooled in cold water. Then, a 10-mL aliquot of the infusion is shaken for 10 min with 10 mL of isobutylmethyl ketone and the two layers are allowed to separate. Next, 2 mL of aliquot of the upper layer, 4 mL of ethanol and 2 mL of Flavognost reagent (2% w/v diphenylboric acid-2-aminoethyl ester in ethanol) are mixed well. The absorbance is then read at 625 nm after 15 min standing at room temperature; a mixture of isobutylmethyl ketone and ethanol (1:1, v/v) is used as a blank. The content of theaflavins in black tea is calculated from the following formula:

$$\text{Theaflavin } (\mu\text{mol/g}) = E_{625} \times 47{,}900/\text{DM}$$

Where E_{625} is optical density and DM the dry matter of tea sample.

Determination of Iron-Binding Phenolic Groups

Brune et al. (1991) developed a spectrophotometric method for determination and evaluation of iron-binding phenolic groups in foods. This method is based on the findings of Mejbaum-Katzenellenbogen and Kudrewicz-Hubica (1966), who observed that Fe-galloyl and Fe-catechol complexes are colored differently. Ferric ions react with phenolic compounds containing three adjacent hydroxyl groups (galloyl groups) to form a blue-colored complex, while phenolics with two adjacent

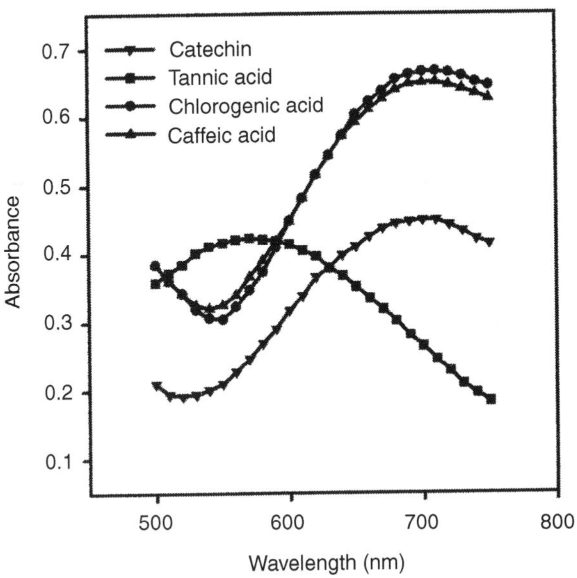

FIGURE 10.5 Absorbance spectra for equimolar (1.378 μmol/mL) concentrations of chlorogenic acid 500 μg/mL (●—●); caffeic acid 248 μg/mL (▲—▲); catechin 400 μg/mL (▼—▼); and tannic acid 200 μg/mL (■—■). (Adapted from Brune, M. et al., 1991, *J. Food Sci.*, 56:128–131; 167.)

hydroxyl groups (catechol groups) produce a green-colored complex. The absorption maximum for iron–tannin complex is around 580 nm, while that for complexes of iron with catechin, pyrocatechol, caffeic acid, and chlorogenic acid is around 680 nm (Figure 10.5). The iron-binding reagent does not lend itself for determination of monohydroxy phenolics and dihydroxy phenolics with a meta- or parahydroxylation pattern.

This procedure involves a 16-h extraction of phenolic compounds in the dark from dry or freeze-dried food samples with a 50% dimethylformamide solution in 0.1 M acetate buffer (pH 4.4) at room temperature. A ferric ammonium sulfate (FAS) reagent consisting of 1 M of HCl solution containing 5% ferric ammonium sulfate and 1% gum Arabic is added to the filtered extract. The resulting color is read spectrophotometrically against the blank at 578 and 680 nm, which corresponds to the absorption maxima of iron–galloyl (blue color) and iron–catechol (green color) complexes, respectively. The absorbances of the blank are also read and subtracted from those of the food sample at the two given wavelengths. Following this, the contents of catechol groups (expressed as catechin equivalents) and galloyl groups (expressed as tannic acid equivalents) are calculated from four standard curves obtained by using tannic acid and catechin at the two specified wavelengths values.

The iron-binding phenolic groups are assayed colorimetrically. To 2.0 mL of the extract, 8 mL of fresh FAS reagent is added and mixed well. The absorbance is then read at 578 and 680 nm after 15 min standing at room temperature. The mixture of extraction solvent and FAS reagent is used as a blank. A food blank is prepared by mixing 2 mL of the extract with 8 mL of the food blank reagent (obtained by mixing 89 parts of 50% dimethylformamide solution in 0.1 M acetate buffer, pH of 4.4, 10 parts of 1% gum Arabic solution, and 1 part of 1 M HCl). The absorbances of the food blank at 578 and 680 nm are read vs. those of the blank prepared by mixing 2 mL of the extraction solvent and 8 mL of food blank reagent. These absorbances (measurement of food sample base color) are subtracted from those of the food sample. The content of iron-binding phenolics is expressed as tannic acid or catechin equivalents.

UV SPECTROPHOTOMETRIC ASSAY

A number of approaches have been used to develop a simple and satisfactory UV spectrophotometric assay. Each group of phenolic compounds is characterized by one or several UV absorption maxima (Table 10.9). Simple phenolics have absorption maxima between 220 and 280 nm (Owades et al., 1958b); however, closely related phenolics show quite wide variations in their molecular absorptivity. Absorption is affected by the nature of solvent and pH; moreover, the possibility of interference by UV-absorbing substances such as proteins, nucleic acids and amino acids should be considered — a rather cumbersome and difficult task. Therefore, suitability of the UV assay depends on the material to be analyzed. Phenolics in tea and beer (Owades et al., 1958a, b), tea, coffee and other beverages (Hoff and Singleton, 1977), as well as those in cereals, legumes (Sharp et al., 1977; Davis, 1982) and oilseeds

TABLE 10.9
UV Absorption Patterns of Various Phenolic Compounds

Class of Compounds	UV Absorption, λ_{max}
Benzoic acids	270–280
Hydroxycinnamic acids	290–300[a]; 305–330
Anthocyanic pigments	270–280, 315–325[b]
Flavonols	250–270; 330[b]; 350–380
Flavan-3-ols	270–280
Coumarins	220–230; 310–350
Flavones	250–270; 330–350
Flavanones, Flavanonols	270–295; 300–330[a]
Chalcones	220–270; 300–320; 340–390
Aurones	240–270
Isoflavones	245–270; 300–350

Note: Solvent is methanol, except for anthocyanin pigments in which methanol containing HCl 0.01% is used.

[a] Shoulder.

[b] In the case of acylation by hydroxycinnamic acids.

Source: Adapted from Macheix, J.-J. et al., 1990, *Fruit Phenolics,* CRC Press Inc., Boca Raton, FL.

such as canola meals (Blair and Reichert, 1984; Naczk et al., 1992b; Shahidi and Naczk, 1990), have been estimated by this method. Determination of the absorbancy at 325 nm has also been proposed for estimating the hydroxycinnamic acid derivatives (chlorogenic acid) in pears (Billot et al., 1978) and coffee (Moores et al., 1948).

UV and visible spectroscopic techniques are often used for identification of isolated phenolic compounds, particularly flavonoids (Mabry et al., 1970). The spectra of phenolics in methanol after addition of Schift reagents are recorded. For example, the addition of sodium hydroxide or hydrochloric acid is helpful in determining the phenolic and carboxylic acid groups present in the molecules (Jurd, 1957). On the other hand, the absorption spectra of total phenolic extracts can be used to identify the presence of groups of predominant phenolic compounds (Macheix et al., 1990).

CHEMOMETRIC METHODS

Traditional spectroscopic assays may lead to overestimation of polyphenol contents of crude extracts from plant materials due to overlapping spectral responses; however, these problems can be overcome by using a chemometric technique to analyze the spectra. Partial least squares (PLS) or principal component analysis (PCA) is commonly used to resolve these types of problems (Kramer, 1998; van der Voort, 1992). Chemometric technique uses information (such as a spectrum) and chemical indices (such as concentration of a component) to establish a mathematical relation between the two. It assumes that the chemical index (concentration) is correct and attributes weightings of the spectral information accordingly. The set-up of the model, correlating the information with a chemical index, is known as calibration (Beebe and Kowalski, 1987).

Monedero et al. (1999) developed a chemometric technique for controlling the content of phenolic aldehydes and acids during production of wine subjected to accelerated aging. The wine was aged by adding oak wood extracts obtained by maceration of charred oak shavings. Charring time and the interactions between charring temperature and time were essential factors to control the content of 10 of the 11 phenolic compounds studied. Edelmann et al. (2001) developed a rapid method of discrimination of Austrian red wines based on midinfrared spectroscopy of phenolic extracts of wine. The samples of wine were cleaned up by solid-phase extraction (SPE) before collection of spectra using FTIR spectrophotometry. These authors also reported that the use of UV-vis spectroscopy was limited to the authentication of the Burgundy species Pinot Noir. Subsequently, Brenna and Pagliarini (2001) employed a multivariate analysis for establishing a correlation between polyphenolic composition and antioxidant power of red wines.

Briandet et al. (1996) applied PCA to differentiate between Arabica and Robusta in instant coffees based on their FTIR spectra. Spectra used in this study were obtained by using the diffuse reflection infrared Fourier transform and attenuated total reflection sampling techniques. According to these authors, the discrimination between species of coffees is based on the different chlorogenic acid and caffeine contents. Later, Downey et al. (1997) successfully applied factorial discriminant analysis and PLS to develop a mathematical model for varietal authentication of lyophilized samples of coffee based on near- and midinfrared spectra.

Schulz et al. (1999) used a near-infrared reflectance (NIR) spectroscopic method for prediction of polyphenols in the leaves of green tea (*Camelia sinensis* (L.) O. Kuntze). The contents of gallic acid and catechins were determined using a reversed-phase HPLC methodology. The PLS method was used to calibrate NIR spectra with the contents of gallic acid and catechins in tea. The models predicted the contents of catechins and gallic acid with good accuracy. On the other hand, Mangas et al. (1999) used the linear discrimination analysis for discrimination of bitter and nonbitter cider apple varieties. The most discriminant variables were chlorogenic acid, phloretin 2-xyloglucoside and an unidentified phenolic acid derivative with a maximum absorption at 316.7 nm. Using this model, 91.3 and 85.7% correct classification was obtained for internal and external evaluation of the model, respectively.

Recently, Naczk et al. (2001c) employed a multivariate model for the prediction of soluble condensed tannins in crude extracts of polyphenols from canola and rapeseed hulls. PLS was used to correlate the spectral data of the crude polyphenols in methanol between 265 and 295 nm with the tannin content in hulls. The tannin contents were determined by using proanthocyandin and vanillin assays. The ratio of the standard deviation of the data to the standard error of calibration indicated that the predictive ability of the models was good. Subsequently, Naczk et al. (2002) used UV spectrophotometry to develop a chemometric procedure for the prediction of total phenolic acids in crude polyphenol extracts from defatted rapeseed and canola meals. The phenolic acids were isolated from meals as described by Krygier et al. (1982) and quantified using the Folin-Denis assay. PLS was employed to correlate the spectral data of the crude extract of phenolics in methanol with the total phenolic acid contents. The best correlation was obtained in the 320 to 355 nm region that includes the absorption maxima specific for canola and rapeseed phenolic acids.

NUCLEAR MAGNETIC RESONANCE SPECTROSCOPY

Various nuclear magnetic resonance (NMR) spectroscopic techniques have been employed for structural elucidation of complex phenolics isolated from foods (Porter, 1989). These include ^{1}H NMR and ^{13}C NMR, two-dimensional homonuclear (2 ^{1}H-^{1}H) correlated NMR spectroscopy (COSY), heteronuclear chemical shift correlation NMR (C-H HECTOR), totally correlated NMR spectroscopy (TOCSY), nuclear Overhauser effect in the laboratory frame (NOESY) and rotating frame of reference (ROESY) (Bax, 1985; Bax and Grzesiek, 1996; Belton et al., 1995; Derome, 1987; Ferreira and Brandt, 1989; Kolodziej, 1992; Hemingway et al., 1992; Newman and Porter, 1992). Combinations of high-resolution spectroscopic techniques with novel mathematical treatments of data now provide greater insight into structural elucidation of mixtures of compounds without their prior separation into individual components (Gerothanassis et al., 1998; Limiroli et al., 1996; Sacchi et al., 1996).

TITRATION METHODS

Titration is one of the official AOAC (1980) methods for determination of tannins in tea, cloves and spices. The assay is based on the oxidation of water extracts of phenolics using potassium permanganate in the presence of indigo as an indicator. Permanganate is not specific for phenolics and also readily oxidizes other substances present in the extract. Indigo not only serves as an end point indicator, but also slows down the oxidation reaction because it is less oxidizable than polyphenols.

Tannins in clove and allspice are assayed according to the AOAC (1980) method: a 2-g sample is extracted with anhydrous diethyl ether for 20 h. The residue is then boiled with 300 mL of water for 2 h, cooled, diluted to 500 mL and filtered. A 25-mL sample of this infusion is transferred to a porcelain dish and 20 mL of indigo solution and 750 mL of water added to it. Following this, a standardized solution of potassium permanganate is added, 1 mL at a time, until the blue color of the solution changes to green; then a few drops are added at a time until the solution becomes golden yellow. Similarly, a mixture of 20 mL of indigo solution and 750 mL of water is titrated. The difference between these two titrations is multiplied by a constant factor to obtain tannic acid equivalents or the amount of oxygen absorbed. The potassium permanganate solution is prepared by dissolving 1.333 g of $KMnO_4$ in 1 L of water and standardizing it against a 0.1 M oxalic acid solution. Indigo solution is prepared by dissolving 6 g of sodium indigotin disulfonate in 500 mL of water by heating. After cooling, 50 mL of sulfuric acid is added and the solution diluted to 1 L and then filtered.

PROTEIN PRECIPITATION ASSAYS

Evaluation of biological activity of polyphenolics is based on their ability to form insoluble complexes with proteins as well as bind certain essential minerals. A number of methods have been developed for quantification of biological activity of phenolics. Most of these methods are based on the ability of phenolics to precipitate proteins such as gelatin, bovine serum albumin, and hemoglobin as well as different enzymes (Table 10.10). Protein precipitation methods are highly correlated with the biological value of high tannin-content foods and feeds (Hahn et al., 1984).

Phenolic compounds have been identified as potent inhibitors of iron absorption, presumably by forming insoluble complexes with iron ions in the gastrointestinal lumen and thus making the iron unavailable for absorption (Brune et al., 1989). Therefore, evaluation of biological activity of food phenolics should also include measurement of the content of iron-binding phenolic groups.

Bovine Serum Albumin Assays

Several methods have been published that use bovine serum albumin to determine the ability of polyphenols to precipitate proteins (Asquith and Butler, 1985; Hagerman and Butler, 1978; Hoff and Singleton, 1977; Makkar et al., 1987, 1988; Marks et al., 1987; Naczk et al., 2001a; Silber et al., 1998; Verzele et al., 1986). The first step in these procedures is the formation of a protein–polyphenolic complex.

TABLE 10.10
Some Methods to Determine Tannins by Protein Precipitation

Tannins Measured	Remarks	Reference
Hemoglobin precipitation	Requires fresh blood; interference by plant pigments, saponins etc.	Bate-Smith, E.C. (1973b); Schultz, J.C. et al. (1981)
AOAC	Accuracy ± 10%; not suitable for sorghum grains; suffers from disadvantages associated with FD, FC, PB methods	AOAC (1980)
β-Glucosidase inhibition	Relationship between enzyme activity and insoluble complex formation not known; cumbersome, expensive	Goldstein, G.L. and Swain, T. (1965)
Immobilized protein	Time-consuming, expensive, difficult to handle large number of samples at a time	Hoff and Singleton (1977)
Dye-labeled BSA precipitation	Simple but rather insensitive; preferentially forms soluble complexes	Asquith, T.N. and Butler, L.G. (1985)
Radial diffusion	Insensitive to acetone; simple but involves an element of subjectivity	Hagerman, A.E. (1987)
BSA precipitation	Simple method; measures phenolics in complex; nonspecific binding of phenols to complex can introduce error; not suitable for comparing tannins from different sources	Hagerman, A.E. and Butler, L.G. (1978)
	Indirect method; protein estimations by Lowry or Bradford assays; takes longer time; less sensitive	Martin, J.S. and Martin, M.M. (1982, 1983)
	Indirect method; protein determination by Kjeldahl method; less sensitive	Amory, A.M. and Schubert, C.L. (1987)
	Indirect method; protein determination by HPLC; more sensitive compared to indirect methods above	Verzele, M. et al. (1986)
	Ninhydrin method used to measure protein in complex after its alkaline hydrolysis; takes only 20 min for hydrolysys; assay a bit messy because of ninhydrin reagent	Makkar, H.P.S. et al. (1987)
	Same as in method of Makkar, H.P.S. et al. (1987) except that hydrolysis of the complex is under acidic conditions and takes longer (22 h)	Marks, D. et al. (1987)
	Measures both tannins and proteins	Makkar, H.P.S., et al. (1988)
	Highly sensitive; requires microquantity of tannins; allows quantitation of hydrolyzable and condensed tannins in plant extract	Dawra, R.K. et al. (1988)

Source: Adapted from Makkar, H.P.S, 1989, *J. Agric. Food Chem.*, 37:1197–1202.

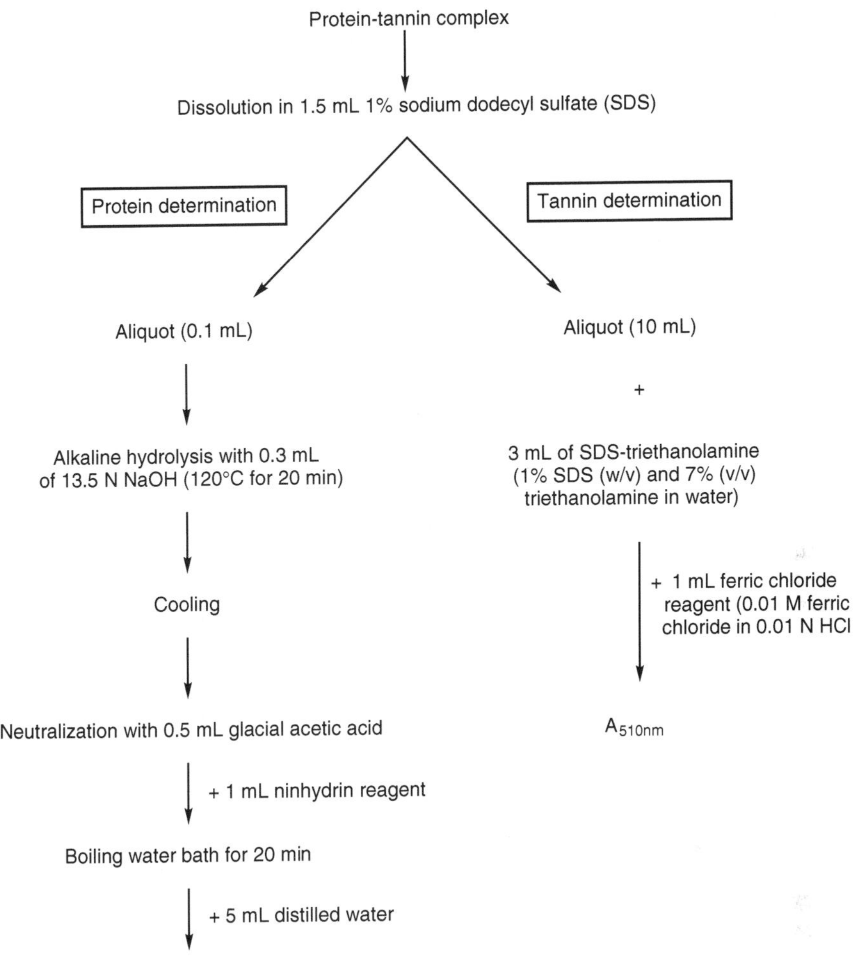

FIGURE 10.6 Estimation of protein and tannin in a tannin–protein complex. (Adapted from Makkar, H.P.S. et al., 1988, *J. Agric. Food Chem.*, 36:523–525.)

Complete precipitation of this complex depends on the structure and quantity of polyphenols and may take from 15 min to 24 h (Hagerman and Robbins, 1987; Makkar, 1989); therefore, the precipitation time should be standardized. The formation of a protein–polyphenol complex is affected by the presence of small amounts of residual organic solvents such as acetone used for the extraction of polyphenols (Dawra et al., 1988). Other factors that may influence the assay include the concentration and nature of proteins used in the test mixture as well as pH and ionic strength of the reaction mixture (Makkar, 1989).

Some methods only determine the quantity of precipitable phenolics (Hagerman and Butler, 1978) while others may also determine the precipitation capacity of phenolics (Figure 10.6) (Asquith and Butler, 1985; Dawra et al., 1988; Hagerman and Butler, 1980; Makkar et al., 1987, 1988; Marks et al., 1987). Naczk et al. (2001a) proposed to express protein precipitating potential of crude tannin extracts as a slope of lines (titration curves) reflecting the amount of tannin–protein precipitated as a function of the amount of tannins added to the reaction mixture. The titration curves are obtained using the protein precipitation assay (Hagerman and Butler, 1978) and the dye-labeled protein assay (Asquith and Butler, 1985). The determination of slope values is based on the statistical analysis (linear regression) of experimental data involving the precipitation of tannin–protein complexes at a minimum of four different tannin levels. Determination of the slopes under standardized conditions (type and concentration of protein, pH and temperature) affords meaningful differences in the slope values for crude tannin extracts from various canola and rapeseed hulls (Naczk et al., 2001a), as well as for condensed tannins of beach pea, evening primrose and faba bean (Naczk et al., 2001b).

Asquith and Butler (1985) developed a simple method to determine the amount of protein precipitated by tannins. In this assay, a bovine serum albumin (BSA) covalently bound with Remazol brilliant R was used as a substrate. The protein-labeling procedure was adapted from Rinderknecht et al. (1968). The dye reacts irreversibly with BSA under mild conditions without apparent denaturation. The blue BSA can be stored at 6°C for several months without loss of precipitating activity. In the presence of excess blue BSA, the amount of precipitated protein is proportional to the amount of tannin added. The plant pigments do not interfere with the assay because the maximum absorption of dye-labeled BSA is at longer wavelengths. Because the assay detects all protein precipitants, it should be corroborated by assays for protein-precipitable phenols before the formation of precipitate can be interpreted as indicative of the presence of tannins.

Recently, Silber et al. (1998) described a novel dye-labeled protein assay based on utilization of a Coomassie brillant blue (G-250)–BSA complex. Coomassie blue pigment exists in a cation form (red with maximum absorbance at 470 nm), an anionic form (blue with maximum absorbance at 595 nm) and a neutral form (green with maximum absorbance at 650 nm) (Matejovicova et al., 1997). Binding Coomassie blue to protein markedly enhances the absorbance of light (Matejovicova et al., 1997; Silber et al., 1998). The complex absorbs light intensely at 600 to 620 nm, but the mechanism of its binding is still not well understood (Atherton et al., 1996; Silber et al., 1998). It takes 3 h of standing at refrigerated temperature to completely precipitate the dye-labeled protein–tannin complexes.

Hagerman (1987) developed a radial diffusion assay for tannin analysis in which tannins were quantified by forming a visible ring precipitation in BSA-containing agarose gel. The tannin-containing solution is placed in a well in a BSA-containing agarose slab. Tannins readily diffuse through the gel to form precipitatble BSA–tannin complexes. The carbohydrate medium of the gel does not form precipitable complexes. The acetone extracts of tannins can be analyzed directly because the acetone does not affect this assay. The area of the visible ring of precipitation is linearly correlated to the amount of tannin placed in a well. The results obtained with this assay are highly correlated with the results of the protein precipitation method of Hagerman and Butler (1978). The detection limit of the assay is 0.025 mg of tannin and depends on the concentration of protein in the plates. Nontannin phenolics such as flavonoids, hydroxybenzoic, and hydroxycinnamic acids do not interfere with the assay.

Hoff and Singleton (1977) developed a method for determination of tannins in foods by employing immobilized bovine serum albumin on a Sepharose matrix. They found that immobilized proteins bind tannins selectively at pH 4. Monomeric polyphenolic substances such as chlorogenic acid, catechol and gallic acid do not form complexes with the immobilized proteins under these conditions. Thus, this method may be used as a means for separating tannins from nontannin materials. Tannin–protein complexes are then dissociated in organic solvents such as methanol or dimethylformamide and the content of tannins determined spectrophotometrically. However, this method is cumbersome and time-consuming, and does not allow handling a large number of samples at a time.

The polyphenols precipitating BSA are assayed as described by Hagerman and Butler (1978): to 2 mL of a standard solution of BSA (1 mg/mL) 1.0 mL of crude extract of polyphenols is added, followed by mixing and standing at room temperature for 15 min. The mixture is then centrifuged at 5000 × g for 15 min and the supernatant is discarded. The pellet is rinsed with 0.2 M acetate buffer (pH 5.0) and then dissolved in 4 mL of an SDS–triethanolamine solution containing 1% SDS and 5% triethanolamine in distilled water. Then 1 mL of a 0.01 M ferric chloride in 0.01 M HCl solution is added and the solution mixed immediately. After 15 to 30 min standing at room temperature, the absorbance of the solution is read at 510 nm. A blank consisting of SDS–triethanolamine and ferric chloride solution is used and the results expressed as A_{510} per gram of tannins.

The assay developed by Asquith and Butler (1985) is: to 4 mL of a standard solution of blue BSA (2 mg/mL) 1.0 mL of a methanolic solution of crude polyphenol extract is added. After 5 min of vigorous mixing at room temperature, the solution is centrifuged at 5000 × g for 15 min and the supernatant discarded. The pellet is then dissolved in 3.5 mL of an SDS–triethanolamine solution containing 1% SDS, 5% triethanolamine and 20% isopropanol in distilled water. The absorbance of the solution is read at 590 nm against a blank consisting of SDS–triethanolamine–isopropanol solution and methanol. The results are expressed as percent of blue BSA precipitated.

Collagen Assay

The collagen assay is one of the official AOAC (1980) methods for determination of tannins. This method is based on the precipitation of tannins with gelatin, hide powder or kaoline, and subsequent oxidation of tannins in an acidic solution with potassium permanganate in the presence of indigo as an indicator. Polyphenols are slowly oxidized in acid solutions by permanganate. Therefore, a large excess of indigo is used to determine the end point of the titration and limit the rate of oxidation reaction in order to avoid oxidation of nonphenolic substances. However, the method has a limitation because different phenols at the same concentration do not reduce the same quantity of permanganate.

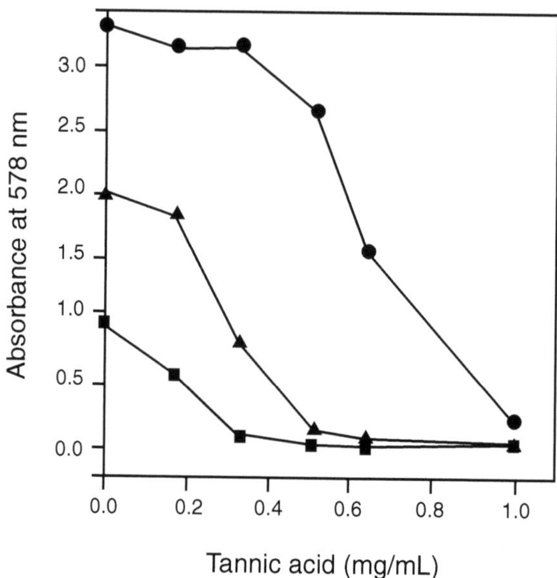

FIGURE 10.7 Optical density of the supernatant at 578 nm after precipitation with tannic acid. Three initial hemoglobin concentrations are shown. (Adapted from Schultz, J.C. et al., 1981, *J. Agric. Food Chem.*, 29:823–829.)

Hemoglobin Assay

The hemoglobin assay is based on the use of hemoglobin as a binding substrate for polyphenols (Bate-Smith, 1973b). This assay requires freshly drawn blood because commercial preparations do not give satisfactory results (Asquith and Butler, 1985). The residual hemoglobin can be measured directly by spectrophotometry. Schultz et al. (1981) found that the relationship between hemoglobin absorbancy and tannin concentration is linear over a large range of tannic acid concentration; however, the linearity range depends on the concentration of hemoglobin (Figure 10.7). At higher initial concentrations of hemoglobin, a specific threshold level of tannin is necessary to initiate precipitation. Thus, the initial concentration of hemoglobin must be adjusted for optimal sensitivity over the range of tannin content to be assayed. Saponins and some enzymes present in plant materials may interfere with this assay, however (Bate-Smith, 1977).

The assay given by Bate-Smith (1977) is: a 1.0-mL sample of freshly drawn blood is diluted with 50 mL of distilled water. To 1 mL of this aliquot, 1 mL of phenolic extract is added; the mixture is rapidly mixed and centrifuged and the content of residual hemoglobin measured at 578 nm (Bate-Smith, 1977). The percentage of hemoglobin precipitation is calculated from the difference between the absorption of the sample and the control consisting of 1 mL of diluted blood with 1 mL of water. The content of phenolics is expressed as tannic acid equivalents and calculated from the relationship describing tannic acid content as a function of precipitation percentage. According to Bate-Smith (1977), tannic acid equivalent contents obtained this way serve as a measure of effective astringency.

Competitive Binding Assays

Interactions between tannins and proteins are often measured using assays based on competition of a test protein and labeled protein for binding and coprecipitation with tannins (Asquith and Butler, 1985; Bacon and Rhodes, 1998; Hagerman and Butler, 1981). These assays permit a direct comparison of protein affinities for condensed taninns.

Hagerman and Butler (1981) developed a competition assay based on the assumption that proanthocyanidin–tannin and antigen–antibody interactions are similar. The similarities include comparable sizes of ligand and binding agent that are multivalently bound to form soluble and insoluble complexes. This procedure utilizes an iodine 125-labeled bovine serum albumine prepared, as described by Hagerman and Butler (1980), as the labeled ligand. In this assay, the proanthocyanidins are simultaneously exposed to a mixture of the competitor (unlabeled protein) and labeled protein. The insoluble tannin–protein complex is separated by centrifugation and the amount of unprecipitated labeled protein is determined in supernatant by counting using the Gamma 300 gamma counter.

Asquith and Butler (1985) developed a competition assay based on the dye-labeled protein assay. A dye-labeled protein for this assay was prepared by covalently binding BSA with Remazol brilliant blue R as described by Rinderknecht et al. (1968). The competition procedure used in this assay is similar to that described by Hagerman and Butler (1981). A standard solution of dye-labeled BSA is mixed with a methanolic solution containing enough tannins to precipitate 70 to 80% of dye-labeled proteins. Varying amounts of competitor protein are mixed with a standard solution of dye-labeled BSA before tannins are added and the amount of precipitated dye-labeled BSA measured spectrophotometrically. Results are presented on a semi-log plot. Following this, the amount of competitor required to inhibit 50% precipitation of dye-labeled protein due to tannin is established and proposed as a measure of relative affinity of competitor protein for tannins.

Bacon and Rhodes (1998) developed a novel competitive tannin–salivary protein-binding assay. This assay was developed based on the principle of a competitive ELISA and competition between an enzyme-labeled tannin (HPR-EGC) and a tested tannin to bind to parotid salivary protein immobilized in the well of microtiter plate. (–)-Epigallocatechin (EGC) conjugated to horseradish peroxidase via a linker molecule, 1,4-butanediol diglycidyl ether, is used as an enzyme-labeled tannin. Coated with parotid salivary proteins and loaded with a known quantity of HPR–EGC, different amounts of the tannin under test are added to the wells. The plates are incubated for 2 h at 37°C and then the peroxidase activity in each well is determined. Following this the amount of tannin required to displace 50% of HRP–EGC bound to the protein coating on microtiter plate is calculated and proposed as a measure of tannin affinity for proteins. A control sample (no test tannin added) is used to measure the amount of HRP–EGC conjugated to proteins coating the microtiter plate.

Enzymatic Assays

Enzymatic methods are based on the ability of tannins to form complexes with enzymes, thus inhibiting their activities. Phenolic compounds can affect enzymes by forming (1) insoluble protein–phenolic complexes and lowering the concentration of enzymes in the reaction mixture or (2) inactive soluble protein–phenolic complexes. Thus, competitive and noncompetitive reaction kinetics as well as mixed kinetics for the inhibition of enzymes of phenolic compounds must be considered. Hagerman and Butler (1978) have suggested, however, that the results of enzymatic assay are difficult to interpret because the relationship between enzymatic activity and formation of insoluble complexes is not fully understood.

Later, Dick and Bearne (1988) studied the inhibition of β-galactosidase activity by polyphenolic compounds. They demonstrated that *trans*-3-(4-hydroxyphenyl)-propenoyl group, common to 4′-hydroxyflavones and *p*-hydroxyphenolic acid, is important for effective inhibition of enzymatic activity. Mole and Waterman (1987c) found that adding tannic acid to a standard trypsin solution leads to formation of an insoluble complex. Trypsin remaining in the solution has an enhanced rate of autolysis, however; conformational changes in enzyme brought about by the formation of soluble complexes with tannic acid might be responsible. This phenomenon has not been observed when excess bovine serum albumin is added to the system containing trypsin and tannic acid.

The inhibition of α-amylase by sorghum polyphenolics has been assayed according to Earp et al. (1981): ground sorghum grains are extracted with water and filtered. A Phadebas tablet, which is a water-insoluble, cross-linked starch polymer with a blue dye bound to it, is added to 4 mL of the extract. Upon hydrolysis the dye is released; 1 mL of α-amylase (0.17 units/mL) is added and the sample incubated for 10 min at 37°C. Adding 1 mL of 0.5 M NaOH to the mixture stops the reaction. After addition of 4 mL of distilled water and filtration, the absorbance of the solution is read at 620 nm. The standard curve of enzyme units of activity vs. absorbance is prepared in order to calculate the percentage inhibition of enzyme activity.

Ittah (1991) developed an enzymatic assay for quantitation of tannins. The assay utilizes the ability of tannins to form complexes with more than one protein and is based on determining the activity of bound alkaline phosphatase. Tannin in this assay is sandwiched between two BSA molecules and alkaline phosphatase. A linear relationship between tannin concentrations and the activity of bound alkaline phosphatase has been found. The procedure involves immobilization of BSA on polystyrene microplates; the quantity of immobilized BSA is low in comparison with the amount of tannin applied. Unbound tannins are washed out and then the microplates are incubated with alkaline phosphate solutions. Following this the unbound enzyme is washed out and the activity of the bound alkaline phosphatase is determined using *p*-nitrophenylphosphate as a substrate.

ELECTROCHEMICAL METHODS

Electrochemical methods such as cyclic voltammetry and liquid chromatography/electrochemistry are very useful tools for the analysis of food phenolics. Most phenolics are electrochemically active at modest oxidation potentials. These methods can be used to determine redox potentials of phenolics, identify the mechanism of oxidation, identify a flavonoid based on comparison with a standard and determine redox potentials for unknown phenolics. The knowledge of redox potential is of great interest for the food industry because oxidized phenolics may have detrimental effects on the quality of wine and beer, grape juice and other beverages (Lunte and Lunte, 1988).

Cyclic voltammetry is used to determine the redox potential as well as the mechanism of oxidation for a pure phenolic compound. The voltammogram (Figure 10.8) is obtained by scanning the potential of the working electrode as a function of time in positive and negative directions. The peak potential for a specific phenolic compound is defined at the point at which the current is diffusion limited. The greater the positive peak potential, the more difficulty a phenolic substance has in oxidizing (Kissinger and Heineman, 1984). Thus, knowledge of peak potentials enables one to estimate the ease of oxidation of food phenolics. Oxidation potentials for some phenolics are given in Table 10.11.

Voltammetry also enables one to estimate the stability of oxidized phenolics. Presence of stable quinone species formed during a voltammetric scanning of potential in the positive direction can be detected when the potential is scanned in the negative direction. This is due to the production of a cathodic current as a result of reduction of quinones to corresponding phenols (Lunte and Lunte, 1988). Voltammetric information also makes it possible to choose the potential that will provide the optimum information for a chromatographic component of interest (Lunte et al., 1988a).

Fernando and Plambeck (1988) demonstrated the ability to measure theaflavin content in tea liquor using a voltammetric method. Theaflavins undergo a completely irreversible reduction during acyclic voltammetry on a hanging mercury drop electrode. The theaflavin content in tea liquor was simultaneously determined by Flavognost method (Hilton, 1973) and by differential pulse polarography. A very highly correlated relationship between the peak currents and the Flavognost absorbance was found with a correlation coefficient equal to 1.

Liquid chromatography/electrochemistry also serves as a useful tool to detect, classify, and identify phenolics in complex samples. It provides information that cannot be obtained using liquid chromatography and a UV detector. According to Chaviari et al. (1988), electrochemistry is a useful alternative to UV detection owing to its better selectivity. The phenolics eluted from the column are detected using a dual electrode detector in the series or parallel configuration.

In the series configuration the upstream electrode is set up at a potential where the flavonoid of interest is oxidized, while a downstream electrode is set at the reducing potential. In cases of reversible oxidation the quinone produced at the upstream electrode may be detected at the downstream electrode. The combination of dual electrodes with a diode array detector can be used to classify phenolics based on their conjugation pattern and hydroxyl substitution (Lunte et al., 1988). In the parallel configuration one of the electrodes is set at the peak potential for the phenolic of interest whereas the other is set at the potential where the current vs. potential responses are changing rapidly. The ratio of these two responses is specific for a

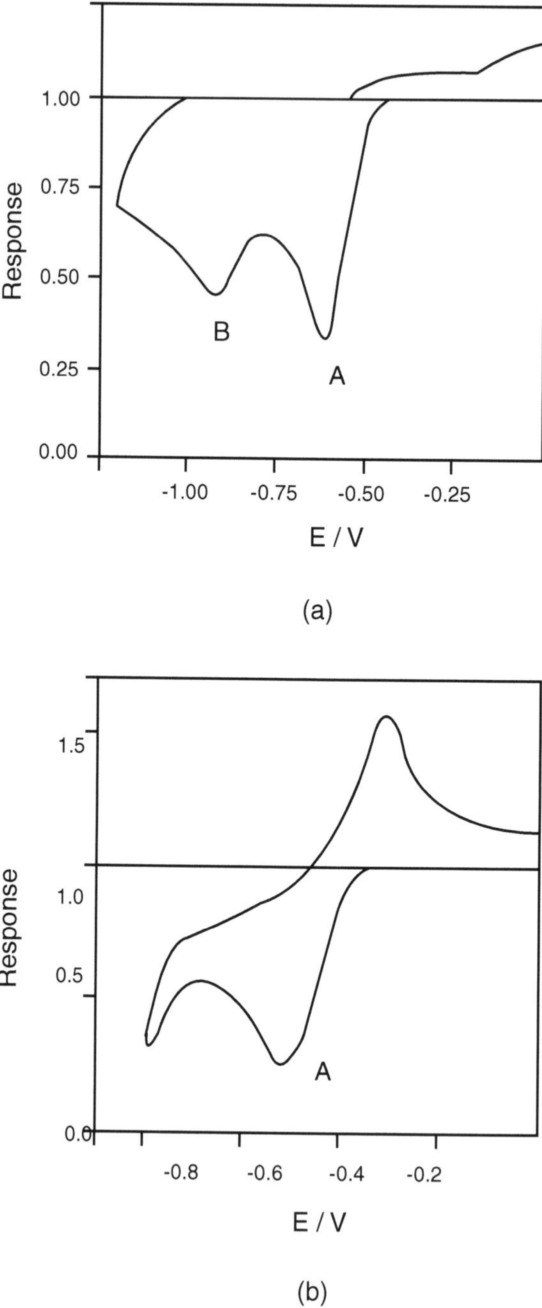

FIGURE 10.8 Cyclic voltammograms of epicatechin; (a) oxidation of the phenolic groups on ring B; and (b) oxidation of the phenolic groups on ring A. (Adapted from Lunte, S.M. et al., 1988b, *Analyst*, 113:99–102.)

TABLE 10.11
Peak Potentials for Oxidation of Selected Phenolic Compounds at Glass Carbon Electrode

No.	Compound	V[a]
1	Eriodictyol	+0.51; +1.20
2	Catechin	+0.53; +0.88
3	Caffeic acid	+0.53
4	Gallic acid	+0.54; +0.87
5	Hydroquinone	+0.57
6	Pyrocatechin	+0.61
7	Syringic acid	+0.75
8	Guaiacol	+0.77
9	Syringaldehyde	+0.81
10	Ferulic acid	+0.82
11	*p*-Coumaric acid	+0.85
12	Vanillic acid	+0.88
13	Acetovanillone	+0.89
14	Vanillin	+0.92
15	Resorcinol	+0.94
16	Phenol	+0.98
17	*o*-Hydroxyphenylacetic acid	+0.98
18	*m*-Hydroxyphenylacetic acid	+1.02
19	*p*-Hydroxyphenylacetic acid	+1.02
20	*p*-Hydroxybenzoic acid	+1.15
21	2,4-Dihydroxyacetophenone	+1.15
22	2-Hydroxyacetohenone	+1.25

Note: All samples are 0.5-mM solutions in methanol — 0.1% perchloric acid in water (15 + 85).

[a] Volts vs. saturated calomel electrode at 20 mVs^{-1}.

Source: Adapted from Chiavari, G. et al., 1988, *Analyst*, 113:91–94.

particular phenolic; thus, the comparison of retention time and current ratios of an unknown with the standard allows one to identify a compound with greater certainty (Lunte and Lunte, 1988; Lunte et al., 1988a).

A combination of HPLC technique and voltammetry has been successfully applied for detection, identification, and quantification of flavonoid and nonflavonoid phenolics in wine. Positive identification may be obtained by comparing the capacity factor (k′) and electrochemical behavior of wine phenols with those of standard solutions containing pure phenolics (Lunte et al., 1988a; Mahler et al., 1988; Woodring et al., 1990).

The voltammetric–amperometric detector has also been applied to detect food phenolics; it operates by scanning the potential of the upstream (voltammetric) electrode and detecting the electrolysis product (amperometric) electrode at a constant potential downstream. This allows one to obtain voltammograms without the charging current problems typical of a single voltammetric detector. Voltammetric–amperometric detection results in a three-dimensional data array of current response vs. time and potential. The voltammetric–amperometric detector also has the capability of detecting each component of phenolic mixture at its optimum potential because each chromatovoltammogram contains the current responses for a range of potentials. Moreover, the chromatographic-voltammetric data array obtained in the course of analysis can also be used for postexperimental optimization of detection potentials (Lunte et al., 1988a).

CHROMATOGRAPHIC TECHNIQUES

Various chromatographic techniques have been employed for separation, preparative isolation, purification and identification of phenolic compounds (Karchesy et al., 1989; Jackman et al., 1987; Merken and Beecher, 2000). Chromatographic procedures have also been used to study the interaction of phenolics with other food components (Oh et al., 1985).

LIQUID CHROMATOGRAPHY

Many liquid chromatographic methodologies have been described in the literature for fractionation of tannins (proanthocyanidins) using Sephadex G-25 (McMurrough and McDowell, 1978; Michaud and Margail, 1977; Somers, 1966), Sephadex LH-20 (Asquith et al., 1983; Boukharta et al., 1988; Davis and Hoseney, 1979; Lea and Timberlake, 1974; Strumeyer and Malin, 1975; Thompson et al., 1972); Sepharose CL-4B (Hoff and Singleton, 1977), Fractogel (Toyopearl) TSK-HW 40(s) (Derdelinckx and Jerumanis, 1984; Mateus et al., 2001; Ricardo da Silva et al., 1991), Fractogel (Toyopearl) TSK 50(f) (Labarbe et al., 1999; Meirelles et al., 1992), inert glass microparticles (Labarbe et al., 1999) and C_{18} Sep-Pak cartridge (Guyot et al., 2001; Jaworski and Lee, 1987; Salagoity-Auguste and Bertrand, 1984; Sun et al., 1998b).

Hoff and Singleton (1977) developed a chromatographic procedure for separating tannins from nontannin materials using bovine serum albumin immobilized on Sepharose CL-4B as a column packing material. Nontannin polyphenolics are first separated by washing the column with an acetate buffer. The protein–tannin complexes are then dissociated by elution of the column with organic solvents such as methanol or dimethylformamide to release and recover tannin substances.

The preparative isolation of proanthocyanins is, however, most commonly achieved by employing Sephadex LH-20 column chromatography (Thompson et al., 1972; Strumeyer and Malin, 1975; Davis and Hoseney, 1978; Asquith et al., 1983). The crude extract is applied to the column, which is then washed with ethanol to elute the nontannin substances. Following this, proanthocyanins are eluted with acetone–water or alcohol–water. For purification of proanthocyanins, the crude extract of phenolic substances is applied to a Sephadex LH-20 column that is then eluted with water containing increasing proportions of methanol (Nonaka et al., 1983).

Derdelinckx and Jerumanis (1984) employed Fractogel (Toyopearl TSK HW-40(s) gel) to separate malt and hop proanthocyanidin dimers and trimers after chromatography of polyphenols on Sephadex LH-20 with methanol. The proanthocyanidin fractions are applied to the Fractogel column, which is then eluted with methanol. The four major peaks of hops polyphenolics correspond to the B3 and B4 procyanidin dimers and two unidentified procyanidin oligomers. On the other hand, malt polyphenolics are separated into a mixture of four proanthocyanidin trimers, two procyanidin trimers and an unknown procyanidin oligomer. According to Derdelinckx and Jerumanis (1984), Fractogel (Toyopearl TSK HW-40(s) gel) allows one to obtain the proanthocyanidins in an advanced state of purity.

Recently, Labarbe et al. (1999) applied an inert glass powder (Pyrex microparticles, 200 to 400 μm) for the fractionation of grape (seed or skin) proanthocyanidins according to their degree of polymerization. Proanthocyanidins are extracted from grape with acetone–water (60:40, v/v). Following this, the nontannin phenolics and proanthocyanidins are fractionated on a Fractogel (Toyoperl TSK HW-50(f) gel) column (35 × 8 cm) as described by Souquet et al. (1996). The nontannin phenolics are washed out from the column with 2 bed volume, of water followed by 5 bed volumes of ethanol–water–trifluoroacetic acid (55:44.05:0.05, v/v/v). Subsequently, proanthocyanidins are washed from the column with 3 bed volumes of acetone–water (60:40, v/v). Purified proanthocyanidins are dissolved in methanol and then applied onto the column filled with glass microparticles and equilibrated with methanol–chloroform (25:75, v/v) and massively precipitated on top of the column with chloroform. Proanthocyandins are sequentially eluted from the column by increasing proportions of methanol in a methanol–chloroform solvent system.

Fulcrand et al. (1999) fractionated wine phenolics into simple (phenolic acids, anthocyanins, flavonols and flavanols) and polymeric phenolics using a Fractogel (Toyopearl) HW-50(f) column (bed 12 × 120 mm). Simple phenolics are eluted from the column with ethanol–water–trifluoroacetic acid (55:45:0.005, v/v/v) while polymeric phenolics are recovered with 60% (v/v) acetone. Later, Mateus et al. (2001) employed a Fractogel (Toyopearl) HW-40(s) column for fractionation of anthocyanin-derived pigments in red wines. Two liters of wine are directly applied onto the Toyopearl gel column (200 × 16 mm i.d.) at a flow rate of 0.8 mL/min. The anthocyanins are subsequently eluted from the column with water–ethanol (20%, v/v). The elution of wine phenolics from Toyopearl column yields malvidin 3-glucoside and three derived pigments, namely, malvidin 3-glucoside pyruvic adduct, malvidin 3-acetylglucoside pyruvic adduct and malvidin 3-coumarylglucoside pyruvic adduct. These glucosides account for 60% of total monoglucoside content.

Oh et al. (1980, 1985) employed a gel chromatographic technique to study tannin–protein interaction. Tannins are immobilized on Sepharose-4B via epoxy activation. Protein is then applied to the column at pH 4 and the elution of protein from the column achieved using a pH gradient or by anionic and nonionic detergents.

Salogoity-Auguste and Bertrand (1984) as well as Jaworski and Lee (1987) demonstrated that a C_{18} Sep-Pak cartridge can be used to separate grape phenolics into acidic and neutral fractions. Later, Sun et al. (1998b) successfully used a C_{18} Sep-Pak cartridge for fractionation of grape proanthocyanidins according to their polymerization degree. The procedure involves passing the extract of grape phenolics through two preconditioned neutral C_{18} Sep-Pak cartridges connected in series. The phenolic acids are then washed out with water; catechins and oligomeric proanthocyanidins are subsequently eluted with ethyl acetate and anthocyanidins and polymeric proanthocyanidins with methanol. The ethyl acetate fraction is redeposited on the same C_{18} Sep-Pak cartridges; catechins are first eluted with diethyl ether and then oligomeric proanthocyanidins with methanol.

High-Performance Liquid Chromatography

High performance liquid chromatographic (HPLC) techniques are now widely used for separation and quantitation of phenolic compounds. Various supports and mobile phases are available for the analysis of anthocyanins, procyanidins, flavonones and flavonols, flavan-3-ols, procyanidins, flavones and phenolic acids (Senter et al., 1989). Introduction of reversed phase columns has considerably enhanced the HPLC separation of different classes of phenolic compounds (Hostettmann and Hostettmann, 1982) and several reviews have been published on the application of HPLC methodology for the analysis of phenolics (Daigle and Conkerton, 1983, 1988; Karchesy et al., 1989; Robards and Antolovitch, 1997; Merken and Beecher, 2000).

Table 10.12 through Table 10.17 summarizes some modern HPLC procedures applied for the analysis of various classes of food phenolics. Food phenolics are commonly detected using UV-Vis and photodiode array (DAD) detectors (Carando et al., 1999; Edenharder et al., 2001; Siess et al., 1996; Tomas-Barberan et al., 2001; Wang et al., 2000a, b; Zafrilla et al., 2001). Other methods used for the detection of phenolics include electrochemical coulometric array detector (EC) (Mattila et al., 2000; Sano et al., 1999), chemical reaction detection technique (de Pascual-Teresa et al., 1998), and fluorimetric detector (Arts and Hollman, 1998; Carando et al., 1999).

Mass spectrometry (MS) detectors coupled to high-performance liquid chromatography (HPLC–MS tandem) have been commonly employed for structural characterization of phenolics. Electrospray ionization mass spectrometry (ESIMS) has been employed for structural confirmation of phenolics in plums, peaches, nectarines (Tomas-Barberan et al., 2001), grape seeds (Peng et al., 2001), soyfoods (Barnes et al., 1998) and cocoa (Hammerstone et al., 1999). Satterfield and Brodbelt (2000) demonstrated that complexation of flavonoids with Cu^{+2} enhances the detection of flavonoids by ESIMS. Mass spectra obtained for metal–flavonoid complexes are more intense and simpler for interpretation than those of corresponding flavonoids. Identification of phenolics collected after HPLC analysis has also been carried out using fast atom bombardment mass spectrometry (FABMS) (Bakker et al., 1992; Edenharder et al., 2001; Sano et al., 1999) and electron impact mass spectrometry (Edenharder et al., 2001). Matrix-assisted laser desorption/ionization mass spectrometry (MALDI-MS) has been employed for qualitative and quantitative analysis of anthocyanins in foods (Wang and Sporns, 1999).

High-Speed Countercurrent Chromatography

High-speed countercurrent chromatography (centrifugal partitioning chromatography; HSCCC) is an all-liquid chromatographic technique very suitable for preparative

TABLE 10.12
Some HPLC Procedures to Determine Isoflavones in Soy Foods

Food	Sample Preparation	Stationary Phase	Mobile Phase	Reference
Seed	Extraction with 80% MeOH; centrifugation	Aquapore C_8; (220 × 4.6 mm)	A: 10% acetonitrile in H_2O with 0.1% TFA; B: 90% acetonitrile in H_2O with 0.1% TFA; linear gradient: 100% A, 0% B, 0 min; 70% A, 30% B, 30 min; step gradient to 0% A, 100% B	Simonne, A.H. et al. (2000)
Soy foods	Extraction with 80% MeOH; centrifugation, evaporation of MeOH, extraction of lipids with hexane	Brownlee Aquapore C_8 reverse phase; (300 × 4.5 mm)	A: 0.1% TFA in H_2O; B: acetonitrile; linear gradient: 100% A, 0% B, 0 min; 53.6% A, 46.4% B, 20.6 min	Coward, L. et al. (1993)
Seeds, soy foods	Extraction with 0.1%HCl-acetonitrile (1:5), filtration	YMC-pack ODS–AM-303 (250 × 4.6 mm, 5µm)	A: 0.1% acetic acid in water; B: 0.1% acetic acid in acetonitrile; 85% A, 15% B, 0 min; 85% A, 15% B, 5 min; 71% A, 29 %B, 31 min; 65% A, 35% B, 8 min; 85% A, 15% B, 3 min	Murphy, P.A. et al. (1997)
Seeds	Extraction with 80% MeOH	YMC-Pack ODS–AQ 303 (250 × 4.6 mm; 5 µm)	A: 0.1% acetic acid in H_2O; B: 0.1% acetic acid in acetonitrile; modified gradient: 85% A, 15% B, 0 min; 69% A, 31% B, 45 min	Wang, J. et al. (2000a)
Soy sauce	Direct injection	Wakosil II 5C18 HG (250 × 4.6 mm) fitted with a precolumn (30 × 4.6 mm) packed with the same material	A: 0.05% TFA in H_2O; B: 90% acetonitrile with 0.05% TFA; 100% A, 0% B, 0 min; 100% A, 0% B, 20 min; 75% A, 25% B, 270 min, linear gradient	Kinoshita, E. et al. (1997)
Soy foods	Extraction with 80% MeOH, centrifugation	NovaPak C_{18} reverse phase (150 × 3.9 mm, 4 µm) coupled to Adsorbosphere C_{18} column guard	A: 10% acetic acid in H_2O; B: MeOH-acetonitrile–dichloromethane (10:5:1 v/v/v). 95% A, 5% B, 0 min; 95% A, 5% B, 5 min; linear gradient: 45% A, 55% B, 20 min; 30% A, 70% B, 6 min; 95% A, 5% B, 3 min	Franke, A.A. et al. (1999)
Soy foods	Extraction with 96% EtOH, centrifugation, filtration	NovaPak C_{18} (150 × 3.9 mm; 4 µm)	Isocratic: acetonitrile–water (33:67, v/v)	Hutabarat, L.S. et al. (2001)

Source: Adapted from Merken, H.M. and Beecher, G.R., 2000, *J. Agric. Food Chem.*, 48:577–599.

TABLE 10.13
Some HPLC Procedures to Determine Catechins and Proanthocyanidins (PA) in Selected Foods

Food	Sample Preparation	Stationary Phase	Mobile Phase	Reference
Grape seed	Extraction with EtOH; fractionation of PA using Sephadex LH-20	Exsil 100 ODS C_{18}, reverse phase (250 × 4.6 mm, 5 µm) coupled to C_{18} column guard	A: 0.2% phosphoric acid (v/v); B: 82% acetonitrile with 0.4% phosphoric acid; gradient: 100% A, 0% B, 0 min; 85% A, 15% B, 15 min; 84% A, 16% B, 25 min; 83% A, 17% B, 5 min; 57% A, 43% B, 3 min; 48% A, 52% B, 1 min; isocratic 52% B, 7 min; 57% A, 43% B, 1 min; 83% A, 17% B, 1 min; 100% A, 0% B, 2 min	Peng, Z. et al. (2001)
Beverages	Direct injection	Spherisorb ODS2 (150 × 46 mm, 3 µm)	A: H_2O; B: MeOH; C: 4.5% aqueous formic acid; D: 4.5% aqueous formic acid–MeOH (90:10, v/v); gradient: 0–10 min, 100% A to 100% C; 10–20 min, 0–15% D in C; 20–30 min, 15% D in C, isocratic; 30–40 min, 15–35% D in C; 40–45 min, 35% D in C, isocratic; 45–60 min, 35–45% D in C; 60–75 min, 45–100% D in C; 75–175 min, 0–50% B in D; 175–180 min, 50–80% B in D	de Pascual-Teresa, S. et al. (1998)
Apples, grapes, beans	Extraction with 90% MeOH (apples, grapes) or 70% MeOH (beans); filtration	Inertsil ODS–2 (150 × 4.6 mm, 5 µm) coupled with Opti-Guard PR C18 Violet A guard	A: 5% acetonitrile in 0.025 M phosphate buffer pH 2.4; B: 25% acetonitrile in 0.025 M phosphate buffer pH 2.4; isocratic, 0–5 min, 10% B; 5–20 min, linear gradient: 5–20 min, 10–80% B; 20–22 min, 80–90% B; isocratic 22–25 min, 90% B; linear gradient, 25–28 min, 10% B; isocratic, 28–37 min, 10% B	Arts, I.C.W. and Hollman, P.C.H. (1998)

Wine	Dealcoholized under vacuum; fractionation of polymeric fraction using Tyopearl TSK gel HW-50 (f), thiolysis	Nucleosil 120, (125 × 4 mm, 3 µm)	A: 2% HCOOH in H_2O; B: CH_3CN–H_2O–HCOOH (80:18:2, v/v/v); linear gradient: 15–75% B, 0–15 min; 75–100% B, 15–20 min	Fulcrand, H. et al. (1999)
Wine	Dealcoholized under vacuum; fractionation of procyanidins and catechins using two C_{18} Sep Pak cartridges in series	Superspher 100 RP18 (250 × 4 mm; 4 µm)	A: H_2O; B: H_2O–acetic acid (90:10, v/v); **catechins**: 10–80% B, 0–5 min; 80–100% B, 5–29 min; 100% B, 29–45 min; **procyaidins**: 10–70% B, 0–40 min; 70–85% B, 40–55 min; 85–100% B, 55–74 min	Sun, B.S. et al. (1998b)
Wine	Filtration, direct injection	Nucleosil 100 C_{18}, (250 × 4 mm, 5 µm) coupled to C_{18} column guard	A: 2 mM $NH_4H_2PO_4$, adjusted to pH 2.6 with H_3PO_4; B: 20% A with acetonitrile; C: 0.2 M H_3PO_4 adjusted to pH 1.5 with ammonia; gradient: 100% A, 0–5 min; 0–4% B, 5–15 min; 4–8% B, 15–25 min; 8% B, 92% C, 25.1 min; 8–20% B, 25.1–45 min; 20–30% B, 45–50 min; 30–40%, 50–55 min; 40–80% B, 55–60 min	Carando, S. et al. (1999)
Cocoa	Extraction of defatted seeds with 70% acetone (v/v), followed by extraction with 70% MeOH (v/v), fractionation of phenolic from nonphenolics on column packed with Baker octadecyl for flash chromatography	Phenomenex Luna (250 × 4.6 mm, 5 µm)	A: dichloromethane; B: MeOH; C: acetic acid–H_2O (1:1,v/v); gradient: 0 min, 14% B in A; 0–30 min, 14–28.4% B in A; 28.4–50% B in A, 30–60 min; 50–86% B in A, 60–65 min; isocratic 65–70 min	Hammerstone, J.F. et al. (1999)
Oolong tea	Extraction with acetonitrile-water (1:1, v/v)	Devolosil PhA-5 (250 × 46 mm)	0.1 M NaH_2 PO_4 buffer (pH 2.5)–acetonitrile (85:15, v/v) with 0.1 mM EDTA2 Na, isocratic	Sano, M. et al. (1999)

(continued)

TABLE 10.13 (CONTINUED)
Some HPLC Procedures to Determine Catechins and Proanthocyanidins (PA) in Selected Foods

Food	Sample Preparation	Stationary Phase	Mobile Phase	Reference
Green tea beverage	Washed with chloroform; extraction with ethyl acetate, ethyl acetate layers combined, evaporated, residue dissolved in 50% acetonitrile	μ–Bondapak C_{18} (300 × 3.9 mm)	Acetonitrile–ethyl acetate-0.05% phosphoric acid (12: 2:86, v/v/v), isocratic	Wang, J. et al. (2000b)
Green tea	Extraction with hot water, filtration, washing with chloroform, extraction of water layer with ethyl acetate, ethyl acetate layers combined and evaporated, residue dissolved in water	Hypersil ODS (250 × 4.6 mm, 5 μm)	0.05% H_2SO_4 aqueous–acetonitrile–ethyl acetate (86:12:2, v/v/v), isocratic	Chen, Z.-Y. et al. (1998)

Source: Adapted from Merken, H.M. and Beecher, G.R., 2000, *J. Agric. Food Chem.*, 48:577–599.

TABLE 10.14
Some HPLC Procedures to Determine Anthocyanins and Anthocyanidins in Selected Foods

Food	Sample Preparation	Stationary Phase	Mobile Phase	Reference
Red onions	Extraction with MeOH containing 0.1% HCl, filtration	Prodigy ODS2 (250 × 4.6 mm, 5 µm)	A: 10% formic acid in H_2O (v/v); B:MeOH-H_2O-formic acid (50:40:10, v/v/v); isocratic: 80% A, 20% B, 0–4 min; linear regression: 20–80% B in A, 4–26 min	Gennaro, L. et al. (2002)
Red wine fruit juices	Dealcoholization of wine under vacuum; dilution of fruit juice with water; separation of anthocyanins using C_{18} Sep-Pak	SPLC-18-DB (250 × 10 mm, 5 µm) preparative reverse phase coupled with preinjection C_{18}saturator with silica-based packing (75 × 4.5 mm, 12 µm) and guard with Supelco LC-18 reverse-phase packing (50 × 4.6 mm, 20–40 µm)	A: 5% formic acid in H_2O; B: formic acid–H_2O–MeOH (5:5:90, v/v/v); linear gradient: 5–20% B in A, 0–1 min; 20–25% B in A, 1–12 min; 25–32% B in A, 12–32 min; 32–55% B in A, 32–38 min; 55–100% B in A, 38–44 min; 100% B, 44–46 min; 100–5% B in A, 46–47 min	Wang, J. and Sporns, P. (1999)
Red wine	Direct injection	Ultrasphere (C18) ODS (250 × 4.6 mm)	A: H_2O–formic acid (9:1, v/v); B: CH_3CN–H_2O–formic acid (3:6:1, v/v/v); gradient: 20–85% B in A, 0–70 min; 85–100% B in A, 70–75 min; isocratic: 100% B, 75–85 min	Mateus, N. et al. (2001)
Red Wine	Centrifugation, addition of formic acid to 1.5%, filtration	Supelcosik LC-18, (250 × 2.1 mm), reverse–phase	A: 5% formic acid in H_2O (v/v); B: MeOH; gradient: 5% B in A, 0–5 min; 5–65% B in A, 5–55 min; 65–100% B, 55–58 min; 100–5% B in A, 58–60 min; 5% B in A, 60–64 min	Mazza, G. et al. (1999)
Red blood orange juice	Homogenization with (acetone–EtOH–hexane, 25:25:50,v/v/v), centrifugation, concentration of acetone–EtOH layer, separation of anthocyanins using C_{18} Sep-Pak	Prodigy ODS3 (150 × 4.6 mm, 5 µm)	A: 0.1% phosphoric acid in H_2O; B: 0.1% phosphoric acid in acetonitrile; gradient: 10% B in A, 0–2 min; 10–50% B in A, 2–32 min; 50% B in A, 32–37 min; 50–70% B in A, 37–57 min	Lee, H.S. (2002)

Source: Adapted from Merken, H.M. and Beecher, G.R., 2000, *J. Agric. Food Chem.*, 48:577–599.

TABLE 10.15
Some HPLC Procedures to Determine Flavones and Flavonols in Selected Foods

Food	Sample Preparation	Stationary Phase	Mobile Phase	Reference
Red onions	Extraction with MeOH stabilized with BHT; dilution with MeOH	Supelcosil LC-18 (250 × 4.6 mm, 5 μm) coupled with a Spherisorb Supelguard LC-18	A: 0.01 M sodium phosphate adjusted to pH 2.5 with H_3PO_4; B: MeOH; linear gradient: 87–60% A in B, 0–13.5 min; 60–10% A in B, 13.5–39 min; 10–0% A in B, 39–42 min; 0–87% A in B, 42–46 min	Gennaro, L. et al. (2002)
Yellow and green french beans	Extraction with chloroform to remove chlorophyll and carotenoids, drying, extraction with 70% MeOH, evaporation of MeOH, purification of phenolics using polyamide cartridge, filtration	LiChrospher 100 RP-18 (250 × 4, 5μm) coupled with guard (4 × 4 mm) packed with the same stationary phase	A: acetonitrile; B: 2% acetic acid in H_2O; gradient: 10–30% A in B, 0–35 min; 30–45% A in B, 35–37 min; 45% A in B, 37–42 min; 45–10% A in B, 42–44 min	Hempel, J. and Bohm, H. (1996)
Red bean seed coats	Extraction with MeOH, separation of tannins using Sephadex LH-20, flavonoid fraction rechromatographed on Sephadex LH-20	Shiseido Capcell Pak C_{18} preparative reverse phase (250 × 10 mm, 5μm)	Isocratic: acetonitrile–H_2O (30:70, v/v)	Beninger, C.W. and Hosfield, G.L. (1999)
Buckwheat	Extraction with 80% MeOH, filtration, evaporation, dissolving in MeOH-H_2O-oxalic acid (13:36:1, v/v/v) filtration	Capcell Pak C_{18}–SG 120, (100 × 4.6 mm, 3 μm)	A: MeOH-H_2O-acetic acid (13:36:1, v/v/v); B: MeOH–H_2O–acetic acid (73:25:2, v/v/v); gradient: 10–50% B in A, 0–20 min; 50% B in A, 20–25 min; 50–10% B in A, 25–30 min	Oomah, B.D. and Mazza, G. (1996)
Tomatoes, onions, lettuce, celery	Extraction with 1.2M HCl in 50% MeOH for 2 h at 90°C; extract adjusted to pH 2.5 with TFA, filtration	C_{18} Symmetry (150 × 3.9 mm, 5μm) reversed-phase, coupled with C_{18} symmetry guard	A: acetonitrile; B: H_2O adjusted to pH 2.5 with TFA; gradient: 15–35% A in B, 0–20 min	Crozier, A. et al. (1997)
Edible tropical plants	Extraction with 1.2 M HCl in 50% MeOH for 2 h at 90°C, filtration	Nova Pak C_{18} (150 × 3.9; 4 μm)	Isocratic: MeOH–H_2O (1:1, v/v) adjusted to pH 2.5 with TFA	Miean, K.H. and Mohamed, S. (2001)

Source: Adapted from Merken, H.M. and Beecher, G.R., 2000, *J. Agric. Food Chem.*, 48:577–599.

TABLE 10.16
Some HPLC Procedures to Determine Other Classes of Phenolics in Selected Foods

Food	Phenolics	Sample Preparation	Stationary Phase	Mobile Phase	Reference
Finger millet	Free phenolic acids	Extraction with 70% EtOH, centrifugation, concentration, adjusting pH to 2–3, extraction with ethyl acetate, evaporation, dilution in MeOH	Shimpak C_{18} (250 × 4.6 mm) reverse phase	Isocratic: H_2Oacetic acid–MeOH (80:5:15, v/v/v)	Subra Rao, M.V.S.S.T. and Muralikrishna, G. (2002)
Barley	Phenolic acids	Extractions: hot H_2O; acid hydrolysis; acid and α-amylase hydrolysis; acid, α-amylase and cellulase hydrolysis; centrifugation	Supelcosil LC-18 (150 × 4.6 mm, 5 μm)	A: 0.01 M citrate buffer pH 5.4 adjusted with 50% acetic acid; B: MeOH; gradient: 2–4% B in A, 0–12 min; 4–13% B in A, 12–20 min; 13% B in A, 20–26 min; 13–2% B in A, 26–30 min	Vasanthan, J.Y.T. and Temelli, F. (2001)
Citrus fruits	Coumarins	Extraction with acetone, filtration, evaporation dissolving in MeOH–acetone (1:1, v/v), filtration	Hypersil ODS (125 × 4 mm, 5 μm)	Isocratic: MeOH–H_2O (75:25, v/v)	Ogawa, K. et al. (2000)
Rice bran oil	γ-Oryzanol	Solubilization of oil in hexane–ethyl acetone (9:1, v/v), removal of lipids using silica column (250 × 25 mm)	Microsorb-MV C_{18}, (250 × 4.6 mm)	Isocratic: MeOH-acetonitrile–dichloromethane-acetic acid (50:44:3:3, v/v/v/v)	Xu, Z. and Golber, J.S. (1999)
Flaxseed flour, defatted	Lignans	Extraction with 1,4-dioxane–95%EtOH (1:1,v/v), centrifugation, evaporation, alkaline hydrolysis, acidification to pH 3, removal of salt using C 18 reversed SPE	Econosil RP C_{18} (250 × 4.6 mm, 5μm)	A: 5% acetonitrile in 0.01 M phosphate buffer, pH 2.8; B: acetonitrile; gradient: 100–70% A in B, 0–30 min; 70–30% A in B, 30–32 min	Johnsson, P. et al. (2000)
Peanuts	Resveratrol	Extraction with 80% EtOH, centrifugation, semipurification Al_2O_3 silica gel 60R_{18} (1:1)	Vydac C_{18} (150 × 4.5 mm) reverse phase	A: acetonitrile; B: 0.1% TFA; gradient: 0%A in B, 0–1 min; 0–15% A in B, 1–3 min; 15–27% A in B, 3–23 min; 27–100% A in B, 23–28 min	Sanders, T.H. et al. (2000)

Source: Adapted from Merken, H.M. and Beecher, G.R., 2000, *J. Agric. Food Chem.*, 48:577–599.

TABLE 10.17
Some HPLC Procedures to Determine Multiple Classes of Phenolics in Selected Foods

Food	Phenolics	Sample Preparation	Stationary Phase	Mobile Phase	Reference
Lingoberry, cranberry, onions, broccoli	Catechins, flavanones, flavones, flavonols	Extraction with 1.2 M HCl in 50% MeOH for 2 h at 90°C; filtration	Inertsil ODS (150 × 4 mm; 3 µm) coupled with C-18 guard	A: 50 mM H_3PO_4 pH 2.5; B: acetonitrile; **catechins**: 86% A in B, isocratic; **other flavonoids**: gradient: 95% A in B, 0–5 min; 95–50% A in B, 5–55 min; 50% A in B, 55–65 min; 50–95% A in B, 65–67 min	Mattila, P. et al. (2000)
Nectarines, peaches, plums	Phenolic acids, catechins, flavonols, procyanidins	Extraction with 80% MeOH containing 2 mM NaF; centrifugation, filtration	Nucleosil C_{18} (150 × 4.6 mm, 5 µm) reverse phase coupled with guard containing the same stationary phase	A: 5% MeOH in H_2O; B: 12% MeOH in H_2O; C: 80% MeOH in H_2O; D: MeOH; gradient: 100% A, 0–5 min; 0–100% B in A, 5–10 min; 100% B, 10–13 min; 100–75% B in C, 13–35 min; 75–50% B in C, 35–50 min; 50–0% B in C, 50–52 min; 100% C, 52–57 min; 100% D 57–60 min	Tomas-Barberan, F.A. et al. (2001)

Red raspberry	Ellagic acids, flavones	Extraction with MeOH, filtration, addition of H_2O, evaporation, semipurification of phenolics using Sep-Pak C18, filtration	Lichrocart 100 RP-18 (250 × 4 mm, 5 μm) reverse phase	A: 5% formic acid in H_2O; B: MeOH; gradient: 10–15% B in A, 0–5 min; 15–30% B in A, 5–20 min; 30–50% B in A, 20–35 min; 50–90% B in A, 35–38 min	Zafrilla, P. et al. (2001)
Propolis	Phenolic acids, flavones, flavonones flavonols	Dilution with EtOH, alkaline hydrolysis, acidification, extraction of phenolic with ethyl acetate, evaporation, dissolving in EtOH	Lichrosorb RP18 (200 × 3, 7 μm) coupled with C-18 guard	A: H_2O adjusted to pH 2.6 with H_3PO_4; B: acetonitrile; gradient: 0–9% B in A, 0–12 min; 9–13% B in A, 12–20 min; 13–40% B in A, 20–40 min; 40–70% B in A, 40–60 min; 70% B in A, 60–85 min	Siess, M.-H. et al. (1996)
Spinach	Flavonols, flavanones	Extraction with 70% MeOH, removal of carotenoids and chlorophyll using ODS-C_{18} packing material, centrifugation, concentration	YMC ODS–AQ (250 × 4.6, 5 μm)	A: H_2O containing 0.01% TFA; B: acetonitrile containing 0.01% TFA; gradient: 100% A, 0–10 min; 100–50% A in B, 10–40 min; 50–0% A in B, 40–50 min	Edenharder, R. et al. (2001)

Source: Adapted from Merken, H.M. and Beecher, G.R., 2000, *J. Agric. Food Chem.*, 48:577–599.

isolations of pure compounds (Degenhart et al., 2000a, b). Separation of compounds is based on their partitioning between two immiscible liquids (Conway and Petrovski, 1995).

Degenhart et al. (2000c) used HSCCC for preparative isolation of anthocyanins from red wines and grape skins. Anthocyanins were fractionated based on their polarities into four solvent systems:

- Solvent I, consisting of *tert*-butyl methyl ether–n-butanol–acetonitrile-water (2:2:1:5, v/v/v/v) containing 0.1% trifluoroacetic acid (TFA), was used as a medium for fractionation of monoglucosides and acylated diglucosides.
- Solvent II, consisting of ethyl acetate–n-butanol–water (2:3:5, v/v/v) and 0.1%TFA, was used as a medium for separation of visitins and diglucosides.
- Solvent III, consisting of ethyl acetate–water (1:1, v/v) and 0.1% TFA, was used as a medium for extraction of coumaryl and caffeoyl monoglucosides.
- Solvent IV, consisting of ethyl acetate–n-butanol–water (4:1:5, v/v/v) was employed as a medium for fractionation of acetylated anthocyanins.

Vitrac et al. (2001) have also applied HSCCC for fractionation of red wine phenolics. Phenolics are extracted first from red wine into ethyl acetate. Subsequently the phenolic extract is chromatographed using a 1.5- × 60-cm cation-exchange Dowex (Sigma) column. Nonphenolic constituents are washed out from the column with water and then phenolics are eluted with aqueous methanol (75%, v/v). Afterwards, the phenolic extract is fractionated using centrifugal partition chromatography in ascendant and descendant modes. The solvent system's water–ethanol–hexane–ethyl acetate in ratios of 3:3:4:5 (v/v/v/v) and 7:2:1:8 (v/v/v/v) are used, at a flow rate of 3 mL/min, for elution of phenolics in ascendent and descendent modes, respectively.

Degenhart et al. (2000d) demonstrated that HSCCC can be used for isolation of theaflavins, epitheaflavic acids, and thearubigins from black tea using hexane–ethyl acetate–methanol-water (2:5:2:5 and 1.5:5:1.5:5, v/v/v/v). Prior to HSCCC, theaflavins were extracted from tea infusion with ethyl acetate and then cleaned up using a Sephadex LH-20 column to avoid coelution of catechins and theaflavins. On the other hand, isolation of thearubigins required cleaning up tea infusion on an Amberlite XAD-7 column prior to HSCCC to remove all nonphenolic compounds.

Baumann et al. (2001) have developed a simple and efficient method for separating catechin gallates from spray-dried tea extract. Tea phenolic extract is first subjected to liquid–liquid partitioning between ethyl acetate and water. The organic layer containing catechins is then submitted to high-speed centrifugal countercurrent chromatography operating in an ascending mode. Using n-hexane–ethyl acetate–water (1:5:5, v/v/v) or ethyl acetate–methanol–water (5:1:5 and 5:2:5, v/v/v) achieves favorable partitioning. Sephadex LH-20 column with methanol as a mobile phase is used for final purification of catechin gallates.

Other Chromatographic Techniques

Other chromatographic techniques have also been employed for purification and separation of food phenolics. Of these, paper chromatographic (PC) and thin-layer chromatographic (TLC) techniques are still widely used for purification and isolation of anthocyanins, flavonols, condensed tannins and phenolic acids using different solvent systems (Chu et al., 1973; Durkee and Harborne, 1973; Fenton et al., 1980; Forsyth, 1955; Francis et al., 1966; Francis, 1967; Harborne, 1958, 1967; Jackman et al., 1987; Leung et al., 1979; Mabry et al., 1970).

PC on Whatman no. 3 has been employed for separation of anthocyanins using butanol–acetic acid–water, chloroform–acetic acid–water, or butanol–formic acid–water as possible mobile phases (Jackman et al., 1987). On the other hand, two-dimensional PC has been used for the analysis of procyanidin oligomers. Chromatograms are developed using 6% acetic acid as a mobile phase in the first direction and 2-butanol–acetic acid–water as mobile phase in the second direction (Haslam, 1966; Thompson et al., 1972).

Azar et al. (1987) identified phenolics of bilberry juice (*Vaccinium myrtillus*) using a two-dimensional TLC. Phenolic acids were chromatographed on a 0.1 mm cellulose layer with solvent I: acetic acid–water (2:98, v/v), and solvent II: benzene–acetic acid–water (60:22:1.2, v/v/v). However, TLC analysis of flavonols was carried out on silica gel plates using ethyl acetate–ethyl methylketone–formic acid–water (5:3:1:1, v/v/v/v) or on cellulose plates using solvent I: t-butanol–acetic acid–water (3:1:1, v/v/v) and solvent II: acetic acid–water (15:85, v/v). The phenolic acids were detected by first spraying the chromatograms with deoxidized *p*-nitroaniline and then with a 15% solution of sodium carbonate in water; flavonols were detected by spraying with a 5% aluminum chloride solution in methanol.

Two-dimensional cellulose TLC plates have also been employed to separate procyanidins. t-Butanol–acetic acid–water (3:1:1, v/v/v) was used for development in the first direction while 6% acetic acid was used for development in the second direction. Detection of polyphenols on TLC was carried out using ferric chloride, potassium ferricyanide or vanillin–HCl solutions (Karchesy et al., 1989). TLC on silica using ethyl acetate–formic acid–water (90:5:5, v/v/v) or toluene–acetone–formic acid (3:3:1, v/v/v) has been used for monitoring the isolation of procyanidins by column chromatography (Karchesy and Hemingway, 1986) and by HPLC (Carando et al., 1999), respectively. On the other hand, phenolic acids have been separated on silica TLC plates using n-butanol–acetic acid–water (40:7:32, v/v/v) as a mobile phase (Dabrowski and Sosulski, 1984).

Various gas chromatographic (GC) methodologies have been employed for separation and quantitation of phenolic acids (Dabrowski and Sosulski, 1984; Krygier et al., 1982), isoflavones (Liggins et al., 1998), capsaicinoids (Thomas et al., 1998), phenolic aldehydes (Friedman et al., 2000), and condensed tannin monomers (Hemes and Hedges, 2000; Matthews et al., 1997). Novel high-temperature gas chromatographic columns, electronic pressure controllers and detectors have significantly improved the resolution and have also led to an increase in the upper range of molecular weights of substances that could be analyzed by GC. Preparation of samples for GC may include removing lipids from the extract, liberating phenolics from ester and glycosidic bonds by alkali (Dabrowski and Sosulski, 1984; Krygier et al., 1982), acid (Zadernowski, 1987) and enzymatic hydrolysis (Liggins et al., 1998) or depolymerizing tannin acid in the presence of nucleophiles such as phloroglucinol (Hemes and Hedges, 2000; Matthews et al., 1997) or benzyl mercaptan (Guyot et al., 2001; Labarbe et al., 1999; Matthews et al., 1997). Prior to chromatography, phenolics were usually transformed to more volatile derivatives by methylation (Jurenitsch et al., 1979; Jurenitsch and Leinmuller, 1980; Kosuge and Furuta, 1970), trifluoroacetylation (Chassagne et al., 1998; Sweeley et al., 1963), conversion to trimethylsilyl derivatives (Dabrowski and Sosulski, 1984; Krygier et al., 1982) or derivatization with N-(*tert*-butyldimethylsilyl)-N-methyltrifluoroacetamide (Hemes and Hedges, 2000).

CAPILLARY ELECTROPHORESIS

Capillary electrophoresis is a novel analytical tool for separating many classes of compounds (Cikalo et al., 1998; Zeece, 1992) based on the electrophoretic migration of charged analytes. Small internal diameter capillary columns minimize the ohmic heating problems that may have an adverse effect on bandwidths. In addition, small sample sizes can be used and separations require little or no organic solvents (Cikalo et al., 1998).

Hall et al. (1994) used capillary electrophoresis for separation of food antioxidants such as butylated hydroxyanisole (BHA) and butylated hydroxytoluene (BHT). Later, Andrade et al. (1998) utilized capillary zone electrophoresis to evaluate the effect of grape variety and wine aging on the composition of noncolored phenolics in port wine. Noncolored phenolics were extracted from wine into diethyl ether, then concentrated to dryness and redissolved in methanol. Subsequently, phenolics were separated on a fused-silica capillary column with 0.1 M sodium borate (pH 9.5) at 30°C and voltage of 20 kV producing a current of 90 μA.

Chu et al. (1998) separated pure forms of *cis*- and *trans*-resveratrol isomers from wine using capillary electrophoresis in micellar mode. Direct separation of resveratrols in wine samples was performed with fused silica capillaries in 25 mM sodium borate, 25 mM sodium phosphate and 75 mM SDS (pH 9.3) at 16 kV and 20°C using a UV detector set at 310 nm. The detection limit of the method was 1.25 mM. To determine rutin content in different fractions of buckwheat flour and bran, Kreft et al. (1999) utilized capillary electrophoresis with a UV detector. The extraction of rutin from buckwheat fractions was carried out using 60% ethanol containing 5% ammonia in water. Identification of resveratrol and rutin was confirmed by spiking the samples with standards.

Moane et al. (1998) utilized capillary electrophoresis for direct detection of phenolic acids in beer. Separation of phenolic acids was performed with fused silica capillary in 25 mM phosphate buffer, pH 7.2 at 25 kV. The sample was injected to capillaries using a reversed-polarity injection technique to remove nonphenolic cationic and neutral compounds. (These substances interfere with electrochemical detection of phenolic acids by passivation the electrode surface.) Recently, Pan et al. (2001) developed a method for determining protocatechuic aldehyde and protocatechuic acid by capillary electrophoresis with amperometric detection. Under optimum conditions these two analytes were completely separated in 8 min with detection limits of 0.10 g/mL for protocatechuic aldehyde and 0.25 μg/mL for protocatechuic acid.

REFERENCES

Adams, J.B. 1972. Changes in polyphenols of red fruits during processing: the kinetics and mechanism of anthocyanin degradation. Campden Food Pres. Res. Assoc. Tech. Bull. P22, Chipping Campden, Gloucestershire, U.K., 1–185.

Amory, A.M. and Schubert, C.L. 1987. A method to determine tannin concentration by measurement and quantification of protein–tannin interactions. *Oecologia*, 73:420–424.

Andrade, P., Seabra, R., Ferreira, M., Ferreres, F., and Garcia-Viguera, C. 1998. Analysis of noncoloured phenolics in port wines by capillary zone electrophoresis. Influence of grape variety and ageing. *Z. Lebensm. Unters Forsch A.*, 206:161–164.

Antolovich, M., Prenzler, P., Robards, K., and Ryan, D. 2000. Sample preparation in the determination of phenolic compounds in fruits. *Analyst*, 125:989–1009.

AOAC, 1980. *Official Methods of Analysis*, 12th ed., Association of Official Analytical Chemists, Washington, D.C.

Arts, I.C.W. and Hollman, P.C.H. 1998. Optimization of quantitative method for the determination of catechins in fruits and legumes. *J. Agric. Food Chem.*, 46:5156–5162.

Asquith, T.N., Izuno, C.C., and Butler, L.G. 1983. Characterization of the condensed tannin (proanthocyanidin) from group II sorghum. *J. Agric. Food Chem.*, 31:1299–1303.

Asquith, T.N. and Butler, L.G 1985. Use of dye-labeled protein as spectrophotometric assay for protein precipitants such as tannins. *J. Chem. Ecol.*, 11:1535–1544.

Atherton, B.A., Cunningham, E.L., and Splittgerber, A.G. 1996. A mathematical model for the description of the Coomassie blue protein assay. *Anal. Biochem.*, 233:160–168.

Austin, F.L. and Wolff, I.A. 1968. Sinapine and related esters in seed of *Crambe abyssinica*. *J. Agric. Food Chem.*, 16:132–135.

Azar, M., Verette, E., and Brun, S. 1987. Identification of some phenolic compounds in bilberry juice, *Vaccinium myrtillus*. *J. Food Sci.*, 52:1255–1257.

Bacon, J.R. and Rhodes, M.J.C. 1998. Development of competition assay for the evaluation of the binding of human parotid salivary proteins to a dietary complex phenols and tannins using a peroxide-labeled tannin. *J. Agric. Food Chem.*, 46:5083–5088.

Bakker, J., Bridle, P., and Koopman, A. 1992. Strawberry juice color: the effect of some processing variables on the stability of anthocyanins. *J. Sci. Food Agric.*, 60:471–476.

Barnes, S., Coward, L., Kirk, M., and Sfakianos, J. 1998. HPLC-mass spectrometry analysis of isoflavones *P.S.E.B.M.* 217:254–262.

Bate-Smith, E.C. 1973a. Tannins in herbaceous leguminosae. *Phytochemistry,* 12:1809–1812.

Bate-Smith, E.C. 1973b. Haemanolysis of tannins: the concept of relative astringency. *Phytochemistry,* 12:907–912.

Bate-Smith, E.C. 1977. Astringent tannins of Acer species. *Phytochemistry,* 16:1421–1427.

Baumann, D., Adler, S., and Hamburger, M. 2001. A simple isolation method for the major catechins in green tea using high-speed countercurrent chromatography. *J. Nat. Prod.*, 64:353–355.

Bax, A. 1985. A simple description of 2D NMR spectroscopy. *Bull. Magn. Reson.*, 7:167–183.

Bax, A. and Grzesiek, S. 1995. ROESY, in *Encyclopedia of NMR*, Grant, D.M., and Harris, R.K., Eds., Wiley, Chichester, U.K., 4157–4166.

Beart, J.E., Lilley, T.H., and Haslam, E. 1985. Polyphenol interactions. Covalent binding of procyanidins to proteins during acid catalyzed decomposition: observation on some polymeric proanthocyanidins. *J. Chem. Soc. Perkin Trans.*, 2:1439–1443

Beebe, K.R. and Kowalski, B.R. 1987. Introduction to multivariate calibration and analysis. *Anal. Chem.*, 59:1007A–1017A.

Belton, P.S., Delgadillo, I., Gil, A.M., and Webb, G.A., Eds. 1995. *Magnetic Resonance in Food Science.* The Royal Society of Chemistry, Cambridge, U.K.

Beninger, C.W. and Hosfield, G.L. 1999. Flavonol glycosides from Montcalm dark red kidney bean: implications for the genetics of seed coat color in *Phaseolus vulgaris* L. *J. Agric. Food Chem.*, 47:4079–4082.

Billot, J., Hartmann, C., Macheix, J.J., and Rateau, J. 1978. Les composes phenoliques au cours de la croissance de la poire, Passe-Crassane (Phenolic compounds during Passe-Crassane pear growth). *Physiol. Veg.*, 16:693–714.

Blair, R. and Reichert, R.D. 1984. Carbohydrates and phenolic constituents in a comprehensive range of rapeseed and canola fractions: nutritional significance for animals. *J. Sci. Food Agric.*, 35:29–35.

Boukharta, M., Girardin, M., and Metche, M. 1988. Procyanidines galloylees du sarment de vigne (*Vitis vinifera*). Separation et identification par chromatographie liquide haute performance and chromatographie en phase gazeuse. *J. Chromatogr.*, 455:406–409.

Brenna, O.V. and Pagliarini, E. 2001. Multivariate analysis of antioxidant power and polyphenolic composition in red wines. *J. Agric. Food Chem.*, 49:4841–4844.

Briandet, R., Kemsley, E.K., and Wilson, R.H. 1996. Discrimination of Arabica and Robusta in instant coffee by Fourier transform infrared spectroscopy and chemometrics. *J. Agric. Food Chem.*, 44:170–174.

Broadhurst, R.B. and Jones, W.T. 1978. Analysis of condensed tannins using acidified vanillin. *J. Sci. Food Agric.*, 29:788–794.

Brune, M., Hallberg, L., and Skanberg, A.B. 1991. Determination of iron-binding phenolic groups in foods. *J. Food Sci.*, 56:128–131; 167.

Brune, M., Rossander, L., and Hallberg, L. 1989. Iron absorption and phenolic compounds. Importance of different phenolic structures. *Eur. J. Clin. Nutr.*, 43:547–558.

Burns, R.E. 1971. Methods for estimation of tannin in grain sorghum. *Agron. J.,* 63:511–512.

Butler, L.G., Price, M.L., and Brotherton, J.E. 1982. Vanillin assay for proanthocyanins (condensed tannins): modification of the solvent for estimation of the degree of polymerization. *J. Agric. Food Chem.*, 30:1087–1089.

Carando, S., Teissedre, P.-L., Pascual-Martinez, L., and Cabanis, J.-C. 1999. Levels of flavan-3-ols in French wines. *J. Agric. Food Chem.*, 47:4161–4166.

Chassagne, D., Crouzet, J., Bayonove, C.L., and Baumes, R.L. 1998. Identification of passion fruit glycosides by gas chromatography/mass spectrometry. *J. Agric. Food Chem.*, 46:4352–4357.

Chavan, U.D., Shahidi, F., and Naczk, M. 2001. Extraction of condensed tannins from beach pea (*Lathyrus maritimus* L.) as affected by different solvents. *Food Chem.*, 75:509–512.

Chaviari, G., Concialini, V., and Galletti, G.C. 1988. Electrochemical detection in the high-performance liquid chromatographic analysis of plant phenolics. *Analyst*, 113:91–94.

Chen, Z.-Y., Zhu, Q.Y., Wong, Y.F., Zhang, Z., and Chung, H.Y. 1998. Stabilizing effect of ascorbic acid on green tea catechins. *J. Agric. Food Chem.*, 46:2512–2516.

Chu, N.T., Clydesdale, F.M., and Francis, F.J. 1973. Isolation and identification of some fluorescent phenolic compounds in cranberries. *J. Food Sci.*, 38:1038–1042.

Chu, Q., O'Dwyer, M., and Zeece, M.G. 1998. Direct analysis of resveratrol in wine by micellar electrokinetic capillary electrophoresis. *J. Agric. Food Chem.*, 46:509–513.

Cikalo, M.G., Bartle, K.D., Robson, M.M., Myers, P., and Euerby, M.R. 1998. Capillary electrochromatography. *Analyst*, 123:87R–102R.

Clandinin, D.R. 1961. Effect of sinapine, the bitter substance in rapeseed oil meal, on the growth of chickens. *Poultry Sci.*, 40:484–487.

Clifford, M.N. and Wright, J. 1973. Meteperiodate — a new structure — specific locating reagent for phenolic compounds. *J. Chromatogr. A*, 86:222–224.

Clifford, M.N. and Wright, J. 1976. The measurement of feruloylquinic acids and caffeoylquinic acids in coffee beans. Development of the technique and its preliminary application to green coffee beans. *J. Sci. Food Agric.*, 27:73–84.

Conway, W.D. and Petrovski, R.J., Eds. 1995. *Modern Countercurrent Chromatography,* ACS Symposium Series 593, American Chemical Society, Washington, D.C.

Corse, J.W., Sondheimer, E., and Lundin, R. 1962. 3-Feruloylquinic acid. A 3′-methylether of chlorogenic acid. *Tetrahedron*, 18:1207–1210.

Coward, L., Barnes, N.C., Setchell, K.D.R., and Barnes, S. 1993. Genestein, daidzein, and their -glucoside conjugates: antitumor isoflavones in soybean foods from American and Asian diets. *J. Agric. Food Chem.*, 41:1961–1967.

Crozier, A., Lean, M.E.J., McDonald, M.S., and Black, C. 1997. Quantitative analysis of the flavonoid content of commercial tomatoes, onions, lettuce and celery. *J. Agric. Food Chem.*, 45:590–595.

Dabrowski, K.J. and Sosulski, F.W. 1984. Quantification of free and hydrolyzable phenolic acids in seeds by capillary gas-liquid chromatography. *J. Agric. Food Chem.*, 32:123–127.

Daigle, D.J. and Conkerton, E.J. 1983. Analysis of flavonoids by HPLC. *J. Liq. Chromatogr.*, 6:105–118.

Daigle, D.J. and Conkerton, E.J. 1988. Analysis of flavonoids by HPLC: an update. *J. Liq. Chromatogr.*, 11:309–325.

Davis, K.R. 1982. Effects of processing on composition and *Tetrahymena* relative nutritive value of green and yellow peas, lentils, and white pea beans. *Cereal Chem.*, 58:454–460.

Davis, A.B. and Hoseney, R.C. 1979. Grain sorghum condensed tannins. I. Isolation, estimation, and selective absorption by starch. *Cereal Chem.*, 56:310–314.

Dawra, R.K., Makkar, H.P.S., and Singh, B. 1988. Protein binding capacity of microquantities of tannins. *Anal. Biochem.*, 170:50–53.

Degen, A.A., Becker, K., Makkar, H.P.S., and Borowy, N. 1995. Acacia saligna as a fodder tree for a desert livestock and the interaction of its tannins with fibre fractions. *J. Sci. Food Agric.*, 68:65–71.

Degenhart, A., Knapp, H., and Winterhalter, P. 2000a. Separation and purification of anthocyanins by high speed countercurrent chromatography and screening for antioxidant activity. *J. Agric. Food Chem.*, 48:338–343.

Degenhart, A., Engelhardt, U.E., Lakenbrink, C., and Winterhalter, P. 2000b. Preparative separation of polyphenols from tea by high-speed countercurrent chromatography. *J. Agric. Food Chem.*, 48:3425–3430.

Degenhart, A., Hofmann, S., Knapp, H., and Winterhalter, P. 2000c. Preparative isolation of anthocyanins by high-speed countercurrent chromatography and application of the color activity concept to red wine. *J. Agric. Food Chem.*, 48:5812–5818.

Degenhart, A., Engelhardt, U.E., Wendt, A.-S., and Winterhalter, P. 2000d. Isolation of black tea pigments using high-speed countercurrent chromatography and studies on properties of black tea polymers. *J. Agric. Food Chem.*, 48:5200–5205.

Delcour, J.A. and de Verebeke, J.D. 1985. A new colourimetric assay for flavonoids in Pilsner beers. *J. Inst. Brew.*, 91:37–40.

de Pascual-Teresa, S., Treutter, D., Rivas-Gonzalo, J.C., and Santos-Buelga, C. 1998. Analysis of flavanols in beverages by high-performance liquid chromatography with chemical reaction detection. *J. Agric. Food Chem.*, 46:4209–4213.

Derdelinckx, G. and Jerumanis, J. 1984. Separation of malt and hop proanthocyandins on Fractogel TSK HW-40 (S). *J. Chromatogr.*, 285:231–234.

Derome, A.E. 1987. *Modern NMR Techniques for Chemistry Research*, Pergamon Press, Oxford, U.K.

Deshpande, S.S. 1985. Investigation of dry beans (*Phaseolus vulgaris* L.): microstructure, processing and antinutrients. Ph.D. thesis. University of Illinois, Urbana Champaign, cited in Salunkhe, D.K., Chavan, J.K., and Kadam, S.S. 1989. *Dietary Tannins: Consequences and Remedies,* CRC Press Inc., Boca Raton, FL, 92.

Deshpande, S.S. and Cheryan, M. 1987. Determination of phenolic compounds of dry beans using vanillin, redox and precipitation assays. *J. Food Sci.*, 52:332–334.

Deshpande, S.S., Cheryan, M., and Salunkhe, D.K. 1986. Tannin analysis of food products. *CRC Crit. Rev. Food Sci. Nutr.*, 24:401–449.

Deshpande, S.S. and Cheryan, M. 1985. Evaluation of vanillin assay for tannin analysis of dry beans. *J. Food Sci.*, 50:905–910.

Dick, A.J. and Bearne, L. 1988. Inhibition of β-galactosidase of apple by flavonoids and other polyphenols. *J. Food Biochem.*, 12:97–108.

Downey, G., Briandet, R., Wilson, R.H., and Kemsley, K. 1997. Near- and mid-infrared spectroscopies in food authentication: coffee varietal identification. *J. Agric. Food Chem.*, 45:4357–4361.

Du, C.T. and Francis, F.J. 1975. Anthocyanins of garlic (*Allium sativum* L.): microstructure, processing and antinutrients. Ph.D. thesis. University of Illinois, Urbana Champaign, cited in Salunkhe, D.K., Chavan, J.K., and Kadam, S.S. 1989. *Dietary Tannins:Consequences and Remedies*, CRC Press Inc., Boca Raton, FL.

Durkee, A.B. and Harborne, J.B. 1973. Flavonol glycosides in *Brassica* and *Sinapis. Phytochemistry,* 12:1085–1089.

Earp, C.F., Akingbala, J.O., Ring, S.H., and Rooney, L.W. 1981. Evaluation of several methods to determine tannins in sorghum with varying kernel characteristics. *Cereal Chem.*, 58:234 238.

Edelmann, A., Diewok, J., Schuster, K.C., and Lendl, B. 2001. Rapid method for the discrimination of red wine cultivars based on mid-infrared spectroscopy of phenolic wine extracts. *J. Agric. Food Chem.*, 49:1139–1145.

Edenharder, R., Keller, G., Platt, K.L., and Unger, K.K. 2001. Isolation and characterization of of structurally novel antimutagenic flavonoids from spinach (*Spinacia oleracea*). *J. Agric. Food Chem.*, 49:2767–2773.

Eskin, N.A.M. 1980. Analysis of rapeseed by $TiCl_4$, in *Proceedings of Analytical Chemistry of Rapeseed and its Product Symposium*, Winnipeg, Manitoba, 171–173.

Fenton, T.W., Leung, J., and Clandinin, D.R. 1980. Phenolic components of rapeseed meal. *J. Food Sci.*, 45:1702–1705.

Fernando, A.R. and Plambeck, J.A. 1988. Determination of theaflavin in tea liquors using voltammetric methods. *Analyst*, 113:479–482.

Ferreira, D. and Brandt, E.V. 1989. New NMR experiments applicable to structure and conformation analysis, in *Chemistry and Significance of Condensed Tannins*, Hemingway, R.W. and Karchesy, J.J., Eds., Plenum Press, New York, 153–173.

Folin, O. and Denis, W. 1912. On phosphotungstic-phosphomolybdic compounds as color reagents. *J. Biol. Chem.*, 12:239–243.

Forsyth, W.G.C. 1955. Cacao polyphenolic substances. *Biochem. J.,* 60:108–111.

Francis, F.J. 1967. Criteria for identification of anthocyanins. *HortSci.*, 2:170–171.

Francis, F.J., Harborne, J.B., and Barker, W.G. 1966. Anthocyanins in the low bush blueberry, *Vaccinium angustifulium*. *J. Food Sci.*, 31:583–587.

Franke, A.A., Hankin, J.H., Yu, M.C., Maskarinec, G., Low, S.-H., and Custer, L.J. Isoflavone levels in soy foods consumed by multiethnic populations in Singapore and Hawaii. *J. Agric. Food Chem.*, 47:977–986.

Friedman, M., Kozukue, N., and Harden, L.A. 2000. Cinnamaldehyde content of foods determined by gas chromatography–mass spectrometry. *J. Agric. Food Chem.*, 48:5702 5709.

Fulcrand, H., Remy, S., Souquet, J.-M., Cheynier, V., and Moutounet, M. 1999. Study of wine tannin oligomers by on-line liquid chromatography electrospray ionization mass spectrophotometry. *J. Agric. Food Chem.*, 47:1023–1028.

Fuleki, T. and Francis, F.J. 1968a. Quantitative methods for anthocyanins. 1. Extraction and determination of total anthocyanin in cranberries. *J. Food Sci.*, 33:72–77.

Fuleki, T. and Francis, F.J. 1968b. Quantitative methods for anthocyanins. 2. Determination of total anthocyanin and degradation index for cranberry juice. *J. Food Sci.*, 33:78–83.

Fuleki, T. and Francis, F.J. 1968c. Quantitative methods for anthocyanins. 3. Purification of cranberry anthocyanins. *J. Food Sci.*, 33:266–274.

Gennaro, L., Leonardi, C., Esposito, F., Salucci, M., Maiani, G., Quaglia, G., and Fogliano, V. 2002. Flavonoid and carbohydrate contents in *Tropea* red onions: effects of home-like peeling and storage. *J. Agric. Food Chem.*, 50:1904–1910.

Gerothanassis, I.P., Exarchou, V., Lagouri, V., Troganis, A., Tsimidou, M., and Boskou, D. 1998. Methodology for identification of phenolic acids in complex phenolic mixtures by high resolution two-dimensional nuclear magnetic resonance. Application to methanolic extracts of two oregano species. *J. Agric. Food Chem.*, 46:4185–4192.

Goldstein, J.L. and Swain, T. 1963a. Changes in tannins in ripening fruits. *Phytochemistry,* 2:371 383.

Goldstein, J.L. and Swain, T. 1963b. The quantitative analysis of phenolic compounds, in *Methods of Polyphenol Chemistry*, Pridham, J.B., Ed., MacMillan, New York, 131–146.

Goldstein, J.L. and Swain, T. 1965. The inhibition of enzymes by tannins. *Phytochemistry,* 4:185 192.

Govindarajan, V.S. and Mathew, A.G. 1965. Anthocyanidins from leucoanthocyanidins. *Phytochemistry,* 4:985–988.

Gupta, R.K. and Haslam, E. 1980. Vegetable tannins: structure and biosynthesis, in *Polyphenols in Cereals and Legumes*, Hulse, J.H., Ed., International Development Research Center, Ottawa, Canada, 15–24.

Guyot, S., Marnet, N., and Drilleau, J.-F. 2001. Thiolysis-HPLC characterization of apple procyanidins covering large range of polymerization states. *J. Agric. Food Chem.*, 49:14–20.

Hagerman, A.E., Zhao, Y., and Johnson, S. 1997. Methods for determination of condensed and hydrolyzable tannins, in *Antinutrients and Phytochemicals in Food,* Shahidi, F., Ed., ACS Symposium Series 662, American Chemical Society, Washington, D.C., 209–222.

Hagerman, A.E. 1987. Radial diffusion method for determining tannins in plant extracts. *J. Chem. Ecol.*, 13:437–449.

Hagerman, A.E. and Robbins, C.T. 1987. Implications of soluble tannin-protein complexes for tannin analysis and plant defense mechanisms. *J. Chem. Ecol.*, 13:1243–1259.

Hagerman, A.E. and Butler, L.G. 1981. The specificity of proanthocyanidin–protein interaction. *J. Biol. Chem.*, 256:4494–4497.

Hagerman, A.E. and Butler, L.G. 1980. Determination of protein in tannin–protein precipitates. *J. Agric. Food Chem.*, 28:944–947.

Hagerman, A.E. and Butler, L.G. 1978. Protein precipitation method for the quantitative determination of tannin. *J. Agric. Food Chem.*, 26:809–812.

Hahn, D.H., Rooney, L., and Earp, C.F. 1984. Tannin and phenols in sorghum. *Cereal Foods World*, 29:776–779.

Hall, C.A., Zhu, A., and Zeece, M.G. 1994. Comparison between capillary electrophoresis and high-performance liquid chromatography separation of food grade antioxidants. *J. Agric. Food Chem.*, 42:919–921.

Hammerstone, J.F., Lazarus, S.A., Mitchell, A.E., Rucker, R., and Schmitz, H.H. 1999. Identification of procyanidins in cocoa (*Theobroma cacao*) and chocolate using high-performance liquid chromatography/mass spectrometry. *J. Agric. Food Chem.*, 47:490–496.

Harborne, J.B. 1967. *Comparative Biochemistry of Flavonoids*, Academic Press, London.

Harborne, J.B. 1958. The chromatographic identification of anthocyanin pigments. *J. Chromatogr.*, 1:473–488.

Hartzfeld, P.W., Forkner, R., Hunter, M.D., and Hagerman, A.E. 2002. Determination of hydrolyzable tannins (gallotannins and ellagitannins) after reaction with potassium iodate. *J. Agric. Food Sci.*, 50:1785–1790.

Haslam, E. 1965. Galloyl esters in the Aceraceae. *Phytochemistry,* 4:495–498.

Haslam, E. 1966. *Chemistry of Vegetable Tannins*, Academic Press, New York, 14–30.

Hemes, P.J. and Hedges, J.J. 2000. Determination of condensed tannin monomers in environmental samples by capillary gas chromatography of acid depolymerization extracts. *Anal. Chem.*, 72:5115–5124.

Hemingway, R.W., Ohara, S., Steynberg, J.P., Brandt, E.V., and Ferrera, D. 1992. C-H HETCOR NMR studies of proanthocyanidins and their derivatives, in *Plant Polyphenols: Synthesis, Properties, Significance,* Hemingway, R.H. and Laks, P.E., Eds., Plenum Press, New York, 321–338.

Hempel, J. and Bohm, H. 1996. Quality and quantity of prevailing flavonoid glycosides of yellow and green French beans (*Phaseolus vulgaris* L.). *J. Agric. Food Chem.*, 44:2114 2116.

Hillis, W.E. and Swain, T.J. 1959. The phenolic constituents of *Prunus domestica.* II. The analysis of tissues of the victoria plum tree. *J. Sci. Food Agric.*, 10:135–144.

Hilton, P.J. 1973. Tea, in *Encyclopedia of Industrial Chemical Analysis*, vol. 18, John Wiley & Sons, New York, 455–518.

Hoff, J.F. and Singleton, K.I. 1977. A method for determination of tannin in foods by means of immobilized enzymes. *J. Food Sci.*, 42:1566–1569.

Hong, V. and Wrolstad, R.E. 1990. Characterization of anthocyanin-containing colorants and fruit juices by HPLC/photodiode array detector. *J. Agric. Food Chem.*, 38:698–708.

Hostettmann, K. and Hostettman, M. 1982. Isolation techniques for flavonoids, in *The Flavonoids: Advances in Research,* Harborne, J.B. and Mabry, T.J., Eds., Chapman & Hall, New York, 1–18.

Hutabarat, L.S., Greenfield, H., and Mulholland, M. 2001. Isoflavones and coumestrol in soybeans and soybean products from Australia and Indonesia. *J. Food Compos. Anal.*, 14:43 58.

Inoue, K.H. and Hagerman, A.E. 1988. Determination of gallotannin with rhodanine. *Anal. Biochem.*, 169:363–369.

Ismail, F. and Eskin, N.A.M. 1979. A new quantitative procedure for determination of sinapine. *J. Agric. Food Chem.*, 27:917–918.

Ittah, Y. 1991. Titration of tannin via alkaline phosphatase. *Anal. Biochem.*, 192:277–280.

Jackman, R.L., Yada, R.Y., and Tung, M.A. 1987. A review: separation and chemical properties of anthocyanins used for their qualitative and quantitative analysis. *J. Food Biochem.*, 11:279 308.

Jaworski, A.W. and Lee, C.Y. 1987. Fractionation and HPLC determination of grape phenolics. *J. Agric. Food Chem.*, 35:257–259.

Johnsson, P., Kamal-Eldin, A., Lundgren, L.N., and Aman, P. 2000. HPLC method for analysis of secoisolariciresinol diglucoside in flaxseeds. *J. Agric. Food Chem.*, 48:5216–5219.

Jurd, L. 1957. The detection of aromatic acids in plant extracts by ultra-violet absorption spectra in their ions. *Arch. Biochem. Biophys.*, 66:284–288.

Jurenitsch, J. and Leinmuller, R. 1980. Quantification of nonylic acid vanillylamide and other capsaicinoids in the pungent principle of *Capsicum* fruits and preparations by gas-liquid chromatography on glass capillary columns. *J. Chromatogr.*, 189:389–397.

Jurenitsch, J., David, M., Heresch, F., and Kubelka, W. 1979. Detection and identification of new pungent compounds in fruits of *Capsicum. Planta Med.*, 36:61–67.

Karchesy, J.J., Bae, Y., Chalker-Scott, L., Helm, R.F., and Foo, L.Y. 1989. Chromatography of proanthocyanidins, in *Chemistry and Significance of Condensed Tannins*, Hemingway, R.W. and Karchesy, J.J., Eds., Plenum Press, New York, 139–152.

Karchesy, J.J. and Hemingway, R.W. 1986. Condensed tannins: (4β→8; 2β→0→7)-linked procyanidins in *Arachis hypogea* L. *J. Agric. Food Chem.*, _34:966–970.

Khanna, S.K., Viswanatham, P.N., Krishnan, P.S., and Sanwai, G.G. 1968. Extraction of total phenolics in the presence of reducing agents. *Phytochemistry,* 7:1513–1517.

Kinoshita, E., Ozawa, Y., and Aishima, T. 1997. Novel tartaric acid isoflavone derivatives that play key roles in differentiating Japanese soy sauces. *J. Agric. Food Chem.*, 45:3753–3759.

Kissinger, P.T. and Heineman, W.R. 1984. Laboratory techniques, in *Electrochemical Chemistry,* Marcel Dekker, New York, 78–124.

Kolodziej, H. 1992. ^{1}H NMR spectral studies of procyanidins derivatives: diagnostic ^{1}H NMR parameters applicable to the structural elucidation of oligomeric procyanidins, in *Plant Polyphenols: Synthesis, Properties, Significance*, Hemingway, R.H. and Laks, P.E., Eds., Plenum Press, New York, 295–320.

Kosuge, S. and Furuta, M. 1970. Studies on the pungent compounds in fruits of *Capsicum.* Part XIV. Chemical constitution of the pungent principle. *Agric. Biol. Chem.*, 34:248–256.

Kozlowska, H., Rotkiewicz, D.A., Zadernowski, R., and Sosulski, F.W. 1983. Phenolic acids in rapeseed and mustard. *J. Am. Oil Chem. Soc.*, 60:1119–1123.

Kozlowska, H., Naczk, M., and Zadernowski, R. 1990. Phenolic acids and tannins in rapeseed and canola, in *Canola and Rapeseed: Production, Chemistry, Nutrition and Processing Technology*, Shahidi, F., Ed., Van Nostrand Reinhold, New York, 193–210.

Kramer, R. 1998. *Chemometric Techniques for Quantitative Analysis*, Marcel Dekker, New York, NY.

Kreft, S., Knapp, M., and Kreft, I. 1999. Extraction of rutin from buckwheat (*Fagopyrum esculentum* Moench) seeds and determination by capillary electrophoresis. *J. Agric. Food Chem.*, 47:4649–4662.

Krygier, K., Sosulski, F., and Hogge, L. 1982. Free, esterified and bound phenolic acids. 1. Extraction and purification. *J. Agric. Food Chem.*, 30:330–333.

Labarbe, B., Cheynier, V., Brossaud, F., Souquet, J.-M., and Moutounet, M. 1999. Quantitative fractionation of grape proanthocyanidins according to their degree of polymerization. *J. Agric. Food Chem.*, 47:2719–2723.

Lea, A.G.H. and Timberlake, C.F. 1974. The phenolics of cider. I. Procyanidins. *J. Sci. Food Agric.*, 25:471–477.

Lee, H.S. 2002. Characterization of major anthocyanins and the color of red-fleshed budd blood orange (*Citrus sinensis*). *J. Agric. Food Chem.*, 50:1243–1246.

Leung, J., Fenton, T.W., Mueller, M.M., and Clandinin, D.R. 1979. Condensed tannins of rapeseed meal. *J. Food Sci.*, 44:1313–1316.

Li, Y.-G., Tanner, G., and Larkin, P. 1996. The DMACA-HCl protocol and the threshold proanthocyanidin content for bloat safety in forage legumes. *J. Sci. Food Agric.*, 70:89–101.

Liggins, J., Bluck, L.J.C., Coward, A., and Bingham, S.A. 1998. Extraction and quantification of daidzein and genistein in food. *Anal. Biochem.*, 264:1–7.

Limiroli, R., Consonni, R., Ranalli, A., Bianchi, G., and Zetta, L. 1996. ^{1}H NMR study of phenolics in the vegetation water of three cultivars of *Olea europea*. Similarities and differences. *J. Agric. Food Chem.*, 44:2040–2048.

Little, A.C. 1977. Colorimetry of anthocyanin pigmented products: changes in pigment composition with time. *J. Food Sci.*, 42:1570–1574.

Lunte, C.E., Wheeler, J.F., and Heineman, W.R. 1988a. Determination of selected phenolic acids in beer extracts by liquid chromatography with voltammetric-amperometric detection. *Analyst*, 113:94–95.

Lunte, S.M., Blankenship, K.D., and Read, S.A. 1988b. Detection and identification of procyanidins and flavonols in wine by dual-electrode liquid chromatography–electrochemistry. *Analyst*, 113:99–102.

Lunte, S.M. and Lunte, C.E. 1988. Electrochemical methods of flavonoid analysis. *Bull. Liason Groupe Polyphenols*, 14:141–144.

Mabry, T.J., Markham, K.R., and Thomas, M.B. 1970. *The Systematic Identification of Flavonoids,* Springer-Verlag, New York.

Macheix, J.-J., Fleuriet, A., and Billot, J. 1990. *Fruit Phenolics,* CRC Press Inc., Boca Raton, FL.

Mahler, S., Edwards, P.A., and Chisholm, M.G. 1988. HPLC identification of phenols in Vidal Blanc wine using electrochemical detection. *J. Agric. Food Chem.*, 36:946–951.

Makkar, H.P.S., Dawra, R.K., and Singh, B. 1987. Protein precipitation assay for quantification of tannins: determination of protein in tannin-protein complexes. *Anal. Biochem.*, 166:435 439.

Makkar, H.P.S., Dawra, R.K., and Singh, B. 1988. Determination of both tannin and protein in a tannin–protein complex. *J. Agric. Food Chem.*, 36:523–525.

Makkar, H.P.S. 1989. Protein precipitation methods for quantification of tannins: a review. *J. Agric. Food Chem.*, 37:1197–1202.

Makkar, H.P.S. and Singh, B. 1991. Distribution of condensed tannins (proanthocyanidins) in various fibre fractions in young and mature leaves of some oak species. *Anim. Feed Sci. Technol.*, 32:253–260.

Makkar, H.P.S., Bluemmel, M., and Becker, K. 1997. *In vitro* rumen apparent and true digestibilities of tannin-rich forages. *Anim. Feed Sci. Technol.*, 67:245–251.

Makkar, H.P.S., Gamble, G., and Becker, K. 1999. Limitation of butanol-hydrochloric acid-iron assay for bound condensed tannins. *Food Chem.*, 66:129–133.

Mangas, J.J., Rodriguez, R., Suarez, B., Picinelli, A., and Dapena, E. 1999, Study of the phenolic profile of cider apple cultivars at maturity by multivariate techniques. *J. Agric. Food Chem.*, 47:4046–4052.

Markham, K.R. 1975. Isolation techniques for flavonoids, in *The Flavonoids,* Harborne, J.B., Mabry, T.J., and Mabry, H., Eds., Chapman & Hall, London, 1.

Marks, D., Glyphis, J., and Leighton, M. 1987. Measurement of protein in tannin–protein precipitation using ninhydrin. *J. Sci. Food Agric.*, 38:55–61.

Martin, J.S. and Martin, M.M. 1983. Tannin essays in ecological studies: precipitation of ribulose-1,5-bisphosphate carboxylase/oxygenase by tannic acid, quebracho, and oak foliage extracts. *J. Chem. Ecol.*, 9:285–294.

Martin, J.S. and Martin, M.M. 1982. Tannin assay in ecological studies: lack of correlation between phenolics, proanthocyanidins and protein-precipitating constituents in nature foliage of six oak species. *Oecologia*, 54:205–211.

Mateus, N., Silva, A.M.S., Vercauteren, J., and de Freitas, V. 2001. Occurrence of anthocyanin-derived pigments in red wines. *J. Agric. Food Chem.*, 49:4836–4840.

Matejovicova, M., Mubagwa, K., and Flameng, W. 1997. Effect of vanadate on protein determination by the Coomassie brillant blue microassay procedure. *Anal. Biochem.*, 245:252–254.

Mathews, S., Mila, I., Scalbert, A., Pollet, B., Lapierre, C., Harve du Penchoal, C.L.M., Rolando, C., and Donelly, D.M.X. 1997. Method for estimation of proanthocyanidin based on their acid depolymerization in the presence of nucleophiles. *J. Agric. Food Chem.*, 45:1195–1201.

Mattila, P., Astola, J., and Kumpulainen, J. 2000. Determination of flavonoids in plant material by HPLC with diode-array and electro-array detections. *J. Agric. Food Chem.*, 48:5834 5841.

Maxson, E.D. and Rooney, L.W. 1972. Evaluation of methods for tannin analysis in sorghum grain. *Cereal Chem.*, 49:719–729.

Mazza, G., Fukumoto, L., Delaquis, P., Girard, B., and Ewert, B. 1999. Anthocyanins, phenolics and color of Cabernet Franc, Merlot, and Pinot Noir wines from British Columbia. *J. Agric. Food Chem.*, 47:4009–4017.

McMurrough, I. and McDowell, J. 1978. Chromatographic separation and automated analysis of flavanols. *Anal. Biochem.*, 91:92–100.

Mejbaum-Katzenellenbogen, W. and Kudrewicz-Hubicka, Z. 1966. Application of urea, ferric ammonium sulfate and casein for determination of tanning substances in plants. *Acta Biochim.*, Polonica 13(1):57.

Merken, H.M. and Beecher, G.R. 2000. Measurements of food flavonoids by high-performance liquid chromatography: a review. *J. Agric. Food Chem.*, 48:577–599.

Merrit, M.C. and Proctor, B.E. 1959. Effect of temperature during the roasting cycle on selected components of different types of whole bean coffee. *Food Res.*, 24:672–680.

Michaud, M.J. and Margail, M.A. 1977. Etude analitique des tanins catechiques. I. Les oligomeres flavanoliques de l'*Actinidia chinesis. Planchon. Bull. Soc. Pharm.*, Bordeaux 116:52–64.

Miean, K.H. and Mohamed, S. 2001. Flavonoid (myrecitin, quercetin, kaempferol, luteoilin, and apigenin) content in edible tropical plants. *J. Agric. Food Chem.*, 49:3106–3112.

Meirelles, C., Sarni, F., Ricardo da Silva, J.M., and Moutounet, M. 1992. Evaluation des procyanidines galloylees dans les vins rouges issue de differents modes de vinification, in *Proc. Int. Polyphenolic Group Conv.*, Lisboa, Vol. 16 (II), 175–178.

Moane, S., Park, S., Lunte, C.E., and Smyth, M.R. 1998. Detection of phenolic acids in beverages by capillary electrophoresis with electrochemical detection. *Analyst*, 123:1931 1936.

Mole, S. and Waterman, P. 1987a. A critical analysis of techniques for measuring tannins in ecological studies. I. Techniques for chemically defining tannins. *Oecologia*, 72:137–147.

Mole, S. and Waterman, P. 1987b. A critical analysis of techniques for measuring tannins in ecological studies. II. Techniques for biochemically defining tannins. *Oecologia*. 72:148–156.

Mole, S. and Waterman, P. 1987c. Tannic acid and proteolytic enzymes: enzyme inhibition or substrate deprivation? *Phytochemistry*, 26:99–102.

Monedero, L., Olalla, M., Martin-Lagos, F., Lopez, H., and Lopez, M.C. 1999. Application of chemometric techniques in obtaining macerates with phenolic compound content similar to that of wines from the Jerez-Shery region subjected to oxidative aging. *J. Agric. Food Chem.*, 47:1836–1844.

Moore, A.B., Francis, F.J., and Clydesdale, F.M. 1982a. Changes in the chromatographic profile of anthocyanins of red onion during extraction. *J. Food Protect.*, 45:738–743.

Moore, A.B., Francis, F.J., and Jason, M.E. 1982b. Acylated anthocyanins in red onions. *J. Food Protect.*, 45:590–593.

Moores, R.G., McDermott, D.L., and Wood, T.R. 1948. Determination of chlorogenic acid in coffee. *Anal.Chem.*, 20:620–624.

Moskowitz, A.H. and Hrazdina, G. 1981. Vacuolar contents of fruit subepidermal cells from *Vitis* species. *Plant Physiol.*, 68:686–692.

Murphy, P.A., Song, T.T., Buseman, G., and Barua, K. 1997. Isoflavones in soy-based infant formula. *J. Agric. Food Chem.*, 45:4635–4638.

Naczk, M., Pink, J., Zadernowski, R., and Pink, D. 2002. Multivariate model for the prediction of total phenolic acids in extracts of polyphenols from canola and rapeseed meals: a preliminary study. *J. Am. Oil Chem Soc.*, 79:759–762.

Naczk, M., Amarowicz, R., Zadernowski, R., and Shahidi, F. 2001a. Protein-precipitating capacity of crude condensed tannins of canola and rapeseed hulls. *J. Am. Oil Chem. Soc.*, 78:1173–1178.

Naczk, M., Amarowicz, R., Zadernowski, R., and Shahidi, F. 2001b. Protein precipitating capacity of condensed tannins of beach pea, canola hulls, evening primrose and faba bean. *Food Chem.*, 73:467–471.

Naczk, M., Pink, J., Amarowicz, R., Pink, D., and Shahidi, F. 2001c. Multivariate model for prediction of soluble condensed tannins in crude extracts of polyphenols from canola and rapeseed hulls. *J. Am. Oil Chem. Soc.*, 78:411–414.

Naczk, M., Amarowicz, R., Pink, D., and Shahidi, F. 2000. Insoluble condensed tannins of canola/rapeseed. *J. Agric. Food Chem.*, 48:1758–1762.

Naczk, M., Shahidi, F., and Sullivan, A. 1992a. Recovery of rapeseed tannins by various solvent systems. *Food Chem.*, 45:51–54.

Naczk, M., Wanasundara, P.K.J.P.D., and Shahidi, F. 1992b. A facile spectrophotometric quantification method of sinapic acid in hexane-extracted and methanol–ammonia–water treated mustard and rapeseed meals. *J. Agric. Food Chem.*, 40:444–448.

Naczk, M. and Shahidi, F. 1991. Critical evaluation of quantification methods of rapeseed tannins. Presented at 8th International Rapeseed Congress, Saskatoon, Canada, July 9–11, 1991.

Naczk, M. and Shahidi, F. 1989. The effect of methanol–ammonia–water treatment on the content of phenolic acids of canola. *Food Chem.*, 31:159–164.

Newman, R.H. and Porter, L.J. 1992. Solid state ^{13}C-NMR studies on condensed tannins, in *Plant Polyphenols: Synthesis, Properties, Significance*, Hemingway, R.H. and Laks, P.E., Eds., Plenum Press, New York, 339–348.

Nonaka, G., Morimoto, S., and Nishioka, I. 1983. Tannins and related compounds. Part 13. Isolation and structures of trimeric, tetrameric, and pentameric proanthocyanidins from cinnamon. *J. Chem. Soc. Perkin Trans.*, 1:2139–2146.

Ogawa, K., Kawasaki, A., Yoshida, T., Nesumi, H., Nakano, M., Ikoma, Y., and Yano, M. 2000. Evaluation of auraptene content in citrus fruits and their products. *J. Agric. Food Chem.*, 48:1763–1769.

Oh, H.I., Hoff, J.E., and Haff, L.A. 1985. Immobilized condensed tannins and their interaction with proteins. *J. Food Sci.*, 50:1652–1654.

Oh, H.I., Hoff, J.E., Armstrong, G.S., and Haff, L.A. 1980. Hydrophobic interaction in tannin protein complexes. *J. Agric. Food Chem.*, 28, 394–398.

Oomah, B.D. and Mazza, G. 1996. Flavonoids and antioxidative activities in buckwheat. *J. Agric. Food Chem.*, 44:1746–1750.

Owades, J.L., Rubin, G., and Brenner, M.W. 1958a. Determination of tannins in beer and brewing materials by ultraviolet spectroscopy. *Proc. Am. Soc. Brew. Chem.*, 6:66–73.

Owades, J.L., Rubin, G., and Brenner, M.W. 1958b. Food tannins measurement, determination of food tannins by ultraviolet spectrophotometry. *J. Agric. Food Chem.*, 6:44–46.

Pan, Y., Zhang, L., and Chen, G. 2001. Separation and determination of protocatechuic aldehyde and protocatechuic acid in *Salivia miltorrhrza* by capillary electrophoresis with amperometric detection. *Analyst*, 126:1519–1523.

Peng, Z., Hayasaka, Y., Iland, P.G., Sefton, M., Hoj, P., and Waters, E.J. 2001. Quantitative analysis of polymeric procyandins from grape (*Vitis vinifera*) by reverse phase high performance liquid chromatography. *J. Agric. Food Chem.*, 49:26–31.

Porter, L.J. 1989. Tannins, in *Methods in Plant Biochemistry*, vol. 1, Harborne, J.B., Ed., Academic Press, San Diego, CA, 389–420.

Porter, L.J., Hrtstich, L.N., and Chan, B.G. 1986. The conversion of procyanidins and prodelphinidins to cyanidins and delphinidins. *Phytochemistry,* 25:223–230.

Price, M.L. and Butler, L.G. 1977. Rapid visual estimation and spectrophotometric determination of tannin content of sorghum grain. *J. Agric. Food Chem.*, 25:1268–1273.

Price, M.L., Van Scoyoc, S., and Butler, L.G. 1978. A critical evaluation of the vanillin reaction as an assay for tannin in sorghum. *J. Agric. Food Chem.*, 26:1214–1218.

Prior, R.L., Lazarus, S.A., Cao, G., Muccitelli, H., and Hammerstone, J.F. 2001. Identification of procyanidins in bluberries and cranberries (*Vaccinium* Spp.) using high-performance liquid chromatography/mass spectrometry. *J. Agric. Food Chem.*, 49:1270–1276.

Reeves, S.G., Gone, F., and Owuor, P.O. 1985. Theaflavin analysis of black tea — problems and prospects. *Food Chem.*, 18:199–210.

Ricardo da Silva, J.M., Rigaud, J., Cheynier, V, Cheminat, A., and Moutounet, M. 1991. Procyanidin dimers and trimers from grape seeds. *Phytochemistry,* 4:1259–1264.

Rinderknecht, H., Geokas, M.C., Silverman, P., and Haverback, B.J. 1968. A new ultrasensitive method for determination of proteolytic activity. *Clin. Chim. Acta*, 21:197–203.

Robards, K. and Antolovitch, M. 1997. Analytical chemistry of fruits bioflavonoids. *Analyst*, 122:11R–34R.

Robertson, A. and Hall, M.N. 1989. A critical investigation into the Flavognost method for theaflavin analysis in black tea. *Food Chem.*, 34:57–70.

Sacchi, R., Patumi, M., Fontanazza, G., Barone, P., Fiordiponti, P., Mannina, L., Rossi, E., and Segre, A.L. 1996. A high-field ^{1}H nuclear magnetic resonance study of minor components in virgin olive oils. *J. Am. Oil Chem. Soc.*, 73:747–757.

Salagoity-Auguste, M.H. and Bertrand, A.J. 1984. Wine phenolics. Analysis of low molecular weight components by high performance liquid chromatography. *J. Sci. Food Agric.*, 35:1241–1247.

Salunkhe, O.K., Chavan, J.K., and Kadam, S.S. 1989. *Dietary Tannins: Consequences and Remedies*. CRC Press, Boca Raton, FL.

Sanders, T.H., McMichael, R.W., and Hendrix, K.W. 2000. Occurrence of resveratrol in edible peanuts. *J. Agric. Food Chem.*, 48:1243–1246.

Sano, M., Suzuki, M., Miyase, T., Yoshino, K., and Maeda-Yamamoto, M. 1999. Novel antiallergic catechin derivatives isolated from oolong tea. *J. Agric. Food Chem.*, 47:1906 1910.

Sarkar, S.K. and Howarth, R.E. 1976. Specifity of vanillin test for flavonols. *J. Agric. Food Chem.*, 24:317–320.

Satterfield, M. and Brodbelt, J.S. 2000. Enhanced detection of lavonoids by metal complexation and electrospray ionization mass spectrometry. *Anal. Chem.*, 72:5898–5906.

Scalbert, A., Monties, B., and Janin, G. 1989. Tannins in woods: comparison of different estimation methods. *J. Agric. Food Chem.*, 37:1324–1329.

Scalbert, A. 1992. Quantitative methods for estimation of tannins in plant tissues, in *Plant Polyphenols: Synthesis, Properties, Significance*, Hemingway, R.W. and Laks, P.S., Eds., Plenum Press, New York, 259–280.

Schultz, J.C., Baldwin, I.T., and Nothnagle, P.J. 1981. Hemoglobin as a binding substrate in the quantitative analysis of plant tannins. *J. Agric. Food Chem.*, 29:823–829.

Schulz, H., Engelhardt, U.H., Wegent, A., Drews, H.-H., and Lapczynski, S. 1999. Application of near-infrared reflectance spectroscopy to the simultaneous prediction of alkaloids and phenolic substances in green tea leaves. *J. Agric. Food Chem.*, 47:5064 5067.

Senter, S.D., Robertson, J.A., and Meredith, F.I. 1989. Phenolic compounds of the mesocarp of cresthaven peaches during storage and ripening. *J. Food Sci.*, 54:1259–1260; 1268.

Shahidi, F. and Naczk, M. 1989. Effect of processing on the content of condensed tannins in rapeseed meals. A research note. *J. Food Sci.*, 54:1082–1083.

Shahidi, F. and Naczk, M. 1990. Contribution of sinapic acid to the phenolic constituents of solvent extracted cruciferae oilseeds. *Bull. Liason Groupe Polyphenols*, 15:236 239.

Shahidi, F., Chavan, U.D., Naczk, M., and Amarowicz, R. 2001. Nutrient distribution and phenolic antioxidants in air-classified fractions of beach pea (*Lathyrus maritimus* L.). *J. Agric. Food Chem.*, 49:926–933.

Sharp, R.N., Sharp, C.Q., and Kattan, A.A. 1977. Tannin content of sorghum grain by UV spectrophotometry. *Cereal Chem.*, 55:117–118.

Siess, M.-H., Le Bon, A.M., Canivenc-Lavier, M.-C., Amiot, M.-J., Sabatier, S., Aubert, S.Y., and Suschetet, M. 1996. Flavonoids of honey and propolis: characterization and effects on hepatic drug-metabolizing enzymes and benzo[α] pyrene-DNA binding in rats. *J. Agric. Food Chem.*, 44:2297–301.

Silber, M.L., Davitt, B.B., Khairutdinov, R.F., and Hurst, J.K. 1998. A mathematical model describing tannin–protein association. *Anal. Biochem.*, 263:46–50.

Simonne, A.H., Smith, M., Weaver, D.B., Vail, T., Barnes, S., and Wei, G.I. 2000. Retention and changes of soybean isoflavones and carotenoids in immature soybean seeds (Edamame) during processing. *J. Agric. Food Chem.*, 48:6061–6069.

Singleton, V.L. and Rossi, J.A. 1965. Colorimetry of total phenolics with phosphomolybdic-phosphotungstic acid. *Am. J. Enol. Vitic.*, 16:155–158.

Snell, F.D. and Snell, C.T. 1953. *Colorimetric Methods of Analysis*, 3rd ed., vol. III, Van Nostrand, Princeton, NJ, 104.

Somers, T.C. 1966. Wine tannins-isolation of condensed flavanoid pigments by gel filtration. *Nature*, 209:368–370.

Sondheimer, E. and Kertesz, Z. 1948. Anthocyanin pigments. Colorimetric determination in strawberries and strawberry preserves. *Anal. Chem.*, 20:245–248.

Souquet, J.M., Cheynier, V., Brossaud, F., and Moutounet, M. 1996. Polymeric proanthocyanidins from grape skins. *Phytochemistry*, 43:509–512.

Spiro, M. and Price, W.E. 1986. Determination of theaflavins in tea solution using the Flavognost complexation method. *Analyst*, 3:331–333.

Strumeyer, D.H. and Malin, M.J. 1975. Condensed tannins in grain sorghum: isolation, fractionation, and characterization. *J. Agric. Food Chem.*, 23:909–914.

Subra Rao, M.V.S.S.T. and Muralikrishna, G. 2002. Evaluation of antioxidant properties of free and bound phenolic acids from native and malted finger millet (*Ragi, Elusine coracana* Indaf-15). *J. Agric. Food Chem.*, 50:889–892.

Sun, B., Ricardo-da-Silva, J.M., and Spranger, I. 1998a. Critical factors of vanillin assay for catechins and proanthocyanidins. *J. Agric. Food Chem.*, 46:4267–4274.

Sun, B.S., Leandro, M.C., Ricardo-da-Silva, J.M., and Spranger, M.I. 1998b. Separation of grape and wine proanthocyanidins according to their degree of polymerisation. *J. Agric. Food Chem.*, 46:1390–1396.

Swain, T. and Hillis, W.E. 1959. Phenolic constituents of *Prunus domestica*. I. Quantitative analysis of phenolic constituents. *J. Sci. Food Agric.*, 10:63–68.

Sweeley, C.C., Bentley, R., Makita, M., and Wells, W.W. 1963. Gas-liquid chromatography of trimethylsilyl derivatives of sugars and related substances. *J. Am. Chem. Soc.*, 85:2497–2507.

Tempel, A.S. 1982. Tannin-measuring techniques. A review. *J. Chem. Ecol.*, 8:1289–1298.

Terrill, T.H., Rowan, A.M., Douglas, G.B., and Barry, T.N. 1992. Determination of extractable and bound condensed tannin concentrations in forage plants, protein concentrate meals and cereal grains. *J. Sci. Food Agric.*, 58:321–329.

Thies, M. and Fischer, R. 1971. New reaction for microchemical detection and the quantitative determination of catechins. *Mikrochim. Acta*, 1:9–13.

Thomas, B.V., Schreiber, A.A., and Weisskopf, C.P. 1998. Simple method for quantitation of capsaicinoids in peppers using capillary gas chromatography. *J. Agric. Food Chem.*, 46:2655–2663.

Thompson, R.S., Jacques, D., Haslam, E., and Tanner, R.J.N. 1972. Plant proanthocyanidins. Part 1. Introduction: the isolation, structure, and distribution in nature of plant procyanidins. *J. Chem. Soc. Perkin Trans.*, 1, 1387–1399.

Timberlake, C.F. and Bridle, B. 1967. Flavylium salts, anthocyanidins and anthocyanins. 1. Structural transformations in acid solutions. *J. Sci. Food Agric.*, 18:473–478.

Tomas-Barberan, F.A., Gil, M.I., Cremin, P., Waterhouse, A.L., Hess-Pierce, B., and Kader, A.L. 2001. HPLC–DAD–ESIMS analysis of phenolic compounds in nectarines, peaches, and plums. *J. Agric. Food Chem.*, 49:4748–4760.

Treutter, D. 1989. Chemical reaction detection of catechins and proanthocyanidins with 4-dimethylcinnamaldehyde. *J. Chromatogr.*, 467:185–193.

Tzagoloff, A. 1963. Metabolism of sinapine in mustard plants. I. Degradation of sinapine into sinapic acid and choline. *Plant Physiol.*, 38:202–206.

Vasanthan, J.Y.T. and Temelli, F. 2001. Analysis of phenolic acids in barley by high performance liquid chromatography. *J. Agric. Food Chem.*, 49:4352–4358.

Vitrac, X., Castagnino, C., Waffo-Teguo, P., Deleunay, J.-C., Vercauteren, J., Monti, J.-P., Deffieux, G., and Merillon, J.-M. 2001. Polyphenols newly extracted in red wine from southwestern France by centrifugal partition chromatography. *J. Agric. Food Chem.*, 49:5934–5938.

van der Voort, F.R. 1992. Fourier-transform infrared spectroscopy applied in food analysis. *Food Res. Int.*, 25:397–403.

Verzele, M., Delahaye, P., and Damme, F.V. 1986. Determination of the tanning capacity of tannic acids by high performance liquid chromatography. *J. Chromatogr.*, 362:363–374.

Wang, C., Sherrard, M., Pagala, S., Wixon, R., and Scott, R.A. 2000a. Isoflavone content among maturity group 0 to II soybeans. *J. Am. Oil Chem. Soc.*, 77:483–487.

Wang, L.-F., Kim, D.-M., and Lee, C.Y. 2000b. Effects of heat processing and storage on flavanols and sensory qualities of green tea beverages. *J. Agric. Food Chem.*, 48:4227 4232.

Wang, J. and Sporns, P. 1999. Analysis of anthocyanins in red wine and fruit juice using MALDI–MS. *J. Agric. Food Chem.*, 47:2009–2015.

Waterman, P.G. and Mole, S. 1994. *Analysis of Plant Phenolic Metabolites,* Blackwell, Oxford, U.K.

Willis, R.B. and Allen, P.R. 1998. Improved method for measuring hydrolyzable tannins using potassium iodate. *Analyst,* 123:435–439.

Wilson, T.C. and Hagerman, A.E. 1990. Quantitative determination ellagic acid. *J. Agric. Food Chem.*, 38:1678–1683.

Woodring, P.J., Edwards, P.A., and Chisholm, M.G. 1990. HPLC determination of non flavonoid phenols in Vidal Blanc wine using electrochemical detection. *J. Agric. Food Chem.*, 38:729–732.

Xu, Z. and Goldber, J.S. 1999. Purification and identification of components of γ-oryzanol in rice bran oil. *J. Agric. Food Chem.*, 47:2724–8.

Zadernowski, R. 1987. Studies on phenolic compounds of rapeseed flours. *Acta Acad. Agric. Olst. Technol. Aliment.*, Suppl. F, 21:1–55.

Zafrilla, P., Ferreres, F., and Tomas-Barberan, F.A. 2001. Effect of processing and storage on the antioxidant ellagic acid derivatives and flavonoids of red raspberry (*Rubus idaeus*) jams. *J. Agric. Food Chem.*, 49:3651–3655.

Zeece, M. 1992. Capillary electrophoresis: a new analytical tool for food science. *Trends Food Sci. Technol.*, 3:6–10.

Index

F

G

H

N

O

P

Q

R

S

W

X

Z